DUKE'S HANDBOOK OF Medicinal Plants OF THE Bible

James A. Duke
with
Peggy-Ann K. Duke
Judith L. duCellie

CRC Press
Taylor & Francis Group
Boca Raton London New York

CRC Press is an imprint of the
Taylor & Francis Group, an **informa** business

CRC Press
Taylor & Francis Group
6000 Broken Sound Parkway NW, Suite 300
Boca Raton, FL 33487-2742

No claim to original U.S. Government works
Printed in the United States of America on acid-free paper
10 9 8 7 6 5 4 3 2 1

International Standard Book Number-13: 978-0-8493-8202-4 (Hardcover)

Library of Congress Cataloging-in-Publication Data

Duke, James A., 1929-
Duke's handbook of medicinal plants of the Bible / James A. Duke.
p. cm.
Includes bibliographical references and index.
ISBN 978-0-8493-8202-4 (alk. paper)
1. Herbs in the Bible 2. Plants in the Bible. 3. Herbs--Therapeutic use. I. Title. II. Title: Handbook of medicinal herbs of the Bible.

BS665.D85 2007
220.8'58163--dc22 2007020177

Visit the Taylor & Francis Web site at
http://www.taylorandfrancis.com

and the CRC Press Web site at
http://www.crcpress.com

Contents

Acknowledgments

For mysterious reasons, this third of my biblical undertakings has required the patience of Job. And for their patience, I am indeed grateful to Barbara Norwitz, my publisher, Jill Jurgensen, senior project coordinator, and Karen Simon, project editor, for tolerating some idiosyncrasies I introduced in 2007, 30 years after my first publication with CRC Press. But Jobian patience has been required by Judi duCellier, my associate for those same 30 years; Peggy Duke, my wife and illustrator, as we creep and creak to our 50th wedding anniversary; and Mary Jo Bogenshutz-Godwin, another long-time associate and frequent co-author. Mary Jo has patiently helped in many ways, especially in getting occasional updates to my USDA phytochemical database, and moving us closer to a *Duke's Handbook of Medicinal Herbs of Latin America*.

Faith-Based Farmaceuticals: Introduction

With the accelerating decline of my long-enjoyed good health, I lose faith more and more each year in the pharmaceutical firms of America. It is clear that they wield incredible and potentially corrupting power. I have little faith in the pharmaceutical firms' desire to see a healthy America. I see them promoting multiple-pills-a-day regimens, to be taken for life, and not necessarily promoting health. Their side effects sometimes outweigh any health advantages they offer. But I gain faith each year in the natural farmaceuticals in the Green Farmacy Garden. While I consider myself both a naturalist and a spiritualist, I have strayed from the Southern Baptists who introduced me to religion before I was old enough to have a questioning mind. Mine is still a questioning mind! Turning 78 on April 4, 2007, I still envy those who believe, without doubt. I wish I did! Surely I so wish, if they are right! Maybe even if they are wrong! They are less likely to worry themselves to death. I am also impressed with figures suggesting that 90 to 95% of Americans resort to prayer in illness. If they believe, they are much more liable to be helped. If they believe more in the healing power of their acupuncturist, chiropractor, herbalist, homeopath, naturopath, orthopedist, or even their minister or shaman, than they do in their allopathic physician, there is a good chance that, with the power of mind/body synergy, those non-allopathic modalities are more likely to succeed than the allopathic modality. Yes, I am saying that with many people, believing is more liable to cure than is allopathy. And those same people, reading that our many Biblical foods are menus of medicinal phytochemicals, from which their bodies can selectively mine those needed by the ailing body, are more liable to be helped by using the biblical food farmacy.

Many of us have come to believe my earlier published statistics that pharmaceuticals kill a thousand times more Americans than herbs. The famous Gary Null and a few others even say that pharmaceuticals are bigger killers than cancer, cardiopathy, diabetes, or stroke. And, it is easy to see that the newest, most expensive, and most advertised pharmaceuticals are among the most dangerous. Yes, shortly after Phase 3 clinical trials and final FDA approval, the pharmaceutical firms get their marketing efforts into gear with the permeating message, "Ask your physicians if New Drugs A, B, C and W, X, Y are right for you," simultaneously giving free samples to the allopaths to encourage you to take the new drug. And off go their drug reps to reward the allopaths with more than free samples — training seminars in romantic places, where they will learn most of the positives about the new drug and few of the negatives. Before you know it, you are taking Drug Z for the side effects of Drug A. You, the American Public, are induced into what I call Phase 4 "Cynical Trials," in which half of the new pharmaceuticals will be relabeled, with stronger warnings, or partially or completely recalled within a decade. Meanwhile, more expensive pharmaceuticals will continue to cause many more deaths than are caused by the safe herbs we are led to believe are dangerous. They are not! Check the Bextra, Celebrex, and Vioxx stories, and soon-to-be-heard statin stories (three close friends of mine, too old to be worried about cholesterol, have been hospitalized because of using statins), and head counts of iatrogenic fatalities. The Null numbers: The total number of annual iatrogenic deaths in America is 783,936 (Null et al., 2003).

Remember that pharmaceuticals have been with us less than 150 years. If our ancestors left Africa via the Holy Land 2000 years ago (for faith-based literalists), maybe a million years ago (for the less literal), then our genes, tracing back to our African/Holy Land ancestors, have had at least ten times more temporal experience with biblical herbs (e.g., cinnamon, coriander, cumin, dill, garlic, grape, mint, milk thistle, myrrh, olive, onion, pomegranate, saffron, turmeric, and the like). Pharmaceuticals and synthetic food additives are relatively new to our genes. Our bodies have had thousands, perhaps millions, of years of evolutionary experience with the thousands of phytochemicals in these edible species. Our bodies may even require many of them. In many cases, I will wager but cannot prove, the body has evolved homeostatic mechanisms for maintaining homeostatic balances for these phytochemicals. Our body can sequester them from our dietary milieus if we need them, excreting them if we do not. We can prove this for simple elemental chemicals such as selenium and zinc. I will also wager that homeostatic balancing activities exist for hundreds of many, long-familiar dietary components. We just assigned an RDA for choline in the past decade. The further we get from our paleolithic diet and, more importantly, the more synthetic pharmaceuticals and food additives we ingest, the more liable we are to suffer imbalances. It is not only food additives that hurt us; it is the subtractives as well. The subtractive phytochemicals are those important nutrients reduced or lost in food processing. Most of the value in whole grains is lost in the processing (e.g., some of the minerals such as copper, iron, manganese, selenium, and zinc, and vitamins such as B_1, B_2, B_3, B_5, B_6, E, and folic acid).

Restoring chemical balance may require getting back to basics, those primitive paleolithic foods rich in phytonutrients. At the same time, we should reduce our consumption of over-processed, nutrient-poor junk foods, avoiding additives and even pharmaceuticals where possible and plausible. I am not saying that there is no place for pharmaceuticals, but I will say that in many cases there are balanced Biblical foods that are pharmacologically competitive with unbalancing pharmaceuticals, and these food farmaceuticals should be drugs of first resort, the pharmaceutical of last resort. And if you believe more in me and my Biblical food farmaceutical shotgun than you believe in your allopath and her/his expensive pharmaceutical silver bullets, then there is a better chance that my natural approach will help you. Believing is half the cure. Can you believe in a company whose 2-billion-dollar-a-year drug was shown in the *Journal of the American Medical Association* (*JAMA*) back in 2002 to be no better than a placebo for major depression. Can you believe that, now 4 years later, that company still has the premier lead-off ad page for *JAMA* touting the $2-billion-a-year drug as so trusted, so reliable, so efficacious? I suspect you would be better off with Biblical walnut oil and Biblical saffron, nourishing and medicating your body, attenuating the depression with few, or no, consequential side effects. Ask your congress person to insist on a new third-arm trial comparing Zoloft with walnut oil/hypericum/saffron, a triple-whammy food-farmaceutical antidepressant. If you count all the possible side effects reported in the fine print of that ad for the $2-billion-a-year pharmaceutical, you will count more than a hundred. When that study was published back in 2002 showing the pharmaceutical as no better than placebo, almost nobody heard that the drug failed, too. The news was instead blaring out that "St. John's wort no better than placebo." True St. John's wort (SJW) fared no better than placebo in this clinical comparison of SJW, Zoloft, and the placebo. But that is the half of the story that Joan Q. Public heard a thousand times, maybe once or twice hearing that the pharmaceutical also failed. Do I think there is a pharmaceutical/FDA/press conspiracy? I will say that they are all singing the same song, and the song is wrong, thus hurting Americans. Their monotonous song drives American consumers from the safer food, herb, and spice farmaceuticals to the more expensive, more dangerous synthetic pharmaceuticals — all this at the expense of our health and the health of our planet. Even our rivers and lakes — and consequently our water supply— are now cocktails of pharmaceutical residues.

That bit of antidepressant news was so depressing that I resorted to song writing as alternative therapy. I like to remind people of the failure of the press to report the real news back in 2002.

About The Author

James A. "Jim" Duke, Ph.D., economic botanist and ethnobotanist, retired after a full career with the United States Department of Agriculture (USDA, Beltsville, MD) in 1995. After "retiring," he served 5 years as senior science adviser with Nature's Herbs. Although he has been retired for more than a decade, the USDA still maintains his Phytochemical Database online at http://www.ars-grin.gov/duke. He has published more than 30 books dealing with herbs, economic botany, and ethnobotany. For the past 5 years, he has taught medical botany as distinguished lecturer with the master of sciences program in herbal medicine at the Tai Sophia Institute, Laurel, Maryland. With an aggregate of 6 years in Latin America, he still leads ethnobotanical trips. In 2007, for example, he took one group to the Amazonian rain forest, accompanied by his *Amazonian Ethnobotanical Dictionary* (Duke, J.A. and Vasquez Martinez, R., 1994. CRC Press, Boca Raton, FL, 215 pp.). He also led a group to Costa Rica in 2007, accompanied by his *Tico Ethnobotanical Dictionary* (online at the USDA Phytochemical Database and an updated version of his *Isthmian Ethnobotanical Dictionary,* first published, 1971). For the 16th year, he led a week-long field course in ethnobotany in coastal Maine. He often hosts tours of his own Green Farmacy Garden in suburban Maryland, with some 300 medicinal herbs.

A Phi Beta Kappa graduate of the University of North Carolina (Chapel Hill), where he also attained his Ph.D., Duke was elected as distinguished alumnus 50 years later. He was appointed honorary president of the Herb Society of America for 2007.

Bibliographic Abbreviations

The following is a listing of the books and journals frequently consulted for this book.

AAB = Arvigo and Balick (1993)
AAH = Allen and Hatfield (2004)
ABS = Abstract
ACT = *Alternative & Complementary Therapies*
AEH = De Smet et al. (1997)
AH2 = McGuffin et al. (2000)
AHL = Liogier (1974)
AHP = McGuffin et al. (1997)
AKT = Tillotson et al. (2001)
APA = Peirce (1999)
AUS = Austin (2004)
AVP = Arsene (1971)
BAT = Batanouny (1981)
BEJ = Bejar (2001)
BGB = Blumenthal et al. (2000)
BI2 = Duke (1999)
BIB = Duke (1983)
BIS = Bisset (1994)
BMD = Montague-Drake (1997)
BNA = Balick et al. (2000)
BO2 = Boik (2001)
BOI = Boik (1995); and for fuller citation, *Cancer & Natural Medicine*, Oregon Medical Press, Princeton, MN, 315 pp. (= BOI)
BOU = Boulos (1983)
BOW = Bown (2001)
BRU = Bruneton (1999)
BUR = Jacobs and Burlage (1958)
CAN = Newall et al. (1996)
CEB = Erichsen-Brown (1989)
CJE = Earle (http://www.botanik.uni-bonn.de/conifers/)
COX = Newmark and Schulick (2000)
CR2 = Duke et al. (2002[[a)
CRC = Duke (1985)
CRH = *Consumer Reports on Health*
CWW = Wright (2002)
DAA = Duke and Ayensu (1985)
DAD = Duke and duCellier (1993)
DAS = Davies and Stewart (1990)
DAV = Duke and Vasquez (1994)
DAW = Duke and Wain (1991)
DEM = Moerman (1998)
DEP = Watt (1889–1892)
DLZ = De Lucca and Zalles (1992)
EAS = Eastman (1992, 1995, 2003)

EB	=	*Economic Botany*
EFS	=	Steinmetz (circa 1957)
EGG	=	Egg (1999)
EMP	=	*Economic and Medicinal Plant Research*
FAC	=	Facciola (1998)
FAD	=	Foster and Duke (1990)
FAH	=	Foster and Hobbs (2002)
FAY	=	Foster and Yue (1992)
FEL	=	Felter and Lloyd (1898)
FFJ	=	*Flavor and Fragrance Journal*
FNF	=	*Father Nature's Farmacy* online database: http://www.ars-grin.gov/duke/
FP1	=	Zohary (1966)
FP2	=	Zohary (1972)
FP3	=	Feinbrun-Dothan (1978)
FT	=	*Fitoterapia*
GAZ	=	Skenderi (2003)
GHA	=	Ghazanfar (1994)
GMH	=	Grieve (1931)
GMJ	=	Grenand et al. (1987)
GOO	=	Google search
HC	=	Herb Clips (American Botanical Council)
HDN	=	Neuwinger (1996)
HEG	=	Hegnauer (1962–1997)
HG	=	*HerbalGram*
HH2	=	Hansel et al. (1992, 1993, 1994)
HH3	=	Blaschek et al. (1998, 1998)
HHB	=	List and Hohammer (1969–1979)
HJP	=	Philips (1958)
HOC	=	Tyler (1994)
HOE	=	Duke (1983)
HOO	=	Tyler (1985)
HOS	=	Duke et al. (2002[[b)
IED	=	Duke (1986)
IHB	=	Burkill (1966)
JAC	=	*The Journal of Alternative and Complementary Medicine*
JAD	=	James A. Duke, personal communication
JAF	=	*Journal of Agricultural and Food Chemistry*
JAH	=	*Journal of the American Herbalists Guild*
JAR	=	*The International Journal of Aromatherapy*
JBH	=	Harborne and Baxter (1983)
JEB	=	*Journal of Ethnobotany*
JFM	=	Morton (1977, 1981)
JLH	=	Hartwell (1982)
JMF	=	*Journal of Medicinal Food*
JN	=	*Journal of Nutrition*
JNP	=	*Journal of Natural Products*
JNU	=	Joseph et al. (2001)
KAB	=	Kirtikar and Basu (Reprint 1975)
KAL	=	Koch and Lawson (1996)
KAP	=	Kapoor (1990)
KC2	=	Huang (1999)

KJV	=	King James Version of the Bible
KOM	=	Blumenthal et al. (1998)
LAF	=	Leung and Foster (1995)
LAW	=	Lawson (1996)
LEG	=	Duke (1981)
LIB	=	Libster (2002)
LMP	=	Perry (1980)
LRNP	=	*Lawrence Review of Natural Products, looseleaf*; periodically updated
LWW	=	Little et al. (1974)
M29	=	Martindale; *The Extra Pharmacopoeia* (29th edition). The Pharmaceutical Press, London (1989), 1930 pp.
MAB	=	Mills and Bone (1999)
MAM	=	Miller and Murray (1998)
MAX	=	Martinez (1969)
MKK	=	Kaul (1997)
MPB	=	Mors et al. (2000)
MPG	=	Gupta (1995)
MPI	=	ICMR (Indian Council of Medical Research) (1976, 1987)
MZN	=	Martin et al. (1961)
NAD	=	Nadkarni (1976)
NH	=	Barbara Grant or Grace Lyn Rich, Nature's Herbs; personal communication
NMH	=	Humphrey (2003)
NP	=	*Natural Products*
NPM	=	Manandhar (2002)
NR	=	*Nutrition Reviews*
NUT	=	Duke (1989)
NWT	=	*New World Translation of the Bible*
ORAFTI	=	Newsletter Number 9 of ORAFTI
PAM	=	Pizzorno and Murray (1985)
PAY	=	Palevitch and Yaniv (2000)
PCS	=	Standley (1920–1926)
PDB	=	Phytochemical Database, http://www.ars-grin.gov/duke/
PED	=	Pedersen (1998)
PER	=	Mark Dafforn, personal communication
PH2	=	Gruenwald et al. (2000)
PHM	=	*Phytomedicine*
PHR	=	Fleming et al. (1998)
PIP	=	Schilcher (1997)
PJB	=	*Protocol Journal of Botanical Medicine*
PM	=	*Planta Medica*
PNC	=	Williamson and Evans (1989)
POR	=	Porcher's Australian Nomenclature Database, http://www.plantnames.unimelb.edu.au
PR	=	*Phytotherapy Research*
PST	=	Press et al. (1994)
RAR	=	Rutter (1990)
RCP	=	*Revista Cubana de Plantas Medicinales*
RIN	=	Rinzler (1990)
RIZ	=	Rizk (1986)
ROE	=	Roersch (1994)
RSV	=	*Revised Standard Version of the Bible*
RyM	=	Roig y Mesa (1928)

SAR = Schultes and Raffauf (1990)
SAY = Stamets (and Yao) (2002)
SF = Foster (1996)
SHT = Schulz et al. (1998)
SKJ = Jain and deFilipps (1991)
SKY = Lininger et al. (1998)
SOU = Soukup (1970)
SPI = Charalambous (1994)
SUW = Suwal (1976)
TAD = Tucker and Debaggio (2000)
TAN = Tanaka (1976)
TGP = Duke (1997)
TIB = Kletter and Kriechbaum (2001)
TMA = *Time-Life*, Editors (1996)
TOM = Tommy Bass; in Crellin and Philpott (1990)
TRA = Germosén-Robineau (TRAMIL) (1997)
ULW = Coe and Anderson (1999)
UPH = Uphof (1968)
UPW = Burkill (1985–1995)
USN = USDA Nomenclature Database, http://www.ars-grin.gov/npgs/tax/taxgenform.html.
VAD = García et al. (1998) (Vadmecum)
VOD = Beauvoir et al. (2001)
VVG = Van Wyk et al. (1997)
WAF = White and Foster et al. (2000)
WAM = White and Mavor (1998)
WBB = Watt and Breyer-Brandwijk (1962)
WHO = World Health Organization (1999)
WI3 = Wichtl (2004)
WIC = Wichtl (1984)
WIN = Kindscher and Hurlburt (1998)
WOI = CSIR (1948–1976)
WO2 = CSIR (1985)
WO3 = CSIR (2000)
X as prefix followed by several numbers = PubMed reference citation (e.g., X123456)
YAB = Yaniv and Bachrach (2005)
ZOH = Zohary (1982)
ZUL = Hutchings et al. (1996)

Chemical and Medical Abbreviations

Most of the miscellaneous abbreviations relate to chemistry, medicines, and dosages, but there are some other abbreviations used in a similar table in Duke et al. (2002a).

ADD = Attention deficit disorder
AHPA = American Herbal Products Association
ALA = Alpha-linolenic acid
APA = American Pharmaceutical Association
BO = Body odor
BPC = British Pharmacopoeia
BPH = Benign prostatic hypertrophy
CDC = Centers for Disease Control
cf. = Compare with
CNS = Central nervous system
COPD = Chronic obstructive pulmonary disorder
COX = Cyclooxygenase
COX-I = Cyclooxygenase-inhibitor (sometimes, more specifically COX-1 or COX-2-inhibitor)
cv = Cultivar
CVI = Chronic venous insufficiency
ED = Erectile dysfunction
ED50 = Effective dose at which 50% of subjects are "cured," "effected," "affected," or "altered"
e.g. = For example
EO = Essential oil
ERT = Estrogen replacement therapy
etc. = Et cetera
f = Folklore, not yet substantiated (following parentheses in "Activities" and "Indications" sections)
g = Gram
GI = Gastrointestinal
GLA = Gamma-linolenic acid
GMO = Genetically modified organism
GRAS = Generally recognized as safe
h (as an abandoned score for an activity or indication) = Homeopathic
H2O2 = Hydrogen peroxide
HCN = Hydrocyanic acid
hmn = Human
HPS = Hepatopulmonary syndrome
IBD = Inflammatory bowel disease
IBS = Irritable bowel syndrome
IC = Inhibitory concentration
IKKbeta = IkappaB-kinase-beta
iNOS = Inducible nitric oxide synthase
ipr = Intraperitoneal
ivn = Intravenous
l = Liter

LD50 = Lethal dose at which 50% of experimental population is killed
LDlo = Lowest reported lethal dose
MAOI = Monoamine oxidase inhibitor
MBC = Minimum bactericidal concentration
MDR = Multi drug resistant
mg = Milligram
MIC = Has been used by different sources differently = Minimum Inhibiting Concentration or Mean Inhibiting Concentration
ml = Milliliter
MLD = Minimum lethal dose; mean lethal dose
mM = Millimolar
MRSA = Methicillin-resistant *staphylococcus aureus*
mus = Mouse
NCI = National Cancer Institute
NO = Nitric oxide
ODC = Ornithine-decarboxylase
OPC = Oligomeric procyanidins
ORAC = Oxygen radical absorbance capacity
orl = Oral
oz = Ounce
PMS = Premenstrual syndrome
ppm = Parts per million
PSA = Prostate-specific antigen
rbt = Rabbit
scu = Subcutaneous
SF = Stephen Foster
SHBG = Sexual hormone binding globulin
SLE = Systemic lupus erythematosus
SOD = Superoxide dismutase
Tbsp = Tablespoon
TCM = Traditional Chinese medicine
tsp = Teaspoon
μl = Microliter
μM = Micromolar
UTI = Urinary tract infection
viz. = Videlicet
Vol = Volume
X = Solitary X in the title line of the herb following the scientific name means "don't take it without advice from an expert." Think of it as a skull-and-crossbones; X followed by serial number = PMID (PubMed ID number)
ZMB = Zero moisture basis

Hushpuppy

The Sad Saga of St. John

— **Jim Duke**

(Performed at Tai Graduation, June 2004; AHG, October 9, 2004)

I remember that sad day
In the year 2002
When I heard the TV say
St. John ain't good for you
I reckon they forgot
What you really oughta know
Two billion bucks of Zoloft
Placed second to placebo

And they also forgot
The good Doctor Cott
The first one to outline
The study design
But after Cott was gone
The design it was redrawn
With no redeeming graces
They took on basket cases.

They forgot the good St. John
Was the German's number one
With better Deutsche direction
They take John for their depression
What's the story we were fed
By our US Institute
They misled us instead
Saying herbs they ain't "sehr gut"

Our NInstitutes of Health
Misleads us local yuppies
They rob health to pay off wealth
Their studies are "hushpuppies"
And the press in all its wisdom
Missed one important score
St. John reduced orgasm
But Zoloft reduced it more

Hushpuppy
A most depressing tune
Hushpuppy

Keep howling at the Moon
St. John
Ain't you groaning in your grave
Pray John
Make the NIH behave

They muted the real news
The placebo beat the pill
But the news gave me the blues
Like liars always will

They forgot to tip us off
What I think we all should know
Two-billion-dollar Zoloft
Was poorer than placebo

That's what really was the news
Hope that everybody knows
Zoloft did really lose
Outdone by mere placebos

So I'm plowing up my herbs
They're much too hard to grow
Gonna move out to the suburbs
And grow me some placebo

Hushpuppy
A most depressing tune
Hushpuppy
Keep howling at the Moon
Hushpuppy
Did you want the herb to lose?
Your study
Was really just a ruse
St. John
Ain't you groaning in your grave
Pray John
Make the NIH behave

— Anonymous, 2002

Format of This Book

It is only natural that a believer in evolution would have an evolving format. The current format has evolved from my CRC *Handbook of Medicinal Herbs*, 2nd edition, which had evolved from my public domain Father Nature's Farmacy, online at the USDA. There are some new features here.

Lead Line: The lead line for each species remains pretty much the same. Common name — (*Scientific name* Author) followed by an X, a +, ++, or +++ representing the rather subjective safety scores, as in the past (X = don't take it, + = OK but probably not as safe as coffee, ++ = OK and probably as safe as coffee, and +++ = OK and probably safer than coffee); then the taxonomic family to which the species belongs. Family names are always in capital letters and end in "ACEAE." Like allopaths, health announcers, and reporters, I reserve the right to change my mind as new information comes in, positive or negative. I assembled this information, based on the published literature — no prescription implied or intended.

Synonyms: The next line may list some outdated synonyms, scientific names that at some time in the past also have been applied to this species.

Notes: The NOTES paragraph almost always begins with biblical quotes from various translations of the Bible (KJV = King James Version; RSV = Revised Standard Version, NWT = New World Translation). It is amazing how many uncopyrighted versions of the Bible in many languages are available on the Internet. And it is amazing how often the plant names (usually underlined in the quotation) are differently translated in the various versions. After these quotes follows a concise paragraph or two commenting on points of interest.

Common Names: Here I have compiled many, but by no means all, common names, often flagged as to language or country of origin. First comes a name in alphabetical order with a parenthetical citation of the country and/or language name or abbreviation. The country/language names/abbreviations always have the initial letter capitalized and subsequent letters in lowercase. These are followed by three-letter abbreviations (all capital letters) of the source(s), sometimes supplemented by journal citations or PubMed abstracts preceded by an X, to tell readers where I found these names. Sometimes one of the references, most frequently KAB, will list more than a hundred common names, from various parts of India and elsewhere, including dozens of Sanskrit names. In some such cases, I took at least one name from that source from each country or language. Few users will want to study all these common names unless it is a country they plan to visit. With an electronic version of the database, they could generate the names pertinent to the country they plan to visit. Often, the name itself will tell something about the plant or its medicinal uses. I have elected to use the standardized common names (Scn.) endorsed by the American Herbal Products Association (AH2) as the pivotal common name in the lead line for the entry. Occasionally, AH2 would offer an optional alternative common name, which I have abbreviated Ocn. (= other common name). Where there was no standardized common name, I often use the abbreviation Nscn. (= no standardized common name). In such cases, not uncommon in this biblical edition, I have consulted the USDA nomenclaturists and their database, trying to ensure that they and I will agree, and this might later influence the American Herbal Products Association should they decide to add some of these to a revised edition of their standardized common name book. With these common names flagged with geographic and linguistic handles, skillful database managers can readily print out mini-medicinal floras for many countries.

Activities: Following the common name paragraph are the activities reported for the herb, followed by a parenthetical scoring of the level of the efficacy of that activity. I have a subjective four-score evaluation of the efficacy of the activities f = strictly folklore; 1 = some animal, epidemiological, *in vitro*, or phytochemical studies support the efficacy (I actually feel that in many cases f may be better than 1); 2 = extracts of plant approved by Commission E, by the TRAMIL Commission, or demonstrated by human clinical trials; and a very rare 3 = herb itself clinically proven in human trials. If there is folkloric data (f), and animal or phytochemical support (1), and clinical proof for extracts or the rare clinical proof for the herb itself (3), as occasionally happens, e.g. with garlic, the efficacy score would read f123. Our computer programs can then print out the best scoring herbs for a given activity or indication. These efficacy scores are referenced like the common name, by three-letter abbreviations of my major sources in capital letters, and/or PubMed citation numbers preceded by X, and/or occasional shorthand journal citations.

Indications: Following the ACTIVITIES are the INDICATIONS reported for the herb, followed by a parenthetical scoring of the level of the efficacy of that indication. I have the same subjective efficacy scores f = folklore; 1 = some supporting animal, epidemiological, *in vitro*, or phytochemical studies; 2 = approved by Commission E, or the TRAMIL Commission, or proved in human clinical trials for simple herbal extracts; and a very rare 3 = herb itself clinically proven in human trials. Combinations of these four scores can appear, especially when many sources have been consulted. Not all sources consulted are cited but I attempt to cite my new source succinctly when the score goes up or down. These scores are referenced by three-letter abbreviations of my major sources, and/or PubMed serial citation numbers preceded by X, and/or an occasional shorthand journal citation. Occasionally, trying to reference every activity and indication, I have to resort to bibliographic shorthand; hence, there will be a cryptic journal citation (especially of journals not covered by PubMed), with an abbreviation for the journal, followed by the volume number and the page number, as in my CRC *Handbook of Medicinal Herbs*. Some examples are:

EB12:368 = *Economic Botany*, Vol. 12, p. 368.

FT67:215 = *Fitoterapia*. Vol. 67, p. 215.

ACT9:251 = *Journal Alternative & Complementary Medicine*, Vol. 9, p. 251.

Dosages: This entry has evolved significantly since CR2, the CRC *Handbook of Medicinal Herbs* (2nd edition, 2002). First, I have added a third scoring element for the food farmacy potential of the plant. FNFF stands for Father Nature's Food Farmacy. Here is the FNFF scoring pattern:

FNFF = X = I found nothing credible suggesting the plant as food.

FNFF = ? = Very questionable survival food.

FNFF = ! = Survival food or little known but locally important; not in United States supermarkets.

FNFF = !! = Important food in some parts of world; not in major supermarkets.

FNFF = !!! = Important enough in the world to be in many United States supermarkets.

Following the food farmacy score, there will be dosages from various sources using the same reference citations. Then there will be folkloric bullets suggesting how various countries and ethnic groups report using the plant. With this new FNFF scoring, my computer can rank the herbs for safety, efficacy, and food farmacy potential. In these litigious days, I feel safer recommending food farmacy to friends and family. I think food farmacy should be the first line of attack when a simple new medical problem arises.

Downsides: Under this heading I often report contraindications, interactions, and side effects, just as in the CRC *Handbook of Medicinal Herbs* (Edition 2, 2002).

Natural History: Because of my increasing interest in zoopharmacognosy and natural history, I added this cubbyhole to permit inclusion of pertinent facts on the natural history. Which animals are using it besides us?

Extracts: Here I try to include news on chemicals or extracts of the plant that have proven effects.

Catalog of "Faith-Based" Farmaceuticals

CILICIAN FIR (*ABIES CILICICA* (ANTOINE & KOTSCHY) CARRIÈRE) ++ ABIETACEAE

NOTES (CILICIAN FIR):

The beams of our house are cedar, and our rafters of fir.

Song of Solomon 1:17 (KJV)

The beams of our house are cedar, our rafters are pine.

Song of Solomon 1:17 (RSV)

The beams of our grand house are cedars, our rafters juniper trees.

Song of Solomon 1:17 (NWT)

Off to a taxonomic bad start. What do you think? What timber was used in the rafters, fir as in KJV, pine as in RSV, or juniper as in the NWT? In this exceptional case, each version has its own rendering for the plant name underlined. No one can say for sure which version is correct. There are no voucher specimens. You will hear me lament that fact many times. And even today, the names fir, juniper, and pine mean different things to different people. If I include them all, I will have more tentative species in my faith-based herbal here. Zohary (1982) lists *Abies cilicica, Cupressus sempervirens,* and *Juniperus excelsa* as candidates for the word *berosh*, found more than 30 times in the scriptures, and interpreted to mean coniferous trees with small scale-like or short linear leaves rather than pine-like needles. Amazingly, he concludes by considering *berosh* a collective name for all three. Perhaps not so amazing; we have popular scrub oak and scrub pine concepts that are suprageneric here in the United States, and our English word *conifer* embraces more kinds of gymnosperms than Zohary's *berosh*. When Zohary encountered *berosh* associated with the word for Lebanon or *erez,* he thinks they mean *Abies cilicica,* which grows in Lebanon mixed with cedar. "The great timber negotiations between King Solomon and Hiram of Tyre undoubtedly included this outstanding species of Lebanese tree, whose southernmost limit of distribution is today the village of Slenfe (at a latitude of about 34° North)". Jane Philips (1958) noted that the tree still occurred near Beirut and Tripoli. Private growers of the trees said they were used for medicine, the resin used for cough medicines and salves. Twigs or dried leaves were boiled up in cough syrups. To prepare the salve, leaves were ground up in a mortar to apply to wounds. Algerians are said to sprinkle powdered leaves in butter as a vulnerary (Philips, 1958).

COMMON NAMES (CILICIAN FIR):

Berosh (Eng.; BIB); Cilician Fir (Eng.; USN); Cilicische Tanne (Ger.; USN); Nscn.

Activities (Cilician Fir):

Antiseptic (1; X11962214); Bactericide (1; X10548751); Vulnerary (f1; BIB; X10548751).

Indications (Cilician Fir):

Bacillus (1; X10548751); Bacteria (1; X10548751); Cough (f1; BIB; X10548751); Enterobacter (1; X10548751); Escherichia (1; X10548751); Infection (1; X11962214); Klebsiella (1; X10548751); Listeria (1; X10548751); Mycobacterium (1; X10548751); Proteus (1; X10548751); Pseudomonas (1; X10548751); Staphylococcus (1; X10548751); Wound (f1; BIB; X10548751).

Dosages (Cilician Fir):

FNFF = ?

- Algerians are said to sprinkle powdered leaves in butter as a vulnerary (BIB).
- Lebanese use the resin or leaf/twig decoction in cough syrups and wound salves (BIB).

INDIAN GUM ARABIC TREE (*ACACIA NILOTICA* (L.) WILLD. EX DELILE) ++ FABACEAE

Synonyms:

Acacia arabica (Lam.) Willd.; *Acacia vera* Willd.; *Mimosa arabica* Lam.; *Mimosa nilotica* L.

Notes (Indian Gum Arabic Tree):

And the angel of the Lord appeared unto him in a flame of fire out of the midst of a bush; and he looked, and behold, the bush burned with fire, yet it was not consumed.

Exodus 3:2 (KJV)

And the angel of the Lord appeared to him in a flame of fire out of the midst of a bush; and he looked, and lo, the bush was burning, yet it was not consumed.

Exodus 3:2 (RSV)

Then Jehovah's angel appeared to him in a flame of fire in the midst of a thornbush. As he kept looking; why, here the thornbush was burning with the fire, and the thornbush was not consumed.

Exodus 3:2 (NWT)

In my first Bible book, I side with the Moldenkes, who concluded that the burning bush was an *Acacia*. But the NWT renders it thornbush. Zohary (1982) argues convincingly that the plant called *sneh* should be translated *Senna*, not *Acacia nilotica* as Moldenke and Tristam concluded, or *Rubus* as other scholars had concluded. Why? Because *Acacia nilotica* and *Rubus* do not occur on the Sinai. He concludes that "the most plausible explanation for *sneh* is *Cassia senna* named sene in all Arabic-speaking countries." The taxonomic name has changed to *Senna alexandrina,* but *sene* persists in most Arabic quarters. However, I still include the *Acacia*. If the burning bush was thorny, as inferred only in the NWT above, then it cannot be *Cassia senna*.

At the crossroads of the African, Asian, and European continents, the Holy Land serves as the bridge from Africa through Egypt, to Asia and Europe. Thus, the medicinal wisdom of the early Arabs, Copts, Hebrews, and Moslems assume great importance because of their empirical antiquity. The scriptures tell us that Abraham's grandson Jacob immigrated into Egypt when he learned that his long-lost son Joseph was prospering there. And that led to what I will call the Jacobian exchange (introduction of middle-eastern species to Egypt), almost two millennia before the Colombian exchange (mixing the flora of the old and new worlds). Jacob's stay was said to be a 400-year odyssey, so he took with him saplings of certain *Acacias* not native to Egypt. The upright frames of the tabernacles were to be built of *Acacias* (BI2). Which *Acacia* can be argued for eons. Thus, man has been moving species around for millennia, sometimes obscuring their true nativity. Some scholars equate this *Acacia* with the thornbush of Exodus, equating the fire with the parasite *Loranthus acaciae*. Roasted seed kernels provide a dye for black strings worn by Nankani women.

Common Names (Indian Gum Arabic Tree):

Abadonui (Dahomey; KAB); Acacia (Eng.; CR2); Acacia d'Egypte (Fr.; BOU); Ajabaksha (Sanskrit; KAB); Akakia (Arab.; Iran; NAD); Ammughilam (Arab.; NAD); Amraya (Mauritania; UPW); Amur (Mauritania; Sahara; KAB); Amura (Mali; UPW); Australian Wattle (Eng.; NPM); Babal (Guj.; NAD); Babbar (Sin.; KAB); Bablia (Guj.; KAB); Babboola (Sanskrit; MPI); Babbuli (Kan.; KAB); Babhul (Mar.; KAB); Babhula (Bom.; Mar.; Sin.; KAB; NAD); Babla (Beng.; Hindi; Pun.; KAB; MPI; NAD); Babli (Mun.; KAB); Babola (Mal.; KAB; NAD); Babul (Eng.; Mah.; Nepal; Kum.; Kon.; Urdu; CR2; KAB; NAD; SUW); Abadanui (Dahomey; UPW); Babul Acacia (Eng.; Ocn.; AH2); Babulla (Ayu.; AH2); Babulo (Oriya; KAB); Babur (Nasirabad; Sibi; KAB); Bagana (Bambara; Ivo.; Malinki; KAB; UPW); Bagaruwa (Hausa; Kano; Sokoto; KAB); Bambolero (Lambadi; KAB); Bamura (Jubbulpore; KAB); Bani (Baraba; Surai; KAB); Bara Na (Fulah; KAB); Barbara (Sanskrit; NAD); Barbaramu (Tel.; KAB; NAD); Bauni (Kan.; NAD); Baval (Porebunder; KAB); Bebned (Wolof; KAB); Black Babul (Eng.; KAB; WO2); Boina (Wassula; KAB); Bois d'Arariba Rosa (Fr.; KAB); Diabbe (Sarakolet; KAB); Egyptian Acacia (Eng.; BOU); Egyptian Thorn (Eng.; BOU; UPW); Espinheira Preta (Port; Guinea-Bissau; UPW); Gabaruwa (Zaria; KAB); Gabur (San.; KAB); Gambia Pods (Eng.; UPW); Gaodi (Peuhl; Tuculor; KAB); Garad (Niger; Nig.; UPW); Gaudi (Gambia; UPW); Gobli; (Mysore; KAB); Gomma da India (Por.; KAB); Gommier d'Egypte (Fr.; BOU); Gommier Rouge (Fr.; BOU; UPW); Gorzia (Ghana; UPW); Gum Arabic (Eng.; SUW); Gurti (Arab.; BOU); Huanlongkyain (Burma; KAB); Indian Gum Arabic Tree (Eng.; Scn.; AH2); Iramangandam (Tam.; KAB); Jali (Kan.; NAD); Kala Babli (Mah.; NAD); Kalikikar (Dec.; NAD); Kaloababal (Guj.; MPI; NAD); Kambani (Bobo; Uper Volta; UPW); Kara (Togo; UPW); Karat (Arab.; GHA); Karemugilan (Iran; NAD); Karijali (Kan.; MPI; NAD); Karuvael (Tam.; NAD); Karuvelakam (Mal.; KAB); Karuvelum (Mal.; Tam.; NAD); Kharemughilam (Iran; KAB); Kikar (Beng.; Hindi; Pun.; Rendli; KAB; NAD); KôBè (Guinea; UPW); Kommi (Greek; KAB); Kusatregon (Gurma; KAB); Nallatumma (Tel.; MPI; NAD); Qarad (Arab.; BOU); Qarat (Arab.; GHA); Red Thorn (Eng.; UPW); Sak (Kas.; Pun.; NAD); Sake (Marke; KAB); Sant (Arab.; BOU); Scorpion Mimosa (Eng.; UPW); Shameeruku (Kon.; MPI; NAD); Shittim (Heb.; KAB); Shoka Masrya (Arab.; BOU); Shoka Qibttya (Arab.; BOU); Siludi (Fulah; KAB); Tamak (Tuareg; KAB); Tulh (Arab.; Dho.; Oman; GHA); Tuma (Tel.; NAD); Unmughilam (Arab.; KAB; NAD); Vabboola (Sanskrit; NAD); Vabbula (Sanskrit; NAD).

Activities (Indian Gum Arabic Tree):

Alexiteric (f; KAB); Algicide (1; ZUL); Amebicide (1; ZUL); Analgesic (1; X8982438); Anthelmintic (f; KAB); Antiaggregant (1; X9251908); Anticarcinogenic (1; WO3; X12616620); Antiedemic (1; X8982438); Antihepatitic (1; PR14:510); Antihistaminic (1; ZUL); AntiHIV (1; X10189947); Antihypertensive (1; X10594935); Anti-inflammatory (f1; X8982438); Antimalarial (1; X10479756); Antimutagenic (1; WO3; X12616620; X11850969); Antiplasmodial (1; X10479756); Antiplatelet (1; X9251908); Antioxidant (1; X11837686); Antiseptic (1; WO3); Antispasmodic (1; X10594935); Antitussive (f; BIB); Aphrodisiac (f; KAB; MPI; ZUL); Astringent (f; GMH; PH2; SUW); Bactericide (1; ZUL; X15476301); Calcium-Antagonist (1; X9251908); Chemopreventive (1; X11850969); Decongestant (f; BIB; EB22:173); Demulcent (f; BIB; DEP; SUW); Expectorant (f; KAB; MPI;

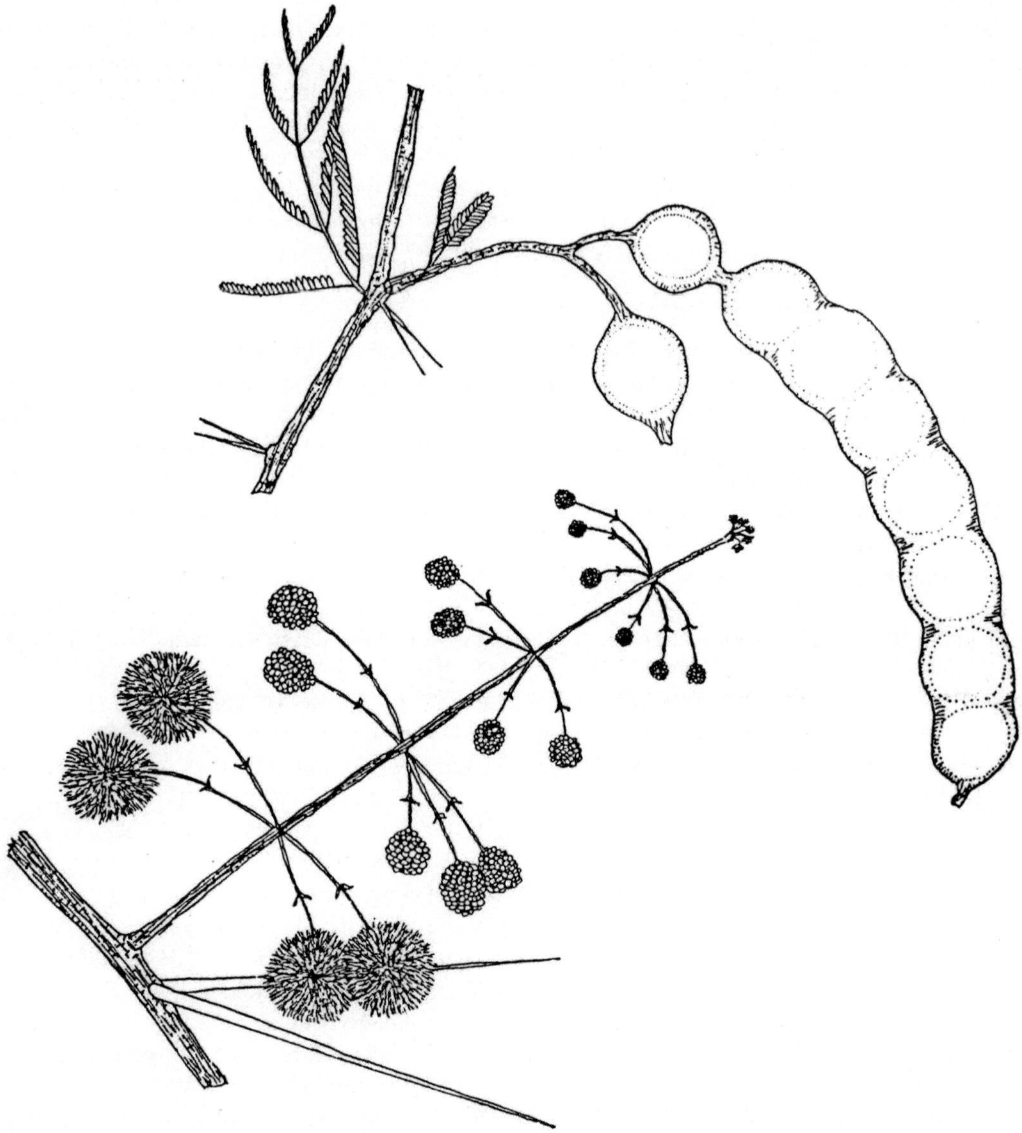

FIGURE 1.1 Indian Gum Arabic Tree (*Acacia nilotica*).

NAD); Fungicide (1; WO3); HCV-Protease Inhibitor (1; PR14:510); Hemostat (f; DEP; NAD); Hepatotonic (f; KAB); Hypertensive (1; X10594939); Hypoglycemic (1; ZUL); Hypotensive (f1; BOU; ZUL; X10594935); Lactagogue (f1; BIB; UPW; 15283686); Mastogenic (1; X15283686); Molluscicide (1; ZUL); Neurostimulant (f; BIB; UPW); p-Glycoprotein Inhibitor (1; X12748979); Plasmodicide (1; X10479756); Protease Inhibitor (1; X11054840); Protisticide (1; ZUL); Spasmogenic (1; X10594939); Stimulant (f; BIB); Taenicide (1; ZUL); Teratologic (f; ZUL); Tonic (f; DEP; SUW); Vasoconstrictor (1; X10594939).

Indications (Indian Gum Arabic Tree):

Alopecia (f; WO3); Ameba (f1; BOU; ZUL); Aphtha (f; NAD); Ascites (f; KAB); Asthma (f; WO3); Bacteria (1; X15476301); Biliousness (f; KAB); Bleeding (f; BIB; NAD); Boil (f; GHA); Bronchosis (f; KAB; WO3); Burn (f; SKJ; WO3); Cancer (f; BIB; JLH); Cancer, ear (f; JLH); Cancer, eye (f; JLH); Cancer, liver (f; JLH); Cancer, spleen (f; JLH); Cancer, testes (f; JLH); Cataract (f; GHA); Catarrh (f; GHA; HH2); Childbirth (f; DEP); Chill (f; ZUL); Cholecystosis (f; BIB; EB22:173); Cholera (f; SKJ; WO3); Cold (f; GHA); Colic (f; KAB); Condyloma (f; BIB); Congestion (f; BIB); Conjunctivosis (f; DEP; NAD); Cough (f; DEP; KAB; NAD); Cramp (f; BOU); Cystosis (f; DEP); Dermatosis (f; BOU; WO3); Diabetes (f1; BOU; DEP; GHA; SUW; WO3; ZUL); Diarrhea (f; GHA; GMH; PH2; SUW); Dysentery (f; BIB; DEP; SUW); Dyslactea (1; X15283686); Dyspepsia (f; ZUL); Dysuria (f; KAB); Edema (1; X8982438); Enterosis (f1; DEP; X15476301); Fever (f; BIB; BOU; UPW); Flu (1; FNF); Fracture (f; KAB); Fungus (1; WO3); Gastrosis (f; DEP); Gingivosis (f; BOU; DEP; PH2); Gonorrhea (f1; DEP; KAB; ZUL); Hemorrhoid (f; BIB; KAB; PH2); Hepatosis (f1; BIB; WO3; PR14:510; X11054840); High Blood Pressure (f1; BOU; ZUL); HIV (1; X10189947); Hypersalivation (f; DEP); Impotence (f; NAD; UPW); Induration (f; BIB; JLH); Infection (1; WO3; ZUL; X15476301); Inflammation (1; PH2; X8982438); Insanity (f; KAB); Leprosy (f; KAB); Leukoderma (f; KAB); Leukorrhea (f; DEP; NAD); Menorrhagia (f; DEP); Micromastia (1; X15283686); Mucososis (f; PH2); Mycosis (1; WO3); Odontosis (f; PNC); Ophthalmia (f; BIB; JLH; KAB); Orchosis (f; BIB); Otosis (f; BIB; JLH); Pain (1; X8982438); Pharyngosis (f; KAB; PH2); Pneumonia (f; ZUL); Prolapse (f; NAD); Proctosis (f; DEP; UPW); Puerperium (f; DEP); Pulmonosis (f; ZUL); Salmonella (1; X15476301); Sclerosis (f; BIB; JLH); Smallpox (f; BIB); Snakebite (f; DEP); Sore (f; DEP; UPW); Sore Throat (f; DEP; SUW; WO2); Spermatorrhea (f; KAB); Splenosis (f; JLH); Staphylococcus (f; ZUL); Stomachache (f; UPW); Stomatosis (f; DEP; PH2; UPW); Strangury (f; KAB); Swelling (f1; GHA; X8982438); Syphilis (f; BIB; WO3); Toothache (f; GHA; UPW; ZUL); Tuberculosis (f; BIB; UPW); Typhoid (f; BIB); Urethrosis (f; KAB); Urogenitosis (f; NAD); Uterosis (f; DEP; KAB); Vaginosis (f; KAB; PH2); Venereal Disease (f1; DEP; NAD; X11483371); Virus (1; X11054840); Worm (1; ZUL); Wound (f; UPW).

Dosages (Indian Gum Arabic Tree):

FNFF = !!

Tender young pods eaten as vegetable; ripe seed kernels roasted and eaten, made into wine (TAN); gum used in confectionary (TAN). Tender pods and shoots used as vegetable, and to stimulate milk production. Roasted seed kernels sometimes used for flavoring; the raw seed is good animal feed. (BIB). Konkani make candy by drying the gum with butter, spices, and balling up with sugar (KAB).

- Arabs inhale smoke from burning pods for cold (GHA).
- Arabs soak crushed seeds overnight in water or fresh milk and drink for diabetes (GHA).
- Asian Indians use bark juice in mother's milk as eyedrops for conjunctivosis (NAD).
- Asian Indians fry gum in ghee for impotence (NAD).
- Asian Indians consume the gum (not converted to sugar) for diabetes (NAD).
- Asian Indians gargle the leaf decoction for gingivosis, sore throat (NAD).
- Asian Indians take pulped leaves for diarrhea and dysentery, anally or orally (NAD).
- Ayurvedics consider the bark alexipharmic, anthelmintic, astringent, and use it for ascites, biliousness, bronchosis, burning sensations, cough, diarrhea, dysentery, dysuria, leukoderma, and piles (KAB).
- Egyptian Nubians believe diabetics can take high carbohydrate foods as long as they regularly take powdered pods (BOU).
- Guinea natives take gum/resin for chest and throat ailments, dysentery, and eye problems (KAB).
- Konkani take one tola of leaves with four mashas cumin, two tolas sugar, eaten or drunk with milk for bloody spermatorrhea (KAB).
- Masai use bark as aphrodisiac and neurotonic (UPW).
- Nigerians suck the gum for oral ulcers (UPW).
- Omani mix resin with egg white as collyrium for cataracts (GHA).

- Rajputanans bruise the leaves to apply to sore eyes in children (KAB).
- Senegalese chew antiscorbutic bark and take bark tea for diarrhea, dysentery, and toothache (UPW).
- Unani consider the leaves astringent, cerebrotonic, febrifuge, hepatotonic, and useful for gonorrhea, leukoderma, and strangury (KAB).
- Unani consider all parts of the plant aphrodisiac (KAB).

Downsides (Indian Gum Arabic Tree):

None covered (AHP, KOM). Large internal doses may lead to constipation and dyspepsia (PH2).

Natural History (Indian Gum Arabic Tree):

Older shrubs are very important in diet of impala and kudu (X15278425). Lac insects often occur on the tree, with resultant production of lac resin and shellac (BIB).

TALH (*ACACIA SEYAL* DELILE.) ++ FABACEAE

Notes (Talh):

I will plant in the wilderness... the Shittah tree.

Isaiah 41:19 (KJV)

I tend to side with Zohary, who identified the shittah as *Acacia tortilis* (q.v.). Still, according to Walker (1957), the Shittah tree is mentioned in the Bible only once but its wood is referred to many times as *shittim*, which is the plural of *shittah* in Hebrew. It was natural for Moses to turn to shittim to build the Ark of the Covenant and the Tabernacle. No one is sure which species of *Acacia* was meant.

Common Names (Talh):

'Alk (Arab.; BOU); Abre à Gomme (Fr.; BOU); Buffalo Thorn (Eng.; UPW); Bulkia (Gambia; UPW); Daci (Niger; UPW); Dedera èl Beida (Arab.; Sen.; UPW); Goga (Ghana; UPW); Gomme Friable (Fr.; UPW); Gum Talha (Eng.; Trade.; AH2); Mimosa Epineux (Fr.; UPW); Sadra Bed (Arab.; Mauritania; UPW); Sasé (Upper Volta; UPW); Seyal (Arab.; BOU); Seyal Acacia (Eng.; FAC); Shittah Tree (Eng.; BOU); Shittim Tree (Eng.; Ocn.; AH2; BOU); Sittim Wood (Eng.; FAC); Suakim (Trade; UPW); Talca (Ocn.; AH2); Talakh (Ber.; BOU); Talh (Scn.; AH2); Tamat (Ber.; BOU); Tefi (Ber.; BOU); Teleh (Arab.; Mali; UPW); Thala (Ber.; BOU); Thirsty Thorn (Eng.; BOU; UPW); Thirty Thorn (Eng.; USN); Tutampolel (Nig.; UPW); Whistling Tree (Ocn.; AH2); Whistling Wood (Eng.; USN); White Galled Acacia (Eng.; UPW); White Whistling Thorn (Eng.; UPW).

Activities (Talh):

Analgesic (f; BOU); Aphrodisiac (f; UPW); Astringent (f; UPW); Diuretic (f; UPW); Emollient (f; UPW); Febrifuge (f; UPW); Hemostat (f; UPW); Insectifuge (f; UPW); Pediculifuge (f; UPW); Stimulant (f; UPW).

Indications (Talh):

Biliousness (f; UPW); Aphrodisiac (f; UPW); Bleeding (f; UPW); Burn (f; UPW); Cancer (f1; UPW); Cold (f; BOU); Conjunctivosis (f; UPW); Diarrhea (f; UPW); Dysentery (f; UPW); Enterosis (f; DAW); Fever (f; BOU; UPW); Gastrosis (f; BOU); Gonorrhea (f; UPW); Headache (f; UPW); Impotence (f; UPW); Inflammation (f; BOU); Jaundice (f; UPW); Leprosy (f; UPW); Lice (f; UPW); Ophthalmia (f; UPW); Pain (f; BOU; UPW); Puerperium (f; BOU); Respirosis (f; BOU); Rheumatism (f; BOU; UPW); Snakebite (f; UPW); Syphilis (f; UPW); Ulcer (f; BOU); Venereal Disease (f; UPW).

Dosages (Talh):

FNFF = !!

The tree yields a good quality gum (but inferior to that of *A. senegal*); said to be edible (BIB; BOU; FAC; UPW).

- Ivory Coastals mix with *Acacia sieberana* for intestinal ailment (BIB).
- Masai consider the bark a stimulant, giving bark infusion to feverish children (UPW).
- Nigerians use the wood smoke against insects, lice, etc. (UPW).
- North Africans use wood smoke as a fumigant for rheumatic pain, and to protect mothers against colds and fever 2 weeks after parturition (BOU).
- North Africans use the edible gum for respiratory inflammations and rheumatism, the bark and leaves for gastric ulcers (BOU).
- Senegalese mix powdered root with hedgehogs' ventral parts as an aphrodisiac (UPW).
- South Africans mention the gum's use as an emollient and astringent for colds, diarrhea, hemorrhage, and ophthalmia (WBB).
- Sudanese direct smoke from the heartwood toward rheumatic pain; the women appreciating the aroma and reddish color it imparts to their skin (UPW).
- Sudanese mix concentrated bark decoction with butter for conjunctivitis and headache (UPW).
- Tanganyikans use the bark as a stimulant (BIB), taking the root for gonorrhea (UPW).

Natural History (Talh):

Phlebotomus orientalis, the vector of visceral leishmania in the Sudan, is typically associated with *Acacia seyal* and *Balanites aegyptiaca* vegetation (X11370250). As with many *Acacias*, the sweet-scented flowers attract bees (UPW). Swellings at the base of the thorns are called ant-galls, and, when hollow, are invaded by ants; the hollowed growths sometimes whistle in the wind (UPW). Like many *Acacia* species, seeds of this one are subject to predation by bruchid beetles (X8169432). Birth seasons of some monkeys appear to be timed to availability of surplus energy and protein. Patas monkeys' high locomotive ability may enable them to obtain more energy from seeds of *Acacia seyal* and gums of *A. sieberiana*, and more protein from grasshoppers and seeds of *A. seyal* in the mid-dry season than the tantalus monkeys. Availability of seeds during the dry season may exert the dominant influence on timing of birth not only in patas, but also in savanna monkeys (*Cercopithecus aethiops*), which include the tantalus monkeys (X11132111).

Extracts (Talh):

Ethanolic extract reduces tumors (UPW).

BIBLICAL ACACIA (*ACACIA TORTILIS* SSP. *RADDIANA* (SAVI) BRENAN) ++ FABACEAE

Synonyms:

Acacia raddiana Savi

Notes (Biblical Acacia):

> **And you shall make upright frames for the tabernacle of acacia wood.**
>
> **Exodus 26:15 (RSV)**

Zohary is convinced that the "common acacia" is the most correct identification of the shittah whose wood was used in constructing the Tabernacle. He argues that the other native species — *A. albida, A. laeta, A. negevensis,* and *A. tortilis* — neither conform as well to the scriptural text, nor are they so suitable for construction. According to Zohary, *A. albida* and *A. laeta* are absent in the Sinai, where the Israelites wandered. *A. negevensis* and *A. tortilis* are either unsuitable for construction or are rare in the Sinai. Of 24 biblical references to *shittim,* 19 are to the acacia tree and 5 to places associated with the tree. The Arabic word *sunt* designates acacia in Arabia, Egypt, and southern Israel. Zohary argues that *sunt* is the linguistic equivalent of the Hebrew *shittah* (ZOH).

Common Names (Biblical Acacia):

Abser (Ber.; BOU); Abzac (Ber.; BOU); Aluki (Mali; UPW); Cilluki (Upper Volta; UPW); Faux Gommier (Fr.; UPW); Gommier de Tunisie (Fr.; BOU); Hares (Arab.; BOU); Samr (Nig.; UPW); Sayal (Arab.; BOU); Seyal (Niger; UPW); Shittah (Heb.; ZOH); Somer (Arab.; Yemen; X15890471); Sunt (Arab.; Egypt; ZOH); Tadjdjart (Ber.; BOU); Tahi (Arab.; Sen.; UPW); Talh (Arab.; BOU); Talha (Arab.; Mauritania; UPW); Tamat (Ber.; BOU); Tihi (Ber.; BOU).

Activities (Biblical Acacia):

Antidote (f; UPW); Antiedemic (f; UPW); Antiseptic (f; BOU); Aphrodisiac (f; UPW); Astringent (f; BOU); Febrifuge (f; UPW); Vulnerary (f; BOU; UPW).

Indications (Biblical Acacia):

Allergy (f; UPW); Avitaminosis (f; UPW); Convulsion (f; UPW); Dermatosis (f; UPW); Diarrhea (f; BOU); Edema (f; UPW); Enterosis (f; X15890471); Fever (f; UPW); Gastrosis (f; X15890471); Hepatosis (f; BOU); Impotence (f; UPW); Infection (f; BOU); Jaundice (f; BOU); Ophthalmia (f; BOU); Pulmonosis (f; BOU); Wound (f; BOU).

Dosages (Biblical Acacia):

FNFF = !
In times of scarcity, pods and seeds may be eaten by humans (UPW).

- Bambara in Mali make a draught from young plants with *Piliostigma* for convulsions (UPW).
- Guinea and Senegal natives use powdered bark to dust on dermatosis and as vermifuge (UPW).
- North Africans dissolve the gum in water for jaundice, ophthalmia, and pulmonosis (BOU).
- Senegalese take bark infusion for fever (UPW).
- Senegalese poultice the leaves with cowpea leaves onto allergic skin conditions and edema (UPW).

Natural History (Biblical Acacia):

All the PubMed abstracts relate not to medicine, but to the symbiotic rhizobial bacteria associated with the roots. In Israel, the tree is host to a bruchid beetle *Caryodon gonagra*, which is a pest of stored groundnuts in Africa (UPW).

CORN COCKLE (*AGROSTEMMA GITHAGO* L.) + CARYOPHYLLACEAE

Synonyms:

Lychnis githago Scop.

Notes (Corn Cockle):

If my land cry against me … Let thistles grow instead of wheat and cockle instead of barley.

Job 31:40 (KJV)

FIGURE 1.2 Corn Cockle (*Agrostemma githago*).

This weed, rarely encouraged for its attractive flower, is more often regarded as a poisonous weed of wheat. In the United States, it is declared a noxious weed seed in Alabama, Arkansas, Delaware, Florida, Georgia, Indiana, Kentucky, Louisiana, Massachusetts, Maryland, Maine, Mississippi, North Carolina, New Hampshire, New Jersey, New York, Ohio, Oklahoma, Pennsylvania, Rhode Island, South Carolina, Tennessee, Texas, Vermont, and West Virginia.

Common Names (Corn Cockle):

Bolderik (Dutch; EFS); Corn Campion (Eng.; GMH); Corn Cockle (Eng.; EFS); Corn Pink (Eng.; GMH); Darnell (Eng.; GMH); Gerzeau (Fr.; EFS); Gith (Eng.; GMH); Githage (Eng.; GMH); Gittaione (It.; EFS); Karamuk (Tur.; EFS); Kornrade (Ger.; EFS); Lolium (Eng.; JLH; GMH); Mazzettone (It.; EFS); Neguillon (Sp.; EFS); Nielle des Blés (Fr.; EFS); Nigela dos Trigos (Por.; EFS); Nigella (Eng.; GMH); Raden (Ger.; JLH); Ray (Eng.; GMH); Tare (Eng.; GMH); Yetón (Arg.; EFS); Zizany (Eng.; GMH).

Activities (Corn Cockle):

Anesthetic (1; CRC); Antileukemic (1; X11408934); Antimitogenic (1; X11255109); Antimycotic (1; PH2); Antiproliferant (1; X11408934); Antiviral (1; X11408934); Apoptotic (1; X11408934); Bacteriostat (1; CRC; WO2); Cytotoxic (1; X14648395); Diuretic (f; BIB; CRC; EFS); Embryotoxic (1; X11453369); Emmenagogue (f; BIB; CRC; EFS); Expectorant (f; BIB; CRC; EFS); Hemolytic (1; HH2); Hypotensive (1; WO2); Narcotic (1; WO2); Ribosome-inactivator (1; X11408934); Toxic (f1; PH2; WO2); Vermifuge (f; BIB; CRC; EFS); Vulnerary (1; FNF).

Indications (Corn Cockle):

Aposteme (f; CRC; JLH); Cancer (f; CRC); Cancer, uterus (f; JLH); Cough (f; PH2); Dermatosis (f; PH2); Dropsy (f; CRC; GMH); Edema (f; JLH; PH2); Exanthemata (f; CRC; WO2); Infection (1; HH2); Fungus (1; HH2); Gastrosis (f; BIB; CRC; PH2); Hemorrhoid (f; CRC; WO2); Induration (f; JLH); Jaundice (f; CRC; EFS; GMH; WO2); Leukemia (1; X11408934); Mycosis (1; HH2); Paralysis (f; BIB; CRC); Swelling (f; JLH; PH2); Tumor (f; CRC); Uterosis (f; JLH); Virus (1; X11408934); Wart (f; JLH); Worm (f; BIB; PH2); Wound (1; FNF).

Dosages (Corn Cockle):

FNFF =?
Young leaves "used as a vegetable with vinegar and bacon for emergency food" (FAC). Homeopathic only (HH2; PH2).

- Germans burned seeds until black and applied them to cancer (JHL).

Downsides (Corn Cockle):

No health hazards known in conjunction with proper administration of designated homeopathic dosages (PH2). Seed toxic, 2–3 g considered harmless to humans, over 5 g potentially lethal; signs of intoxication include colic, conjunctivosis, cramps, delirium, diarrhea, dizziness, headache, lachrymation, mucositis, and restlessness (PH2). The FDA Poisonous Plant database listed 95 citations as of November 2004.

Natural History (Corn Cockle):

The large purple flowers have pale streaks, the "honey guides." The long calyx teeth seem to serve as landing strips for butterflies and moths, which pollinate the flower. Nectar is secreted at the bottom of the tube, too deep for bees. Anthers shed their pollen before the stigmata mature (GMH).

Extracts (Corn Cockle):

LD50 (saponin mix) = 750 mg/kg orl mus HH2. LD50 (saponin mix) = 2.3 mg/kg ivn rat HH2. LD50 (saponin mix) = 50 mg/kg orl rat HH2. Sprouts contain allantoin.

HOLLYHOCK (*ALCEA ROSEA* L.) + MALVACEAE

SYNONYMS:

Althaea ficifolia (L.) Cav.; *Althaea rosea* (L.) Cav.

NOTES (HOLLYHOCK):

Can that which is tasteless be eaten without salt, or is there any taste in the slime of the purslane?

Job 6:6–7 (RSV)

As Zohary notes, biblical scholars suggest both *Alcea* and *Malva* as the best candidates for the Hebrew *halamuth*. Both genera are common Israeli herbs, in early winter, used as edible potherbs. In toto, Zohary argues stronger for *Malva* than *Alcea*, and who knows, so many thousands years later, what was meant by these non-botanists of days gone by. Zohary's picture leads me to believe that he is talking about the same hollyhock that keeps coming up back by my greenhouse, which once was called *Althea rosea*. Anthropologist Jane Philips says it is one of the most widely used folk cures, both in Lebanon and America (BIB). It is in Zohary's book that I first saw "the slime of the *purslane*" (ZOH), and I have personally seen what I call purslane (*Portulaca oleracea*) on the streets of Tel Aviv.

COMMON NAMES (HOLLYHOCK):

Alcea (Sp.; USN); Alcée (Fr.; KAB); Alcée Rose (Fr.; KAB); Altaia (Greek; KAB); Altea (Peru; EGG); Alteia Rosada (Ma.; JFM); Althée (Fr.; KAB); Althée Rose (Fr.; KAB); Augenpappel (Ger.; KAB); Bastun ta San Giusepp (Malta; KAB); Bâton de Saint Jacques (Fr.; KAB); Baummalve (Ger.; EFS; KAB); Baumrose (Ger.; KAB); Binafsa (Arab.; GHA); Black Hollyhock (Eng.; FAC); Bourdon de Saint Jacques (Fr.; KAB); Brandrose (Ger.; KAB); Braunrose (Ger.; KAB); Chernaya Roja (Rus.; KAB); Ehrenrose (Ger.; KAB); Felriss (Ger.; KAB); Feuerbluete (Ger.; KAB); Garden Hollyhock (Eng.; GMH); Gartenmalve (Ger.; KAB); Glochenpappel (Ger.; KAB); Glockrose (Ger.; KAB); Guimauve Rose Trémière (Fr.; EFS; KAB); Gülhatmi (Tur.; EFS); Halsrose (Ger.; KAB); Hanna Aoi (Japan; TAN); Herbstrose (Ger.; KAB); Herzleuchte (Ger.; KAB); Hochleuchte (Ger.; KAB); Hock Herb (Eng.; KAB); Hollyhoke (Eng.; JLH); Hollyhock (Eng.; Scn.; AH2; CR2; USN); Kohlrose (Ger.; KAB); Malva (Peru; Sp.; EFS; EGG); Malva Arborea (Sp.; KAB); Malva de India (Por.; EFS); Malva de Jardin (Sp.; JFM); Malva de San Jose (Sp.; JFM); Malvaííso (Por.; EFS); Malva Jaspeada (Chile; JLH); Malvaloca (Sp.; JFM; KAB; USN); Malva Real (Chile; Peru; Ven.; EGG; EFS; JFM; JLH; KAB); Malvarrosa (It.; Malta; Sp.; EFS; JFM; KAB; USN); Malvavisco (Sp.; EFS); Malvone (It.; EFS; KAB); Mályvarózsa (Hun.; EFS); Mauve (Eng.; JLH); Mauve Arborée (Fr.; KAB); Mauve des Jardines (Fr.; KAB); Mauve Rose (Fr.; KAB); Mirame Lindo (Sp.; JFM); Mundrose (Ger.; KAB); Nachrose (Ger.; KAB); Nalba de Gradina (Rom.; KAB); Pappelrose (Ger.; EFS); Passe Rose (Fr.; EFS; KAB); Roemische Malve (Ger.; KAB); Rosa di Mare (It.; EFS); Rose Alcée (Fr.; KAB); Rose _ Băton (Fr.; KAB); Rose de Mer (Fr.; KAB); Rose d'Outre Mer (Fr.; KAB); Rose Mallow (Eng.; EFS); Rose Papale (Fr.; KAB); Rose Trémière (Fr.; EFS; USN); Rosen Eibisch (Ger.; EFS); Rosenpappel (Ger.; KAB); Rosoni (It.; EFS); Round Dock (Eng.; KAB); Schwartz Pappelblüüte (Ger.; EFS); Schwartzmalve (Ger.; EFS; KAB); Shtok Rosa (Rus.; KAB); Shu K'uei (China; EFS; KAB); Shu Kui Hua (Pin.; DAA); Siegmarsblume (Ger.; KAB); Stangenrose (Ger.; KAB); Stockmalve (Ger.; EFS; KAB; USN); Stockrose (Ger.; USN); Stokroos (Dutch; EFS); Stokrose (Den.; EFS); Stokross (Dutch; KAB); Thuc Quy (Ic.; KAB); Ungerblume (Ger.; KAB); Varita de San Jose (Cuba; Dr.; AHL; RyM); Varra de San Jose (Sp.; JFM); Weinrose (Ger.; KAB); Wetterrose (Ger.; KAB); Winterrose (Ger.; KAB); Zhanba (Mongolia; X12795226).

ACTIVITIES (HOLLYHOCK):

Analgesic (f1; BRU; X2504193); Antiaromatase (1; X12502186); Antidiaphoretic (f; BIB); Antiedemic (f1; X2504193); Antiestrogenic (f1; X12502186); Antiinflammatory (f1; X2504193);

FIGURE 1.3 Hollyhock (*Alcea rosea*).

Aromatase Inhibitor (1; X12502186); Astringent (f; DEP); Circulostimulant (f; DAA); Demulcent (f; DEP; GMH); Diuretic (f; DEP; EGG); Emollient (f1; BRU; GMH); Expectorant (f; EFS); Febrifuge (f; DEP; LMP); Fungistat (f; FNF); Gastroprotective (1; FNF); Hypoglycemic (1; WO3); Refrigerant (f; DEP); Stomachic (f; LMP); Vasoprotective (1; FNF).

Indications (Hollyhock):

Abscess (f; BIB; GHA; JFM); Aphonia (f; JFM); Aposteme (f; JLH); Bite (f; DAA); Bleeding (f; DAA); Boil (f; WO3); Bronchosis (f; JFM); Bruise (f; BIB); Burn (f; DEP); Cancer (f; BIB; JFM); Childbirth (f; LMP); Cold (f; BIB); Colosis (1; BRU); Constipation (f; DAA); Cough (f1; BIB; BRU; DEP; GHA; JFM; PH2); Cramp (f1; BIB; BRU); Cystosis (f; DEP); Dermatosis (f1; BRU; JFM; PH2; WO3); Diabetes (1; WO3); Dysentery (f; BIB; DEP; KAB); Dysmenorrhea (f; DAA; LMP); Earache (f; EGG); Edema (f1; X2504193); Enterosis (f; DEP; PH2); Fever (f; BIB; PH2); Gastrosis (f; EGG; PH2); Goiter (f; WO3); Gravel (f; BIB); Hematemesis (f; EGG); Hematuria (f; DAA); Hemorrhoid (f; DAA); Inflammation (f1; JFM; PH2; X2504193); Itch (1; BRU); Jaundice (f; WO3); Laryngosis (f; JFM); Malaria (f; DAA); Miscarriage (f; BIB); Ophthalmia (f; BIB); Pain (f1; BRU; X2504193); Pharyngosis (f1; BRU; PH2); Pulmonosis (f; GMH; PH2); Respirosis (f; JFM; PH2); Rheumatism (f; BIB); Scirrhus (f; JLH); Snakebite (f; DEP); Sore (f; DAA; DEP; PH2); Sore Throat (f; BIB; JFM); Stomatosis (f1; BRU; PH2); Strangury (f; JFM); Sweating (f; BIB); Swelling (f1; EGG; X2504193); Tenesmus (f; DEP); Thirst (f; PH2); Tumor (f; JLH); Urethrosis (f; JLH; PH2); Vaginosis (f; EGG).

Dosages (Hollyhock):

FNFF = !!
Young mucilaginous leaves cooked and eaten; flower petals, raw or cooked, and cooked flower buds also eaten; black hollyhock petals useful for imparting deep colors to teas and wines. Roots yield an edible starch (FAC; TAN) 1–2 oz mucilage (DEP). 1–2 g fl/teacup (PH2). 1.5 g fl/100 ml water as mouthwash (PH2).

- Arabians drink the sweetened floral tea for cough (GHA).
- Arabians use leaves in collyria and poulticed onto abscesses (GHA).
- Brazilians apply the leaves to inflammation (JFM).
- Chileans apply the root decoction to tumors (JLH).
- Costa Ricans take floral infusion for cough and sore throat (JFM).
- Gypsies pound roots with honey, taking 2x per day to prevent miscarriage (BIB).
- Lebanese mix juice with powdered coffee, soot, spiderwebs, or sugar to stop bleeding (BIB).
- Lebanese take tea of dried plants and/or flowers for colds, cramps, sweating, and sore throat (BIB).
- Middle Easterners apply mucilage to bruises, or with dough and olive oil to tumors (BIB).
- Peruvians poultice leaves, cooked in oil or milk, on swellings (EGG).
- Peruvians use sudorific decoction for cough and bloody vomiting (EGG).
- Peruvians use fresh leaf decoctions in douches (EGG).
- Peruvians use steam from floral decoction for earache (EGG).
- Punjabi use flowers for rheumatism, the root for dysentery (KAB).
- Venezuelans take floral/foliar decoction/tea for aphonia, bronchosis, laryngosis, and strangury (JFM).

Downsides (Hollyhock):

Class 1 (AHP). No health hazards known in conjunction with the proper administration of designated therapeutic dosages (PH2).

CAMEL'S THORN (*ALHAGI MAURORUM* MEDIK.) ++ FABACEAE

Synonyms:

Alhagi camelorum Fisch.

Notes (Camelthorn):

> **We have sent you money to buy burnt-offerings, and sin offerings, and incense, and prepare ye <u>manna</u>.**
>
> **Baruch 1:10 (KJV)**

Because the Baruch manna was for sale, it was probably the resinous gum from some tree of the Levant. During the heat of the day, a sweet gummy substance oozes from the leaves and stems. This hardens upon contact with the air and is then collected by shaking over drop-cloths.

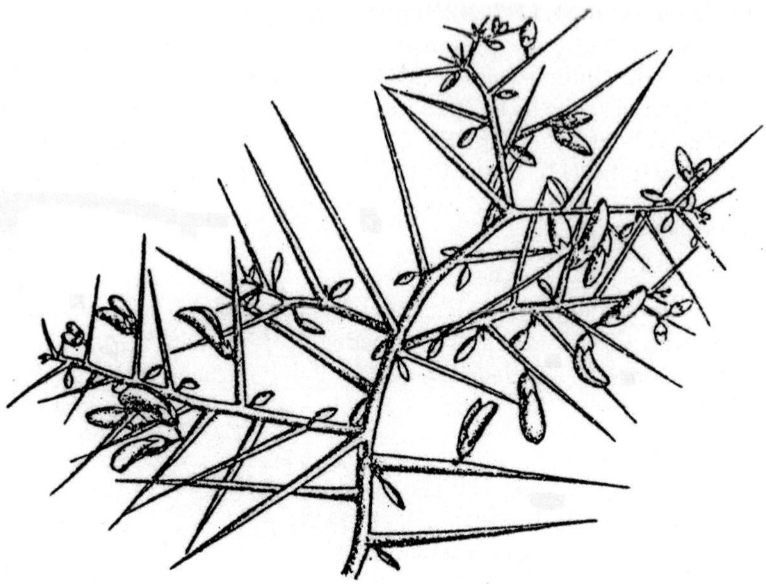

FIGURE 1.4 Camel's Thorn (*Alhagi maurorum*).

Common Names (Camelthorn):

Aagul (Arab.; KAB); Adhikantaka (Sanskrit; KAB); Agoul (Fr.; KAB); Agul (Arab.; Mali; UPW); Ahagul (Arab.; Mali; UPW); Al Heef (Arab.; GHA); Alhaju (Arab.; KAB); Ananta (Sanskrit; KAB); Aqul (Oman; Qatar; Saudi; GHA); Arabian Manna Plant (Eng.; KAB); Athariyun (Urdu; KAB); Bahukantaka (Sanskrit; KAB); Ballituruche (Kan.; KAB); Borellia (Mali; UPW); Camel's Thorn (Eng.; Scn.; AH2); Dhirghamula (Sanskrit; KAB); Dulallabha (Beng.; KAB); Duralabba (Ayu.; AH2); Duramula (Sanskrit; KAB); Durlabha (Sanskrit; KAB); Durlambha (Nepal; KAB); Dusparsha (Sanskrit; KAB); Elbo (Niger; UPW); Farakiyun (Urdu; KAB); Gandhari (Sanskrit; KAB); Ghaz (Kila Saifulla; KAB); Girikarnika (Sanskrit; Tel.; KAB); Girikarnika Yavasa (Sanskrit; DEP); Hagah Matsui (Isr.; PAY); Haj (Arab.; KAB); Halhagi (Fr.; KAB); Igol (Arab.; GHA); Javansa (Hindi; DEP; KAB); Javasa (Hindi; Urdu; KAB); Javasha (Ayu.; AH2); Javaso (Guj.; KAB); Jawas (Mar.; KAB); Jawassa (Bom.; KAB); Jawassi (Bom.; KAB); Junwasa (Hindi; KAB); Juwasa (Hindi; DEP; KAB); Kachchura (Sanskrit; KAB); Kahribuz (Bal.; KAB); Kag (Oman; GHA); Kandar (Gandava; KAB); Kandeira (Gandava; KAB); Kandera (Sibi; KAB); Kantakaluka (Sanskrit; KAB); Kantaki (Sanskrit; KAB); Kantechumbaka (Mar.; KAB); Kappattumpa (Mal.; KAB); Kas (Mar.; Sin.; KAB); Kaskhandero (Sin.; KAB); Kharebuz (Iran; KAB); Kharibuz (Iran; NAD); Kharishutr (Iran; KAB); Kshudrengudi (Sanskrit; KAB); Makhe (Kila Saifulla; KAB); Marudbhava (Sanskrit; KAB); Persian Manna Plant (Eng.; KAB); Rodanika (Sanskrit; KAB); Samudranta (Sanskrit; KAB); Shinz (Kharan; Laleji; Las Bela; Ormara; KAB); Shoukuljamal (Arab.; DEP; KAB); Shutarkhar (Iran; DEP; KAB); Shutharkhar (Bal.; Iran; DEP; KAB); Sukshmapatra (Sanskrit; KAB); Tandan (Kohlu; KAB); Tanwan (Loralai; KAB); Tella (Tel.; DEP); Tellaginiya (Tel.; KAB); Tikshnakantaka (Sanskrit; KAB); Tindan (Sibi; KAB); Toreyingalu (Kan.; KAB); Triparnika (Sanskrit; KAB); Tz'u Mi (China; KAB); Ushtarkhar (Iran; KAB); Usturkhar (Sin.; KAB); Vanadarbha (Sanskrit; KAB); Vasanta (Sanskrit; KAB); Vishaghna (Sanskrit; KAB); Vivarnaka (Sanskrit; KAB); Yas (Sanskrit; KAB); Yavasa (Hindi; Sanskrit; DEP; KAB); Yavasaka (Sanskrit; KAB); Yawas (Mar.; KAB); Zoz (Sibi; Zhob; KAB); Zuwasha (Cutch; KAB).

ACTIVITIES (CAMELTHORN):

Alexiteric (f; BIB); Antiatherosclerotic (1; WO3); Antibilious (f; BIB; DEP); Antiemetic (f; WO2); Antihypercholesterolemic (1; WO3); Antinociceptive (1 × 15507342); Antiseptic (f1; WO2); Aperient (f; BIB); Aphrodisiac (f; BIB; PAY); Cholagogue (f; BIB); Demulcent (f; BIB); Depilatory (f; WO2); Depurative (f; BIB; KAP); Diaphoretic (f; KAB); Diuretic (f; BIB; WO2); Ergogenic (f1; WO3); Expectorant (f; BIB); Febrifuge (f; KAB); Hypolipidemic (1; WO3); Hypotensive (1; WO3); Laxative (f; BIB; WO2); Orexigenic (f; KAB; SAY); Proteolytic (1; WO2); Refrigerant (f; KAB); Suppurative (f; BIB); Sympathomimetic (1; WO2); Tonic (f; KAB; SAY).

INDICATIONS (CAMELTHORN):

Abscess (f; BIB; KAB); Adenopathy (f; JLH; UPW); Anorexia (f; BIB; SAY); Arthrosis (f; GHA); Asthma (f; BIB); Atherosclerosis (1; WO3); Bacteria (1; WO2); Biliousness (f; DEP); Bleeding (f; BIB); Bronchosis (f; BIB); Cancer (f; JLH); Cancer, abdomen (f; JLH); Cancer, gland (f; JLH); Cardiopathy (1; X1305866); Cataract (f; GHA; PAY); Catarrh (f; PAY); Cerebrosis (f; BIB); Constipation (f; PAY); Corneosis (f; BIB); Cough (f; DEP; PAY); Dermatosis (f; BIB); Diarrhea (f1; SAY; X15138016); Enterosis (f; UPW); Epistaxis (f; BIB); Fever (f; PAY); Gastrosis (f; PAY); Gingivosis (f; PAY); Halitosis (f; PAY); Headache (f; BIB; WO3); Hematachezia (f; PAY); Hemicrania (f; BIB; KAB); Hemorrhoid (f; BIB; KAB); High Blood Pressure (1; WO3); High Cholesterol (1; WO3); High Triglycerides (1; WO3); Impotence (f; PAY); Infection (f1; PAY; WO2); Jaundice (f; GHA); Leprosy (f; BIB; KAB); Migraine (f; BIB); Nephrosis (f; PAY); Obesity (f1; BIB; KAB; WO3); Odontosis (f; PAY); Opacity (f; BIB); Ophthalmia (f; KAB); Pain (f1; GHA; WO3; X15507342); Polyp (f; JLH); Pulmonosis (f; JLH; PAY; WO3); Rheumatism (f; PAY; WO2); Smallpox (f; BIB); Sore (f; BIB; PAY); Splenosis (f; PAY); Stomachache (f; PAY); Stomatosis (f; PAY); Swelling (f; BIB; KAB; WO2); Thirst (f; KAB); Tumor (f; UPW); Water Retention (f; KAP).

DOSAGES (CAMELTHORN):

FNFF = !!

Roots dug and consumed in the Sahara (UPW). The sugary secretion (manna) is edible, occurring in small round grains, consisting mostly of sugars: melizitose, 47.1; sucrose, 26.4; and invert sugar, 11.6% (BIB); 1–2 g herb (KAP); 48–96 ml herb decoction (KAP).

- Ayurvedics regard the plant for anorexia, bronchosis, cerebrosis, constipation, dermatosis, epistaxis, fever, leprosy, obesity, and thirst (KAB).
- Israelis boil root, steep overnight, strain, and drink 2 to 3 × day for kidney sand (PAY).
- Israelis boil root until the water is half gone, drinking it to stop bloody diarrhea (PAY).
- Israelis expose rheumatic pain to the crushed root steam (PAY).
- Israelis take seed tea for constipation, hemorrhoids, spleen infections, and stomachache (PAY).
- Konkani smoke the plant with ajwan seed, black datura, and tobacco for asthma (KAB).
- Ormara natives apply root decoction topically on abscesses and swelling (KAB).
- Romans used the plant for nasal polyps (UPW).
- Unani view the plant as alexiteric, aperient, using for corneal opacities, hemicrania, and hemorrhoids; they used the manna as aperient, aphrodisiac, cholagogue, depurative, expectorant, using it for asthma, eruptions, hemorrhoids, nausea, and smallpox (KAB).

ONION *(ALLIUM CEPA* L.) +++ LILIACEAE

NOTES (ONION):

We remember the fish, which we did eat in Egypt freely; the cucumbers, and the melons, and the leeks, and the onions, and the garlick: But our soul is dried away: there is nothing at all, beside this manna, before our eyes.

Numbers 11:5–6 (KJV)

We remember the fish we ate in Egypt for nothing, the cucumbers, the melons, the leeks, the onions, and the garlic; but now our strength is dried up, and there is nothing at all but this manna to look at.

Numbers 11:5–6 (RSV)

How we remember the fish that we used to eat in Egypt for nothing, the cucumbers and the water melons, and the leeks and the onions, and the garlic; But now our soul is dried away. Our eyes are on nothing at all except the manna.

Numbers 11:5–6 (NWT)

Although widely used in biblical times, the onion is only mentioned once, in Numbers 11, true also of the leek and the garlic that we hear built the pyramids. Medicinally, I view the leek and onions as dilute garlic (BIB).

COMMON NAMES (ONION):

Akaakai (Hawaii; LIB); Albasa (Hausa; KAB); Albassa-Haoussa (Sudan; AVP); Azalim (Ber.; BOU); Azlim (Ber.; BOU); Basal (Arab.; Malta; GHA; KAB); Basar (Sin.; NAD); Basl (Arab.; DEP; EFS; NAD); Basla (Malta; KAB); Baslim (Ber.; BOU); Bassal (Arab.; BOU); Bawang (Mal.; Malaya; DEP; EFS; NAD); Besla (Arab.; BOU); Bhazal (Heb.; KAB); Bolle (Ger.; KAB); Btsong (Tibet; NPM; TIB); Ceapa (Rom.; KAB); Ceba (Lan.; KAB); Cebo (Lan.; KAB); Cebola (Mad. Por.; AVP; EFS); Cebola Comun (Por.; AVP); Cebola las Hortas (Por.; KAB); Cebola Ordinario (Por.; KAB); Cebolla (Bel.; Sp.; AVP; EFS); Cebolla Cabezona (Sp.; AVP); Cebula (Pol.; KAB); Ceola (It.; KAB); Cepa (Lan.; KAB); Chabura (Uvosha; Peru; SOU); Ciboria (Peru; Shipibo/Conibo; EGG); Ciboule (Fr.; KAB); Cipolla (It.; EFS; KAB); Dirghapatra (Sanskrit; KAB); Dungari (Guj.; Sin.; DEP; KAB); Dungari Kandu (Guj.; Sin.; NAD); Durgandha (Sanskrit; NAD); Echte Zwiebel (Ger.; EFS); Eerulli (Mal.; NAD); Fara Albassa (Sudan; AVP); Gabu (Sokoto; KAB); Gemein Zwiebel (Ger.; EFS); Guda (Kano; KAB); Gudagi (Sokoto; KAB); Hagyma (Hun.; KAB); Hui Hui Ts'ung (China; EFS); Hu Ts'ung (China; EFS); Ira-Vengayam (Tam.; DEP; KAB); Irulli (Tam.; DEP; KAB; NAD); Jaman Pullu (Aym.; Bol.; DLZ); Kanda (Bom.; Mah.; Mar.; DEP; KAB; NAD); Kando (Kon.; KAB); Kesunni (Burma; DEP; KAB; NAD); Khtim (Cam.; KAB); Kiska (Tur.; EB54:155); Krommyon (Greek; KAB); Ku Kut (Mex.; AVP); Kunbali (Kan.; DEP; KAB); Kyetthwonni (Burma; DEP; KAB; NAD); Lasona (Ilo.; KAB); Lasuna (Tag.; KAB); Lauch (Ger.; AVP); Lawashi (Sokoto; KAB); Lebsal (Ber.; BOU); Loegen (Den.; KAB); Log (Den.; EFS); Lok (Swe.; EFS); Luk (Rus.; KAB); Lunu (Sin.; KAB; NAD); Makakanda (Sanskrit; KAB); Makkhang (Limbu; NPM); Mi (Sur.; AVP); Neermulli (Tel.; NAD); Neerulli (Kan.; NAD); Nirulli (Kan.; Tel.; DEP; KAB); Nripakanda (Sanskrit; KAB); Nripapriya (Sanskrit; KAB); Nripavhaya (Sanskrit; KAB); Nripeshtha (Sanskrit; KAB); Ochong (Lepcha; NPM); Ognon (Haiti; AVP); Oignon (Fr.; Haiti; AVP; EFS); Oignon Comun (Fr.; KAB); Oignon de Cuisine (Fr.; KAB); Onion (Eng.; CR2; NAD); Onyan (Ulwa; ULW); Palandu (Beng.; Sanskrit; EFS; NAD); Piaj (San.; KAB); Piau (Kon.; KAB); Piyaj (Beng.; DEP; NAD); Piyang (Beng.; NAD);

FIGURE 1.5 Onion (*Allium cepa*).

Piyas (Assam; Hindi; India; Iran; DEP; EFS; KAB; NAD); Piyaz (Hindu; Iran; Nepal; Urdu; EFS; KAB; NAD; SUW); Pulantic (Beng.; NAD); Puyaj (Bom.; KAB); Pyaj (Beng.; Bhojpuri; Chepang; Gurung; Magar; Mooshar; Nepal; Newari; Sunwar; Tamang; Tharu; NAD; NPM); Rajapalandu (Sanskrit; KAB); Rajapriya (Sanskrit; KAB); Rajeshtha (Sanskrit; KAB); Raktakanda (Sanskrit; KAB); Ri-Sgog (Tibet; NPM); Rochaka (Sanskrit; KAB); Roedloek (Swe.; KAB); Safa (Sokoto; KAB); Seba (Cat.; KAB); Sebúya (Garifuna; IED); Seigola (It.; AVP); Shaja (Kano; KAB); Sibojo (Ma.; JFM); Sibuyas (Pam.; Tag.; KAB); Siepel (Dutch; EFS); Siwulla (Aym.; Bol.; Que.; DLZ); Sochan (Tur.; KAB); Sogan (Tur.; EFS; EB49:406); Sommer-Kuchen Zwiebel (Ger.; EFS); Tamanegi (Japan; TAN); Tibsal (Ber.; BOU); Tongolo (Hova; KAB); Tongolobe (Hova; KAB); Tongolovazaha (Hova; KAB); Ts'ong Tse (China; KAB); Ui (Dutch; Ma.; EFS; JFM); Uigen (Dutch; KAB); Ullegaddi (Tam.; NAD); Vella-Vengayam (Tam.; DEP; KAB); Vengayam (Kan.; Tam.; DEP; NAD); Voroshagyma (Hun.; EFS); Vulli-Gaddalu (Tam.; Tel.; DEP; KAB); Xonocatl (Mex.; AVP); Yangoa (Korea; TAN); Yavanestha (Sanskrit; KAB); Yerragadda (Tel.; NAD); Zalim (Ber.; BOU); Zipolle (Ger.; KAB); Zippel (Ger.; KAB); Z'Oignon (Haiti; AHL); Zongnon (Haiti; AVP); Zonyon (Creole; Haiti; VOD); Zwiebel (Ger.; AVP; DEP; KAB).

Activities (Onion):

Allergenic (f1; APA); Amebicide (f1; X10594976); Anthelmintic (f; WHO); Antianthrax (1; LIB); Antiaggregant (12; KOM; MPI; SHT; WHO; WO3); Antiallergic (1; BRU; PHR; WHO); Antianaphylactic (1; X3932203); Antiasthmatic (f1; PHR; PNC); Antiatherosclerotic (f_{12}, APA; WO1); Antibiotic (1; PNC); Anticarcinogenic (1; WO3); Anticariogenic (1; LIB); Anticystitic (1; X11272677); Antidermatophytic (1; X7600010); Antidiabetic (f12; X15582196); Antidote (Tobacco); (f; NAD); Antiedemic (f1; APA; WHO); Antihistamine (1; WHO); Antihypercholesterolemic (1; MPI); Antihyperglycemic (12; GHA; WHO; X15582196); Antihyperlipidemic (1; BGB; WHO); Antihypertensive (1; DAD); Antiinflammatory (f12; PHR; VOD); Antimitotic (1; WHO); Antimutagenic (1; X9838070); Antioxidant (1; WO2; X15582196); Antiplatelet (1; BGB; WHO); Antiproliferant (1; X15506817; X15890236); Antiseptic (f12; APA; BGB; JFM; PH2); Antispasmodic (f; PNC); Antithrombotic (1; X15342218); Antithromboxane (1; WHO); Antitoxigenic (1; X7600010); Antitumor (f1; APA; BGB; JLH); Aphrodisiac (f1; DAD; SKY; WHO; JAC7:405); Apoptotic (1; JAF51:208; X15890236); Bacillus (1; LIB); Bactericide (12; KOM; PH2; SHT; WHO); Candidicide (f1; WHO); Cardiotonic (f; DAD; JFM); Carminative (f; APA; PNC; WHO); Chemopreventive (1; X15506817); Choleretic (f; JFM); Collyrium (f; GHA); Copper Chelator (1; WO2); Cyclooxygenase Inhibitor (1; WHO); Decongestant (f1; APA); Demulcent (f; NAD); Deobstruent (f; KAP); Diaphoretic (f; JFM); Diuretic (f1; BGB; PNC; SKJ; VOD; WHO); Emmenagogue (f; DAD; DEM; KAB; WHO); Expectorant (f1; APA; GHA; PNC; SKJ; WHO); Febrifuge (f; GHA); Fibrinolytic (12; APA; MPI; SHT; WHO); Fungicide (1; APA; WHO; X7600010; X15113089); Gram(+)-icide (1; WO3); Gram(−)-icide (1; WO3); Hepatoprotective (f; X15582196); Hypocholesterolemic (12; BGB; JNU; MPI; WHO; X15539326); Hypoglycemic (f12; APA; MPI; PNC; WHO; X15738612); Hypotensive (12; DAD; KOM; SHT); Immunodepressant (1; X9103661); Lipolytic (2; BGB; KOM; SHT); Lipoxygenase Inhibitor (1; WHO); Mast Cell Stabilizer (1; X3932203); Nephroprotective (f; X15582196; X15539326); Orexigenic (2; BGB; PH2; WHO); Parasiticide (f1; X10594976); Pectoral (f; KAB); Peristaltic (f; KAB); Phospholipase Inhibitor (1; WHO); Propecic (f; EGG); Protein Kinase Inhibitor (1; WHO); Protisticide (1; X10594976); Rubefacient (f; JFM); Sedative (f; KAB); Soporific (f; DEP); Spermagenic (1; JAC7:405); Stimulant (f; JFM; NAD); Stomachic (f; KAB); Streptococcus (1; LIB); Thrombolytic (1; X15342218); Tonic (f; WHO); Vermifuge (f; APA); Vulnerary (f; VOD).

Indications (Onion):

Abscess (f; EB49:406); Adenopathy (f; JLH); Allergy (f1; BRU); Alopecia (f; EGG); Ameba (f1; X10594976); Anaphylaxis (1; X3932203); Angina (f; BGB; PHR); Anorexia (2; BGB; KOM; PH2;

WHO); Anthrax (1; LIB); Aphonia (f; DLZ); Apoplexy (f; DEP); Atherosclerosis (2; APA; KOM; PH2; SHT); Asthma (f1; APA; BRU; JFM; PHR; PH2); Atherosclerosis (1; JFM; WO2); Bacillus (1; X4064797); Bacteria (1; JFM; PH2); Bite (f; DEP); Biliousness (f; KAB); Bite (f; BOU; NAD); Bleeding (f; KAB); Blister (f1; EGG; SKJ); Boil (f1; NAD; SKJ); Bronchosis (2; BGB; PHR; PH2; WHO); Bruise (f; EGG; PHR; WHO); Bugbite (f; PHR); Burn (f; JLH; PHR); Cancer (1; APA; JLH); Cancer, breast (f1; FNF; JLH; JNU); Cancer, colon (f1; JNU); Cancer, esophagus (f1; JNU); Cancer, gland (f1; FNF; JLH); Cancer, liver (f1; FNF; JLH); Cancer, lung (f1; JNU); Cancer, rectum (f1; FNF; JLH); Cancer, stomach (f1; APA; BRU; FNF; JLH; JAC7:405); Cancer, uterus (f1; FNF; JLH); Candida (f1; X10594976); Carbuncle (f; KAB; LIB); Cardiopathy (f1; APA; JFM; JNU); Caries (1; X9354029); Cataract (f; BOU); Catarrh (f; KAB); Chest Cold (f; JFM); Chilblain (f; KAP; X15664457); Cholecocystosis (f; JFM; PHR); Cholera (f; DEP; WHO); Circulosis (f; EGG); Cold (f12; DEM; GHA; PHR; PH2); Colic (f; EGG; PHR; PH2; WHO); Colosis (f; KAP); Congestion (f1; APA; BGB; JFM); Convulsion (f; KAB; LIB); Corn (f; JLH; LIB); Cough (f12; BGB; FNF; GHA; JFM; PHR; PH2); Cramp (f; GHA); Cystosis (1; X11272677); Deafness (f; GHA; JFM); Depressed Immune System (2; PHR); Dermatosis (f1; GHA; SKY; X7600010); Diabetes (f12; APA; WHO; X15582196; X15738612); Dropsy (f; BGB; DAD; DLZ; GMH); Dysentery (f; BGB; DAD; JNU); Dysgeuzia (f; KAB); Dysmenorrhea (f; PHR); Dyspepsia (f12; JFM; PHR; PH2); Dyspnea (f; BGB); Dysuria (f; KAB); Earache (f1; APA; DEM; DEP); Edema (f; JFM; LIB); Enterosis (f; KAP); Epilepsy (f; JFM); Epistaxis (f; KAB; LIB); Escherichia (1; PH2; X4064797); Felon (f; JLH); Fever (f2; DEM; GHA; PHR; PH2; WHO); Flu (f; DEM); Fracture (f; EB51:195); Fungus (1; X10594976); Furuncle (f; DLZ; PHR); Gallstone (f; EGG); Gas (f; DAD; JFM; SKJ); Gastrosis (f; EGG; GHA); Giardia (1; X10594976); Gingivosis (1; X9354029); Gravel (f; BGB; DAD; GMH); Headache (f; LIB; VOD); Hemorrhoid (f; LIB); Hepatosis (f; KAB); High Blood Pressure (2; PH2); High Blood Pressure (2; PHR; SHT; WHO); High Cholesterol (2; APA; SHT; WHO);High Triglycerides (1; WHO); Hyperlipidemia (f; SKJ); Hysteria (f; DEP; KAB); Impotence (f; KAB); Induration (f; JLH); Infection (2; JNU; PHR; PH2); Inflammation (2; PHR); Insomnia (f; KAB); Interstitial Cystosis (1; X11272677); Jaundice (f; LIB; WHO); Laryngitis (f; DLZ); Leishmania (1; FT75(1):9); Lymphangites (f; KAB); Malaria (f; DEP; GHA; JFM); Mange (f; JFM); Migraine (f; KAB); Mycosis (1; X10548758); Nephrosis (f; BGB; GHA); Neuralgia (f; JFM); NIDDM (2; WHO); Nyctalopia (f; KAB); Obesity (f1; BGB; LIB; SKJ); Odontosis (f; KAB); Ophthalmia (f; SKJ); Osteoporosis (1; ACT5:330; JNU); Otosis (f; SKJ; WHO); Pain (f; GHA; JFM); Paralysis (f; DLZ); Parasite (f1; JFM; ULW; X10594976); Periodontosis (1; X9354029); Pertussis (f; PHR); Pharyngosis (2; PHR); Phthisis (f; DEP); Pimples (f; JFM; WHO); Proctosis (f; KAB); Prolapse (f; KAB); Protozoa (1; X10594976); Pulmonosis (f; ULW); Rabies (f; KAB); Rash (f; GHA); Respirosis (f; ULW); Rheumatism (f; JFM); Salmonella (1; PH2); Scabies (f; JFM; LIB); Scurvy (f1; NAD); Sinusosis (f; LIB); Sores (f1; JNU; WHO); Sore Throat (f; DEP; GHA; LIB); Splenosis (f; DEP; LIB); Splinter (f; X15664457); Staphylococcus (1; LIB); Sting (f; JFM); Stomachache (f; PHR); Stomatosis (2; PHR); Strangury (f; KAP); Streptococcus (1; X9354029; X4064797); Sunstroke (f; DEP); Swelling (f; JFM); Syncope (f; DEP; KAB); Tenesmus (f; BGB); Thorn (f; X15664457); Thrombosis (f; JFM); Tinnitus (f; JFM); Tonsilosis (f; JFM); Toothache (f; JNU); Tuberculosis (f; JFM; LIB); Varicosis (f; DLZ); Venereal Disease (f; LIB); Vertigo (f; KAB); Virus (1; X10594976); Vision (f; GHA); Wart (f; PHR); Whitlow (f; JLH); Worm (f; JFM); Wound (f; NAD; PHR); Yeast (1; WHO; X10594976).

Dosages (Onion):

FNFF = !!!
Bulbs and leaves widely eaten; flowers and vivapoarous plantlets also eaten (FAC; TAN; EB54:155); 0.25–1 onion (2–5 oz) (APA); 1 tsp onion juice 3 to 4 ×/day (APA); 1 onion per day (JAD); 10–20 ml bulb or leaf infusion (KAP); 1–3 g powdered seed (KAP); 50 g fresh onion or 20 g dry onion (KOM; SHT; WHO); 4–5 tsp tincture/day (PHR); 4–5 Tbsp onion syrup (PHR).

- Grated bulbs heated and mixed with grated soap and applied to abscess (EB49:406).
- Grated bulbs mixed with albumen, pine tar, and soap, and applied topically in setting fractures (EB51:195).
- Arabs use honey extract as expectorant and for stomach cramps; they rub it on the face to remove dark spots or rash (GHA).
- Arabs cook red onions with eggs and sesame oil for cold and cough (GHA).
- Arabs apply crushed bulb juice as eardrops for deafness and infections (GHA).
- Arabs eat fresh onion to clear irritated throat (GHA).
- Bahamans put a slice of onion in the shoe near the heel to treat a cold (JFM).
- Bolivians eat the bulbs for respiratory problems, and for bladder and kidneystones and inflammation (DLZ).
- Bolivians take candied onions for cough, dropsy, insomnia, pertussis, strangury, onion soup for cramps, paralysis, rheumatism, varices (DLZ).
- Curacaons plug an earache or ringing ear with the onion neck (JFM).
- Dominicans use diced bulbs with honey for bronchitis and catarrh (AHL).
- Nicaraguan Garifuna take juice orally for respiratory-pulmonary disorders, worms, and intestinal parasites (IED).
- Haitians apply sliced onion to head for headache (VOD).
- Italians use onions for chilblains, splinters, and thorns (X15664457).
- Japanese put a cut onion under the pillow for insomnia (LIB).
- Peruvians and Bolivians report onion skin tea for laryngitis with loss of voice (DLZ; SOU).
- Peruvians apply the outer layers of the onion to burns to prevent blistering (EGG).
- Russians boil bulbs in vinegar and apply to corns (JLH).
- Spaniards recommend onion juice for buzzing in the ears, and even deafness (JLH).
- Trinidadans take onion decoction for chest cold, cough, and tuberculosis (JFM).
- Yucatanese take 3 tsp onion juice per day for coronary thrombosis, edema, proteinuria, strangury, adding lemon juice for flu, rheumatism, and tonsilitis (JFM).

Downsides (Onion):

Class 1. Some idiopathic allergies (JAD). Allergic rhinoconjunctivitis and contact dermatosis reported (WHO).

Extracts (Onion):

I have long believed that string beans and raw onions have helped me avoid diabetes. Studies in 2005 — some positive, some negative — have not weakened my belief (X15582196). El-Demerdash et al. (X15582196) strengthened my case, demonstrating hypoglycemic activities of both onion and garlic juice (at 10 ml/kg body weight or 4 g/kg, a huge dose for me, equivalent to 400 g onion juice). The antioxidant and antihyperglycemic activities of onion and garlic may protect against liver and renal damage (X15582196). Conversely, Jelodar et al. (2005) found no hypoglycemic activity for onion, as compared to much more potent garlic (X15738612). Wetli et al. (2005) say that osteoporosis costs America $17 billion economically, over and beyond the pain and suffering. Adding 7% dried onion bulbs to diets decreases bone resorption and increases bone mineral content in growing rats. Rutin had been thought to contribute some of this activity but Wetli et al. (2005) ruled out flavonoids as the major contributor(s) and suggested instead gamma-L-glutamyl-*trans*-S-1-propenyl-L-cysteine sulfoxide (GPCS), adding this to the long list of dietary phytochemicals that support bone health, Ca, K, Mg, vit. C, D, and K, phytoestrogens (coumestrol, humulone, isoflavones, zearelenol), possibly other flavonoids (hesperidin, rutin) and monoterpenes (X15853380). Chang et al. (2005d) demonstrated a growth inhibitory effect of alk(en)yl thiosulfates from onion and garlic on tumor cell lines. Sodium n-propyl thiosulfate and sodium 2-propenyl thiosulfate (natural constituents of onion and garlic, respectively) were originally deemed to cause hemolytic anemia in dogs. They inhibit *in vitro* proliferation of three human tumorigenic cell lines in a dose-dependent manner. Both induced apoptosis (X15890236). Wetli et al. (2005) showed that a gamma-glutamyl peptide (gamma-L-glutamyl-trans-S-1-propenyl-L-cysteine sulfoxide) inhibited bone resorption by osteoclasts. One gram onion added to rat food significantly inhibits bone resorption at 2 mM (X15853380). Corea et al. (2005) found four new antispasmodic compounds (furostanol saponins); high concentrations of quercetin, quercetin 4(I)-glucoside, taxifolin, taxifolin 7-glucoside, and phenylalanine were also isolated (X15713001). (=) SMCS proved a better antioxidant than glibenclamide and insulin, but the drugs were better for antidiabetic activity (X12587728). Feeding rats 1 g powdered onion/day/month boosted bone mineral content 17%, bone thickness more than 15%, performing better than calcitonin (=) (JNU). In the study by Park and Shin (2005), cinnamon and onion oil followed garlic and clove bud oils in lethality to the Japanese termite, *Reticulitermes speratus* Kolbe. Diallyl trisulfide was most toxic, then diallyl disulfide, eugenol, diallyl sulfide, and beta-caryophyllene (X15913300).

LEEK (*ALLIUM PORRUM* L.) +++ LILIACEAE

Synonyms:

A. ampeloprasum L. *fide* some Auct.; *Allium ampeloprasum* var. *porrum* (L.) J. Gay; *Allium laetum* Salisb.; *Allium porrum* var. *maximim* Schweinf.; *Porrum comune* Reichb.; *Porrum sativum* Mill.; *Porrum sectile* Schult. *fide* HH2

Notes (Leek):

We remember the fish, which we did eat in Egypt freely; the cucumbers, and the melons, and the leeks, and the onions, and the garlick: But our soul is dried away: there is nothing at all, beside this manna, before our eyes.

Numbers 11:5–6 (KJV)

We remember the fish we ate in Egypt for nothing, the cucumbers, the melons, the leeks, the onions, and the garlic; but now our strength is dried up, and there is nothing at all but this manna to look at.

Numbers 11:5–6 (RSV)

How we remember the fish that we used to eat in Egypt for nothing, the cucumbers and the water melons, and the leeks and the onions, and the garlic; But now our soul is dried away. Our eyes are on nothing at all except the manna.

Numbers 11:5–6 (NWT)

Zohary notes that the leek (*hatzir*) is widely cultivated in Israel and is "indeed the most precious" of the few cultivated species of *Allium*. Some authors seem to think of *A. ampeloprasum* as the plant when grown for its bulb, *A. porrum* as the leek. More from Cornucopia than Pharmacopeia, i.e., from a culinary point of view, Facciola groups them all under *Allium ampeloprasum:* (1) the Levant Garlic, *Allium ampeloprasum,* cultivated for its large roots, and including the Argentine garlic, elephant garlic, Levant garlic, multiplier leek, Perennial Sweet Leek, Persian chives, and Yorktown onion; (2) *Allium ampeloprasum* var. *babingtonii,* the British leek or Welsh leek; (3) *Allium ampeloprasum* KURRAT group, the Salad leek or Kurrat Nabati, believed to be the leek of ancient Egypt (leaves found in Egyptian tombs); (4) *Allium ampeloprasum*, Porrum group, the leek (FAC). Kirtikar and Basu lump them both under *Allium ampeloprasum* (KAB).

Common Names (Leek):

Ail à Tuniques (Fr.; KAB); Ail d'Orient (Fr.; TAD); Ajet (Sp.; EFS); Ajo Porro (Cuba; Sp.; AVP; USN); Àlbásàà Mai Kara (Hausa; Nig.; UPW); Alho Porró (Por.; USN); All Porret (Cat.; KAB); Argentine Garlic (Eng.; FAC); Basal (Arab.; EFS); Breitlauch (Ger.; HH2); British Leek (Eng.; FAC); Burri (Ger.; EFS); Cebollín (Sp.; AVP); Currat (Malta; KAB); Currat Salvagg (Malta; KAB); Dungali (India; EFS); Giant Garlic (Eng.; TAD); Goondina (Iran; EFS); Great Headed Garlic (Eng.; TAD); Hatzir (Heb.; Isr.; BIB); He (Vn.; EB42:413); Jumbo Garlic (Eng.; TAD); Kânda (India; EFS); Karâts (Arab.; JLH); Kiras (Arab.; DEP); Kirath (Arab.; DEP; EFS); Kourât (Arab.; JLH); Krachhai (Cam.; KAB); Kurrat Nabati (?; FAC); Kyet Thoon (Burma; DEP); Lauch (Ger.; KAB; USN); Leek (Eng.; FAC; EB42:413); Levant Garlic (Eng.; FAC); Look (Dutch; KAB); Mama LaSalle's French Garlic (Eng.; TAD); Multiplier Leek (Eng.; FAC); Paru (Beng.; DEP; KAB); Pearl Onion (Eng.; TAD); Perennial Sweet Leek (Eng.; FAC); Persian Chives (Eng.; FAC); Pferdknobauch (Ger.; TAD); Pirasa (Tur.; EFS); Platandu (Sanskrit; EFS); Poireau (Fr.; Haiti; AVP; JLH; TAN; USN); Porei (Rus.; KAB); Poro (Sen.; UPW); Porre (Den.; EFS); Porreau (Fr.; Haiti; AVP; EFS); Porree (Ger.; USN); Porret (Eng.; EFS); Porreta (Por.; UPW); Porrey (Ger.; KAB); Porrina (It.; HH2); Porro (Cat.; It.; Peru; Sp.; EGG; USN); Porro Hortense (Por.; UPW); Pourrat (Lan.; KAB); Pourriole (Fr.; KAB); Prasium (JLH); Prei (Dutch; Sur.; AVP; EFS); Praz (Rom.; KAB); Puerro (Dr.; Peru; Sp.; AHL; EGG; TAN; USN); Puerro Agreste (Sp.; KAB); Puerro de Vina (Sp.; KAB); Pulantu (India; EFS); Purjo (Swe.; EFS); Purlok (Den.; JLH); Purret (Eng.; HH2); Pyaz (India; EFS); Salad Leek (Eng.; TAN); Shan Chiu (China; EFS); Siboyas sa Taal (Pi.; KAB); Spanischer Lauch (Ger.; HH2); Suppenlauch (Ger.; HH2); Taree Irani (FAC); Ti l'Onion (St. Lucia; AVP); Toi tay (Vn.; EB42:413); Tsung (China; EFS); Verruga (Sp.; AVP); Welch Leek (Eng.; FAC); Welschlauch (Ger.; HH2); Winterlauch (Ger.; HH2); Winter leek (Eng.; TAN); Yorktown Onion (Eng.; FAC).

Activities (Leek):

Antitumor (1; WO3); Aphrodisiac (f; DAW); Apoptotic (1; JAF51:208); Constipation (f; UPW); Digestive (f; DAW; EFS); Discutient (f; DAW; EFS); Diuretic (f; BIB; EFS; KAB; UPW); Emollient (f; KAB); Expectorant (f; EFS; KAB); Hypoglycemic (f; EB49:406); Hypotensive (f; BIB); Laxative (f; UPW); Nephrotonic (f; KAB); Stimulant (f; KAB); Stomachic (f; DAW); Suppurative (f; KAB); Vermifuge (f1; FNF; HH2).

Indications (Leek):

Abscess (f; DAW); Angina (1; FNF); Arthrosis (f; DAW; JLH); Ascariasis (f; DAW); Bladderstone (f; KAB); Boil (f; KAB); Calculus (f; KAB); Callus (f; JLH); Cancer (f1; FNF; JLH; WO3); Carcinoma (f1; FNF; JLH); Chafing (f; KAB); Chilblain (f; KAB); Cold (f; DAW); Conjunctivosis (f; KAB); Corn (f; JLH); Cough (f; DAW); Cystosis (f; KAB); Diabetes (1; FNF; EB49:406); Diarrhea (f; DAW); Dysentery (f; DAW); Enterosis (f; JLH); Gastrosis (f; JLH); Hepatosis (f; JLH); High Blood Pressure (f1; BIB; FNF); Impotence (f; DAW); Induration (f; JLH); Infection (f1; DAW; FNF); Mycosis (f1; DAW; FNF); Nephrosis (f; DAW); Ophthalmia (f; KAB); Otosis (f; DAW); Polyp (f; JLH); Pulmonosis (f; DAW); Rhinosis (f; DAW; JLH); Ringworm (f1; DAW; FNF); Scrofula (f; DAW); Stone (f; KAB); Tuberculosis (f1; DAW; FNF); Tumor (f; JLH); Uterosis (f; JLH); Wart (f1; FNF; JLH); Wen (f; JLH); Worm (f1; DAW; FNF; HH2); Wound (f; BIB).

Dosages (Leek):

FNFF = !!

Whether you take the narrow or broad taxonomic interpretation, all parts of most varieties are eaten somewhere. Of the stricter interpretation of leek, Tanaka says, all parts, except root, are used in soups and stews (TAN); Facciola says of the stricter Porrum leek group, leaves and stalks are eaten, boiled, braised, steamed, or raw, in "salads. … Widely used in gourmet French cuisine, especially *vichyssoise. Glamorgan sausages,* popular in Wales, are meatless sausages made from leeks, cheese, and breadcrumbs. Sprouted seeds are eaten like those of onion or garlic" (FAC).

- Asian Indians suggest leek juice with cream for chafing, chilblains, and sore eyes (KAB).
- Brits say, "to eat leeks in Lide and Ramsins in May and all the year physicians may play (AAH)."
- Cambodians use the plant as a diuretic and emollient (KAB).
- Danish suggest boiled leek for induration of the abdomen and stomach (JLH).
- French suggest crushed leek for tumors of the joints (JLH).
- Germans plastered the plant or its juice on cancers (JLH).
- Lebanese tie bruised leaves around cuts (BIB).
- Lebanese with HBP are encouraged to eat more of the diuretic leek (BIB).

Extracts (Leek):

Viewing this as generic with garlic, but dilute, I would use larger doses for any of the indications for which garlic has proven successful, including antisepsis, both prophylactic and curative, and as an immunostimulant.

GARLIC (*ALLIUM SATIVUM* L.) +++ LILIACEAE

Synonyms:

Porrum sativum Mill.

Notes (Garlic):

> **We remember the fish, which we did eat in Egypt freely; the cucumbers, and the melons, and the leeks, and the onions, and the garlick: But our soul is dried away: there is nothing at all, beside this manna, before our eyes.**
>
> **Numbers 11:5–6 (KJV)**

> **We remember the fish we ate in Egypt for nothing, the cucumbers, the melons, the leeks, the onions, and the garlic; but now our strength is dried up, and there is nothing at all but this manna to look at.**
>
> **Numbers 11:5–6 (RSV)**

> **How we remember the fish that we used to eat in Egypt for nothing, the cucumbers and the water melons, and the leeks and the onions, and the garlic; But now our soul is dried away. Our eyes are on nothing at all except the manna.**
>
> **Numbers 11:5–6 (NWT)**

To me, garlic is biblical food farmacy for cancer, cardiopathy, cerebropathy, diabetes, and sepsis. I often ponder a paradox. While many health writers will tell you not to chronically tweak your immune systems with such things as boneset, echinacea, elderberry, huangchi, I have not heard them say the same thing about garlic, one of the best immune boosters, with at least a dozen compounds reported to boost the immune system. Back during the anthrax scare, I spoke of garlic (seriously, although most people assumed I was talking frivolously) as the "herbal duct tape" because of its multiple antiseptic *and* immune-stimulating phytochemicals. The Centers for Disease Control (CDC), Food and Drug Administration (FDA), and the National Institutes of Health (NIH) are quick to say that people with depressed immune systems are more likely to get the "disease of the year"

FIGURE 1.6 Garlic (*Allium sativum*).

(my flippant term for, as examples, anthrax, bird flu, SARS, or West Nile virus). However, I have never heard them recommend boosting the immune system with antiseptic garlic to improve one's odds against these very diseases. That is a ponderous paradox. Back during our first anthrax scare, there was no clinical proof that Cipro or garlic was useful against anthrax. Garlic is a broad-spectrum antibiotic, not liable to lead to multiple drug resistance. And now garlic has been shown *in vitro* to help slow anthrax (X14598920). Will the phytochemical politicians again advise us not to suggest that garlic can help with the "disease of the year?" I can get garlic immediately in case of anthrax attack — but I might need several days, a doctor's appointment, a prescription, and a few dollars to get Cipro. Garlic would help; Cipro would help; resistance will emerge to pure Cipro; resistance would be less likely to develop to the synergic mix of dozens of antiseptic compounds in whole garlic. The year 2004 saw garlic identified as the herb of the year. If I were told I could have only one medicinal herb, it would certainly be the biblical garlic. I think of it as the medicine of the millennium. I confess that I have not reviewed all the PubMed citations — 120 alone in the first quarter of 2005.

Common Names (Garlic):

Aglidion (Greek; KAB); Aglio (It.; Malta; AVP; EFS; KAB); Ail (Fr., Haiti; AHL; AVP; EFS); Ail Comun (Fr.; AVP); Ail de Cuisine (Haiti; AHL); Aiu (Rom.; KAB); Ajo (Bel.; Sp.; Dr.; AHL); Ajo comun (Sp.; EFS); Ajusa (Aym.; Bol.; Que.; DLZ); Alho (Mad.; Por.; AVP); Alho Manso (Por.; JFM); Alho Ordinario (Por.; AVP; EFS; KAB); All (Cat.; KAB); Arishtha (Sanskrit; KAB); Asna Pullu (Que.; DLZ); Ay (Cat.; KAB); Ayo (Ga; KAB); Banag (Ilo.; KAB); Baoang (Vis.; KAB); Bauang (Vis.; KAB); Baratchouria (Bas.; AVP); Bauangpoti (Jolo; KAB); Bawang (Java; Tag.; KAB; TAN); Bawang Puteh (Mal.; EFS); Belluli (Can; KAB); Bhutabhna (Sanskrit; KAB); Chesnok (Rus.; KAB); Churl's Treacle (Eng.; KAB); Czosnek (Pol.; AVP; KAB); Cu Toi (Annam; KAB); Da Suan (Pin.; JAD); Dawang (Malaya; KAB); Dirghapatraka (Sanskrit; KAB); Foom (Arabic; KAB); Fokhagyma (Hun.; EFS); Ganda (Vis.; KAB); Garlic (Eng.; CR2); Gartenlauch (Ger.; AVP); Gogpa (Tibet; NPM); Gogpa (Sherpa; NPM); Grinjana (Sanskrit; KAB); Hsiao-suan (Chi.; EFS); Hvildog (Den.; AVP); Hwitolk (Swe.; AVP); Ivimba (Xosa; KAB); Kanchai (Cam.; KAB); Katukanda (Sanskrit; KAB); Kesumphiu (Burma; KAB); Knofflook (Dutch; AVP); Knoblauch (Ger.; EFS); Knoflook (Dutch; EFS); Knoflook (Sur.; AVP); Knoplook (Dutch; KAB); Kyalic (Miskito; ULW); Kyatthoubega (Burma; KAB); Kyetthwunbya (Burma; KAB); Lahasun (Bhojpuri; NPM); Lahsan (Hindi; KAB); Lai (Haiti; AHL); Lasan (Guj.; Hindi; India; EFS; KAB); Lashan (Beng.; KAB); Lashuna (Sanskrit; KAB); Laso (Vis.; KAB); Lasun (Beng.; Danuwar, Mooshar, Sunwar, Tharu; KAB; NPM); Lasuna (Sanskrit; Malaya; Mar.; EFS; KAB); Lasunas (Mar.; KAB); Lauch (Ger.; KAB); Lava (Newari; NPM); Lay (Creole; Haiti; TRA; VOD); Lye (Haiti; AVP); Lehsun (Urdu; KAB); Lha-ming-khrag (Tibet; NPM); Look (Dutch; KAB); Lossun (Kon.; KAB); Mahakanda (Sanskrit; KAB); Mahaushana (Sanskrit; KAB); Mahusudha (Sanskrit; EFS); Manmuk (Rai; NPM); Manul (Korea; TAN); Mlechhakanda (Sanskrit; KAB); 'Mpunzi (Xosa; KAB); Naharu (Assam; KAB); Ninniko (Japan; TAN); Noh (Tamang; NPM); Poor Man's Treacle (Eng.; KAB); Q'achi (Callawaya; DLZ); Rahushhishta (Sanskrit; KAB); Rasona (Sanskrit; KAB); Rasonaka (Sanskrit; KAB); Rasun (Beng.; San.; KAB); Samanachupang (Krobo; KAB); Samsak (Tur.; KAB); Sarimsak (Tur.; EFS; EB49:406); Sarmisak (Tur.; EB54:155); Sar Moussak (Tur.; AVP); Saum (Arab.; KAB; TAN); Scorodon (Greek; KAB); Seer (Iran; AVP); Sgog Skya (Tibet; TIB); Shuklakandha (Sanskrit; KAB); Shunuan (Dec.; KAB); Sir (Iran; EFS; KAB); Som (Arab.; EFS); Soom (Arab.; AVP); Suan (China; KAB); Suan t-eou (China; KAB); Sudulunu (Sin.; KAB); Tafarnuwa (Hausa; KAB); Tai-Tsoua (China; AVP); Ta Suan (China; KAB); Taum (Arab.; KAB); Tchesnoks (Rus.; AVP); Teum Tal Ichell (Malta; KAB); Theriague du Paysan (Fr.; KAB); Thom (Arab.; GHA); Thoum (Arab.; AVP; BOU); Thujsa Pullu (Aym.; Bol.; DLZ); Tiskert (Ber.; BOU); Tissert (Ber.; BOU); Tongologasy (Hova; KAB); Tongolonkova (Hova; KAB); Toum (Arab.; BOU); Ugragandha (Sanskrit; KAB); Usturoiu (Rom.; KAB); Vatari (Sanskrit; KAB); Vellaippundu (Tam.; KAB); Vellullitellagadda (Tel.; KAB); Vitlok (Swe.; EFS); Yavaneshta (Sanskrit; KAB).

Activities (Garlic):

Acarifuge (1; KAL); ACE Inhibitor (1; NP6:1); Adaptogenic (1; X15881870); Alterative (f; KAP; PED); Alexiteric (f; KAB); Amebicide (f1; APA; X11101670); Analgesic (f1; BGB; ULW); Angiotensin Receptor Blocker (1; NP6:1); Anodyne (f; DAD); Androgenic (1; KAL); Anthelmintic (f1; KAL; WHO); Antiaflatoxin (1; X1394115); Antiaggregant (f123; APA; FNF; KOM; PH2; SHT; VOD); Antiallergic (f1; AKT; JN131:1075s); Antiandrogenic (1; DAD; JN131:1075s); Antianemic (f1; JN131:1016s); Antiangiogenic (f; BO2); Antiarteriosclerotic (f12; KAL); Antiarthritic (f1; KAL); Antiatherogenic (2; BGB; WHO); Antibiotic (f12; AKT; PNC; PED); Anticancer (f1; KAL; SKY); Anticholinesterase (1; KAL); Anticollagenase (f; BO2); Antidiabetic (f1; KAL; PNC); Antidote (f; VOD; WO2); Antieicosanoid (1; BO2); Antifertility (1; KAL; WO2); Antifungal (f12; KAL; SKY); Antigiardal (1; X11101670); Antihepatotoxic (1; CAN; KAL); Antihypercholesterolemic (1; WHO); Antihyperglycemic (1; KAP); Antihyperlipidemic (1; WHO); Antihypertensive (f1; SKY; VOD; WHO); Antiinflammatory (f1; APA; BGB); Antiintegrase (1; KAL); Antiisoprene (1; BO2); Antileukemic (1; JN131:1027s); Antilymphomic (1; BO2); Antimycobacterial (1; PR14:303); Antimycotic (f12; BGB; KAL); anti-NF-kB (1; BO2; JN131:1020s); Antinitrosaminic (1; VOD; JN131:1027s); Antioxidant (1; AKT; KAL; PH2; SHT; WO3); Antiplatelet (1; WHO); Antiproliferant (1; X15890236); Antiprostaglandin (1; WHO); Antipyretic (f1; WHO); Antiradicular (1; VOD); Antirheumatic (f1; KAL); Antiseptic (f123; AKT; APA; GHA; PH2; PNC; SKY; VOD); Antisickling (1; JN131:1016s); Antispasmodic (f1; DLZ; PED; WHO); Antistress (1; KAL); Antithrombotic (f1; FAY; PH2; PNC); Anti-TNF (1; BO2); Antithyroid (1; KAL); Antitubercular (1; PR14:303); Antitumor (f1; BGB; PNC); Antiulcer (f1; X11238826); anti-VEGF (1; BO2); Antiviral (1; AKT; APA; KAL; SKY); Aphrodisiac (f1; DAD; WHO); Apoptotic (1; BO2; X15890236); Bactericide (2; AKT; FAD; KOM; SKY; WHO); Beta-Blocker (1; NP6:1); Calcium-Channel Blocker (1; NP6:1); Cardioprotective (f1; FNF; VOD; X15881870); Cardiotonic (f1; AKT; JFM); Carminative (f1; PED; RIN; WHO); Choleretic (f1; MAM); Decongestant (f1; FAY); Detoxicant (f; AKT; FAY); Diaphoretic (f; JFM; PED; PNC); Digestive (f1; AKT; PED); Diuretic (f1; FAD; WHO); Edemagenic (1; WO3); Emmenagogue (f1; JFM; WHO); Estrogenic (1; KAL); Expectorant (f; PED; PNC; WOI); Fibrinolytic (123; APA; GHA; KAL; KOM; PH2); Fungicide (f12; FAD; KOM; MAM); Gastroprotective (f1; FNF; VOD); Gastrotonic (f; KAB); Glutathiogenic (1; BO2; PH2; JN131:1010s); Hepatoprotective (1; BGB; JFM; WO3); Hyperglycemic (1; PNC); Hypocholesterolemic (12; AKT; DAD; FAD; PH2; SHT); Hypoglycemic (f1; DAD; KAL; PED; PNC; X15738612); Hypolipidemic (f1; BGB; DAD; PED; PNC); Hypoperistaltic (2; WHO); Hypotensive (12; AKT; BGB; FAD; SHT); Hypotriglyceridemic (1; AKT; VOD); Hypouricemic (f; JFM); Immunostimulant (1; AKT; BGB; CAN; FAY; PED); Insectifuge (f1; KAL); Insulin-sparing (1; PNC); Interferonigenic (1; X11238818); Interleukenogenic (1; WO3); Lactogogue (f; NMH); Larvicide (1; WO2); Lipogenic (f; KAB); Lipolytic (12; KOM; PH2; SHT; WHO); Lymphocytogenic (1; AKT); Memorigenic (f1; GHA; JN131:1016s); Mucolytic (1; MAB); Myocontractant (1; CAN); Myorelaxant (1; CAN); Nervine (PED); NKC Enhancer (1; AKT; PH2); NO-genic (1; KAL); Orexigenic (f; KAB); Ovicide (1; WO3); Oxytocic (1; WO2); Parasiticide (f1; AKT); Phagocytotic (1; AKT; JN131:989s); Prooxidant (1; BO2); Protisticide (1; KAL); Radioprotective (JN131:1010s); Rubefacient (f; JFM); Sedative (1; WHO); Spermicide (1; KAL); Tonic (f; KAB); Vasodilator (1; SHT; WHO); Vermifuge (f1; AKT; APA; VOD); Vulnerary (f1; PED). (For much more information, see Koch and Lawson's excellent *Garlic Book* (KAL): I do not have time to read and rescore, but if they tabulate clinical trials demonstrating the efficacy of whole garlic, it should get a 3.)

Indications (Garlic):

Abscess (f1; DAA; PNC); Acne (f; FAD); Adenopathy (f; JLH); Aegilops (f; JLH); Aging (f1; PH2); Allergy (f1; AKT; EGG); Alopecia (f1; WHO; WO2); Alzheimers (1; JN131:1010s); Amebiasis (f12; FAY; PNC); Amenorrhea (f1; BGB; JFM; LIB); Anemia (f; DAD); Angina (f1; FNF); Anorexia (f; FAY); Anthrax (1; X14598920); Appendicitis (1; FAY; PNC); Aphthae (1; KAL); Arrhythmia (f; EGG); Arteriosclerosis (12; BGB; BIS; FAD; KAL; PH2); Arthrosis (f1; FAD; KAL; PHR; PH2; EB49:406); Asthma (f1; PNC; WHO); Atherosclerosis (f123; AKT; APA; PHR; SHT: WHO); Athlete's Foot (f12;

LIB; TGP); Bacillus (1; LAW; X10548758); Bacteria (1; JFM; PH2); Bite (f; FAY; JFM); Boil (f1; DAA); Bronchiestasis (1; KAL); Bronchosis (f12; FAD; PHR; PH2; BOD; WHO); Burn (f12; KAL); Callus (f; JFM; PH2); Cancer (f12; AKT; FAD; PH2); Cancer, abdomen (f1; AKT; FNF; JLH); Cancer, bladder (f1; FNF; JLH; X11341051; X11238811); Cancer, breast (f1; BRU; JN131:989s); Cancer, colon (f1; AKT; (f1; FNF; JLH)); Cancer, esophagus (f1; JN131:1075s); Cancer, gland (1; X11238818); Cancer, liver (f1; BO2; PR14:564); Cancer, lung (f1; BRU; FNF; JLH; JN131:989s); Cancer, prostate (f1; X11102955); Cancer, skin (f1; FNF; JLH); Cancer, stomach (f1; AKT; VOD; X11238811); Cancer, uterus (f1; FNF; JLH); Candidiasis (f12; CAN; KAL; TRA; VOD); Carbuncle (f; FAY); Cardiopathy (f123; BGB; EGG; FAD; SKY; VOD); Caries (f1; FNF; KAB); Catarrh (f1; AKT; BGB); Celiac (1; KAL); Chilblain (f; EGG); Childbirth (f; JFM; KAB); Cholecocystosis (f; APA); Cholera (f1; PNC; TRA); Chronic Fatigue (f; JFM); Circulosis (f; DLZ); Coccidiosis (1; KAL); Cold (f12; AKT; FAD; GHA; PHR; PNC); Colic (f1; GHA; WHO); Colosis (1; KAL; LAW); Congestion (f1; FAY); Conjunctivosis (f; GHA); Constipation (f; JFM; PH2); Convulsions (f; KAB; PHR); Corn (f; EGG; JLH; LIB; PHR); Cough (f12; APA; FAD; PHR); Cramp (f; PH2); Cryptococcus (1; DAA); Cystosis (f; JFM); Cytomegalovirus (1; KAL); Dandruff (f; EGG; GHA); Deafness (f; LAW); Debility (f; PH2); Dementia (1; X11238823); Dermatosis (f1; AKT; DAA; DAD; KAL; PNC; VOD); Diabetes (f1; GHA; MAM; PH2; PNC; X15738612); Diarrhea (f1; AKT; GHA; PNC); Diptheria (f; DAA; DAD; EGG); Dropsy (f; KAB); Dyspepsia (f1; AKT; JFM; KAL); Dysentery (f12; AKT; DAD; FAD; PNC); Dysmenorrhea (f; PHR; PH2); Dyspepsia (f1; BIS; PNC; WHO); Dyspnea (1; FAD; FAY); Earache (f1; FAD; TRA); Edema (f; JFM; PNC); Enterosis (f12; AKT; APA; FAD; GHA; PH2; VOD; WHO); Epigastrosis (2; WHO); Epilepsy (f; AKT; FAY); Escherichia (1; LAW; WO2); Felon (f; JLH); Fever (f12; FAD; PHR; PH2); Fibrinolytic (SHT); Fibroid (f; DAD; JLH); Filaria (1; KAL); Flatulence (f1; GHA; WHO); Flu (f1; AKT; APA; EGG; KAL; PNC; TRA); Fungus (f1; AKT; JFM); Gangrene (f; EGG; KAP); Gas (f1; DAD; GHA; JFM; PH2; TRA; VOD); Gastroenterosis (f12; BIS; DAD; FAD); Gastrosis (f12; AKT; FAD; FAY; PH2; WHO); Giardia (f1; KAL; X11101670); Gout (f; DEP; FAD; JFM); Headache (f; JFM); Helicobacter (1; AKT; X11238826); Hemorrhoid (f; JFM); Hepatosis (f12; APA); Hepatotoxicity (Acetaminophen); (2; MAM); Herpes (f1; KAL; TRA); High Blood Pressure (f12; AKT; FAD; PH2; SHT; ULW; WHO); High Cholesterol (123; AKT; APA; KAL; PH2; SHT); High Triglycerides (123; AKT; APA; KAL; SHT); HIV (1; KAL); Hookworm (f1; AKT; KAL; WHO); HPS (1; X15833681); Hyperlipidemia (123; SHT; WHO); Hyperperistalsis (2; WHO); Hypoglycemia (f; FAY); Hypotension (f; DAD); Hysteria (f; JFM); Immunosuppression (12; PHR; SKY); Impotence (f1; AKT; X112388); Infection (f123; AKT; APA; EGG; GHA; PH2; PNC; SKY); Insomnia (f; JFM); Intermittent Claudication (12; BGB; SHT; TGP); Itch (f1; TRA; VOD); Keratosis (1; KAL); Lambliasis (1; KAL); Laryngosis (f1; KAL; KAP); Lead Poisoning (1; PNC); Leishmania (1; X11119248); Leprosy (f; JFM); Leukemia (f; JLH); Leukoderma (f; EGG; KAB); Lumbago (f; PH2); Lupus (f; KAL); Lymphoma (1; BO2; JLH); Malaria (f; DAD; EGG; JFM); Malnutrition (f; VOD); Mange (f; JFM); Melancholy (f; JFM); Melanoma (1; JN131:1027s); Meningosis (f; DAA); Menopause (f; JFM); Mucososis (1; KAL); Myalgia (f; PHR; PH2); Mycosis (f1; AKT; PNC; TRA); Myofascitis (f; DAA); Nausea (f1; TRA; WHO); Nephrosis (f1; DLZ; GHA; KAL); Neuralgia (f1; KAL; PHR); Neuroblastoma (1; JN131:1027s); Nicotinism (1; KAL); Odontosis (f; KAB); Osteoporosis (1; X15173999); Otitis (f1; BOU; FAD; SKY); Pain (f; GHA; JFM; PH2; EB49:406); Palpitation (f; JFM); Paradentosis (1; KAL); Paralysis (f; KAB); Parasite (f1; AKT; ULW); Paratyphoid (f; KAP); Paratyphus (f; LAW); Periodontosis (1; LAW; X15892950); Pertussis (f12; DAD; FAD; FAY; PNC); Pharyngosis (f12; PHR); Pinworm (f1; AKT; FAY); Pneumonia (f1; DAD; KAL; VOD); Poliomyelitis (1; KAL; LAW); Polyp (f; JLH); Porphyromonas (1; X15892950); Pulmonosis (f; KAP; VOD); Pulposis (1; LAW); Raynaud's Disease (2; TGP); Respirosis (f1; AKT; BGB; KAL; PH2; WHO); Rheumatism (f1; FAD; KAL; PH2; VOD; EB49:406); Rhinosis (f12; BGB); Ringworm (f1; APA; DAA; WHO); Roundworm (f1; KAL; WHO); Salmonella (1; WO2); Scabies (f1; DAA; JFM); Sciatica (f; PHR; PH2); Senile Dementia (1; KAL; X11238823); Sepsis (f1; KAL); Shigella (1; LAW; TRA; WO2); Sinusosis (f1; FAY); Snakebite (f; FAD; FAY; GHA); Sore (f1; FAD; JFM); Sore Throat (f1; KAL); Soroche (f; KAL); Splenosis (f; EGG; KAB); Sporotrichosis (1; KAL); Staphylococcus (1; LAW); Stomachache (f; FAY);

Stomatosis (f12; PHR); Streptococcus (12; X9354029; X15892950); Stroke (1;JN131:1010s); Swelling (f; AKT; FAD; FAY; GHA; JFM); Syncope (f; KAB); Tapeworm (f; JFM); Thirst (f; KAB); Thrombosis (f123; APA; EGG; GHA; KAL; KOM; PH2); Toothache (f1; TRA); Tonsilosis (f1; LAW); Trachoma (f; DAA); Trichomonaisis (f1; DAA); Trypanosomiasis (1; KAL); Tuberculosis (f1; APA; EGG; GHA; JFM; KAL; TRA); Typhoid (f; DAA); Typhus (f1; DAD; KAL); Ulcer (f1; AKT; X11238826); Ulcus cruris (12; KAL); UTIs (f1; WHO); Vaginosis (f12; APA; DAA; KAL); Varicosity (f; DLZ; JFM); Virus (1; PH2; TRA); Vitiligo (f; EGG); Wart (f; EGG; PHR; PH2); Wen (f; JLH); Whitlow (f; JLH); Worm (f1; AKT; APA; EGG; JFM; VOD); Wound (f; GHA; PHR); Yeast (f12; APA; CAN; JAD; WO2).

Dosages (Garlic):

FNFF = !!!

Bulbs and leaves widely eaten more as spice than vegetable; seeds, seedlings, and vivaparous sprouts also eaten (FAC, TAN; EB54:155); 1–5 cloves/day (APA); 2–4 g, 3 ×/day (CAN); 2–4 ml tincture (1:5 in 45% ethanol) 3 ×/day (CAN); 0.03–0.12 ml garlic oil/day (CAN); 2–8 ml garlic syrup (CAN); 2–4 ml garlic juice (CAN); 9–15 g fresh bulb (FAY); 1.5–6 g fresh tuber (KAP); 1–2 minims garlic oil (KAP); 4 g fresh garlic/day (KOM); one 400-mg StX/day; 3–4 550-mg capsules 3 ×/day (NH). One enteric-coated 400-mg tablet (StX to contain at least 3 mg allicin potential) 1 ×/day at mealtime (NH); 1/4–1/2 cup fresh bulb (PED); 6–12 g dry bulb (PED); 9 g dry bulb/45 ml alcohol/45 ml water (PED); 2–4 ml garlic juice (PNC); 2–8 ml garlic syrup (PNC); 600–900 mg/day coated garlic (SHT); 4 g garlic or one average clove; 5000 µg allicin/day (SKY); 8–12 g bulb, 3 ×/day (TRA).

- Arabs inhale steam from boiling water with garlic for tuberculosis (GHA).
- Arabs suggest eating cloves daily for colic, diabetes, diarrhea, enteralgia, and swellings (GHA).
- Arabs warm skinned cloves in cow's milk, add fat, heat until thickened; add ground ginger, laurel, pepper, and senna; take for colds, cough, nephrosis, and improving memory (GHA).
- Arabs use ashes from burnt cloves to treat eye infections (GHA).
- Bolivians suggest a clove garlic for arteriosclerosis, cerebral congestion, high arterial pressure; garlic soup for circulatory and kidney problems (DLZ).
- Bolivians suggest three cloves garlic mashed in one cup of milk with honey for arteriosclerosis, gastric, respiratory, or pulmonary problems (DLZ).
- Bolivians suggest a glass of water with 20 drops garlic tincture for hemorrhage, high blood pressure, palpitations, varicose veins (DLZ) (maybe a martini with garlic instead of onion? (JAD).
- Dominicans suggest rubbing garlic onto rheumatism (AHL).
- Haitians take bulb decoctions, juices, or teas for bronchitis, dermatoses, gas, itch, pneumonia, and other pulmonary problems; they eat the bulbs for worms and hypertension (VOD).
- Mexicans suggest taking three cloves in milk each morning to prevent malaria and tuberculosis (JFM).
- Mexicans take a leaf decoction as emmenagogue (JFM).
- Peruvians eat the cloves for arrhythmia, arteriosclerosis, cardiopathy, catarrh, diabetes, embolism, hysteria, hypertension, malaria, menopause, smoker's syndrome, splenosis, and thrombosis (EGG).
- Peruvians suggest rubbing cut garlic on corns, dandruff, infection, itch, rheumatism, sores, vitiligo, and warts (EGG).
- Trinidadans take garlic decoction for dyspepsia, enterosis, hypertension, stomachache, strangury, and rubbing the crushed cloves on the belly to facilitate childbirth (JFM).
- Yucatanese suggest three chipped bulbs in milk or tomato juice for diabetes, malaria, or rheumatism, or steeping in alcohol 5 days and taking up to 20 drops a day for asthma, atherosclerosis, or hypertension (JFM).

Downsides (Garlic):

Class 2c (AHP, 1997). Some thiol-bearing compounds in garlic and onion and their relatives can cause acantholysis *in vitro* (Brenner et al., 1995) and possibly pemphigus *in vivo*. "More than 5 cloves a day may induce flatulence and heartburn (Castleman, 1996) and 'thin blood'" (people taking blood thinners may overthin their blood thereby). Some people are very allergic to garlic. Contraindicated in hyperthyroid (TRA); Commission E reports rare GI disturbances, allergic reactions, and change of odor of skin and breath (Commission E). Allergic reactions of contact dermatosis and severe asthmatic attacks (from inhalation of garlic powder) may occur. Topical application of garlic or garlic oil may cause local irritating effects. Nausea, vomiting, and diarrhea may occur following ingestion of fresh garlic bulbs, extracts, or oil (AEH1). Sulfides may irritate the GI tract or cause dermatosis (CAN). Fresh garlic is reportedly dangerous to children (AHP). Use sparingly with children under 2 years of age; may irritate mouth or stomach if used too liberally (WAM). Then there is Miller and Murray's extremely cautious but not critical review (MAM). Although possibly "useful for mild hypertension … routine use is not recommended (MAM)" After informing us that no drug-drug interactions have been reported for garlic, they feed us a long list of potential drug-drug interactions. Hasty readers, especially medical doctors, will take this as a proven drug-garlic interaction, "Avoid concomitant use … with NSAIDS, anticoagulants and drugs that inhibit liver metabolism" (e.g., cimetidine (Tagamet), ciproflaxin (Cipro), clarithromycin, diltiazem (Cardizem), enoxacin, erythromycin, fluoxetine (Prozac), fluvoxamine, itraconazole, ketoconazole, nefazodone, paroxetine (Paxil), ritonavir) "may at least additively and perhaps synergistically interact with garlic." Watch also drugs extensively metabolized by the liver (alprazolam, amitriptyline, astemizole, carbamazepine, cisapride, clozapine, corticosteroids, cyclosporine, desipramine, diazepan (Valium), imipramine, phenytoin (Dilantin), propranolol, terfenadine (Seldane), theophylline, triazolam, warfarin (Coumadin), and drugs that may be affected by liver inhibition (e.g., propranolol, diazepam) (MAM). Miller and Murray (1998) tabulate allergic contact dermatosis, burning GI sensations, diaphoresis, diarrhea, light-headedness, menorrhagia, metrorrhagia, nausea, spinal epidural hematoma, and vomiting as side effects of garlic. They even try to attribute a case of spinal hematoma to garlic (in an 87-year-old male ingesting 2 g daily "to prevent heart disease." Perhaps they are too eager to accentuate the negative, attributing the problem "to garlic's ability to inhibit normal platelet function." Still accentuating the negative, they talk about rats given massive doses (50 mg/day garlic powder) developing degenerative changes in 45 days and severe testicular lesions after 70 days (MAM). The credibility of their uncritical data is questionable; for example, on one page (i.e., p. 144), they talk about 0.75 mg garlic essential oil divided in three doses a day causing anorexia, nausea, severe vomiting, diarrhea, marked weight loss, metrorrhagia, and menorrhagia; yet on the following page (i.e., p. 145), they casually discuss a dosage more than three orders of magnitude higher for 20 days lowering platelet aggregation from 30.37% to 21.21% (MAM). I feel I have to report this to my readers, although I consider it mostly hyperbolic. At 3 × 300 mg/day coated garlic powder tablet dosages, GI discomfort was the most frequent side effect (also bloating, dizziness, headache, hypotensive circulatory reactions, outbreaks of sweating); daily doses of 900 to 1200 mg were associated with garlic odor. "May potentiate the effect of antihypertensive and anticoagulant medications" (SHT). No known contraindications during pregnancy and lactation (SKY).

Extracts (Garlic):

Of the 120 new PubMed abstracts for garlic in the first quarter of 2005, some were disappointing. Iranian scientists (e.g., Jelodar et al., 2005) found, contrary to my expectations, that garlic, but not onion and fenugreek, is hypoglycemic in experimental rats. I think they should have also compared the mix of the three biblical herbs, anticipating synergy or additivism, all recommended in Persian folklore medicine as good for diabetes (X15738612). Bakri and Douglas (2005) extended the well-known antiseptic activity of garlic to bacteria involved in periodontitis (X15892950). In general, the minimal inhibitory and minimum bactericidal concentrations for Gram-negative strains (garlic MIC range 35.7–1.1mg/ml;

allicin mean MIC 4.1 μg/ml; mean MBC 7.9 μg/ml) were lower than those for the Gram-positive strains tested (garlic MIC range 142.7–35.7 mg/ml; allicin mean MIC 27.5 μg/ml; mean MBC 91.9 μg/ml). The putative periodontal pathogens had among the lowest MICs (17.8–1.1 mg/ml garlic) and MBCs (35.7–1.1 mg/ml garlic) (X15892950). Verma et al. (2005) demonstrated adapotogenic activity of garlic oil on exercise tolerance in coronary patients. Thirty patients were given garlic oil for 6 weeks. The 6-week treatment reduced heart rate at peak exercise and resultant workload on the heart (X15881870). Kim et al. (2005) showed that too much diallyl disulfide could be cytotoxic to neuronal cells. Levels of free radicals and membrane lipid peroxidation increased dose dependently at levels higher than 25 μM (X 15950962). Chang et al. (2005) found that sodium 2-propenyl thiosulfate had cyclooxygenase inhibitory as well as antiaggregant activity in canine platelets (X15850716). Akyuz and Kaymakoglu (2005) suggest garlic and lamivudine in combination as a natural/chemotherapy for hepatopulmonary syndrome (HPS), one I never heard of previously. HPS is characterized by abnormalities of arterial oxygenation in patients with chronic liver disease, with or without portal hypertension. There is no definitive treatment except liver transplantation. One HPS patient with liver cirrhosis and HPS received garlic and lamivudine for 3 years. Signs of liver failure and hypoxemia gradually improved, indicating that lamivudine may improve the functional reserve of the liver, while garlic may help to reduce the signs and symptoms of HPS (X15833681). Chang et al. (2005) suggest that garlic oil's anticarcinogenic activities may be due to (1) antioxidant activity, (2) induction of apoptosis, (3) inhibition of DNA-adduct formation, (4) modulation of immune function, and/or (5) modulation of xenobiotic-metabolizing enzyme activities (X15796590). I can suggest dozens of other phytochemical reasons. For several other useful phytochemical activities in whole garlic, consult the multiple-activity-menu site at the USDA (http://www.ars-grin.gov/duke/dev/all.html) — and one might well be overwhelmed by the 19-page printout. Active hypoglycemic compounds may have insulin-sparing activity, the thiol groups competing for insulin with the inactivating compounds (PNC). Ajoene is antiaggregant, antilipoxygenase, antiprostaglandin (CAN; PNC) synergizes the antiaggregant activity of dipyramidole, forskolin, indomethacin, and prostacyclin. Garlic (or allicin) is antiseptic to *Actinobacter, Aeromonas, Aspergillus, Bacillus, Candida albicans, Citrobacter, Corynebacterium, Cryptococcus, Epidermophyton, Escherichia coli, Hafnia,* Herpes, Influenza, *Klebsiella, Microsporum, Mycobacterium, Pasturella, Proteus, Providencia, Pseudomonas, Rhodotorula, Salmonella paratyphi, Salmonella typhi, Shigella dysenterica, Staphylococcus aureus, Torulopsis, Trichomonas* sp., *Trichophyton, Trichosporum,* and *Vibrio cholera* (CAN; PNC); LD50 = 60 mg/kg ivn mus (SHT); 120 mg/kg scu mus M11 (SHT) might be a good way to cut back on your grocery bill (except for garlic), if you believe this quote: "Rats fed up to 2000 mg/kg garlic extract for 6 months showed no weight loss but did show a slightly reduced food intake relative to controls. There were no changes in renal function, hematologic parameters, or selected serologic parameters; and there was no evidence of any pathologic changes in organs or tissues." Experimentally antiaggregant, bactericidal, diuretic, fungicidal, hypocholesterolemic, hypotensive (FAD; FNF). Clinical studies suggest utility in arteriosclerosis, cardiopathy, GI disorders, high cholesterol, and high blood pressure (FAD). Commission E approvals differ: Blumenthal et al. (1998) approve 4 g fresh garlic or equivalent preparations "supportive to dietary measures at elevated levels of lipids in blood" and preventive measures for age-dependent vascular changes," while Gruenwald et al. (1998) approve garlic for almost the same things for which they approve echinacea, viz. arteriosclerosis, bronchosis, cold, cough, fever, pharyngosis, stomatosis, and "tendency to infection."

ALOE (*ALOE VERA* (L.) BURM. F.) (GEL) ++ ALOEACEAE (LILIACEAE)

SYNONYMS:

Aloe barbadensis Mill; *Aloe indica* Royle, nom. nud.; *Aloe perfoliata* var. *vera* L. [basionym]; *Aloe vulgaris* Lam., nom. illeg. *fide* USN

FIGURE 1.7 Aloe (*Aloe vera*).

NOTES (ALOE):

And there came also Nicodemus, which at the first came to Jesus by night, and brought a mixture of myrrh and aloes, about an hundred pound weight. Then took they the body of Jesus, and wound it in linen clothes with the spices, as the manner of the Jews is to bury.

John 19:39–40 (KJV)

Nicode'mus also, who had at first come to him by night, came bringing a mixture of myrrh and aloes, about a hundred pounds' weight. They took the body of Jesus, and bound it in linen cloths with the spices, as is the burial custom of the Jews.

John 19:39–40 (RSV)

Nicode'mus also, the man that came to him in the night the first time, came bringing a roll of myrrh and aloes, about a hundred pounds [of it]. So they took the body of Jesus, and bound it up with bandages with the spices, just the way the Jews have the custom of preparing for burial.

John 19:39–40 (RSV)

Early authors, mostly American or English, on biblical botany, and even I, concluded that the biblical aloe of the New Testament was *Aloe perryi* or *Aloe socotrina*. Knowing how taxonomically difficult this genus of some 250 to 300 species is, I am relieved to accept the opinion of Israeli scientist Michael Zohary, who concludes that the aloe of John 19 was "probably an oil extracted from the succulent leaves of *Aloe vera* (in its broader sense also including *A. succotrina* and *A. barbadensis*)." It was widely used in that part of the world for embalming and for medicine (ZOH). It is generally agreed that the aloes and myrrh, not cheap even then, were to facilitate Jesus' embalming and wrapping in linen (BMD). As mentioned in my second edition (CR2), there is still much confusion resulting from varying interpretations by various scientists. I will abide by AHPA's decision to treat the scientific name as the standardized common name, whether I like it or not.

COMMON NAMES (ALOE):

Acibar (Sp.; USN); Adala (Sanskrit; KAB); Aloe (Creole; Guy.; Ocn. Sp.; AH2; GMJ; USN); Aloès (Haiti; USN); Aloès de Jardins (Fr.; Haiti; AHL); Aloès Vulgaire (Fr.; USN); Aloe Vera (Scn.; AH2); Aloi (Greek; KAB); Angani (Tam.; KAB); A'oe (Ger.; AVP); A'ona (Pol.; AVP); Azvre (Por.; AVP); Babosa (Mad.; Por.; AVP); Babosa Commun (Por.; AVP); Bamboo (Usa.; AVP); Bamboo key (Fla.; AVP); Barbados Aloe (Eng.; Ocn.; AH2; USN); Brahmi (Kan.; KAB); Chenninayakam (Mal.; KAB); Chinnakalabanda (Tel.; KAB); Curaçao Aloe (Eng.; Ocn.; AH2; USN); Darakhte Sibr (Iran; DEP); Darakhtesinn (Iran; KAB); Echte Aloe (Ger.; USN); Eliya (Dec.; DEP); Ghiguvara (Hindi; KAB); Ghigvar (Hindi; DEP); Ghikanuar (Hindi; KAP); Ghikavar (Hindi; DEP); Ghikuanri (Oriya; KAB); Ghiqwara (Urdu; KAB); Ghiu Kumari (Nepal; SUW); Ghrita Kumari (Beng.; India; AH2; DEP; KAB); Girta Kunvar (Beng.; DEP); Kadvikunar (Guj.; KAB); Kalabanda (Tel.; DEP); Kanyá (Sanskrit; DEP); Kanyrasara (India; AH2); Katalai (Sri.; Tam.; AVP; DEP); Kattala (Mal.; DEP); Key Bamboo (Fla.; AVP); Komarisa (Singh.; DEP); Koraphada (Mar.; DEP; KAB); Kumári (Hindi; Sanskrit; DEP); Kunvar (Tel.; DEP); Laloi (Haiti; AVP); Lalwa (Creole; Haiti; VOD); Lankhu Laphi (Aym.; DLZ; SOU); Lephaee (Sin.; DEP); Lolu Sora (Kan.; DEP); Lu Hui (Pin.; DAA; KAB); Musabar (Arab.; KAB); Nha Dam Luu Hoi (Ic.; AVP); Patte laloi (Haiti; AVP); Pet'k'kin'ki (Maya; AVP); Sabar (Arab.; GHA); Sabara (Arab.; DEP); Sábila (Bol.; Peru; Sp.; DLZ); Sabur Obiknovennoi (Rus.; AVP; KAB); Sambour (Tur.; AVP); Saqal (Arab.; GHA); Sávila (Peru; Mdd.; Sp.; DAV; USN); Sávila Penca (Ecu.; BEJ); Sawila (Aym.; Que.; DLZ); Semper Vive

(Sp.; AVP); Sempervivum (Ma.; JFM); Sentebibu (Ma.; JFM); Shazaon Lepa Burma; DEP); Sibhir (Iran; AVP); Sibr (Iran; DEP); Simple Bible (Ma.; JFM); Sinkle Bible (Ma.; JFM); Singwanaka Luih (Ulwa; ULW); Tazavon-Lepa (Burma; DEP); True Aloe (Eng.; USN); West Indian aloe (Eng.; USN); Yerba de Gomas (Ma.; JFM); Zabbara (Malta; KAB); Zabila (Mex.; Sp.; AVP; DLZ).

Activities (Aloe):

Abortifacient (f1; GMJ; WO3); Analgesic (f1; EGG; PH2); Angiogenic (1; X14517429); Anthelmintic (1; MPI); Antiaggregant (f; CRC); Antiaging (f; WO3); Antialcoholic (1; WO3); Antiapoptotic (1; X15613791); Antibradykinin (1; X15182910); Anticancer (f1; JLH; X15531293); Antidote (alcohol) (f1; DLZ; WO3; X8937458); Antiedemic (f1; ALH; CAN; WHO); Antigliomic (1; X15747063); Antiherpetic (1; AAB; PH2; RCP7(1)); Antihistamine (1; APA); Anti-inflammatory (f1; AAB; CAN; PH2; WHO; WO3; X15751795; X15182910); Antilipoxygenase (1; X15751795); Antioxidant (1NP9(2):8); Antiplaque (f; WO3); Antiproliferant (1; X15531293); Antiprostaglandin (1; WHO; NP9(2):8); Antiradicular (1NP9(2):8); Antiretroviral (1; NP9(2):8); Antiseptic (f1; CRC; PH2; X15615409); Antithromboxane (1; PH2; WHO; NP9(2):8); Antiulcer (f1; AAB; PH2); Antiviral (1; AAB; PH2; WO3); Antiwrinkle (f; WO3); Aperient (f; DAA; DEP); Aphrodisiac (f; NP9(2):8); Apoptotic (1; X15531293; X15747063); Arylamine-N-Acetyltransferase Inhibitor (1; PH2); Bactericide (1; APA); Bitter (FED); Cholagogue (f; CRC); Cicatrizant (f1; DLZ; EGG; RCP7(1)); Collagenic (1; PH2; NP9(2):8); Collagenase Inhibitor (1; X12479983); COX-2-I (1; X12677534); Cyclooxygenase Inhibitor (1; PH2); Decongestant (f; DLZ); Demulcent (f1; WAM); Depurative (f; CRC); Digestive (f1; CRC; WAM); Emmenagogue (f1; DAA; JFM; MPI); Emollient (f1; GMJ; WAM); Fibroblastogenic NP9(2):8; Fungicide (1; APA; PH2; NP9(2):8); Gram(+)-icide (1; X15615409); Gram(–)-icide (1; X15615409); Hemostat (1; WO3); Hypocholesterolemic (1; WO3; X14598919); Hypoglycemic (1; CAN; JAC7:405; X15117555); Hypotensive (1; X11731923); Hypotriglyceridemic (1; JAC7:405); Immunomodulator (1; PH2); Immunostimulant (1; NP9(2):8); Insecticide (f; CRC); Larvicide (f; CRC); Laxative (f1; PH2; WAM; NP9(2):8); Metalloproteinase Inhibitor (1; X12479983); Microcirculatory Stimulant (CAN); Mitogenic (1; WO3); Moisturizer (f1; CRC); Nematicide (1; CRC); Osteogenic (1; RCP7(1); Phagocytotic (1; CAN; PNC); Propecic (f; KAP); Purgative (f; DLZ); Radioprotective (1; MPI; X15613791); Stimulant (f; CRC); Stomachic (f; CRC; MPI); Thromboxane-A-2-Synthase Inhibitor (1; X12677534); TNF-genic (1; NP9(2):8); Tonic (f; MPI); Tyrosine Kinase Inhibitor (1; PH2); Vermifuge (f; IED); Vulnerary (f1; CAN; EGG; WAM; WHO; WO3; X14517429).

Indications (Aloe):

Abrasions (f1; WHO); Abscess (f; CAN; DLZ); Acne (f; CRC; WHO); Acrochordons (f; CRC); Adenopathy (f; DEP); Alopecia (f; CRC; DAV; KAP); Amenorrhea (f; CRC; PH2); Anemia (f; WHO); Apoplexy (f; DEP); Arthrosis (f1; CAN; EGG; WO3); Asthma (f12; CAN; DLZ; EGG; KAP; PNC; RCP9(1); NP9(2):8); Bacteria (1; APA; PH2; NP9(2):8); Baldness (f; CRC); Bite (f; ULW); Bleeding (f; CRC); Blindness (f; WHO); Boil (f; AAB); Bronchosis (1; CAN; WO3); Bruise (f1; JFM; WHO); Bugbite (f1; APA; ULW); Burn (f1; AAB; CAN; ULW; VOD; WAM; WHO; X15751795); Cancer (f1; FNF, JAD; JLH; PH2); Cancer, anus (f1; CRC; JLH); Cancer, breast (f1; CRC; JLH); Cancer, larynx (f1; CRC; JLH); Cancer, lip (f1; CRC; JLH); Cancer, liver (f1; 1 CRC; JLH); Cancer, lymph (f; DEP); Cancer, nose (f1; CRC; JLH); Cancer, skin (f1; CRC; JLH); Cancer, stomach (f1; CRC; JLH; PH2); Cancer, tongue (f; EGG; JLH); Cancer, uterus (f; CRC; JLH); Cerebrosis (f; DLZ); Childbirth (f; CRC; DAA); Cholecystosis (f; JFM); Cold (f; CRC; JFM); Colic (f; DEP; KAP; PH2); Colitis (f12; JFM; X15199891; X15043514); Condyloma (f; CRC; JLH); Congestion (f; DLZ); Conjunctivosis (f; EGG; DEP; PHR); Constipation (f12; DAA; PH2; WAM; adult only); Consumption (f; KAP); Convulsion (f; CRC; NP9(2):8); Cough (f; APA; CRC; KAP); Cystosis (f; JFM); Decubitis (f; AAB); Dermatosis (f1; PH2; ULW; WHO; WO3; NP9(2):8); Diabetes (f1; APA; CAN; VOD; JAC7:405; NP9(2):8); Dysmenorrhea (f; AHL; KAP); Dyspepsia (f; AHL; CRC); Eczema (f; CAN; CRC); Edema (f1; CAN; JFM; VOD; WHO); Enterosis (f; CRC; VOD;

WO3); Epilepsy (f; KAP); Erysipelas (f; CRC; EGG); Erythema (f12; X15857459); Escherichia (1; NP9(2):8); Fever (f; DEP; GHA; VOD; NP9(2):8); Flu (1; NP9(2):8); Fracture (1; RCP7(1)); Frostbite (f12; APA; PH2; WHO; NP9(2):8); Fungus (1; AAB; APA; MPI; PH2); Gastrosis (f; CRC; VOD; WO3); Gingivosis (f; WO3); Glaucoma (f; WHO); Glioma (1; X15747063); Glossosis (f; JLH); Gonorrhea (f; JFM); Headache (f; GHA; VOD); Hemorrhoid (f; APA; CRC; WHO); Hepatosis (f1; BEJ; CRC; DEP; RCP6(1)); Herpes (12; AAB; PH2; NP9(2):8; RCP7(1)); High Cholesterol (1; WO3; X14598919); High Triglycerides (1; JAC7:405); HIV (1; WO3); Hysteria (f; CRC; VOD); IBD (12; X15043514); Immune Deficiency (1; CAN; PNC); Impotence (f; NP9(2):8); Indigestion (1; WAM); Infection (f1; APA; BEJ; PH2; NP9(2):8) Infertility (1; CRC; MPI); Inflammation (f1; CAN; CRC; GHA; PH2; VOD; WHO; NP9(2):8; X15182910); Intoxication (1; X8937458); Ischemia (1; PH2); Itch (f; DAA); Jaundice (f; CRC; KAB); Klebsiella (1; NP9(2):8); Leprosy (f; KAB); Leukemia (f; CRC; JLH); Leukorrhea (f; JFM); Lumbago (f; KAB); Measles (1; NP9(2):8); Mouth Sores (f1; CAN; EGG); Myalgia (f; KAB); Mycosis (1; FNF; PH2; WHO); Nephrosis (f; JFM); Ophthalmia (f; DEP; KAB); Pain (f; EGG; GHA; PH2); Peptic Ulcers (f1; CAN; CRC); Periodontosis (f; EGG); PMS (f; APA); Pneumonia (f; JFM); Proctosis (f; CRC); Pseudomonas (1; NP9(2):8); Psoriasis (f12; CAN; PH2; WHO; NP9(2):8; X15857459); Pulmonosis (f; JFM); Radiation Burns (1; CRC; DAA; WHO); Rash (f1; AAB; IED); Rheumatism (f; WO3); Ringworm (f1; APA); Salmonella (1; CRC); Seborrhea (f; WHO); Shock (1; X15566601); Snakebite (f; IED); Sore (f1; IED; X15751795); Sore Throat (f; JFM); Splenosis (f; DEP; KAB); Sprain (f; JFM); Staphylococcus (1; CRC; PH2); Sting (f; IED); Stomatosis (f; JLH); Strangury (f; KAB); Streptococcus (1; CRC); Sunburn (f1; AAB; PNC; VOD; WAM); Swelling (f; GHA; JFM); Syphilis (f; PHR); Toothache (f; DAV); Trachosis (f; WO3); Tuberculosis (1; DAA); Tumors (f1; CRC); Ulcer (f1; APA; JFM; PH2; VOD; WAM; WHO; WO3); Uterosis (f; CRC); Vaccinia (1; NP9(2):8); Vaginosis (f; APA); Venereal Disease (f; CRC; JFM); Virus (1; PH2; NP9(2):8); Wart (f; CRC; JLH); Weaning (f; VOD); Worm (f1; IED; MPI; PH2); Wound (f1; APA; CAN; CRC; NP9(2):8).

Dosages (Aloe):

FNFF = !
I would not myself think of it as food but TAN calls it a vegetable and FAC says the gel is often added to juices and jellies; bitter extracts are used in some beverages and candies and Dr. Samst Swedish bitters (FAC; TAN); 50–300 mg powder in a single dose at bedtime (AHP); 50–200 mg powder (APA); 1 Tbsp gel 3 ×/day (APA); 25 mg in 701 mg soybean oil, 1–2 softgels 3 ×/day; apply topically; or 1 tsp juice after meals (SF).

- Arabs rub fresh leaves or juice over the body to cool fevers (GHA).
- Ayurvedics regard the plant as alexiteric, alterative, anthelmintic, aphrodisiac, and useful for asthma, bronchitis, dermatitis, erysipelas, fever, hepatosis, jaundice, leprosy, ophthalmia, splenomegaly, and tumors (KAB).
- Caribbeans eat the "jelly" for constipation, cough, and sore throat (JFM).
- Chinese as early as 100 A.D. used aloe for convulsions, dermatosis, fever, and sinusosis (NP9(2):8).
- Curacaons take the sap for gallbladder ailments (JFM).
- Greeks in Dioscorides' time used the sap for boils, dermatosis, itch, sores, and took it internally for infections and stomach disorders (NP9(2):8).
- Haitians make a hot aqueous extract of dried leaves as an antidiabetic, purgative, and vermifuge (VOD).
- Jamaicans take the "jelly" in decoction for biliousness and cold (JFM).
- Kenyans taught me how to use the gel as an efficacious sunscreen (JAD).
- Latinos eat jelly-like flesh to relieve sore throat (JFM).
- Omani apply the gel to swollen eyes (GHA).

- Peruvians apply the gel to burns, conjunctivitis, erysipelas, inflammation, and sores (EGG).
- Trinidadans steep the flesh in stout, for jaundice; in rum, for pneumonia (JFM).
- Unani consider the plant antiinflammatory, carminative, digestive, purgative, tonic, and useful for biliousness, hemorrhoids, lumbago, myalgia, ophthalmia, splenitis (KAB).
- Yucatanese apply heated leaves to abscesses, bruises, erysipelas, and gum boils (JFM).

Downsides (Aloe):

Gel Class 1 Internally; Gel Class 2d Externally (AHP); Class 2b, 2c, 2d for the powder. Dermatosis; Diarrhea; Intestinal Cramps; Nephrosis; Ulcers. Newall, Anderson, and Phillipson caution that anthraquinones may be purgative, and an irritant to GI tract. Because of its cathartic and reputed abortifacient actions, its use in pregnancy and lactation should be avoided. However, they suggest that topical, but not oral, application might be okay during pregnancy and lactation (CAN). The latex can be a drastic cathartic. Contraindicated for pregnant women and children (LRNP). Do not use internally in pregnancy (WAM). Do not use with undiagnosed abdominal pain (WAM). Do not use internally for more than 10 days (WAM). Epidemiological studies in Germany reveal that abusers of anthranoid laxatives have three times higher rate of colon carcinoma (AEH). May cause allergic dermatosis. Taken in excess it may result in ulcers or irritated bowels (TMA, 1996). Hypoglycemic. Commission E reports contraindications, adverse effects, and interactions of anthranoid laxatives (AEH). Naturopaths Yarnell and Meserole (1996) state that people allergic to aloe may develop a severe rash following its application. Alcoholic extract at 100 mg/kg for 3 months is toxic in mice (AAB). While not indicting this species, Neuwinger (1996) notes several fatalities from other species of Aloe. Few botanists can identify the species of aloe with nothing but the latex there, no leaf, no flower, etc. (JAD).

Extracts (Aloe):

Aloe gel, Barbados aloe, and aloe powder contain, respectively, 4.87, 4.65, and 4.21% aloin, and 2.2, 2.1, and 2.03% aloe emodin. Gel preparations are reportedly effective against peptic ulcers (unless stress induced), radiation burns, and skin ulcers, and ineffective against stress-induced gastric and peptic ulcers. Healing wounds (10 mg/kg scu rat; 100 mg/kg scu mus); Bezakova et al. (X15751795), finding antilipoxygenase activity for aloe extracts, further rationaliz use of aloe extracts in acute inflammation, especially minor burns and skin ulcers (X15751795). Paulsen et al. (2005), in a double-blind, placebo-controlled study of commercial *Aloe vera* gel, found it useful in psoriasis vulgaris. In 41 patients with stable plaque psoriasis, erythema, infiltration, and desquamation decreased in 72.5% [but 82.5% improved on placebo, making it better than the aloe. Sounds like placebo and Zoloft (X15857459).]. Mijatovic et al. (2005) reported antigliomic action of aloe emodin, a chemical found in aloe and many other unrelated species. The antigliomic activity involves induction of both apoptosis and autophagy, as well as differentiation of glioma cells (X15747063). Biswas and Mukherjee (2003) proved vulnerary activity of several folkloric wound healers, including aloe, found effective in experimental models (X15866825).

JOINTED ANABASIS (*ANABASIS ARTICULATA* (FORSSK.) MOQ.) + CHENOPODIACEAE

Notes (Jointed Anabasis):

The sons of Gad: Ziphion, Haggi, <u>Shuni</u> …

Genesis 46:16 (KJV)

Zohary, equating Shuni with Anabasis, believes that "there must have been Hebrew names for a species predominant through hundreds of miles of the Judaean Desert, the Negev, Edom and elsewhere. … These names, while not finding a place among the flora of the Bible, have survived as proper names in Hebrew and as plant names in Arabic" (ZOH). It is one of the most common desert dwarf shrubs, often dominating the vegetation type, enduring extreme drought for several consecutive years. Several of the 25 species in the genus contain the alkaloid anabasine. This species, rich in potassium and saponins, is used as a detergent. Dried plants are collected as fuel.

Common Names (Jointed Anabasis):

Agram (Arab.; ZOH); 'Ajram (Arab.; ZOH); Ashna (Heb.; ZOH); Agram. (Leb.; HJP); Balbal (Leb.; HJP; ZOH); Hagarmi (Heb.; ZOH); Jointed Anabasis (Eng.; ZOH); Shenan (Arab.; ZOH); Shuni (Heb.; ZOH); Tatir (Atab.; HJP); Ushnan (Arab.; ZOH).

Activities (Jointed Anabasis):

Vulnerary (f; HJP).

Indications (Jointed Anabasis):

Sore (f; HJP); Wart (f; HJP).

Dosages (Jointed Anabasis):

FNFF = !
Stems produce an edible gum or manna (UPH).

- Lebanese goat herders smear mashed plant, with blood and mud, on goat lesions (HJP).
- Lebanese use fresh leaf tea, or dry plant ashes, on running sores (HJP).

Natural History (Jointed Anabasis):

The plants are heavily grazed by camels and goats (BIB; UPH). The fat sand rat (*Psammomys obesus*), a diurnal gerbillid, wholly herbivorous rodent, is able to survive consuming only this halophytic chenopod (X11073792).

Extracts (Jointed Anabasis):

Contains a complex mix of saponins and prosaponins, involving glucose, glucuronic acid, and oxytriterpenic acid derived from anabasic acid (HEG; HOC). Thus far I find no reports of the interesting insecticidal alkaloid anabasine in this species, but I suspect it could well be there. Anabasine, if present, is reported to have antismoking, insecticidal, myorelaxant, respirastimulant, sialogogue activities, and is fairly potent as a rodenticide [Ldlo = 10 mg/kg (orl rat)] (PDB).

ROSE-OF-JERICHO (*ANASTATICA HIEROCHUNTICA* L.) ++ BRASSICACEAE

Notes (Rose-of-Jericho):

O my God, make them like a wheel; as the stubble before the wind.

Psalms 83:13 (KJV)

Folklore has it that Mary clenched this in her hand when birthing Jesus (GHA). A typical tumbleweed and resurrection plant, it occurs in sand depressions or wadis in hot deserts, like the Judaean and Negev. It is probably what was mentioned as the "wheel" in Psalms 83. Zohary calls it the true Rose of Jericho. Dry fruiting branches hygroscopically expand into the "Jericho Rose" when placed in water, even if dry for several years. The rose is sold in Middle Eastern markets as "Kaff Mariam," (e.g., Qatar and other Arabian States and Emirates). The plant is soaked in water and, when it has unfurled, the water is drunk by the expectant mother, perhaps in hopes that the offspring will fill out as readily as the Jericho Rose, a symbol of resurrection (BAT; RIZ).

Common Names (Rose-of-Jericho):

Akaraba (Ber.; BOU); Bint en Nebi (Arab.; BOU); El Kemsha (Arab.; BOU); Elkmisha (Arab.; BOU); Haddaq (Arab.; BOU); Jericho Rose (Eng.; BIB); Jerose (Fr.; BOU); Kaff al-'Adhra (Arab.; GHA); Kaff el-'adra (Arab.; BOU); Kaff e Maryam (Arab.; Qatar GHA); Kaff Lella Fatma (Arab.; BOU); Kaff Maryam (Arab.; BOU); Kamaash (Arab.; BOU); Kamché (Arab.; Mauritania; UPW); Kemshe (Arab.; BOU); Kemshet en Nebi (Arab.; BOU); Kershoud (Arab.; BOU); Kmisa (Arab.; BOU); Komechi en Nebi (Mail; UPW); Kufayfah (Arab.; BOU); Main de Fathma (Alg.; Fr.; BOU; UPW); Mary's Flower (Eng.; BIB); Qebad (Arab.; BOU); Rose de Jericho (Fr.; BOU); Rose of Jericho (Eng.; BOU); Shagaret Maryam (Oman; Qatar; Saudi; GHA); Palestinian Tumbleweed (Eng.; BIB); Shagret Mariam (Arab.; BOU); Shajaret et Talq (Arab.; BOU); St. Mary's Flower (Eng.; BOU); Tamkelt (Ber.; BOU); True Rose of Jericho (Eng.; USN); Wheel (Eng.; BIB); Yedd Fatma (Arab.; BOU).

Activities (Rose-of-Jericho):

Analgesic (f; BOU); Emmenagogue (f; BOU); Hepatoprotective (1; X12643908).

Indications (Rose-of-Jericho):

Amenorrhea (f; BOU); Childbirth (f; BOU; GHA); Cold (f; BOU); Epilepsy (f; BOU); Hepatosis (1; X2643908); Pain (f; BOU).

Dosages (Rose-of-Jericho):

FNFF = ?

- Arabians, near birth, soak the dry plant in water and drink to avoid birthing pain (BAT; BOU; GHA; RIZ).
- Rabat Arabs take the infusion for colds and epilepsy (BOU).

Extracts (Rose-of-Jericho):

Anastatins A and B have hepatoprotective effects (on D-galactosamine-induced cytotoxicity in primary cultured mouse hepatocytes) even more potent than commercial silybin (X12643908).

WINDFLOWER (*ANEMONE CORONARIA* L.) + RANUNCULACEAE

Notes (Windflower):

> **Consider the lilies of the field, how they grow; they toil not, neither do they spin: And yet I say unto you, that even Solomon in all his glory was not arrayed like one of these.**
>
> **Matthew 6:28 (KJV)**

Traditionally, says Zohary, this is believed to be the lilies of the field of Matthew and Luke. Common, colorful, showy in much of the open areas of the Middle East, coming in blue, pink, purple, red, scarlet, white, violet (FP1; ZOH). This spectacular ornamental is said to still be abundant on the Mount of Olives, as it was in Jesus' day (BIB). Today this is the most common anemone in the

FIGURE 1.8 Windflower (*Anemone coronaria*).

florist trade. Mythology has it that the anemone sprung from the tears of Venus pining for Adonis: where a tear dropped, a windflower grew. Even Pliny swore that anemones only flowered when the wind was blowing. Magicians gathered them in his day as a remedy against disease, tying the flowers around the neck or arm of the patient as a charm to cure all illness (BIB).

Common Names (Windflower):

Crown Anemone (Eng.; ZOH); Garden Anemone (Eng.; BIB); Lily (Eng.; BIB); Lily of the Field (Eng.; BIB); Poppy Anemone (Eng.; BIB); Sakkir (Arab.; Syria; HJP); Shaka'ik un Za'man (Arab.; Syria; HJP); Windflower (Eng.; BIB; HJP).

Activities (Windflower):

Vulnerary (f; HJP).

Indications (Windflower):

Leprosy (f; HJP); Malaria (f; HJP); Rhinosis (f; HJP); Sore (f; HJP); Tuberculosis (f; HJP); Tumor (f; JLH).

Dosages (Windflower):

FNFF = X

- Arabs used the flowers for treating tumors (JLH).
- Lebanese immigrants claim that the plant is good for malaria (HJP).
- Middle Easterners still use classically to cleanse the nose and sores, using the root for tuberculosis and the plant for leprosy (HJP).

Downsides (Windflower):

As of July 2004, the FDA Poisonous Plant Database listed titles alluding to the toxicity of this species.

Natural History (Windflower):

Although not producing nectar, the flowers, opening by day closing by night, attract insects that feed on the copious pollen. The flowers are mostly cross-pollinated and the copious seeds distributed by the wind.

DILL (*ANETHUM GRAVEOLENS* L.) ++ APIACEAE

Synonyms:

Anethum sowa Kurz; *Peucedanum graveolens* L.; *Peucedanum sowa* Kurz.

Notes (Dill):

> **Woe unto you, scribes and Pharisees, hypocrites! for ye pay tithe of mint and anise and cummin, and have omitted the weightier matters of the law, judgment, mercy, and faith: these ought ye to have done, and not to leave the other undone. Ye blind guides, which strain at a gnat, and swallow a camel.**
>
> **Matthew 23:23–24 (KJV)**

> **Woe to you, scribes and Pharisees, hypocrites! for you tithe the mint and dill and cummin, and have neglected the weightier matters of the law, justice and mercy and faith; these you ought to have done, without neglecting the others. You blind guides, straining out a gnat and swallowing a camel.**
>
> **Matthew 23:23–24 (RSV)**

> **Woe to you, scribes and Pharisees, hypocrites! because you give the tenth of the mint and the dill and the cumin, but you have disregarded the weightier matters of the Law, namely justice and mercy and faithfulness, These things it was binding to do, yet not to disregard the other things. Blind guides, who strain out the gnat but gulp down the camel**
>
> **Matthew 23:23–24 (NWT)**

FIGURE 1.9 Dill (*Anethum graveolens*).

Although some scholars think this "biblical anise," only in the KJV, might be Pimpinella, (ZOH) Zohary argues that it is unlikely that Pimpinella "has ever been grown in the biblical countries." Dill is more characteristically a plant of oriental cultivation than anise. In postbiblical literature, dill is named *shiveth,* which is identical to the arabic *sabth* (ZOH). Dill was grown by the ancient Greeks and Romans. Lebanese have a legend that Abraham taught his people about dill when he came from the north. The Talmud records that its seeds, stems, and leaves were subject to tithe. The essential oil has shown inhibitory effects on various organisms, such as *Bacillus anthracis.*

Common Names (Dill):

Adas (Malaya; EFS); Adas Manis (Malaya; EFS); Anega (Ger.; Sp.; AVP; EFS); Anet (Eng.; USN); Aneta Odorant (It.; EFS); Aneth (Fr.; BOU); Aneth Odorante (Fr., Fwi.; AVP); Aneth Denouil (Fr.; EFS); Aneto (It.; Por.; AVP; EFS); Ayn-Jaradi (Arab.; Syria; HJP); Baluntshep (Bom.; DEP); Baston do Diale (Belgium; JLH); Buzzalchippet (Arab.; AVP); Chebit (Syria; AVP); Chibith (Syria; AVP); Dereotu (Tur.; EBS; EB54:155); Dild (Den.; EFS); Dilkfruid (Dutch; EFS); Dill (Eng.; Swe.; CR2; EFS); Dilla (Nor.; EFS); Dillenkraut (Ger.; EFS; HHB); Dillkraut (Ger.; EFS; HHB); Dillsmen (Ger.; AVP); Ecarlade (Fr.; Fwi.; AVP); Endro (Por.; AVP; EFS); Eneldo (Peru; Sp.; AVP); Faux Anis (Fr.; Fwi.; EFS; AVP); Fenouil Batard (Fr.; Fwi.; AVP; BOU); Fenouil Puant (Fr.; Fwi.; AVP; BOU); Finnochio Fetido (It.; EFS); Funcho (Mad.; JAD); Gurkenkraut (Ger.; HHB); Hinojo Hedionodo (Sp.; EFS); Hulwa (Oman; GHA); Keper (Pol.; AVP); Keraonia (Arab.; AVP); Kerti Kapor (Hun.; EFS); Koper (Pol.; AVP); Misreya (Sanskrit; DEP); Misroya (Sanskrit; EFS); Neto (It.; EFS); Sabth (Arab.; ZOH); Sadhab el Barr (Arab.; Syria; HJP); Sadhan al Barr (Oman; GHA); Sa Myiet (Burma; DEP); Sata Kuppi (Tam.; DEP); Satapushpi (Sanskrit; DEP); Shabat (Qatar; GHA; RIZ); Shamar (Arab.; BOU; HJP); Shebat (Arab.; Syria; HJP); Shebet (Arab.; BOU); Sheveth (Heb.; ZOH); Shibit (Arab.; BOU); Shibith (Oman; Syria; GHA; HJP); Shi Luo (China; DAA); Shi Luo Zi (Pin.; DAA); Shubit (Arab.; EFS); Shulupa (Beng.; DEP); Shumar (Arab.; Syria; HJP); Shumra (Arab.; Syria; HJP); Soi (Kas.; DEP); Sowa (India; AVP; EFS); Soya (Kum.; Nepal; Pun.; DEP; SUW); Spinet (Oman; GHA); Sulpa (Beng.; DEP); Sulpha (Beng.; DEP); Surva (Guj.; DEP); Sutopsha (Hindi; DEP); Suva (India; EFS); Tebs (Arab.; BOU); Tere Otou (Tur.; AVP); Venkel (DWI; JFM).

Activities (Dill):

Allergenic (1; X12868972); Antidote (f; BOU); Antiemetic (f; BOU); Antihyperlipidemic (1; X11409638); Antihypercholesterolemic (1; X11409638); Antioxidant (1; X15364640); Antiradicular (1; X15364640); Antisecretory (1; X12493079); Antiseptic (f1; ZUL; X12797755); Antispasmodic (f12; APA; BOU; KOM; PH2; TRA); Antitoxic (1; TRA); Antitumor (f1; TRA); Bacteriostat (12; APA; KOM; PH2; TRA; ZUL); Candidicide (1; X12797755); Cardiodepressant (1; APA); Carminative (f1; BOU; JFM; PNC; ZUL); Collyrium (f; JFM); Deobstruent (f; DEP); Detergent (f; CRC); Digestive (f; BOU; CRC); Diuretic (f1; APA; BOU; CRC; DEP; JFM); Emmenagogue (f; DEP); Emollient (f; BOU); Estrogenic (f1; FNF); Fungicide (1; TRA); Gastroprotective (f1; X12493079); GST Inducer (1; X1438594); Hypoglycemic (1; TRA); Hypotensive (1; APA; FNF; TRA); Insecticide (1; X15125529); Insectifuge (1; ZUL); Lactagogue (f1; APA; BOU; CRC; JFM; NMH; TRA); Larvicide (1; X15125529); Laxative (f; CRC); Mutagenic (1; X7411385); Myorelaxant (1; APA); Narcotic (f; CRC); Orexigenic (f; APA); Psychadelic (f; CRC); Respirostimulant (f; APA); Sedative (f1; BOU; TRA); Soporific (f; JFM); Stimulant (f; CRC); Stomachic (f1; BOU; PNC; ZUL); Vasodilator (1; TRA).

Indications (Dill):

Anorexia (f12; APA; PHR); Aposteme (f; CRC; JLH); Bacillus (1; TRA); Bacteria (12; APA; ZUL); Bronchosis (12; PHR); Bruise (f; CRC); Cancer (f; CRC; JLH); Cancer, abdomen (f; CRC; JLH); Cancer, anus (f; CRC; JLH); Cancer, breast (f; JLH); Cancer, colon (f; CRC; JLH); Cancer, liver (f; CRC; JLH); Cancer, mouth (f; CRC; JLH); Cancer, stomach (f; CRC; JLH); Cancer, throat (f; CRC; JLH); Candida (1; X12797755); Cardiopathy (1, TRA) Childbirth (f; JFM); Cholecocystosis (2; PHR); Cold (f12; PHR); Colic (f1; CRC; GHA; PNC; ZUL); Condylomata (f; CRC); Cough (f12; APA; CRC; PHR); Cramp (f1; BOU; PHR; PH2; TRA); Diabetes (1; TRA); Dropsy (f; AHL; CRC); Dyslactea (f1; APA; BOU; CRC; JFM; NMH; TRA); Dyspepsia (f12; APA; CRC; KOM); Enterosis (f12; APA; GHA; JLH; PHR; PH2); Escherichia (1; TRA); Fever (f12; PHR); Fibroid (f; JLH); Fungus (1; X12797755); Gas (f1; JFM; TRA; ZUL); Gastralgia (f1; TRA); Gastrosis (f12; APA; PHR;

PH2; TRA; X12493079); Halitosis (f1; APA; PH2); Heart Problems (1; TRA); Hemorrhoid (f1; APA; CRC); Hepatosis (f2; JLH; PHR); High Blood Pressure (1; APA; FNF; TRA); High Cholesterol (1; X11409638); Induration (f; CRC; JLH); Infection (f12; APA; PHR); Insomnia (12; APA; CRC; PHR; PH2); Jaundice (f; AHL; CRC); Mastosis (f; CRC; JLH); Mycosis (1; TRA; X12797755); Nephrosis (f; APA; PH2); Neuroses (1; APA); Obesity (1; X11409638); Pain (f; BOU; DEP); Pharyngosis (12; PHR); Respirosis (f; PH2); Salmonella (1; TRA); Sclerosis (f; CRC); Scirrhus (f; JLH); Scurvy (f1; CRC); Shigellosis (1; TRA); Salmonella (1; TRA); Sore (f; CRC); Spasms (f; PHR); Splenosis (f; JLH); Staphylococcus (1; CRC); Stomachache (f1; APA; BOU; CRC; JFM; TRA); Stomatosis (12; PHR); Syphilis (f; PH2); Tumor (f; CRC; JLH); Urethrosis (1; APA; PH2); Uterosis (f; JLH); UTI's (f; APA; PHR); Venereal Disease (f; PH2); Worm (f; PH2); Yeast (1; X12797755).

Dosages (Dill):

FNFF = !!!

Dill is used primarily as a condiment. Dried fruits (seeds) are used in pickles, soups, spiced beets, fish, and fish sauces, with eggs, and in potato salads. Roasted fruits serve as a coffee substitute. Hot fruit extracts are used to make jams and liqueurs; the dill oil is also used in liqueurs. Fresh leaves are used in salads, with cottage cheese, cream cheese, steaks, chops, avocado, cauliflower, green beans, squash, tomatoes, and tomato soup, zucchini, and shrimp. Dried leaves, known as dill weed, are also used to season various foods. Ethiopians use tender plant parts, dried fruits, and flowers in flavorings, especially alcoholic beverages. Oil from the seed is used chiefly as a scent in soaps and perfumes, and in the pickle industry. Weed oil, from the aboveground parts of the plant, is used in the food industry because of its characteristic dill herb smell and flavor (FAC, TAN). Chew 1/2–1 tsp seed (APA); 2 tsp crushed seed/cup water (APA); 1 g seed (HHB); 3 g seed (KOM; PHR); 0.1–0.3 g EO (KOM; PHR); 0.3–1 ml concentrated dill water (PNC); 0.05–2 ml (I would have said 0.05–0.2 ml) essential oil (PNC); (cf. celeryseed, close kin) (PNC).

- Asian Indians use the seed for abdominal tumors (JLH).
- Bahamans use the plant decoction for stomachache in children (JFM).
- Belgians apply the leaves to tumors (JLH).
- Dominicans suggest bruised leaves for bruises and sores (AHL).
- Dominicans suggest the antiscorbutic stimulant root as a diuretic in dropsy and jaundice (AHL).
- Dutch Antilleans use the cooled plant decoction as a collyrium (JFM).
- Lebanese suggest the use of the tea alone, with or interchangebly with true anise, for baby colic (HJP).
- Peruvians use the seeds for cancerous indurations and as lactagogues (JLH).
- Peruvians use foliar or seed infusion as a diuretic (EGG).
- Peruvians use the plant juice as an antiseptic and carminative (EGG).
- Venezuelans use the decoction as carminative, diuretic, lactagogue, and soporific (JFM).

Downsides (Dill):

Class 1 (AHP, 1997). None known (KOM). No health hazards or side effects in conjunction with proper administration of designated therapeutic dosages (PH2). Contact photodermatosis possible as in most umbellifers. Fresh juice may possibly cause photodermatosis (PHR; X12868972). We might extend to all apiaceous oils Bisset's comments on celeryseed oil, "The drug is contraindicated in inflammation of the kidneys, (BIS)" because apiaceous essential oils may increase the inflammation as a result of epithelial irritation (BIS). Tramil (TRA) conservatively cautions for dill and fennel; contraindicated in reproductive women; limit dosage to 3 days. Doses of 175 mg/kg aqueous extract gastrically intubated in pregnant rats may be embryotoxic and or teratogenic (TRA).

Extracts (Dill):

Proestos et al. (2005) checked the species for flavonoid phenolics and their antioxidant and antimicrobial activity, finding circa 360 ppm quercetin, 445 ppm (–)-epicatechin, and 16 ppm vanillic acid in dill (X15713039). LD50 (Hydroethanolic Fruit Extract) = 1000 mg/kg ipr rat (TRA).

MAYWEED (*ANTHEMIS COTULA* L.) + ASTERACEAE

Synonyms:

Anthemis abyssinica J. Gay; *Anthemis foetida* L.; *Anthemis heterophylla* Wallr.; *Anthemis psorosperma* Ten.; *Anthemis ramosa* Link.; *Chamaemelum cotula* (L.) All.; *Chamaemelum foetidum* Baumg.; *Marula cotula* (L.) DC.; *Marula foetida* (Lam.) S.F. Gray; *Marula vulgaris* Bluff & Fingerh. *fide* HH2

Notes (Mayweed):

> **But go into a field of flowers, where no house is builded, and eat only the flowers of the field; taste no flesh, drink no wine, but eat flowers only.**
>
> **4 Ezra Apocrypha 9:24 (KJV)**

Zohary notes that the yellow-white, diminutive daisy-like flowers (more than 20 species of *Anthemis* being found in Israel) ensure it a primary place among the flowers of the field. Wisely, he designates dog fennel as *Anthemis* sp. (ZOH).

Common Names (Mayweed):

Amarusca (Eng.; JLH); Arbiyan (Arab.; Syria; HJP); Bald Eye Grow (Eng.; BUR); Camomilla Mezzana (It.; HH2); Camomille des Chiens (Fr.; USN); Camomille Puante (Fr.; USN); Chiggy Weed (Eng.; BUR); Cotula Fetida (It.; EFS); Dillweed (Eng.; BUR); Dilly (Eng.; BUR); Dillweed (Eng.; BUR); Dog Daisy (Eng.; BUR); Dog-fennel (Eng.; AH2; USN); Doggis Fenkel (Eng.; JLH); Echhuan (Arab.; Syria; HJP); Fetid Chamomile (Eng.; BUR); Fieldweed (Eng.; BUR); Fieldwort (Eng.; BUR); Foul Wormwood (Eng.; JLH); Fuss ul Kilab (Arab.; Syria; HJP); Hashish el Knanzir (Arab.; Syria; HJP); Hay Fennel (Eng.; BUR); Horse Daisy (Eng.; BUR); Hunds Kamille (Ger.; HH2); Ikhawan (Arab.; ZOH); Ironwort (Eng.; AAH); Kahwan (Arab.; Syria; HJP); Kuhdill (Ger.; HH2); Macela Fetida (Por.; EFS; HH2); Madder Mayweed (Eng.; BUR); Magarza (Sp.; HH2); Manzanilla Fetida (Sp.; EFS); Manzanilla Hedionda (Sp.; HH2; USN); Manzanilla Malagata (Sp.; HH2); Margaça (Mad.; Por.; JAD); Margarzuela (Sp.; EFS); Mawth (Eng.; JLH); Maydewode (Eng.; JLH); Mayflower (Eng.; BUR); Mayweed (Eng.; Scn.; AH2; USN); Mayweed Chamomile (Eng.; USN); Northern Dogfennel (Eng.; BUR); Poison Daisy (Eng.; BUR); Ribyaan (Arab.; ZOH); Stinkende Kamille (Dutch; EFS); Stinking Chamomile (Eng.; HH2; USN); Stinking Mayweed (Eng.; HH2); Stinkende Hundskamille (Ger.; USN); Stinkkamille (Ger.; HH2); Wild Chamomille (Eng.; EFS; HH2).

Activities (Mayweed):

Anodyne (f; BUR); Antiinflammaory (f; WO2); Antispasmodic (f; HH2); Bitter (f; BUR); Carminative (f; EFS); Collyrium (f; BUR); Diaphoretic (f; HJP; TOM); Emetic (f; EFS; HH2); Emmenagogue (f; EFS; HH2); Febrifuge (f; EFS); Insecticide (f; WO2); Insectifuge (f; AAH); Sedative (f; HJP); Tonic (f; EFS; HH2; TOM); Unguent (f; EB28:316); Vermifuge (f; EFS); Vesicant (f; BUR); Vulnerary (f; EFS).

Indications (Mayweed):

Amenorrhea (f; UPH); Arthrosis (f; BUR); Asthma (f; BUR; HJP); Cancer (f; JLH); Conjunctivosis (f; HJP); Dermatosis (f; HJP); Diarrhea (f; HJP; EB31:353); Dropsy (f; BUR); Dysentery (f; HJP); Dyspepsia (f; EB31:353); Epilepsy (f; HJP); Fever (f; EFS); Gas (f; EFS); Headache (f; HH2);

Hyperkinesis (f; HJP); Hysteria (f; HH2; HJP); Insecticide (f; UPH); Neurosis (f; HH2); Ophthalmia (f; HJP); Pain (f; BUR); Pulmonosis (f; BUR); Rheumatism (f; BUR); Scrofula (f; BUR); Sore (f; AAH; EB28:316); Spasm (f; EFS); Sting (Bee) (f; UPH); Worm (f; EFS).

Dosages (Mayweed):

FNFF = !
Used for teas, and in Peru for flavoring (FAC).

- British use in salves for cancer (JLH).
- Irish use with crowfoot for cancer (JLH).
- Lebanese make floral tea to calm adolescents and hystericals (HJP).
- Mapuche Indians use for hysteria and neuroses (HH2).
- Russians use as a vermifuge (HH2).

Downsides (Mayweed):

As of November 2004, the FDA Poisonous Plant Database listed 39 titles alluding to the toxicity of this species.

Natural History (Mayweed):

The flowers attract many insect pollinators, which collect the pollen and/or sip the nectar. When flowering, the plant turns the outer flowers down in the evening and spreads them out in the morning (ZOH).

AGARWOOD (*AQUILARIA MALACCENSIS* LAM.) + THYMELAEACEAE

Synonyms:

Agallochum malaccense O.K.; *Aquilaria agallocha* Roxb.; *Aquilaria malaicense* Rumph.; *Aquilaria ovata* Cav.; *Aquilaria secundaria* DC.; *Aquilariella malaccensis* Van Thiegh *fide* HH2

Notes (Agarwood):

All thy garments smell of myrrh, and aloes...

Psalm 45:8 (KJV)

This is the aloe of the Old Testament; that of the New Testament is generally considered to be *Aloe vera*. The darker of several types of eaglewood, especially when partially diseased and decayed, is highly valued in perfumery, as incense, and for fumigation. The treasured unhealthy wood is so much more valuable than the healthy, that the healthy tree is too often destroyed in the quest for patches of unhealthy wood. Called agar, it is powdered and used as a flea and lice repellent. The soft and fragrant inner wood, itself worth its weight in gold, was molded and used as a setting for precious stones. Myth has it in the East that eaglewood is the only tree to have descended to Man from the Garden of Eden, all others having perished. Supposedly, Adam brought one of its shoots and transplanted it to the land where he settled, other eaglewoods having sprung from this shoot. That is why it is sometimes called Shoot of Paradise and Paradise Wood. Malayans made cloth from the pounded bark. Annamese make a paper substitute from the bark, having presented 30,000 rolls of aloe paper to a Chinese emperor in A.D (BIB).

Common Names (Agarwood):

Adlerholz (Ger.; USN); Agallochon (Greek; DEP); Agallochum (Eng.; DEP); Agalugi (HH2); Agar (Bom.; Guj.; Hindi; Iran; Jap. Tam.; Urdu; DEP; HH2; KAB); Agare Hindi (Arab.; DEP); Agarwood

FIGURE 1.10 Agarwood (*Aquilaria malaccensis*).

(Eng.; Scn.; AH2; USN); Agaru (Ayu.; Beng.; Kan.; Sanskrit; AH2; DEP); Aggalichandana (Tam.; DEP; KAB; WO2); Aggar (Arab.; HH2); Aggur (Arab.; HH2); Aghal (Sanskrit; ZOH); Aghalukhi (Arab.; DEP); Agnikashtha (Sanskrit; KAB); Agre Hindi (Iran; DEP); Agru (Tel.; DEP); Agulugin (HH2); Aguru (Sanskrit; HH2); Ahalim (Heb.; IHB); Ahaloth (Heb.; DEP; IHB; ZOH); Akatau (Mal.; KAB); Akyan (Burma; NAD); Akyau (Burma; DEP); Akyaw (Eng.; DEP); Aloes Wood (Eng.; Ocn.; AH2; USN); Anaryaka (Sanskrit; KAB); Aquilaire (Fr.; KAB); Asara (Sanskrit; KAB); Aud Hindi (Arab.; DEP); Aude Hindi (Arab.; DEP); Belanjirj (Iran; NAD); Bois d'Aigle (Fr.;

HH2); Bois d'Aloes (Fr.; HH2); Bringhaja (Sanskrit; KAB); Calambac (Eng.; DEP); Chan Krasna (Cam.; KAB); Cheng Ch'en Chi (China; KAB); Ch'en Hsiang (China; KAB); Chen Xiang (Pin.; AH2); Eaglewood (Eng.; USN); Gaharu (Malaya; IHB); Garu (Mal.; DEP); Hindiagara (Bom.; DEP); Indian Aloewood (Eng.; USN); Jinkoh (Malaya; HH2); Jishvarupa (Sanskrit; KNAD); Jonk (Sanskrit; KAB); Kanankoh (Malaya; HH2); Karas (Mal.; IHB); Kashthaka (Sanskrit; KAB); Kayagahru (Mal.; NAD); Kayu (Mal.; DEP); Kĕkaras (Malaya; IHB); Kihay (Sin.; DEP; KAB); Klim (Malaya; Semang; IHB); Krasne (Cam.; KAB); Krimigandha (Sanskrit; KAB); Krimija (Sanskrit; KAB); Krishna (Sanskrit; KAB); Krishnagaru (Kan.; Tam.; Tel.; NAD); Kyara (Malaya; HH2); Laghu (Sanskrit; KAB); Lign-aloes (Eng.; JLH; USN); Lignum Aloes (Eng.; DEP); Loha (Sanskrit; KAB); Lohakhya (Sanskrit; KAB); Malacca Eaglewood (Eng.; WO2); Malayan Aloewood (Eng.; WOI); Mi Hsiang (China; KAB); Nwahmi (Thai; DEP); Nyaw Chah (China; DEP); Ood Hindi (Arab.; DEP); Owd (Iran; HH2); Owd Hindi (Iran; HH2); Paradise Wood (Eng.; HH2); Pataka (Sanskrit; KAB); Pichhila (Sanskrit; KAB); Pravara (Sanskrit; KAB); Rajarah Kalijya (Sanskrit; NAD); Rajarha (Sanskrit; KAB); Sam Chit (Malaya; KAB); Sási (Assam; DEP); Shoot of Paradise (Eng.; BIB); Sinnah (Sin.; DEP; KAB); Tabak (Sakai; IHB); Tĕbal (Malaya; Pangan; IHB); Tĕngkaras (Malaya; IHB); Ud (Mal.; KAB); Ud el Juj (Arab.; NAD); Ud Hindi (Arab.; DEP); Ude Hindi (Arab.; DEP); Ugar (Beng.; DEP); Uggor (Arab.; HH2); Vanshika (Sanskrit; KAB); Varnaprasadana (Sanskrit; KAB); Xylaloe (Greek; ZOH); Yelunjooj (;? HH2); Yogasha (Sanskrit; KAB).

Activities (Agarwood):

Anodyne (f; BIB; NAD); Antianaphylactic (1; X9324002); Anticancer (1; X7320738); Antihistaminic (1; X9324002); Aphrodisiac (f; BIB); Astringent (f; KAB); Cardiotonic (1; WO2); Carminative (f; BIB); Cerebrotonic (f; KAB); Cholagogue (f; NAD); CNS Depressant (1; X8441779; WO3); Cytotoxic (1; X7320738); Deobstruent (f; NAD); Diuretic (f; BIB); Febrifuge (1; X8441779); Insectifuge (f1; WO2); Pulifuge (f1; WO2); Sedative (1; X8441779); Stimulant (f; BIB); Stomachic (f; BIB); Tonic (f; BIB).

Indications (Agarwood):

Allergy (1; X9324002); Anaphylaxis (1; X9324002); Anuria (f; HH2); Asthma (f1; BIB; WO2; X9324002); Bleeding (f; DEP); Bronchosis (f; BIB; NAD); Cancer (f; BIB); Cancer, colon (f; BIB); Cancer, liver (f; BIB); Cancer, lung (f; BIB); Cancer, stomach (f; BIB); Cancer, thyroid (f; BIB); Cardiopathy (f; IHB); Childbirth (f; BIB; IHB); Cholera (f; HH2); Colic (f; BIB; DEP); Congestion (f; BIB); Cough (f; HH2); Dermatosis (f; BIB); Diarrhea (f; BIB); Dropsy (f; BIB); Dysgeusia (f; KAB); Enteralgia (f; BIB); Enterosis (f; BIB); Fever (f1; BIB; DEP; X8441779); Gastrosis (f; BIB); Gout (f; BIB; DEP); Headache (f; NAD); Hepatosis (f; KAB); Hiccup (f; BIB); Impotence (f; KAB); Induration (f; BIB); Leukoderma (f; BIB; KAB); Malaria (f; BIB; HH2); Nausea (f; BIB; DEP); Nephrosis (f; BIB); Ophthalmia (f; BIB); Otosis (f; BIB); Pain (f; BIB); Palpitation (f; IHB); Palsy (f; BIB; DEP); Paralysis (f; BIB); Pulmonosis (f; BIB); Rheumatism (f; BIB; DEP); Smallpox (f; IHB); Thirst (f; KAB); Tumor (f; JLH); Vertigo (f; BIB; DEP); Wound (f; BIB).

Dosages (Agarwood):

FNFF = ?

- Asian Indians apply a paste of *Agaru* and *Isvari* in brandy to the chest in bronchitis and to the head in headache (NAD).
- Asian Indians use the plant for abdominal tumors (JLH).
- Ayurvedics use wood for ear and eye ailments, asthma, dermatosis, hiccup, and leucoderma (KAB).
- Chinese consider the wood aphrodisiac, carminative, stimulant, and tonic (KAB).

- Malayans, considering it carminative, stimulant, and tonic, take it in childbirth, for female ailments, and for coronary palpitations (IHB).
- Unani use wood for asthma, bronchitis, diarrhea, enteritis, gastritis, hepatitis, nausea, and to stabilize the fetus in the uterus (KAB).

Natural History (Agarwood):

Only fungus-infected wood provides the prized agarwood of commerce. Species of *Aspergillus, Fusarium, Penicillium,* and *Fungi imperfecti.*

HORSERADISH (*ARMORACIA RUSTICANA* P. GAERTN. ET AL.) ++ BRASSICACEAE

Synonyms:

Armoracia lapathifolia Gilib.; *Cardamine armoracia* O. Ktze.; *Cochlearia armoracia* L.; *Cochlearia rusticana* Lam.; *Nasturtium armoracia* (L.) Fr.; *Radicula armoracia* (L.) Robinson; *Rorippa armoracia* (L.) Hitchc.; *Rorippa rusticana* Gren. & Godr.

Notes (Horseradish):

> **The fourteenth day of the second month at even they shall keep it, and eat it with unleavened bread and bitter herbs.**
>
> **Numbers 9:11 (KJV)**

Although most books on medicinal plants of the Bible, including my own, do not mention the horseradish, it seems to be the most important passover herb and is mentioned in the Torah. Under several orthographic variants (marror, maror, mohror, moror, morror), the Torah mentions eating the morror (marror, maror, moror), or the bitter herb. Many of my Jewish friends consider it the bitter herb mentioned in Numbers 9:10–11 (Helen Metzman; Wayne Silverman, separate personal communications, 2007).

I find it a great spice, especially with ketchupy seafood cocktail sauces like I enjoyed last night with shrimp cocktail. I was taken aback when the first title crossing my desk as I settled back into compiling today was "Deodorization of Swine Manure using Minced Horseradish Roots and Peroxides." Pennsylvania scientists Govere et al. (2005) removed all offensive phenolics without recurrence for 72 hours, but human panels considered the odor reduced 50% in intensity, dare I call it IC50 = 1 part horseradish to 10 parts manure + calcium peroxide (26 or 34 mM) + hydrogen peroxide (34, 52, or 68 mM). The authors conclude that using horseradish "as enzyme carriers and peroxides as electron acceptors emerges as an effective approach to phenolic (*p*-cresol- *p*-ethylphenol) and phenol odor control in animal manure," skillfully avoiding the issue of the nonphenolic contributors (volatile fatty acid-like n-butyric, n-caproic, isobutyric, isocapoic, isovaleric, propionic, and n-valeric acids) and indoles (indole, scatole). "More work is required to find ways to increase the removal of indolic odorants and volatile fatty acids." Govere et al. (2005)

Common Names (Horseradish):

Aed-mädarõigas (Estonia; POR); Balsamita Jaramago (Sp.; EFS); Barbaforte (It.; EFS; HH2); Bayírtupu (Tur.; EFS); Boereradijs (Dutch; POR); Chren (Rus.; HH2); Chrzan (Pol.; HH2); Chrzan pospolity (Pol.; POR); Cran (It.; EFS); Cran de Bretagne (Fr.; EFS); Cranson (Fr.; USN); Cren (It.; HH2); Grand Raifort (Fr.; POR); Great Raifort (Eng.; GMH); Hoosu radiishu (Japan; POR); Horseradish (Eng.; Scn.; AH2; CR2); Khren (Rus.; POR); Közönséges Torma. (Hun.; POR); Kren (Ger.; POR); Křen (HH2); Kreno (Dutch; POR), La Gen (China; POR); Mädarõigas (Estonia; POR); Makatakak (Hocak; WIN); Ma luo po (China; TAN); Meeretisch (Ger.; POR); Meerradi (Ger.; POR); Meerrettich (Ger.; POR; USN); Mérédic (Fr.; HH2); Meredik (Dutch; POR), Mierik (Dutch;

FIGURE 1.11 Horseradish (*Armoracia rusticana*).

POR), Mierikswortel (Dutch; EFS); Morror (Heb.; ?); Mountain Radish (Eng.; GMH); Moutarde des Allemands (Fr.; GMH); Moutarde des capucins (Fr.; POR); Moutardelle (Fr.; POR); Peberrod (Den.; POR); Pepparot (Swe.; POR); Piparjuuri (Fin.; POR); Rábano Forte (Por.; HH2); Rábano Picante (Sp.; Por.; USN); Rábano Picanto (Por.; USN); Rábano Rusticana (Sp.; EFS; USN); Rábano Silvestre (Por.; POR); Rafano (It.; EFS); Raifort Cran (Fr.; USN); Raifort Sauvage (Fr.; EFS); Raiz Forte (Brazil; Por.; POR; USN); Red Cole (Eng.; GMH); Seiyo wasabi (Japan; TAN); Seiyou wasabi (Japan; POR); Taramago (Sp.; POR).

Activities (Horseradish):

Abortifacient (f; DEM; FEL; LIB); Allergenic (1; CAN); Antiedemic (f; BGB); Antiinflammatory (f1; BGB; COX; X15231456); AntiMRSA (1; X17260672); Antimutagenic (1; X16250249); Antioxidant (1; FNF; PED; X15231456); Antisarcomic (1; WO2); Antiseptic (f12; APA; KOM; SKY; X17260672); Antispasmodic (f1; HHB; PHR; PH2); Antitumor (f1; FAD); Bactericide (1; BGB; FAD; X17260672); Bitter (f1; PED; WO2); Carcinostatic (f1; PHR; PH2); Carminative (f; EFS;

WO2); Chemopreventive (1; X15231456); Circulatory Stimulant (f; CAN); COX-2 Inhibitor (1; X15231456); Counterirritant (f1; PED); Decongestant (f1; APA); Depurative (f; DEM); Diaphoretic (f; PNC); Digestive (f; APA; CAN; DEM); Diuretic (f; APA; FEL; LIB; PNC); Emmenagogue (f; LIB); Expectorant (f1; APA; FAD; PED); Fungicide (f1; HHB); Gram(+)-icide (1; X17260672); Gram(–)-icide (1; X17260672); Hyperemic (2; KOM; PHR; PH2); Hypotensive (1; BGB; CAN); Hypothyroid (1; CAN); Immunostimulant (f; LIB; PED); Insecticide (1; X16786497); Larvicide (1; X16786497); Laxative (f; LIB); Mucolytic (f; MAB); Orexigenic (f; DEM; EFS); Pectoral (f; EFS); Rubefacient (12; APA; SKY); Sialogogue (f; FEL; WO2); Stimulant (f; PNC); Vesicant (f1; FAD); Vulnerary (f; CAN; LIB).

Indications (Horseradish):

Abrasion (f; HOO); Allergy (f1; LIB; PED); Alzheimer's (1; COX; X15231456); Anorexia (f; APA; DEM); Arthrosis (f1; APA; BGB; CAN; COX; X15231456); Asthma (f1; BGB; DEM; FNF); Atony (f; FEL); Bacillus (1; X10548758); Bacteria (12; HHB; HH2; KOM; X17260672; X10548758); Bronchosis (f12; APA; PHR; PH2; SKY; X16618018); Bruise (f; HOO); Cancer (1; FNF; JLH); Cancer, abdomen (f1; FNF; JLH); Cancer, breast (f1; FNF); Cancer, colon (f1; FNF; JLH; X15231456); Cancer, liver (f1; FNF; JLH); Cancer, nose (f1; FNF; JLH); Cancer, spleen (f1; FNF; JLH); Cancer, stomach (f1; FNF; JLH); Cancer, skin (f1; FNF; JLH; WO2); Catarrh (1; KOM; PHR; X17260672); Chilblain (f; GMH); Cholecystosis (f; PHR; PH2); Cold (f1; DEM; SKY); Colic (f; APA; PH2); Congestion (f1; APA); Cough (f12; GMH; PHR; PH2); Cramp (f1; HHB; WIN); Cystosis (1; LIB; PHR); Debility (f; BOW); Dental Plaque (f; FAD); Diabetes (f; DEM; LIB); Dropsy (f; FEL; GMH; HHB); Dysmenorrhea (f; DEM); Dyspepsia (f; PHR; PH2; SKY); Dysuria (CAN; PED; fi PHR); Edema (f; BGB; CAN); Enterosis (1; PH2; WO2); Epistaxis (f; HOO); Escherichia (1; HH2; X17260672; X10548758); Fever (f; BOW); Flu (f1; GMH; PHR; PH2; X17260672); Freckle (f; FEL; HOO); Fungus (1; HHB; X10548758); Gastrosis (f; LIB); Glossosis (f; DEM); Gout (f; BGB; GMH; HHB; PHR; WO2); Gravel (f; DEM); Haemophilus (1; X17260672); Headache (f; HOO); Hepatosis (f; HHB; PHR; PH2); High Blood Pressure (1; LIB); Hoarseness (f; FEL; GMH; HOO; WO2); Induration (f; JLH); Infection (f12; HH2; PH2; X17260672; X16618018); Inflammation (f1; CAN; PH2); Lumbago (f; SKJ); Moraxella (1; X17260672); MRSA (1; X17260672); Myalgia (f12; BGB; KOM; PH2); Mycosis (f1; HHB; HH2; X10548758); Neuralgia (f; DEM; GMH); Pain (f1; DEM; PH2); Pericardosis (f; BOW); Pertussis (f; GMH; LIB); Pleurisy (f; BOW); Pseudomonas (1; X17260672); Pulmonosis (f12; APA; DEM; KOM; PHR; PH2; X17260672); Respirosis (f12; APA; DEM; KOM; PHR; PH2; X17260672); Rheumatism (f; DEM; HHB; PHR); Rhinosis (f1; JLH; PED); Sciatica (f; APA; BGB; GMH); Sinusitis (f12; LIB; SKY; WAF; X16618018); Sore (f; LIB; WIN); Sore Throat (f; LIB; SKY); Splenosis (f; GMH; WO2); Staphylococcus (1; HH2; X17260672); Stomachache (f; HOO; LIB); Stomatosis (f; DEM); Stone (f1; CAN; LIB); Streptococcus (1; X17260672); Swelling (f; BGB; JLH); Toothache (f; DEM; LIB); Typhoid (f1; WO2); Urethrosis (12; KOM; PH2); UTIs (urinary tract infections) (12; APA; BGB; KOM; PH2; X17260672; X16618018); Worm (f; APA; GMH); Wound (f; APA; BOW; HOO); Yeast (1; X10548758).

Dosages (Horseradish):

FNFF = !!!

Roots (and occasionally young leaves) widely eaten as spice (as pickle, potherb, or salad ingredient); sliced roots cooked and eaten like parsnips (FAC; TAN), 2–4 g fresh root before meals (CAN); 1–2 drachms grated root (FEL); 20 g fresh root (KOM); 1–2 Tbsp fresh root (PED); 1.5–3 g dry root (PED); 2 g dry root: 10 ml alcohol/10 ml water (PED); 0.5–1 tsp root 3 ×/day (SKY); 2–3 ml tincture 3 ×/day (SKY).

- Hoosiers apply bruised leaves to the forehead for headache, and the stomach for stomachache (HOO).
- Hoosiers apply leaves wet with vinegar to abrasions, bruises, sprains, and wounds (HOO).

- Hoosiers take sweetened vinegar decoction of horseradish for hoarseness (HOO).
- Hoosiers wash freckles regularly with a sour milk infusion (5 hours) of grated horseradish (HOO).
- Hoosiers sniff powdered root for nosebleed (Tyler strongly discourages such) (HOO).

Downsides (Horseradish):

Class 2d. Contraindicated with gastrosis or GI mucososis, and nephrosis. Not for children under 4 years old (AHPA, 1997; AEH; KOM). No health hazards or side effects known in conjunction with proper therapeutic dosages (PH2). Newall, Anderson, and Phillipson (1996) caution that glucosinolates are allergenic and an irritant. Because of the irritant oil, excessive ingestion should be avoided during pregnancy and lactation. May depress thyroid function, an action "common to all members of the cabbage and mustard family." "The oil is one of the most hazardous of all essential oils and is not recommended for either external or internal use" (CAN). Excessive doses may lead to diarrhea or night sweats. "One case of a heart attack has been recorded — the patient survived" (TAD).

Extracts (Horseradish):

Horseradish peroxidase hypotensive (ivn cat) stimulates arachidonic acid metabolites (CAN). Germans have a commercial preparation (Angocin Anti-Infekt N), mixing shoots of *Tropaeolum major* and roots of horseradish. It is used to treat respiratory and urinary tract infections. The volatile mustard oils are activated in the GI tract after ingestion. Both oils have antibacterial activity. (*Haemophilus influenzae, Moraxella catarrhalis, Escherichia coli, Pseudomonas aeruginosa, Streptococcus pyogenes*, methicillin-susceptible [MSSA] and -resistant *Staphylococcus aureus* [MRSA]). Combining the two herbs led to synergy against *Pseudomonas* (X17260672).

WHITE WORMWOOD (*ARTEMISIA HERBA-ALBA* ASSO) ++ ASTERACEAE

Synonyms:

Artemisia inculta Del.; *Artemisia sieberii* Besser; *Seriphidium herba-album* (Asso) Soják *fide* GHA & RIZ

Notes (White Wormwood):

But her end is bitter as wormwood, sharp as a two-edged sword.

Proverbs 5:4 (KJV)

Zohary maintains that this is most likely to be the biblical species in the Holy Land (ZOH). *A. judaica* is similarly used. Bedouins sell both species in Cairo markets. Herba-alba is used in native perfumery. The drink absinthe is made from a similar species, and thousands of gallons were once consumed annually, especially in France. It is said at first to produce pleasant sensations, inspiring the mind with grandiose ideas, making one " drunken with wormwood." Others compare the effects of the smoke from various species of *Artemisia* with smoke from *Cannabis*. Bedouins believe that the fumes will keep away the cosmopolitan evil eye, perhaps a parallel with Amerindian smudging. Such smudgings are occasionally visited by police officials because the *Artemisia* smoke also smells like *Cannabis* smoke. Bedouins use dry wooly galls from the wormwood for tinder to start fires (BIB). In Lebanon, Philips maintains that *A. arborescens, A. herba-alba, A. judaica,* and *A. maritima* are used interchangeably.

Common Names (White Wormwood):

Abelbel (Ber.; BOU); 'Alala (Arab.; BOU); Apsinthos (Greek; ZOH); Armoise (Fr.; BOU); Armoise blanche (Fr.; BOU); biblical Wormwood (Eng.; BIB); Desert Wormwood (Eng.; RIZ); Ghoreird

(Arab.; BOU); Ghurayrah (Arab.; Syria; HJP); Ifsi (Ber.; BOU); Izri (Ber.; BOU); Laanah (Heb.; ZOH); Laehnah (Heb.; BI2); Odessir (Ber.; BOU); Shih (Arab.; Qatar; Syria; BI2; BOU; GHA; HJP); Thym des steppes (Fr.; BOU); White Wormwood (Eng.; HJP; USN); Wormwood (Eng.; BOU); Zeri (Ber.; BOU); Zerzeri (Ber.; BOU).

Activities (White Wormwood):

Abortifacient (f; HJP); Anthelmintic (f1; HJP; X7162537); Antienteritic (f; DAW); Antigastritic (f; DAW); Antihemolytic (1; FNF); Antileishmanic (1; X11346978); Antiophidic (1; FNF); Antiseptic (1; BOU; DAW; X8786657); Antispasmodic (f1; CWW; DAW; EB22:165); Antiulcer (1; CWW); Ascaricide (1; GHA); Bactericide (1; Yashphe et al., 1995); Decongestant (f; DAW); Diuretic (f; HJP); Emmenagogue (f; BOU; DAW); Febrifuge (f; BOU; DAW); Fungicide (1; CWW); Gastro-acidulant (1; X10189949); Gastroprotective (1; X10189949); Hemostat (f; DAW); Hyperglycemic (1; X8786657); Hypoglycemic (1; X8786657); Hypolipidemic (1; CWW); Hypotensive (1; CWW); Insecticide (1; BOU); Parasiticide (1; BOU); Stimulant (f; BOU); Tonic (f; BOU); Vermifuge (f; BOU; DAW; HJP; X7162537).

Indications (White Wormwood):

Abscess (f; BOU; DAW); Amenorrhea (f; DAW); Bacteria (1; X458619); Bedbug (f; HJP); Bleeding (f; BOU; DAW); Bronchosis (f; DAW); Burn (f; BOU); Cold (f; HJP); Congestion (f; DAW); Cough (f; BOU; DAW); Cramp (f; DAW); Dermatosis (f; HJP); Diabetes (1; X9324004; X8786657); Diarrhea (f; BOU); Dysphoria (f; DAW); Enterobius (1; X3256117); Enterosis (f; CWW; DAW; ZOH; EB22:165); Epilepsy (f; HJP); Escherichia (1; CWW); Fever (f; BOU; DAW); Fungus (1; CWW); Gastrosis (f1; DAW; EB22:165; HJP; X10189949; X3747566); Gingivosis (f; HJP); Headache (f; BOU; DAW); Head Cold (f; HJP); High Blood Pressure (f1; CWW; HJP); Infection (1; DAW; X8786657; X458619); Leishmania (1; X11346978); Malaria (f; HJP); Mycosis (1; CWW); Neurosis (f; BOU; DAW); Ophthalmia (f; BOU; DAW); Respirosis (f; CWW); Rheumatism (f; DAW); Rhinosis (f; DAW); Salmonella (1; CWW); Shigella (1; CWW); Sore (f; DAW); Spasm (f1; CWW; DAW; EB22:165); Stomachache (f; HJP); Toothache (f; CWW: DAW; FP3); Ulcer (1; CWW); Worm (f1; BOU; DAW; GHA; ZOH); Wound (f; BOU; HJP).

Dosages (White Wormwood):

FNFF = !
Dried leaves used as tea by Negev and Sinai Bedouin (ZOH).

- Bedouins tamp leaves in the nostrils as a nasal decongestant for cold (BIB).
- Bedouins drink leaf tea, in water or milk, for colds and cough (BIB).
- Bedouins, when newly born, inhale smoke from burning leaves to insure good health (similar aromatic species of Artemisia are used by Orientals and Amerindians in moxibustion and smudging, respectively) (BIB).
- Egyptians use wormwood for tapeworms (BI2).
- Israelis and Palestinians use the leaves for toothache (FP3; HJP).
- Lebanese apply leaf macerate with shea and olive oil to cuts, dermatosis, and lesions (HJP).
- Lebanese make a strong tea from dried whole herb to expel worms (HJP).
- North Africans use the plant to calm bleeding, cough, emotions, fever, headache, nerves, ophthalmia, and stomach (BOU).
- Saudi Bedouins inhale the smoke for medicinal purposes (GHA).

Natural History (White Wormwood):

Although bitter, the plant is grazed by goats (ZOH). Camels that graze on the plant are said to be spared certain skin diseases. Snakes are reported from the shade of the herb (BIB). Farmers in the Nile Valley fumigate poultry with smoke from burning leaves. They view the herb as a snake repellent.

Extracts (White Wormwood):

Aqueous extract is antileshmanial at 4 µg/ml (X11346978). Stomach acid content increased 251% by ethanol extracts (X10189949).

GIANT REED (*ARUNDO DONAX* L.) + POACEAE

Synonyms:

Arundo bengalensis Retz.; *Donax arundinaceus* Beauv.

Notes (Giant Reed):

A reed shaken with the wind.

Matthew 11:7 (KJV)

Many scholars have identified this as the reed of the Bible, while Zohary has instead picked the very similar *Phragmites communis*, which see, from four species of reed in Israel with long rootstocks and tall stems, hollow and jointed, entirely covered with large leaves, each ending in tassels of flowers. But Zohary admits to a "collective" colloquial concept of reeds much broader than the taxonomist's narrower concepts (ZOH). Presenting us with some interesting information on *Arundo donax*, as well as several common names, the USDA's R.E. Perdue (now retired) also noted that many of these common names are applied to both *Arundo* and *Phragmites*. With no voucher specimens, who knows what was really in the hands or minds of early biblical writers and fisherfolk? Various authors state that both the names and uses are almost interchangeable. (But only *Arundo* of these seems to contain the rather dangerous compound bufotenine.) What follows is some of what I wrote in 1985, hoping that it accrues to *Arundo*. Although believed by Egyptians to be a Syrian introduction, *Arundo* apparently is rare or absent as a truly wild plant or seedling. It is cultivated along water courses, but usually above the water level. Even around big lakes, it seems to be exclusively propagated by root divisions, usually by fishermen. Stems serve as support for vines and similar climbing plants, and for making trellises and the like for climbing cultivated plants. In Egypt, the reeds are also used for fencing and roofing. Reeds are also used as measuring rods, walking sticks, arrow shafts, fishing poles, musical instruments (e.g., clarinets and bagpipes in Europe), baskets, and mats. Romans used such reeds for pens. It makes good-quality paper, and Italians use it in manufacturing rayon (BIB).

Common Names (Giant Reed):

Adavi Keekasa Gaddi (Tel.; WO2); Alo Kyu (Burma; WOI); Arrow Reed (Eng.; EFS); Baktusing (Limbu; NPM); Bamboo Reed (Eng.; NPM); Bambou Indien (Guad.; AVP); Bambou Mare (Guad.; AVP); Bansi (Pun.; WOI; WO2); Bara Nal (Hindi; WO2); Bhok (Chepang; NPM); Calami (It.; EB12:368); Cana (Por.; EFS); Caña Bambua (Pr.; Sp.; AVP); Caña Brava (Pr.; Sp.; AVP); Caña Brava de Castilla (Col.; Sp.; AVP); Caña Común (Sp.; EFS); Caña de Castilla (Cuba; Dr.; Sp.; AVP; EFS); Caña Guana Carrizo (Sp.; AVP); Caña Hueca (Sp.; Cuba; AVP); Caña India (Pr.; Sp.; AVP); Cañaveral (Mex.; Sp.; AVP); Cana-vieira (Mad.; JAD); Canna (It.; EB12:368); Canna Berganena (It.; EB12:368); Canna Commune (It.; EB12:368); Canna da Rocchi (It.; EB12:368); Canna di Cannitu (It.; EB12:368); Canna di Stenniri (It.; EB12:368); Canna do Reino (Por.; AVP);

FIGURE 1.12 Giant Reed (*Arundo donax*).

Canna Domestica (It.; EB12:368); Canna du Provenza (It.; EFS); Canna Montana (It.; EFS); Canna Zagariddara (It.; EB12:368); Canne de Province (Fr.; EFS); Carrizo (Mex.; Peru; EGG; EB12:368); Carruzo (Peru; Sp.; EFS); Ceneviera (It.; EB12:368); Ciane Gergane (It.; EB12:368); Ckingkan (Aguaruna; Peru; EGG); Cow Cane (Eng.; AVP); Cuin (Cuba; AVP); Danubian Reed (Eng.; NPM); Dexne

(Peru; Uvosha; EGG); Donaci (It.; EB12:368); Donax Cane (Eng.; NPM); Foocañunhyta (Ocaina; Peru; EGG); Gaba Nal (Assam; WO2); Gaha Nal (Beng.; WO2); Giant Reed (Eng.; NPM); Grand Roseau (Fr.; Mart.; AVP); Great Reed (Eng.; Scn.; AH2; NPM); Guajana (Pr.; Sp.; AVP); Gutamu (It.; EB12:368); Honrkandma (Rai; NPM); Italian Reed (Eng.; NPM); Kadi Lak (Sanskrit; WO2); Kalamos (Greek; EB12:368); Kamiş (Tur.; EFS); Kasab (Egypt; EB12:368); Khen (Laos; AVP); Korukkai (Tel.; WO2); Lata (Col.; Sp.; AVP); Lenrakut (Tharu; NPM); Manglong (Lepcha.; NPM); Nal (India; WO2); Naldura (Gharwal; WO2); Narhal (Delhi; WO2); Narkat (Bhojpuri; Danuwar; Gurung; Nepal; Sunwar; Tamang; NPM); Narkul (India; WO2); Narsal (Delhi; WO2); Nyapan kathi (Newari; NPM); Panache (Mart.; AVP); Pfahlrohr (Ger.; AVP); Pfeilrohr (Ger.; EFS); Pijlriet (Dutch; EFS); Provence Cane (Eng.; NPM); Quassab (Arab.; AVP); Racine de Canne (Fr.; AVP); Riesenschilf (Ger.; EB12:368); Riet (Dutch; AVP); Rohr (Ger.; AVP); Roseau (Fr.; EB12:368); Roseau des Jardines (Haiti; AVP); Say Lau (Vn.; AVP); Spanish-Bambooreed (Eng.; WO2); Sukana (Pun.; WO2); Taquara (Por.; AVP); Taquari (Por.; AVP); Těběrau Gading (Malaya; IHB); Thagal (Nepal; NPM); Uaichi (Aguaruna; Peru; EGG); Virtamu (It.; EB12:368); Wasserrohr (Ger.; EFS); Z'herbe Roseau (Haiti; AVP).

Activities (Giant Reed):

Analgesic (f; EGG); Antifeedant (1; X10803974; X8254352); Antiacetylcholinesterase (1; WO2); Antigalactagogue (f; DAW; MPI; SKJ; WO2); Antihistaminic (1; WO2); Antispasmodic (f; MPI); Cyanogenic (1; BIB); Depurative (f; AHL; DAW); Diaphoretic (f; DAW); Diuretic (f; BIB; DAW; NPM); Emmenagogue (f; MPI; NPM; SKJ); Emollient (f; DAW); Hallucinogenic (1; BIB); Hypertensive (1; DAW); Hypotensive (1; DAW; WO2); Insecticide (1; WO3); Myorelaxant (1; MPI); Respirodepressant (1; MPI); Sudorific (f; DAW); Vasopressor (1; FNF).

Indications (Giant Reed):

Cancer (f; BIB); Condyloma (f; BIB); Dropsy (f; BIB); Dysmenorrhea (f; NPM); Headache (f; NPM); High Blood Pressure (1; DAW); Induration (f; BIB); Low Blood Pressure (1; DAW); Mastosis (f; BIB); Pain (f; EGG); Syphilis (f; AHL); Venereal Disease (f; AHL); Wart (f; BIB); Wound (f; JAD).

Dosages (Giant Reed):

FNFF = !
Young shoots occasionally eaten.

- Dominicans suggest the roots as antilactagogue, antisyphilitic, depurative, and diuretic (AHL).
- Egyptians use *Arundo* and *Phragmites* as diaphoretic and diuretic.
- Italians boil roots in wine with honey for various cancers (JLH).
- Peruvians consider the rhizomes analgesic and diuretic (EGG).

Downsides (Giant Reed):

The bufotenine alkaloids could be problematic.

Natural History (Giant Reed):

Agriculture Handbook No. 165 lists the following as affecting this species: *Armillaria mellea* (root rot), *Leptostroma donacis, Papularia sphaerosperma, Puccinia coronata* (crown rust), and *Selenophoma donacis* (stem speckle).

Extracts (Giant Reed):

Ethanolic rhizome extracts show antiacetylcholinesterase, antihistaminic, antispasmodic, and hypotensive activities (WO2). Leaves emit isoprene (WO3).

Said to raise the blood pressure in high doses, lower it in lower doses (AHL). Leaves may contain bufotenine, which is reportedly antialzheimeran, anticholinesterase (1/30th physostigmine), aphrodisiac, cardioactive, CNS-active, hallucinogenic, hypertensive, mydriatic, respiraparalytic, and serotoninergic (http://www.ars-grin.gov/duke).

TRAGACANTH MILKVETCH (*ASTRAGALUS GUMMIFER* LABILL.) +++ FABACEAE

Synonyms:

Astracanthus gummifer (Labill.) Podlech; *Astragalus adpressus* Ehrenb ex Walp.; *Astragalus erianthus* Willd.; *Astragalus noemiae* Eig. *fide* HH2

Notes (Tragacanth Milkvetch):

> **And their father Israel said unto them, If it must be so now, do this; take of the best fruits in the land in your vessels, and carry down the man a present, a little balm, and a little honey, spices, and myrrh, nuts, and almonds.**
>
> **Genesis 43:11 (KJV)**

> **Then their father Israel said to them, "If it must be so, then do this: take some of the choice fruits of the land in your bags, and carry down to the man a present, a little balm and a little honey, gum, myrrh, pistachio nuts, and almonds.**
>
> **Genesis 43:11 (RSV)**

> **So Israel, their father, said to them, "If then, that is the case, do this: Take the finest products of the land and carry them down to the man as a gift, a little balsam, and a little honey, labdanum, and resinous bark, pistachio nuts, and almonds.**
>
> **Genesis 43:11 (NWT)**

Working with the RSV, the only version of the three to mention *gum*, clearly differing botanically from the KJV and NWT above, Zohary concludes that the word *nekhoth*, mentioned only twice in the Hebrew Bible, should be translated *tragacanth*. The KJV seems to have substituted spices, the NWT labdanum, for the gum in the RSV. As a mediocre botanist, not a historian or biblical scholar, I do not know which, if any, version is correct. Further complicating life, the AH2 specifies that the Standardized Common Name of the dried gummy exudate of the stems of *A. gummifer* is gum tragacanth. Zohary notes that the genus *Astragalus* has about 1800 species, "many of them producing the tragacanth gum." (ZOH) Zohary says tragacanth is "a dried gum exuded from certain spiny and shrubby species of *Astragalus* occurring in several countries of the Middle East, including Israel" (ZOH). Zohary lists 50 species of *Astragalus*, but not *A. gummifer*, in the *Flora of Palestine* (FP2). It is difficult for trained taxonomists who have devoted their lives to the study of legumes to identify flowering and fruiting specimens of the shrub; who, then, can identify the "gum," perhaps mentioned in Genesis? It could have been the gum of many species of *Astragalus*, perhaps even admixtures of gum from several species or even other gum- or resin-bearing shrubs. Before his death, Howard Scott Gentry, who himself had studied tragacanth in the Middle East, tried to interest me in just such a study. There was much anticancer interest in the various species of *Astragalus*. I declined, frightened by the taxonomic complexity. I fear no one man in his lifetime could resolve the taxonomic difficulties of this complex but useful genus. Hungry early man doubtless used tragacanth as a survival food, after seeing that ants, goats, and sheep relish the sweeter gums. Tragacanth gum is one of the oldest natural emulsifiers known to man, extensively used in vaginal jellies and creams, low-calorie syrups, toothpastes, and hand lotions. The gum is used in such foods as salad dressings, sauces, ice creams, confections, syrups, milk powder stabilizers, citrus oil emulsions, and cheeses. Together with guar, tragacanth may make the cheaper ice creams healthier, substituting soluble cholesterol-free polysaccharides for the unhealthy cream. Some of these polysaccharides are active against ascites tumor types in mice. Among the species taxonomically confused with or called tragacanth, or some derivative thereof, Hartwell notes "folk-cancer" remedies for chronic indurations of the liver (including, probably, cirrhosis), nasal polyps, non-ulcerated cancers, superfluous flesh, and tumors of the eyes, fauces, and liver. It was once used to stiffen calico, crape, etc. Need for fuelwood in poor desert areas may jeopardize many natural stands of this shrub (e.g., in Iran, source of the best tragacanth, where it is largely used in medicine and confectionary) (BIB).

Common Names (Tragacanth Milkvetch):

Adragante (Fr.; EFS); Alquitira (Sp.; USN); Angira (India; EFS); Astragale à Gomme (Fr.; USN); Besem (Heb.; BI2); Dragante (It.; EFS); Goat-Thorn Bush (Eng.; AH2); Gomme Adragante (It.; EFS); Gum (Eng.; BIB); Gum Tragacanth (Eng.; Scn.; AH2; USN); Gummitragant (Ger.; USN); Katâd (Arab.; Syria; HJP); Kathira Bayda (Arab.; Syria; HJP); Kitre Givenne (Tur.; EFS); Nakaa (Arab.; ZOH); Nakaath (Arab.; ZOH); Nekhoth (Heb.; ZOH); Tragacanth (Eng.; FAC); Tragacanth Milkvetch (Eng.; AH2); Tragacanthstruik (Dutch; EFS); Tragacanto (Sp.; USN); Tragant (Den.; Ger.; USN); Traganthstrauch (Ger.; EFS).

Activities (Tragacanth Milkvetch):

Anticancer (1; BIB); Aphrodisiac (f; BIB); Demulcent (f; BIB; EFS; HJP); Diuretic (f; BIB; HJP); Emollient (f; BIB; EFS); Laxative (1; BIB; EFS; PH2); Peristaltic (1; PH2); Vulnerary (f; HJP).

Indications (Tragacanth Milkvetch):

Burn (f; BIB); Cancer (1; BIB; JLH); Cancer, eye (1; BIB; JLH); Cancer, liver (1; BIB; JLH); Cancer, throat (1; BIB; JLH); Cirrhosis (f; JLH); Cough (f; BIB); Diarrhea (f; BIB); Hepatosis (f; JLH); Induration (f; JLH); Ophthalmia (f; JLH); Polyp (f; BIB; JLH); Rhinosis (f; BIB; JLH); Wound (f; BIB; HJP).

Dosages (Tragacanth Milkvetch):

FNFF = !!
More food additive than food, the FDA defines GRAS gum tragacanth as "the exudate from one of several species of *Astragalus gummifier* Labillardiere," permitting its use up to 2000 ppm in baked goods and baking mixes, 7000 ppm in condiments and relishes, 13,000 ppm with fats and oils, 8000 ppm with gravies and sauces, 2000 ppm with meat products, 2000 ppm with processed fruits and fruit juices, and up to 1000 ppm in all other food categories. 1 tsp (circa 3 g) granulated drug added to 250–300 ml liquid (PH2).

- Lebanese apply powdered tragacanth to cuts and wounds (HJP).

Downsides (Tragacanth Milkvetch):

None at therapeutic dosages except for rare allergy. Should be taken with plenty of water, as with all bulking agents (PH2).

SEA ORACHE (*ATRIPLEX HALIMUS* L.) + CHENOPODIACEAE

NOTES (SEA ORACHE):

Who cut up mallows by the bushes, and juniper roots for their meat.

Job 30:4 (KJV)

They pick mallow and the leaves of bushes, and to warm themselves the roots of the broom.

Job 30:4 (RSV)

They were plucking the salt herb by the bushes, And the root of broom trees was their food.

Job 30:4 (NWT)

Zohary suggests that this difficult passage, like so many from Job, might have been better rendered as, "They pick the leaves of the orache and the wormwood." The translation of maluah as "orache," and not true mallow, is better because Job is alluding to the desert, where orache, also called salt bush (if not salt herb as in NWT), is common. It is a strong-growing bush or shrub, 5 feet or more tall, with gray foliage and inconspicuous flowers. One of the most common desert plants in salt flats, it is used as survival food and as a salt source. The closely related *Atriplex rosea*, which occurs in the biblical area as well, has been a folk remedy for such cancerous conditions as corns, hard lumps, and indurations (JLH). Smoke from burning seed is used to treat skin ailments and sores. Lebanese doctors are said to extract anodynes, emetics, hypnotics and purgatives from the plants. According to Boulos, the seeds are in small doses emetic, in large doses poisonous (BOU). Ashes of the plant are taken for gastric acidity, the roots for dropsy (BIB).

COMMON NAMES (SEA ORACHE):

Abougboug (Ber.; BOU); Alimo (It.; HH2); Álimo (Sp.; USN); Aramès (Ber.; BOU); Arams (Ber.; BOU); Armuelle Glauco (Sp.; USN); Arroche (Fr.; BOU); Arroche Halime (Fr.; HH2; USN); Halime (Fr.; BOU); Kattaf (Arab.; Syria; HJP); Mallow (Eng.; BIB); Maluh (Arab.; BOU); Malûhh (Arab.; HJP); Marisma (Sp.; USN); Mediterranean Saltbush (WO2); Meermelde (Ger.; HH2); Meerportulak (Ger.; HH2); Orzaga (Sp.; USN); Osagra (Sp.; USN); Pourpier de Mer (Fr.; USN); Qataf (Arab.; BOU); Qataf Bahhari (Arab.; HJP); Qtout (Arab.; BOU); Roghata (Arab.; BOU); Rughat (Arab.; HJP); Rughl (Arab.; GHA); Salgada (Sp.; USN); Salgadeira (Por.; USN); Salguiera (Madiera; Por.; JAD); Sea Orache (Eng.; BOU; FAC; USN; WO2); Sea Purslane (Eng.; BOU; FAC; USN); Shrubby Orache (Eng.; USN); Staudenmelde (Ger.; HH2); Strauchmelde (Ger.; USN); Tall Shrubby Orache (Eng.; HJP).

ACTIVITIES (SEA ORACHE):

Anodyne (f; HJP); Antacid (f; BOU); Antidiabetic (f1; HH2); Emetic (f; BOU; GHA); Hypnotic (f; HJP); Hypoglycemic (1; HH2); Insulin Potentiator (1; HH2); Insecticide (1; X9617055); Larvicide (1; X9617055); Mosquitocide (1; X9617055); Purgative (f; HJP); Toxic (f; BOU); Vermifuge (f; HH2); Vulnerary (f; HJP).

INDICATIONS (SEA ORACHE):

Dermatosis (f; HJP); Diabetes (f1; HH2); Gastrosis (f; BOU); Hyperacidity (f; BOU); Infection (f; HH2); Sore (f; HJP); Worm (f; HH2); Wound (f; HJP).

Dosages (Sea Orache):

FNFF = !!

Described as famine food in North Africa (WO2). Leaves and young shoots have served as a pot-herb. These "mallows" are commonly eaten by the poor between Aleppo and Jerusalem (e.g., eaten by hungry Palestinian shepherds like spinach). According to the Talmud, Jews working on the reconstruction of the Temple in 520–516 B.C. ate these "mallows" (BIB; FP1). Facciola makes it sound a bit more enticing. Young leaves and shoot tips eaten raw or in casseroles, pastas, quiches, soups, etc. Plant produces an edible manna (FAC).

- Arabians view the seed decoction as emetic (GHA).
- Lebanese apply dried leaf powder to sores and wounds (HJP).
- Lebanese use seed infusion or direct seed smoke to dermatoses and sores (HJP).
- Negev Bedouins take the leaf tea for diabetes (HH2).
- North Africans cut the root into long narrow pieces used as a toothbrush (BOU).
- North Africans take the alkaline ashes of the plant for gastric acidity (BOU).

Downsides (Sea Orache):

North Africans view small doses of the seeds as emetic, large doses poisonous (BOU).

Natural History (Sea Orache):

The fat sand rat *(Psammomys obesus)*, a diurnal gerbillid rodent, is herbivorous and able to thrive while consuming only sea orache, a plant relatively low in energy content and high in ash and water content. Before consuming *A. halimus* leaves, the fat sand rats scrape off the outer layer with their teeth, thereby removing many of the electrolytes but increasing the gross energy and organic matter content of the leaves by only about 3.1%. There are advantages to this diet: (1) it provides a more stable diet throughout the year than do seeds; (2) fat sand rats have no competition for this food resource from other rodents; and (3) their burrows are at the base of the plants, requiring little energy to forage (X8679505).

DESERT DATE (*BALANITES AEGYPTIACUS* (L.) DELILE.) ++ BALANITACEAE

Notes (Desert Date):

> **Is there no balm in Gilead; is there no physician there? why then is not the health of the daughter of my people recovered?**
>
> **Jeremiah 8:22 (KJV)**

> **Is there no balm in Gilead? Is there no physician there? Why then has the health of the daughter of my people not been restored?**
>
> **Jeremiah 8:22 (RSV)**

> **Is there no balsam in Gilead? Or is there no healer there? Why is it, then, that the recuperation of the daughter of my people has not not come up?**
>
> **Jeremiah 8:22 (NWT)**

Moldenke and Moldenke note that the Douay Bible of 1609 renders Jeremiah 8:22 to read, "Is there no rosin in Gilead?" resulting in this edition being termed the Rosin Bible. The Bishop's Bible of 1568 reads, "Is there no tryacle in Gilead?" and is termed the Treacle Bible. Should we call the

FIGURE 1.13 Desert Date (*Balanites aegyptiacus*).

Jehovah's Witness version (NWT) the balsam version, and the KJV and RSV the balm versions? Jericho monks are said to regard Balanites as the balm or balsam, selling an oily gum from the fruit as such to travelers. Both *Balanites* and *Pistacia* are common in Palestine and commonly called balm. *Balanites* is also revered by the Mohammedans in western India. Israeli Zohary rules out *Balanites* in favor of *Commiphora*, or myrrh, favoring the assumption that the Queen of Sheba brought it as a gift from Arabia to King Solomon (BIB). But with so many opinions, I include *Balanites* in case the Jericho monks are right and Zohary wrong. Here in America, the balm of Gilead is yet something else, a poplar with propolis-like resin. Scholars, not me, will probably argue about the Balm of Gilead for millennia. In these days of pharmaceutically inspired allopathic shenanigans, I would worry more about there being no balm in Gilead than there being no physician in Gilead. Physicians kill more people than guns, which kill more people than herbs.

Common Names (Desert Date):

Addoua (Ber.; BOU); Adua (Hausa; KAB; UPW); Alo (Ber.; BOU); Angarvriksha (Sanskrit; KAB); Angulidala (Sanskrit; KAB); Anilantaka (Sanskrit; KAB); Balah Harara (Arab.; BOU); Balanite d'Egypte (Fr.; BOU); Betu (India; USN); Bhallakivriksha (Sanskrit; KAB); Bupapabu (Niger; UPW); Ciatt (Tigrinia; KAB); Daruparnaphala (Sanskrit; KAB); Dattier du Désert (Fr.; BOU; KAB); Dattier Sauvage (Fr.; USN); Desert Date (Eng.; BOU; TAN); Ebora (Ber.; BOU); Egorea (Guj.; KAB; NAD); Egyptian Balsam (Eng.; BOU); Egyptian Myrobalan (Eng.; FAC); Elheglyg (Arab.; KAB); Gak (Tigrinia; KAB); Gara (Tel.; KAB); Garachetti (Tel.; NAD); Garbaïe (Mali; UPW); Garrah (Gond.; KAB); Gaurtvaka (Sanskrit; KAB); Gongo (Upper Volta; UPW); Gouasa (Tigrinia; KAB); Gouritivac (Sanskrit; NAD); Guasa (Tigrinia; KAB); Gungo (Ghana; UPW); Guri (Tel.; WOI); Hadjlidj (Arab.; Sen.; UPW); Haugeleg (Fr.; BOU); Heghelig (Arab.; KAB); Heglig (Arab.; Nig.; BOU; UPW); Heglik (Sudan; KAB); Héglik (Fr.; USN); Hingan (Beng.; Dec.; Hindi; KAB; NAD; WOI); Hinganbet (Bom.; Cutch; Dec.; Mar.; KAB; WOI); Hingane (Mar.; KAB); Hingar (Ulwar; KAB); Hingen (Hindi; KAB); Hinger (Bom.; Guj.; KAB; NAD); Hingol (Beng.; Dec.; Hindi; NAD); Hingon (Beng.; KAB); Hingoriyun (Porebunder; KAB); Hingot (Gwalior; Hindi; Urdu; KAB; NAD; WOI); Hingota (Hindi; KAB); Hingotia (India; USN); Hingu (Hindi; KAB); Hinguputra (Sanskrit; KAB); Ibororhen (Ber.; BOU); Ingalarade (Kan.; KAB); Ingalare (Kan.; KAB); Ingalukke (Kan.; KAB); Ingua (Hindi; KAB); Inguda (Sanskrit; KAB); Ingudi (Sanskrit; Tel.; KAB; WOI); Ingudihala (Uriya; KAB); Ingul (Sanskrit; KAB); Inguni (Sanskrit; KAB); Jalajantu (Sanskrit; KAB); Kantaka (Sanskrit; KAB); Krisharaka (Sanskrit; KAB); Kroshtuphala (Sanskrit; KAB); Kunja (Togo; UPW); Kuosa (Tigrinia; KAB); Lalo (India; USN); Maghe (Tigrinia; KAB); Manchuta (Mal.; NAD); Mirobalano de Egipto (Sp.; USN); Mjunju (Swahili; KAB); Mogha (Tigrinia; KAB); Munipadapa (Sanskrit; KAB); Murtoki (Gambia; UPW); Myrobalan d'Egypt (Fr.; UPW); Nanjunda (Tam.; NAD; WOI); Nanjundan (Tam.; KAB); Nanjunta (Mal.; KAB); Ol Ngoswa (Masai; KAB); Putigandha (Sanskrit; KAB); Putripatra (Sanskrit; KAB); Regorea (Guj.; WOI); Ringri (Tel.; KAB; NAD); Selibatico (Bogos; KAB); Sereno (Ivo.; UPW); Shulari (Sanskrit; KAB); Soapberry Bush (Eng.; BOU); Soapberry Tree (Eng.; FAC; TAN; USN); Soump (Wolof; KAB); Taboraq (Ber.; BOU); Tailabija (Sanskrit; KAB); Tanupatra (Sanskrit; KAB); Tapasamudrama (Sanskrit; KAB); Tapasataru (Sanskrit; KAB); Teïchat (Arab.; Mauritania; UPW); Teisset (Ber.; BOU); Thorn Tree (Eng.; BOU); Tiktaka (Sanskrit; KAB); Tiktarnajja (Sanskrit; KAB); Toruvattu (Tam.; KAB); Tugga (Ber.; BOU); umNulu (Tonga; KAB); Vinashaka (Sanskrit; KAB); Vishakantaka (Sanskrit; KAB); Zachum Oil Tree (Eng.; BOU); Zachun (India; USN); Zachumbaum (Ger.; USN); Zaqqoum (Arab.; BOU).

Activities (Desert Date):

Alexipharmic (f; KAB); Alterative (f; KAB); Analgesic (f; KAB); Anthelmintic (f; BOU; HDN; KAB; X10904170); Antidiabetic (1; WO3); Antidote (Curare) (f; BOU); Antiedemic (1; X15763372); Antifeedant (1; HDN; WO3); Antifertility (1; X12179631); Antiinflammatory (1; X15763372); Antinociceptive (1; X15763372); Antioxidant (1; X15763372); Antischistosomal (1; HDN; X15664459); Antiseptic (f; HDN); Antispasmodic (1; HDN); Aperient (f; KAB); Aphrodisiac (fl; UPW); Bactericide (1; HDN); Burn (f; BIB); Cathartic (f; DEP); Cholagogue (1; HDN); Choleretic (1; HDN); CNS Depressant (1; HDN); Curare (1; HDN); Emetic (f; BOU; KAB); Expectorant (f; DEP; WO2); Fungicide (1; HDN); Hemolytic (1; HDN); Hemostat (f; HDN); Hepatoprotective (1; PR15:598); Hypotensive (1; HDN); Insectifuge (f; HDN); Lactagogue (f; UPW); Larvicide (1; WO3); Laxative (f; BOU); Molluscicide (1; HDN); Nematicide (1; WO3); Piscicide (1; HDN); Purgative (f; DEP; HDN; NAD); Stimulant (f; UPW); Vermifuge (f; BOU; NAD).

Indications (Desert Date):

Abscess (f; UPW); Angina (f; UPW); Anxiety (f; HDN); Asthma (f; HDN); Bacillus (1; HDN); Bacteria (1; HDN); Bilharzia (f; HDN); Bite (f; KAB); Bleeding (f; HDN); Blennorrhea (f; UPW); Boil (f; BOU; KAB); Bronchosis (f; UPW); Bubo (f; HDN); Burn (f; NAD; WO2); Carbuncle (f; UPW); Caries (f; UPW); Catarrh (f; HDN); Childbirth (f; WO2); Circumcision fi (BOU); Cold (f; DEP; HDN); Colic (f; BIB; KAB; NAD; UPW); Conjunctivosis (f; HDN); Cough (f; BIB; DEP; KAB; NAD); Cramp (f; HDN); Dermatosis (f; KAB); Diabetes (1; WO3); Diarrhea (f; HDN); Dysentery (f; KAB; UPW); Edema (1; X15763372); Fasciolaris (1; X10904170); Fever (f; BOU; HDN); Freckle (f; NAD; WO2); Fungus (1; HDN); Gingivosis (f; UPW); Guinea Worm (1; WO3); Hemorrhoid (f; UPW); Hepatosis (f1; HDN; UPW; PR15:598); Herpes (1; BIB; HDN); High Blood Pressure (1; HDN); Impotence (f; UPW); Infection (f; BIB); Infection (1; HDN); Infertility (f; HDN); Inflammation (f1; HDN; X15763372); Insanity (f; HDN; UPW); Jaundice (f1; UPW; PR13:439; X10441790); Leprosy (f; UPW); Leukoderma (f; BOU; KAB); Malaria (f1; BIB; BI2 BOU); Mycosis (1; HDN); Pain (f1; BOU; HDN; X15763372); Paralysis (f; UPW); Pertussis (f; WO2); Pneumonia (f; WO2); Pulmonosis (f; WO2); Rheumatism (f; BIB; UPW); Schistosomiasis (1; HDN; 15664459); Shingle (1; HDN); Sleeping Sickness (f; KAB); Smallpox (f; HDN); Snakebite (f; HDN); Sore (f; KAB); Splenosis (f; BOU; UPW); Stomachache (f; HDN); Stomatosis (f; UPW); Swelling (f1; UPW; X15763372); Syphilis (f; BIB; BOU; UPW); Urethrosis (f; HDN); Venereal Disease (f; BIB); Virus (1; HDN); Worm (f; BI2; BOU; HDN; NAD); Wound (f; BI2; BOU; HDN); Yaws (f; UPW); Yellow Fever (fl; UPW).

Dosages (Desert Date):

FNFF = !!

Fruits eaten fresh, dried, in alcoholic beverages (e.g., the Hausa *kango*), and syrups; seeds eaten raw or dried, in breads or soups, source of edible oil; flowers and leaves also eaten as vegetables or in soups (e.g., in Chad, Ghana, Nigeria, Sudan) (BI2; FAC; UPW).

- African Arabs use the fruit pulp as detergent, the bark to poison fish (KAB).
- Asian Indian suggest 2 to 30 g seed as expectorant (DEP).
- Asian Indians suggest 1 to 20 grains fruit as purgative (DEP).
- Ayurvedics use the fruits as alexipharmic, alterative, analgesic, anthelmintic, antidermatitic, and antidysenteric (KAB).
- Ethiopians use bark as an antiseptic, the leaf to dress wounds, and the fruit as an anthelmintic laxative (BIB).
- Ghanans use smoke from stem to heal circumcision wounds, leaves as vermifuge (BI2).
- Lebanese apply the oil to sores, treating dermatosis and rat bites with fruits (BI2).
- Libyans use the leaves to clean infected wounds, and root for herpes and malaria (BI2).
- Nigerians consider the plant abortifacient (BI2).
- Nigerians eat the unopened flower buds as an aphrodisiac (UPW).
- Nigerian Yoruba take the floral decoction for sore throat (UPW).
- Saharans take powdered bark for angina and bronchosis (UPW).
- Sudanese use the oil as a purgative (KAB).
- Turks suggest this as one of the best stomachics, and great for curing wounds.
- Ugandans use the oil to treat sleeping sickness (BI2).
- Unani use fruits for boils, dermatoses, and leukoderma (KAB).

Natural History (Desert Date):

In the biblical desert, there are often strings of oases with about a dozen tropical tree species, including *Acacia, Balanites, Moringa, Salvadora*, etc. Needing high temperature and humidity, they adhere to the mouth of tributaries (e.g., in the Aravah and Jordan Valleys). In Arabia and Somaliland, *Balanites* often occurs in thornbush vegetation under arid tropical conditions, with *Commiphora, Maerua, Ziziphus*, etc.

Extracts (Desert Date):

The active principle, probably a saponin, is lethal to cercaria, fish, miracidia, mollusks, and tadpoles. One fruit weighing 25 g has enough active ingredient to kill the bilharzial mollusks in 30 l water.

FRANKINCENSE (*BOSWELLIA SACRA* BIRDW.) ++ BURSERACEAE

Synonyms:

Boswellia carteri Birdw.

Notes (Frankincense):

And the LORD said unto Moses, Take unto thee sweet spices, stacte, and onycha, and galbanum; these sweet spices with pure frankincense: of each shall there be a like weight:

Exodus 30:34 (KJV)

And the LORD said to Moses, "Take sweet spices, stacte, and onycha, and galbanum, sweet spices with pure frankincense (of each shall there be an equal part).

Exodus 30:34 (RSV)

And Jehovah went on to say to Moses, "Take to yourself perfumes; stacte drops and onycha, and perfumed galbanum, and pure frankincense. There should be the same portion of each.

Exodus 30:34 (NWT)

It is nice when all three versions translate it as frankincense, all three actually being pretty consistent in the formula for a holy incense. On reading Zohary (1982), I see he had a change in plans just before publication, eliminating one species of *Boswellia*, for he says "the resins from the two species listed above" (but he only listed one). He even hints, without clearly stating, that frankincense may involve resins from as many as 24 species of *Boswellia*. *Boswellia carteri* has been reduced to synonymy with *Boswellia sacra*. Frankincense came to the Holy Land via the famous spice route across southern Arabia and some of the littoral stations of East Africa, the same caravan highway used also for goods from India and points farther east (Zohary, 1982). Zohary rationalizes his conclusions by pointing to similarities between the Arabic *luban* and the Hebrew *levonah*. Today, the Catholic Church may be the major consumer, often using frankincense in ceremonial incenses.

Another true scholar, botanical historian John W. Thieret (1996), seems to agree with Zohary, noting that a main source of frankincense is *Boswellia sacra*. "Herodotus (born 484 BC) wrote that the frankincense trees were guarded by vast numbers of small winged serpents; he was wrong." (ZOH) Most frankincense comes from Somalia (following bananas and cattle as leading export) where it provides work for some 10,000 Somali families, but some is gathered in Arabia. Most goes to Saudi Arabia, Yemen, and Egypt, the major markets, but it heads off in lesser quantities in all directions. Early botanist Theophrastus, some three centuries before Christ, said that most frankincense came from Saba (southwestern Arabia, once ruled by the famed Queen Sheba). That ancient country became rich from the incense trade. Approximately 333 B.C., Alexander the Great's army captured Gaza, plundering its frankincense and sending it to Greece. Tons of incense were buried in the temples of Babylon and Nineveh. And in King Tut's tomb, 3000-year-old balls of frankincense were recovered. "In today's churches, frankincense is an ingredient in the incense that sometimes nearly suffocates the faithful.... Because frankincense and myrrh no longer enjoy the esteem that they did two millennia ago, I wonder what the Wise Men would bring today. Perhaps gold, dates, and oil" (Thieret, 1996). I would have guessed saffron.

FIGURE 1.14 Frankincense (*Boswellia sacra*).

Import statistics are difficult to come by. Scholarly historian Thieret (1996) suggests total yearly production of myrrh is perhaps 500 tons, of frankincense 1000 tons. Recently, U.S. imports run 5 to 20 tons. The United Kingdom imports circa 30 tons frankincense each year, one perfume manufacturer alone consuming 5 tons annually (Thieret, 1996).

Ghazanfar (1994) notes that in southern Arabia, *luban* trees occur in wadis extending to the coast on the lower slopes of the gullies and runoffs. The gum exuding from cuts is the major medicinal incense, being burned to give a perfumed smoke, used to improve the aroma of clothing, hair, and residences. The resin, used to stimulate digestion, to treat mastitis, and strengthen the teeth is also mixed into hair products. Soot collected from burning the resin is used as *kohl* memory device collyrium for soothing sore eyes. Pregnant Yemenis chew the gum, and it is also chewed for emotional and psychological problems. Arabians often chew it as a masticatory, believed to improve the memory, or add it to coffee. The resin is presumed to be diuretic and purgative. Thieret (1996) adds that in Greco-Roman medicine, frankincense was prescribed for abscesses, bruises, chest ache, hemorrhage, hemorrhoids, paralysis, and ulcers. In northern Africa, it is used for back problems, chest congestion, chronic coughs, poliomyelitis, and venereal ailments (Thieret, 1996). I fear that much of the information I have compiled should be viewed as generic rather than specific. I have focused on *Boswellia sacra* and *carteri* (HHB and WO2 data below; however, refer to *B. serrata*, so-called Indian Olibanum, syn. *B. glabra* Roxb.). I doubt that there are many people who can swear on a stack of Bibles as to whether a resin is frankincense, myrrh, or one or the other species or genera or a mixture of many species. Let the buyer beware. The frankincense problem is a frankensteinian taxonomic nightmare. I think the common name probably conveys, in this case, as much accuracy as the scientific name, for those resins for which there is no voucher specimen. I have, with that in mind, added the EFS common names for frankincense below, finding much congruence with names derived from other sources.

Frankincense was mentioned 22 times in the Bible, 16 times as an item of worship, 3 times as a product of the garden of Solomon, twice as a tribute of honor, and only once as an item of merchandise. It is chiefly used in incense as a perfume, especially in Catholic ceremonies. Recent authorities maintain that the "incense" used in the service of the Tabernacle was a mixture, in definite proportions of frankincense, galbanum (*Ferula gumosa*), onycha (*Styrax benzoin*), and stacte (*Styrax officinalis*), and the use of any incense not composed of these four ingredients in the proper proportions was strictly forbidden. Frankincense was highly regarded by Egyptians for embalming and fumigating. The gum is used as a masticatory, to clean the mouth. Oil of olibanum is used in high-grade perfumes, especially for oriental and floral types, and was once used as a depilatory. Resin is imported into Lebanon, primarily as incense, but secondarily as a cosmetic and medicine (BIB).

Common Names (Frankincense):

Árbol del Incensio (Sp.; USN); Arbre à Encens (Fr.; USN); Baga ul Bân (Syria; HJP); Bakhor (Arab.; GHA); Encens Mâle (Fr.; EFS); Frankincense (Eng.; CR2; FAC); Hsün lu Hsiang (China; EFS); Incenso (It.; USN); Incienso (Sp.; EFS); Ju Hsiang (China; EFS); Kapitthaparni (Sanskrit; EFS); Levonah (Heb.; ZOH); Lobhan (India; EFS); Lubân (Arab.; Yemen; EFS; GHA; ZOH; X15890471); Lubân Dhakar (Syria; HJP); Magher (Arab.; USA); Menjan Arab (Malaya; EFS); Mogar (Arab.; USA); Moxor (Somalia; USN); Mughur (Arab.; USA); Mustikim (Malaya; EFS); Oliban (Fr.; USN); Olibán (Sp.; USN); Olibano (It.; Sp.; EFS; USN); Olibanum (Eng.; CR2; FAC); Olibanum Tree (Eng.; USN); Ru Xiang (Pin.; DAA); Ru Xiang Shu (Pin.; AH2; USN); Salai (India; EFS); Saleh (India; EFS); Weihrauchbaum (Ger.; USN); Weihrauchpflanze (Ger.; USN); Wierookboom (Dutch; EFS).

Activities (Frankincense):

Abortifacient (f; EFS); Alterative (f; BIB; EFS); Analgesic (f; HHB); Anticomplementary (1; PH2); Antidote (hemlock) (f; BIB); Antielastase (1; X12244881); Antihepatitic (1; PR14:510); Antiinflammatory (1; X12244881); Antileukotriene (1; X12244881); Antiseptic (1; PH2); Antitussive (f; X15890471); Astringent (f; BIB); Carminative (f1; BIB; EFS; PH2); Decongestant (f; BOW); Depilatory (f; BIB); Digestive (f; GHA; HAD); Diuretic (f; BIB; EFS; GHA); Ecbolic (f; EFS); Emmenagogue (f; EFS); Expectorant (f; BIB; BOW); Fumigant (f; BIB); HCV-Protease Inhibitor (1; PR14:510); Irritant (1; PH2); 5-Lipoxygenase Inhibitor (1; X12244881); Memorigenic (f; BIB; GHA); Purgative (f; GHA); Sedative (f; BIB; EFS); Stimulant (f; BIB; EFS); Tonic (f; BIB; EFS).

Indications (Frankincense):

Abscess (f; HAD); Alzheimer's (1; COX; FNF); Anxiety (f; BOW); Arthrosis (1; COX; FNF); Asthma (f1; HHB; X12244881); Backache (f; HAD); Bilharzia (f; BIB); Bleeding (f; BIB; HAD); Boil (f; DEP); Bronchosis (f1; BIB; DEP; X12244881); Bruise (f; HAD); Callus (f; BIB); Cancer (1; COX; FNF; JLH); Cancer, anus (1; BIB; COX); Cancer, breast (1; BIB; COX); Cancer, eye (1; BIB; COX); Cancer, penis (1; BIB; COX); Cancer, spleen (1; BIB; COX); Cancer, teat (1; BIB; COX); Cancer, testicle (1; BIB; COX); Carbuncle (f; DEP; JLH); Cerebrosis (1; X12244881); Chest ache (f; BIB); Colitis (1; FNF; X12244881); Congestant (f; HAD); Corn (f; JLH); Cough (f; HAD); Crohn's Diseases (1; X12244881); Dermatosis (f; GMH); Dysentery (f; BIB); Dysmenorrhea (f; BOW); Dyspepsia (f; HAD); Edema (1; FNF); Fever (f; BIB); Gingivosis (f; BOW); Gonorrhea (f; BIB); Hemorrhoid (f; HAD); Hepatosis (1; PR14:510); Infection (f; BOW); Laryngitis (f; BIB; DEP); Leprosy (f; BIB); Leukemia (1; FNF); Mastosis (f; GHA; JLH); Meningioma (1; FNF); Myelosis (f; HAD); Neurosis (f; BIB; GHA; HAD); Ophthalmia (f; GHA; JLH); Orchosis (f; JLH); Pain (f; HHB); Paralysis (f; BIB); Pharyngosis (f; X15890471); Polio (f; HAD); Polyp (f; JLH); Proctosis (f; JLH); Psychosis (f; HAD); Respirosis (f; PH2); Rheumatism (1; BIB; FNF; X12244881); Side ache (f; BIB); Sore (f; DEP); Spermatorrhea (f; BIB); Splenosis (f; JLH); Stomachache (f; BIB); Stomatosis (f; BOW); Swelling (f1; BIB; FNF; HJP); Syphilis (f; BIB); Ulcer (f; HAD); Ulcerative colitis (1; X12244881); Urogenitosis (f; BIB); Uterosis (f; HHB); UTI (f; BOW); Vaginosis (f; BOW); Venereal Disease (f; BIB); Vomiting (f; HAD); Wound (f; PH2).

Dosages (Frankincense):

FNFF = !!
Aromatic frankincense chewed or essential oil extracted as flavoring used in baked goods, candy, gelatin, ice cream, and soft drinks (FAC).

- Asian Indians use frankincense for nervous disorders and rheumatism (BIB).
- Arabians use the digestive resin as hair oil, to stimulate digestion, and to strengthen teeth (GHA).
- Arabians use soot from burning frankincense to sooth sore eyes (GHA).
- Chinese use for urogenital ailments (BIB).
- East Africans use bark exudate as a tonic and diuretic, and to treat syphilis (BIB).
- Lebanese chew the frankincense to cleanse the mouth (HJP).
- Saudis chew it or add to coffee as diuretic, memorigenic (35 grains), and purgative (BIB; GHA).
- Swahili use the gum as a diuretic (BIB).
- Yemeni chew the gum for pregnancy and emotional and psychological problems (GHA).
- Tanganyikans boil resin with sesame oil, taking daily for bilharzia (BIB).

Extracts (Frankincense):

Class 1 (AHP). No health hazards or side effects known with proper therapeutic dosages (PH2).

BLACK MUSTARD (*BRASSICA NIGRA* (L.) W. D. J. KOCH ++ BRASSICACEAE

Notes (Black Mustard):

Another parable put he forth unto them, saying, The kingdom of heaven is like to a grain of mustard seed, which a man took, and sowed in his field: Which indeed is the least of all seeds: but when it is grown, it is the greatest among herbs, and becometh a tree, so that the birds of the air come and lodge in the branches thereof.

Matthew 13:31–32 (KJV)

FIGURE 1.15 Black Mustard (*Brassica nigra*).

Another parable he put before them, saying, "The kingdom of heaven is like a grain of mustard seed which a man took and sowed in his field; it is the smallest of all seeds, but when it has grown it is the greatest of shrubs and becomes a tree, so that the birds of the air come and make nests in its branches."

Matthew 13:31–32 (RSV)

Another illustration he set before them, saying, "The kingdom of the heavens is like a mustard grain which a man took and planted in his field; which is, in fact, the tiniest of all the seeds, but when it has grown it is the largest of the vegetables and becomes a tree, so that the birds of heaven come and find lodging among its branches."

Matthew 13:31–32 (NWT)

Zohary favors *Brassica nigra*, the black mustard, as the grain of mustard seed in the Bible. As an Israeli botanist, he would certainly know better than I that *Brassica nigra* is the tallest of the local species of *Brassica* and closely related *Sinapis*, and that its seeds are small (circa 1 mm). But he also admits that "The Greek *sinapis* is undoubtedly 'mustard'." No need to quibble about whether it is the least of seed or not. (I suspect a mustard seed is a hundred times heavier than orchid seed, even than *Artemisia annua* seed.) I suspect that in biblical times, as in Maryland fields in spring, few but botanists make distinctions between the *Brassica* and *Sinapis*. Many writers do not distinguish between black, brown (Indian), and white mustard (*Brassica nigra, Brassica juncea,* and *Sinapis alba,* respectively) but the spice and medicine trades seem to favor the white. And the canola and rapeseed varieties have been hopelessly manipulated, even in the GMO field. Few if any taxonomists and agronomists can distinguish all *Brassica* varieties and species, and probably fewer chemists, herbalists, pharmacists, and physicians can be sure of the variety or species. My entries can be no more reliable than their sources. Seems as though the group might better be treated generically than specifically or varietally. Black mustard is cultivated for its seeds, one source of commercial table mustard, and used as a condiment and medicine. Seeds also contain both a fixed and an essential oil, used as a condiment, lubricant, and soap constituent. Black mustard is mixed with white mustard (*Sinapis alba*) to make mustard flour, used in various condiments as "English Mustard" when mixed with water, and "Continental Mustard" with vinegar. The leaves are eaten as a potherb. Mustard flowers are good honey producers. In agriculture, mustard is also used as a cover crop. Smoke from burning plants may repel flies and mosquitoes (BIB).

Common Names (Black Mustard):

Aslrai (Hindi; KAB); Bilesasive (Kan.; KAB); Black Mustard (Eng.; CR2; WIN); Ch'ing Cheih (China; EFS); Chou Noir (Fr.; BOU); Fekete Mustár (Hun.; EFS); Gruener Senf (Ger.; KAB); Kalirai (Guj.; KAB); Kalo Tori (Nepal; NPM); Kalorai (India; EFS); Khardal (Arab.; BOU); Khardal Aswad (Arab.; BOU; HJP); Khurdal (Arab.; KAB); Kuro Garashi (Japan; TAN); Libdan (Arab.; BOU); Libsan (Arab.; BOU); Lifsan (Arab.; BOU); Mostarda (Mad.; JAD); Mostarda Negra (Por.; EFS); Mostarda Ordinario (Por.; EFS); Mostaza Negra (Sp.; EFS); Mosterd (Dutch; KAB); Moutarde Noir (Fr.; BOU); Napi (Greek; KAB); Navuce Rouge (Fr.; KAB); Raisarisha (Beng.; KAB); Rayo (Nepal; SUW); Sansonv (Kon.; KAB); Sar Shaf (Iran; EFS); Sarsan (Bom.; KAB); Sarshapa (Sanskrit; EFS); (Ger.; EFS); Schwartzer Senf (Ger.; EFS); Senapa (It.; KAB); Senapa Nera (Malta; KAB); Senape Nera (It.; EFS); Senep (Den.; KAB); Sénevé Noir (Fr.; BOU); Senfkohl (Ger.; EFS); Siya Hardal (Tur.; EFS); Sort Sennep (Den.; EFS); True Mustard (Eng.; KAB); Tzu Cheih (China; EFS); Xawipanatek (Hocak; WIN); Yungs-nag (Tibet; NPM); Zwarte Mosterd (Dutch; EFS).

Activities (Black Mustard):

Abortifacient (f; CEB; DAW); Anodyne (f; DAW); Antidote (Narcotics) (f; CEB; FEL; HJP); Antiedemic (f; KAB); Antifertility (f; WOI); Antiinflammatory (f; KAB); Antiscorbutic (f1; BOU; FNF); Carminative (f; DAW); Cordial (f; EFS); Diuretic (f; DAW; FEL; HHB); Emetic (f; BOU; DAW); Febrifuge (f; BOU); Gastrotonic (f; CEB); Insectifuge (f; KAB); Laxative (f; BOU; DAW); Memorigenic (f; HJP); Orexigenic (f; DAW); Pancreatonic (f; CEB); Rubefacient (f; DAW; EFS; SUW); Stimulant (f; DAW; WOI); Stomachic (f; DAW; EFS; HHB; WOI); Vesicant (f; DAW).

Indications (Black Mustard):

Abscess (f; DAW); Adenopathy (f; JLH); Ague (f; DEM); Alopecia (f; DAW); Ameba (f; DAW); Amenorrhea (f; FEL); Angina (f; FEL; HHB); Anorexia (f; DAW; DEM; KAB); Apoplexy (f; FEL); Arthrosis (1; FNF); Asthma (f; DEM); Bite (f; CEB); Bloat (f; DAW); Boil (f; DAW); Bronchosis (f; HHB; PH2); Cancer (1; FNF; JLH); Cancer, colon (1; FNF; JLH); Cancer, liver (1; FNF; JLH); Cancer, neck (1; FNF; JLH); Cancer, sinew (1; FNF; JLH); Cancer, skin (1; FNF; JLH); Cancer, spleen (1; FNF; JLH); Cancer, throat (1; FNF; JLH); Cancer, uterus (1; FNF; JLH); Cancer, wrist (1; FNF; JLH); Cardiopathy (f; HHB; PH2); Chest cold (f; HJP); Cholera (f; FEL); Circulosis (f; PH2); Cold (f; DAW; DEM; WIN); Congestion (f; DAW; FEL); Constipation (f; DAW; FEL); Cough (f; DAW); Cramp (f; DAW); CVI (f; PH2); Depression (f; HJP); Dermatosis (f; DAW); Dropsy (f; HJP); Dysentery (f; DAW); Dysmenorrhea (f; FEL); Dyspepsia (f; CEB; FEL); Ectoparasite (f; DAW); Edema (f; DAW); Enterosis (f; FEL; PH2); Epilepsy (f; DAW); Fever (f; BOU; DAW; DEM; FEL; HHB); Fibroid (f; JLH); Fibroma (f; JLH); Gastrosis (f; FEL; PH2; WIN); Glaucoma (f; PH2); Headache (f; DAW; FEL; PH2); Head-cold (f; DEM); Heartburn (f; HHB); Hepatosis (f; JLH); Hiccups (f; DAW); Induration (f; JLH); Inflammation (f; DAW; FEL); Itch (f; DAW); Lumbago (f; PH2); Lymphoma (f; JLH); Maculitis (1, FNF); Meningosis (f; FEL); Nervousness (f; DEM); Neuralgia (f; DAW; WOI); Ophthalmia (f; DAW); Otosis (f; DAW); Pain (f; DEM; FEL; HHB); Pharyngosis (f; DAW); Phthisis (f; DEM); Pleurisy (f; DAW; FEL; PH2; WOI); Polyp (f; JLH); Pneumonia (f; DAW; PH2; WOI); Pulmonosis (f; FEL); Respirosis (f; PH2); Rheumatism (1; DAW; FNF; PH2; WOI); Rhinosis (f; DAW); Sciatica (f; PH2); Sclerosis (f; JLH); Sinusosis (f; PH2); Snakebite (f; DAW; SUW); Splenosis (f; DAW; JLH; KAB); Spine (f; FEL); Stomachache (f; DAW); Sore Throat (f; DAW; KAB); Stroke (f; HJP); Syncope (f; HJP); Toothache (f; CEB; DAW; DEM); Tuberculosis (f; DEM); Tumor (f; KAB); Typhus (f; FEL); Uterosis (f; JLH); Worm (f; DAW).

Dosages (Black Mustard):

FNFF = !!!

Seeds widely eaten as spice, or sprouted; also an oil source; leaves eaten raw or cooked; young flower clusters cooked like broccoli (FAC; TAN).

- Ayurvedics suggest the plant for anorexia, cough, dermatosis, fever, splenomegaly, itch, parasites, throat, tumors, and worms (KAB).
- Balkans take black mustard early in the morning to prevent fainting spells and stroke, to cheer the mind and help the memory (HJP).
- Iranians use mustard as an emetic for narcotic poisoning (HJP).
- Lebanese boil the seed with juniper berries for dropsy (HJP).
- Lebanese poultice the seeds, with or without flaxseed, for chest cold and counterirritant (HJP).
- Syrians use mustard for indurations of the spleen (JLH).
- Unani view seeds as antiedemic, antiinflammatory, antitussive, bechic, laxative, orexigenic, stomachic, using for boils, rhuematism, splenomegaly, and toothache (KAB).

Downsides (Black Mustard):

Class 1 (Internal, ingestion of too much can be irritating); Class 2b (External; duration not to exceed 2 weeks); not for children under 6 years of age. Severe burns can occur with long-term topical use (AHP). Contraindications: children younger than 6 years; renal disease (mustard oil is absorbed through the skin). Even external poultice should be limited to 5 to 10 minutes pediatrically, 10 to 15 minutes for adults, less for sensitive patients (KOM). Millspaugh has said "unground seeds ... proved dangerous, as they are liable to become impacted in the bowel and set up a fatal inflammation" (CEB), 15 to 30 minutes plaster can cause severe burns (AHP). Adverse effects: skin and nervous damage (prolonged use). Should not be used for more than 2 weeks (AEH). Avoid taking with ammonia-containing products as ammonia with mustard oil yields inactive thiosinamine (PH2). Contraindicated in GI ulcers and nephrosis (PHR). Overdoses internally cause GI distress (PHR). Hyperthyroidism with goiter traced "to the use of the isothiocyanates in mustard" (APA). Delaneyite nitpickologists will doubtless clamber to put the same goitrogenic warning on all members of the mustard family as well as papaya, caper, and nasturtium.

Natural History (Black Mustard):

The plants are fairly high in vitamins, minerals, and protein. The leaves are eaten by ducks, musk-rats, and deer, and serve as shelter for small aquatic animal life. Black mustard is insect pollinated. Bees collect the copious mustard nectar and produce a mild-flavored, light-colored honey. Mildews appear on the leaves, causing malformation of flower heads and pods, a situation often controlled by sulfur dusting or spraying with Bordeaux Mixture. Main insect pest is Mustard Sawfly (*Athalia lugens proxima*), larvae of which feed on the leaves. Nematodes include *Ditylenchus dipsaci, Heterodera crucifera, H. schachtii, Meloidogyne arenaria, M. hapla, Nacobbus aberrans, Xiphinema indicum, Pratylenchus penetrans,* and *P. pratensis* (HOE).

BOXWOOD (*BUXUS SEMPERVIRENS* L.) + BUXACEAE

Synonyms:

Buxus wallichiana Baillon *fide* DEP

Notes (Boxwood):

I will set in the desert the fir tree, and the pine, and the box tree together.

Isaiah 41:19 (KJV)

As in my out-of-print 1985 book, I have cited here those uses usually attributed to the European box (*Buxus sempervirens*) because the biblical box (*Buxus longifolia*) is reported to have similar properties. Balfour does not even distinguish them, referring to the box of Isaiah as "the *Buxus sempervirens*" of botanists (BIB). Indeed, some authors consider this only a variant of the European box, *Buxus sempervirens*. Zohary fails to list either (ZOH). The wood, hard and taking a fine polish, is valued wherever a hardwood is needed, for carvings, combs, mathematical instruments, spoons, and turnery, etc. Cultivated by the Romans for the hard wood, inlaid with ivory for cabinet work and jeweled caskets. The wood was also used for tablets that were covered with wax and used for writing. It was said to be used both as a substitute for hops and as a green manure for hops. Boiled with lye it was supposed to tint the hair auburn. It is prized in the Holy Land as an ornamental evergreen. Leaves sometimes used to adulterate uva-ursi (BIB).

Common Names (Boxwood):

'Athaq (Arab.; BOU); Azazzer (Ber.; BOU); Bakas (Arab.; KAB); Baqs (Arab.; BOU); Beuqs (Ber.; BOU); Bois Bénit (Fr.; KAB); 'Boix (Cat.; BOU); Boj (Sp.; EFS); Boje (Sp.; EFS); Bojeo (Sp.; EFS);

FIGURE 1.16 Boxwood (*Buxus sempervirens*).

Bosso (It.; EFS); Bossolo (It.; EFS); Boxboom (Dutch; KAB); Boxwood (Eng.; Scn.; AH2; CR2; USN); Buchsbaum (Ger.; EFS); Buco (Por.; HH2); Buis (Fr.; BOU); Buis Commun (Fr.; KAB); Buis Toujours Vert (Fr.; BOU; EFS); Buje (Sp.; HH2); Bujo (Sp.; HH2); Buksus (Rus.; KAB); Bush Tree (Eng.; KAB); Busso (It.; KAB); Buxbom (Swe.; KAB); Buxo (It.; Por.; EFS; KAB); Bwès Beni (Belgium; JLH); Chikri (India; Kas.; Pun.; DEP; EFS; KAB; WOI); Cimisu (Rom.; KAB); Cimsu (Rom.; KAB); Common Box Tree (Eng.; BOU; EFS); Common Boxwood (Eng.; USN); Dudgeon (Eng.; GMH); Duolgeon (Eng.; KAB); Dwarf Box (Eng.; KAB); European Box (Eng.; USN); Evergreen Box (Eng.; KAB); Huang Yang Mu (China; EFS); Ibiqis (Ber.; BOU); Merisor Turcesc (Rom.; KAB); Mortella (It.; KAB); Ozanne (Fr.; KAB); Palm (Dutch; KAB); Palmboom (Dutch; KAB); Palmboompje (Dutch; EFS); Pampri (India; EFS); Papar (Pun.; KAB); Papri (Jaunsar; Pun.; KAB); Papur (Pun.; KAB); Poppar (Him.; Pun.; KAB); Pyxos (Greek; KAB); Samshit (Rus.; KAB); Sansadu

(Jaunsar; KAB); Shamshad (Arab.; Him.; Iran; Pun.; Urdu; BOU; KAB; WOI); Shanda Laghune (Afg.; DEP; KAB); Shumaj (Pun.; DEP; KAB); Şimşir (Tur.; EFS); Turkish Boxwood (Eng.; USN).

Activities (Boxwood):

Allergenic (1; HH2); Alterative (f; CRC); Analgesic (f; KAB); Antiacetylcholinesterase (1; X15036468); Antibutylcholinesterase (1; X15036468); Anticholinesterase (1; X15036468); Antioxidant (f; X15143441); Antiseptic (1; X11926538); Antitumor (f; CRC); Astringent (f; KAB); Bactericide (1; PH2; X11926538); Cardiotonic (f; KAB); Cathartic (f; BIB; CRC); Cerebrotonic (f; KAB); Cholagogue (f; CRC; EFS); Cytotoxic (1; HH2; PH2); Depurative (f; CRC; PHR); Dermatitigenic (1; HH2); Diaphoretic (f; BIB; CRC; DEP; PH2); Diuretic (f; EFS; CRC); Emetic (f; BIB; CRC); Febrifuge (f; CRC; DEP; EFS); Hepatotonic (f; KAB); Herbicide (1; X11926538); Hypotensive (1; HH2; PH2; X1720559); Laxative (f; CRC; EFS; HHB); Narcotic (f; BIB; CRC; EFS); Phytotoxic (1; X10346940); Propecic (f1; EFS; HH2); Purgative (f; DEP; KAB); Sedative (f; BIB; CRC; EFS); Sudorific (f; CRC; EFS); Teratogenic (f; HH2); Tonic (f; EFS); Toxic (f; EFS); Vermifuge (f; CRC; EFS).

Indications (Boxwood):

Alopecia (f1; BIB; EFS; PH2); Bacteria (1; PH2; X11926538); Bite (f; GMH); Bot (f; GMH); Cancer (f; HHB; JLH); Constipation (f; PH2); Dermatosis (f; HHB; PH2); Epilepsy (f; BIB; CRC; GMH; HHB); Fever (f; CRC; DEP; EFS; KAB); Gout (f; CRC; HHB; PH2); Headache (f; KAB); Hemorrhoids (f; BIB; CRC); Hepatosis (f; KAB); Infection (1; X11926538); Insomnia (f; EFS); Leprosy (f; CRC); Malaria (f; CRC; EFS; HHB; PH2); Pain (f; KAB); Paralysis (1; PH2); Pneumonia (f; PH2); Proctosis (f; KAB); Prolapse (f; KAB); Pulmonosis (f; HH2); Rash (f; PH2); Rheumatism (f; BIB; BOU; CRC; HHB; PH2); Stomatosis (f; KAB); Syphilis (f; BOU; CRC); Tetanus (1; PH2); Toothache (f; BIB; CRC; HH2); Tuberculosis (1; PH2); Venereal Disease (f; BIB; BOU; CRC); Worm (f; CRC; EFS).

Dosages (Boxwood):

FNFF = ?
I do not find anything in my usual edible plant books, but at least four sources suggest that the leaves and bark are used as a substitute for tea in Gahrwal (WO2). One drachm powdered leaves as purgative (GMH), 10 to 20 g powdered leaves as vermifuge (GMH); 1 to 2 oz wood in decoction as sudorific (GMH).

- Belgians use the wood oil for cancers (JLH).
- Unani consider the seeds astringent, cardiotonic, cerebrotonic, hepatotonic, the leaves good for headache, pain, and prolapsed anus (KAB).

Downsides (Boxwood):

Class 3 (AHP). No health hazards known with proper therapeutic dosages (PH2). The FDA Poisonous Plant Database yielded 62 abstracts as of November 2004. Narcotic and sedative in full doses, emeto-cathartic, possibly fatally so, and convulsant in overdoses (BIB). Leaves have caused fatalities in grazing animals. Toxic symptoms include collapse, convulsions, cramps, dermatosis, diarrhea, nausea, paralysis, shakes, vertigo, vomiting, and possibly death due to asphyxiation (CRC; PH2). The lethal dose of the alkaloid mix called "Buxin" is 800 mg per dog, or 100 mg/kg body weight, which speculatively estimated to humans would be 5 to 10 g/kg (HH2).

Natural History (Boxwood):

Leaves, occasionally browsed by goats and other animals, have proven fatal to cattle (WOI).

Extracts (Boxwood):

Most of the data above apply to *Buxus sempervirens*. Atta-ur-Rahman et al. reported new steroidal alkaloids from *Buxus longifolia*: (+)-cyclovirobuxeine F, N-benzoyl-O-acetylbuxalongifoline, buxasamarine, (+)-cyclobuxamidine, and two known steroidal bases (16 alpha-acetoxybuxabenzamidienine and *trans*-cyclosuffrobuxinine) from the leaves of *Buxus longifolia*. The new alkaloids showed significant antibacterial activity (X9358638).

GIANT MILKWEED (*CALOTROPIS PROCERA* (WILLD.) AIT. X ASCLEPIADACEAE

Synonyms:

Asclepias procera Ait.; *Asclepias procera* Willd.; *Asclepias heterophylla* Decaisne.; *Calotropis hamiltonii* Wight.; *Calotropis heterophylla* Wall.; *Calotropis inflexa* Chiovenda; *Calotropis mudari* Ham.; *Calotropis syriaca* (Gmel.) Woodson; *Calotropis wallichii* Wight.

Notes (Giant Milkweed):

For their vine is of the vine of Sodom, and of the fields of Gomorrah: their grapes are grapes of gall, their clusters are bitter.

Deuteronomy 32:32 (KJV)

For their vine comes from the vine of Sodom, and from the fields of Gomor'rah; their grapes are grapes of poison, their clusters are bitter.

Deuteronomy 32:32 (RSV)

For their vine is from the vine of Sodom, and from the terraces of Gomor'rah; their grapes are grapes of poison, their clusters are bitter.

Deuteronomy 32:32 (NWT)

Only Zohary, among the biblical scholars I have read, suggests this species as a biblical species, noting, as I have noted, that it is very common in the Holy Land; wherever it is found, it is called by its Arabic name *osher*. The apple of Sodom, he continues, is more a symbol than a botanical, gently criticising other writers, like perhaps even me, who have equated the Apple of Sodom with such species as *Citrullus colocynthis*, *Solanum incanum*, or *Solanum sodomaeum*, "unjustifiably" see the latter species). Symbolizing the evil Gomorrah and Sodom, its cursed fruit is devoid of flesh and full of hairs (the silk). The juice is poisonous. Sodom was once happy, famed for its fruits and wealth. It was burned by lightning but there are still remainders of that divine fire, and traces of the five cities are still there. The *Calotropis* fruits may look fit to eat, but when picked they dissolve into smoke and ashes. That certainly could apply to our local milkweed fruits and could apply equally well to the ripe fruits of the giant milkweed, better than those of *Citrullus* or *Solanum*. This is how Josephus interpreted the Apple of Sodom. Zohary accepts Josephus's interpretation. Having seen giant milkweed in Israel, Egypt, Kuwait, and wherever I go in the arid tropics, I am inclined to agree.

C. gigantea and *C. procera* are confused, with good reason, in the field and in the herbarium. They have very similar activities and indications (WO2). My Haitian source (AVP) includes *Asclepias gigantea* as a synonym for *Calotropis procera*, implying to me that AVP was not distinguishing the species. Vice versa for EFS. I have consequently entered all their common names below. Both are called giant milkweed by the USDA, but that common name is preferred for the namesake *C. gigantea*. AH2 further confounded the issue by giving them both the same standardized name, *mudar*, but

FIGURE 1.17 Giant Milkweed (*Calotropis procera*).

some different and some shared common names for the two species. Here I reluctantly accede to the *Herbal PDR* (*Edition 2*), which treats the species separately. KAB keys the species as follows:

- Corolla lobes narrow, shorter than the staminal column with 2 obtuse auricles just below the entire apex: *C. gigantea*
- Corolla lobes broader, as long as or longer than the staminal column without auricles below the 2-cleft apex: *C. procera*

Common Names (Giant Milkweed):

Akanda (India; EFS); Alabara (Sudan; AVP); Alarka (Sanskrit; EFS); Algodón de Arbol (Ma.; JFM); Algodón de Seda (Cuba; Dr.; Sp.; AVP; RyM); Algodón de Sierra (Ma.; JFM); Algodón Extranjero (Dr.; Sp.; AHL; JFM); Apple of Sodom (Eng.; ZOH); Arbole de Seda (Sp.; AVP; EFS; JFM); Arbre à Soie (Fr.; Haiti; AVP; EFS; LWW); Arka (Ayu.; Sanskrit; AH2; EFS); Asclépiade Géante (Fr.; AVP); Asclépias Géant (Guad.; AVP); Ashkar (Arab.; GHA); Auk, Auricula Tree (Ma.; JFM); Babambi (Cameroon; HDN); Bambango (Togo; HDN); Baramadha Aka (?; Nepal); Baranbakh (Arab.; BOU); Biduri (Malaya; EFS); Bo (Somalia; HDN); Bomba (Ma.; Pr.; JFM; LWW); Bowstring Hemp (Eng.; Ocn.; AH2); Calotrop (Ma.; JFM); Calotrope (Fr.; BOU); Cazuela (Cuba; Sp.; AVP; JFM); Ciúme (Brazil; MPB); Coton de France (Mart.; LWW); Coton Soie (Haiti, AHL; AVP; JFM); Cow Heel (Ma.; JFM); Crown Flower (Eng.; AVP); Dindare (Eth.; HDN); Dumb Cotton (Ma.; JFM); Estrella de Hollanda (Sp.; AVP); Etethero (Kenya; HDN); Fafetone (Fr.; BOU); Faftan (Sen.; HDN); Faftan Calotrope (Ma.; JFM); Flôr da Sêda (Brazil; MPB); Flor de Seda (Brazil; MPB); French Cotton (Bar.; Jam.; BOU; JFM; LWW); French Jasmine (Ma.; JFM); Ganganpi (Ivo.; HDN); Giant Milkweed (Eng.; Ocn.; AH2; AVP); Giant Swallowwort (Eng.; Ocn.; AH2; AVP); Gigantic Swallowort (Eng.; EFS); Hortênsia (Brazil; MPB); Huré Go (Burkina Faso; HDN); Jester's Cap (Eng.; AVP); Katuna di Seda (Dwi.; Ma.; JFK; LWW); Khok (Iran; EFS); Kihuta (Uganda; HDN); Kipanpango (Gambia; HDN); Kpampuug (Ghana; HDN); Krenka (Arab.; BOU); Lechoso (Ma.; JFM); Lembega (Malaya; EFS); Leta-Unde (Malwai; Yao; HDN); Liberty Tree (Ma.; Wi.; JFM; LWW); Madar (Ocn.; AH2); Mal Casada (Dr.; Sp.; AVP); Mandara (Sanskrit; EFS); Mata de Seda (Pr.; Sp.; AVP); Mercurio Vegetal (Ma.; JFM); Mudar (Fr.; Eng.; Scn.; AH2; EFS; ZOH); Mudar Pflanze (Ger.; EFS); Nfogo (Sudan; AVP); Negyi (Ber.; BOU); Niogo (Sudan; AVP); Oshaar (Egypt; HDN); Oshar (Arab.; GHA); Osher (Arab.; ZOH); Oshur (Arab.; GHA); Palomitas de Seda (Ma.; JFM); Palu di Leche (Ma.; JFM); Pomme de Sodom (Fr.; Ma.; BOU; JFM); Pompompogolo (Sudan; AVP); Rembega (Malaya; EFS); Remiga (Malaya; EFS); Safed (Hindi); Sano Chad; HDN); St. Thomas Bush (Bah.; Ma.; JFM; LWW); Sodom Apple (Eng.; HJP); Sprainleaf (Ma.; JFM); Swallowwort (Ocn.; AH2); Torcha (Ber.; BOU); Totafia (Sudan; AVP); Tounfafiya (Sudan; AVP); Tourza (Ber.; BOU); Tula (Ma.; Pr.; JFM; LWW); Tumfafia (Hausa; Mali; Niger; Nig.; HDN); Turdja (Maur.; HDN); Turuturu (Tanzania; HDN); Ushar (Arab.; Sudan; Syria; EFS; HDN; HJP); Ushayr (Arab.; Syria; HJP); Ushr (Arab.; Syria; HJP); Wild Cotton (Bah.; Ma.; JFM; LWW); Wild Down (Bah.; Ma.; JFM; LWW); Zijdkkatoen (Dwi.; Ma.; JFM; LWW).

Activities (Giant Milkweed):

Abortifacient (f; HDN); Acaricide (1; X14620053); Alexiteric (f; KAB); Alterative (f; DEP; KAP); Analgesic (f1; KAB; MPB; X15848031); Anesthetic (f; HDN); Anthelmintic (1; HDN; KAB; X16085379); Anticoagulant (1; HDN); Anticonvulsant (1; X16446065); Antidiabetic (1; X16054794); Antiedemic (1; X15643548; X16192673); Antihistaminic (1; X16192673); Antiinflammatory (1; HDN; KAB; MPB; X15643548; X16192673); Antinociceptive (1; X15848031); Antioxidant (1; X16054794); Antiseptic (f; HDN); Antispasmodic (f1; DEP; GHA; KAP; X15752643); Antitumor (1; JFM; PH2; X15689169; X16688796); Anti-VEGF (1; X16688796); Anxiolytic (1; X16446065); Bactericide (1; HDN); Bitter (f; KAP); Cardioactive (1; X3086679); Cardiotonic (f1; BOU; KAP; PH2); Curare (1; HDN); Depilatory (f; DEP; HDN; KAB; MPB); Diaphoretic (f; DEP; PH2; SUW); Digestive (f; KAB); Diuretic (f; HJP); Emetic (f; DEP; HJP; KAP; PH2; SUW); Expectorant (f; BOU; DEP; KAP; PH2; SUW); Febrifuge (1; HDN; X16106390); Fibrinolytic (1; HDN; X15922393); Fungicide (f; HJP); Gastrotonic (f; KAB); Hemostat (f; X15922393); Hepatoprotective (1; HDN; X16688796); Hypertensive (1; HDN); Insecticide (1; HDN; JFM); Insectifuge (f; HDN); Larvicide (1; HDN; X12557934); Molluscicide (1; HDN); Nervine (f; DEP); Myorelaxant (1; X15752643); Neurotonic (f; GHA); Ovicide (1; HDN); Proteolytic (1; HDN; KAB; MPB; X15922393); Purgative (f; GHA; KAP; SUW); Respirastimulant (1; HDN); Rubefacient (f; DEP); Sedative (1; X16446065); Stimulant (f; KAB); Stomachic (f; EFS; KAB); Sudorific (f; EFS; KAP); Tonic (f; DEP; KAP); Tranquilizer (f; MPB); Uterotonic (1; HDN; KAP); Vasoconstrictor (1; HDN); Vermifuge (f1; HDN; X16085379); Vulnerary (f; HJP).

Indications (Giant Milkweed):

Abscess (f; HDN); Amenorrhea (f; HDN); Anasarca (f; DEP; KAB; PH2); Ancylostomiasis (f; HDN); Anorexia (f; DEP); Aphtha (f; DEP); Apoplexy (f; BOU); Arthrosis (f1; DEP; HDN; HJP); Ascites (f; DEP; PH2); Asthma (f; BOU; DEP; KAB; SUW); Bacillus (1; HDN); Bacteria (1; HDN); Bite (f; KAB); Bleeding (f; X15922393); Bronchosis (f; DEP; KAP); Cachexia (f; DEP); Cancer (f1; JLH; PH2; X15689169); Cancer, abdomen (f1; JLH; X15689169); Cancer, liver (f1; JLH; PH2; X15689169; X16688796); Cancer, ovary (f1; JLH; X15689169); Cancer, skin (1; PH2; X15689169); Cardiopathy (1; FNF; HDN); Caries (f; HDN); Catarrh (f; DEP; KAB); Chancre (f; HDN); Cold (f; SUW); Colic (f; HDN); Constipation (f; DEP); Convulsion (f1; SEP; PH2; X15752643); Cough (f; GHA; KAB; PH2; SUW); Cramp (f1; DEP; KAP; X15752643); Dermatosis (f; DEP; JFM; SUW); Diabetes (1; X16054794); Diarrhea (f; SUW); Dropsy (f; DEP; HJP; KAB); Dysentery (f; BOU; DEP; HJP; KAP; PH2; SUW); Dysmenorrhea (f; HDN); Dyspepsia (f; PH2; SUW); Dyspnea (f; GHA); Dystocia (f; HDN); Earache (f; HJP); Edema (f1; HDN; X16192673); Elephantiasis (f; BOU; DEP; SUW); Enterosis (f; KAB; PH2; SUW); Epididymitis (f; DEP); Epilepsy (f; DEP; HDN); Escherichia (1; HDN); Fever (1; DEP; HDN; KAB; X16106390); Fungus (f; HJP); Gastrosis (f; HDN); Gonorrhea (f; DEP); Headache (f; HDN; HJP; JFM); Hemorrhoid (f; DEP; HJP; KAB); Hepatosis (f; DEP; KAB); High Blood Pressure (f; HDN); Hysteria (f; DEP); Infection (1; HDN); Inflammation (1; HDN; KAB; MPB; X16192673); Insanity (f; HDN); Insomnia (f; HDN); Leprosy (f; BOU; DEP; HJP; JFM; PH2; SUW); Leukoderma (f; KAB); Malaria (f; DEP; HJP; KAB); Mycosis (f; HJP); Obesity (f; PH2); Pain (f1; KAB; MPB; X15848031); Paralysis (f; DEP; KAB); Parasite (f; HDN); Pertussis (f; HDN); Pharyngosis (f; KAP; PH2); Phthisis (f; DEP); Psychosis (f; HDN); Rheumatism (f; DEP; MPB; SUW); Rhinosis (f; KAP; PH2); Ringworm (f; KAB); Scabies (f; BOU; KAB); Sinusosis (f; HDN); Snakebite (f; DEP; KAB); Sore (f; BOU; JLH); Spasm (f1; DEP; KAP; X15752643); Splenosis (f; DEP; KAB); Staphylococcus (1; HDN); Stomachache (f; HDN); Stomatosis (f; DEP); Swelling (f1; DEP; JLH; KAB; SUW; X16192673); Syphilis (f; BOU; JFM; PH2; SUW); Toothache (f; BOU; DEP; JFM; PH2); Tumor (f; JLH); Tuberculosis (f; HDN); Venereal Disease (f; DEP; HJP; PH2); Vomiting (f; PH2; SUW); Wart (f; JLH); Whitlow (f; HDN); Worm (f1; PH2; SUW; X16085379); Wound (f; DEP; GHA; KAB).

Dosages (Sodom's Milkweed):

FNFF = X
200–600 mg bark as a diaphoretic and expectorant, 2–4 g as emetic (HHB; PHR). Boil seven leaves for 1 hour in 50 ml water; drink 4 ×/day (HDN).

- Arabians mix leaves with clove oil, Indian almond fruit, and black cumin seeds, heat, and rub onto paralyzed limbs (GHA).
- Asian Indians use the roots or leaves for cancer, especially abdominal tumors (JLH).

- Ayurvedics consider the flowers antisialogogue, orexigenic, stomachic, and tonic, using them for asthma and piles; viewing the plant as anthelmintic, its ashes as expectorant, they use the hot leaves to treat enteralgia (KAB).
- Baluchistani use the bark as an expectorant, roots as purgative, and dried juice as antispasmodic and neurotonic (GHA)
- Barbadans apply latex or leaves to rheumatic pains or swelling (JFM).
- Colombians consider the latex antisyphilitic, diaphoretic, emetic, vermifuge, and insert it into aching teeth (JFM)
- Curacaons use fresh leaves, upper surface out, for headaches, and for sprains and swellings, especially when a cold is coming on (JFM).
- Egyptian Bedouins use fresh leaf cataplasm for sunstroke, the latex for scabies in camels and goats; they apply the latex to loosen them or alleviate toothache (BOU).
- Lebanese use latex as abortifacient, fungicide, vulnerary, for arthrosis, dermatosis, dropsy, earache, headache, venereal disease, and wounds (HJP).
- North Africans smoke dry leaves for asthma (BOU).
- North Africans use stomachic sudorific root bark for dysentery, elephantiasis, and syphilitic ulcers (BOU).
- Omani oil painful spots on back or joints and then apply heated *Calotropis* leaves (GHA).
- Pathans use the root as a toothbrush, hoping to cure toothache (KAB).
- Venezuelans use the latex to treat dermatoses and leprosy (JFM).
- Yemeni use dry leaf/twig decoction for breathing difficulties (GHA).

Downsides: (Giant Milkweed):

No health hazards or side effects known with proper therapeutic dosages (PH2). Dangerous plant. Very high doses may cause death (PH2), following bradycardia, convulsion, diarrhea, and vomiting (PH2). Traditional use in India may cause severe bullous dermatosis, leading occasionally to hypertrophic scars. Perkins and Payne note convulsions, diarrhea, vomiting, slowed but stronger heartbeat, labored respiration, increased blood pressure, and possible death (CRC).

Extracts (Giant Milkweed):

Calotropine effective *in vitro* against epidermoid tissue cultures of the rhinopharynx (PH2). Many extract activities listed in HDN.

CAPER BUSH(*CAPPARIS SPINOSA* L.) +++ CAPPARACEAE

Synonyms:

Capparis rupestris Sm. *fide* USN

Notes (Caper Bush):

Also when they shall be afraid of that which is high, and fears shall be in the way, and the almond tree shall flourish, and the grasshopper shall be a burden, and desire shall fail: because man goeth to his long home, and the mourners go about the streets.

Ecclesiastes 12:5 (KJV)

They are afraid also of what is high, and terrors are in the way; the almond tree blossoms, the grasshopper drags itself along and desire fails; because man goes to his eternal home, and the mourners go about the streets.

Ecclesiastes 12:5 (RSV)

FIGURE 1.18 Caper Bush (*Capparis spinosa*).

> **Also they have become afraid merely at what is high, and there are terrors in the way. And the almond tree carries blossoms, and the grasshopper drags itself along and the caper berry bursts; because man is walking to his long-lasting house, and the wailers have marched around in the street.**
>
> **Ecclesiastes 12:5 (NWT)**

While KJV and RSV call it "desire," the NWT graciously, and I hope accurately, calls it caper. I know that the capers of commerce belong to two taxonomic species: *Capparis decidua* (Scn. = karira, AH2) and *Capparis spinosa* (Scn. = caper bush, AH2; the biblical *desire,* ZOH), both of which occur in the Holy Land. The karira is almost leafless, like a broom, with corymbose red flowers, while the caper bush is a leafy clambering shrub with white flowers solitary or in fascicles of two or three (KAB; ZOH). Sharing some of the healthy virtures with the mustard family (Brassicaceae), this might be considered another healthy component of the Mediterranean diet.

Common Names (Caper Bush):

Abiyonah (Heb.; BI2); Alcaparra (Por.; USN); Alcaparro (Sp.; EFS; USN); Amseilih (Ber.; BOU); Asef (Arab.; BOU); Assaaf (Arab.; BOU); Avionah (Heb.; ZOH); Azuf (Arab.; KAB); Bandar

(Pun.; KAB); Barar (Pun.; KAB); Barari (Pun.; KAB); Bassar (Pun.; KAB); Bauri (Jaunsar; Pun.; KAB); Belachem (Ber.; BOU); Ber (Hindi; Pun.; KAB); Bussar Ultakanta (Kum.; KAB); Cabriola (Sp.; EFS); Caper (Eng.; CR2); Caper Bush (Eng.; Scn.; AH2); Cappara (Malta; KAB); Capparo (It.; KAB); Cappero (It.; TAN); Câprier (Fr.; BOU; USN); Câprier Commun (Fr.; BOU); Câprier Epineux (Fr.; BOU); Common Caper (Eng.; HJP); Edible Caper (Eng.; DEP); El Azaf (Arab.; ZOH); Ezov (Heb.; KAB; ZOH); Fakouha (Arab.; GHA); Felfel el-djebel (Arab.; BOU); Kabar (Arab.; Bom.; Iran; Syria; Urdu; DEP; KAB); Kabar Karak (Iran; EFS); Kabarish (Tur.; KAB); Kabarit el Weli (Arab.; Syria; HJP); Kabarra (Pun.; KAB); Kabawa (Afg.; KAB); Kabbar (Arab.; BI2; BOU); Kabbara (Afg.; KAB); Kabra (Hindi; India; Lad.; Pun.; Tibet; EFS; KAB; MKK); Kabur (Arab.; KAB); Kachra (India; EFS); Kakadani (Sanskrit; EFS); Kakri (Pun.; KAB); Kalo Kantharo (Guj.; WO2); Kalvari (Sin.; KAB); Kander (Pun.; KAB); Kantharo (Guj.; WO2); Kaparis (Greek; KAB); Kapernbaum (Ger.; USN); Kapernstrauch (Ger.; USN); Kapersovyi Kust (Rus.; KAB); Kapertjessstruik (Dutch; EFS); Kapperboom (Dutch; KAB); Karika (Sanskrit; EFS); Katrimullinagidda (Kan.; WO2); Kaur (Pun.; KAB); Kebir (Iran; KAB); Kebre (Tur.; EFS); Keri (Pun.; KAB); Khafkhander (Jhalawan; KAB); Khakandir (Jhalawan; KAB); Khawarg (Bal.; KAB); Kiabara (Arab.; EFS); Kiari (Jaunsar; Pun.; KAB); Kirap (Las Bela; Ormara; KAB); Kokilakshamu (Tel.; KAB; WO2); Krap (Las Bela; Ormara; KAB); Kronbeiser (Arab.; BOU); Kulvari (Sin.; DEP); Kurak (Iran; KAB); Lasafa (Arab.; BI2; GHA); Lassaf (Arab.; BOU; ZOH); Marattamaggu (Kan.; WO2); Mullukattari (Kan.; KAB); Ouailoulou (Ber.; BOU); Pahinro (Diwana; Las Bela; KAB); Panetero (Las Bela; KAB); Qabbar (Arab.; Mali; UPW); Ouailoulou (Ber.; BOU); Shafallah (Arab.; Oman; Qatar; Saudi; GHA); Shalem (Arab.; BOU); Shwak el Hhimâr (Arab.; Syria; HJP); Tailouout (Ber.; BOU); Taker (Pun.; KAB); Taparera (Cat.; KAB); Tapenier (Fr.; KAB); Taybout (Ber.; BOU); Tsailaloul (Ber.; BOU); Ul Aşaf (Arab.; Syria; HJP); Ulta Kanta (Kum.; DEP).

Activities (Caper Bush):

Alterative (f; EFS); Analgesic (f; BIB; SKJ); Anthelmintic (f1; BIB); Antiallergic (1; X15799005); Anticystic (1; WO2); Antidote (Beryllium) (f; WO2); Antierythemic (1; X15799005); Antihistaminic (1; X15799005); Antiparalytic (f; HAD); Antiseptic (f; BOU); Antiviral (1; WO2); Aphrodisiac (f; BIB; BOU); Astringent (f; EFS; WO2); Bactericide (1; BIB; WO2); Capillariprotective (f; BOW); Deobstruent (f; DEP); Diuretic (f; EFS; SKJ; WO2); Emmenagogue (f; BIB; DEP); Expectorant (f; BIB; EFS; GHA; WO2); Fungicide (1; BIB; WO2); Hemostat (f1; WO2); Molluscicide (1; X15287187); Orexigenic (f; BIB; BOU; HOS; WO2); Purgative (f; HAD); Stimulant (f; BIB; EFS); Tonic (f; EFS; GHA; SKJ).

Indications (Caper Bush):

Adenopathy (f; BIB; JLH); Aging (f; BIB); Allergy (1; X15799005); Amenorrhea (f; DEP); Arteriosclerosis (f; BIB; BOU); Arthrosis (f; BIB); Bleeding (f; BOW); Cancer (f1; BIB); Cancer, abdomen (f1; FNF; JLH); Cancer, bladder (f1; FNF; JLH); Cancer, colon (f1; FNF; JLH); Cancer, groin (f1; FNF; JLH); Cancer, head (f1; FNF; JLH); Cancer, kidney (f1; FNF; JLH); Cancer, liver (f1; FNF; JLH); Cancer, neck (f1; FNF; JLH); Cancer, spleen (f1; FNF; JLH); Cancer, uterus (f1; FNF; JLH); Cataract (f; BIB); Chill (f; BIB; BOU); Cirrhosis (f; WO2); Cold (f; BIB); Conjunctivosis (f; BOW); Cough (f; BOW; GHA); Cramp (1; HOS); Cystosis (f; JLH); Dengue (f; BIB; HJP); Diabetes (f; GHA); Diarrhea (f; BOW; BOU); Dropsy (f; BIB; BOU; DEP); Dysentery (f; BIB); Dysmenorrhea (f; BIB); Earache (f; BI2; GHA); Enterosis (f; BOW); Erythema (1; X15799005); Fever (f; BOU); Fracture (f; BIB); Ganglion (f; BOU); Gastrosis (f; BOW; GHA); Gout (f; DEP; SKJ; WO2); Headache (f; BIB); Hepatosis (f1; JLH; HOS; WO2); Hyperacidity (f; MKK); Impotence (f; BOU); Induration (f; JLH); Infection (f; BOW; BOU); Infertility (f; BIB; BOU); Inflammation (f1; HOS); Malaria (f; BIB; HJP); Malta Fever (f; BIB); Nephrosis (f; BOU; JLH; WO2); Ophthalmia (f; BIB; BOU); Otosis (f; BIB); Pain (f; BIB); Palsy (f; DEP); Paralysis (f; HAD); Rheumatism (f; HJP;

WO2); Sclerosis (f; BIB); Sciatica (f; BIB); Scrofula (f; BIB; BOU); Scurvy (f1; DEP; WO2); Snakebite (f; BIB); Sore (f; DEP); Splenomegaly (f; BIB); Splenosis (f; BIB; BOU; WO2); Toothache (f; BIB); Tuberculosis (f1; BIB; WO2); Tumors (f1; BIB; HOS); Ulcer (f; BIB); Uterosis (f; JLH); Wart (f; BIB; JLH); Worm (f; GHA).

Dosages (Caper Bush):

FNFF = !!
Food farmacy. Young pickled buds, the capers of commerce, imparted the biblical "desire" or relish to food. Mediterranean capers are gathered and steeped in vinegar for an appetizer. Flower buds used to flavor canapes, gravies, salads, and sauces, after being cooked and pickled. Raw capers are all but unpalatable and much improved in the pickling process. Bay leaves, black pepper, and/or tarragon are good in the pickling vinegar. Some favor capers pickled in sea salt. Some French sauces graced with capers include ravigote, remoulade, tartare, and some vinaigrettes. Sprouts sometimes eaten like asparagus, as well as the buds and shoots. Pickled fruits eaten in Arabia, in Cyprus, and Punjab; branch tips are also pickled (FAC; HOS; TAD; TAN).

- Algerians boil whole plant in oil as a puerperal hydragogue (HOS).
- Arabians use leaves with leaves of *Rhazya stricta* for diabetes (GHA).
- Arabians use powdered leaves in cooking oil as eardrops for earache (GHA).
- Asian Indians apply the expressed caper juice to earache (BI2).
- Asian Indians use capers for burns, pulmonosis, sores, tuberculosis, and wounds (BI2).
- Bedouins use, with *Teucrium pilosum,* as a cold inhalant; they boil chopped or powdered leaves, inhaling the vapors for headache; they apply decoction in cloth poultices to arthritic pains; barren women are covered with ground leaves of *Capparis* and *Tamarix* to inhale the vapors (HOS).
- Hoggar use the plant to treat rheumatism (UPW).
- Iranians use the plant for intermittent fever and rheumatism (HOS).
- Latinos take brandy, honey, or wine decoctions for liver ailments (JLH).
- Lebanese take plant decoction for dengue, malaria, and Malta fever, regarding the root specific for malaria or splenomegaly following malaria (HJP),
- North Africans use flower buds in compresses for the eyes (BOU).
- Spaniards think eating capers staves off old age (BI2).
- Unani consider the root bark aperient, analgesic, emmenagogue, expectorant, and vermifuge, using it in adenopathy, paralysis, rheumatism, splenomegaly, and toothache; they use the juice to kill worms (KAB).

Downsides (Caper Bush):

Isothiocyanates in overdose can cause problems.

Natural History (Caper Bush):

Leaves and fruits are favored foods of goats and sheep (DEP). The nocturnal flowers are pollinated by hawk moths, and the seeds are disseminated by birds (ZOH).

Extracts (Caper Bush):

Stachydrine, in animals, hastens coagulation. Shirwaikar et al. (1996) found antihepatotoxic activity in alcoholic, ether, ethyl acetate, and petrol extracts of root bark. All reduced elevated serum transaminases (orally in rats up to 2000 mg/kg with no evident toxicity) (FT67(3):200).

SAFFLOWER (*CARTHAMUS TINCTORIUS* L.) +++ ASTERACEAE

SYNONYMS:

Cnicus indicus fide EFS

NOTES (SAFFLOWER):

Thy plants are an orchard of pomegranates, with pleasant fruits; camphire, with spikenard, Spikenard and saffron; calamus and cinnamon, with all trees of frankincense; myrrh and aloes, with all the chief spices:

Song of Solomon 4:13–14 (KJV)

Your shoots are an orchard of pomegranates with all choicest fruits, henna with nard, nard and saffron, calamus and cinnamon, with all trees of frankincense, myrrh and aloes, with all chief spices.

Song of Solomon 4:13–14 (RSV)

Your skin is a paradise of pomegranates, with the choicest fruits, henna plants along with spikenard plants; spikenard and saffron, cane and cinnamon, along with all sorts of trees of frankincense, myrrh and aloes, along with all the finest perfumes.

Song of Solomon 4:13-14 (NWT)

Israeli authors such as Zohary, more familiar with the Israeli Flora and the Holy Land than am I, should be better equipped to speculate as to which herbs were really meant in some elusive passages. I am both pleased and displeased to note that Zohary, too, leaves a few problems unresolved, including one involving two major medicinal plants (saffron and turmeric) and one minor medicinal (safflower), all sources of yellow dyes. All three can be grown in the warmer regions of Israel, but the turmeric would be difficult. Saffron and safflower would both be easy to grow. Here are points that Zohary makes: "Saffron (in Hebrew, *karkom*) is mentioned only once in the Bible." (ZOH) Some commentators identify it with turmeric, which "was never grown" (ZOH) in Israel, others with saffron, which was probably grown but only in postbiblical times. There is linguistic support for both possibilities. "There is no doubt that the sown karkom fields mentioned in the Mishnah (of the Talmud) refer to *Crocus sativus*." (ZOH) More data he presents point "to the identification of biblical *karkom* as turmeric and not as crocus…. But doubt arises when one considers another widely cultivated annual yielding numerous heads of orange flowers" (ZOH) (safflower, *Carthamus tinctorius*). Zohary notes that safflower was cultivated in Egypt as early as 3500 B.C., first for its flowers and dyes, later for its oilseed potential. Where does this leave me? Should I include just one or all three candidates for the one mention of saffron in the Bible? From the medicinal point of view, turmeric seems even more important than saffron, which appears even more important than safflower. Ditto from the likelihood point of view, based on the views of Zohary. In reading Zohary, I think he would place his bets first on safflower, then on turmeric, and finally on crocus, as the biblical saffron.

COMMON NAMES (SAFFLOWER):

Açafrão (Mad.; Por.; EFS; PST); Açafrão Bastardo (Por.; USN); African Saffron (Eng.; KAB); Agnishikha (Sanskrit; KAB); Agnisikha (Tel.; KAB); Akhariza (Arab.; KAB); Alazar (Sp.; EFS); Alazor (Spain; VAD); American Saffron (Eng.; EFS); Azafrán (Sp.; Ven.; JLH); Azafrán e la Tierra (Pi.; KAB); Azafrán Romí (Sp.; KAB; EFS); Barre (Hindi; Nwp.; DEP; KAB); Bastard Safran (Ger.; EFS); Bastard Saffron (Eng.; HJP); Basterd Saffraan (Dutch; KAB); Biri (Tag.; KAB); Brandusa de Tvvamna (Rom.; KAB); Bundi (Rajputana; KAB); Cachumba (Pam.; KAB); Cartamo (It.; EFS);

FIGURE 1.19 Safflower (*Carthamus tinctorius*).

Cártamo (Sp.; EFS; USN); Carthame des Teinturiers (Fr.; EFS); Carthamine Dye (Eng.; DEP); Casabha (Vis.; KAB); Chendurukam (Tam.; KAB); Cnigue (Fr.; KAB); Croco Bastardo (It.; EFS); Croco Hortense (It.; KAB); Crocus (Eng.; JLH); Daccam (Ic.; KAB); Dyer Saffron (Eng.; ZOH); Farberdistel (Ger.; EFS); Farber Saflor (Ger.; EFS); Faux Safran (Fr.; EFS); Galapmachu (Manipur; KAB); Gartensafran (Ger.; KAB); Ghosfor (Malta; KAB); Graine de Perroquet (Fr.; EFS); Grano de Perrouget (Lan.; KAB); Gulekafshah (Iran; KAB); Habb et Quirthim (Arab.; JLH); Heboo (Burma; KAB); Hong Fah (Malaya; KAB); Hong Hua (China; Pin.; AH2; EFS); Hong Lang Hoa (China; KAB); Ihhrid (Arab.; Syria; HJP); Kadaya (Mar.; KAB); Kagireh (India; EFS); Kajirah (Beng.; KAB); Kamalottara (Sanskrit; DEP; EFS); Kar (Hindi; Nwp.; DEP; KAB); Karada (Guj.; KAB); Kardai (Bom.; DEP; KAB); Karha (Urdu; KAB); Karkom? (Heb.; ZOH); Kartam (Tur.; EFS); Kasembar (India; EFS); Kasumba (Malaya; EFS); Kazhirah (Iran; DEP); Kazirah (Iran; EFS); Kesumba (Malaya; EFS); Khasaddana (Iran; EFS); Khoinbo (Sin.; KAB); Knikos (Greek; KAB); Kurdi (Mar.; DEP); Kurtam (Pun.; DEP); Kurthum (Arab.; EFS); Kurtim (Egypt; KAB); Kusambe (Kan.; KAB); Kusanbe (Kan.; DEP); Kusbo (Kon.; KAB); Kushumba Vittulu (Tel.; DEP); Kusum (Beng.; DEP); Kusumba (Cutch; India; DEP; EFS); Kusumbha (Ayu.; AH2); Mexican Saffron (Eng.; FAC); Parrot Seed (Eng.; HJP); Powari Jo Bij (Sin.; DEP); Qirtim (Arab.; DEP); Qurtum (Arab.; Syria; HJP); Saffler (Swe.; KAB); Saffloer (Dutch; EFS); Safflower (Eng.; Scn.; AH2; HJP; USN; ZOH); Safir (Pun.; KAB); Saflor (Rus.; KAB); Safra Bort Saafrano (Cat.; KAB); Safran Bâtard (Fr.; DEP; EFS); Sendurgam (Tam.; DEP); Su (Burma; DEP); Usfar (Arab.; EFS); 'Usfur (Arab.; Syria; HJP); Wilder Safran (Ger.; EFS); Wilde Saffraan (Dutch; EFS); Wild Saffron (Eng.; EFS); Za'farân (Arab.; Syria; HJP); Zafferano Falso (It.; EFS).

Activities (Safflower):

Abortifacient (AHP; HHB; PH2); Analgesic (f1; EFS; WO3); Anodyne (f; EFS); Antiacne (1; JAR12:99); Antiaggregant (1; AHP; PNC); Antiaging (f; KAB); Antiedemic (1; WO3); Antifertility (f; DAA); Antihydrotic (f; HHB; PH2); Antiinflammatory (1; PNC; WO3; X8987908); Antiischemic (1; X8425843; X12802724); Antimelanogenic (1; X15577216); Antimelanomic (1; X15577216); Antioxidant (1; X15706901); Antipyretic (f; PED); Antitumor (1; PH2; X8604239); Aphrodisiac (f; KAB); Bactericide (1; WO2); Bechic (f; KAB); Bitter (PED); Calcium Blocker (1; WO3; X8281577); Cardiotonic (1; WO3); Carminative (f; KAB); Cathartic (1; WO2); Cytotoxic (1; X11090999); Decongestant (f; DAA); Diaphoretic (f; DAA; HJP; PED); Diuretic (f; KAB; PNC); Emmenagogue (f1; AHP; EFS; HHB; PH2; WO2); Expectorant (f; HHB; PH2); Febrifuge (f; DAA); Fungicide (1; DAA; VAD); Hypocholesterolemic (1; HHB); Hypoglycemic (f1; VAD); Laxative (f1; DEP; HHB; HJP; PH2; PNC; WO2); Lipolytic (f; VAD); Nematicide (1; VAD; WO2); Neuroprotective (1; X8425843; X12802724); Orexigenic (f; KAB); Phototoxic (f; DAA); Purgative (f; EFS; PH2); Sedative (f1; KAB; WO2); Stimulant (f1; HHB; PH2; WO2); Teratogenic (1; X11090999); Thrombolytic (1; X15806964); Tyrosinase Inhibitor (1; X15577216); Uterotonic (1; AHP; X7646782); Vermifuge (f; DAA).

Indications (Safflower):

Acne (1; JAR12:99); Amenorrhea (f1; AHP; DAA; DEP; PH2); Anorexia (f; KAB); Arteriosclerosis (f; VAD); Arthrosis (f; HJP; PH2); Bacteria (1; DAA); Boil (f; DAA); Bronchosis (f; KAB; PH2); Bruises (f; PNC); Cancer (f; JLH); Cancer, liver (f; JLH); Cancer, skin (f1; PH2; X8604239); Cancer, stomach (f; PH2); Cancer, uterus (f; JLH); Candida (1; DAA); Cardiopathy (f; PNC); Catarrh (f; KAB); Chickenpox (f; PED); Childbirth (f; DAA; DEP); Cold (f; KAB); Comedon (1; JAR12:99); Congestion (f; DAA); Constipation (f; DEP; HJP; VAD); Cough (f; PH2); Dermatosis (f; KAB; PNC; VAD); Dysmenorrhea (f; DAA; PNC); Dystocia (f; DAA); Dysuria (f; KAB); Edema (1; WO3); Embolism (f1; VAD; X15806964); Enterosis (f; VAD); Escherichia (1; DAA); Fever (f; DAA; PNC); Fungus (f1; DAA; KAB; VAD); Hepatosis (f; JLH; KAB; PED); High Cholesterol (f; VAD);

Impotence (f; KAB); Induration (f; JLH); Infection (1; DAA; VAD); Inflammation (1; JLH; WO3); Insomnia (1; WO3); Ischemia (1; X8425843); Itch (f; KAB); Jaundice (f; DEP; KAB); Leprosy (f; KAB); Leukoderma (f; KAB); Measles (f; DEP; KAB; PNC); Metrorrhagia (f; HHB); Mycosis (f1; DAA; KAB; VAD); Neurosis (1; X8425843); Ophthalmia (f; KAB); Pain (f1; PH2; VAD; WO3); Paralysis (f; DEP; HHB); Parasite (f; VAD); Pemphigus (f; DAA); Pneumonia (f; DAA; HHB; PH2); Pulmonosis (f; KAB); Rheumatism (f; DEP; HHB); Ringworm (f; KAB); Scabies (f; KAB; PH2); Scarlatina (f; DEP; KAB); Sore (f; DEP; KAB); Sore Throat (f; KAB); Sprain (f; HJP); Strangury (f; KAB); Swelling (1; WO3); Thrombosis (f1; VAD; X11243195); Uterosis (f; DAA; DEP); Wound (f; DAA; PH2; PNC); Yeast (1; DAA).

Dosages (Safflower):

FNFF = !!

Seeds roasted and eaten as vegetable; leaves cooked like spinach; flowers used as poor man's saffron, often cooked with rice or used as food dye; seed oil commercially available in the United States (DEP; FAC); 1 g in tea; to 3 ×/day (HHB); 1–2 tsp fresh flower (PED); 2–3 g dry flower (PED); 3 g dry flower:15 ml alcohol/15 ml water (PED); 10 g achene/day as laxative (VAD).

- Ayurvedics suggest laxative flowers for leprosy, strangury; seeds aphrodisiac; leaves diuretic, laxative, orexigenic, for dysuria and ophthalmia (KAB).
- Bengali apply seed oil, three to six applications, for itch (KAB).
- Indochinese use flowers as emmenagogue and tonic, for dysmenorrhea and paralysis (KAB),
- Iranians use seed oil in salves for rheumatism and sprains (HJP).
- Jamaicans mash the seed in sweet water as laxative, flowers for jaundice (DEP).
- Lebanese often give children the diaphoretic laxative floral tea (HJP).
- Lebanese use safflower oil in liniments (e.g., for rheumatism) (HJP).
- Philippinos use flowers to treat jaundice (KAB).
- Unani view flowers as diuretic, expectorant, hepatotonic, hypnotic, for boils, bronchitis, complexion, leukoderma, piles, ringworm, scabies; the seeds or seed oil good for old folk, analgesic, aphrodisiac, bechic, carminative, purgative, for arthrosis, catarrh, hepatosis, leukoderma, scabies, sore throat (KAB).
- Venezuelans apply the plant in cataplasms to tumors (JLH).

Downsides (Safflower):

Class 2b-2d. Contraindicated in patients with hemorrhagic disease or peptic ulcers. Reportedly abortifacient, emmenagogue, and uterotonic. May prolong blood coagulation time (AHP). No health hazards or side effects known with proper therapeutic dosages (PH2).

Natural History (Safflower):

Safflower is self-pollinated with some cross-pollination. Pollen and nectaries are abundant with insects working the flowers. Safflower is attacked by many fungi: *Alternaria carthami* (leaf spot and bud rot), *A. zinniae, Bremia lactucae, Cercospora carthami, Cercosporella carthami, Chaetomium globosum, Collectorichum capsici, Corticium solani, Ectoctroma carthami, Epicoccum nigrum, Erysiphe cichoracearum, Fusarium acuminatum, F. solani, Gloeosporium carthami, Glomerella cingulata, Leveillula compositarum, L. taurica, Macrophomina phaseoli, Macrosporium carthami, Marsonia carthami, Oidium carthami, Oidiopsis taurica, Phyllosticta carthami, Phytophthora drechsleri* (root rot), *Ph. palmivora, Ph. parasitica, Puccinia carthami* (rust), *Pythium debaryanum, P. oligandrum, Ramularia carthami, Sclerotinia sclerotiorum, Septoria carthami, Verticillium albo-atrum* (wilt). It is parasitized by *Orobanche cernua* and *Striga lutea,* and is attacked by the Chilli mosaic and Cucumber mosaic viruses. *Pseudomonas solanacearum*, a bacterium, attacks it. Among the nematodes, the following have been isolated from safflower: *Meliodogyne incognita acrita, M. javanica*. Insect pests include Lygus bugs, wireworms, aphids, leaf hoppers, thrips, and sunflower moth larvae (HOE).

CEDAR OF LEBANON (*CEDRUS LIBANI* A. RICH.) ++ PINACEAE

Synonyms:

Cedrus cedrus Huth.; *C. libanotica* Link; *C. libanitica* (Trew) Pilger; *C. libanensis* Juss. ex Mirb.; *C. patula* K. Koch, *Larix cedrus fide* CJE and EFS

Notes (Cedar of Lebanon):

Let fire come out of the bramble, and devour the cedars of Lebannon.

Judges 9 (KJV)

This marvelous handsome tree once forested many slopes of Lebanon, but many were cut because of their excellent qualities. Both the First and the Second Temples in Jerusalem were made of cedar. Biblical scholars (e.g., Moldenke and Moldenke) lament the "history of the Holy Land," which has been transformed from a land of palms, flowing with milk and honey, to its present inhospitable desert condition. How ironic that this plundering of the forests was begun by men who wanted to build a temple to God, a place where men might worship their Creator. William Cullen Bryant beautifully praised the cedars, clearly predicting my sentiments:

"The groves were God's first temples,
Ere man learned
To hew the shaft …"

Solomon, in one of the first recorded "botany lectures," spoke of trees "from the cedar that is in Lebanon even unto the hyssop that springeth out of the wall." Ironically, he is one of the first recorded plunderers of the forest as well. He is said to have raised a group of 30,000 Israelites and sent them in turns, some 10,000 a month, to fell the trees, with the assistance of 150,000 slaves, under the supervision of 3300 officers. They labored seven years to build a "temple for the glory of God" and thirteen more to build the fabulous House of the Forest of Lebanon that so impressed the Queen of Sheba. Adding conservational insult to injury, goats were permitted to browse the deforested slopes, finishing off the once moist forests, converting it to "one of the most impoverished" areas in the world. The cedar exudes a balsam that makes the durable wood very aromatic. Resistant to decay and insects, the timber is excellent for such building. Lebanese correctly believe the tree purifies the air (BIB).

Common Names (Cedar of Lebanon):

Arz (Arab.; Syria; HJP); Arz Libnan (Arab.; Syria; HJP); Arz ur Rabb (Arab.; Syria; HJP); Cedar (Eng.; ZOH); Cedar of Lebanon (Eng.; CJE; CR2; HJP); Ceder van de Libanon (Dutch; EFS); Cèdre du Liban (Fr.; EFS); Cedro del Libano (It.; Sp.; EFS); Doedar (India; EFS); Devadaru (Sanskrit; EFS); Erez (Heb.; ZOH); Glory of Lebanon (Eng.; ZOH); Ibhûl (Arab.; Syria; HJP); Kedros (Greek; JLH); Lebanese Cedar (Eng.; CJE), Libanon Zeder (Dutch; EFS); Sanobar i Hind (Iran; EFS); Stnûb (Arab.; Syria; HJP).

Activities (Cedar of Lebanon):

Antiseptic (1; X11962214); Bactericide (1; X10548751); Diuretic (f; BIB; EFS); Expectorant (1; HHB; PH2); Fungicide (1; FNF); Insecticide (f; BIB).

FIGURE 1.20 Cedar of Lebanon (*Cedrus libani*).

Indications (Cedar of Lebanon):

Asthma (f; BIB; HJP); Bacteria (1; X10548751); Bacillus (1; X10548751); Blenorrhagia (f; BIB); Boil (f; BIB; HJP); Bronchosis (f; BIB); Burn (f; BIB; JLH); Cancer (f; BIB; JLH); Catarrh (1; PH2); Cough (1; FNF; HHB); Dermatosis (f; BIB); Enterobacter (1; X10548751); Fungus (1; FNF); Gastrosis (f1; × 10473175); Helicobacter (1; X10473175); Induration (f; BIB; JLH); Infection (f1; BIB; HJP; X11962214); Klebsiella (1; X10548751); Listeria (1; X10548751); Mycobacterium (1; X10548751); Phthisis (f; BIB); Proteus (1; X10548751); Pseudomonas (1; X10548751); Pulmonosis (f; HJP); Rash (f; BIB); Respirosis (f; BIB; HJP); Staphylococcus (1; X10548751); Tuberculosis (1; BIB; HHB).

Dosages (Cedar of Lebanon):

FNFF = ?
Tree sometimes produces an edible manna (TAN).

- Algerians use deodar cedar pitch for boils and dislocations (HJP).
- Egyptians use sap or oil from the wood for burns, cancer, and indurations (JLH).
- Lebanese inhale the smoke from burning branches for asthma-like pulmonary difficulties (HJP).
- Lebanese use the pitch and ashes in salves to poultice onto deep infections (HJP).
- Turks use the cones, active against *Helicobacter*, for various stomach ailments (X10473175).

Downsides (Cedar of Lebanon):

No health hazards or side effects known with proper therapeutic dosages (not given) (PH2).

Extracts (Cedar of Lebanon):

Essential oil fungitoxic at 1000 ppms (FFJ4(1):1).

PURPLE STAR THISTLE (*CENTAUREA CALCITRAPA L.*) + + ASTERACEAE

Synonyms:

Calcitrapa stellata Lam.

Notes (Purple Star Thistle):

Thorns also and thistles shall it bring forth to thee.

Genesis 3 (KJV)

Worst of weeds, some species head high, some thistles are said to make nutritious vegetables, potherbs, and depuratives. Back in 1929, Temple, a London writer, in his *Flowers and Trees of Palestine*, said, probably of thistles in general, "Many of these thistles, which grow from 10 to 15 feet high (*C. calcitrapa* rarely attains 1 m in height; MPG), and briers are decided obstructionists to travelers who leave the beaten track, and are consequently described as noxious thorny plants." (Temple, 1929). Zohary does not specifically mention this species nor is it reported in the *Flora of Palestine* (FP3; ZOH). Hence, I fear I may have been too eager to accept earlier suggestions that this is one of the "thistles" of the Holy Land. I will bet it is there in Israel, if not mentioned in the Bible. Zohary, who surely would know more about the matter than I, suggests *Centaurea iberica* instead.

Common Names (Purple Star Thistle):

Aboujoulj (Ber.; BOU); Abre Puño (Arg.; Sp.; EFS); Abrojo (Sp.; EFS); Aceb (Ber.; BOU); Ad Dardarriyah (Arab.; Syria; HJP); Aourmela (Ber.; BOU); Bou Neggar (Arab.; BOU); Bou Shweika (Arab.; BOU); Calcatreppolo (It.; EFS); Calcitrapa (Por.; EFS; MPG); Caltrops (Eng.; BOU); Cardo Estrelado (Por.; EFS); Cardo Estrellado (Por.; Sp.; EFS; MPG; VAD); Ceceprete (It.; EFS); Centaurée Chausse-Trape (Fr.; USN); Chardon Étoilé (Fr.; BOU; EFS); Chaussetrape (Fr.; BOU); Common Star Thistle (Eng.; EFS; HJP); Corn Flower (Eng.; HOC); Garnanzos del Cura (Sp.; EFS); Hassak (Arab.; BOU); Ippofesto (It.; EFS); Kalketrap (Dutch; EFS); Mouse Thorn (Eng.; BOU); Murrâr (Arab.; Syria; BOU; HJP); Murrayr (Arab.; Syria; HJP); Nowar Bellaremj (Arab.; BOU); Purple Star Thistle (Eng.; USN); Red Star Thistle (Eng.; USN); Star Thistle (Eng.; USN); Stern-Flockenblume (Ger.; USN); Stern-Flodkenblume (Ger.; EFS); Trepa Caballos (Sp.; EFS).

FIGURE 1.21 Purple Star Thistle (*Centaurea calcitrapa*).

Activities (Purple Star Thistle):

Antioxidant (1; X12203269); Antiseptic (1; MPG); Aperitive (f; VAD); Bactericide (1; MPG); Cholagogue (f; BIB); Choleretic (f; VAD); Cytostatic (1; MPG); Depurative (f; BIB); Digestive (f; VAD); Diuretic (f; EFS); Emmenagogue (f; BIB; BOU); Febrifuge (f; EFS; VAD); Hypoglycemic (f1; MPG; VAD); Hypotensive (1; MPG); Orexigenic (f; BOU; VAD); Stimulant (f; EFS); Stomachic (f; BOU); Sudorific (f; BIB; EFS); Tonic (f; BIB; EFS); Vermifuge (f; BOU); Vulnerary (f; BOU).

Indications (Purple Star Thistle):

Amenorrhea (f; BOU); Anorexia (f; BOU; VAD); Bacteria (1; MPG); Brucella (1; MPG); Cancer (f; BIB; JLH); Cold (f; VAD); Corns (f; JLH); Diabetes (f1; MPG; VAD); Dyskinesia (f; VAD); Fever (f; BIB; EFS); Fistula (f; BIB; WO2); Flu (f; VAD); Gallstone (f; HJP); Gravel (f; BIB; WO2); Headache (f; BOU); High Blood Pressure (1; MPG); Hyperglycemia (f; VAD); Infection (1; MPG); Jaundice (f; BIB); Kidney stone (f; HJP); Malaria (f; BOU); Mycobacterium (1; MPG); Nephrosis (f; BOU; HJP); Ophthalmia (f; BOU); Pain (f; BOU); Pseudomonas (1; MPG); Salmonella (1; MPG); Staphylococcus (1; MPG); Stone (f; BIB; BOU; WO2); Wound (f; BOU); Worm (f; BOU).

Dosages (Purple Star Thistle):

FNFF = !
Young shoots consumed raw in Egypt, as a potherb elsewhere (e.g., among Albanians in North Italy) (X12203269). Bedouins harvest the oleiferous seed and grind it for food (HJP). Fresh plant suspension 125–600 g per day (VAD). 20–40 g herbs infused 10 minutes per liter of water, 3–5 cups/day (VAD). Decoction of 30 g plant per liter of water, 3–5 cups/day (MPG; VAD).

- Argentines use the juice to treat corns (JLH).
- Europeans take seed powder in wine for stones (WO2).
- Europeans use powdered roots for fistula and gravel (WO2).
- Lebanese eat boiled stems for jaundice (HJP).
- Lebanese take (with dilating herbs such as belladonna) to help pass gall and kidney stones (HJP).
- North Africans consider the seeds anodyne, antilithic, febrifuge, and vulnerary, using the whole plant for malaria and ophthalmia, and the leaves for headache (BOU).
- Portuguese use flowers and leaves as febrifuge and vulnerary, and the roots and fruits as diuretics (MPG).
- Spaniards suggest the plant for anorexia, cold, diabetes, hepatobiliary dyspepsia, hyperglycemia, hyposecretory dyspepsia, flu, and wounds (VAD).

Downsides (Purple Star Thistle):

Contraindicated in pregnant and lactating women (VAD). Diabetics should watch insulin levels (VAD).

Natural History (Purple Star Thistle):

Although grazed by animals, it can produce intoxication (e.g., encephalomalacia in horses that consume it) (VAD).

SPANISH THISTLE (*CENTAUREA IBERICA* SPRENG.) ++ ASTERACEAE

Notes (Spanish Thistle):

Thorns also and <u>thistles</u> shall it bring forth to thee; and thou shalt eat the herb of the field.

Genesis 3:18 (KJV)

Thorns and <u>thistles</u> it shall bring forth to you; and you shall eat the plants of the field.

Genesis 3:18 (RSV)

And thorns and thistles it will grow for you; and you must eat the vegetation of the field.

Genesis 3:18 (NWT)

Thorn and thistles shall grow up on their altars.

Hosea 10:8 (RSV)

Zohary notes that the word *dardar* shows up only twice in some versions of the Bible. He suggests that *dardar* may refer to species of *Centaurea,* possibly *C. iberica,* called *dardar* by the Arabs in the Holy Land. Even Zohary is not sure that *C. iberica* was intended in the Bible, especially the citation in Genesis 3:18. (See *Silybum* for my equally weak second opinion.) Nowhere does *C. iberica* occur as a weed in the fields. These Arab and Hebrew names may mean potherb in general rather than *Centaurea* specifically. This type of thistle is distinguished from other thistles and thorns by the flat, whorled rosette hugging the ground in winter. The Arabs gather the Iberian *dardar* (meaning whorls), also called *murrier* (*merorim* in Hebrew). These terms may mean potherb in general rather than *Centaurea* specifically. In India, the flower heads are used for eczema.

Common Names (Spanish Thistle):

Dardar (Arab.; ZOH); Dardur (Arab.; ZOH); Iberian Knapweed (Eng.; GOO); Iberian Star Thistle (Eng.; USN); Merorim (Heb.; ZOH); Mureir (Heb.; ZOH); Spanish Thistle (Eng.; ZOH); Thistle (Eng.; BIB).

Activities (Spanish Thistle):

Antidiabetic (1; X14750205); Insulinogenic (1; X14750205).

Indications (Spanish Thistle):

Diabetes (1; X14750205); Eczema (f; BIB).

Dosages (Spanish Thistle):

FNFF = !
Arabs gather the leaves of young plants of this and several other species as a potherb.

SYRIAN SCABIOUS (*CEPHALARIA SYRIACA* (L.) ROEM. & SCHULT.) + DIPSACACEAE

Notes (Syrian Scabious):

The kingdom of heaven may be compared to a man who sowed good seed in his field; but while men were sleeping, his enemy came and sowed weeds among the wheat, and went away.

Matthew 13: 24–5 (RSV)

Zohary equates this with sowing weed seeds that resemble the seeds of the crops. "Darnels and scabious are both noxious weeds that grow only among crops, and damage them. … The Syrian scabious resembles wheat only by virtue of its seed", (ZOH) while the darnel plant and seed alike resemble wheat. Its grains are closely associated with certain wheat varieties and are harvested accidentally with the wheat. As a result, the weed seeds are sown with the following wheat crop, in some cases overwhelming the crop, in others contributing their bitterness to the resultant flour. I gather from Zohary's conclusion that the seed must be somewhat edible. "Sometimes the weed overwhelms the wheat, so that the farmer is forced to harvest it instead of the sown plant." (ZOH) According to the USDA Nomenclature Database, the seeds are human food (valued in Georgia, previously of the USSR, but not elsewhere, when contaminating wheat and ground into flour for specially flavored bread) (USN).

Common Names (Syrian Scabious):

Abrepuños (Sp.; USN); Céphalaire de Syrie (Fr.; USN); Escabiosilla (Sp.; USN); Makhobeli (Rus.; USN); Taradan Shalam (Arab.; ZOH); Zuwan Aswad (Arab.; ZOH).

Dosages (Syrian Scabious):

FNFF = !
Grown as an oilseed in Russia (FP3); sometimes made into wheat flour as the seeds are threshed with the wheat in which it is a weed.

CAROB (*CERATONIA SILIQUA* L.) +++ FABACEAE (CAESALPINIACEAE)

Notes (Carob):

> **John had his raiment of camels hair, and a leathern girdle about his loins; and his meat was locusts and wild honey.**
>
> **Matthew 3:4 (KJV)**

This is one of the most important tree species in the Middle Eastern vegetation type known as maquis, with *Pistacia*. Arab proverbs hint that the carob tree is cursed, and one should not sleep under it. And yet, carob is widely cultivated for its fruit and seeds, both high in sugar and calcium, low in protein and fats. It is used in "health foods" as a chocolate substitute. A single carob tree may yield 800 pounds of fruit. The seeds are said to be the ancient weight used by goldsmiths, and instituted from early times as carat weight. Seeds commanded a high price; singers imagined they cleared the voice. Carob is also used in textile printing, synthetic resins, insecticides, and fungicides. American imports were mostly used in tobacco flavorings and cosmetics, but have now moved into the health food arena. In the Prodigal Son story, the younger son asked for his inheritance, which he then spent quickly and unwisely. When he had no money or food left, he hired out as a swine tender. Because he had no bread, he wanted to eat the carob pods being fed to swine. On April 25, 1982, in Burtonsville, Maryland, carob-coated walnuts sold for $2.89 a pound, more fitting for a profligate son. The price is more than twice that today.

Common Names (Carob):

Abernid (Ber.; BOU); Albero della Carroba (It.; HH3); Alfarrobeira (Mad.; Por.; JAD; KAB); Algaroba; (Sp.; DEP); Algarroba; (Sp.; Peru; EGG; RAR; VAD); Algarroba del Libano (Sp.; EFS); Algarroba Europeo (Sp.; EFS); Algarrobeira (Por.; USN); Bockshornbaum (Ger.; HH3); Bread Bean (Eng.; KAB); Carob (Eng.; Scn.; AH2; CR2); Carob Bean (Eng.; HH3); Carob Tree (Eng.; BOU); Caroba (Sp.; USN); Carobe (It.; KAB); Carobo (It.; KAB); Carobole (It.; KAB); Caroube (Fr.; USN); Caroubier (Fr.; BOU; EFS; KAB); Caroubier a Siliques (Fr.; TAN); Carouge (Fr.; HH3; KAB); Caruba (Ger.; KAB); Carrubio (It.; DEP; KAB); Carrubo (It.; Malta; EFS; KAB); Ceratonia (Peru; EGG); Chechire Gum (Eng.; HH3); Curenecillo (Sp.; SOU); Garrofe (Cat.; KAB); Garrofero (Sp.; EFS; KAB); Goma de Garrofín (Sp.; VAD); Guainella (It.; HH3; KAB); Hansbrod (Den.; EFS); Harruba (Malta; KAB); Haruv (Heb.; ZOH); Haruvim (Heb.; ZOH); Ikidou (Ber.; BOU); Inago Mame (Japan; TAN); Jans Broodboom (Dutch; KAB); Johannisbrod (Swe.; KAB); Johannisbrotbaum (Ger.; KAB; USN); Johannisbrotkehrmehl (Ger.; HH3); Johnsbread Tree (Enf.; EFS);

FIGURE 1.22 Carob (*Ceratonia siliqua*).

Karobbaum (Ger.; KAB); Karobenbaum (Ger.; HH3); Karobenboom (Dutch; KAB); Keciboynuzu agaci (Tur.; EFS); Keratia (Greek; KAB); Keration (Heb.; BI2); Keratonia (Greek; KAB); Kharnub (Arab.; Pun.; BOU; DEP; KAB); Kharnub Nupti (Arab.; Pun.; DEP); Kharnub Shami (Arab.; DEP); Kharroub (Arab.; BOU); Kharrouba (Arab.; BOU); Kharrub (Arab.; Syria; HJP); Kharruh (Arab.; BI2); Khirnub (Arab.; KAB); Locust Bean (Eng.; Ocn.; AH2; WO2); Locust Tree (Eng.; BOU; WO2); Meccaniamli (Pun.; WO2); Pain de Jean Baptiste (Fr.; EFS); Pane di Santo Giovanni (It.; HH3); Qarmatit (Yemen; GHA); Ribba (Arab.; BOU); Rojkovvi (Rus.; KAB); Roscov (Rom.; KAB); Selarwa (Ber.; BOU); Sodbrodbaum (Ger.; KAB); St. John's Bean (Eng.; SKJ; ZOH); St. John's Bread (Eng.; Ocn.; AH2; BOU; SKJ); Tikherroubt (Ber.; BOU); Tikida (Ber.; BOU); Tikidat (Ber.; BOU); Tikidit (Ber.; BOU); Tisliwha (Ber.; BOU).

Activities (Carob):

Antacid (f; DEP; KAB); Antibacterial (12; SKY; WO2); Anticancer (f1; X12490228); Anticarcinoma (f1; X12490228); Anticoagulant (1; PH2); Antiexudative (1; HH3; PH2); Antioxidant (1; X11782210); Antiproliferant (1; X12490228); Antiseptic (f1; WO2); Antitoxic (f1; SKY); Antitussive (f; BIB); Antiviral (1; HH3; PH2); Apoptotic (1; X12490228); Astringent (f1; BOU; SKY; WO2); Bechic (f; BOU); Caspase-3 Inducer (1; X12490228); Demulcent (f; BIB); Digestive (f1; PH2); Diuretic (f; BOU); Fungicide (f; VAD); Hemolytic (1; WO2); Hypocholesterolemic (1; BIB; LAF); Hypoglycemic (1; HH3; LAF); Hypoinsulemic (1; LAF); Hypolipidemic (1; PH2); Laxative (f; BOU; LAF); Pancreatonic (1; HH3); Pectoral (f; BIB; DEP; EFS); Purgative (f; BIB; DEP; EFS); Resolvent (f; BIB).

Indications (Carob):

Asthma (f; BIB); Atherosclerosis (f; VAD); Cancer (f1; X12490228); Carcinoma (f1; X12490228); Catarrh (f; BIB; IED; PNC); Celiac (f; PH2); Childbirth (f; PH2); Colitis (f; PH2); Constipation (f; IED; HJP); Cough (f; BIB; DEP; HHB; PH2; WO2); Dehydration (f; WO2); Diabetes (f1; LAF; VAD; WO2); Diarrhea (f12; GHA; HHB; SKY); Duodenosis (f; VAD); Dyspepsia (f12; SKY; WO2); Enterosis (f; PH2); Gastrosis (f; VAD); Heartburn (f1; SKY); Hepatosis (1; X12490228); High Cholesterol (1; BIB; LAF); Hyperglycemia (1; LAF); Hyperlipemia (f; VAD); Hyperperistalsis (f; WO2); Induration (f; JLH); Infection (f; VAD); Mononucleosis (f; IED); Mycosis (f; VAD); Obesity (f1; LAF; VAD); Sprue (f; PH2); Steatorrhea (f; HHB); Ulcer (f; VAD); Voice (f; PNC); Vomiting (f; PH2); Wart (f; JLH).

Dosages (Carob):

FNFF = !!!

The fruit, food for the poor in Jewish folklore, also appears in the Christian tradition as "St. John's Bread," eaten by St. John the Baptist in the wilderness. Powdered pulp used as chocolate substitute; ripe fruits made into molasses like *dibs* (e.g., in Palestine); scorched seeds a coffee substitute; tragasol, a gum extracted from the seeds, is used in dressings, ice cream, pickles, and sauces. Alcoholic beverages are made from the pods (BIB; FAC; FP2; TAN). One story in the Talmud tells that the Jewish sage Rabbi Shimeon Bar Yohai, hiding from the Romans in Galilean caves with his son, sustained life with carob alone for 12 years (ZOH). In Cyprus, a brittle candy known as "pasteli" is made from the pods (BIB). 15–20 g carob mixed in applesauce (SKY); 50 drops fluid extract (1:1) goma garrofin 3 ×/day (VAD); 500–1500 mg capsule 1/2 hour before meal, with plenty of juice or water (VAD).

- Arabians take the dry seed kernels with sugar for diarrhea (GHA).
- Egyptians apply carob to warts (JLH).
- Lebanese grind pods into a laxative treacle called "dibbis kharub" (HJP).
- Southern Europeans used the pods for asthma and cough (BIB).
- Spaniards take the flour for diarrhea, gastritis, gastroduodenal ulcers, and infantile nausea (VAD).
- Spaniards take the seed starch as a laxative, and to help obesity and prevent atherosclerosis (VAD).
- Spaniards use green fruit as a fungicide (VAD).

Downsides (Carob):

Class 1 (AHP, 1997). No health hazards or side effects known with proper therapeutic dosages (PH2). Infant diarrhea must be monitored by a professional to ensure proper hydration with high electrolyte fluid during acute diarrhea (SKY, 1998). Avoid with intestinal obstructions or stenoses

(VAD). Best to monitor insulin in diabetics (VAD). Can interfere with intestinal absorption of other medicines (e.g., penicillin and lithium salts) (VAD).

Extracts (Carob):

Papagiannopoulos et al. (2004) quantified the polyphenols in carob fruits, reporting 41 individual phenolics (448 mg/kg extractable polyphenols comprising gallic acid, hydrolyzable and condensed tannins, flavonol-glycosides, and traces of isoflavonoids) (X15186098). While no clear competitor with green tea, carob should share in many of the purported health benefits of green tea in the major killers, cancer, cardiopathy, and diabetes. Carob tannins inactivate toxins by binding with them; inhibit bacteria (SKY; X14170956). I suspect that is why I was given carob when I suffered salmonella in Panama nearly half a century ago. By making stomach contents more viscous, fibers and sugars may interfere with acid reflux in the esophagus (SKY). Rats fed a fiber-free diet containing 1% cholesterol exhibited a small increase in serum cholesterol and a fivefold increase in liver cholesterol. Addition of 10% pectin or 10% locust bean gum kept the increase in liver cholesterol down (BIB). LD50 (Carob gum) = 9100 mg/kg orl dog HH3; LD50 (Carob gum) = 10,000 mg/kg orl hamster HH3; LD50 (Carob gum) = 13,000 mg/kg orl mus HH3; LD50 (Carob gum) = 13,000 mg/kg orl rat HH3.

JUDAS TREE (*CERCIS SILIQUASTRUM* L.) ++ FABACEAE

Notes (Judas Tree):

And he cast down the pieces of silver in the temple, and departed, and went and hanged himself.

Matthew 27:5 (KJV)

A handsome ornamental tree, most probably not the tree on which Judas hanged himself — if he hanged himself at all. Tradition, more than 200 years old, has it that the red color of the flowers results from the trees blushing or burning with shame when Judas selected it. Because Zohary does not mention it in his plants of the Bible, I suppose he doubts that Judas hanged himself on the Judas tree. But he does illustrate it well in the *Flora of the Palestine* (FP2). Hence, it is certainly a handsome tree of the Holy Land although not mentioned per se in the Bible. Other authors (GMH) think Judas hanged himself on an elderberry, *Sambucus* (GMH), but there are no elderberries listed for the native *Flora of Palestine*. *Sambucus nigra* is cultivated in that part of the world and could support the weight of a hanging man. I saw a substantial *Sambucus nigra* tree near Hildegarde von Bingen's. The Judas tree of Palestine is also said to be an ornamental timber tree, the flower buds of which are pickled and eaten (FP2).

Common Names (Judas Tree):

Albero di Guida (It.; EFS); Arból de Amor (Sp.; EFS); Arból de Judea (Sp.; EFS); Arbre de Judie (Fr.; EFS); Arjorán (Sp.; EFS); Ching P'i (China; EFS); Echter Judasbaum (Ger.; EFS); Erguvan (Tur.; EFS); Gainier (Fr.; EFS); Judas Boom (Dutch; EFS); Judas Tree (Eng.; USN); Love Tree (Eng.; USN); Namdumbu (Sp.; EFS); Olaia (Por.; EFS); Siliquastro (It.; EFS); Tzu Ching (China; EFS); Arból de Judea (Sp.; EFS).

Activities (Judas Tree):

Aromatic (f; EFS); Bitter (f; EFS).

Indications (Judas Tree):

Catarrh (f; EFS); Headache (f; EFS).

Dosages (Judas Tree):

FNFF = !
Flowers used in pickles and salads (BIB; FP2).

CHICKPEA (*CICER ARIETINUM* L.) +++ FABACEAE

Notes (Chickpea):

The oxen likewise and the young asses that ear the ground shall eat clean provender, which hath been winnowed with the shovel and with the fan.

Isaiah 30:24 (KJV)

And the oxen and the asses that till the ground will eat salted provender, which has been winnowed with shovel and fork.

Isaiah 30:24 (RSV)

And the cattle and the full-grown asses cultivating the ground will eat fodder seasoned with sorrel, which was winnowed with the shovel and with the fork.

Isaiah 30:24 (NWT)

Provender, perhaps a mistranslation, was mentioned six times in the Bible (KJV). In my first *Medicinal Plants of the Bible*, I did not include the chickpea; none of the scholars I had read seemed to stress that as a possibility. Since then, however, I have read that the word *provender* of the Bible might better have been translated as hummus, and indeed may have been the chickpea. And yes, the leaves of the provender are eaten, more by animals than by humans. But humans do ingest cooked young leaves like spinach. According to Zohary, the biblical *hamitz*, cognate with the Arabic *humus* and the Aramaic *himtza,* means chickpea, and today *himtza* is modern Hebrew for chickpea ("The RSV translation as 'provender' is mistaken.")(ZOH). I will accept Zohary's conclusions and include chickpea. But there are still version variations; it is "clean" in KJV, it is "salted" in RSV, and "seasoned with sorrel" in NWT (Zohary did not index Rumex = sorrel in ZOH, but lists more than 15 varieties and species of Rumex in FP1). Chickpea, apparently, was originally domesticated in Turkey and some neighboring countries. It has been found in pre-pottery Neolithic levels of some prehistoric sites, in Early Bronze Age deposits of Jericho, in Iraq, as well as elsewhere. Earliest records are from Turkey "dating from 5000 BC" (ZOH). Allegedly, the *cicer* of the Romans, Horace referred to parched seed as food for the poor. Supposedly cultivated in Egypt since the earliest times of the Christian era, it was perhaps considered common and/or unclean (DEP). Today, I believe that chickpea was one of several pulses, including *Pisum*, *Vicia*, even *Trigonella* (if you consider that a pulse) already cultivated in the biblical world in biblical times, more than 2000 years ago.

Common Names (Chickpea):

Badam (Mooshar; NPM); Bagolyborsó (Hun.; POR); Balabhaishajya (Sanskrit; KAB); Balabhoyja (Sanskrit; KAB); Becudo (Lan.; KAB); Beiqa (Arab.; BOU); Bengal Gram (Eng.; KAB; MPI); Black Chickpea (Eng.; POR); Black Desi (Eng.; POR); Black-seeded Chickpea (Eng.; POR); Booto (Oriya; WO2); Bukkeert (Nor.; POR); But (Beng.; Hindi; San.; Urdu; KAB; WO2); Butakala (Beng.; POR); Butkalai (Beng.; KAB; NAD); Butma (Assam; POR; WO2); Cabuli (Nwp.; DEP); Café Francais (Fr.; KAB); Cana (Nepal; POR); Canaa (Nepal; POR); Cece (It.; DEP; KAB); Ceci (It.; TAN); Ceseron (Fr.; KAB); Ceze (Lan.; KAB); Cezerous (Lan.; KAB); Cezes (Lan.; KAB); Chahna (Sin.;

KAB); Chana (Beng.; Guj.; Hindi; Kas.; Pun.; Urdu; KAB; MPI; WO2); Chanaa (Guj.; POR); Chanaabatulaa (Beng.; POR); Chanabartula (Beng.; POR); Chanaka (Sanskrit; MPI; WO2); Chanakamulu (Tel.; NAD); Chania (Guj.; KAB; WO2); Chaniaa (Guj.; POR); Chano (Kon.; MPI); Chenna (Dec.; KAB); Chennuka (Sanskrit; DEP); Chickpea (Eng.; CR2); Chola (Beng.; Iran; Pun.; DEP; KAB; NAD; WO2); Chole (Hindi; Pun.; WO2); Chono (Kon.; KAB); Chot Abut (Beng.; MPI); Chotobata (Beng.; POR); Chotobut (Beng.; POR); Chunna (Hindi; DEP; KAB); Chunnaa (Hindi; POR); Cicer (Roman; DEP); Cicérole (Fr.; BOU; KAB); Cicerolle (Fr.; KAB); Cicri (Malta; KAB); Ciecierzyca Pospolita (Pol.; POR); Ciserole (Fr.; KAB); Ciseron (Fr.; KAB); Cizrna Beraní (Czech; POR); Common Gram (Eng.; IHB; KAB); Csicseri borsó (Hun.; POR); Da Zi Ji Dou (China; POR); Djelbane (Ber.; BOU); Dhal (India; LEG); Echte Kicher (Ger.; POR); Ee Chip T'eu Kong (Korea; POR); Erebinthos (Greek; DEP); Ervanço (Por.; KAB; POR); Gairance (Fr.; BOU); Gairoutte (Fr.; KAB); Garavance (Malta; KAB); Garbance (Fr.; KAB); Garbanzo (Eng.; Sp.; KAB; LEG; POR); Garbanzos (Dr.; Sp.; AHL; DEP); Garoutte (Fr.; KAB); Garvance (Fr.; KAB); Garvane (Belgium; Fr.; JLH; KAB); Gram (Eng.; CR2); Grno de Bico (Por.; KAB; POR; USN); Grno Gravanço (Por.; POR); Grauwe Erwt (Dutch; POR); Hamaz (Ber.; BOU); Harbara (Bom.; Mar.; KAB; MPI; WO2); Harbarchana (Bom.; NAD); Harimandhakam (Tel.; DEP; KAB); Harimandhakama (Tel.; POR); Harimantha (Sanskrit; KAB; POR; WO2); Hei Xiao Zi Ji Dou (China; POR); Hei Zi Ji Dou (China; POR); Hhimmass (Arab.; POR); Himtza (Aramaic; Heb.; ZOH); Hiyoko-Mame (Japan; POR; TAN); Hommos (Arab.; BOU; POR); Hommos Malana (Arab.; POR); Homos (Egypt.; DEP; KAB); Horse Bean (Eng.; IHB); Hui Hui Dou (China; POR); Humug (Arab.; NAD); Humus (Arab.; ZOH); Ikiker (Ber.; BOU); I Chip T'eu Kong (Korea; POR); Ji Dou (China; POR); Ji Tou Dou (China; POR); Jivana (Sanskrit; KAB); Jumez (Arab.; KAB; WOI); Kabuli Chickpea (Eng.; POR); Kachang Arab (Malaya; POR); Kachang Kuda (Dei.; Malaya; IHB; POR); Kadala (Kan.; Mal.; MPI; WO2); Kadalai (Tam.; POR; WO2); Kadalakka (Mal.; WO2); Kadale (Kan.; MPI; WO2); Kadle (Tel.; NAD); Kadli (Carnatic; KAB); Kaffeärt (Swe.; POR); Kahviherne (Fin.; POR); KalabPh. (Burma; POR); Kalai (Beng.; DEP; WOI); Kalapai (Burma; DEP; KAB); Kanchuki (Sanskrit; KAB); Karikadale (Kan.; KAB; POR); Katala (Mal.; MPI); Keker (Dutch; KAB; POR); Kempukadale (Kan.; KAB); Kichererbse (Ger.; KAB; POR); Kichererbsen (Ger.; POR); KikFrt (Den.; POR); Kikärt (Swe.; POR); KikerFrt (Den.; POR); Kodala (Sin.; POR); Kondakkadala (Sin.; Tam.; KAB; MPI); Kondi (Sin.; POR); Kreukerwt (Dutch; POR); Krishnachanchuka (Sanskrit; KAB); Kudole (Kan.; KAB); Kulopan (Burma; NAD); Large-seeded Chickpea (Eng.; POR); Makhud (Punj.; MPI); May Bu Ri Kong (Korea; POR); Moroejang (Mun.; KAB); Morujang (Mun.; WO2); Nakhud (Iran; Pun.; DEP; NAD); Naut (Rom.; KAB); Nahot (Rom.; KAB); Nochut (Tur.; POR); Nohud (Tur.; POR); Nohut (Tur.; EB54:155); Nut Baranii (Rus.; POR); Ovetche Harokh (Rus.; KAB); Pajito (Mad.; Por.; JAD); Pesette (Fr.; KAB); Pisello Cece (It.; POR); Pisello Cornuto (It.; POR); Poischi (Fr.; NAD); Pois Becu (Fr.; KAB); Pois Blanc (Fr.; KAB); Pois Chiche (Fr.; BOU; MPI; TAN); Pois ciche (Fr.; KAB); Pois Cornu (Fr.; KAB); Pois Gris (Fr.; KAB); Pois Pointu (Fr.; KAB); Provender (Bib.; Eng.; BIB; ZOH); Ram's Head Chickpea (Eng.; POR); Revithi (Greek; POR); Revithia (Greek; POR); Sakalapriya (Sanskrit; KAB); Salealpriya (Sanskrit; POR); Sanagalu (Tel.; WO2); Sannaagalu (Tel.; DEP); Sanuagalu (Tel.; KAB); Senagalu (Tel.; MPI); Sigro (Cat.; KAB); Sisér (Arm.; POR); Sisiér (Arm.; POR); Sissererwt (Dutch; POR); Sissererwten (Dutch; POR); Sisyr (Arm.; POR); Siuro (Cat.; KAB); Smooth-seeded Chickpea (Eng.; POR); Sugandha (Sanskrit; KAB); Tete de bellier (Fr.; KAB); Thua Hua Chaang (Thai; POR); Turetskii Gorokh (Rus.; POR); Vajibhakshya (Sanskrit; KAB); Vajimantha (Sanskrit; KAB; POR); Wrinkled-seeded Chickpea (Eng.; POR); Xiao Zi Ji Dou (China; POR); Ying Zui Dou (China; POR); Zweigerbse (Ger.; MPI).

Activities (Chickpea):

Allergenic (1; X15662964); Anthelmintic (f; KAB; WO2); Antianemic (1; FNF); Antiatherogenic (1; MPI); Antibilious (f; DEP; MPI); Anticervisotic (1; FNF); Anticheilitic (1; FNF); Anticoronary (1; FNF); Antidementic (1; FNF); Antidepressant (1; FNF); Antigingivitic (1; FNF); Antiglossitic

(1; FNF); Antigout (1; FNF); Antihyperlipidemic (1; WO3); Antiinfertility (1; FNF); Antiinflammatory (f; KAB); Antileukemic (1; WO3); Antimetaplastic (1; FNF); Antimyelotoxic (1; FNF); Antineuropathic (1; FNF); Antiperiodontotic (1; FNF); Antiplaque (1; FNF); Antipolyp (preventive) (1; FNF); Antipsychotic (1; FNF); Anti-spina-bifida (1; FNF); Antistress (1; WO3); Antiviral (1; X11848297); Aphrodisiac (f; BOU; DEP; MPI; WO2); Astringent (f; WO2); Bifidogenic (1; X15850967); Cardioprotective (1; WO3); Depurative (f; KAB); Diuretic (1; WO2); Estrogenic (f; WO2); Flatugenic (f; KAB); Fungicide (1; WO2; X12895650); Hematopoietic (1; FNF); Hypocholesterolemic (2; MPI; WO2); Hypolipidemic (1; WO2); Immunostimulant (1; FNF); Lactagogue (f; LEG); Laxative (f; MPI); Lipolytic (2; MPI); Mitogenic (1; X11848297); Orexigenic (f; KAB); Proteolytic (1; X11791480); Refrigerant (f; WO2); Reverse-Transciptase Inhibitor (1; X11848297); Soporific (f; LEG); Stimulant (f; WO2); Stomachic (f; MPI); Tonic (f; DEP; WO2); Trypsin Inhibitor (1; X11791480); Xanthine-Oxidase Inhibitor (1; FNF); Uricosuric (1; FNF).

Indications (Chickpea):

Alactea (f; JAD); Anemia (1; FNF); Anorexia (f; KAB); Atheroclerosis (1; MPI; WO3); Biliousness (f; DEP; MPI; WO2); Bite (f; DEP; SKJ); Blood (f; KAB); Bronchosis (f; DEP; WO2); Calculus (f; KAB); Cancer, colon (f1; JLH; X15517915; X15298756); Cancer, penis (f; JLH); Cancer, testicle (f; JLH); Cardiopathy (1; MPI; WO3); Catarrh (f; DEP); Cervicosis (1; FNF); Cheilosis (1; FNF); Cholera (f; JAD); Cirrhosis (1; FNF); Constipation (f; SKJ); Cough (f; WO2); Cutamenia (f; DEP; LEG); Dandruff (f; WO2); Dementia (1; FNF); Depression (1; FNF); Dermatosis (f; BOU; WO2); Diarrhea (f; NAD; WO2); Dislocation (f; NAD; WO2); Dysentery (f; WO2); Dysmenorrhea (f; KAB; NAD); Dyspepsia (f; DEP; WO2); Edema (f; WO2); Fever (f; NAD); Fracture (f; WO2); Fungus (1; WO2); Gas (1; JAD; NAD); Gingivosis (1; FNF); Glossosis (1; FNF); Gout (1; FNF); Headache (f; WO2); Hepatosis (1; FNF); High Cholesterol (2; MPI; WO2); HIV (1; X11848297); Impotence (f; DEP; KAB; WO2); Infection (1; WO2; X12895650); Infertility (1; FNF); Inflammation (f; KAB); Itch (f; BOU); Leprosy (f; BOU; DEP; WO2); Mycosis (1; WO2; X12895650); Nausea (f; DEP); Neuropathy (1; FNF); Obesity (1; WO2; WO3); Orchosis (f; JLH); Ozoena (f; KAB); Pain (f; KAB); Periodontosis (1; FNF); Pharyngosis (f; KAB; WO2); Plaque (1; FNF); Polyp (1; FNF); Pulmonosis (f; KAB); Smallpox (f; BOU); Snakebite (f; DEP; KAB; SKJ); Sore Throat (f; WO2); Spina Bifida (1; FNF); Splenosis (f; DEP); Sprain (f; NAD; WO2); Stress (1; WO3); Stroke (f; DEP; SKJ); Thirst (f; KAB); Toothache (f; WO2); Vomiting (f; DEP); Wart (f; JAD); Worms (f; KAB).

Dosages (Chickpea):

FNFF = !!!
Food farmacy at its best (JAD). Seeds, sprouts, young pods, young leaves, and the vinegar off the leaves are all eaten by humans. Great for the heart (say I) is hummus with plenty of garlic and olive oil; sesame and parsley optional but good. Let us hope the Tibetans are right, "Skinny people will gain weight, while fat people will lose weight when consuming sesame oil" (TIB). Juice of fresh leaves is used as hair tonic (NPM).

- Ayurevedics, deeming the seeds aphrodisiac, flatugenic, orexigenic, stimulant, and tonic, use them for bronchitis, dermatosis, fever, leprosy, ozoena, pharyngosis, thirst, and worms (KAB).
- Ayurvedics deem the tart leaves astringent, flatugenic, orexigenic, using for bronchitis and enterosis (KAB).
- Ayurvedics use the acid leaf exudate for constipation and dyspepsia (KAB).
- Belgians and Italians applied in cataplasm to testicular cancers (JLH).
- Deccanese with dysmenorrhea sit over a steaming decoction of the plant (KAB).
- French applied the plant to warts (JLH).

- North Africans use decoction (seed?) for itch, leposy, and smallpox (BOU).
- Unani, deeming the seeds anthelmintic, aphrodisiac, and tonic, use for fever, halitosis, hepatosis, inflammation, otitis, pharyngosis, pulmonosis, and splenosis (KAB).

Downsides (Chickpea):

The oxalic acid may be contraindicated in people with calculus (DEP). Boulos notes that inadequately cooked chickpeas can cause paralysis, like lathyrus peas can cause lathyrism (BOU).

Extracts (Chickpea):

Queiroz-Monici et al. (2005) found chickpea and pea better as bifidogenics than beans and lentils. Chickpea-fed groups consumed more food and showed better food conversion efficiency. Animals fed leguminous diets showed higher counts of Bifidobacterium, and lower *Enterobacter* and *Bacteroides* (X15850967). It also seems to be as "health-giving" and "heart-friendly" as soy without the high fat. The germinated seeds ("sprouts") contain the flavonoids, daidzein, formononetin, pratensin, liquiritgenin, isoliquiritigenin, and its 4′-glucoside, 4′,7-dihydroxyflavonol, garbanzol, biochanin-7-glucoside, and *p*-coumaric acid. Kaufman et al. (1997) reported 45 ppm genistein in chickpea seeds; cf. 25 for soy in comparing circa 75 legume accessions for isoflavone content (X9395689). In USA Patent No. 6599536, "Premenopausal Uses of Clover-Derived Isoflavones" claimed uses for the composition are for treatment or amelioration of premenopausal, benign disorders associated with an abnormally high activity of steroidal estrogen, for example, cyclical acne, endometriosis and endometrial hyperplasia, mastalgia, ovarian cysts, polycystic ovarian disease, and uterine fibroids. The patent cites specific "clovers (*Trifolium* spp.)" and "chick peas" as sources of isoflavones. Remember, my reader, that being cited in a patent does not necessarily make chickpea good for these ailments, but were I suffering any, I might increase my intake of hummus and/or chickpeas [USA Patent No. 6599536 (2003)].

ENDIVE (*CICHORIUM ENDIVIA* L.) +++ ASTERACEAE

Synonyms:

Cichorium divaricatum Schousb.; *Cichorium endivia* subsp. *divaricatum* (Schousb.) P.D. Sell; *Cichorium intybus* subsp. *pumilum* (Jacq.) Ball; *Cichorium pumilum* Jacq.

Notes (Endive):

> **The fourteenth day of the second month at even they shall keep it, and eat it with unleavened bread and bitter herbs.**
>
> **Numbers 9:11 (KJV)**

> **In the second month on the fourteenth day in the evening they shall keep it; they shall eat it with unleavened bread and bitter herbs.**
>
> **Numbers 9:11 (RSV)**

> **In the second month on the fourteenth day between the two evenings, they should prepare it. Together with unfermented cakes and bitter greens they should eat it.**
>
> **Numbers 9:11 (NWT)**

Some scholars regard endive as the bitter herb of Moses. Apparently Zohary does not (ZOH), more convinced that *C. pumilum* Jacq. was the bitter herb. Not to worry; that is just a variety or subspecies

FIGURE 1.23 Endive (*Cichorium endivia*).

of endive. The *Flora of Palestine*, not noted for including alien cultivars, lists neither endive nor chicory but does list the dwarf chicory (which see). Jane Philips (HJP) implies that the uses of the three pertinent taxa are interchangeable medicinally, at least in Lebanon. The chemistry and synonymy reported in *Hager's Handbook* (HH2) seem to support Philips's implications. I accept the HH2 and USN view, accepting *C. pumilum* and *C. endivia* as subspecies of the same species, *C. endivia.*

Zohary does comment that the dwarf chicory, like several of the nine Mediterranean species of *Cichorium*, is eaten by cattle and humans alike. A weedy progenitor is more likely to have been eaten by biblical peasants than the cultivated chicory or endive, which Zohary excludes from his account. However, neither Zohary or I now can guarantee that any or all species were or were not eaten in the biblical setting. I will wager that all that occurred there in those days served as bitter herbs. Zohary notes that chicory (and *Reichardia*) are only possible representatives of a large group of so-called edible bitter herbs in the Middle East, almost like our spring tonic greens down South. Not necessarily chicory, nor dandelion, nor endive, nor fenugreek were necessarily the most important of the biblical *merorim* (Arabic *mureir*), a general term to embrace many different species of bitter edible herbs.

Common Names (Endive):

Achicoria (Peru; Sp.; EGG); Andijvie (Dutch; EFS); Chicorée des Jardins (Fr.; EFS; HH2); Chicorée Frisée (Fr.; EFS); Chiccoria (Por.; EFS); Endive (Eng.; Fr.; CR2; EFS); Endive Chicory (Eng.; HJP); Endivia (It.; Sp.; EFS); Endivia Sallat (Swe.; EFS); Endivie (Ger.; Den.; EFS); Escarola (Peru; Sp.; EFS; EGG); Escarolla (Por.; HH2); Handabah (Arab.; Syria; HJP); Hindiba (Tur.; EFS); Garden Endive (Eng.; WOI); Induba (Arab.; EFS); Kashini (Hindi; WO2); Kasini (Hindi; WO2); Kasni Virai (Tam.; WO2); Kassin (Beng.; WO2); Koshae Vittulu (Tel.; WO2); Koshi (Tel.; WO2); Ku Chü (China; EFS); Memorim (Heb.; ZOH); Mureir (Arab.; ZOH); Radicchio (It.; Sp.; EGG); Scariola (It.; EFS); Shikuryah (Arab.; Syria; HJP); Winterendivie (Ger.; HH2).

Activities (Endive):

Allergenic (f1; FNF); Antibilious (f; BIB; DEP; EFS); Antioxidant (1; FAH; X12137499); Antiradicular (1; X12137499); Aphrodisiac (f; HJP); Bitter (f; BIB); Carminative (f; BIB; EFS); Choleretic (f; BIB); Decongestant (f; HJP); Emetic (f; HJP); Gout (f; HJP); Demulcent (f; BIB); Depurative (f; HJP); Gout (f; HJP); Digestive (f; HJP); Diuretic (f; BIB); Febrifuge (f; DEP; WO2); Laxative (f; BIB); Orexigenic (f; HJP); Refrigerant (f; BIB); Resolvent (f; BIB); Sedative (f; HJP); Stimulant (f; DEP; EFS); Tonic (f; BIB; HJP); Vermifuge (f; HJP); Vulnerary (f; HJP).

Indications (Endive):

Anorexia (f; HJP); Biliousness (f; WOI); Cancer (f; JLH); Cancer, liver (f; JLH); Cancer, spleen (f; JLH); Cancer, throat (f; JLH); Cancer, uterus (f; JLH); Catarrh (f; HJP); Cramp (f; HJP); Diarrhea (f; HJP); Dropsy (f; BIB; HJP); Dyspepsia (f; BIB); Enterosis (f; HJP); Fever (f; BIB; WO2); Gastrosis (f; HJP); Gout (f; HJP); Headache (f; BIB); Hemorrhoid (f; HJP); Hepatosis (f; BIB); Impotence (f; HJP); Induration (f; BIB); Inflammation (f; JLH); Jaundice (f; BIB); Pharyngosis (f; BIB); Pulmonosis (f; HJP); Splenosis (f; BIB); Swelling (f; JLH); Toothache (f; HJP); Uterosis (f; BIB); Wart (f; HJP); Water (f; BIB); Worm (f; HJP); Wound (f; HJP).

Dosages (Endive):

FNFF = !!!
Food farmacy (JAD); now widely available and about as increasingly boring as head lettuce; interpreted by some as one of the bitter herbs of the Bible. Leaves eaten raw in salads or as a potherb. Javans pickle the leaves. Seeds are used in sherbets. The Jews of Alexandria, who translated the

Pentateuch, would have known better than I which herbs were eaten with the Paschal Lamb. Fresh endive leaves smell bitter but full-grown, blanched leaves are pleasant. To blanch, tie the outer leaves in a cone, or place a big flower pot over each plant to curb the light. Blanching takes about 3 weeks (BIB; FAC; KAB; WO2).

- Lebanese used endive as choleretic and in the same way as their many uses for chicory (HJP).

EXTRACTS (ENDIVE):

Those activities and indications of chicory based on cichoric acid or inulin also might apply to the endive, if it has as much of the same compounds.

CHICORY (*CICHORIUM INTYBUS* L.) +++ ASTERACEAE

NOTES (CHICORY):

And they shall eat the flesh in that night, roast with fire, and unleavened breads; and with <u>bitter herbs</u> they shall eat it.

Exodus 12:8

The children of Israel may have learned to eat bitter herbs from the Egyptians. Ancient Egyptians used to place healthy green herbs on the table, mixed with mustard, and then dunk their healthy whole-grain bread in the mixture, like my dad and I used to dunk cornbread in turnip green broth. Moldenke and Moldenke believed that *Cichorium endivia, Cichorium intybus, Lactuca sativa, Nasturtium officinale, Rumex acetosella*, and *Taraxacum officinale* were among the green herbs of the Bible. By contrast, local Israeli botanist Michael Zohary lists none of these in his *Plants of the Bible* (ZOH), and only the watercress is listed as occurring in the *Flora of Palestine.* Zohary figures instead that chicory and the poppy-leaved *Reichardia* (which looks like dandelion) as more promising candidates. Regarding bitter herbs, Zohary says, "Many plants, especially those belonging to the Mustard and Daisy families, are frequently collected and used as potherbs and salad plants" (ZOH).

COMMON NAMES (CHICORY):

Achicoria (Peru; Sp.; EGG); Achicoria Amarga (Sp.; EFS); Achicoria Para Ensalada (Sp.; EFS); Almeirão (Por.; EFS; HH2); Almiron Amarga (Sp.; EFS); Ambuboia (Iran; NAD); Arhilon (Ber.; BOU); Barbe de Capucin (Fr.; BOU; EFS); Basarula (Arab.; NAD); Bitterste Cichory (Dutch; KAB); Blaue Wegwarte (Ger.; EFS); Blausamenwirbel (Ger.; KAB); Bois de Corde (Fr.; KAB); Brussel Witloof (Eng.; TAN); Bunk (Eng.; KAB); Camaroja (Cat.; KAB); Camarroya (Sp.; EFS); Cheveux de paysan (Fr.; BOU; KAB); Chicorea Brava (Por.; KAB); Chicoree (Fr.; BOU); Chicoree Amere (Fr.; BOU; KAB); Chicorée Sauvage (Fr.; EFS; KAB; TAN); Chicória (Por.; HH2); Chicória Brava (Por.; HH2); Chicoria do Café (Por.; EFS); Chicory (Eng.; CR2; HH2); Chicourey (Arab.; GMH); Cichorei (Dutch; EFS); Cicoare (Rom.; KAB); Cicorea (It.; KAB); Cicoria (It.; Malta; HH2; KAB); Cicoria Radicchio (It.; EFS); Cicoureio (Provence; KAB); Cicueira (Malta; KAB); Cikoria (Swe.; EFS); Cikorie (Den.; EFS); Citlik (Tur.; EB54:155); Djouldjoulan (Arab.; BOU); Ecoubette (Fr.; KAB); Gewöhliche Wegwarte (Ger.; HH2); Gul (Pun.; DEP; KAB); Hand (Pun.; KAB); Handhiposh (Kas.; MKK); Hendiban (Arab.; Saudi; GHA); Hendibeh (Arab.; GMH); Herbe a Café (Fr.; KAB); Herbe Amere (Fr.; KAB); Hindeg (Ger.; KAB); Hindiba (Arab.; DEP; GHA); Hinduba (Beng.; Bom.; Guj.; Hindi; NAD; WO2); Hindubar (Arab.; KAB); Indian Endive (Eng.; DEP); Indyba (Arab.; KAB); Inthybe (Fr.; KAB); Ju Qu (Pin.; Jule Salat (Den.; EFS); DAA); Kasani (Guj.; Iran; Urdu; DEP; KAB); Kashin (Loralai; KAB); Kashini (Tam.; KAB); Kashini

FIGURE 1.24 Chicory (*Cichorium intybus*).

Virai (Tam.; DEP); Kasini (Tel.; KAB); Kasini Virai (Tam.; NAD); Kasini Vittilu (Tel.; DEP); Kasni (Beng.; Iran; Hindi; Mal.; Pun.; DEP; EFS; KAB; WO2); Kichora (Greek; KAB); Kikori (Greek; KAB); Kikorion (Greek; KAB); Kiku Nigana (Japan; TAN); Ku Ku (China; NAD); Ku Tsai (China; NAD); Lishkan (Jammu; WO2); Memorim (Heb.; ZOH); Mersag (Ber.; BOU); Munnchatti Kizhangu (Mal.; WO2); Mureir (Arab.; ZOH); Peewortel-plant (Dutch; EFS); Podroznik (Pol.; KAB); Radicchio (It.; Malta; HH2; KAB); Radicheta (Sp.; EFS); Ragged sailor (Eng.; KAB); Rau Di'ep Dang (Vn.; TAN); Seris (Arab.; BOU); Seris Picris (Greek; KAB); Shikoriah (Arab.; NAD); Shikouria (Arab.; BOU); Succory (Eng.; ERS; HH2; TAN); Suchal (Pun.; KAB); Suikerij (Dutch; EFS); Tilfaf (Arab.; BOU); Timerzouga (Ber.; BOU); Timizagt (Ber.; BOU); Tsalina (Ber.; BOU); Tsikorie (Rus.; KAB); Tukhm-e-kasani (Iran; EFS); Verfluchte Jungfer (Ger.; KAB); Waegwarda (Swe.; KAB); Wegenwachter (Dutch; EFS); Weglunge (Ger.; KAB); Wegwarte (Ger.; KAB); Wegweisse (Ger.; KAB); Wild Bachelor's Button (Eng.; KAB); Wild Chicory (Eng.; BOU); Wild Endive (Ger.; KAB); Wild Succory (Eng.; EFS); Wilde Cichory (Dutch; KAB); Wilde Zichorie (Ger.; EFS; HH2); Witloof (Dutch; EFS); Xicoina (Cat.; KAB); Xicoira (Cat.; KAB); Yabani Hindiba (Tur.; EB54:155); Zichorie (Ger.; KAB; HH2); Zikorifa (Rus.; EFS); Zral (Bal.; KAB).

Activities (Chicory):

Alexiteric (f; KAB); Allergenic (f1; FAH; WO3); Alterative (f; NAD); Analgesic (f; KAB); Antibilious (f; WO2); Antiexudative (1; HH2; PHR; PH2); Antifeedant (1; WO3); Antiinflammatory (f1; APA; FAD; PNC; X15649409); Antilipogenic (f1; ORAFTI9); Antimalarial (f1; X15507374); Antimetastatic (1; X15877900); Antimutagenic (1; APA); Antioxidant (1; FAH; X12137499; X15800389); Antiradicular (1; X12137499); Antiseptic (1; FAD); Antispermatogenic (1; WO2); Antisteatotic (1; ORAFTI9); Bactericide (1; FAD; WO2; X15567253); Bifidogenic (1; FNF); Bitter (f1; FAH; HHB); Bradycardic (1; WO2); Cerebroprotective (1; X16218660); Cardiodepressant (1; FAD; PNC); Cardiotonic (1; FAD); Carminative (f; DEP; WO2); Cerebrotonic (f; WO2); Chemopreventive (1; X15877900); Cholagogue (f1; BOU; PHR; PH2; VVG; WO2); Choleretic (f12; ABA; BOU; KOM; PH2; VVG); COX-2 Inhibitor (1; X15649409); Decongestant (f; DEP); Demulcent (f; DEP; GHA; WBB); Depurative (f; BOU; EGG; HHB; VVG); Digestive (f1; APA; FAH; NAD; WBB; WO2); Diuretic (f1; BOU; FAD; FAH; PNC; VVG; WO2); Emmenagogue (f; DEP; KAB; WBB; WO2); Febrifuge (f; BOU; GHA); Hepatoprotective (1; APA; VVG; WO2); Hypocholesterolemic (1; FAH; PHR; PH2); Hypoglycemic (1; FAD); Hypotensive (1; VAD); Laxative (f1; APA; BOU; FAD; FAH; GHA; PNC); Lipolytic (f1; FAH); Negative Chronotropic (1; PH2); Negative Inotropic (1; PH2); Nervine (f; DEM; FAH); Neuroprotective (1; X16218660); Orexigenic (f1; APA; BOU; FAH; VAG; VVG); Peristaltic (1; WO2); Phytoalexin (1; WO3); Prebiotic (1; X15649409); Refrigerant (f; DEP); Sedative (f1; APA; FAD; GMH); Stomachic (f; HHB; WO2); Syndrome X (1; ORAFTI9); Tonic (f; APA; BOU; FAH; PNC; VVG); Xanthine-Oxidase Inhibitor (1; X12203269).

Indications (Chicory):

Acne (f; WO2); Adenopathy (f; JLH); Ague (f; GMH); Alzheimer's (1; COX; X15649409); Amenorrhea (f; DEP; KAB); Anorexia (f12; APA; BOU; FAD; FAH; KOM; PH2; VVG); Arrhythmia (1; APA); Arthrosis (f1; COX; KAB; X15649409); Asthma (f; WO2); Atony (f; BOU); Bacteria (1; FAD; X15567253; X15567253); Biliousness (f; HJP; KAB; WBB); Cancer (f1; APA; JLH); Cancer, breast (f1; JLH); Cancer, colon (f1; COX; X15649409; X15877900); Cancer, face (f1; JLH); Cancer, gum (f1; JLH); Cancer, liver (f1; JLH); Cancer, lung (f1; X7569446); Cancer, mouth (f1; JLH); Cancer, spleen (f1; JLH); Cancer, stomach (f1; JLH); Cancer, tongue (f1; JLH); Cancer, uterus (f1; JLH); Cardiopathy (1; APA; FAD); Catarrh (f; HHB; HJP; WBB); Chancre (f1; DEM); Cholecystosis (2; HHB; PHR); Conjunctivosis (f; GMH); Constipation (f1; FNF); Cramp (f; JHP; PH2); Cystosis (f; VAD); Deafness (f; PH2); Dermatosis (f; APA; FAD; PH2); Diabetes (1; FAD); Diarrhea (f; PH2; WO2); Dropsy (f; NAD); Dysmenorrhea (f; WBB; WO2); Dyspepsia (f12; APA; BOU; FAD; GHA;

KOM; PH2; VVG); Edema (f; VAD); Enterosis (f; PH2); Epilepsy (f; WO3); Fever (f; BOU; DEP; DEM; FAD; GHA; WO2); Gallstone (f; FAD; FAH); Gastrosis (f; HHB; JLH; WBB); Gingivosis (f; JLH); Glossosis (f; JLH); Gout (f1; PNC; WO2; X12203269); Gravel (f; GMH; NAD); Headache (f; GHA; PH2; WO2); Heartburn (f; GAZ); Hemorrhoid (f; HJP; PH2; WBB); Hepatosis (f12; DEP; FAD; FAH; JLH; PHR; PNC; VVG); High Blood Pressure (1; VAD); Hypercholesterolemia (1; FAH; PHR); Hyperglycemia (1; FAD); Hypertriglyceridemia (1; ORAFTI9); Induration (f; JLH); Infection (1; FAD); Inflammation (f1; APA; FAD; GMH; WO2; X15649409); Insomnia (f; GMH); Jaundice (f; FAD; GHA; GMH; VVG; WO2); Lachrymosis (f; JLH); Lumbago (f; KAB); Malaria (f1; X15507374); Melancholy (f; PH2); Nausea (f; DEP; WBB); Nephrosis (f; VAD; VVG); Obesity (f1; FAH; VAD); Oliguria (f; VAD); Ophthalmia (f; DEM); Pain (f; KAB); Pharyngosis (f; WO2); Pseudomonas (1; X15567253); Pulmonosis (f; GMH); Pyelonephrosis (f; VAD); Respirosis (f; HHB); Rash (f; PH2); Rheumatism (f; GMH; PNC; WO2); Sclerosis (f; JLH); Sore (f; DEM); Sore Throat (f; PH2; WO2); Splenomegaly (f; NAD; WO2); Splenosis (f; JLH); Stone (f; VAD); Swelling (f1; APA; GMH); Tachycardia (1; APA); Toothache (f; HJP); Tuberculosis (f; GMH; PH2); Urethrosis (f; VAD); Urolithiasis (f; VAD); Uterosis (f; JLH); Vomiting (f; PH2; WO2); Wart (f; JLH); Worm (f; HJP); Wound (f; HJP).

Dosages (Chicory):

FNFF = !!!
Long eaten, for example, by the Romans as vegetable or salad, as mentioned by Horace, Ovid, Pliny, and Virgil (GMH). Leaves grown as vegetable; roots eaten as vegetable or roasted as coffee substitute; flowers eaten fresh, pickled, or in broth; Turks prepare a chewing gum (*dagsakizi*) from the latex (EGG; FAC; TAN). 3 g root (KOM); 3–5g powdered root (PHR); 2–4 g /150–250 ml water PH2.; 30–0 drops fluid extract, 1–3 ×/day, before or after meals (VAD); 50–100 drops 1:1 tincture 1–3 ×/day (VAD).

- Arabians eat the leaves or take decoction to lower fever (GHA).
- Arabians eat the seeds for headache, or take decoction for jaundice (GHA).
- Asian Indians use for acne, diarrhea, fever, ophthalmia, pharyngitis, splenomegaly, and vomiting (WO2).
- Egyptians value the root for tachycardia (BIB).
- Iranians take as a refrigerant, resolvent, medicines, for biliousness (HJP).
- Lebanese, considering the plant aphrodisiac, calmative, choleretic, depurative, laxative, and tonic, take it for dyspepsia and jaundice (HJP).
- Peruvians suggest eating the leaves or taking the plant juice to purify the blood (EGG).
- Ukranians take the plant for catarrh, diarrhea, enterosis, gastrosis, hemorrhoids, pulmonosis, and worms (HJP).
- Unani use the wild bitter type for asthma, biliousness, and inflammation (BIB).

Downsides (Chicory):

Class 1 (AHP). No health hazards or side effects known with proper therapeutic dosages (PH2). Commission E reports contraindications of hypersensitivity to chicory and other Asteraceae and adverse effects of rare allergic skin reactions. Patients with bile-stones or gallstones should first consult a physician (AEH; KOM). Cadot et al. (1996) report a first case of occupational allergy to chicory (*Cichorium intybus*) in a vegetable wholesaler. Symptoms occurred after oral, cutaneous, or inhalatory exposure. The patient also reported reactions after ingestion of botanically related endive (*Cichorium endivia*) and lettuce (*Lactuca sativa*). They identified the responsible allergen by SDS-PAGE and immunoblot to be a 48-kDa protein, confined to the nonilluminated parts of the plants. No cross-reactivity was found with mugwort (*Artemisia vulgaris*), ryegrass (*Lolium perenne*), or birch (*Betula verrucosa*) pollen, which suggests that the vegetable is the primary allergenic material. Carcinogens, viz. 1,2-benzoperylene; 3,4,benzopyrone; and floranthene occur in chicory, increasing on roasting, especially above 175 F° (WOI). Ingestion of 100 g may cause slight bradycardia.

Extracts (Chicory):

Industrial source of fructose, inulin, oligofructose, and the sugar enhancer, maltol (APA; JAD). Cichoric acid was recently heralded as very important in immunomodulation and antiviral activity, even anti-HIV. The compound lactucin and to a lesser extent lactucopicrin are CNS-sedative and can antagonize the stimulation of caffeine (BIB). Lactucin and lactucopicrin proved to be antimalarial compounds in chicory, folklorically regarded for malaria in Afghanistan (X15507374). The guaianolide 8-deoxylactucin is a key inhibitor of COX-2 expression (X15649409). Italians are producing circa 250,000 tons of red chicories a year, 100-g servings of which can provide as much as 130 mg anthocyanins and 650 mg total phenolics, scavenging highly reactive oxidants in the stomach, benefitting age-associated oxidative stress, and improving neuronal and cognitive brain function (X16218660).

CASSIA (*CINNAMOMUM AROMATICUM* NEES) +++ LAURACEAE

Synonyms:

Cinnamomum cassia Auct.; *Cinnamomum obtusifolium* var. *cassia* Perr. & Eb.; *Laurus cassia* Nees.; *Laurus cinnamomum* Andr. *fide* HH2

Notes (Cassia):

> **Take thou also unto thee principal spices, of pure myrrh five hundred shekels, and of sweet cinnamon half so much, even two hundred and fifty shekels, and of sweet calamus two hundred and fifty shekels.**
>
> **Exodus 30:23 (KJV)**

> **And of cassia five hundred shekels, after the shekel of the sanctuary, and of oil olive an hin.**
>
> **Exodus 30:24 (KJV)**

> **and of cassia five hundred, according to the shekel of the sanctuary, and of olive oil a hin.**
>
> **Exodus 30:24 (RSV)**

> **and cassia five hundred units, by the shekel of the holy place, and olive oil a hin.**
>
> **Exodus 30:24 (NWT)**

It is nice to see that all versions cite the cassia, and most scholars agree that it is the cassia of today's spice trade. Although the spice trade tends to lump cassia and cinnamon in the same spice jar, Israeli botanist Michael Zohary and the Bible itself treat them as separate items. Although both are mentioned many times in the Bible, nowhere are they both mentioned in the same verse; but it is very close in the case quoted above, where cinnamon was mentioned in verse 23 and cassia in verse 24. Of the cassia, citing Exodus 30:24, Job 42:14, and Psalms 45:7–8, Zohary says, "In the quoted passages, the Hebrew *ketziah* and *kiddah* are translated as 'cassia'; the former is also applied as a personal name. The question as to whether *ketziah* and *kiddah* are synonymous or refer to different plants or drugs will never be resolved" (ZOH). Moses and Solomon probably obtained cassia, via trade, from Sri Lanka, where it is only cultivated or more possibly from China, where it is native (BIB; USN).

FIGURE 1.25 Cassia (*Cinnamomum aromaticum*).

Common Names (Cassia):

Bastard Cinnamon (Eng.; EFS); Bunga Lawang (Java; IHB); Canéficier (Fr.; EFS); Canela de China (Cuba; RyM); Canela de la China (Sp.; USN); Canelero Chino (Sp.; USN); Canella de Coromandel (It.; HH2); Canella di China (It.; KAB); Canelle de Chine (Fr.; EFS); Canelle de Cochinchine (Fr.; USN); Cannelier Casse (Fr.; USN); Casia (Sp.; EFS); Casse Ligneux (Fr.; EFS); Cassia (Eng.; It.;

Scn.; AH2; CR2; EFS); Cássia (Por.; USN); Cassia Lignea (It.; HH2); Cassis (Fr.; EFS); Chinazimpt (Ger.; USN); Chinese Cassia (Eng.; BIB); Chinese Cassiaboom (Dutch; EFS); Chinese Cinnamon (Eng.; Ocn.; AH2; BIB); Chinese Kaneelboom (Dutch; EFS); Chinesischer Kassia (Ger.; EFS); Chinesischer Zimptbaum (Ger.; EFS; HH2); Chinesischer Zimtstrauch (Ger.; HH2); Ch'ün Kuei (China; EFS; KAB); Çin Tarçini (Tur.; EFS); Dalchini (India; EFS); Darasini (Arab.; EGS); Fahej (Hun.; KAB); Gudatvak (Sanskrit; EFS); Gui Zhi (Pin.; AH2); Holzkassia (Ger.; EFS); Holzzimpt (Ger.; EFS); Hout Cassia (Dutch; EFS); Kanel Kassia (Den.; EFS); Kashia Keihi (Japan; USN); Kassienzimpt (Ger.; EFS); Kayu Manis China (Malaya; EFS; IHB); Ketziah (Heb.; ZOH); Kiddah (Heb.; ZOH); Kitaiskaya Koritsa (Rus.; KAB); Koritsa (Rus.; KAB); Koui Chou (China; KAB); Kuei (China; EFS); Kulit Manis (Malaya; EFS); Kwa P'i (China; EFS); Kwai Phee (China; EFS); Laurier Casse (Fr.; EFS); Lauro Cassia (It.; EFS); Malabar Leaf (Eng.; JLH); Malabathron (Greek; JLH); Mou Kuei (China; EFS); Rou Gui (Pin.; AH2; DAA); Saila Myah (Iran; EFS); Salikha (Arab.; KAB); Taj (Iran; Urdu; KAB); Zimtbaum (Ger.; HH2); Zimtkassie (Ger.; EFS; HH2); Zimtstrauch (Ger.; HH2).

Activities (Cassia):

Aldose-Reductase Inhibitor (1; X12553890); Analgesic (f; WO2); Anesthetic (f1; DAA; WO2); Antiaging (f; DAA); Antiaggregant (1; CAN); Antiallergic (1; WO2); Anticomplement (1; CAN); Antidiarrheic (f1; CAN); Antidote (f; WO2); Antiemetic (f1; CAN); Antifertility (f1; DAA); Anti-inflammatory (f1; X15710356); Antimutagenic (1; X11506812); Antioxidant (1; X12916067); Antiproteinuric (1; WO2); Antipyretic (f; WO2); Antiseptic (f1; CAN; DAA; WO2); Antispasmodic (f1; CAN); Antitumor (f1; CAN); Antiulcer (1; BGB; CAN; PH2; WO2); Antiviral (1; BGB; DAA; LAF); Apoptotic (1; X14587878); Astringent (f1; AHP; WO2); Bactericide (1; BGB; LAF; PH2; X12423924); Cardiotonic (f1; DAA; WO2); Carminative (f1; BGB; CAN; DAA; WO2); COX-2 Inhibitor (1; X12413723); Diaphoretic (f; AHP); Digestive (f; BGB); Diuretic (f; WO2); Expectorant (f; WO2); Febrifuge (f1; DAA); Fungicide (1; BGB; HH2; LAF; PH2); Hepatotonic (f; WO2); Hypotensive (1; DAA; WO2); Immunostimulant (1; PH2); Insectifuge (1; X15264623); Larvicide (1; BGB; LAF; X15796573); Metalloproteinase-9 Inhibitor (1; X15652283); Mosquitofuge (1; X15264623); NF-kappa B Inhibitor (1; X15710356); NO Inhibitor (1; X15710356); iNOS Inhibitor (1; X12475291); Parasiticide (1; X12847923); Purgative (f; WO2); Sedative (f1; DAA; WO2); Stomachic (f; AHP; BGB); Tranquilizer (f; WO2); Vasodilator (1; KC2); Xanthine-Oxidase Inhibitor (1; X11025157).

Indications (Cassia):

Allergy (1; WO2); Alzheimer's (1; COX; X12413723); Amenorrhea (f1; DAA; PH2; WO2); Anesthetic (f1; WO2); Anorexia (12; BGB; KOM; PH2); Arthrosis (f1; COX; DAA; X12413723); Ascites (f; WO2); Asthenia (f; BGB); Asthma (1; BGB; WO2); Bacillus (1; X12423924); Bacteria (1; X12423924); Bloating (2; BGB; KOM); Bronchosis (1; BGB); Cancer (f1; JLH; X15652283; X12860272); Cancer, bladder (f1; JLH; X15652283); Cancer, colon (f1; COX; JLH; X12413723); Cancer, diaphragm (f1; JLH; X15652283); Cancer, kidney (f1; JLH; X15652283); Cancer, liver (f1; JLH; X15652283); Cancer, rectum (f1; JLH; X15652283); Cancer, spleen (f1; JLH; X15652283); Cancer, stomach (f1; JLH; X15652283); Cancer, vagina (f1; JLH; X15652283); Cancer, uterus (f1; JLH; X15652283); Chills (f; DAA); Circulosis (f; X15796573); Cold (f; BGB; CAN); Colic (f1; BGB; CAN; DAA; PH2); Condyloma (f; JLH); Cough (f; BGB; DAA); Cramps (f1; BGB); Cystosis (f; JLH); Diabetes (f; DAA); Diaphragmosis (f; JLH); Diarrhea (f1; BGB; CAN; DAA; PH2); Dysmenorrhea (f; DAA); Dyspepsia (f12; BGB; CAN; KOM; PH2); Dysuria (f; DAA; WO2); Edema (f; WO2); Enteralgia (f; BGB); Enterosis (f; BGB; PH2; WO2); Enuresis (f; PH2); Epilepsy (f; WO2); Escherichia (1; X12423924); Exhaustion (f; PH2); Fever (f1; BGB; DAA; WO2; X15796573); Fungus (1; HH2); Gas (f1; BGB; PH2); Gastrosis (f; BGB; DAA; PH2; WO2); Goiter (f; DAA); Gout

(1; X11025157); Gray Hair (f; WO2); Hepatosis (f; JLH); Hernia (f; PH2); Impotence (f; PH2); Induration (f; JLH); Infection (f1; HH2; X15796573); Inflammation (f1; X15710356; X15796573); Insomnia (f; DAA); Jaundice (f; DAA); Listeria (1; X12423924); Lumbago (f; DAA); Menopause (f; PH2); Mycosis (1; HH2); Nephrosis (1; BGB; WO2); Neuralgia (f1; WO2); Neurasthenia (f; PH2); Ophthalmia (1; WO2); Orchosis (f; PH2); Pain (f1; WO2); Parasite (1; X12847923); Pharyngosis (f; WO2); Salmonella (1; X12423924); Sore (f; JLH); Splenosis (f; JLH); Stomachache (f; DAA); Staphylococcus (1; X12423924); Streptococcus (1; HH2); Tracheosis (1; WO2); Trypanosoma (1; X15567249); Urethrosis (f; WO2); Uterosis (f; WO2); Vaginosis (f; JLH); Vertigo (f; DAA); Vomiting (f; PH2); Wart (f; JLH).

Dosages (Cassia):

FNFF = !!!
Dried green fruits are the cassia buds of commerce, which resemble cloves. Cassia bark is also an important spice. All parts of the plant possess an essence, cinnamic aldehyde, which may be distilled for export. Buds of the tree are used in place of cloves to season dishes (BIB; FAC; TAN). 2–4 g ground bark/day (BGB; PH2); 0.7–1.3 g bark in 150 ml water 3 ×/day (BGB); 0.5–1 g bark, as tea, 3 ×/day (CAN); 0.05–0.2 ml cassia oil 3 ×/day (CAN); 0.3–1.2 ml flower tincture (1:5 in 90% ethanol) 3 ×/day (CAN).

- Chinese suggest that prolonged use improves the complexion, making it more youthful (DAA).
- Chinese use the plant for amenorrhea, arthritis, cancer, chills, cold, colic, cough, diabetes, diarrhea, dizziness, dysmenorrhea, dyspepsia, dysuria, fever, goiter, headache, jaundice, lumbago, rheumatism, and stomachache (DAA).
- Egyptians use the leaves for cancer of the womb, the "grains" for condylomata, vaginosis, and warts (JLH).
- Indonesians use the plant for tumors (JLH).
- Iranians use bark tea for excessive salivation (BIB).
- Javan brides must drink a potion containing two Bunga Lawang (cassia buds) (IHB).
- Malayans use the imported bark in decoction with other herbs for chest complaints and cough (IHB).
- Unani, considering the bark carminative, emmenagogue, hematotonic, and tonic, use it for headache, inflammation, piles, and pregnancy (KAB).

Downsides (Cassia):

Class 2b, reportedly abortifacient (AHP, 1997). Newall, Anderson, and Phillipson (1996) caution that the cinnamaldehyde in the volatile oil is allergenic and an irritant (CAN). May interfere with absorption of tetracycline (AHP, 1997). No health hazards or side effects known with proper therapeutic dosages (PH2). Prolonged use of the essential oil should be restricted during pregnancy (AHP, 1997). Commission E reports contraindications for bark; hypersensitivity to cinnamon or Peruvian balsam; pregnancy, and adverse effects often allergic reactions of skin and mucosae. Flower not permitted for therapeutic use. Contraindications for hypersensitivity to cinnamon or Peruvian balsam, pregnancy, and adverse effects allergic skin reactions and mucosal reactions (AEH).

Extracts (Cassia):

He et al. (2005) note that cinnamaldehyde (83% or bark essential oil, 65% or twig essential oil) has antifungal, antioxidant, antipyretic, antiseptic, cytotoxic, and larvicidal activities, inhibiting the production of lymphocytes and modulating T-cell differentiation. In TCM, cassia is used for circulatory disorders, dyspepsia, gastritis, and inflammation (X15796573). EO LD50 = 320 mg/kg dermal (CAN) (should not be used on skin at levels >0.2%); Aqueous extracts of cassia deemed as effective as cimetidine in preventing ulcers (BGB; WO2). *Trans*-cinnamaldehyde (IC50 = 3 μg/ml) and weakly cinnamyl alcohol, *trans*-cinnamic acid, and eugenol inhibited aldose reductase (but quercitrin was 6 times more potent than cinnamaldehyde) (X12553890). *Cinnamomum cassia* inhibited epimastigote forms of *Trypanosoma cruzi*, (IC50 = 3.9 μg/ml) (X15567249). Butanol extracts inhibit metalloproteinase-9 (IC > 90 = 100 μg/ml) (X15652283); LD50 (cinnamaldehyde) = 2200–3350 mg/kg orl rat HH2; LD50 (cinnamaldehyde) = 200 mg/kg ipr mus HH2; LD50 (cinnamaldehyde) = 132 mg/kg ivn mus HH2; LD50 (cinnamaldehyde) = 2225 mg/kg orl mus HH2; LD50 (EO) = 5200 mg/kg orl rat HH2.

CINNAMON (*CINNAMOMUM VERUM* J. PRESL) +++ LAURACEAE

SYNONYMS:

Cinnamomum zeylanicum Blume; *Laurus cinnamomum* L.

NOTES (CINNAMON):

> **And the merchants of the earth shall weep and mourn over her; for no man buyeth their … cinnamon, and odours, and ointments.**
>
> **Revelation 18 (KJV)**

And on a more romantic tone:

> **I have perfumed my bed with myrrh, aloes, and cinnamon. Come, let us take our fill of love until the morning: let us solace ourselves with loves. For the goodman is not at home, he is gone a long journey.**
>
> **Proverbs 7:17–19 (KJV)**

> **I have perfumed my bed with myrrh, aloes, and cinnamon. Come, let us take our fill of love till morning; let us delight ourselves with love. For my husband is not at home; he has gone on a long journey.**
>
> **Proverbs 7:17–19 (RSV)**

> **I have besprinkled my bed with myrrh, aloes, and cinnamon. Do come, let us drink our fill of love until the morning; do let us enjoy each other with love expressions. For my husband is not in his house; he has gone traveling on a way of some distance.**
>
> **Proverbs 7:17–19 (NWT)**

There is near unanimity in the version regarding the intended use of these aromatic spices. And they are consistent in translating them as myrrh, aloes (in this case *Aquilaria*), and cinnamon, the latter more appealing to my olfactories. Zohary confirms that the long-discussed identification of the biblical *kinnamon* as *Cinnamomum* has been confirmed by various scholars. Alien to the Holy Land, and native to Sri Lanka and coastal India, it must have followed the old trade routes for drugs, incenses, perfumes, and spices.

COMMON NAMES (CINNAMON):

Like spice dealers, I once aggregated cinnamon and cassia, so many of the common names, activities, and indications below could well apply to either species. I would not hesitate to use one or the other for the indication of one or the other. Many of the reported studies were, in fact, performed on purchased materials that may have been one or the other or a mixture of the two. Who really knows which they studied if the material they studied had already been reduced to powdered bark (JAD).

FIGURE 1.26 Cinnamon (*Cinnamomum verum*).

Aakerpatri (Tel.; WO2); Arbol de la Canela (Sp.; KAB); Bahugandha (Sanskrit; KAB); Bois de Cannelle (Fr.; KAB); Bojevar (Bom.; KAB); Canalleira da India (Por.; AVP); Canela (Cr.; Sp.; AVP; USN); Canela de Ceilán (Cuba; Dr.; Sp.; AVP); Canela de Ceylán (Sp.; AVP); Canela de Ceylán (Cuba; RyM); Canela de India (Por.; AVP); Canaleiro (Por.; USN); Canelero de Ceilán (Sp.; USN); Canelo (Sp.; AVP); Cannalavangapattai (Tam.; KAB; WOI); Cannelier (Guad.; Seychelles; AVP; KAB); Cannelier de Ceylan (Fr.; USN); Cannella (It.; EFS); Cannella di Ceylon (It.; KAB); Cannelle (Fwi.; Haiti; AVP); Cannelle de Ceylan (Fr.; EFS); Cannelle Aromatique (Guad.; AVP); Ceylonzimt (Ger.; USN); Ceylonzimtbaum (Ger.; USN); Ceylon Cinnamon (Eng.; Ocn.; AH2; EFS); Ceylon Kaneel (Dutch; EFS); Ceylon Kanel (Den.; Swe.; EFS; KAB); Ceylon Cimtbaum (Ger.; EFS); Ceylonzimt (Ger.; KAB); Cheriyilaivannam (Mal.; DEP); Chinese Cassia (Eng.; MPI); Chitari (Peru; Shipibo/ Conibo; EGG); Cinnamon (Eng.; Scn.; AH2; CR2; VOD); Cirfah (Arab.; Syria; HJP); Cynamon (Pol.; HH2); Daichini (Guj.; DEP); Dalachini (Kan.; DEP); Dalchini (Beng.; Dec.; Hindi; Mar.; Nepal; DEP; SUW); Dalochini (Oriya; WO2); Darachini (Mar.; KAB); Darchini (Urdu; KAB); Darasini (Arab.; NAD); Darchini (Pun.; DEP); Darchinisailaniyah (Iran; KAB); Darsini (Arab.; Syria; DEP; HJP);

Daruchini (Beng.; Guj.; Hindi; Kan.; Kas.; Mah.; NAD); Darushila (Sanskrit; SKJ); Dasamchakkaluk (Tel.; MPI); Echter Ceylonzimt (Ger.; USN); Gudatvak (Sanskrit; DEP); Gudatwoko (Oriya; KAB); Hmanthin (Burma; KAB); Ijin (Tulu; KAB); Ilayanngam (Tam.; WO2); Kalphah (Bom.; NAD); Kanèl (Creole; Haiti; VOD); Kanèl dès Indes (Creole; Haiti; VOD); Karitsa (Rus.; KAB); Karruwa (Tam.; DEP); Kaya Manis (Malaya; EFS); Kevei (China; NAD); Kinnamomum (Greek; NAD); Kinnamon (Bib.; ZOH); Kirfa (Pun.; KAB; NAD); Korica (Rus.; HH2); Kukhi taj (Nepal; SUW); Kulit Manis (Malaya; NAD); Kurundo (Sin.; NAD); Kuruva (Sri.; KAB); Lavanga (Kadir; KAB); Lavanga Patta (Mal.; DEP); Lavangachakke (Kan.; KAB); Lowangapatta (Mal.; Tam.; Tel.; NAD); Lulingyaw (Burma; DEP); Malabar Leaf (Egypt; JLH); Qalamidarchini (Beng.; Dec.; Hindi; KAB); Qirfahesailaniyah (Arab.; KAN); Qualami (Dec.; NAD); Rassu Kurundu (Singh.; DEP; KAB); Saila Myah (Iran; NAD); Salikhah (Arab.; Syria; HJP); Sanalinga (Tel.; DEP); Sannalavanga (Tel.; KAB); Scortisoare (Rom.; KAB); Seylan Tarçina (Tur.; EFS); Skoricovnik Ceylonsky (Che.; HH2); Taj (Bom.; Guj.; DEP); Tali Khahe (Iran; DEP); Tamalapatra (Sanskrit; WOI); Tarcin (Tur.; KAB); Timbootikyoobo (Burma; NAD); Tiqui (Kon.; KAB); True Cinnamon (Eng.; Ocn.; AH2; VOD); Tvach (India; JLH); Tvak (Ayu.; AH2); Twak (Sanskrit; MPI); Varangam (Sanskrit; NAD); Vazhana (Mal.; WO2); Zimmt (Ger.; NAD); Zimtbaum (Ger.; USN); Zimtlorbeerbaum (Ger.; EFS).

Activities (Cinnamon):

Adipogenic (1; X15468836); Allergenic (1; APA; X15186386); Analgesic (f1; APA; CAN; ZUL); Anesthetic (f1; APA; CAN; TRA; ZUL); Anthelmintic (f; LIB); Antiaggregant (1; X10632089); Anticonvulsant (f1; LIB; TRA); Antidiarrheal (f1; CAN; TRA); Antiemetic (f1; HOS; WO2); Antifungal (f; CRC); Antihistamine (1; WHO); Antiinflammatory (f1; HOS; LIB); Antileukemic (1; TRA; WO2); Antilymphomic (1; WO2); Antimycobacterial (1; PR14:303); Antioxidant (1; CRC; X10641152; X14585184); Antipeptic (f1; HOS); Antiplatelet (1; X10632089); Antiprostaglandin (1; HH2; TRA); Antipyretic (f1; CRC; TRA); Antiseptic (f1; APA; CAN; TRA; WO2; ZUL); Antisialogogue (f; HOS); Antispasmodic (f1; CAN; MPI; TRA; WHO); Antitubercular (1; PR14:303); Antitussive (2; HOS); Antiulcer (f1; APA; WHO); Antiviral (1; APA; BGB; CAN; TRA); Aphrodisiac (f; LIB; KAB); Aromatic (f; CRC); Astringent (f1; CAN; TRA; ZUL); Bactericide (12; APA; KOM; PH2; TRA; WHO); Candidicide (1; APA; CAN; WO2; JAR12:83); Carminative (f1; APA; CAN; TRA; VOD; WHO); Choleretic (1; APA); Circulostimulant (1; HOS); Cordial (f; CRC); COX-2 Inhibitor (1; HOS; X12444669); Cyclooxygenase Inhibitor (f; ZUL); Cytotoxic (1; CAN; TRA); Demulcent (f1; HOS); Depurative (f; APA); Diaphoretic (f; AHP; LIB); Digestive (f; VOD); Emmenagogue (f; LIB); Emollient (f; JLH); Estrogenic (1; PHR; PH2; TRA); Expectorant (f1; WO2); Febrifuge (f; VOD); Fungicide (1; KOM; PH2; TRA; WHO; JAR12:83); Germicide (f; CRC); Gram(+)-icide (1; WO2); Gram(–)-icide (1; WO2); Hemostat (f; KAB; MPI); Hepatotonic (f; KAB); HMG-CoA Reductase Inhibitor (1; X4585184); Hypocholesterolemic (12; X14633804; X14585184); Hypoglycemic (12; X14633804); Hypotensive (1; ZUL); Hypothermic (f; CRC); Hypotriglyceridemic (12; X14633804); Hypouricemic (1; X11025157); Immunostimulant (1; HOS); Insecticide (1; PHR; PH2; WO2); Insectifuge (1; TRA); Insulin Potentiator (1; JAF52:65); Lactagogue (f; LIB); Larvicide (1; APA; BGB; CAN; TRA); Lipolytic (1; APA; BGB; LAF; TRA; WO2); Lipoxygenase Inhibitor (f; HOS; ZUL); Mutagenic (1; TRA; WHO; WO2); Myorelaxant (1; TRA; WHO); Narcotic (f; NAD); Nematicide (1; TRA); Nervine (f; NAD); Neurotonic (f; MPI); Orexigenic (f; CAN; LIB); Refrigerant (f; CAN); Secretogogue (1; HOS); Sedative (f1; CRC; LIB; TRA); Sialogogue (1; APA); Spasmolytic (1; ZUL); Stimulant (f; CRC; WO2); Stomachic (f; AHP; MPI); Teratogenic (1; WHO); Tonic (f; IED); Tranquilizer (1; HOS); Uterorelaxant (f; APA); Uterotonic (f; APA); Vibriocide (1; WO2); Vulnerary (f1; X13680838); Xanthine-Oxidase Inhibitor (1; X11025157).

Indications (Cinnamon):

Adenopathy (1; HOS); Alzheimer's (1; COX; FNF; HOS); Amenorrhea (f1; CRC; KAB; WHO; WO2); Amnesia (f; ZUL); Anorexia (f12; CAN; GAZ; KOM; PH2; WHO); Aphonia (f; HOS); Arthrosis (f1; COX; CRC; HOS); Asthenia (1; BGB); Asthma (f; CRC; LIB); Bacillus (1; X10548758); Bacteria (1;

WO2); Biliousness (f; KAB); Bleeding (f; KAB); Bloating (f1; BGB); Bronchosis (f12; CRC; KAB; PHR); Cancer (f1; COX; CRC; HOS); Cancer, abdomen (f1; COX; HOS; JLH); Cancer, bladder (f1; COX; HOS; JLH); Cancer, breast (f1; COX; HOS; JLH); Cancer, colon (f1; COX; HOS; JLH); Cancer, diaphragm (f1; COX; HOS; JLH); Cancer, ear (f1; COX; HOS; JLH); Cancer, gum (f1; COX; HOS; JLH); Cancer, kidney (f1; COX; HOS; JLH); Cancer, liver (f1; COX; HOS; JLH); Cancer, mouth (f1; COX; HOS; JLH); Cancer, neck (f1; COX; HOS; JLH); Cancer, rectum (f1; COX; HOS; JLH); Cancer, sinus (f1; COX; HOS; JLH); Cancer, spleen (f1; COX; HOS; JLH); Cancer, stomach (f1; COX; HOS; JLH); Cancer, vagina (f1; COX; HOS; JLH); Cancer, uterus (f1; COX; HOS; JLH); Candida (f1; CRC; LIB; JAR12:83); Cardiopathy (f1; EGG; KAB; LIB; X14633804); Cerebrosis (f; KAB); Childbirth (f; LIB); Chill (f; PHR; PH2); Cholera (f1; CRC; SKJ; WO2); Cold (f12; CAN; GAZ; PHR; ZUL); Colic (f1; APA; CAN; EGG; TRA); Condylomata (f; JLH); Conjunctivosis (f; WHO); Convulsion (f; LIB); Cough (2; CRC; PHR); Cramp (f1; APA; DEP; VOD; ZUL); Dandruff (1; JAR12:83); Debility (f; LIB); Depression (f; LIB); Dermatosis (1; JAR12:83); Diabetes (f12; TGP; X14633804; JAF52:65); Diarrhea (f1; DEP; EGG; PHR; TRA; WHO); Dropsy (f; NAD); Dysentery (f; CRC; DEP; WO2); Dysmenorrhea (f1; APA; DEP; WHO); Dyspepsia (f12; CAN; IED; KOM; PH2; WHO); Dyspnea (f; WHO); Earache (f; LIB); Edema (f1; HOS); Enteralgia (f1; WHO); Enterosis (f; JLH; VOD); Enterospasm (2; KOM; WHO); Epidermophyton (1; JAR12:83); Escherichia (1; CRC; X10548758); Exhaustion (f; LIB); Fatigue (f; GAZ); Fever (f12; AHP; PHR; TRA; VOD); Fistula (f; CRC; SKJ); Flatulence (f12; KOM; VOD; WHO); Flu (f; PHR; PH2); Frigidity (f; LIB; WHO); Fungus (1; GAZ; LIB; X10548758); Gas (f1; APA; DEP; TRA; VOD); Gastrosis (f; DEP; HOS; VOD; WO2); Gastrospasm (f12; KOM; VOD); Gingivosis (f; JLH); Glossosis (f; DEP; HOS; WO2); Gonorrhea (f; LIB; NAD); Gout (1; X11025157); Halitosis (f; PH2); Headache (f1; DEP; WO2; ZUL); Heart (f; CRC); Hemorrhage (f1; APA); Hemorrhoid (f; KAB); Hepatosis (f; JLH; NAD); Hiccup (f; KAB); High Blood Pressure (f; LIB; ZUL); High Cholesterol (12; X14633804); Hydrocele (f; KAB); Hyperglycemia (f12; X4585184); Hypertriglyceridemia (f12; X4585184); Immunodepression (1; HOS); Impotence (f; LIB; WHO); Infection (2; PHR; WO2); Inflammation (f1; HH2; HOS; LIB); Itch (f; KAB); Leukemia (1; TRA; WO2); Leukorrhea (f; WHO); Listeria (1; X12380758); Lumbago (f; CRC); Lungs (f; CRC); Lupus (f; LIB); Lymphoma (1; WO2); Malassezia (1; JAR12:83); Mastosis (f; JLH); Melancholy (f; NAD); Menorrhagia (f; CRC; LIB); Mycosis (1; ZUL; JAR12:83); Nausea (f; CRC; EGG; TRA; ZUL); Nephrosis (f; CRC; LIB); Neuralgia (f; DEP; WHO; WO2); Obesity (12; X4585184); Oketsu Syndrome (f; LIB); Otosis (f; LIB); Pain (f1; KAB; WHO; WO2); Paralysis (f; DEP; HOS; WO2); Pharyngosis (2; PHR); Phthisis (f; CRC); Phymata (f; JLH); Proctosis (f; JLH; KAB); Prolapse (f; CRC; SKJ); Pseudomonas (1; HH2); Psoriasis (f; CRC); Rheumatism (f; APA; WHO; WO2; ZUL); Salmonella (1; WO2); Sinusosis (f; JLH); Sore (f; JLH); Spasm (f; CRC); Splenosis (f; JLH); Staphylococcus (1; CRC; HH2); Stomachache (f; EGG); Stomatosis (2; CRC; JLH; PHR); Stress (f; LIB); Syncope (f; WO2); Tension (f; LIB); Thirst (f; SKJ); Thrush (f1; LIB); Toothache (f; DEP; PH2; WHO); Tuberculosis (1; LIB; PR14:303); Tumor (f; CRC; JLH); Typhoid (f; LIB; NAD); Ulcer (f1; HOS; WHO); Vaginosis (f; CRC; JLH; WHO); Venereal Disease (f; LIB); Virus (f; LIB); Vomiting (f; CRC; PH2); Wart (f; CRC; JLH); Wen (f; JLH); Worm (f; PHR; PH2); Wound (f1; PHR; PH2; WHO; X13680838); Xerostomia (f; KAB); Yeast (f1; APA; WO2; X10548758; JAR12:83).

Dosages (Cinnamon):

FNFF = !!!

The bark is one of the world's premier spices. Cinnamon leaves used also as spice (e.g., in Jamaica's jerked pork) (FAC). 1 tsp bark/cup water/2–3 ×/day with meals (APA); 0.5–1 g bark, as tea, 3 ×/day (CAN); 0.5–1.0 ml liquid extract (1:1 in 70% ethanol) 3 ×/day (CAN); 2–4 ml cinnamon tincture (CAN, PNC); 20 grains bark for dysentery (DEP); 2–4 g bark (KOM); .05–0.2 g EO (KOM); 0.05–0.2 ml cinnamon oil (PNC); 0.3–1.2 ml spirit of cinnamon (PNC); 0.3–1 g powdered bark (PNC); 2–4 g bark/day (WHO); 0.05–0.2 g essential oil/day (WHO); 1 tsp bark/cup water 2–3 ×/day (WIC).

- Asian Indians use the bark in bolmes, enemas, or ghees for abdominal cancers (JLH).
- Asian Indians use a spicy triad *trijataka* (cardamom, cinnamon, and "tejapatra," possibly cassia) for lengual paralysis, stomach cramps, and toothache (HOS).
- Ayurvedics consider the bark aphrodisiac and tonic, using for biliousness, bronchitis, diarrhea, itch, parched mouth, worms, and cardiac, rectal, and urinary diseases (HOS).
- Belizeans for snoring suggest 1 cup of cinnamon tea with two teaspoons grated ginger and honey and milk added. Drink at bedtime each night until cured (or until death do us part!) (AAB).
- Caribbean Tramileños take the bark infusion for diarrhea and nausea (TRA).
- Dominicans take bark decoction, with or without cilantro, for enterosis and fever (VOD).
- Egyptians use the leaves for uterine cancer, the seeds for venereal warts (JLH).
- Haitians use the bark decoction as carminative, digestive, and febrifuge (VOD).
- Haitians use the essence topically for rheumatism, internally (dilute I presume) for enteric or gastric gas and spasms (VOD).
- Lebanese use cinnamon as a stimulant, for colds, rheumatism, halitosis, and slobbering (HJP; HOS).
- Pakistanis chew the bark for dysmenorrhea (DEP).
- Peruvians suggest the bark infusion for the heart, the decoction for colic (EGG).
- Ukrainians give raw grated carrots with cinnamon for anemia (HJP).
- Unani consider the oil carminative, emmenagogue, and as a tonic to the liver, using it for abdominal pains, bronchitis, head colds, and inflammation (HOS).

Downsides (Cinnamon):

Class 2b,2d; "Not for long-term use; do not exceed recommended dose (2–4 g bark/day; 50–200 mg essential oil daily). May overstimulate the vasomotor center" (AHP, 1997). Commission E reports bark contraindications: hypersensitivity to cinnamon or Peruvian balsam and adverse effects: often allergic reactions of skin and mucosae. Tramil warns against continued use because of mutagenicity (TRA). Other sources report contraindications: GI ulcer, pregnancy (AEH). Newall, Anderson, and Phillipson (1996) caution that the cinnamaldehyde in the volatile oil is allergenic and an irritant. The allergenic oil should not be taken internally (CAN). "There are no known problems with the use of cinnamon during pregnancy and lactation, provided that doses do not greatly exceed the amounts used in foods. … May cause some people to break out in a rash" (Castleman, 1996). Regrettably, I was unable to read the article on allergic contact dermatitis from cinnamon used as an odor-neutralizing agent in shoe insoles (X15186386). High doses caused vomiting in experiments with dogs, corresponding with reported side effects in humans. Cinnamaldehyde 5% in petrolatum is a skin irritant. Prolonged contact with cinnamon oil on skin may cause burns. Cinnamaldehyde in cosmetics or perfumes may cause allergic reactions. Allergic reactions (i.e., swollen lips or tongue, itching, burning sensation, blistering of the oral mucosa, and urticaria) reported from contact with ointments, toothpaste, mouthwash, or foods containing cinnamon oil or cinnamaldehyde (AEH1). Sensitized and sensitive justifiable chemophobes may develop dermatosis using mouthwash, perfume, soap, or toothpaste flavored with camphor, cassia, or cinnamon (FNF; RIN). May reduce the activity of tetracycline (WHO). Extracts and cinnamaldehyde reported mutagenic in some studies, nonmutagenic in others.

Toxicity: Following ingestion of cinnamon, contact dermatosis may flare up. Eugenol has been reported to be an irritant and a weak tumor promoter. Cinnamic aldehyde in perfumes can cause dermatosis. In toothpaste it can cause sensitivity (DAD).

Extracts (Cinnamon):

In a study by Park and Shin (2005), cinnamon and onion oil followed garlic and clove bud oils in lethality to the Japanese termite, *Reticulitermes speratus* Kolbe. Diallyl trisulfide was most toxic, then diallyl disulfide, eugenol, diallyl sulfide, and beta-caryophyllene (X15913300). Chericoni et al. (2005)

found that eugenol was, by far, the most potent antioxidant in cinnamon's essential oil, recounting its use as antioxidant, antiperoxidant, antiradicular, antiseptic, hepatoprotective, and sedative. Oral eugenol is rapidly absorbed, reaching blood plasma levels of 5 μM, significantly antioxidant levels, 2 hours after 150 mg of the eugenol, but almost completely excreted in the urine by 24 hours (X15941312). Pakistani scientists (X14633804) found that cinnamon improves glucose and lipids of type-2 diabetics. Cinnamon, consumed (1, 3, or 6 g/day) for 40 days followed by a 20-day washout period, reduced mean fasting serum glucose (18–29%), triglyceride (23–30%), LDL cholesterol (7–27%), and total cholesterol (12–26%) levels; compared to placebo (X4585184). Korean scientists (Lee et al., 2003) showed that cinnamate, a phenolic in the bark, enhances hepatic lipid metabolism and antioxidant defense systems in high cholesterol–fed rats. Cinnamate supplementation resulted in higher catalase and glutathione peroxidase activities. Lee et al. (2003) suggested that dietary cinnamate inhibits hepatic HMG-CoA reductase activity, resulting in lower hepatic cholesterol (X14585184). LD50 (EO) = 690 mg/kg der (CAN); perhaps second only to some varieties of clove (up to 20% eugenol), cinnamon (to 3.8%) is a major source of eugenol, which has all sorts of biological activities. Analgesic; anesthetic 200–400 ppm; antiaggregant IC50 = 0.3 μM (PR4:93); antiarachidonate; anticonvulsant; antiedemic, 100; antiinflammatory (11 μM); antimitotic; antimutagenic; antinitrosating; antioxidant, IC65 = 30 ppm; antiprostaglandin, 11 μM, IC50 = 9.2 mM; antiradicular, EC50 = 2 μl/l; antiseptic (3 ml/man/day); antithromboxane; antitumor; antiulcer; apifuge; bactericide, 500 ppm; calcium antagonist, IC50 = 224 μM; cancer preventive; candidicide; carminative; choleretic; CNS depressant; cytochrome-p450 inhibitor; enterorelaxant; febrifuge (3 ml/man/day); fungicide; hepatoprotective, 100 ppm; larvicide; motor depressant; sedative; spasmolytic; trypsin enhancer; and vermifuge (FNF).

LADANUM (*CISTUS CRETICUS* L.) + CISTACEAE

SYNONYMS:

Cistus incanus auct.; *Cistus incanus* var. *creticus*; *Cistus ladinifera*; *Cistus villosus* L., *Cistus villosus* var. *creticus*. *fide* EFS

NOTES (LADANUM):

> **A company of Ishmeelites came from Gilead with their camels, bearing spicery and balm and myrrh.**
>
> **Genesis 37 (KJV)**

The generally recognized sources of ladanum have been *Cistus creticus, C. cyprius, C. ladanifer,* and *C. monspeliensis*. Some of the EFS data below may indeed apply to *C. ladanifer.* All VAD data below accrue to *C. ladinifer.* Because ladanum used to be combed from the fur of sheep, or the beards of goats that had been grazing the Rock Rose, it often represented a mixture of species. While I suspect goats and sheep are discriminating grazers, I doubt that they were restricted to grazing one of the many species in so many Mediterranean areas. It is doubtful that all ladanum has been scientifically or taxonomically verified, so the accumulated literature may apply to various species. The biblical ladanum was probably not *C. ladaniferus,* and more probably *C. creticus* or *C. incanus.*

COMMON NAMES (LADANUM):

Ciste à Gomme (Fr.; EFS); Ciste de Crète (Fr.; EFS); Ciste d'Espagne (Fr.; EFS); Ciste Ladinifère (Fr.; EFS); Cisto Canescente (It.; EFS); Cisto di Creta (It.; EFS); Cisto di Spagno (It.; EFS); Cisto Ladinifero (It.; EFS); Cistus Ladon (Eng.; JLH); Cretan Rock Rose (Eng.; EFS); Girit Ladeni (Tur.; EFS); Gum Cistus (Eng.; JLH); Jara (Sp.; VAD); Kretische Cistrose (Ger.; EFS); Labdanum (Eng.; BIB); Ladanum (Dutch; Eng.; EFS; JLH; ZOH); Ladum (Eng.; JLH); Ledum (Eng.; JLH); Lot (Heb.?; ZOH); Manna de Hasta (Sp.; EFS); Rock Rose (Eng.; JLH); Spanische Cistrose (Ger.; EFS); Nscn.

FIGURE 1.27 Ladanum (*Cistus creticus*).

Activities (Ladanum):

Alanyl-aminopeptidase Inhibitor (1; X11801385) Allergenic (1; X3365959); Antiaggregant (1; X15325737); Antileukemic (1; X9581515); Antioxidant (1; X10917563); Antiperoxidant (1; X10917563); Antiproliferant (1; X11801385); Antiradicular (1; X10917563); Antiseptic (1; JAD; X8134413); Antiulcer (1; X7784302); Astringent (f; EFS); Bactericide (1; JAD; X8134413); Candidicide (1; JAD; X8134413); Cardioprotective (1; X15325737); Cytotoxic (1; X8134413); Dermoprotective (1; X10917563); Dipeptidylpeptidase-IV Inhibitor (1; X11801385); Diuretic (f; JAD); Emmenagogue (f; GMH); Expectorant (f; GMH; ZOH); Fungicide (1; JAD); Gastroprotective (1; X7784302); Gram(+)-icide (1; X9342956); Gram(–)-icide (1; X9342956); Hemostat (f; EFS); Myorelaxant (1; X15138007); Nervine (f; EFS); Purgative (f; EFS); Revulsive (f; EFS); Sedative (f; VAD): Spasmolytic (1; X15138007); Stimulant (f; GMH; ZOH); Vulnerary (f; BIB).

Indications (Ladanum):

Anxiety (f; VAD); Arthrosis (f; VAD); Asthma (f; BOU); Bacteria (1; JAD; X8134413); Bronchosis (f; JAD); Cancer, anus (f; JLH); Cancer, uterus (f; JLH); Candida (1; JAD); Cardiopathy (1; X15325737); Catarrh (f; BIB; BOU; ZOH); Corn (f; JLH); Diarrhea (f; JAD); Duodenosis (f; VAD); Dysentery (f; BIB; BOU; ZOH); Dyspnea (f; BIB); Escherichia (1; JAD); Fracture (f; JAD); Fungus (1; JAD); Gastrosis (f; JLH; VAD); Headache (f; VAD); Hepatosis (f; JLH); Hernia (f; JAD); Hysteria (f; BIB); Induration (f; JLH); Infection (1; JAD; X8134413); Inflammation (f; VAD); Insomnia (f; VAD); Leukemia (1; X9581515); Leukorrhea (f; JAD); Myalgia (f; VAD); Mycosis (1; JAD); Neuralgia (f; VAD); Osteoarthritis (f; VAD); Polyp (f; JLH); Proctosis (f; JLH); Rhinosis (f; JLH); Sore (f; JLH); Spasm (1; X15138007); Splenosis (f; JLH); Staphylococcus (1; JAD); Ulcer (f; VAD); Uterosis (f; JLH).

Dosages (Ladanum):

FNFF = !

Labdanum gum, the oleoresin, essential oil, and absolute are used as GRAS food flavoring elements, usually at levels less than 10 to 20 ppm, and have also been used in cosmetics, creams, detergents, perfumes, and soaps. The absolute is reportedly used in levels up to 4000 ppm, the essential oil up to 8000 ppm but only in perfumery.

Downsides (Ladanum):

The oleoresin of *C. ladanifer* is described as hepatotoxic, nephrotoxic, and neurotoxic (VAD).

Extracts (Ladanum):

The essential oil and the resin are antibiotic against *Candida, Escherichia,* and *Staphylococcus.*

COLOCYNTH (*CITRULLUS COLOCYNTHIS* (L.) SCHRAD.) X CUCURBITACEAE

Synonyms:

Cucumis colocynthis L.

Notes (Colocynth):

Behold, I will feed them, even this people, with wormwood, and give them water of gall to drink.

Jeremiah 9:15 (KJV)

Common Names (Colocynth):

Abur (Ber.; BOU); Aferziz (Arab.; Mauritania; UPW); Ahlandal der Araber (Ger.; KAB); Alkat (Ber.; BOU); Atmaraksha (Sanskrit; KAB); Bitter Apple (Eng.; Ocn.; AH2; BOU; CR2); Bitter Cucumber (Eng.; KAB); Bitter Gourd (Eng.; BOU; CR2); Bitter Melon (Eng.; GHA); Calebasse de Serpent (Fr.; KAB); Castravete Amar (Rom.; KAB); Chicotin (Fr.; BOU); Chittipapara (Tel.; KAB); Cocomero Amaro (It.; KAB); Colocynth (Eng.; Scn.; AH2; BOU; CR2); Coloquinte (Fr.; BOU; EFS; UPW); Coloquintida (Por.; KAB); Coloquintide (It.; EFS); Corin M'bodi (Fulah; KAB); Darwawal (Las Bela; KAB); Dingel (Arab.; BOU); Donkey's Melon (Eng.; GHA); Ebucehilkarpuzu (Tur.; EFS); Endrayani (?; Nepal); Ghorumba (Hindi; KAB); Gorkaya Tikva (Rus.; KAB); Habid (Arab.; BOU); Hadaj (Arab.; BOU); Hadaq (Arab.; BOU); Hadj (Arab.; Mali; UPW); Hadja (Arab.; BOU); Haguellet (Arab.; BOU); Handal (Arab.; Nig.; Oman; Qatar; BOU; GHA; UPW); Hanzal (Arab.; EFS); Hanzul (Arab.; EFS); Hedeg (Dho.; Oman; GHA); Hedej Lehmar (Arab.; BOU); Henzil (Dec.; KAB); Hindavanahetalkh (Iran; KAB); Indrak (Guj.; KAB): Indravaruni (Ayu.; Sanskrit; AH2; JLH; KAB); Indrayan (Hindi; India; Urdu; FS; KAB); Kaddukankri (Bom.; HAB); Kharbuza-Talkt (Iran; Khiasi, Burma; KAB); Kolokvint (Den.; Nor.; Swe.; EFS; KAB); Kolokvinter (Den.; KAB); Kolokwint (Dutch; EFS); Koloquint (Ger.; EFS); Koron Mboddi (Fula; Guinea; UPW); Kurkushta (Bal.; KAB); Kwartowa (Hausa; UPW); Kwintappel (Dutch; KAB); Makhal (Beng.; KAB); Maraghuna (Zhob.; KAB); Marghun (Sharug; KAB); Marghuni (Sibi; KAB); Marhoum (Arab.; BOU); Merraret el Sekhour (Arab.; BOU); Murrah (Oman; GHA); Pavamekkekayi (Kan.; KAB); Peykkommaddi (Sri.; KAB); Peykkommutti (Mal.; KAB); Phidangourgia (Greek; JLH); Picrokolokunthia (Greek; JLH); Pikrangougia (Greek; JLH); Purgierparadiesapfel (Ger.; EFS); Qittat en Na'am (Arab.; BOU); Sartol (Hun.; KAB); Serere (Soussou; KAB); Shary (Qatar; GHA); Shetiputsa (Sin.; KAB); Sikya Pikra (Greek; KAB); Sise (Arab.; BOU); Sisigi (Diola; Sen.; UPW); Suri (Arab.; GHA); Tadjellet (Ber.; BOU); Taferzizt (Ber.; BOU); Thorliyindrayan (Mar.; KAB); Tifersit (Ber.; BOU); Trujogosht (Sin.; KAB); (Indi Tuera (Sp.; KAB); Ubruzi (Ber.; BOU); Verittumatti (Tam.; KAB); Vicala (India; JLH); Visala (India; JLH); Wild Gourd (Eng.; CR2).

Activities (Colocynth):

Abortifacient (f; CRC; EFS; WBB; X14430893); Acaricide (1; BIB); Alterative (f; KAP; MPI); Analgesic (f; GHA); Anthelmintic (f; WO2); Antiaggregant (1; WO2); Antiandrogenic (1; X12660478); Anticholinergic (1; MPI; WO2); Antifertility (1; X12660478); Antihistaminic (1; KAB; MPI; WO2); Antihyperglycemic (1; X10904181); Antioxidant (1; X12484560); Antiperoxidant (1; X12484560); Antiproliferant (f; X14732962); Antitumor (f; X14732962); Bactericide (1; WO2); Bitter (1; KAB); Carcinogenic (1; X6746706); Cardiodepressant (1 WO2); Carminative (f; CRC; KAB; WO2); Cathartic (1; CRC; MPI); Depurative (f; WO2); Diuretic (1; KAP; MPI; WO2); Ecbolic (f; BIB; CRC); Emetic (1; MPI); Emmenagogue (f; CRC; EFS; WBB); Expectorant (1; MPI; WO2); Febrifuge (f; BIB; CRC); Hepatoprotective (1; WO2; X12484560); Herbicide (1; WO2); Hydragogue (f; CRC); Hypoglycemic (1; WO2; X10904181); Immunostimulant (1; X12963136); Insecticide (1; BIB; KAP; WBB; WO2); Insulinotropic (1; X10909260); Irritant (1; PH2); Laxative (f1; GHA; PHR); Mucoirritant (1; PHR); Negative Chronotropic (1; MPI); Negative Inotropic (1; MPI); Nematicide (1; WO2); Poison (1; PHR); Protisticide (1; WO2); Purgative (1; CRC; MPI; PH2; WBB); Refrigerant (f; KAB); Repellant (f; CRC); Ribosome Inactivator (1; X2248976); Toxic (f; EFS); Uterorelaxant (1; WO2); Vermifuge (1; BIB; CRC).

Indications (Colocynth):

Adenopathy (f; CRC; JLH); Alopecia (f; WBB); Amenorrhea (f; BIB; CRC); Anemia (f; CRC; KAB); Arthrosis (f; CRC; GHA); Ascites (f; CRC; KAP; PH2; WBB); Asthma (f; CRC; KAB); Bacteria (1; WO2); Baldness (f; WBB); Biliousness (f; BIB; CRC; KAP); Bite (f; KAP; WBB); Blennorrhagia (f; UPW); Blood (f; WO2); Breast (f; CRC); Bronchosis (f; CRC; KAB); Cancer (f1; CRC; GHA;

KAB); Cancer, abdomen (f; CRC; JLH); Cancer, bladder (f; CRC; JLH); Cancer, breast (f; CRC; JLH); Cancer, colon (f; CRC; JLH); Cancer, eye (f; CRC); Cancer, liver (f; CRC); Cancer, sinew (f; CRC; JLH); Cancer, spleen (f; CRC); Carcinoma (f; CRC); Catarrh (f; HHB); Cerebrosis (f; CRC); Childbirth (f; KAB; PH2); Cholecystosis (f; PHR; PH2); Colic (f; GHA; KAP); Constipation (f1; CRC; GHA; PHR); Corn (f; CRC; JLH); Cough (f; WO2); Cramp (f; HHB; WO2); Cystosis (f; HHB; JLH); Debility (f; CRC); Dermatosis (f; BOU; UPW); Diabetes (f1; BIB; BOU; WO2; X9324004); Dropsy (f; BIB; CRC; KAP); Dysmenorrhea (f; HHB); Dyspepsia (f; CRC; KAB); Dysuria (f; CRC; KAP; WBB); Elephantiasis (f; CRC; KAB; PH2); Endothelioma (f; CRC; JLH); Enterosis (f; KAP; WO2); Epilepsy (f; BOU; CRC; KAP; WBB); Epithelioma (f; JLH); Fetal Atrophy (f; CRC); Fever (f; BIB; CRC; WO2); Frostbite (f; BIB; CRC); Gangrene (f; BIB; CRC; WO2); Gargantosis (f; CRC; KAB); Glaucoma (f; KAP); Gonorrhea (f; BOU; UPW); Gray Hair (f; KAP; WO2); Headache (f; WO2); Hemicrania (f; CRC; KAB); Hemorrhoid (f; WO2); Hepatosis (f; KAP; PHR; PH2; WO2); Induration (f; JLH); Infection (1; WBB); Inflammation (f; CRC; WO2); Itch (f; BOU); Jaundice (f; BIB; CRC; WBB); Leishmania (1; WO2); Leprosy (f; CRC); Leukemia (f; CRC; JLH); Leukoderma (f; BOU; CRC); Migraine (f; KAB; WO2); Nephrosis (f; HHB); Neuralgia (f; HHB; KAP; WO2); Neurosis (f; HHB); Ophthalmia (f; CRC; WO2); Pain (f; GHA; JLH); Paralysis (f; WO2); Parasite (f; KAP); Pediculosis (f; CRC); Protozoa (1; WO2); Rheumatism (f; BIB; CRC; KAB; KAP; WBB); Sarcoma (f; JLH); Sciatica (f; KAP); Scirrhus (f; JLH); Snakebite (f; BOU; CRC); Splenomegaly (f; CRC); Splenosis (f; JLH); Sting (f; WBB); Swelling (f; WO2); Throat (f; CRC); Tumors (f; CRC); Urogenitosis (f; BIB; WO2); Uterosis (f; CRC; KAB); Varicose Veins (f; BIB; CRC; WO2); Venereal Disease (f; UPW); Worm (1; WO2); Wounds (f; CRC; WO2).

Dosages (Colocynth):

FNFF = !

Barely food farmacy; fruits pickled after boiling in several changes of water to remove bitter elements. Seed, removed from the poisonous pulp, is eaten in Central Sahara. Seed kernels eaten with dried dates (BIB; TAN). 120–300 mg individually/day; up to 600 mg/day (HHB). 0.2–0.4 g root powder (KAP); 0.1–0.4 g fruit powder (KAP); 3–10 ml root tea (KAP). Homeopathic dilutions only (JAD); allopathic doses no longer defensible (PH2).

- Algerians use colocynth in gargles and mouthwash, and as a counterirritant in chest cold plasters. They poultice salted rind onto frostbite (BIB).
- Arabians apply crushed leaves with garlic to bites and stings (GHA).
- Arabians apply seeds crushed in water as hair-darkening shampoo (GHA).
- Arabians mix crushed fruits with oil to massage painful arthritis (GHA).
- Asian Indians use the seed oil for bites and stings and epilepsy, and to promote hair growth (UPW).
- Ayurvedics use the root for arthritic pain, breast inflammation, ophthalmia and uterine pain; and the fruit for adenopathy, anemia, ascites, asthma, bronchitis, constipation, dyspepsia, elephantiasis, fetal atrophy, jaundice, leucoderma, splenomegaly, throat diseases, tubercular glands, tumors, ulcers, and urinary discharges (KAB).
- Bedouins tie a slice of fresh gourd onto the heel before retiring for rheumatism (in the Sinai, I was told that one tastes the bitter gourd in the morning as a result) (BIB).
- Guinea natives poultice the leaves onto migraine and neuralgia (KAB).
- Lebanese apply the pulp to open varicose veins, also using it for cancer, gangrene, and wounds (BIB; HJP).
- Mauritanians with gonorrhea insert the penis into cooked fruit for circa 1 hour to treat blenorrhagia (UPW).
- Mauritanians apply baked root powder in butter or camel's milk to head lice (UPW).
- North Africans swallow one unchewed seed per day for 21 days for diabetes (BOU).

- North Africans take root decoction with garlic for snakebite (BOU).
- Unani, considering the fruit abortifacient, carminative, and purgative, use it for brain disorders, epilepsy, hemicrania, inflammation, leprosy, ophthalmia, and weakness of the limbs (KAB).

Downsides (Colocynth):

"The purgative action is so drastic as to have caused fatalities. One woman who took 120 g to induce abortion died in 50 hours. In case of poisoning, stomach evacuation is recommended, followed by oral or rectal administration of tincture of opium, followed by stimulating and mucilaginous beverages" (CRC). Toxic doses (600–1000 mg) may cause colic, diarrhea, hematchezia, nephrosis, and vomiting; lethal doses (as low as 2 g) may cause convulsions, paralysis, and possibly death due to circulatory collapse (PH2).

Natural History (Colocynth):

Despite the bitterness, the fruits are eaten by grazing animals and the seeds gathered by desert rodents. Hungry Bedouins may even eat the seed after soaking in water, able to survive nearly 2 weeks on the seed (although probably with diarrhea). Goats and wild game eat the stem and leaves (BIB); the fruit is eaten only by donkeys, gazelles, and ostriches (UPW). The following fungi affect colocynth: *Colletotrichum bryoniae, Erysiphe cichoracearum, E. polyphaga, E. semitectum, Fusarium oxysporum,* and *Puccinis citrulli*. The Bottle gourd mosaic virus and the nematode, *Meloidogyne* sp., also attack this plant (HOE).

Extracts (Colocynth):

Alpha-elaterin-2-d-glucopyranoside: anticholinergic (WO2); antihistaminic (WO2); cardiodepressant (WO2); purgative (WO2); uterorelaxant (WO2); toxic dose 0.5–1 g orl man; LDlo = 4000 mg orl man. Alpha-spinasterol significantly (circa 1000 X simvastin, a coenzyme-A inhibitor) modulates development and/or progression of diabetic nephropathy. It reduced significantly attendant increases of serum triglycerides, renal weight, and urinary protein excretion in diabetic mice (X15326549).

WATERMELON (*CITRULLUS LANATUS* (THUNB.) MATSUM. & NAKAI.) +++ CUCURBITACEAE

Synonyms:

Citrullus caffer Schrad.; *Citrullus edulis* Spach.; *Citrullus lanatus* (Thunb.) Matsum. & Nakai subsp. *mucosospermus* Fursa; *Citrullus vulgaris* Schrad.; *Colocynthis citrullus* (L.) O. Kuntze; *Cucurbita citrullus* L., *Momordica lanata* Thunb.

Notes (Watermelon):

> **We remember the fish, which we did eat in Egypt freely; the cucumbers, and the melons, and the leeks, and the onions, and the garlick: But our soul is dried away: there is nothing at all, beside this manna, before our eyes.**
>
> **Numbers 11:5–6 (KJV)**

> **We remember the fish we ate in Egypt for nothing, the cucumbers, the melons, the leeks, the onions, and the garlic; but now our strength is dried up, and there is nothing at all but this manna to look at.**
>
> **Numbers 11:5–6 (RSV)**

FIGURE 1.28 Watermelon (*Citrullus lanatus*).

> **How we remember the fish that we used to eat in Egypt for nothing, the cucumbers and the water melons, and the leeks and the onions, and the garlic; But now our soul is dried away. Our eyes are on nothing at all except the manna.**
>
> **Numbers 11:5-6 (NWT)**

Of the three versions, only NWT specifies watermelons. And it really is a life-sustaining treasure in the desert when you have lost your canteen. Many of the common names that follow were taken from Porcher's excellent Australian database (http://www.plantnames.unimelb.edu.au). Porcher's team adopted the group names "Lanatus" for wild and semi wild African watermelons, "Vulgaris" to cover "edible" cultivars, and "Citroides" to cover the preserving melons and fodder cultivars, deeming it a "sensible option" (POR). I have also taken at least one common name for each country from a large colloquial name collection presented in UPW. Noting that watermelon has been known from Egypt since the Bronze Age, Zohary speculates that watermelon was domesticated in Africa during the Neolithic Period.

Common Names (Watermelon):

Anguria (It.; POR); Arbuz (Pol.; Rus.; POR); Arbuz Stolovyj (Rus.; POR); Arpuusi (Fin.; POR); Avatiach (Heb.; POR); Avatiach Pashut. (Heb.; POR); Avatihim (Heb.; ZOH); Bateekh (Arab.; POR); Batia (Ma.; JFM); Battikh (Arab.; Syria; HJP; POR); Belancia (Por.; AVP); Belik Zichi (Arab.; EFS); Betteakh (Arab.; Egypt; POR); Chaya Pula (Sanskrit; EFS); Choei Koa (China; POR); Cimangko (Dwi.; POR); Citron Melon (Eng.; POR); Cocomero (It.; AVP; POR); Coloquinte (Fr.; POR); Da Zi Gua Zi Xi Gua (China; POR); Dinia (Bul.; POR); Dua Do (Vn.; POR); Dua Hao (Vn.; POR); Dua Hao Ruoi Do (Vn.; POR); Dua Hao Ruoi Vang (Vn.; POR); Egusi Melon (Eng.; POR); Eguzui Suika (Japan; POR); Facé (Por.; AVP); Fodder Melon (Eng.; POR); Futtermelone (Ger.; POR); Gewöhnliche Wassermelone (Ger.; POR); Görögdinnye (Hun.; POR); Grosse Wasser Melon (Ger.; EFS); Grote Water Meloen (Dutch; EFS); Gua Zi Xi Gua (China; POR); Han Koa (China; POR); Han Kua (China; EFS); Hia Koa (China; POR); Hinduana (BAL; KAB); Hsi Kua (China; EFS); Jabas (Arab.; Syria; HJP); Jacé (Por.; AVP); Ka-bed (Tibet; NPM); Karpusi (Greek; POR); Karpouzia (Cyprus; POR); Karpusi (Greek; POR); Karpuz (Tur.; POR); Kavun (Ukraine; POR); Kawon (Pol.; POR); Kharbuza (Hindi; POR); Kharmuja (Hindi; POR); Lubenica (Croatia; Macedonia; POR); Lubenica Meloun (Czech.; POR); Lubenice (Slovenia; POR); Lubenitsa (Serbia; POR); Malancia (Por.; EFS); Matao (Thai; POR); Mehal (BAL; KAB); Melancia (Por.; JFM; POR); Melância (Mad.; Por.; POR); Melâo d'Agua (Por.; AVP; JFM); Melone d'Acqua (It.; POR); Melón de Agua (Cuba; Peru; Sp.; EFS; JFM; RyM; SOU); Melon d'Eau (Fr.; Haiti; AVP; EFS); Melon Dlo (Creole; Haiti; AVP; VOD); Melon Fourrager (Fr.; POR); Mendikai (Malaya; POR); Môô (Laos; POR); Mtango (Swahili; POR); Mtikiti (Swahili; POR); Nzara (Sudan; EFS); 'Öö'w Llök (Khmer; POR); Pakwan (Tag.; POR); Pastecca (It.; POR); Pastek (Creole; Haiti; VOD); Pastèque (Fr.; Fr. Guiana; AVP; JFM); Pastèque à Graine Rouge (Fr.; POR); Pastèque Fourragère (Fr.; POR); Patia (Ma.; JFM); Patilla (Dr.; Pr.; Ven.; AVP); Patille (Ven.; EFS; JFM); Pati Yache (Ma.; JFM); Pepene Verde (Rom.; POR); Popone (It.; AVP); Preserving Melon (Eng.; POR); Raqqi? (Iran; POR); Red-Seeded Citron (Eng.; POR); Red-Seeded Preserving Melon (Eng.; POR); Samangka (Sunda; POR); Sandia (Bel.; Bol.; Mex.; Peru; BNA; EGG; JFM; SOU); Sandilla (Peru; RAR); Semangka (Malaya; POR); Shiryou Suika (Japan; POR); Shi Yong Xi Gua (China; POR); Shokuyou Suika (Japan; POR); Si koa (Canton; POR); Si Liao Xi Gua (China; POR); Soo Bahk (Korea; POR); Stock Melon (Eng.; POR); Su Bak (Korea; POR); Suika (Japan; POR); Taeng Chin (Thai; POR); Taeng Moh (Thai; POR); Tarabuja (Nepal; NPM; POR); Tarabuucha (Guj.; POR); Tarabuuja; (Mar.; Pun.; POR); Tarabuujaa (Nepal; POR); Tarabuuza (Hindi; POR); Tarbooz (Hindi; POR); Tarbuj (Hindi; POR); Tarbuz (Hindi; POR); Tarmuj (Hindi; POR); Taramuj (Beng.; POR); Tarbuz (India; EFS); Tèng mô (Laos; POR); Tembikai (Malaya; POR); Ts'ing Teng Koa (China; POR); Vandmelon (Den.; POR); Vannmelon (Nor.; POR); Vattenmelon (Swe.; POR); Vesimeloni (Fin.; POR); Waatlemoen (Afrikan; POR); Wasserkürbis (Ger.; AVP); Wassemelone (Ger.; POR); Wasserzitrulle (Ger.; POR); Watermeloen (Dutch; POR); Watermelon (Eng.; Scn.; AH2; CR2; NPM); Watesan (Java; POR); Wild Melon (Eng.; POR; USN); Xi Gua (Pin.; AH2; DAA); Xi Gua Pi (Pin.; AH2).

Activities (Watermelon):

Antemetic (f; VOD); Anthelmintic (1; HDN; WO2); Antiallergic (1; X8463793); Anticancer (1; FNF); Antioxidant (1; FNF; X15351703); Antiprostatitic (1; FNF); Antiseptic (f; BIB; EFS); Aphrodisiac (f; KAB); Cerebrotonic (f; BIB); Chemopreventive (1; FNF); Curare (1; HDN); Demulcent (f; EFS; WO2); Diuretic (f1; JFM; VOD; WO2); Febrifuge (f; HJP; VOD; WO2); Hypocholesterolemic (1; FNF); Hypotensive (1; BIB; WO2); Litholytic (f1; AHL; VOD; WO2); Nephrotonic (f; HJP); Purgative (f; WO2); Snake Repellant (f; HDN); Vermifuge (f1; JFM; SOU).

Indications (Watermelon):

Bite (f; HDN); BPH (1; FNF); Bronchosis (f; JFM); Cancer (1; FNF; WO2); Catarrh (f; BIB); Cystosis (1; BIB; JFM); Depression (1; BIB; FNF); Dermatosis (f; HDN); Diarrhea (f; BIB; WO2); Dyspepsia (f; BIB); Dysuria (f; BIB; JLH; MAX; WO2); Fever (f; BIB; HDN; VOD); Gas (f; JFM); Gonorrhea (f; BIB); Headache (f; JFM); Hepatosis (f; JFM); High Blood Pressure (1; JFM); Infection (f; WO2); Inflammation (f; BIB); Kidney stone (f1; VOD; WO2); Maculitis (1; FNF); Malaria (f; HDN; JFM); Nephrosis (f; BIB; WO2); Pulmonosis (f; JFM); Sinusitis (f; BIB); Sore (f; BIB; WO2); Sore Throat (f; BIB); Stomatosis (f; BIB); Stone (f1; VOD; WO2); Strangury (f; WO2); Roundworm (1; WO2); Tapeworm (1; WO2); Typhus (f; BIB); Urethrosis (f; MAX); Vaginosis (f; BIB); Venereal Disease (f; BIB); Worm (f1; HDN; JFM; RAR; WO2).

Dosages (Watermelon):

FNFF = !!!
Fruit pulp edible; seeds edible, raw or toasted; toasted seeds served as coffee substitute; pickled young fruits and rinds edible; leaves used as potherb (EGG; FAC).

- Bahamans decoct bruised seeds and drink as diuretic 3 ×/day (JFM).
- Cherokee have adopted the seeds for kidney ailments and enuresis, Cheyenne take seed as diuretic, Chickasaw take seed for bloody urine, Iroquois use them for oliguria, and the Rappahannock take seed infusions for gravel or kidney stones, in men or horses (DEM).
- Curacao Natives bind the rind around head for headache (JFM).
- Haitians, believing the fruit juice diuretic and febrifuge, drink it for kidney stones (VOD).
- Haitians ingest leaves for respiratory problems and nausea (VOD).
- Latinos express 5 to 8 g juice from fresh seeds in 1 cup water for worms (JFM; SOU).
- Mexicans take leaf decoction for malaria (JFM).
- Puerto Ricans eat the diuretic, tonic fruit for bronchitis, catarrh, gas, and lung problems (JFM).
- Venezuelans poultice mashed rind onto liver problems (JFM).

Downsides (Watermelon):

"Eating unripe watermelons causes serious illness, even death" (JFM).

Extracts (Watermelon):

Citrin (cucurbocitrin) hypotensive. Mandel et al. (2005), after finding citrullinaemia and high arginine in a 19-month-old girl with developmental delay, learned she had consumed large quantities of watermelon, a fruit rich in free citrulline and arginine. Then they fed watermelon to six healthy adults. All developed elevated plasma citrulline (386–1069 μmol/l) and moderately elevated plasma arginine (128–251 μmol/l). Citrullinaemia, new to me, is indicated by elevated plasma citrulline and arginine, in the absence of orotic or arginosuccinic aciduria or hyperammonaemia (X15902549).

CITRON (*CITRUS MEDICA* L.) ++ RUTACEAE

Notes (Citron):

> **And ye shall take you on the first day the boughs of <u>goodly trees</u>, branches of palm trees, and the boughs of thick trees, and willows of the brook; and ye shall rejoice before the LORD your God seven days.**
>
> **Leviticus 23:40 (KJV)**

FIGURE 1.29 Citron (*Citrus medica*).

And you shall take on the first day the fruit of goodly trees, branches of palm trees, and boughs of leafy trees, and willows of the brook; and you shall rejoice before the LORD your God seven days.

Leviticus 23:40 (RSV)

And you must take for yourselves on the first day the fruit of splendid trees, the fronds of palm trees, and the boughs of branchy trees, and poplars of the torrent valley; and you must rejoice before Jehovah your God seven days.

Leviticus 23:40 (NWT)

Most translators and exegetes agree that the Hebrew *ethrog* and the "goodly trees" (*etz hadar*) of KJV and RSV (and I assume the *splendid trees* of NWT) above represent *Citrus medica* and that it grew in Israel at the time of the Bible. Even Moldenke agreed that *etz hadar* was *Citrus medica*. However, some scholars argue that it was not a specific tree at all. I will agree with Moldenke and Zohary, and include the citron among my biblical species and try to find a hardy one for the garden. Until then, I will let my *Poncirus* substitute for the ethrog when giving my biblical tours of the garden. Supposedly the first *Citrus* "liberated" from Asia, probably India, citron relics are found in Babylonian excavations of the Sumerian epoch, 4000 B.C. It is thought to have reached the Mediterranean 300 years before Christ, perhaps with Alexander's armies returning from India (ZOH).

COMMON NAMES (CITRON):

Adam's Apple (Eng.; KAB); Amlakeshara (Sanskrit; KAB); Bajauri (Pun.; KAB); Bajauri Nimbu (Pun.; NAD); Balank (Guj.; KAB; NAD); Bara Nimbu (Beng.; Hindi; KAB; WOI); Begpura (Beng.; WOI); Bija Pura (Bom.; KAB); Bijaura (Hindi; KAB; WOI); Bijoru (Guj.; WOI); Bijoura (Guj.; NAD); Bil-ba (Tibet; NPM); Bimara (Tamang; NPM); Bimbiri (Nepal; KAB); Bimiro (Danuwar, Nepal; NPM); Cederno (It.; KAB); Cedraten (Ger.; KAB); Buyag (Tag.; KAB); Céédratier (Fr.; KAB; USN); Chholongo Nebu (Beng.; NAD); Chu Yuan (China; KAB); Cidra (Por.; Sp.; EGG; KAB; USN); Cidraero (Sp.; EGG); Cidreira (Mad.; Por.; KAB); Cidro (Sp.; USN); Citrat (Malta; KAB); Citroenboom (Dutch; KAB); Citron (Eng.; Scn.; Swe.; AH2; KAB; NPM; USN); Citron Lemon (Eng.; HJP); Cytryna (Pol.; KAB); Ethrog (Heb.; HOC); Etrog (Isr.; AH2; BIB); Etz Hadar (Heb. [= Goodly trees]; ZOH); Gilam (Mal.; KAB; WOI); Gou Yuan (Pin.; AH2); Gou Yuan (Pin.; AH2); I-Lemura (Sen.; UPW); Jrako (Ivo.; UPW); Kachchhe (Lepcha; NPM); Kachikung (Lepcha; KAB); Kadaranarathai (Tam.; WOI); Kagdi limbu (Mah.; NAD); Karuna (Sanskrit; NAD); Kitrea (Greek; KAB); Kogilachim (Tam.; KAB); Lamain (Rom.; KAB); Lanagi (Tharu; NPM); Langkok (Magar; NPM); Lebu (Beng.; KAB); Leemáám Mááasar (Hausa; Nig.; UPW); Lemun (Sokoto; KAB); Lemuna (Guinea; UPW); Limon (Pi.; KAB); Limonnow Dyerevo (Rus.; KAB); Lungamu (Tel.; KAB; WOI); Madala (Kan.; WOI); Madalada hannu (Kan.; NAD); Madalanarakam (Mal.; NAD); Madeephalamu (Tel.; NAD); Mahalung (Mah.; NAD); Mahalunga (Mar.; Sanskrit; NAD; WOI); Mahaphala (Kan.; WOI); Mangsai (Chepang; NPM); Mapala (Tulu; KAB); Maphal (Hindi; NAD); Mary Bushukan (Japan; TAN); Matalanarakam (Mal.; WOI); Matulang (Sanskrit; NAD); Matulungga (Ayu.; AH2); Mauling (Kon.; KAB); Mavalinga (Kon.; NAD); Mavalung (Mar.; KAB; WOI); Medischer Apfel (Ger.; USN); Melon Lime (Eng.; KAB); Narotte (Madras; KAB); Punseme (Cat.; KAB); Ruranj (Iran; KAB); Rusaka (Kan.; WOI); Rusakam (Mal.; WOI); Sedaran (Sin.; KAB); Shauktakera (Burma; KAB); Soippa (Limbu; NPM); Sunpekawa (Rai; NPM); Tasi (Newari; NPM); Toronja (Sp.; USN); Trunj (Arab.; Syria; HJP); Turanj (Guj.; Hindi; Kotra; KAB; WOI); Turin (Mach; Rindli; KAB); Utraj (Arab.; KAB); Voamandina (Betsimisaraka; KAB); Voasaty (Malagasy; KAB); Xiang Yuan (Pin.; AH2); Zitronazitrone (Ger.; USN).

Activities (Citron):

Analgesic (f; WO2); Antidote (f; NAD); Antiinflammatory (f; NAD); Antiscorbutic (f; WO2); Antiseptic (f1; HJP); Antispasmodic (f; EGG); Aphrodisiac (f; KAB); Astringent (f; WOI); Bronchoprotective (1; X15598576); Cardiotonic (f; WO2); Digestive (f; NPM); Hypotensive (1; WO2); Laxative (f; EGG); Orexigenic (f; NAD); Refrigerant (f; WOI); Sedative (f; EGG; KAB); Stimulant (f; DAD; EGG); Stomachic (f; WO2); Tonic (f; DAD); Vermifuge (f; EGG; HOC).

Indications (Citron):

Anorexia (f; NAD); Asthma (f1; DAD; X15598576); Biliousness (f; NAD); Bronchosis (f; BIB); Calculus (f; KAB); Cancer (f1; DAD; JLH); Caries (f; KAB); Colic (f; DAD); Constipation (f; EGG); Cough (f; DAD); Cramp (f; EGG); Diarrhea (f; DAD); Dysentery (f; NPM; WOI); Dyspepsia (f; EGG; NAD); Dysuria (f; KAB); Earache (f; KAB); Enterosis (f; DAD); Fever (f; NAD); Gastrosis (f; DAD); Halitosis (f; KAB); Hemorrhoid (f; KAB); Hiccough (f; KAB); High Blood Pressure (1; WO2); Impotence (f; KAB); Infection (f; HJP); Inflammation (f1; NAD; X15598576); Intoxication (f; KAB); Jaundice (f; DAD); Leprosy (f; KAB); Lumbago (f; BIB); Nausea (f; NAD); Odontosis (f; KAB); Ophthalmia (f; HJP); Otosis (f; KAB); Palpitation (f; WO2); Pharyngosis (f; KAB); Rheumatism (f; UPW); Sclerosis (f; DAD); Seasickness (f; BIB); Snakebite (f; NAD); Sore Throat (f; EGG; KAB); Splenosis (f; DAD); Sting (f; NAD); Stomatosis (f; EGG; UPW); Stomachache (f; DAD; HOC); Stomatosis (f; EGG; UPW); Syphilis (f; DAD); Thirst (f; NAD); Tumor (f; DAD); Venereal Disease (f; DAD); Worm (f; EGG).

Dosages (Citron):

FNFF = !!

Diamante citron was the first citrus known to European civilization, and is still cultivated in Calabria, Corsica, Crete, and Israel. The variety "Etrog" is the official citron in the Jewish Feast of the Tabernacle ritual, the entire fruit being eaten. Thick rind used in fruit cakes, salads, sweet rolls, etc. (FAC). The main products are candies and liqueurs, the oil used in flavoring beverages and sweets. Peel candied (TAN).

- Africans deemed the fruits useful in rheumatism and sore mouth (UPW).
- Ayurvedics view the rind as an aphrodisiac, the stimulant tonic seeds as useful for biliousness, hemorrhoids, and inflammation (KAB).
- Ayurvedics view the ripe fruits as stimulant and tonic and good for asthma, cough, earache, hiccup, leprosy, and sore throat (KAB).
- Ayurvedics view the root as anthelmintic, laxative, and useful in calculus, cancer, caries, colic, dysuria, nausea, the flower buds astringent and orexigenic and good for asthma, cough, enterosis, hiccup, intoxication, and nausea (KAB).
- Lebanese use citron, much as did the Choco Indians of Panama use lemon, as pediatric ophthalmic eyedrops (HJP).
- Nepalese use the fruits as a digestive for dysentery (NPM).
- Peruvians use bark tea for cramps, dyspepsia, and distraught nerves, the floral/foliar tea for sore mouth and throat (EGG).
- Peruvians use distilled floral water as antispasmodic (EGG).
- Peruvians regard the seed infusion as laxative and vermifuge (EGG).
- Peruvians suggest the floral tincture as a stimulant (EGG).

Downsides (Citron):

I view it as a primitive citrus, less water, more phytochemicals (hence quite possibly relatively more medicinal).

Natural History (Citron):

This species is being crossed with lemon seeking resistance to the mal secco disease, *Phoma tracheiphila* fungus that lives in vessels in the wood, causing typical tracheomycosis, most serious disease of orchard lemons, affecting up to 100% of orchards of susceptible cultivars (X15941331).

Extracts (Citron):

For those with no other citrus, this species, like my *Poncirus*, can provide many of the phytonutrients common to many citrus species; my *Poncirus* fruits hang on late into autumn. Verzera et al. (2005) analyzed the oil of cultivar "Diamante," reporting 55 components, with the important anticancer compound limonene constituting circa 52% of the oil, gamma-terpinene at 27.7%, circa 2% ocimene, circa 2% alpha-pinene, circa 2% beta-pinene, 1.7% neral, and 2.8% geranial and perillaldehyde. Most of the other compounds were well below 2%, many below 0.1% (X15941331). I think limonene is a significant contribution from citrus. Israeli scientists (Keinan et al., 2005) suggest that limonene (constitutes more than 50% of citron's essential oil) might help asthmatics, especially those aggravated by ozone pollution. Could poor children in our inner cities, where asthma is increasing dramatically, reduce asthma attacks and/or symptoms (especially around ozone pollution, as on school buses in inner cities) by squeezing citrus peels (apparently all contain limonene) and inhaling the pleasant aroma periodically. Because the sedative antiinflammatory limonene is also absorbed through the skin, rather rapidly I might add, I would also consider adding crushed citrus peels to bath water, which might help at the end of the day's commute. Keinan's group predicted that electron-rich olefins, known ozone scavengers, could be used to prevent asthmatic episodes. Volatile, unsaturated monoterpenes, like limonene, could saturate the pulmonary membranes, equipping airways with local chemical protection against ozone. In Keinan's experimental rats, limonene inhalation significantly prevented bronchial obstruction (eucalyptol, alias cineole, saturated and inert to ozone, did not) (X15598576). I doubt I can convince our government to compare citrus inhalation in clinical trials as a third arm against placebo and some expensive pharmaceuticals. (As a gray-haired botanist, I cannot and do not prescribe. But if I had a hundred asthmatic grandchildren, you could bet some of them would be trying citrus peel and others would not, followed by a vice versa crossover, recording the frequency and severity of their attacks for old grandpa.) Faith-based types might best be praying too, and using the peel of the citron, called ethrog or etrog in some versions of the Bible. Without praying to improve my odds, I will still bet the citrus peel would score well along side the pharmaceuticals. I know it would be cheaper and bet it would be safer. That asthmatic possibility is relatively new; it has been known for a decade that limonene can also prevent cancers. Mondello et al. (1995) found that limonene was the most frequent monoterpene component in the citrus oils they examined, (from 50% in lime oil to circa 97% in grapefruit; about 52% in the etrog (X10554196). d-Limonene works in preclinical models of breast cancer, causing more than 80% of carcinomas to regress with little host toxicity (Crowell et al., 1994). Tsuda et al. (2004) singled out promising chemopreventive anticancer phytochemicals: vitamin derivatives, phenolic and flavonoid agents, fatty acids, organic sulfur compounds, isothiocyanates, curcumins, and d-limonene (X15499193). Those last four suggest a biblical chutney: the garlic and onion for organic sulfur compounds, watercress and horseradish for isothiocyanates, turmeric for curcumin, and citrus peels for limonene. Lu et al. (2003) showed that d-limonene exerts a cytotoxic effect on gastric cancer by inducing apoptosis (X12921557). Parija and Das (2003) and Kaji et al. (2001) reiterate the well-known anticarcinogenic activity of d-limonene, mentioning also its chemopreventive nature in hepatocarcinogenesis (X12688534;

X11433412). And during the month that we focused on lung cancer, PubMed only had 11 citations regarding limonene and lung cancer. If I were diagnosed with lung cancer today, I would be ingesting more garlic and more citrus too. I might even be inhaling limonene-containing essential oils as I squeezed citrus peels. Raphael and Kuttan (2003) note that several naturally occurring monoterpenes, such as carvone, limonene, and perillic acid, inhibit experimental lung metastasis. (Limonene at 100 μM/kg body weight, 10 doses ipr, remarkably reduced metastatic tumor nodule formation.) Limonene and perillic acid just might even inhibit the metastatic progression of melanoma cells (X14582701). I doubt that Witschi (2000) would join me in judiciously inhaling citrus essential oils. Under the catchy title, "Successful and Not So Successful Chemoprevention of Tobacco Smoke-Induced Lung Tumors," Witschi notes that none of the following "chemopreventive" agents — green tea, phenethyl isothiocyanate, acetylsalicylic acid, N-acetylcysteine, 1,4-phenylenebis[methylene]selenocyanate, and the d-limonene — reduced lung tumor multiplicity or incidence (X11195468). I would still be eating my garlic, eating my Brazil nuts, and inhaling my citrus, although no clinical trials have indicated their safety or efficacy. I suspect they will do as much good and will do less harm at a much lower cost — economically, mentally, and physically — than the chemotherapeutic cocktails being offered by the allopaths and their covert sponsors, the pharmaceutical firms. Many people believe more in the faith-based botanical herbs than they do in the ACS and NCI poisons. Me too! Such people have a better chance of being helped by these safer food farmacy items, also improving their odds against the other diseases of modern man, cardiopathy and diabetes and iatrogenesis.

AFRICAN MYRRH (*COMMIPHORA AFRICANA* (A. RICH.) ENGL. ++ BURSERACEAE

NOTES (AFRICAN MYRRH):

... there is <u>bdellium</u> and the onyx stone

Genesis 2:12 (KJV)

Bdellium is just another variant on the many myrrhs. The tree that provides the aromatic gum known as bdellium grew in the territory east of Persia. When the bark was incised, gum would ooze out "the bigness of a white olive." Gum removed from the bark of the tree would soon harden, become transparent and waxlike, and resemble a pearl. In Tabore and Ugugo, Africa, the gum was melted with butter as a perfume. Egyptian women carried pouches of bdellium, for a delightful perfume. In West Africa, the resin is used as an insecticide believed to repel termites. The wood is used for beads, the stems as a chewstick (BIB). Bdellium is a folk cancer remedy for indurations of the liver and sinews, tumors of the spleen, polyps, carcinomata, and scirrhus. In tropical Africa, the resinous exudate is sometimes applied as a plaster for fever and spasms. Washed bark, mixed with salt is used for snakebite. Pounded leaves with millet are taken with milk as a stomachic. Ronga use the remedy for stomach troubles. The plant is regarded as a stomachic and collyrium. West Africans hold their face over the steaming pot for eye inflammations (BIB). Because of its readiness to strike root from cuttings, many ethnic groups associate the tree and the myrrh with immortality. For example, African Tuaregs consider the plant a symbol of immortality (UPW).

COMMON NAMES (AFRICAN MYRRH):

I find it confusing that AHP designated myrrh as the Standardized Common Name for *Commiphora africana*, *Commiphora erythracea*, *Commiphora madagascariensis*, *Commiphora molmol*, *Commiphora myrrha*, and *Commiphora schimperi*, while Zohary says myrrh is identical with *Commiphora abyssinica*, which the USDA Nomenclature database equates with *Commiphora habessinica*.

FIGURE 1.30 African Myrrh (*Commiphora Africana*).

Adras (Arab.; Sen.; UPW); Adres (Arab.; Mali; Mauritania; UPW); African Bdellium (Eng.; UPW); African Myrrh (Eng.; UPW); Badadi (Fula; Mali; UPW); Barakanti (Bambara; Upper Volta; UPW); Bdellium (Eng.; JLH); Bdellium d'Afrique (Fr.; UPW); Gafal (Arab.; Niger; UPW); Kuénu (Togo; UPW); Kussum (Chamba; Nig.; UPW); Myrrh Africaine (Fr.; UPW); Narga (Ghana; UPW).

Activities (African Myrrh):

Antiseptic (f; UPW); Collyrium (f; BIB); Emmenagogue (f; UPW); Hemostat (f; UPW); Insecticide (f1; UPW); Purgative (f; UPW); Sedative (f; UPW); Soporific (f; UPW); Stomachic (f; UPW); Taenicide (f; UPW); Termitifuge (f; UPW); Vermifuge (f; UPW).

Indications (African Myrrh):

Arthrosis (f; JLH); Bleeding (f; UPW); Cancer (f; JLH); Carcinoma (f; JLH); Childbirth (f; UPW); Chill (f; UPW); Conjunctivosis (f; UPW); Dermatosis (f; UPW); Diarrhea (f; UPW); Fatigue (f; UPW); Gastrosis (f; UPW); Hepatosis (f; JLH); Induration (f; JLH); Infection (f; UPW); Insanity (f; UPW); Insomnia (f; UPW); Leprosy (f; UPW); Obesity (f; UPW); Ophthalmia (f; UPW); Pain (f; UPW); Polyp (f; JLH); Respirosis (f; UPW); Rhinosis (f; JLH); Snakebite (f; UPW); Sore (f; UPW); Splenosis (f; JLH); Sterility (f; UPW); Stiffness (f; UPW); Sting (f; UPW); Stomachache (f; UPW); Tumor (f; JLH); Worm (f; UPW).

Dosages (African Myrrh):

FNFF = !
Roots are dug and consumed raw in parts of Africa; fruit pulp also eaten.

- Africans chew the stems to cleanse the teeth (BIB).
- Ivory Coastals and Upper Voltans take the decoction for male sterility (UPW).
- Kenyans use the fruit as an oral hemostat and styptic (UPW).
- Nigerians use root decoction as taenicide (60 g powdered seed in water for tapeworm); Nigerian Fula use decoction for insanity (UPW).
- Ronga use the remedy for stomach troubles (UPW).
- Tanganyikans poultice stiff neck caused by chills with root decoction, and take the root decoction for childbirth, diarrhea, and stomachache; they eat pounded bark for malaria (UPW).
- Tropical Africans apply the resinous exudate as a plaster for fever and spasms (BIB).
- West Africans hold their face over the steaming pot for eye inflammations (BIB).

Natural History (African Myrrh):

Foliage readily grazed by cattle, goats, and sheep (UPW).

BALM OF GILEAD (*COMMIPHORA GILEADENSIS* (L.) C. CHR.) + BURSERACEAE

Synonyms:

Amyris gileadensis L.; *Amyris opobalsamum* L. *Commiphora opobalsamum* (L.) Engl.; *Commiphora opobalsamum* var. *gileadensis* Engl.

Notes (Balm of Gilead):

> **[T]hey traded in thy market wheat of Minnith, and Pannag, and honey, and oil, and *balm*.**
>
> **Ezekiel 27:17 (KJV)**

Balm is the gum or thickened juice exuding from the balsam tree, once prolific in Judea, and sometimes called balm of Gilead, an emblem of Palestine. Cultivated shrubs were protected by guards. The tree, native to Yemen, was believed to have been introduced to Palestine by the Queen of Sheba on her famous visit to King Solomon. It was later cultivated in Palestine, mostly around Jericho, where it still existed during the Roman conquest. The soldiers carried balm branches back to Rome as symbols of their having defeated the Hebrews.

Common names (Balm of Gilead):

Balasan (Arab.; Syria; HJP); Balm of Gilead (Eng.; Scn.; AH2; HJP); Balsam (Eng.; HJP); Balsam Makkah (Arab.; Syria; HJP); Gilead Balsam (Eng.; HOC); Jibbali (HOC); Mecca Balsam (Eng.; Ocn.; AH2); Mecca Myrrh (Eng.; Ocn.; AH2); Opobalsamum (Eng.; Ocn.; AH2); Sukof (Arab.; Dho.; GHA).

FIGURE 1.31 Balm of Gilead (*Commiphora gileadensis*).

Activities (Balm of Gilead):

Antioxidant (1; X15702514); Antiseptic (f; BIB); Antiulcer (1; X15814261); Astringent (f; BIB; EFS); Bradycardic (1; X9292417); Carminative (f; DAW; HJP); Demulcent (f; EFS); Digestive (f; DAW); Diuretic (f; EFS); Febrifuge (f; HOC); Fumitory (f; HJP); Gastroprotective (1; X15814261); Hepatoprotective (1; X15702514); Hypotensive (1; X9292417); Sudorific (f; EFS); Urogenital (f; EFS); Vulnerary (f; EFS; X15814261).

Indications (Balm of Gilead):

Bite (f; GHA); Cacoethes (f; BIB); Cancer, bladder (f; BIB); Cancer, breast (f; BIB); Cancer, eye (f; BIB); Cancer, gum (f; BIB); Cancer, kidney (f; BIB); Cancer, liver (f; BIB); Cancer, mouth (f; BIB); Cancer, rectum (f; BIB); Cancer, stomach (f; BIB); Cancer, uterus (f; BIB); Cancer, vagina (f; BIB); Carcinoma (f; BIB; JLH); Cardiopathy (1; X9292417); Cold (f; BIB); Cystosis (f; BIB); Dermatosis (f; HOC); Dyspepsia (f; HOC); Eczema (f; HOC); Fever (f; HOC); Flu (f; HJP); Gas (f; HOC); Gastrosis (f1; HJP; X15814261); Gonorrhea (f; EFS); Hepatosis (f1; JLH; X15702514); High Blood Pressure (1; X9292417); Induration (f; JLH); Infection (f; BIB); Infertility (f; HOC); Mastosis (f; BIB); Nephrosis (f; BIB); Plague (f; HOC); Proctosis (f; BIB); Sclerosis (f; JLH); Shingle (f; HOC); Shivering (f; BIB); Sore (f; HOC); Splenosis (f; BIB); Ulcer (1; X15814261); Urogenitosis (f; EFS); Uterosis (f; BIB); Vaginosis (f; BIB); Wound (f; EFS; HJP; X15814261).

Dosages (Balm of Gilead):

FNFF = !

Stem juice sweet when chewed; ripe fruits edible; leaves used for fodder; underbark used as tea substitute (GHA; HOC).

- Christians dissolved bits of myrrh in wine as a digestive (BIB).
- Dhofari Arabs rub the resin over the body as a deodorant (GHA).
- Dhofari Arabians use the bark and resin as a cosmetic, and a soothing agent for dogbite (GHA).
- Dhofari Arabs use water from boiled bark to lighten skin color (GHA).
- Isfahans swallowed a few fruits, whole, for colds and shivering (BIB).
- Lebanese use the myrrhs similarly, as carminative, fumitory, and vulnerary, using dried fruits for gastric problems and flu. They direct the smoke onto wounds (HJP).

ABYSSINIAN MYRRH (*COMMIPHORA HABESSINICA* (O. BERG) ENGL.) + BURSERACEAE

Synonyms:

Balsamodendrum habessinicum O. Berg; *Commiphora abyssinica* (O. Berg) Engl., orth. var.

Notes (Abyssinian Myrrh):

Your robes are all fragrant with myrrh, and aloes and cassia.

Psalms 45:8 (RSV)

And when they were come into the house, they saw the young child with Mary his mother, and fell down, and worshipped him: and when they had opened their treasures, they presented unto him gifts; gold, and frankincense, and myrrh.

Matthew 2:11 (KJV)

And going into the house they saw the child with Mary his mother, and they fell down and worshiped him. Then, opening their treasures, they offered him gifts, gold and frankincense and myrrh.

Matthew 2:11 (RSV)

And when they went into the house they saw the young child with Mary its mother, and, falling down, they did obeisance to it. They also opened their treasures and presented it with gifts, gold and frankincense and myrrh.

Matthew 2:11 (NWT)

Following those two biblical quotes, Zohary says, "*Mor* is justly translated 'myrrh', which is identical with the species of *Commiphora abyssinica*," (ZOH) now orthographically corrected to *Commiphora habessinica*. But Zohary goes on to say, "Further investigation is required to confirm the identification." (ZOH) Conveniently, this is alphabetically the first of 20 species (there are about 200 species in this confusing genus) covered by the USDA Nomenclature Database, until you change the orthography. Then, *C. africana* is first alphabetically. Neither Zohary nor I know which species is the myrrh of the Bible. But Zohary selected this one; in my first Bible book (BIB), I selected *C. africana* as bdellium and *C. myrrha* as myrrh, after wrestling with the long lists of candidates. With a little lobbying input from my editorial assistants, I honed in on *C. erythraea* in my second Bible book (BI2). I am not embarrassed to side with Zohary's selection of *C. habessinica*. After all, Zohary has lived most of his life and his career as a botanist in the Holy Land, and I have spent no more than a total of 6 weeks in the Holy Land (Egypt, Israel). Zohary notes that myrrh (of undetermined specific identity), most precious of the resins, is connected with both the birth and death of Jesus (ZOH). Because of its readiness to strike root from cuttings, many ethnic groups associate the tree and the myrrh with immortality.

Common Names (Abyssinian Myrrh):

Abyssinian Myrrh (Eng.; USN); Arabian Myrrh (Eng.; USN); Boeotian Myrrh (Eng.; JLH); Maceron (?; JLH); Medigeh (Arab.; GHA); Mirra (Eng.; JLH); Mrr (Arab.; Dho.; GHA); 'Okor (Arab.; GHA); Opopanax (Eng.; JLH); Smyrna Boiotike (Eng.; JLH); Yemen Myrrh (Eng.; USN).

Activities (Abyssinian Myrrh):

Anthelmintic (f; GHA); Antiseptic (f; GHA); Astringent (f; UPH): Stimulant (f; UPH); Stomachic (f; UPH).

Indications (Abyssinian Myrrh):

Adenopathy (f; GHA); Cancer (f; JLH); Carcinoma (f; JLH); Chest Cold (f; GHA); Cold (f; GHA); Cough (f; GHA); Diarrhea (f; GHA); Dyspepsia (f; GHA); Dyspnea (f; GHA); Infection (f; GHA); Pulmonosis (f; GHA); Scirrhus (f; JLH); Sore (f; JLH); Swelling (f; GHA); Worm (f; GHA).

Dosages (Abyssinian Myrrh):

FNFF = !
Arabians peel and chew the roots for their sweet taste; branches sucked for sustenance; ripe berries eaten (GHA).

- Arabians inhale the smoke of burning resin for breathing disorders, chest colds, and swollen glands (GHA).
- Arabians take the resin orally for coughs, diarrhea, dyspepsia, and worms (GHA).
- Arabians use resin for chest and other infections, applying it externally as a disinfectant (GHA).

MYRRH (*COMMIPHORA MYRRHA* (NEES) ENGL. AND OTHER SPP.) ++ BURSERACEAE

Synonyms:

Balsamodendron myrrha Nees; *Commiphora molmol* (Engl.) Engl.; *Commiphora myrrha* var. *molmol* Engl. *fide* USN

Notes (Myrrh):

My beloved put in his hand by the hole of the door, and my bowels were moved for him. I rose up to open to my beloved; and my hands dropped with myrrh, and my fingers with sweet smelling myrrh, upon the handles of the lock.

Song of Solomon 5:4–5 (KJV)

My beloved put his hand to the latch, and my heart was thrilled within me. I arose to open to my beloved, and my hands dripped with myrrh, my fingers with liquid myrrh, upon the handles of the bolt.

Song of Solomon 5:4–5 (RSV)

My dear one himself pulled back his hand from the hole [of the door], and my inward parts themselves became boistrous within me. I got up, even I, to open to my dear one, and my own hands dripped with myrrh and my fingers with liquid myrrh; upon the hollows of the lock.

Song of Solomon 5:4–5 (NWT)

FIGURE 1.32 Myrrh (*Commiphora myrrha*).

All three versions seethe with sex and myrrh, a not uncommon linkage in the Bible. But which myrrh? I doubt that Herbal PDR and Commission-E writers are any wiser than the wise men of the Nativity; or Felter in 1898; or Madaus in 1938; or even me in 1985, 1999, 2005, or today, at knowing which species of *Commiphora* is myrrh, especially when faced only with the imported gum, and not flowering and fruiting herbarium vouchers. Myrrh shows up in at least a dozen places in ten books of the Bible. We will never know which of many species it was, of the complicated genus *Commiphora*. Many authors equate this one with the biblical myrrh, but none with final authority. PH2 entries below were derived from Gruenwald's entry for *C. molmol*. The USDA (USN) and

Philips (HJP) equate *C. molmol* with *C. myrrha*. AH2 treats them as distinct species but with the same standardized common name, myrrh (AH2). Hence, the data below may be attributed to either species, whether or not the source of the data believed that they are one and the same.

Common Names (Myrrh):

African Myrrh (Eng.; USN); Bálim-Tra-Pólam (Tel.; DEP); Baisabole (India; EFS); Balasan (Arab.; Syria; HJP); Balintrap-Dum (Tel.; NAD); Balsam Makkah (Arab.; Syria; HJP); Bhensa (Bom.; NAD); Bhensabol (India; EFS); Bisabole (India; EFS); Ból (Beng.; Bom.; Cutch; Dec.; Guj.; Hindi; Iran; DEP; EFS; NAD); Bóla (Ayu.; Kan.; Sanskrit; AH2; DEP; EFS); Bólam Gandarassa (Singh.; DEP; NAD); Bysabole (India; EFS); Common Myrrh (Eng.; Ocn.; AH2); Gandharas (Beng.; NAD); Gandha-Rasaha (Beng.; Sanskrit; DEP); Habaghadi (Arab.; EFS); Habak (Arab.; EFS); Heerabole (India; DEP; EFS); Herabol Myrrh (Eng.; USN); Hírábói (Beng.; DEP); Hírábol (Cutch; Guj.; Mah.; DEP; NAD); Hirabol Myrrh (Eng.; Ocn.; AH2); Karam (India; EFS); Minaharma (Sanskrit; EFS); Mirha (Tur.; EFS); Mirra (It.; Sp.; EFS; USN); Molmol (Eng.; Ocn.; AH2); Mor (Heb.; DEP); Mo Yao (Pin.; AH2; DAA); Mo Yao Shu (Pin.; AH2); Mukula (Arab.; EFS); Mur (Arab.; DEP); Murr (Arab.; DEP; GHA; NAD); Myrrh (Arab.; Eng.; Scn.; AH2; CR2; GHA); Myrrha (Den.; EFS); Myrrhabaum (Ger.; EFS); Myrrhe (Fr.; Ger.; EFS; USN); Myrrheboom (Dutch; EFS); Rasagandha (Sanskrit; EFS; NAD); Rasagandhaha (Sanskrit; DEP); Saindhava (Sanskrit; NAD); Samudraguggul (Sanskrit; NAD); Somali Myrrh (Eng.; Ocn.; AH2); Subr (Arab.; GHA); Vellaipa-Pòlam (Tam.; DEP; NAD); Vola (Sanskrit; NAD); Mo Yao (Pin.).

Activities (Myrrh):

Abortifacient (f; DEP); Analgesic (1; APA; BGB; WI3); Anesthetic (1; WI3; JNP64:1460); Anticlastogenic (1; X7518189); Antiedemic (1; CAN; X9582001); Antiinflammatory (1; APA; BGB; PNC); Antioxidant (1; X9032627); Antipyretic (1; APA; CAN); Antiradicular (1; X9032627); Antirheumatic (1; BGB); Antiseptic (1; APA; BGB; FEL; PH2; PNC; SKY; WI3); Antispasmodic (f1; APA; DEP; EFS; PNC); Antithrombotic (1; X10353165); Antitumor (1; X7956458); Antiulcer (1; X9032627); Astringent (f12; APA; DEP; KOM; NAD; PH2; SKY; WI3); Bactericide (1; WI3; JNP64:1460); Carminative (f; BGB; PHR; PNC; WI3); Collyrium (f; DEP); Cytotoxic (1; JNP64:1460); Decongestant (f1; APA; DEP); Deodorant (1; BGB; HHB; WI3); Digestive (f; GHA; PH2); Emmenagogue (f; APA; DEP; FEL; NAD); Expectorant (f; DEP; EFS; FEL; NAD; PHR; PH2; PNC; WI3); Fasciolicide (12; X5125536); Fungicide (1; WI3; JNP64:1460); Gastroprotective (1; X9032627); Haematogenic (1; X11390128); Hepatoprotective (1; X15125513); Hypocholesterolemic (1; CAN); Hypoglycemic (1; APA; CAN); Hypotriglyceridemic (1; CAN); Immunostimulant (1; APA; PNC); Lactagogue (f; DEP); Laxative (f; NAD); Larvicide (1; X11478451); Lipolytic (1; CAN); Mitodepressant (1; X7518189); Mosquitocide (1; X11478451); Mucogenic (1; X9032627); Orexigenic (f; PH2); Prostaglandigenic (1; X9032627); Schistosomicide (1; X15125513); Stimulant (f; APA; EFS; FEL; NAD); Stomachic (f; EFS; NAD); Thyrostimulant (1; X9032627); Tonic (f; DEP; EFS); Vermifuge (f; DEP); Vulnerary (f; PNC).

Indications (Myrrh):

Abrasion (1; CAN); Adnexitis (f; MAD); Alopecia (f; MAD; NAD); Amenorrhea (f; BGB; DEP; FEL; MAD; NAD; PH2); Aphthae (1; CAN); Arthrosis (1; GHA); Arteriosclerosis (f; MAD); Asthma (f1; APA; DEP; FEL); Athlete's Foot (1; SKY); Bacteria (1; JNP64:1460); Bedsores (f; APA); Bladder stone (f; BIB); Boils (f; PNC); Bronchosis (f1; APA; BGB; DEP; FEL); Bruise (f; BOW); Cancer (f; APA; PH2); Cancer, abdomen (f; PH2); Cancer, colon (f; PH2); Candidiasis (f1; BGB; NAD); Canker Sore (1; APA; SKY); Carbuncle (f; PH2); Caries (f; FEL; NAD); Catarrh (f; BGB; CAN; FEL); Chilblain (f; BIB); Chlorosis (f; BIB; DEP; NAD); Circulosis (f; BOW); Cold (f1; BGB; CAN; GHA; SKY); Congestion (f1; APA; BGB; DEP); Consumption (f; MAD); Cough (f; PH2); Dandruff (1; JAR12:83); Decubitis (f; BGB; BOW); Dermatosis (1; APA; MAD; PH2; WI3); Diabetes (f1; JNP64:1460; X12506289); Diarrhea (f; MAD; JNP64:1460); Dicrocoeliasis (12; X15287191); Diptheria (f; NAD); Dropsy (f; BIB); Dysentery (f; MAD); Dyslactea (f; DEP); Dysmenorrhea (f1; BGB; NAD; PH2); Dyspepsia (f; APA; DEP; FEL; GHA; NAD); Dysuria (f; MAD); Earache (f; BIB); Edema (1; X9582001); Enterosis (f; DEP; PH2); Epilepsy (f; DEP); Erysipelas (f; MAD); Fascioliasis (12; X5125536); Fever (f; BIB; DEP; GHA; MAD); Fracture (f; GHA); Freckle (f; MAD); Fungus (1; JNP64:1460); Furunculosis (1; CAN; PH2); Gangrene (f; FEL); Gas (f; APA; DEP; MAD); Gastrosis (f; FEL; PH2; PNC; JNP64:1460); Gingivosis (f1; APA; DEP; FEL; PNC; SKY); Gleet (f; FEL); Gonorrhea (f; FEL); Halitosis (f; FEL); Hemorrhoids (f; APA; BGB; BIB; GHA); Hepatosis (f1; MAD; X15125513); Hoarseness (f; APA); Hypothyroidism (1; WAF); Impotence (f; GHA); Infection (f12; DEP; PH2; JNP64:1460); Infertility (f; MAD); Inflammation (f1; BGB; DEP; GHA; PH2; WI3); Itch (f1; WI3); Laryngitis (f; FEL); Leprosy (f; APA); Leukorrhea (f; FEL; MAD); Menopause (1; BGB); Menorrhagia (f; MAD); Mononucleosis (f; BOW); Mucososis (f1; APA; DEP; FEL; PH2; WI3); Mycosis (1; JNP64:1460); Odontosis (f; MAD); Ophthalmia (f; BIB; DEP); Orchosis (f; DEP); Osteoalgia (f; BGB); Otosis (f; BOW); Pain (1; JNP64:1460); Parasite (f12; DEP; X15287191); Pharyngosis (f12; APA; FEL; KOM; MAD; PH2; PNC; WI3); Phthisis (f; NAD); Pruritis (f1; WI3); Pulmonosis (f; DEP; MAD); Respirosis (f; BGB); Rheumatism (f; BGB); Rhinosis (f; APA; BIB); Salpingitis (f; MAD); Schistosomiasis (12; X15287168; X15125513); Sinusitis (1; APA); Snakebite (f; GHA); Sore (f1; APA; DEP; FEL; PNC); Sore Throat (f12; BGB; DEP; FEL: KOM; MAD; SKY); Stitch (f; MAD); Stomatosis (f12; APA; DEP; KOM; MAD; NAD; PH2; PIP; WI3); Swelling (f1; APA; X9582001); Thrombosis (1; X10353165); Tonsilosis (1; APA; BGB; FEL; PNC); Toothache (f; GHA); Tuberculosis (f; MAD); Tumor (f1; DEP; X7956458); Ulcer (f; APA; PH2; X11113992); Uterosis (f; MAD); Uvulosis (f; FEL); Venereal Disease (f; FEL); Water Retention (f; MAF); Weaning (f; GHA); Worm (f; DEP; FEL; MAD); Wound (f; APA; BGB); Wrinkle (f; MAD); Yeast (f1; BGB; NAD).

Dosages (Myrrh):

FNFF = !!

Myrrh, more food additive than food, is used to flavor baked goods, beverages, candy, chewing gums, frozen desserts, gelatins, meat, puddings, soft drinks, Swedish bitters (FAC); myrrh dissolved in water used in Arabia to flavor coffee (GHA). 1 tsp powdered myrrh/cup water/1–2 ×/day (APA); 5–10 drops tincture per glass water (for mouthwash or gargle) (APA); 8–10 drops myrrh extract to 4 ×/day (APA); 2.5–5.0 ml myrrh tincture (CAN; PNC); 0.3–1.2 g resin/day (HHB). 0.3–1.5 g (MAD); 6–10 drops tincture, several times a day (MAD); 1–2 ml tincture 3 ×/day (SKY); 1 g resin 3 ×/day (SKY); 1/8–1/4 tsp myrrh tincture 3 x/daily (WAF).

- Arabians smear resin on a black cloth that, after hardening, is used to bind fractures (GHA).
- Asian Indians dissolve myrrh in mother's or asses' milk as a collyrium (DEP).
- Asian Indians give myrrh with gúr to increase flow of milk (DEP).
- Asian Indians mix borax with myrrh for parasitic stomatitis or thrush (NAD).
- Asian Indians mix myrrh tincture with glycerine for diptheria (NAD).
- Asian Indians suggest myrrh tincture for chlorosis and dysmenorrhea in young girls (NAD).
- Dhofari soak the resin in water and drink it or rub it on the body for fever (GHA).
- Lebanese use the myrrhs similarly, as carminative, fumitory, vulnerary, using dried fruits for gastric problems and flu. They direct the smoke on wounds (HJP).
- Omani apply myrrh to caries heating the myrrh with a hot rod for toothache (GHA).
- Saudi apply the resin to the breast to wean babies (GHA).
- Yemeni paste myrrh on snakebites and wounds; on the penis as an aphrodisiac (GHA).

Downsides (Myrrh):

None known (KOM; PHR). Class 2b. Emmenagogue; uterotonic. Contraindicated in uterorrhagia. Doses >2–4 g may cause diarrhea and nephrosis. French only permit external application (AHP, 1997). Undiluted tincture may produce burning and local irritation (AEH). Newall, Anderson, and Phillipson (1996) caution that because it is reputed to affect the menstrual cycle, its use in pregnancy and lactation should be avoided. May interfere with diabetic therapies. In view of the lack of toxicological data, excessive use should be avoided (CAN). "No adverse effects from myrrh have been reported" (SKY). Apprehension, diarrhea, hiccups, and restlessness have been reported as side effects of related gugulipid administration (CAN).

Extracts (Myrrh):

Hypoglycemic; resin kills germs and stimulates macrophages (SKY); astringent myrrh soothes gingivosis, stomatosis, and sore throat. Extracts of *C. habessinica* stimulate phagocytosis in mice inoculated with *Escherichia coli.* Other species of *Commiphora* have demonstrated antiaggregant, antioxidant, cardioprotective, hypocholesterolemic, and hypotriglyceridic activities, suggesting the utility of these gums, like many other gums, in preventing and moderating heart disease. I kind of like the idea of a gum for a gum disease.

POISON HEMLOCK (*CONIUM MACULATUM* L.) X APIACEAE

Synonyms:

Cicutaria vulgaris Clus; *Conium major* Bauh.; *Conium vulgaris major* Park

Notes (Poison Hemlock):

… Judgment springeth up as <u>hemlock</u> in the furrows of the field …

Hosea 10:4 (KJV)

[F]or you have turned judgment into gall, and the fruit of righteousness into <u>hemlock</u>.

Amos 6:12(KJV)

These are the only citations for hemlock in the King James Version online. The poison that Socrates took, hemlock, is too dangerous for herbal administration by the uninitiated. Some authors (e.g., Walker) relate the biblical hemlock to *Conium*, a truly dangerous medicinal plant, while Moldenke and Moldenke relate it to a more innocuous herb, closer to wormwood. Zohary, admitting that the Hebrew word *rosh*, generally meaning poison, more likely means a "bitter and poisonous drink or food." Without saying definitively that hemlock is intended, Zohary does recount that the plant is poisonous, and quite possibly the poison drink of Socrates, yet praised by Avicennia for breast tumors (BIB).

Common Names (Poison Hemlock):

Anisillo (Ven.; AVP); Ansarinha Malhada (Port.; EFS); Baldiran (Tur.; EFS); Barbousha (Arab.; BOU); Beaver Poison (Eng.; GMH); Bikhe Shoukaran (Arab.; BOU); Bisbis Barri (Arab.; Syria; HJP); Carrot-fern (Austr.; Eng.; USN); Cashes (Eng.; BUR); Cegude (Por.; EFS); Cerfeuil (Fr.; AVP); Cicuta (Braz.; Col.; Peru; Por.; Sp.; Ven.; AVP; EFS; EGG; VAD); Cicuta Magguire (It.; EFS); Cicuta Mayor (Sp.; EFS); Ciguë (Fr.; JLH); Ciguë Officinale (Fr.; EFS); Ciguë Tachée (Fr.; BOU); Conoi (Arg.; JLH); Cow Bane (Eng.; BUR); Culantrillo (Peru; Sp.; EGG); Culén (Peru; Sp.; EGG); Deadly Hemlock (Eng.; BUR); Djerir (Arab.; BOU); Doll Kraut (Ger.; EFS); Dolle Kervel

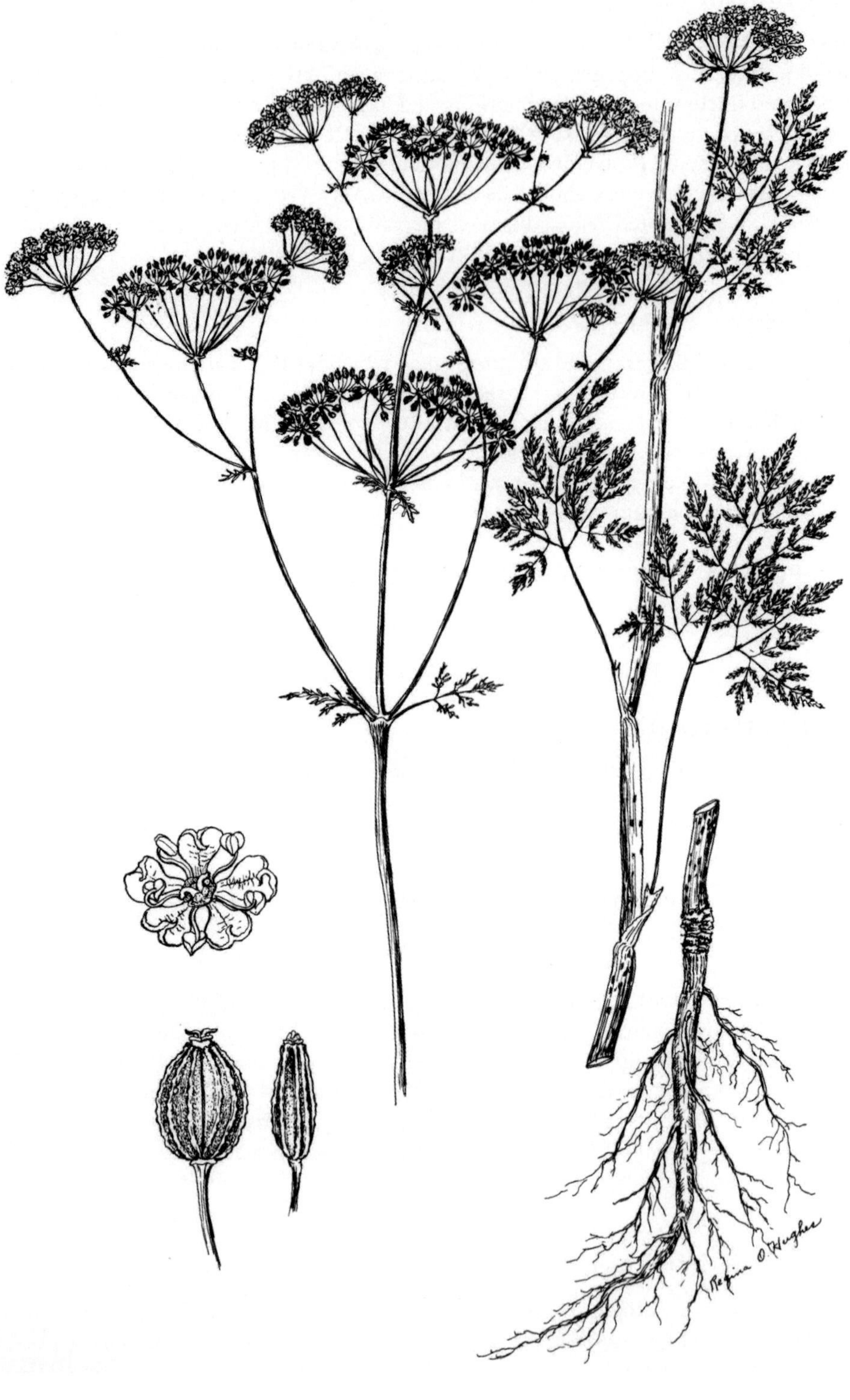

FIGURE 1.33 Poison Hemlock (*Conium maculatum*).

(Dutch; EFS); Echte Schierling (Ger.; EFS); Fool's-parsley (Eng.; USN); Geflechter Giftschierling (Ger.; EFS); Gevlekte Scheerling (Dutch; EFS); Gift Petersilie (Ger.; EFS); Grande Ciguë (Fr.; JLH); Guebaba (Ber.; BOU); Harmal el Djezair (Arab.; BOU); Hemlock (Eng.; USN); Herb Bennet (Eng.; GMH); Herb Bonnet (Eng.; BUR); Isojuuri (Fin.; JLH); Kecksies (Eng.; GMH); Kex (Eng.; GMH); Kill Cow (Eng.; BUR); Kita Anis (Peru; Sp.; EGG); Kurdumana (India; EFS; SKJ); Monte Zanahoria (Peru; Sp.; EGG); Mushquash Root (Eng.; GMH); Odört (Swe.; EFS); Poison Hemlock (Eng.; CR2; USN; ZOH); Poison Parsley (Eng.; BUR); Poison Root (Eng.; BUR); Poison Snakeweed (Eng.; BUR); Rosh (Heb.; ZOH); Sakaran (Arab.; ZOH); Saûvadge Feno (Belgium; JLH); Sellata (Ber.; BOU); Shawkaran (Arab.; BOU); Sikran (Arab.; BOU); Skarntyde (Den.; EFS); Spotted Corobane (Eng.; GMH); Spotted Cowbane (Eng.; BUR); Spotted Hemlock (Eng.; EFS; USN); Spotted-parsley (Eng.; USN); Stinkweed (Eng.; BUR); Tahhmâ (Arab.; Syria; HJP); Viznaga (Eng.; BUR); Wild Hemlock (Eng.; BUR); Wodewhistle (Eng.; BUR); Ziata (Arab.; BOU); Nscn.

Activities (Poison Hemlock):

Alterative (f; CRC); Analgesic (f; BUR; CRC); Anaphrodisiac (f; BIB; CRC); Anodyne (f; BIB; CRC); Antidote (Strychnine) (f; GMH); Antinicotinic (1; PH2); Antispasmodic (f; BIB; CRC; EFS); Aphrodisiac (f; BIB; CRC; EFS; SKJ); Cardiodepressant (f; BUR); Hypertensive (1; PH2); Hypotensive (1; PH2); Insecticide (1; EGG); Negatively Inotropic (1; PH2); Nervine (f; BIB; CRC; EFS); Nicotinic (1; PH2); Paralytic (1; PH2); Poison (12; BIB; DEM; EFS; PH2); Respirodepressant (1; PH2); Respirostimulant (1; PH2); Sedative (f; BIB; BUR; CRC); Teratogenic (1; PHR; PH2).

Indications (Poison Hemlock):

Adenopathy (f; CRC; JLH; PH2); Amenorrhea (f; CRC); Arteriosclerosis (f; BOU); Arthritis (f; BIB; PHR; PH2); Asthma (f; BUR; CRC; PHR; PH2); Backache (f; PHR; PH2); Bite (f; BOU); Bronchosis (f; CRC; PHR; PH2); Burn (f; AAH); Cachexia (f; CRC); Cacoethes (f; CRC); Cancer (f; CRC); Cancer, breast (f; CRC; JLH); Cancer, colon (f; JLH); Cancer, face (f; JLH); Cancer, ganglia (f; JLH); Cancer, gland (f; JLH); Cancer, intestine (f; CRC); Cancer, liver (f; JLH); Cancer, mesentery (f; JLH); Cancer, neck (f; JLH); Cancer, nose (f; CRC); Cancer, pancreas (f; CRC); Cancer, parotids (f; CRC); Cancer, penis (f; JLH); Cancer, scrotum (f; JLH); Cancer, skin (f; CRC; JLH); Cancer, spleen (f; JLH); Cancer, sternum (f; CRC); Cancer, stomach (f; JLH); Cancer, testicle (f; JLH); Cancer, uterus (f; CRC; JLH); Cancer, viscera (f; CRC); Carcinoma (f; CRC); Carditis (f; CRC); Cerebrosis (f; PH2); Chorea (f; CRC); Colic (f; CRC); Cough (f; CRC; PHR); Cramp (f; PHR; PH2); Delirium (f; CRC); Depression (f; PH2); Dermatosis (f; CRC); Dyslactea (f; CRC); Dysmenorrhea (f; CRC); Eczema (f; CRC); Edema (f; JLH); Enterosis (f; JLH); Epilepsy (f; CRC; PHR; PH2); Erysipelas (f; CRC); Heart (f; CRC); Hepatosis (f; JLH); Hiccup (f; HJP); Hyperlactation (f; CRC); Hepatosis (f; CRC; JLH); Icterus (f; CRC); Impotence (f; CRC); Induration (f; JLH); Jaundice (f; CRC); Laryngismus Stridulus (f; CRC); Leprosy (f; CRC); Mania (f; CRC; GMH); Myosis (f; PHR; PH2); Neck (f; PHR); Nervousness (f; HJP); Neuralgia (f; BUR; CRC; PHR; PH2; VAD); Neurosis (f; BIB); Orchosis (f; JLH); Pain (f; BOU; CRC; JLH); Palpitation (f; CRC); Pancreatosis (f; CRC; JLH); Paralysis (f; CRC); Pareses (f; PH2); Parotosis (f; JLH); Pertussis (f; CRC; GMH); Prostatosis (f; BOU); Pylorosis (f; PH2); Rheumatism (f; BOU; BUR; CRC; PHR); Sciatica (f; PHR; PH2); Scirrhus (f; JLH); Sclerosis (f; CRC); Scrofula (f; CRC; GMH); Sore (f; AAH; CRC); Spasm (f; CRC); Splenosis (f; JLH); Stye (f; CRC); Swelling (f; CRC); Syphilis (f; CRC); Tetanus (f; CRC; PHR); Tumor (f; CRC); Ulcer (f; CRC); Vertigo (f; CRC); Wen (f; JLH); Wound (f; AAH).

Dosages (Poison Hemlock):

FNFF = X

Do not use (JAD). Many of the indications are homeopathic. Maximum dose 300 mg; standard dose 100 mg; not to exceed 1500 mg/day (HHB; PHR; PH2).

- Arab and Greek physicians consider hemlock for arthrosis, cancer, dermatosis, hydrophobia, pain, scrofula, sores, and swelling (HJP).
- Asian Indians report using the herb to paralyze the nerves in painful dermatoses (SKJ).
- British claim to have cured cancer by taking leaf decoctions (AAH).
- British (not me) suggest hemlock juice for cramps, epilepsy in dentition, laryngeal spasms, mania, teething in children, and in inhalations for asthma, bronchitis, and pertussis (GMH).
- English Gypsies use hemlock to cure ulcers (HJP).
- Iranians apply fruits externally as an anodyne (BIB).
- Irish poultice the plant onto burns, rheumatism, sores, swelling, and wounds (AAH).
- Irish use the poisonous plant to treat giddiness. (AAH)
- Lebanese suggest the plant for asthma, cancer, hiccups, nervousness, pain, and whooping cough (HJP).
- North Africans suggest the dangerous tincture for arteriosclerosis, cancer, prostatitis, spasms, the dilute infusion for pain, and rheumatism (BOU).
- Rabat natives fumigate insect bites with the roots (BIB).

Downsides (Poison Hemlock):

"Classed appropriately by the FDA as an unsafe herb containing the poisonous alkaloid coniine and other closely related alkaloids." Plant can also cause contact dermatosis. Ingestion may cause debility, drowsiness, nausea, labored respiration, paralysis, asphyxia, and death. Following lethal doses, animals rapidly begin to show symptoms; among them are paralysis of the tongue, mydriasis, head pressure, giddiness, nausea, vomiting, diarrhea, and collapse into central paralysis (first the feet and legs, then the buttocks, arms, then paralysis of the swallowing and speech). With increasing dyspnea and cyanosis, death ensues through central respiratory paralysis. Lethal dose is about 500 to 1000 mg coniine for man (CRC; HHB). (Note that in CRC (1985), I misquoted HHB and said 500 to 100 mg.)

Natural History (Poison Hemlock):

Larks and quail eat the fruit with impunity, but their meat becomes poisonous in the process. Thrush can eat the fruits but ducks are poisoned thereby (GMH).

CORIANDER (*CORIANDRUM SATIVUM* L.) ++APIACEAE

Synonyms:

Coriandrum diversifolium Gilib.; *Coriandrum globosum* Salisb.; *Coriandrum majus* Gouan

Notes (Coriander):

And the house of Israel called the name thereof Manna: and it was like coriander seed, white; and the taste of it was like wafers made with honey.

Exodus 16:31 (KJV)

Now the house of Israel called its name manna; it was like coriander seed, white, and the taste of it was like wafers made with honey.

Exodus 16:31 (RSV)

And the house of Israel began to call its name "manna." And it was like coriander seed, and its taste was like that of flatcakes with honey.

Exodus 16:31 (RSV)

FIGURE 1.34 Coriander (*Coriandrum sativum*).

All three versions call it coriander, which generally tranquilizes me into believing that this is a reasonably certain biblical citation. But Zohary questions it for linguistic, culinary, and ecological reasons. Coriander does not occur in the desert (ZOH) (although I have seen it in irrigated deserts in Egypt (JAD)). And Zohary rightly comments that its green to brown seeds are not whitish. However, I have seen some that looked ivory colored, not too far from cream. He questions the KJV, RSV, and NWT translations of *gad* as coriander. And, the Arabic *gidda,* cognate with *gad,* is a white *Artemisia*, not a greenish coriander. Thus, while leaving coriander in his biblical book as will I, Zohary doubts that it is intended in the quoted passages. AH2 gives two standardized common names for this species: cilantro for the leaves and coriander for the seed. This will be confusing in places like Peru where this is the temperate species used in the Andes, whereas tropical *Eryngium foetidum*, of the same family, bears the name cilantro in the lowlands. AH2 gives culantro as the standardized common name for *Eryngium foetidum* (AH2). I suspect that both will serve well and similarly for the many indications listed below. Coriander was used in love potions, its use as an aphrodisiac being mentioned in *The Thousand and One Nights* (BIB).

COMMON NAMES (CORIANDER):

Ababika (Sanskrit; KAB); Allaka (Sanskrit; KAB); Behan (Pun.; DEP); Bööbberli (Swiss; POR); Bopchukuksun (Rai; NPM); Brasyal (Tamang; NPM); Chamem (Arm.; POR); Chhatra (Sanskrit; KAB); Chinese Parsley (Ocn.; Por.; AH2; POR; USN); Chinesische Petersilie (Ger.; POR); Chrapfechöörnli (Swiss; POR); Cilantro (Eng.; Scn.; Sp.; AH2; EFS; EGG; USN); Coentro (Por.; POR); Coentro (Mad.; Por.; EFS; USN); Conzra (Arab.; GHA); Col (Essex; KAB); Cominos (Pi.; KAB); Coriander (Eng.; Hun.; Scn.; AH2; NPM; POR; USN); Coriandolo (It.; EFS; POR; USN); Coriandre (Fr.; BOU; EFS; USN); Coriandro (It.; Por.; Sp.; EFS; KAB; POR); Coriándru (Rom.; POR); Cosbor (Malta; KAB); Culantro (Cuba; Ocn.; Pan.; Peru; Pi.; Sp.; AH2; EFS; IED; POR; RyM; USN); Culantro Chino (Ma.; JFM); Culantro de Cartagena (Ma.; JFM); Culantro de Castilla (Ma.; JFM); Culantro Domestico (Ma.; JFM); Daaniwal (Kas.; POR); Danga (Nepal; DEP); Danyalu (Tel.; DEP); Daun Ketumbar (Malaya; POR); Debja (Arab.; BOU); Dembilal (Eth.; POR); Dhaanya (Hindi; POR); Dhaanyakam (Hindi; POR); Dhana (Bom.; Mar.; Sanskrit; DEP; KAB; POR); Dhanak Chi (Tur.; DEP; KAB); Dhanayaka (Sanskrit; POR); Dhane (Beng.; DEP; KAB; POR); Dhaneyaka (Sanskrit; KAB); Dhania (Beng.; Hindi; India; Oriya; Pun.; Urdu; EFS; POR); Dhania Saabut (Hindi; POR); Dhanika (Sanskrit; KAB); Dhaniya (Bhojpuri, Danuwar, Gurung, Magar, Mooshar, Nepal, Sunwar, Tam.; Tharu; Urdu; KAB; NPM; POR; SUW); Dhaniyaa (Hindi; Urdu; POR); Dhaniya Dhap (Tamang; NPM); Dhaniyaka (Sanskrit; KAB; POR); Dhaniyaalu (Tel.; POR); Dhano (Sin.; DEP; KAB); Dhanya (Hindi; Mar.; Sanskrit; DEP; KAB; POR); Dhanyabija (Sanskrit; KAB); Dhanyaka (Ayu.; Sanskrit; AH2; DEP; KAB); Dhanyika (Sanskrit; DEP); Dhennika (Sanskrit; KAB); Dhoney (Beng.; POR); Dimbilal (Eth.; POR); Dyovunco (Ocaina; EGG); Falscher Anis (Ger.; KAB); Gad (Heb., KAB; ZOH); Gakaka (Piro; EGG); Gemeiner Coriander (Ger.; Swe.; KAB; NAD); Geshnes (Iran; POR); Goid (Punic; ZOH); Gosangn (Newari; NPM); Gouzbir (Ber.; BOU); Grain(e) du Coriandre (Fr.; Havija; Kan.; DEP; NAD); Haraa Dhania (Hindi; POR); Havija (Kan.; POR); Hiang T''sai (China; KAB); Hom Phak (Ic.; KAB); Hom Pom (Ic.; KAB); Ho Tuy (Ic.; KAB); Hridyagandha (Sanskrit; KAB); Hsiang Hsui (China; EFS); Huang Thai Tu (Ic.; KAB); Hu Sui (Pin.; China; AH2; DAA; POR); Indische Petersilie (Ger; POR); Iuen Siu (China; KAB); Iuen Soui (China; KAB); Janapriya (Sanskrit; KAB); Kabzara (Arab.; GHA); Kajbira (Arab.; KAB); Kambari (Kan.; POR); Karbijar (Arab.; KAB); Kashmirkhuska (Iran; KAB); Kasriza (Iran; KAB); Katumba (Sumatra; IHB); Katumber (Java; IHB); Katunchar (Sunda; IHB); Kerti Koriander (Tur.; EFS); Kesbour (Arab.; BOU); Ketumbar (Malaya; IHB; POR); Khabzara (Arab.; GHA); Khotbir (Mar.; DEP; KAB); Khotmir (Mar.; DEP; KAB); Kishnetz (Rus.; KAB); Kisnis (Tur.; EB49:406); Kisnish (Tur.; POR); Kishniz (Iran; DEP; EFS); Kizniz (Tur.; EFS); Koendoro (Japan; POR; X11776997); Kolendra (Pol.; POR); Kolendra Siewna (Pol.; POR); Koljandra (Rus.; POR); Konphir (Guj.; KAB; POR); Korander (Dutch; KAB); Koriander (Den.; Dutch; Ger.; Hung.; Nor.; Rus.; Swe.; EFS; KAB; POR); Korianderfrucht (Ger.; POR), Koriandr (Czech.; Rus.; POR); Koriandr Posevnoi (Rus.; POR); Koriandrze (Pol.; KAB); Koriannon (Greek; POR); Korijander (Croatia; POR); Korion (Greek; POR); Koriyun (Greek; DEP); Koryander (Pol.; KAB); Ko Soo (Korea; POR); Ko Su (Korea; POR); Kotamalli (Tam.; KAB); Kotambari (Kan.; DEP; KAB); Kotamrbi-beeja (Kan.; DEP; EFS); Kot Bor (Kon.; KAB); Kot Bori (Kon.; KAB); Kothamali (Tam.; POR); Kothamalli (Tam.; POR); Kothambala (Kas.; POR); Kothambalari Kothambri (Kan.; POR); Kothamira (Bom.; DEP; KAB); Kothambri (Kan.; WOI); Kothimber (Mar.; POR; WOI); Kothimbir (Mar.; POR); Kothmir (Mar.; DEP); Kothmiri (Guj.; POR); Kothmiri Bija (Kan.; POR); Kothumpalari (Mal.; POR); Kothumpalati (Mal.; KAB); Kothumpkalari Bija (Mal.; POR); Kotimiri (Tel.; DEP; KAB; NAD; POR); Kottamali (Mal.; Tam.; DEP; EFS; POR); Kottamalli Virae (Tam.; POR); Kottambari,

Kothambari (Kan.; POR); Kottampalari (Tel.; NAD); Kotthamalie (Sinh; KAB; POR); Kottmir (India; EFS); Kottumburi (Kon.; NAD); Koyendoro (Japan; POR); Kunati (Sanskrit; KAB); Kurbusar (Arab.; DEP; KAB); Kusbara (Arab.; Heb.; BOU; DEP; EFS; POR); Kusbarah (Arab.; DEP); Kushniz (Iran; EFS; KAB); Kusthumbari (Kustumburi) (Sanskrit; EFS; KAB; POR); Kuzbarah (Arab.; KAB; POR); Kuzbura (Arab.; POR); Libdhane (Guj.; POR); Nan Nan Bin (Burma; POR); Nan Nan Zee (Burma; POR); Nau-nau (Burma; DEP; NAD); Ngo (Vn.; EB42:413); Ngo Tham (Ic.; KAB); Nisara (Sanskrit; KAB); Ongsay (Pi.; KAB); Pak Chi (Thai; IHB); Penjilang (Java; Malaya; IHB; POR); Persil Arabe (Fr.; POR); Petite Coriandre (Fr.; KAB); Phààk Kaawm (Laos; POR); Phak Hom (Laos; POR); Pucioagna (Rom.; KAB); Rau mùùi (Vn.; KAB; POR); Rüügeliküümmi (Swiss; POR); Saquil (Ma.; JFM); Schwindelkornerr (Ger.; EFS); Shakayogya (Sanskrit; KAB); Silantro (Peru; EGG); Stinkdill (Ger.; KAB); Sukshmapatra (Sanskrit; KAB); Sugandhi (Sanskrit; KAB); Tabel (Arab.; BOU); Tansanaqua (Ma.; JFM); Tumbaru (Sanskrit; KAB); Tunchar (Sunda; IHB); Uchung (Lepcha; NPM); Ushu (Lad.; MKK); Ussu (Bhoti; KAB); U-su (Tibet; NPM); Vedhaka (Sanskrit; KAB); Veshana (Sanskrit; KAB); Vitunakka (Sanskrit; KAB); Wanzendill (Ger.; POR); Wanzenküümmel (Ger.; POR); Wan Swee (China; POR); Xiang Sui (China; POR); Yan sui (China; POR; USN); Yuan Sui (China; Pin.; AH2; EFS); Yuan Sui Zi (Pin.; AH2); Zaub Thwb Qaib (Hmong; EB57:365); Zaub Txib Gab (Hmong; EB57:365).

Activities (Coriander):

Alexiteric (f; BIB); Analgesic (f; KAB); Anaphrodisiac (f; HHB; JFM); Antidote (Mercury) (1; FNF; X15721537; X8914687); Antiedemic (f1; APA); Antiimplantation (1; APA); Antiinflammatory (f1; PNC); Antimutagenic (1; APA; X15451560); Antinitrosaminic (1; X15451560); Antioxidant (1; JAF51:6961; X15364640); Antiperoxidant (1; X15364640); Antiradicular (1; JAF51:6961); Antiseptic (f1; PH2); Antispasmodic (f1; BGB; HHB; PHR; PH2); Anxiolytic (1; X15619553); Aphrodisiac (f; APA; BOU; DEP; SUW); Bactericide (1; APA; PHR; PH2; X15612768); Cardiotonic (f; GHA); Carminative (f1; APA; DEP; GHA; HHB; PHR; PH2); Cerebrotonic (f; KAB); Contraceptive (f; APA); Digestive (f1; BGB; BIB; GHA; IHB); Diuretic (1; APA; BIB; DEP); Emmenagogue (f; BIB; EFS); Febrifuge (f; DEP); Fungicide (1; APA; PHR; PH2); Gastrogogue (1; PH2); Hepatotonic (f; KAB); Hypnotic (f; KAB); Hypocholesterolemic (1; X15462185); Hypoglycemic (1; APA; PNC); Hypolipidemic (1; X15462185) Hypotensive (f; APA); Hypotriglyceridemic (1; X15462185); Lactogogue (f; NMH); Larvicide (f1; APA; PNC; X15623234); Lipolytic (f1; BGB; LAF); Mercury Chelator (1; X15721537; X8914687); Myorelaxant (1; APA; X15619553); Orexigenic (f2; EFS; PH2); Pectoral (f; BIB; DEP); Sedative (f1; BIB; DEP; X15619553); Stimulant (f1; BGB; BIB; DEP); Stomachic (f1; BGB; BIB; HHB); Tonic (f; BIB; DEP; GHA); Vermifuge (f; JFM).

Indications (Coriander):

Adenopathy (f; KAB); Ameba (f; PH2); Amenorrhea (f; JFM); Anorexia (f2; APA; EFS; KOM; PH2); Arthrosis (f; BIB; HHB); Asthenia (f; BOU); Bacillus (1; HH2; X15612768); Bacteria (1; PH2; X15612768); Biliousness (f; BIB; DEP; SUW); Bleeding (f; DEP; EGG; PH2); BO (f; APA); Bronchosis (f; KAB); Burn (f; BOU); Cancer (f; JLH); Cancer, abdomen (f; JLH); Cancer, colon (f; JLH); Cancer, sinew (f; JLH); Cancer, spleen (f; JLH); Cancer, uterus (f; JLH); Carbuncle (f; BOU; DEP); Cardiopathy (f; BIB; GHA); Catarrh (f; BIB); Chickenpox (f; PH2; SKJ); Childbirth (f; IHB: PH2); Cholecocystosis (f; PHR); Cholera (f; BOU); Colic (f; DEP; GHA; HHB); Condyloma (f; JLH); Conjunctivosis (f; DEP; GHA); Coryza (f; KAB); Cough (f; IHB; PHR; PH2); Cramp (f1; BGB; BIB; PH2); Cystosis (f; PH2); Dermatosis (f; PHR; PH2); Diabetes (f; JFM); Diarrhea (f; APA; EGG; HHB); Dysentery (f1; APA; PHR; PH2); Dyspepsia (f12; APA; DEP; GHA; HHB; KOM; PH2); Dysuria (f; PH2); Edema (f; PH2); Enterosis (f2; BGB; JLH; PHR; PH2); Epistaxis (f; EGG; PH2); Erotomania (f; BIB); Erysipelas (f; BIB); Erythema (f; DEP); Escherichia (1; HH2; X15612768); Fever (f; PHR; PH2); Fungus (1;PHR); Gingirrhagia (f; KAB); Gleet (f; KAB); Halitosis (f; APA;

DEP; PHR; PH2); Headache (f1; PH2; X15612768); Gas (f1; APA; BGB; DEP; EGG); Gastrosis (f1; BGB; HHB; BIB; DEP; EGG; PHR; PH2); Hemorrhoid (f; APA; DEP; PH2); Hernia (f; BIB); Hiccup (f; KAB); Hysteria (f; BIB; BOU; JFM); Impotence (f; BIB; BOU); Induration (f; JLH); Infection (f1; HH2; PH2); Inflammation (f; KAB); Intoxication (f; BIB; DEP); Jaundice (f; KAB); Kernel (f; JLH); Leprosy (f; PHR; PH2); Listeria (1; X11929164); Measles (f; APA; HAD; PH2); Mycosis (f1; DEP; HH2; X15612768); Nausea (f; BIB; GHA; IHB); Nervousness (f; BIB); Neuralgia (f; APA; BIB; EGG; NAD); Neurosis (f; BOU); Obesity (1; X15462185); Ophthalmia (f; BOU; DEP; GHA); Orchosis (f; BOU); Otosis (f; BOU); Pain (f; KAB; PH2); Parasite (f; BOU); Pharyngosis (f; PHR; PH2); Ptomaine (f; BIB); Puerperium (f; PHR); Rash (f; PHR; PH2); Rheumatism (f; BOU; HHB; NAD); Salmonella (1; HH2; X15161192); Scabies (f; KAB); Sclerosis (f; BIB); Scrofula (f; PH2); Smallpox (f; DEP); Snakebite (f; BIB); Sore (f; DEP); Soroche (f; EGG); Splenosis (f; BIB); Stomachache (f; BIB; EGG; EB49:406); Stomatosis (f; KAB; PHR; PH2); Swelling (f; DEP; GHA); Syphilis (f; BIB; KAB); Thirst (f; NAD); Thrush (f; DEP); Toothache (f; APA); Tumor (f; JLH); Ulcer (f; BIB); Uterosis (f; JLH); Venereal Disease (f; BIB; KAB); Vertigo (f; HHB; NAD; PH2); Wart (f; JLH); Wen (f; JLH); Worm (f; APA; BOU; JFM); Wound (f; HH2); Yeast (f1; DEP).

Dosages (Coriander):

FNFF = !!!

Seeds, the size of a peppercorn, have a sharp but pleasant aroma, pleasing to many ethnics (e.g., Arabs, Egyptians, some Europeans, Asian Indians), sometimes flavoring breadstuffs, cakes, and confections therewith. Used as early as 1550 B.C., the dried fruits, called coriander seed, combining the taste of lemon peel and sage, is used in pastries, cookies, buns, processed meats (such as sausage, bologna, and frankfurters), pickling spice, and curry powder. Also used to flavor liqueurs, such as gin and vermouth; in the cocoa, chocolate, and cordial industries. Young plants used in salads as a vegetable and in chutneys, sauces, soups, and curries. (Bib.; FAC; TAN). 1–2 tsp crushed fruit/cup water up to 3 ×/day (APA); 0.1 g essential oil, 2–3 ×/day (HH2); 3 g fruit (KOM; PHR); 0.3–1 g powdered fruit (PNC); 0.5–2 ml liquid fruit extract (PNC); 0.05–2 ml (they said 2 ml, I would have said 0.2 ml; cf. celery seed, close kin) essential oil (PNC).

- Asian Indians paste powdered seed on carbuncles, headache, sores, and gargle for thrush (DEP).
- Asian Indians report pulverized roots and leaves in alcohol for measles eruptions (KAB).
- Asian Indians suggest the seed infusion or tincture for biliousness, catarrh, dyspepsia, enterosis, gas, and sore throat (NAD).
- Asian Indians suggest equal parts coriander, cardamom, and caraway (1:1:1) as digestive (NAD).
- Asian Indians suggest powdered seed for colic, dyspepsia, and halitosis (DEP).
- Asian Indians suggest equal parts coriander, cottonseed, poppy seed, 2 parts sugar, and some rose water for vertigo (NAD).
- Ayurvedics recommend for biliousness, bronchitis, dysentery, fever, nausea, and thirst, viewing it as aphrodisiac, aperitif, anthelmintic, antipyretic, diuretic, laxative, refrigerant, stimulant, and stomachic (KAB).
- Cubans suggest the seed decoction for diabetes and neuralgia (JFM).
- Ethiopians chew the leaves for colic and stomachache (BIB).
- Iranians use the leaf for headache (BIB).
- Latinos report the tea 2 ×/day is a female anaphrodisiac (JFM).
- Latinos boil 1 tsp fruit in 0.25 liter wine as emmenagogue and vermifuge (JFM).
- Middle Easterners steep seed in vinegar for one day and drink with sugar as cardiotonic, general tonic (GHA).
- Lebanese use seed decoction as a stimulant or as a narcotic anodyne (HJP).

- Mohammedens use carminative, pectoral, sedative seeds in a collyrium to prevent smallpox from destroying the eyes, as well as for chronic conjunctivosis (DEP).
- Peruvians paste the crushed leaves on the forehead for altitude sickness (EGG).
- Peruvians suggest the leaf tea for gas, headache, neuralgia, pain, and stomachache (EGG).
- Saudis suggest seed decoction for failing vision (GHA).
- Unani used the leaves, considered analgesic and hypnotic, for bleeding gums, eye pains, gleet, hiccup, inflammation, jaundice, piles, scabies, stomatitis, toothache, and tubercular glands. They used the seed to prevent bronchitis and coryza, for biliousness, dyspepsia, headache, syphilis, and ulcers on the penis, viewing the seed as aphrodisiac, cardiotonic, cerebrotonic, hepatotonic (KAB).
- Seed ground with raisins for tumors (JLH); with honey or raisins for burns, carbuncles, orchosis, sores, and sore ears (BOU).

Downsides (Coriander):

Class 1 (AHP, 1997). None known (KOM). "Health hazards or side effects following the proper administration of designated therapeutic dosages are not recorded" (PH2). Leaves may harbor *Listeria*, especially in hot, moist situations.

Extracts (Coriander):

Iranian scientists (Emamghoreishi et al., 2005) demonstrated anxiolytic activity of seed extracts of coriander, used for anxiety and insomnia in Iranian folk medicine (10, 25, 50, 100 mg/kg, ipr mus). It may have myorelaxant and sedative effects (X15619553). Karunasagar et al. (2005) showed that a sorbent prepared from coriander could remove inorganic mercury (Hg^{2+}) and methyl mercury (CH_3Hg^+) from aqueous solutions. Such a sorbent could be used to decontaminate inorganic- and methyl mercury–contaminated waters (X15721537). Earlier Japanese scientists had suggested that coriander could help remove mercury from the human body (X8914687). [My dentist takes coriander following a day drilling in mercurous fillings.] Indian scientists (Harve and Kamath, 2004) report on an interesting case of interspecific synergy. Acetone and petroleum ether extracts of *Murraya koenigii, Coriandrum sativum, Ferula asafoetida,* and *Trigonella foenum-graceum* potentiated synthetic larvicides Fenthion and Temephos as larvicides against *A. aegypti* larvae. They used 0.05 ppm Temephos and Fenthion with 25 ppm of *M. koenigii, F. asafetida, T. foenum graceum,* and 100 ppm of *C. sativum.* All plants showed synergic potential but were weakly larvicidal when tested individually (X15623234). Lo Cantore et al. (2004) demonstarted the antibacterial activity of coriander was better than that of related fennel against 27 phytopathogenic bacterial species (including *Bacillus* and *Escherichia*) and two mycopathogenic ones (X15612768). Essential oil is particularly effective against *Listeria monocytogenes* (X11929164). Kubo et al. (2004) found that coriander's (2E)-l dodecanal was about twice as potent and (2E)-undecenal about equipotent with gentamicin at killing *Salmonella*. They were additive rather than synergic (X15161192). Proestos et al. (2005) checked the species for flavonoids and phenolics and their antioxidant and antimicrobial activity (X15713039). LD50 (EO) = 4130 mg/kg orl rat (HH2).

SAFFRON (*CROCUS SATIVUS* L.) ++ IRIDACEAE

Notes (Saffron):

> **Thy plants are an orchard of pomegranates, with pleasant fruits; camphire, with spikenard, Spikenard and saffron; calamus and cinnamon, with all trees of frankincense; myrrh and aloes, with all the chief spices.**
>
> **Song of Solomon 4:13–14 (KJV)**

Your shoots are an orchard of pomegranates with all choicest fruits, henna with nard, nard and saffron, calamus and cinnamon, with all trees of frankincense, myrrh and aloes, with all chief spices.

Song of Solomon 4:13–14 (RSV)

Your skin is a paradise of pomegranates, with the choicest fruits, henna plants along with spikenard plants; spikenard and saffron, cane and cinnamon, along with all sorts of trees of frankincense, myrrh and aloes, along with all the finest perfumes.

Song of Solomon 4:13–14 (NWT)

In biblical times, saffron was important to oriental people as a condiment and sweet perfume. Still, it was not mentioned by early Sanskrit writers (although KAB present nearly fifty Sanskrit names for the plant; DEP listed only three). Arabian authors speak of its cultivation in the 10th century at Darband and Ispahan. Chinese state that Mohammedens introduced it to China during the Yuen dynasty. Mullahs are said to write charms with a saffron ink (DEP). The stigmas were especially valued as a food colorant (e.g., adding yellows to Indian curries). It was mentioned by Dioscorides. In Pliny's time, benches at public theaters were strewn with saffron, the precious petals placed in small fountains, to diffuse the scent into public halls. One cannot believe the figures one sees for this most expensive of spices until we realize that it was not specified whether they were talking fresh weight (for the flowers stigmata) or perhaps dry weight for the saffron (dried stigmata). From my reading, "It takes 100,000 flowers to yield 1 kg saffron" (one flower yields 10 mg, or 3.33 mg per stigmata; BIB), or "4000 to make an ounce" (circa 3000 mg), or one flower yielding only 0.75 mg, or 0.25 mg per stigmata (DEP). So I asked our garden director, Holly Vogel, to weigh some so we would know how many flowers it would take to give that 30-mg saffron dose. Her sister sent data from an online spice company suggesting that a single stigma would weigh 2 mg. I like that answer; it will take 15 stigmata (there are three in a flower) to attain that 30-mg posology. My calculations suggested 40 flowers or 120 stigmata at the low dry weight, or only three flowers at the high dry weight. Maybe if I munch 15 stigmata next spring, in the sunshine, it will cure my midwinter Seasonal Affective Disorder. If you are foolish enough to wish to check our calculations, buy some saffron; if each orange thread is simple, you may have the real spice; but if, on analysis, each thread turns out to be a small withered tubular flower, you may have the poor man's saffron, *Carthamus*, mentioned previously. If you find strands longer than an inch long, it might be dyed corn silk. This most expensive spice, saffron, is often adulterated. Other than food usage, it is also used in cosmetics for eyebrows and nail polishes, and as incense. Dioscorides comments on its use as a perfume, Harrison, on its use as a deodorant. Dissolved in water, it is applied to foreheads on religious and ceremonial occasions; it is also used as an ink. A Bronze Age (circa 3000–1100 BCE) Aegean wall painting in the building of Xeste 3 at Akrotiri, Thera, features *Crocus*. The frescoes concern saffron and healing, even depicting the harvest of stigmata; and 90 medical indications (and you thought my lists were long?) for saffron, starting in the Bronze Age. The frescoes depict a Theran goddess with her phytotherapy, saffron (X15259204).

Common Names (Saffron):

Açafrão (Por.; EFS; USN); Agnishekhara (Sanskrit; KAB); Asfar (Arab.; BOU); Azafrán (Sp.; EFS; USN); Bhavarakta (Sanskrit; EFS; NAD); Castagnolo (Potenza; KAB); Croco Florito (It.; EFS); Echter Safran (Ger.; HH3); Fan Hung Hua (China; Pin.; AH2; EFS); Giallone (It.; KAB); Grogo (Tuscany; KAB); Jafran (Beng.; KAB; NAD; WOI); Jafrana (Arab.; KAB); Jafranekar (Urdu; KAB); Kamakuma (Malaya; EFS); Kambama (Malaya; EFS); Karkom (Heb., KAB); Kasmirajanma (Sanskrit; DEP); Kecara (Bom.; Mar.; DEP); Kesar (Hindi; DEP; KAB); Kesara (Mar.; KAB; WOI); Keshar (Guj.; DEP; KAB); Keshara (Sanskrit; WOI): Kessar (Bom.; KAB); Koma Koma (India; EFS); Kong (Kas.; DEP; KAB; WOI); Konger (India; EFS); Kormar Romar (India; EFS);

FIGURE 1.35 Saffron (*Crocus sativus*).

Kouzrkour (Ber.; BOU); Krokos (Greek; KAB); Kruku (Arab.; BOU); Kumkuma (Sanskrit; DEP); Kungkumapave (Tel.; KAB; WOI); Kungumapu (Tam.; DEP; KAB; WOI); Kunkuma (Ayu.; AH2); Kunkumakesari (Kan.; Kon.; NAD; WOI); Kunkumapave (Tel.; DEP); Kunkumappu (Mal.; Tam.; NAD); Kunkumapurru (Tel.; NAD); Kunkumapuvva (Tel.; NAD); Kunkumkesarei (Kan.; KAB); Kunkumma Purru (India; EFS); Kurkam (Arab.; ZOH); Kurkum (Arab.; Bhote; DEP; KAB; ZOH); Larkimasa (Iran; KAB); Pewva (India; EFS); Piwva (India; EFS); Sa'faram (Arab.; Syria; HJP); Saferam (Arab.; ZOH); Safferian (Treviso; KAB); Saffron (Eng.; Scn.; AH2; CR2; USN); Saffron Crocus (Eng.; USN); Saffraan (Dutch; EFS); Safra (Cat.; KAB); Safran (Fr.; Den.; Ger.; Tur.; BOU; EFS; KAB; USN); Safran Cultivé (Fr.; BOU); Safran Vrai (Fr.; BOU); Saurab (Sanskrit; DEP; EFS); Schafran (Rus.; KAB); Spanish Saffron (Eng.; Ocn. AH2; KAB); Szafrana (Pol.; HH3; KAB); Thanwai (Burma; DEP; KAB); True Saffron (Eng.; Ocn.; AH2); Xi Hong Hua (Pin.; AH2); Zaafaran (Arab.; Iran; DEP; WOI); Zafar (Tur.; DEP; KAB); Za'faran (Arab.; BOU); Zafferano (It.; EFS); Zaffran (Hindi; India; EFS; NAD); Zafrah (Arab.; Iran; EFS; NAD); Zafran (Hindi; WOI); Zafrane Hor (Alg.; HH3); Zahafaran (Arab.; WOI); Zang Hong Hua (Pin.; DAA); Zipharana (Iran; NAD).

Activities (Saffron):

Abortifacient (f12; PHR; PH2); Anodyne (f; APA; CRC; MAD); Antiaggregant (1; X16180089); Antidepressant (f12; CRC; DAA; PNC; X15852492); Antidote (f; MAD); Antiedemic (f1; APA); Antigenotoxic (1; X11665650; X15331343); Antihysteric (f; CRC); Antiinflammatory (f1; X11914135); Antilymphomic (1; JAC7:405); Antinociceptive (1; X11914135); Antioxidant (1; PR14:149; X15653110); Antiplatelet (1; X16180089); Antiradicular (1; PR14:149; X15653110); Antiseptic (f; CRC; HJP); Antispasmodic (f1; APA; CRC; HHB; HJP); Antitumor (f1; PR14:149); Aphrodisiac (f; APA; BOU; CRC; MAD); Balsamic (f; CRC); Cardiotonic (f; CRC; EFS; MAD); Carminative (f1; CRC; PNC; VAD); Circulostimulant (f; BOW; X12776492); Diaphoretic (f; APA; CRC); Digestive (f; APA; BOW); Ecbolic (f; CRC); Emmenagogue (f; BOU; CRC; HHB; PNC); Emollient (f; APA); Expectorant (f; APA; CRC); Gastrogogue (f; PHR; PH2); Hemostat (f; MAD); Hypocholesterolemic (1; APA); Hypolipemic (1: PR14:149); Hypotensive (1; APA; BOW; X12648816); Immunostimulant (1; X12776492); Lactagogue (f; KAB); Memorigenic (1; X16028982); Myorelaxant (f; APA); Narcotic (f; BOU; CRC; SKJ); Nervine (f; CRC); Neuroprotective (1: PR14:149); Orexigenic (f; VAD); Sedative (f; APA; CRC; HHB); Stimulant (f; BOU; CRC; HHB); Stomachic (f; CRC; HHB); TNF-alpha Inhibitor (1; X11720092); Tonic (f; BOU); Toxic (f; CRC); Uterotonic (1; PHR; PH2).

Indications (Saffron):

Adenopathy (f; JLH; X12776492); Aegilops (f; JLH); Amenorrhea (f1; CRC; MAD; NAD; PH2); Anorexia (f; VAD); Anuria (f; ?); Arthrosis (f; KAB); Asthenia (f; VAD); Asthma (f; BOU; MAD); Bladder Ailments (f; CRC); Bleeding (f; DAA; MAD); Blood Disorders (f; CRC); Bronchosis (f; PH2): Burn (f; JLH); Cacoethes (f; JLH); Cancer (f1; APA; PR14:149; X11582266); Cancer, abdomen (f1; APA; CRC); Cancer, bladder (f1; APA; CRC); Cancer, breast (f1; APA; CRC; JLH); Cancer, colon (f1; APA; JLH); Cancer, diaphragm (f1; APA; JLH); Cancer, ear (f1; APA; CRC); Cancer, eye (f1; APA; JLH); Cancer, kidney (f1; APA; CRC); Cancer, larynx (f1; APA; JLH); Cancer, liver (f1; APA; CRC); Cancer, mouth (f1; APA; CRC); Cancer, neck (f1; APA; CRC); Cancer, spleen (f1; APA; CRC); Cancer, stomach (f1; APA; CRC; JLH); Cancer, testicle (f1 APA; JLH); Cancer, throat (f1; APA; JLH); Cancer, tonsil (f1; APA; CRC); Cancer, uterus (f1; APA; CRC; JLH); Cardiopathy (f1; APA; X12776492); Catarrh (f; CRC; DEP; SKJ); Cerebrosis (f1; APA; KAB); Chemopreventive (f1; X11582266); Chickenpox (f; HOS); Childbirth (f; DAA; PH2); Chlorosis (f; HOS); Cholera (f; CRC; HOS); Chorea (f; HHB; MAD); Circulosis (1; X12776492); Cold (f; CRC); Colic (f; NAD); Condyloma (f; DAA); Conjunctivosis (f; MAD); Cough (f; DAA; MAD); Cramp (f1; DAA; HHB; VAD); Cystosis (f; JLH); Cytotoxic (1; HH3); Debility (f; NAD); Depression (f12; CRC; DAA; PNC; X15852492);

Dermatosis (f; CRC; KAB); Diabetes (f; CRC); Diarrhea (f; NAD); Dysmenorrhea (f; DAA; HHB; MAD; PNC); Dyspepsia (f1; VAD); Edema (f1; APA); Enterosis (f; JLH); Epistaxis (f; MAD); Fear (f; CRC; DAA); Fever (f; CRC; NAD; PH2); Fibroid (f; JLH); Fibrosarcoma (1; HH3); Fractures (f; KAB); Gas (f1; MAD; VAD); Gastrosis (f; JLH); Gingivosis (f1; VAD); Gout (f; MAD); Hangover (f; LIL); Headache (f; KAB; PH2); Hemicrania (f; KAB); Hemoptysis (f; DAA; MAD); Hemorrhoid (f; NAD); Hepatosis (f1; CRC; DEP; JLH; SKJ; X12776492); High Blood Pressure (1; APA; X12648816); High Cholesterol (1; APA); Hysteria (f; BOU; CRC; DAA; MAD); Induration (f; JLH); Infection (f1; HJP); Inflammation (f1; JLH; X11914135); Lachrymosis (f; JLH); Laryngosis (f; JLH); Leukemia (f1; JLH; X12776492); Lochiostasis (f; PH2); Lymphoma (1; APA; JLH; HH3); Measles (f; CRC; DAA; MAD); Melancholy (f12; CRC; DEP; HHB; KAB; X15852492); Menorrhagia (f; HHB; HOS; PH2); Menoxenia (f; CRC); Nausea (f; KAB); Nephrosis (f; JLH; KAB); Neuralgia (f; NAD); Neurosis (1; CRC; FNF); Obesity (f1; VAD; PR14:149); Ophthalmia (f; JLH; HOS); Orchosis (f; JLH); Osteosarcoma (1; HH3); Pain (f1; DAA; X11914135); Papilloma (1; X12776492); Parotosis (f; HOS; JLH); Pertussis (f; BIB; BOU; DAA; MAD); Pharyngosis (f; KAB); Phymata (f; JLH); Plague (f; MAD); Puerperium (f; CRC); Rhabdomyosarcoma (1; X12776492); Sarcoma (1; X12776492); Scabies (f; KAB); Sclerosis (f; CRC); Shock (f; CRC; DAA); Snakebite (f; NAD; SKJ); Sore Throat (f; PH2); Spasms (f; CRC); Splenosis (f; CRC; JLH; KAB); Spermatorrhea (f; NAD); Swelling (f1; APA); Thrombosis (1; X16180089); Tonsilosis (f; JL); Toothache (f; VAD); Twitching (f; MAD); Uterosis (f; CRC; DAA; JLH); Venereal Disease (f; CRC; DAA); Vertigo (f; MAD); Vomiting (f; PH2); Wart (f; CRC).

Dosages (Saffron):

FNFF = !!!

Prominent coloring agent and spice in cookery (especially Spanish), soups, stews, especially chicken dishes, and in confectionery to give color, flavor, and aroma (BIB). 10–15 stigmata/cup water (APA); 0.5–1.5 g day (APA; HHB). 10 g as abortive (HH3); 0.1–1 g powdered saffron (MAD); 15–16 drops tincture (MAD). 0.5–2.5 g saffron (PNC); tea (2g/l) 3 ×/day (VAD).

- Algerians and Gypsies use the saffron infusion as a collyrium (HJP).
- Asian Indians use saffron for bladder, kidney, and liver ailments; also for cholera. Mixed with ghee, it is used for diabetes (BIB).
- Ayurvedics, deeming saffron alexiteric, anthelmintic, laxative, tonic, use it for anuria, biliousness, bronchosis, cerebrosis, dermatoses, headache, pharyngitis, scabies, and vomiting (KAB).
- Iranians view saffron as antispasmodic and stimulant (HJP), and now as antidepressant (X15852492). Iranians report saffron (30 mg/day) as effective as imipramine (100 mg/day) in mild to moderate depression (X15341662).
- Lebanese add a dozen pistils to a large cup of hot water for children coming down with chickenpox, measles, or mumps (HJP).
- Germans take saffron in milk for measles (MAD).
- North Africans suggest eight to ten filaments (stigmata) per cup of tea as a "narcotic for cases of asthma, whooping cough, hysteria" (BOU).
- Spanish suggest massaging the gums with tincture (1:10) for gingivosis or toothache (VAD).
- Unani, deeming the herb antiinflammatory, aphrodisiac, bechic, diuretic, hematinic, lactagogue, laxative, use it for disorders of brain, kidney, liver, and spleen (KAB).

Downsides (Saffron):

Class 2b. Abortifacient; emmenagogue; uterotonic. Severe side effects may result from ingesting 5 g saffron (lethal dose = 20 g) (AHP, 1997). "Health risk or side effects following the proper administration of designated therapeutic dosages are not known" (PH2). Controversial. The 200-mg/kg dose of saffron alleged to extend the life of cancerous mice translates to 22,000 mg or 22 g saffron with a 100-kg rat named Jim Duke. Commission E reports no risks for doses up to 1.5 g; however, 5 g is toxic, 10 g is abortive, and 20 g is lethal (AEH; PHR). Conversely, Tucker and DeBaggio report that "ingesting 0.05 ounce (1.5 g) saffron has resulted in death" (TAD). Paradoxically, the life-saving dose is lethal! It is good that saffron is so expensive, so thatwe will not have too many fools overdosing on it. Preferring to err on the safe side, we can think of saffron only as an expensive spice to be used judiciously. I think these numbers may be alarmist. Of mice and men, the oral LD50 in mice is 20,000 mg/kg according to the German *Hager's Handbuch* (HH3), while for man, 5 to 10 g can prove lethal.

Extracts (Saffron):

Iranians found that saffron at 30 mg/day was as effective as imipramine at 100 mg/day in mild to moderate depression ($F = 2.91$, d.f. $= 1$, $P = 0.09$) (X15341662). Jessie and Krishnakantha (2005) found that saffron extract inhibited platelet aggregation *in vitro* (IC50 = 0.35–0.86 mg) (X16180089). LD50 (saffron) = 20,000 mg/kg orl mus HH3; LD50 (saffron extracts >600 mg/kg ipr mus HH3; LDlo 5,000–10,000 mg total orl hmn HH3). If *Hager's Handbuch* numbers are correct, we had better leave the saffron for the mouse, to whom the saffron seems orders of magnitude safer than to the man, where a few milligrams could be good, a few grams could be bad.

MELON (*CUCUMIS MELO* L.) +++ CUCURBITACEAE

Synonyms:

Bryonia collosa Rottler; *Cucumis collosus* (Rottler) Cogn.; *Cucumis trigonus* Roxb. *fide* USN; *Cucumis utilissimus* Roxburgh *fide* NPM

Notes (Melon):

> **We remember the fish, which we did eat in Egypt freely; the cucumbers, and the melons, and the leeks, and the onions, and the garlick: But our soul is dried away: there is nothing at all, beside this manna, before our eyes.**
>
> **Numbers 11:5-6 (KJV)**

> **We remember the fish we ate in Egypt for nothing, the cucumbers, the melons, the leeks, the onions, and the garlic; but now our strength is dried up, and there is nothing at all but this manna to look at.**
>
> **Numbers 11:5-6 (RSV)**

> **How we remember the fish that we used to eat in Egypt for nothing, the cucumbers and the water melons, and the leeks and the onions, and the garlic; But now our soul is dried away. Our eyes are on nothing at all except the manna.**
>
> **Numbers 11:5-6 (NWT)**

Some think the biblical melon was this; others think, as did the NWT translators, that it was watermelon. Both are cultivated as pleasantly juicy fruits today in the Holy Land and other arid lands, always welcome in dry climates. We will never be sure what the non-taxonomic scholars implied. Even some hard-core taxonomists are loathe to make herbarium specimens of melons. Hence, I include both in my faith-based food farmacy, although the word "melon" shows up only once in my

FIGURE 1.36 Melon (*Cucumis melo*).

computerized version of the Bible. Zohary insists that the Hebrew words *kishuim* and *mikshaw* must equate to muskmelons, arguing that cucumbers did not exist in Egypt in biblical times. That makes the NWT even more interesting if one reads "muskmelon" instead of "cucumber."

Common Names (Melon):

Afqous (Ber.; BOU); Aggour (Arab.; BI2; BOU); Al Bata (Mali; UPW); Ao Uri (Japan; TAN); Bachang (Kon.; NAD); Bambous (Ber.; BOU); Batanyah (Hausa; Nig.; UPW); Battikh Asfar (Arab.; Syria; HJP); Bettikha (Arab.; BOU); Bittikh (Arab.; BOU); Blewah (Java; TAN); Cantaloupe (Eng.; CR2); Cataloup (Fr.; EFS); Chibuda (Bom.; DEP); Chibunda (Mar.; DEP); Cikiire (Fula; Nig.; UPW); Dungra (India; DEP); Esaka (Cameroon; UPW); Feggous (Arab.; BOU); Ghurmi (Mooshar; Satar; NPM); Gidhro (Sin.; DEP); Gidro (Sin.; NAD); Goihmha (Tharu; NPM); Hsiang Inabe (Togo; UPW); Kakari (Nepal; NPM); Kalinga (Sanskrit; EFS; NAD); Kalingada (Kan.; NAD); Kan Kua

(China; EFS); Kanteloep (Dutch; EFS); Kavun (Tur.; EFS; EB54:155); Khad (Sen.; UPW); Kharbazeh (Iran; DEP); Kharbuja (Bal.; Hindi; Nepal; DEP; KAB; NPM); Kharmuj (Beng.; DEP); Kharvuja (Sanskrit; DEP); Khurbuj (India; EFS); Khurbuza (Kangra; DEP); Kishuim (Heb.; ZOH); Kua (China; EFS); Kurubombonmesengo (Gambia; UPW); Makuwauri (Japan; EFS); Ma-ru-tse (Tibet; NPM); Mazarisharif (Afg.; TAN); Melâo (Mad.; Por.; EFS); Melâo Casca de Carvalho (Por.; AVP); Melâo de Mesa (Por.; AVP); Meloen (Sur.; AVP); Melon (Den.; Eng.; Fr.; Swe.; AH2; BOU; CR2; EFS; NPM); Melón (Peru; Sp.; EGG); Melón Almazaleno (Sp.; AVP); Melón de Castilla (Haiti; AVP); Melon d'Espagne (Haiti; AVP); Melone (Ger.; It.; AVP; EFS); Melonegurke (Ger.; NAD); Melon France (Haiti; AVP); Melon Musqué (Fr.; Guad.; Mart.; St. Lucia; AVP); Melon Sucré (Fr.; EFS); Mikshaw (Heb.; ZOH); Mulampandu (Tel.; DEP); Mulkun Noshi (Korea; TAN); Muskmelon (Eng.; AVP; BOU; ZOH); Nashiuri (Japan; TAN); Phut (Nepal; NPM); Popone (It.; AVP); Qawoun (Arab.; Syria; BOU; HJP); Quisu Im (Heb.; BI2); Remó (Naga; DEO); Sakkar Teti (India; EFS); Sardapaliz (Afg.; DEP; KAB); Senat (Sudan; UPW); Shammam (Arab.; Leb.; Syria; BOU; JNP); Shamman (Arab.; Nig.; UPW); Shao Kua (China; EFS); Shima Uri (Japan; TAN); Shiro Uri (Japan; TAN); Sirdapaliz (Afg.; DEP; KAB); Small Gourd (Eng.; NPM); Snap Melon (Eng.; NPM); Sweet Melon (Eng.; EFS); Tarbucha (Guj.; DEP); Tarbuj (San.; DEP); Tarrsimte (Ber.; BOU); Teng Tai (Thai; TAN); T'ien Kua (China; EFS); Turbuch (Guj.; NAD); Urmi (Nepal; NPM); Vaelapalam (Tam.; NAD); Velipandu (Tel.; NAD); Vellari Verai (Tam.; DEP); Yaqtin (Arab.; BOU); Yengani (Ghana; UPW); Yomba (Sen.; UPW); Yüeh Kua (China; EFS); Zaghun (Lad.; DEP); Zuckermelone (Ger.; EFS).

Activities (Melon):

Allergenic (1; X14533664); Antiacne (1; PAM); Antiaging (1; FNF); Antiasthmatic (1; PAM); Anticancer (f1: FNF; JNU; PAM); PAM; Antiemetic (f; BIB; LMP); Antiinflammatory (f1; JLH; X15261965); Antimastitic (1; PAM); Antimutagenic (1; FNF); Antioxidant (1; PAM; X15261965); Antiozenic (1; FNF); Antiphotophobic (30–300 mg man/day) (1; M29); AntiPMS (1; JAF34:409); Antiporphyric (30–300 mg/man/day) (1; MAR); Antipityriasic (1;PAM); Antiproliferant (1; JNU); Antipsoriac (1; PAM); Antistress (1; DAS); Antitussive (f; BOU); Antiulcer (1; PAM); Antivinous (f; BIB); Aphrodisiac (f; BIB; KAB); Cardioprotective (1; FNF); Cardiotonic (f; KAB); Cerebrotonic (f; KAB); Demulcent (f; BIB; EFS; UPW); Digestive (f; BIB; BOU; LMP); Diuretic (f; BIB; UPW; WOI); Emetic (f; BIB; BOU; UPW; WBB); Emmenagogue (f; BIB); Emollient (f; WBB); Expectorant (f; BIB; BOU); Fungicide (1; UPW; WBB); Immunostimulant (1; FNF; PAM); Insectifuge (f; HJP); Interferon synergist (1; FNF); Lachrymatory (f; BIB; KAB); Lactagogue (f; KAB); Laxative (f; KAB); Mucogenic (1; PAM); Phagocytotic (1; PAM); Refrigerant (f; BIB; BOU: DEP); Stomachic (f; BIB; EFS; LMP); Taenifuge (f; BIB); Thymoprotective (1; PAM); Tonic (f; BIB); Vermifuge (f; BIB); Vulnerary (f; KAB).

Indications (Melon):

Acne (1; FNF; MAR); Anasarca (f; BIB); Anuria (f; DEP); Apoplexy (f; BIB); Ascites (f; BIB; KAB); Biliousness (f; BIB); Bronchosis (f; BIB); Bruise (f; BIB); Burn (f; BIB); Cancer (f; BIB); Cancer, bladder (f1; FNF; JLH); Cancer, liver (f1; FNF; JLH); Cancer, stomach (f1; BIB; JLH; JNU); Cancer, uterus (f1; BIB; JNU); Cardiopathy (1; FNF; JNU); Childbirth (f; HJP); Cold (f; BIB); Coryza (f; BIB); Cough (f; BIB; BOU); Cystosis (f; FNF; JLH); Dermatosis (f; BIB); Diabetes (f; BIB); Dyspepsia (f; BIB); Dysuria (f; BIB; DEP; NPM); Eczema (f; BIB; UPW; WOI); Extravasation (f; BIB); Fatigue (f; BIB); Fever (f; BOU); Freckle (f; BIB); Fungus (1; UPW); Gastrosis (f; JLH); Gonorrhea (f; BIB); Hepatosis (f; BIB; JLH); Hyperkeratosis (1; FNF; PAM; DAS); Ichthyosis (1; FNF); Infection (1; UPW); Inflammation (f1; JLH; X15261965); Insanity (f; KAB); Jaundice (f; BIB; LMP); Leukoplakia (1; FNF; PAM); Lupus (1; FNF; MAR); Mastosis (1; FNF); Menorrhagia (f; BIB); Mycosis (1; UPW); Nephrosis (f; BIB); Oliguria (f; BIB); Ophthalmia (f; BIB); Pityriasis (1; FNF);

Photophobia (1; FNF); PMS (1; FNF); Polyp (f; BIB); Porphyria (1; FNF); Prostatosis (f; FAC); Psoriasis (1; FNF); Rhinosis (f; BIB); Sore (f; BIB); Sore Throat (f; KAB); Stomatosis (f; BIB); Stress (1; FNF); Sunburn (f; BIB); Sunstroke (f; BIB); Thirst (f; KAB); Tumor (f; BIB); Tympanites (f; DEP); Ulcer (1; FNF); Uterosis (f; JLH); Venereal Disease (f; BIB); Xerophthalmia (1; FNF).

Dosages (Melon):

FNFF = !!!
Fruit widely eaten raw, pickled, or baked; seeds eaten raw or toasted, or slurried in fruit juices; or the oil is expressed. Roasted charmagaz is a mixture of peeled melon, cucumber, pumpkin, and watermelon seed (FAC; TAN; EB54:155); leaves eaten as a relish, vegetable, and in soups (UPW).

- Algerians give floral tea after surgically removing fetus (HJP).
- Asian Indians apply bruised seeds to children with tympanites (DEP).
- Asian Indians suggest fruit, raw or cooked, as lotion in eczema, freckles, and sunburn (NAD); one fruit a day for eczema (DEP).
- Asian Indians consider their beverage called burfi good for the prostate (FAC).
- Asian Indians consider the seed oil diuretic (NAD).
- Ayurvedics consider the fruit aphrodisiac, diuretic, febrifuge, laxative, and tonic, using it for ascites, biliousness, fatigue, and insanity (KAB).
- Chinese suggest the seed for stomach cancer (JLH).
- Lebanese rub fruit pulp on breast and give it to weaning children to suck (HJP).
- Lebanese believe the melon repels bedbugs (HJP).
- Unani, considering the seed diuretic, lachrymatory, and tonic, use for bronchosis, fever, hepatosis, nephrosis, ophthalmia, sore throat, thirst; the fruit they consider cardiotonic, cerebrotonic, diaphoretic, diuretic, lactagogue, used for dysuria and ophthalmia, the vulnerary rind applied to the hypogastrium as diuretic (KAB).

Extracts (Melon):

Vouldoukis et al. (2004) proved antioxidant and antiinflammatory activities in a melon extract selected for its high superoxide dismutase activity. The extract dose dependently inhibited the production of superoxide anion (maximal at 100 μg/ml). The antiinflammatory properties of the extract were, in part, due to the induction of production of IL-10 by peritoneal macrophages (X15261965). Alpha-spinasterol significantly (circa 1000 x simvastin, a coenzyme-A inhibitor) modulates development and/or progression of diabetic nephropathy. It reduced significantly attendant increases in serum triglycerides, renal weight, and urinary protein excretion in diabetic mice (X15326549).

CUCUMBER (*CUCUMIS SATIVUS* L.) ++ CUCURBITACEAE

Notes (Cucumber):

And the daughter of Zion is left as a cottage in a vineyard, as a lodge in a garden of <u>cucumbers</u>, as a besieged city.

Isaiah 1:8 (KJV)

And the daughter of Zion is left like a booth in a vineyard, like a lodge in a <u>cucumber</u> field, like a besieged city.

Isaiah 1:8 (RSV)

FIGURE 1.37 Cucumber (*Cucumis sativus*).

And the daughter of Zion has been left remaining like a booth in a vineyard, like a lookout hut in a field of cucumbers, like a blockaded city.

Isaiah 1:8 (NWT)

Cucumber is mentioned only three times in the Bible, twice as a garden of cucumbers. And all three versions in my trilogy concur with cucumber. Zohary reminds us, rightly or wrongly, that garden cucumbers did not exist in Egypt in biblical times. And the lodges or lookout huts in the cucumber fields were booths of twigs or mats used then, and now, to guard against theft. But Zohary maintains that the huts were in melon fields rather than in cucumber fields. I would certainly guard my cantaloupes and watermelons more than cucumbers, although they are all pleasingly cool and juicy in hot arid climates. In case Zohary is wrong, perhaps highly unlikely, I will retain the prolific cucumber in my garden, hoping to prevent the wrinkles generated by these preplexing historical puzzles. Like many cultivated, multiple-variety members of the pumpkin family, it has a whole host of common names around the world. The USDA database listed only cucumber, gherkin concombre (Fr.; USN); cornichon (Fr.; USN); Gurke (Ger.; USN); khira (India; USN); pepino (Por.; Sp.; USN); and cohombro (Sp.). But the USDA website links one to the much more extensive index of common names, which I have copied below (those followed by POR are abbreviations for citations from Porcher (porcher@landfood.unimelb.edu.au). Cucumber soap is said to be especially beneficial for windburn. Cucumber juice is said to kill cockroaches, repelling fish moths and wood lice. Even the strewing of the green peel on the floor at night is effective (BIB).

Common Names (Cucumber):

Aggouria (Greek; DEP); Agurk (Den.; AVP; EFS; POR); Agurka (Bohemia; DEP); Airelu Kankro (Nepal; POR); Alpicoz (Sp.; EFS); Asare Kankro (Nepal; POR); Augurk (Dutch; EFS); Bahuphala (Sanskrit; KAB); Bakritaear (Mun.; KAB); Bazarula (Arab.; EFS; NAD); Betbinho (Guinea-Bissau; UPW); Bi Bai (Ic.; KAB); Bonteng (Sunda; IHB; POR); Cabul (Vis.; KAB); Calavaya (Vis.; KAB); Castravete (Rom.; KAB); Cedriuolo (It.; KAB); Cetrinolo (It.; AVP); Cetriola (It.; EFS); Cetriolo (It.; KAB; POR); Cetriuolo (It.; KAB); Chaja (Kalmuck; DEP); Chiar (Arab.; DEP); Citriuolo (It.; KAB); Cocombre (Haiti; AVP); Cocombro (Por.; AVP); Cocomero (It.; EFS); Cogombre (Cat.; KAB); Cohombro (Sp.; AVP; RAR; USN); Concombre (Fr.; Haiti; AVP; EFS; TAN); Concombre Commun (Fr.; KAB); Concombre Cultive (Fr.; KAB); Concombre Ordinaire (Fr.; AVP); Concombre Vrai (Fr.; AVP); Cornichon (Fr.; EFS; USN); Cucumber (Eng.; NPM); Dib (Hmong; EB57:365); Dlib (Hmong; EB57:365); Dobba (Malaya; IHB); Dosakaya (Tel.; WOI); Dosekaya (Tel.; NAD); Dozakaya (Tel.; KAB); Dua Chuot (Ic.; KAB); Dua Gang (Ic.; KAB); Dureagurk (Den.; EFS); Entimum (Malaya; IHB); Faggus (Arab.; Mali; UPW); Gemeine Gurke (Ger.; AVP; TAN); Gherkin (Eng.; EFS); Gnyo (Magar; NPM); Gros Concombre (Fr.; AVP); Gurka (Swe.; AVP; EFS; KAB); Gurke (Ger.; DEP; USN); Haswey (Dho.; Omar; Saudi; GHA); Hiyar (Tur.; EFS); Ho qua (Ic.; KAB); Hoang Qua (Ic.; KAB); Huang Gua (China; POR); Huang Kua (China; DAA; EFS); Huang Kwa (Canton; POR), Hu Kua (China; EFS); Kakari (Guj.; KAB); Kakdi (Hindi; Mah.; NAD; POR); Kakkari (Mal.; NAD); Kaknai (Orissa; KAB); Kakri (Bom.; Hindi; Nepal; Simla; POR; SKJ); Kakrikai (Tam.; WOI); Kakro (Nepal; POR); Kalabaga (Vis.; POR); Kandalu (Sanskrit; KAB); Kankdi (Guj.; KAB); Kankri (Bom.; Dec.; Guj.; Hindi; NAD); Kankro (Nepal; NPM; SUW); Kankur karkti (India; EFS); Kantakilaki (Sanskrit; KAB); Kantakiphala (Sanskrit; KAB); Karan (Arm.; DEP); Katiman (Java; IHB); Katimun (Sunda; IHB); Ketimun (Indonesia; POR); Kheyar (Arab.; GHA); Khiar (Tur.; AVP); Khira (Beng.; Hindi; India; Mar.; NAD; USN; WOI); Khiyar (Iran; EFS); Kiar (Tartar; DEP); Kira (India; EFS); Kokomba (Ghana; UPW); Komkommer (Dutch; POR); Konkommer (Dutch; EFS); Koshaphala (Sanskrit; KAB); Krastavat (Slav.; DEP); Krastave (Slav.; AVP); Kratsavets (Albania; DEP); Kumba (Sierra Leone; UPW); Kunkummer (Ger.; KAB); Kurkku (Fin.; POR) Kusud (Arab.; EFS); Kyu Uri (Japan; POR); Kyuri (Japan; TAN); Langlai (Tamang; NPM); Lathai (Gurung; NPM); Machicho (Par.; AVP); Melon Alon (Sp.; AVP); Mentimun (Java; POR);

Moro Kyu (Japan; POR); Mullanvellari (Tam.; NAD); Mullusavte (Kan.; NAD); Mulluvellari (Tam.; NAD); Mutrulla Sookasa (Sanskrit; EFS); Ngurli (Nig.; UPW); Ogorek (Pol.; KAB); Oguretz (Rus.; KAB); Oh Ee (Korea; POR); Oi (Korea; POR); Penpininho (Por.; EFS); Pepineiro (Por.; KAB); Pepinillo (Sp.; Peru; EFS; EGG); Pepino (Cuba; Mad.; Peru; Por.; Sp.; EGG; RAR; RyM); Pepino de Castilla (Sp.; RAR); Pipingha (Sin. POR); Pipingkai. (Sin.; KAB; POR); Pipingkay (Sin.; Tam.; KAB); Pipino (Tag.; POR); Pipinya (Sing.; POR); Pitapushpa (Sanskrit; KAB); Qing Gua (Canton; POR); Quishooaim (Heb.; KAB); Qitha Bostany (Arab.; AVP); Qitsa (Arab.; AVP); Sakusa (Sanskrit; EFS; NAD); Salatagurk (Den.; EFS); Salatalik (Tur.; EB54:155); Santekayi (Kan.; KAB); Sasa (Beng.; KAB; NAD); Sautekayi (Kan.; NAD); Setimun (Sakai; IHB); Shiyarekhurd (Iran; KAB); Sikua (Greek; DEP); Sikuos (Greek; DEP); Sikya (Greek; KAB); Sma Gurka (Swe.; EFS); Sudhavasa (Sanskrit; KAB); Sukasa (Sanskrit; NAD; SKJ); Sushitala (Sanskrit; KAB); Taear (Munari; KAB); Taeng Kwaa (Thai; POR); Taeng Om (Thai; POR), Taeng Raan (Thai; POR); Tansali (Guj.; KAB); Tavas (Mar.; NAD); Tavase (Mar.; KAB); Tèèng (Laos; POR); Teng Kwa (Thai; IHB); Teng Ran (Thai; IHB); Thagwa (Burma; KAB); Thakhwa (Burma; POR); Thakhwathee (Burma; KAB); Tihu (Hindi; POR); Timun (Java; Malaya; Sunda; IHB; POR); Timun China (Malaya; IHB); Tolombro (Por.; AVP); Touxem (Kon.; KAB); Tovxini (Kon.; KAB); Towshay (Kon.; NAD); Trapukarkati (Sanskrit; KAB); Trapusha (Sanskrit; NAD); Tra sac (Ic.; KAB); Trâsâk (Khmer; POR); Tseng Kwa (China; POR); Tukhmi Khiyarain (India; DEP); Tundilaphala (Sanskrit; KAB); Uborka (Hun.; KAB); Uggurits (Estonia; DEP); Ukkurits (Estonia; DEP); Urits (Estonia; DEP); Vellari (Madras; KAB); Vellarikkai (Tam.; SKJ; WOI); Voantangombazaha (Guj.; KAB).

Activities (Cucumber):

Anthelmintic (f; KAB; UPW); Antiinflammatory (f; EGG; KAB); Antiseptic (1; X15151242); Bactericide (1; X15151242); Demulcent (f; GHA; NAD; SUW); Diuretic (f; DEP; GHA; NAD; SUW); Emetic (f; DAA; GHA); Febrifuge (f; NAD); Insectifuge (f; DAA; UPW); Insecticide (f; UPW); Proteolytic (1; KAB); Purgative (f; DAA; GHA); Tonic (f; SUW); Vermifuge (f; EFS; GHA).

Indications (Cucumber):

Bacillus (1; X15151242); Bacteria (1; X15151242); Beri-beri (f; DAA); Biliousness (f; KAB); Burn (f; DAA); Cancer (f; JLH); Cancer, abdomen (f; JLH); Cancer, bladder (f; JLH); Cancer, liver (f; JLH); Cancer, stomach (f; JLH); Condyloma (f; JLH); Cystosis (f; JLH; NAD); Dermatosis (f; DAA; KAB); Dysentery (f; IHB); Dysuria (f; GHA; NAD); Enterosis (f; GHA; JLH); Escherichia (1; X15151242); Fever (f; NAD); Gastrosis (f; JLH); Hepatosis (f; JLH); Induration (f; JLH); Infection (1; X15151242); Inflammation (f; EGG; NAD); Listeria (1; X15151242); Malaria (f; NAD); Ophthalmia (f; EGG); Salmonella (1; X15151242); Spermatorrhea (f; NAD); Scald (f; DAA); Sore Throat (f; UPW); Sprue (f; IHB); Strangury (f; NAD); Sunstroke (f; DEP; NAD); Swelling (f; DAA); Tapeworm (f; GHA); Throat (f; DEP; NAD); Tumor (f; JLH); Wart (f; JLH); Urethrosis (f; NAD); Worm (f; DAA; UPW).

Dosages (Cucumber):

FNFF = !!!

Fruits widely eaten raw or pickled, oil-roasted, stewed, stuffed, or wrapped in banana leaves and baked; pickles in rice bran essential in Japanese diet; seed edible; young leaves and stems eaten as potherb (FAC; NPM; TAN; EB54:155).

- Africans report that the juice repels fish moths and woodlice; and that peelings left on the floor overnight will kill cockroaches that eat them in three or four nights (UPW).
- Arabians use fruit pulp to soothe and whiten the skin (GHA).
- Ayurvedics consider the fruit aphrodisiac, diuretic, febrifuge, laxative, and tonic, using it for ascites, biliousness, fatigue, and insanity (KAB).

- Indochinese suggest candied cucumber in children's dysentery (BIB).
- Indonesians suggest fruit or its juice for gallstones and sprue (BIB).
- Koreans make cucumber salves for burns, scalds, and skin disorders (BIB).
- Lebanese mix the sap that oozes from scraped cucumber skin with yogurt to treat cold sores (HJP).
- Lebanese believe that cucumbers prevent colds (HJP).
- Lebanese use a favorite salad, *laban* (cucumbers and yogurt), to soften skin, dispel acne, smooth rash, and heal sunburn (HJP).
- Madagascans use the fruit as an anthelmintic, a use showing up in many cultures (KAB).
- Peruvians apply sliced cucumber to eye problems and inflammation (EGG).
- Unani, considering the seed diuretic, lachrymatory, and tonic, use it for bronchosis, fever, hepatosis, nephrosis, ophthalmia, sore throat, and thirst; the fruit they consider cardiotonic, cerebrotonic, diaphoretic, diuretic, and lactagogue; used for dysuria and ophthalmia; the vulnerary rind is applied to the hypogastrium as diuretic (KAB).

Downsides (Cucumber):

Like other members of the cucurbit family, this also contains cucurbitacins.

Extracts (Cucumber):

Buescher et al. (2004) showed that two cucumber volatiles (E,Z)-2,6-nonadienal and (E)-2-nonenal could inactivate or weaken bacterial pathogens (*Bacillus cereus, Escherichia coli, Listeria monocytogenes* (but weakly), *Salmonella typhimurium*) (at 250 and 500 ppm) (X15151242). Alpha-spinasterol significantly (circa 1000 x simvastin, a coenzyme-A inhibitor) modulates development and/or progression of diabetic nephropathy. It reduced significantly attendant increases in serum triglycerides, renal weight, and urinary protein excretion in diabetic mice (X15326549).

CUMIN (*CUMINUM CYMINUM* L.) +++ APIACEAE

Synonyms:

Cuminia cyminum J. F. Gmel.; *Cuminum hispanicum* Bunge; *Cuminum odorum* Salisb.; *Ligusticum cuminum* (L.) Crantz

Notes (Cumin):

> **Woe unto you, scribes and Pharisees, hypocrites! for ye pay tithe of mint and anise and cummin, and have omitted the weightier matters of the law, judgment, mercy, and faith: these ought ye to have done, and not to leave the other undone.**
>
> **Matthew 23:23 (KJV)**

> **Woe unto you, scribes and Pharisees, hypocrites! for ye pay tithe of mint and anise and cummin, and have omitted the weightier matters of the law, judgment, mercy, and faith: these ought ye to have done, and not to leave the other undone. Ye blind guides, which strain at a gnat, and swallow a camel.**
>
> **Matthew 23:23-4 (KJV)**

> **Woe to you, scribes and Pharisees, hypocrites! for you tithe mint and dill and cummin, and have neglected the weightier matters of the law, justice and mercy and faith; these you ought to have done, without neglecting the others. You blind guides, straining out a gnat and swallowing a camel.**
>
> **Matthew 23:23-4 (RSV)**

FIGURE 1.38 Cumin (*Cuminum cyminum*).

> **Woe to you, scribes and Pharisees, hypocrites! because you give the tenth of the mint and the dill and the cumin, but you have disregarded the weightier matters of the Law, namely justice and mercy and faithfulness, These things it was binding to do, yet not to disregard the other things. Blind guides, who strain out the gnat but gulp down the camel.**
>
> **Matthew 23:23-4 (NWT)**

Search the KJV electronically for cumin and you will not find it. In that version of the Bible, it is spelled "cummin." Zohary comments that the plant is undoubtedly native to the Middle East, but so long cultivated that it has not been positively identified in the wild state (ZOH). However, all three versions above seem to agree on the cumin and the mint, such that I confidently include them in my biblical garden. In biblical times, cumin was used as a medicine and an appetite stimulant. Philips notes that cumin (*Cuminum*) was the materia medica of the ancient world, almost paralleling black cumin's (*Nigella*) role in the Muslim world as the medicine for every disease except death. Among Mediterranean people, they used to speak of "cummin-splitters," or avaricious individuals perhaps more greedy than today's hair splitters.

COMMON NAMES (CUMIN):

Acham (Ber.; BOU); Ajaji (Sanskrit; DEP; EFS; KAB); Ajajika (Sanskrit; KAB); Anis Aigre (Fr.; KAB); Anis Faux (Fr.; KAB); Anisacre (Fr.; KAB; NAD); Azkar (Ber.; BOU); Borsos Köményy (Hun.; POR); Camon (Heb.; POR); Camon Tarbuti (Heb.; POR); Carvi (It.; EFS); Cheerakum (Mal.; NAD); Chemmun (Malta; KAB); Chimen (Roumanian; KAB); Cimino (It.; KAB); Cimonagero (It.; KAB); Cominho (Por.; TAD; USN); Cominhos (Por.; EFS; KAB); Comino (It.; Peru; Sp.; CR2; EGG; KAB; TAD; VAD); Comino Blanco (Sp.; POR); Comino Comun (Malta; Sp.; EFS; KAB); Comino Fino (Sp.; EFS); Cominos (Ma.; JFM); Cumi (Cat.; KAB); Cumin (Eng.; Fr.; Malta; Scn.; AH2; BOU; CR2; EFS; KAB; TAD); Cumin de Malte (Fr.; USN); Cumin du Maroc (Fr.; POR); Cumin Officinal (Fr.; KAB); Cuminho (Ma.; JFM); Cumino (It.; EFS; KAB); Cummin (Eng.; Malta; EFS; KAB); Dipaka (Sanskrit; KAB); Dipya (Sanskrit; KAB); Dipyaka (Sanskrit; KAB); Dirghaka (Sanskrit; KAB); Dirghakana (Sanskrit; KAB); Duru (Sin.; KAB); Egyiptomi Köményy (Hun.; POR); Faux Anis (Fr.; KAB); Gee (Nepal), Go-snyod (Tibet; NPM); Guarajaji (Sanskrit; KAB); Guarajiraka (Sanskrit; KAB); Hime Unikyoo (Japan; POR); Hrasvanga (Sanskit; NAD); Ichammen (Ber.; BOU); Ichoumane (Ber.; BOU); Jarana (Sanskrit; KAB); Jeelakara (Tel.; NAD); Jeera (Kon.; Mah.; Nepal; NAD; SUW); Jeeraka (Sanskrit NAD); Jeerakam (Mal.; POR); Jeerigay (Kan.; NAD; POR); Jeerige (Tel.; WOI); Ji (Newari; NPM); Jilakara (Tel.; WOI); Jilakarra (Tel.; POR); Jintan (Dei.; POR); Jintan Puteh (Malaya; EFS; POR); Jintan Putih (Dei.; POR); Jinten (Dei.; Java; POR); Jinten Bodas (Sunda.; IHB); Jinten Puteh (Java; IHB); Jinten Putih (Indonesia; POR); Jira (Beng.; Bhojpuri; Chepang; Danuwar; Magar; Nepal; Sunwar; DEP; KAB; NPM); Jirah (Urdu; KAB); Jiraj (Beng.; Kan.; Hindi; Sanskrit; KAB; WOI); Jiraka (Ayu.; Sanskrit; Tel.; AH2; DEP; KAB; WOI); Jirakam (Mal.; KAB); Jirana (Sanskrit; KAB); Jirautmi (Guj.; DEP; KAB); Jirna (Sanskrit; KAB); Jiregire (Mar.; KAB; WOI); Jiringe (Kan.; KAB); Jiro (Rai; NPM); Jiru (Guj.; DEP; KAB); Jorekam (Mal.; WOI); Jyajyurakat (Lepcha; NPM); Kamon (Heb.; KAB; ZOH); Kammon (Arab.; Heb.; BI2; EFS; POR; TAD); Kammûn (Arab.; Syria; HJP); Kammun Abiad (Arab.; Syria; HJP); Kamon (Heb.; POR); Kamun (Arab.; Egypt; Morocco; EFS; KAB; TAD); Kamuna (Arab.; Sanskrit; KAB; WOI); Kana (Sanskrit; KAB); Kanajira (Sanskrit; KAB); Kanavha (Sanskrit; KAB); Kemmoun (Arab.; BI2; BOU); Kemum (Accadian; ZOH); Kemun (Arab.; Syria; KAB; ZOH); Kimion (Arm.; POR); Kimoon (Arab.; GHA); Kimyon (Tur.; EFS); Kloeftsvoeb (Den.; POR); Kmin (Rus.; Pol.; POR; TAD); Kmin Tminovyj (Rus.; POR); Kmin Tminovyi (Rus.; POR); Kminek (Pol.; POR); Komijn (Dutch; EFS; POR; TAD); Kommoon (Malta; KAB); Komyn (Dutch; KAB); Kreuzkuemmel (Ger.; KAB); Kreuzkümmel (Ger.; EFS; TAD; USN); Kumel (Ger.; HIL); Kumin (Japan; TAD); Kumin Rimskii (Rus.; POR); Kumin Tminovyi (Rus.; POR); Kumina (Fin.; POR); Kuminmag (Hun.; POR); Kumyn (Dutch; KAB); Kunchika (Sanskrit; EFS); Kyminon (Greek; KAB; POR); Kyminos (Greek; KAB); Ma-Ch'in (China; Khmer; POR; TAD); Magadha (Sanskrit; KAB); Maruk (Limbu; NPM); Maustekumina (Fin.; POR); Mitadipya (Sanskrit; KAB); Mitajaji (Sanskrit; KAB); Mohrenkummel (Ger.; EFS); Mutterkuemmel (Ger.; KAB); Mutterkummel (Ger.; EFS); Ou Shi Luo (China; POR); Pepparkummin (Swe.; POR); Ravamuna (Arab.; KAB); Rimskiy Tmin (Russian; KAB); Roemischerkuemmel (Ger.; KAB); Római Köményy (Hun.; POR); Roman Caraway (Eng.; POR); Romersk Kommen (Den.; EFS); Romerks Kummin (Swe.; POR); Romischer Kummel (Ger.; EFS; TAD); Safed Jeera (Beng.; Hindi; India; EFS; NAD); Safed Jiraun (Guj.; NAD); Sanoot (Arab.; GHA); Sannût (Arab.; Syria; HJP; POR); Seerugam (Tam.; KAM); Shiagira (Beng.; Hindi; NAD); Shimai Shombu (Tam.; NAD); Shiragam (Tam.; DEP; KAM); Siragam (Tam.; WOI); Spidskommen (Den.; POR); Spiskummin (Swe.; POR; TAD); Spisskarve (Nor.; POR); Spisskummen (Nor.; POR); Sududuru (Sin.; DEP; KAB; POR); Thian-Khao (Thai; POR); Thien Khaw (Laos; POR); Vahmisakha (Sanskrit; KAB); Venidischer Kummel (Ger.; NAD); Vit Kummin (Swe.; POR); Weißer Kreuzkümmel (Ger.; POR); Yee Raa (Thai; POR); Zeera (Hindi; Iran; Urdu; EFS; POR; WOI); Zero (Sind; DEP; KAB; NAD); Zira (Hindi; Iran; Urdu; DEP; EFS; KAB; NAD; POR; WOI); Ziraa (Urdu; POR); Zi Ran Qin (China; POR); Zireh (Iran; POR); Ziya (Burma; DEP; KAB; POR).

Activities (Cumin):

Abortifacient (f; BIB; PH2); Aldose-Reductase Inhibitor (1/2 quercitrin) (1; X15796577); Alpha-Glucosidase Inhibitor (circa 1/2 acarbose) (1; X15796577); Analgesic (f; HJP); Anesthetic (f1; BIB); Anodyne (f1; BIB; DEP; PHR; PH2); Anthelmintic (f; BIB; KAB; VAD); Antiaggregant (1; PHR; PH2); Antibilious (f; BIB); Antidiabetic (1; X15796577); Antidote (f; BIB); Antifertility (1; PHR; PH2; JAC7:405); Antihyperglycemic (1; JAC7:405; X8569244); Antiinflammatory (f1; BIB); Antioxidant (1; BIB; LAF; X15364640); Antiseptic (1; BIB; PH2); Antispasmodic (f; BIB; EFS; PHR; PH2); Antitumor (f1; BIB); Aphrodisiac (f; BIB; HHB; KAB; PH2); Apoptotic (1; X10675556); Astringent (f; DEP; JFM; SUW; WOI); Bactericide (1; BIB; LAF; X10548758); Candidicide (1; HH2); Carminative (f1; DEP; EFS; NPM; PH2; SUW; VAD); Chemopreventive (1; JAC7:405); Decongestant (f; JFM); Diaphoretic (1; BIB); Digestive (f; EFS; VAD); Diuretic (f; BIB; HHB; JFM; PH2); Emmenagogue (f; BIB; BOU; EFS; PH2); Estrogenic (1; HH2; PHR; PH2; VAD); Fungicide (1; PHR; PH2); Glutathiogenic (1; JAC7:405); Hypoglycemic (1; JAC7:405; X8569244); Lactogogue (f; BOU; EFS; HHB; JFM; KAP; NMH; PH2); Larvicide (1; BIB; HH2); Mutagenic (f1; BIB; PH2; X14531636); Orexigenic (f; BIB; VAD); Photodermatitic (1; LAF); Secretagogue (1; X10812814); Sedative (f; VAD); Stimulant (f; DEP; EFS; NPM; PH2; SUW); Stomachic (f; BIB; BOU; DEP; EFS; SUW); Sudorific (f; BIB; BOU; EFS); Tonic (f; BIB); Uterocontractant (f; GHA); Uterotonic (f; KAB); Vulnerary (f; BIB).

Indications (Cumin):

Adenoma (1; X16608205); Adenopathy (f1; JLH; X16608205); Alzheimer's (1; COX; FNF); Amenorrhea (f1; BOU; HH2; VAD); Anorexia (f; BIB; VAD); Asthma (f; BIB); Atherosclerosis (1; COX; FNF; X16608205); Bacillus (1; X10548758); Bacteria (1; X10548758; X15934015; X15631509); Bite (f; DEP); Boil (f; BIB; KAB); Cancer (f1; JLH; JAC7:405; X16608205); Cancer, abdomen (f1; JLH; JAC7:405); Cancer, colon (f1; JLH; JAC7:405; X16608205); Cancer, ear (f; JLH); Cancer, esophagus (f1; JAC7:405); Cancer, liver (f; JLH); Cancer, pancreas (f1; JNU); Cancer, spleen (f; JLH); Cancer, stomach (f1; JLH; JAC7:405); Cancer, testes (f; JLH); Cancer, throat (f; JLH); Cancer, uterus (f; JLH); Cancer, uvula (f; JLH); Candida (1; HH2); Cardiopathy (f1; BIB; HJP; X16608205); Childbirth (f; HJP); Chills (f; BIB); Cholera (1; HH2); Cold (f; BIB; BOU); Colic (f; BOU; BIB; EGG; GHA; PHR; PH2); Condylomata (f; BIB; JLH); Conjunctivosis (f; BIB); Constipation (f; BIB); Consumption (f; BIB); Corn (f; BIB; JLH); Corneal Opacities (f; BIB; KAB); Cough (f; BIB); Cramps (f; BIB; BOU); Debility (f; NAD); Dermatosis (1; BIB; JAR12:83); Diabetes (1; JAC7:405; X8569244); Diarrhea (f; BIB; GHA; PHR; PH2; SUW; WOI); Diuretic (f; JFM); Dysentery (f1; BIB; HH2); Dyslactea (f; BOU; NAD); Dysmenorrhea (f; VAD); Dyspepsia (f; BIB; DEP; EGG; SUW; WOI); Dysuria (f; JFM); Enterosis (f; JLH; VAD); Epilepsy (1; X16226415); Epistaxis (f; BIB; GHA); Escherichia (1; X10548758); Esophagosis (1; JAC7:405); Fever (f; BIB); Flu (f; BIB); Fungus (1; PH2); Gas (f; BIB; JFM; PH2); Gastrosis (f1; PH2; JAC7:405); Gonorrhea (f; BIB; DEP; KAB; NAD); Headache (f; BIB; PHR; PH2); Helicobacter (1; X15934015); Hematochezia (f; PH2); Hemoptysis (f; BIB); Hepatosis (f1; BIB; JLH; JAC7:405; X16106395); Hiccup (f; BIB; KAB; NAD); Hoarseness (f; NAD); Hypothermia (f; BOU); Hysteria (f; BOU); Impotence (f; BIB; GHA; NAD); Induration (f; JLH); Infection (1; PH2); Inflammation (f1; BIB; X16608205); Klebsiella (1; HH2); Leprosy (f; BIB; KAB; PH2); Leukemia (1; X10675556); Leukoderma (f; BIB; KAB); Leukorrhea (f; NAD); Mastosis (f; SKJ); Microsporum (1; JAR12:83); Mycosis (1; PH2); Obesity (1; X16106395); Ophthalmia (f; PH2); Orchosis (f; JLH; JFM); Pain (f; BIB; DEP; HJP; PH2); Palpitation (f; BIB); Parasite (f; VAD); Parotitis (f; BIB; BOU; JLH); Puerperium (f; BIB); Rheumatism (f; PHR); Salmonella (1; HH2); Scabies (f; BIB; KAB); Sclerosis (f; BIB; JLH); Scorpion Sting (f; BIB); Shigella (1; HH2); Snakebite (f; SUW); Sore (f; BIB; KAB); Spasm (f; BIB; VAD); Splenomegaly (f; BIB; KAB); Splenosis (f; JLH; KAB); Staphylococcus (1; HH2); Stomachache (f; BIB); Stone (f; PH2); Streptococcus (1; HH2); Sty (f; BIB); Syncope (f; BIB); Swelling (f; SKJ); Tachycardia (f; BIB); Tumor (f; BIB); Ulcer (f1; BIB; X15934015); Uterosis (f; BIB; JLH); Venereal Disease (f; JLH); Vibrio (1; HH2); Wart (f; BIB; JLH); Whitlow (f; BIB; JLH); Worm (f; BIB; KAB); Yeast (1; X10548758).

Dosages (Cumin):

FNFF = !!!
The spice seed cumin is crushed and mixed with foods, for example, in biblical times, with fish and meat, especially stews. Egyptians sprinkled the seeds on bread and cakes. Today the seeds flavor breads, cheeses, chutney, meat, pickles, rice, sauerkraut, sausage, and soups. Essential oil used in liqueurs and perfumes (BIB; FAC; TAN); 300–600 mg (HHB); 0.06–0.2 ml EO (HHB; HH2). 5–10 fruits (PHR); "300 to 600 gm" (PHR) (I am sure they did not mean that; that is, 1 to 2 pounds, close to a lethal dose, I would think). Corrected in PH2 to 300–600 mg (=5–10 fruit).

- Algerians use cumin for colic and dyspepsia, with butter for coughs and colds (HJP).
- Ancient Assyrians suggested cumin with garlic for constipation and gas (BIB).
- Arabs take oil of cumin as an aphrodisiac, mixed with honey and pimento, taken 3 ×/day (BIB).
- Arabians drink decoction of ground lime and cumin seed for colic (GHA).
- Arabians steep leaves in vinegar; boil in water; drink warm for diarrhea (GHA).
- Arabians stuff ground seed with vinegar in nose to stop bleeding (GHA).
- Asian Indians smoke seeds coated with ghee in a pipe to relieve hiccup (NAD).
- Asian Indians suggest seeds with lime juice for pregnant ladies with bilious nausea (NAD).
- Asian Indians suggest 10 to 30 grains seed, even in food, for diarrhea, dyspepsia, gonorrhea, and hoarseness (NAD).
- Ayurvedics consider the fruit aphrodisiac, anthelmintic, and alexipharmic, using it for belching, biliousness, consumption, dysentery, eye diseases, fever, leprosy, leucoderma, scorpion stings, and tumors (BIB).
- Ethiopians apply pounded leaves to skin disorders (BIB).
- Iranians suggest the seeds as analgesic for pain following childbirth (HJP).
- Iraqis and Iranians use as carminative (HJP).
- Lebanese use seed oil, with or without orange flower water, for cramps, syncope, and tachycardia (HJP).
- North Africans poultice the seeds on the nape of the neck for mumps (BIB).
- Peruvians suggest the carminative seed tea for dyspepsia and infants with colic (EGG).
- Shi'ites cook it with *kibi* to prevent gas (HJP).
- Unani use the fruit for asthma, boils, corneal opacities, epistaxis, gonorrhea, hemoptysis, hiccup, inflammation, scabies, splenomegaly, styes, and ulcers, considering it abortifacient, astringent, carminative, emmenagogue, and vulnerary (KAB).
- Yemeni use seeds in aphrodisiac and uterocontractant preparations (GHA).

Downsides (Cumin):

AHP Class 1. None noted (PHR). "Health hazards or side effects following the proper administration of designated therapeutic dosages are not known" (PH2). "A very weak oxidative mutagenic action has been revealed by cumin" (X14531636). Spaniards (VAD) are more cautious but it may be generic for essential oils. Except for specified VAD indications, not for pregnant nor lactating women; not for not children less than 6 years old. Not for patients with Crohn's, epilepsy, gastritis, hepatosis, IBS, neuroses, Parkinson's, and ulcers. There is a canned contraindication that could apply to every herb: "Do not prescribe alcoholic tinctures to recovering alcoholics" (VAD).

Extracts (Cumin):

Could the biblical cumin have prevented the diabetes in those 300 million people worldwide who have it; a leading cause of amputation, blindness, heart attack, and kidney failure among adults? Lee (2005) establishes that cuminaldehyde inhibits aldose reductase (IC50 = 0.85 μg/ml) and alpha-glucosidase (IC50 = 500 μg/ml). Lee (2005) optimistically champions cuminaldehyde for its antidiabetic potential (X15796577). Cuminaldehyde was half as powerful at inhibiting alpha-glucosidase as acarbose and quercetin and could serve as an antidiabetic (X15796577).

ITALIAN CYPRESS (*CUPRESSUS SEMPERVIRENS* L.) ++ CUPRESSACEAE

Notes (Italian Cypress):

He heweth him down cedars, and taketh the cypress and the oak, which he strengtheneth for himself among the trees of the forest: he planteth an ash, and the rain doth nourish it.

Isaiah 44:14 (KJV)

Zohary (1982) notes that biblical *berosh* occurs more than 30 times in the Scriptures, but thinks of it as a collective term, meaning merely conifers with scale-like rather than needle-like leaves. He lists *Abies cilicica, Cupressus sempervirens,* and *Juniperus excelsa* as candidates for the word *berosh.* He interpreted it to mean coniferous trees with small scale-like or short linear leaves. He concludes that berosh is a collective name for all three. Amazingly, on page 120 of his excellent book, he identified ash in the HJV quote above as laurel (*Laurus nobilis*). The colloquial taxonomic names are so different in the RSV that I quote here Zohary's quote of the RSV:

He cuts down cedars; or he chooses a holm tree or an oak and lets it grow strong among the trees of the forest; he plants a cedar and the rain nourishes it.

Note that the RSV says cedar, the KJV says ash. Perhaps this is not so amazing; there are many supraspecific and suprageneric terms in the United States, such as scrub oaks and conifer, respectively. The latter embraces more kinds of gymnosperms than Zohary's *berosh.* The cypress was an important biblical timber tree, used by the Egyptians for coffins in olden times, and in Greece more recently. The doors of St. Peter's in Rome and the gates of Constantinople, made of cypress, both survived more than 1000 years. Its timbers were used for house building, ship building (even the ark), and musical instruments. David and all the house of Israel played on musical instruments made of cypress (BIB). Oil of cypress is a valuable perfume ingredient, providing ambergris- and ladanum-like odors. The trees are often planted as ornamentals in cemeteries, gardens, and parks. The Island of Cyprus, where the tree was once worshipped, derives its name from the cypress. Regarded as antiseptic, astringent, diuretic, expectorant, pectoral, styptic, sudorific, vasoconstrictor, vermifuge, and vulnerary, cypress is used in folk remedies for cancer or tumors of the eyes, nose, breast, tests, uterus, and indurations of the liver, spleen, stomach, and testicles. In Palestine, the oil derived from the leaves was used for whooping cough. Lebanese use for cough, dyspepsia, hiccup, inflammation, and ulcers. The berries serve in a cough syrup. Mashed berries are applied to lesions. Algerians ate stewed fruits for dyspepsia. In India, the fruits are described as "an aromatic stimulant" in piles. The cone has been used for bronchitis, cough, diarrhea, enuresis, fever, hemorrhage, and hemorrhoids (BIB).

Common Names (Italian Cypress):

Arella (Ber.; BOU); Bestana (Arab.; BOU); Chiparos (Rom.; KAB); Churam (Mal.; KAB); Ciprés (Cuba; Peru; Sp.; EGG; KAB; RyM); Ciprés Común (Sp.; USN); Ciprés italiano (Sp.; USN); Cipresso (It.; KAB); Cyprès (Fr.; BOU; KAB); Cyprès Commun (Fr.; BOU; USN); Cyprès d'Italie (Fr.; USN); Cyprès Pyramidal (Fr.; KAB); Cypress (Eng.; Swe.; BOU; CR2; KAB); Cypresseboom (Dutch; KAB); Cypreste (Por.; KAB); Echte Zypresse (Ger.; USN); Evergreen Cypress (Eng.; BOU; HJP); Farash

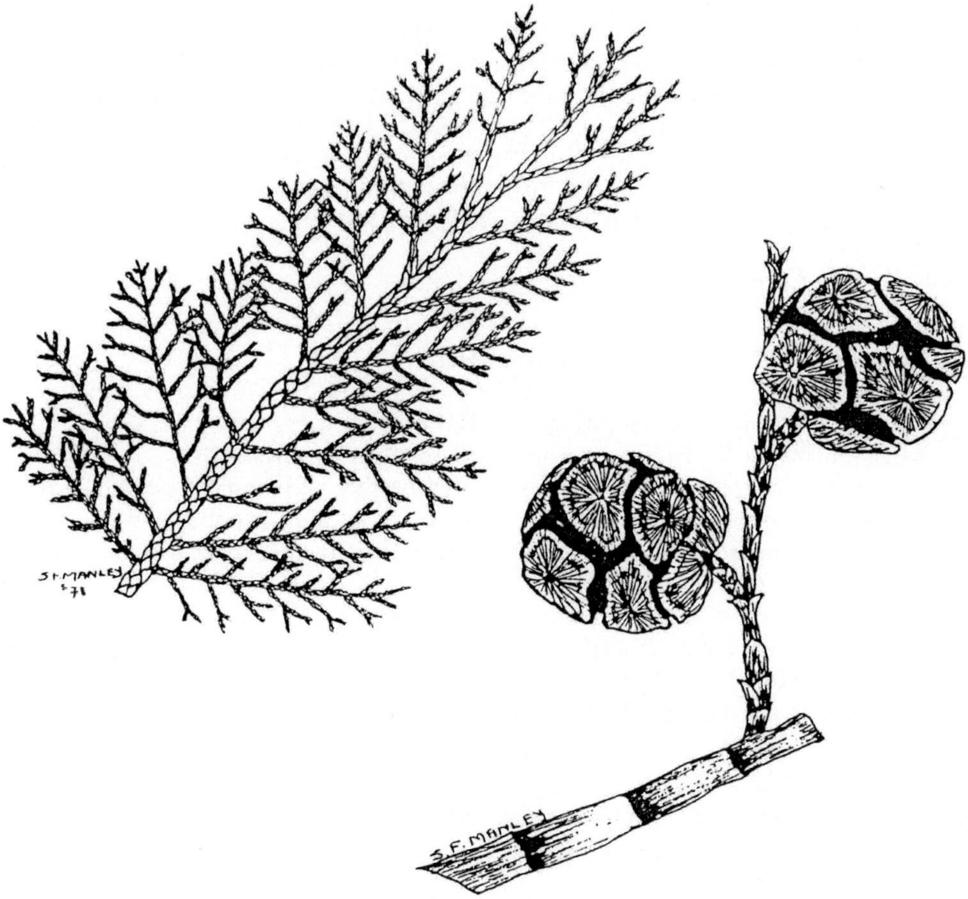

FIGURE 1.39 Italian Cypress (*Cupressus sempervirens*).

(Sin.; KAB); Hayat (Arab.; BOU); Italian Cypress (Eng.; Scn.; AH2; USN); Italienische Zypresse (Ger.; USN); Kyparis (Rus.; KAB); Kyparissi (Greek; KAB); Kyparissios (Greek; KAB); Mediterranean Cypress (Eng.; CR2; USN; VOD); Sara (Hindi; Nwp.; KAB); Sa'ed (Arab.; BOU); Saras (Hindi; Nwp.; KAB); Saru (Arab.; Hindi; Syria; HJP; KAB); Saruboke (Mar.; KAB); Sarw (Arab.; BOU); Serwal (Arab.; BOU); Shagaret el Hayat (Arab.; BOU); Sharbin (Arab.; Syria; HJP); Sipré (Creole; Haiti; VOD); Surahva (Sanskrit; KAB); Suram (Tam.; KAB); Tiddi (Ber.; BOU); Zipressenbaum (Ger.; KAB).

Activities (Italian Cypress):

Anthelmintic (f; KAB; WOI); Antielastase (1; X7884634); Antiseptic (f; EFS); Astringent (f1; BOU; HHB; KAB); Diuretic (f; BOU; EFS); Expectorant (f1; BOU; EFS; PH2); Febrifuge (f; BOU); Hemostat (f; EFS); Insectifuge (f; WOI); Pectoral (f; EFS); Sedative (f; BOU); Stimulant (f; KAB); Sudorific (f; BOU; EFS); Tonic (f; BOU); Vasoconstrictor (f; EFS); Vermifuge (f; VOD); Vulnerary (f; BIB).

Indications (Italian Cypress):

Abortion (f; BIB); Bleeding (f; BOU; EFS; VOD); Bronchosis (f; BOU; HHB; PH2); Cancer, breast (f; JLH); Cancer, eye (f; JLH); Cancer, liver (f; JLH); Cancer, nose (f; JLH); Cancer, spleen (f; JLH); Cancer, stomach (f; JLH); Cancer, testes (f; JLH); Cancer, uterus (f; JLH); Cold (f; PH2);

Convulsion (f; VOD); Cough (f; BOU; PH2; VOD); Diabetes (f; BIB); Diarrhea (f1; BOU; HHB; VOD); Dyspepsia (f; BIB); Enterosis (f; HJP; VOD); Enuresis (f; BOU); Fever (f; BIB); Flu (f; BIB); Gastrosis (f; JLH); Hemorrhoid (f1; BOU; HHB; KAB; VOD); Hepatosis (f; JLH); Hiccup (f; HJP); Infection (f; EFS); Inflammation (f; HJP); Mastosis (f; JLH); Menopause (f; VOD); Menorrhagia (f; VOD); Myofibroma (f; BIB); Neurosis (f; VOD); Orchosis (f; JLH); Pertussis (f; EFS; WOI); Polyp (f; BIB); Prolapse (f; BIB); Pulmonosis (f; VOD); Rheumatism (f; VOD); Rhinosis (f; JLH); Sclerosis (f; BIB); Splenosis (f; JLH); Swelling (f; BIB; HJP); Ulcer (f; HJP); Uterosis (f; VOD); Varicosity (1; HHB); Wart (f; BIB); Worm (f; HHB; VOD); Wound (f; HJP).

Dosages (Italian Cypress):

FNFF = X
0.5–2 g fluid extract; 0.15–0.2 g aqueous extract (HHB).

- Algerians eat stewed fruits for dyspepsia (HJP).
- Asian Indians use the fruits as "an aromatic stimulant" in piles (BIB).
- Haitians use bark, fruit, or wood decoction for diarrhea and chest complaints (VOD).
- Haitians use essential oil for convulsive coughs and intestinal worms (VOD).
- Haitians use the fruit decoction for hemorrhoids, menstrual and nervous disorders, and uterorrhagia (VOD).
- Haitians use the leaf decoction topically for rheumatism (VOD).
- Lebanese use for cough, dyspepsia, hiccup, inflammation, and ulcers. The berries serve in a cough syrup. Mashed berries are applied to lesions (HJP).
- Palestinians used the leaf oil for whooping cough (BIB).

Downsides (Italian Cypress):

"Health risks or side effects following the proper administration of designated therapeutic dosages are not known" (PH2). Large doses may irritate kidney (PH2). More recent PubMed abstracts refer more to allergenicity than to medicinal uses.

TURMERIC (*CURCUMA LONGA* L.) +++ ZINGIBERACEAE

Synonyms:

Curcuma domestica Valeton

Notes (Turmeric):

Thy plants are an orchard of pomegranates, with pleasant fruits; camphire, with spikenard, Spikenard and saffron; calamus and cinnamon, with all trees of frankincense; myrrh and aloes, with all the chief spices.

Song of Solomon 4:13–14 (KJV)

Your shoots are an orchard of pomegranates with all choicest fruits, henna with nard, nard and saffron, calamus and cinnamon, with all trees of frankincense, myrrh and aloes, with all chief spices.

Song of Solomon 4:13–14 (RSV)

Your skin is a paradise of pomegranates, with the choicest fruits, henna plants along with spikenard plants; spikenard and saffron, cane and cinnamon, along with all sorts of trees of frankincense, myrrh and aloes, along with all the finest perfumes.

Song of Solomon 4:13–14 (NWT)

FIGURE 1.40 Turmeric (*Curcuma longa*).

Israeli authors such as Zohary, more familiar with the Israeli flora and the Bible than am I, should be better equipped to speculate as to which herbs were really meant in some elusive passages. I am both pleased and displeased to note that Zohary like me, leaves a few, perhaps insoluble problems unresolved, including one involving two major medicinal plants — saffron and turmeric — and one minor medicinal — safflower— all sources of yellow dyes, and all used culinarily. All three can be grown in warmer regions of Israel, but the turmeric would be most difficult. Saffron and safflower would both be easy. Here are points that Zohary makes. "Saffron (in Hebrew, *karkom*) is mentioned only once in the Bible." (ZOH) Some commentators identify it with turmeric, which "was never grown" in this country, others with saffron, which was probably grown only in postbiblical times. There is linguistic support for both possibilities. "There is no doubt that the sown *karkom* fields mentioned in the Mishnah (of the Talmud) refer to *Crocus sativus*." (ZOH)Other data he presents point "to the identification of biblical *karkom* as turmeric and not as crocus…. But doubt arises when one considers another widely cultivated annual yielding numerous heads of orange flowers" (safflower, *Carthamus tinctorius*; (ZOH). Where does this leave me? Should I include just one or all three of the candidates for the one mention of saffron in the Bible? From the medicinal point of view, turmeric seems even more important than saffron, which appears even more important than safflower. Ditto from the likelihood point of view, based on the views of Zohary. Expansively, I have included all three in this account. Regardless of if and when it is cultivated in Israel, Xia et al. (2005) tell us that turmeric has been in Traditional Chinese Medicine (TCM) since Su Song in 627–649. In China it is commonly used for dysmenorrhea, epilepsy, hepatitis, pain, rheumatism, and traumatic diseases. The essential oil has antitumor and antiviral properties (X16028990).

Common Names (Turmeric):

Açafrao da India (Por.; AVP); Aneshta (Sanskrit; KAB); Angai (Pam.; KAB); Arazó (Chiriguano; DLZ); Arishina (Kan.; DEP); Aurukesafar (Arab.; DEP; B12); Azafrán (Peru; EGG; DAV; MDD); Azafrán Bastardo (Bol.; DLZ); Azafrán Cimarrón (Peru; EGG); Azafrán de la India (Sp.; USN); Azafrán Indio (Peru; RAR); Bahula (Sanskrit; KAB); Banley (Cam.; KAB); Batatilla (Sal.; AVP); Besar (Chepang; Danuwar; Magar; Nepal; Tharu; NPM); Bhadra (Sanskrit; KAB); Calanag (Vis.; KAB); Calavaga (Vis.; KAB); Camagüey Amarillo (Cuba; AVP); Camotillo (Cr.; Sal.; AVP); Chiang Huang (China; KAB); Chichima (St. Lucia; AVP); Chobah (Iran; EFS); Chukurayo (Aym.; Bol.; DLZ); Common Turmeric (Eng.; Ocn.; AH2); Coron (Peru; Shipibo/Conibo; EGG); Culalao (Pam.; KAB); Culiao (Ilo.; KAB); Cucurma (Sp.; SAR); Cunig (Ilo.; KAB); Curcuma (Cr.; Fr.; Mart.; Pr.; AHL; AVP; USN); Cúrcuma (Peru; RAR); Cucurma di Levante (It.; EFS); Cucurma Indiano (It.; EFS); Curcuma Long (Fr.; EFS); Curry (Eng.; EGG); Dar Sard (Iran; DEP); Darzardi (Iran; KAB); Dilao (Tag.; KAB); Dirgharaga (Sanskrit; KAB); Djavé (Tur.; AVP); Dragón (Cuba; AVP); Dulao (Vis.; KAB); Erouqsoir (Arab.; AVP); Gandhapashika (Sanskrit; KAB); Gangamau (Hausa; KAB); Gauri (Sanskrit; KAB); Gauriz Kamal (Sanskrit; EFS); Gelbwurzel (Ger.; AVP; EFS; USN); Gengibre Cimarrón (Dr.; AHL); Gengibrillo (Dr.; AHL); Gharshani (Sanskrit; KAB); Gōnōwe'ka (Siona; SAR); Guisador (Peru; DAV; DAV; MDD); Gurgemeie (Swe.; EFS); Gurkmeje (Den.; AVP); Guskmeja (Swe.; AVP); Halada (Guj.; DEP; KAB); Haladi (Sanskrit; Urdu; KAB); Haldar (Beng.; Pun.; DEP); Halde (India; EFS); Haldi (Hindi; Tamang; DEP; NPM); Halede (Mar.; DEP); Haledo (Nepal; NPM; SUW); Halja (Pun.; DEP); Halu (Nepal; NPM); Halud (Beng.; DEP); Harandi (Limbu; NPM); Hardi (Bhojpuri; Mooshar; Sunwar; NPM); Haridra (Sanskrit; AH2; KAB); Haridra Etta (Sanskrit; EFS); Hardithuli (Rai; NPM); Haridra (Sanskrit; DEP); Harita (Sanskrit; KAB); Hatusasang (Mun.; KAB); Hawa Hawa (Ese'Eja; EGG); Hemaragi (Sanskrit; KAB); Hemaragini (Sanskrit; KAB); Hohomich (Sunwar; NPM); Holodi (India; EFS); Hridivilasina (Sanskrit; KAB); Hsa-nwen (Burma; DEP; KAB); Huang Chiang (China; EFS); Idi (Culina; RAR); Indian Saffron (Eng.; Ocn.; AH2; USN); Jayanti (Sanskrit; KAB); Jengibrillo (Dr.; Pr.; AVP); Jiang huang (Pin.; AH2; DAA); Juquillo (Pr.; AVP); Jvarantika (Sanskrit; KAB); Kaha (Sing.; Ari.; AVP; DEP); Kahu Halu (Nepal; SUW); Kalo Haledo (Nepal; SUW); Kanchani (Sanskrit; KAB); Karkom? (Heb.; ZOH); Keveri (Sanskrit; KAB); Kiaong Hoang (China; DEP); Koenir (Sur.; AVP); Korkoom (Arab.; AVP); Krimighni (Sanskrit; KAB); Kshanada (Sanskrit; KAB); Kshapa (Sanskrit; KAB); Kunjit (Malaya; EFS); Kunyit (Malaya; EFS); Kurkum (Arab.; Heb.; BI2; DEP; ZOH); Kurkuma (Ger.; EFS; USN); Kürküma Uzun (Tur.; EFS); Kurkumel (Pol.; AVP); Lakshmi (Sanskrit; KAB); Lange Curcuma (Dutch; EFS); Lange Kurkuma (Ger.; EFS); Lisangay (Zambalese; KAB); Long Root Turmeric (Jam.; AHL; AVP); Mandiwinshi (Amahuaca; Peru; EGG); Mangalaprada (Sanskrit; KAB); Mangalya (Sanskrit; KAB); Manjal (Tam.; DEP); Mannal (Mal.; DEP; KAB); Marinalu (Mal.; DEP; KAB); Mbaemboyubá (Chiriguano; Iguembo; DLZ); Mehaghni (Sanskrit; JFM); Nghê (Vn.; AVP); Nisa (Sanskrit; DEP); Nisha (Sanskrit; EFS); Nishakhya (Sanskrit; KAB); Nishavha (Sanskrit; KAB); Ollod (Kon.; KAB); Ollodi (Kon.; KAB); Palillo (Peru; DAV); Palillo Cholón (Peru; EGG; RAR); Palillo Chuncho (Peru; RAR); Pampi (Tel.; DEP); Pangas (Pam.; KAB); Pasupu (Tel.; DEP); Pavitra (Sanskrit; KAB); Pinga (Sanskrit; KAB); Pinja (Sanskrit; KAB); Pita (Sanskrit; KAB); Pitavaluka (Sanskrit; KAB); Pitika (Sanskrit; KAB); Pitras (Beng.; DEP; KAB); Ponly (Cam.; KAB); Porenki (Machiguenga; EGG); Pwalojir Pagi (Piro; Yine; EGG); Quinamboy (Vis.; KAB); Rabhangavasa (Sanskrit; KAB); Racine Safrán (Fr.; AHL); Rajani (Sanskrit; EFS); Ranjani (Sanskrit; KAB); Ratrinamika (Sanskrit; KAB); Romiet (Cam.; KAB); Safrán (Haiti; AVP; TRA); Safrán Cooli (Mart.; AHL); Safrán des Antilles (Guad.; AHL); Safrán des Indes (Fr.; AHL; USN); Safrán de St. Domingue (Fr.; AHL); Safrán du Pays (Fr.; AHL); Safrán Pays (Mart.; AHL); Sanae (Burma; KAB); Sanoe (Burma; DEP); Sarsud (Arab.; DEP); Ser-po (Tibet; NPM); Shifa (Sanskrit;

KAB); Shiva (Sanskrit; KAB); Shobhana (Sanskrit; KAB); Shyma (Sanskrit; KAB); Souchet (Fr.; AHL); Souchet Long (Fr.; KAB); Souchet Odorant (Fr.; KAB); Subgavavhaya (Sanskrit; KAB); Suvarna (Sanskrit; KAB); Suvarnavarna (Sanskrit; KAB); Tamasini (Sanskrit; KAB); Tambrick (Jam.; AHL; AVP); Tamotamo (Malagasy; KAB); Tanun (Burma; DEP; KAB); Terre Mérite (Fr.; KAB); Timmer (Egypt; AVP); Tjitjima (St. Lucia; TRA); Tumeric (Jam.; AHL); Tumerico (Pr.; AVP); Turmeric (Eng.; Scn.; Trin.; AH2; AVP; NPM; USN); Turmero (Ven.; AVP); Ukon (Japan; TAN); Uma (Sanskrit; KAB); Unkwisi Manya (Siona; SAR); Urakya (Gurung; NPM); Uruk es Saba Ghin (Arab.; DEP); Uruk es Subr (Arab.; DEP); Urukus Safé (Arab.; EFS); Vara (Sanskrit; KAB); Varangi (Sanskrit; KAB); Varavarnini (Sanskrit; KAB); Varnadatri (Sanskrit; KAB); Varnavât (Sanskrit; EFS); Varnavati (Sanskrit; KAB); Varnini (Sanskrit; KAB); Vishaghni (Sanskrit; KAB); Wat Kam (Malaya; KAB); Wong Keong (Malaya; KAB); Wong Keung (Canton; KAB); Yamini (Sanskrit; KAB); Yellow Ginger (Bel.; Eng.; Ocn.; AH2; BNA); Yoshitapriya (Sanskrit; KAB); Yü chin (China; EFS); Yu Chiu (China; USN); Yu Jin (Pin.; AH2; DAA); Yü Kin (China; EFS); Yü Lin (China; EFS); Yuet Kam (Canton; KAB); Yungs-ba (Tibet; NPM); Yuquilla (Cr.; Cuba; AVP); Yuvati (Sanskrit; KAB); Zafran al Hend (Arab.; AVP); Zard (Iran; EFS); Zard Choba (Iran; DEP); Zard Chubah (Iran; DEP); Zerdé (Tur.; AVP); Zirand (Arab.; EFS); Zirsood (Arab.; AVP).

Activities (Turmeric):

Abeta-Blocker (1; X15974909); Alterative (f; DAD; SUW); Amebicide (1; MPI); Analgesic (f1; BIB; COX; X16028990); Antacid (f; BIB; DAD); Anthelmintic (f; KAB; SUW); Antiaflatoxin (1; X1394115); Antiadenomic (1; X7954412); Antiaggregant (f1; AKT; MAB; SKY; VAD); Antialzheimeran (1; X15974909); Antiamyloid (1; X15974909); Antiangiogenic (1; MAB); Antiarthritic (f1; APA; PED; WHO); Anticholeretic (f1; DAD); Antidiabetic (f1; BOW; JMF8:251); Antidote (arsenic) (f; DAD); AntiEBV (1; X10389986); Antiedemic (f1; WHO); Antifertility (1; PH2; PNC); Antifibrinolytic (1; PR14:443); Antiflatulent (f1; WHO); Antihistaminic (1; MAB; MPI; SKY); AntiHIV (1; MAB); Antihyperlipidemic (12; PHR; JMF8:256); Antiinflammatory (f12; APA; KOM; PH2; TRA; WAM); Antiintegrase (1; MAB; WHO); Antileishmanic (1; X10865470); Antileukemic (1; AKT); Antileukotriene (1; BGB; PR14:443); Antilipoperoxidant (1; X7714712); Antilymphomic (1; APA; JAD; MAB); Antimutagenic (1; BGB; LIB; MAB); Antinitrosation (1; X3054526); Antioxidant (1; PHR; PH2; WAM; WHO); Antipapillomic (1; X8879271); Antiparasitic (f; DAD); Antiplatelet (1; MAB); Antiprostaglandin (1; PH2); Antipsoriatic (1; FNF); Antiscorbutic (f; DLZ); Antiseptic (f1; MAB; PH2; PNC); Antispasmodic (f1; BIB; SHT; VAD); Antithromboxane (1; MAB); Antitumor (f1; APA; MAB; PH2; TRA); Antitumor promoter (1; X7586157); Antiulcer (f1; TRA; WHO); Antivenom (1; JAF51:6802); Antiviral (1; X10389986); Aperitive (f; DLZ); Apoptotic (1; X15356994); Astringent (f; BIB); Bactericide (1; APA; MAB; MPI; VAD); Bitter (f1; AKT); Cardioprotective (1; MAB; X15622377); Carminative (f1; APA; MAB; SUW; WHO); Chemopreventive (1; MAB); Cholagogue (f1; AHL; BGB; SHT; TRA); Choleretic (f12; KOM; SHT; TRA; WHO); Cholecystokinetic (2; KOM; SHT; WHO); Cicatrizant (f; EGG); Circulotonic (1; BOW); COX-2 Inhibitor (1; COX); Cyclooxygenase Inhibitor (1; MAB; PNC); Cytotoxic (1; MAB); Decongestant (f; BIB); Depurative (f; MAB; SUW); Digestive (f1; MAB); Diuretic (f; APA; BIB); Dysgeusia (f; KAB); EGF Inhibitor(1; X15356994); Emmenagogue (f1; AHP; DAD; LIB); Expectorant (f; BIB); Febrifuge (f1; BIB; COX); Fibrinolytic (1; MAB); Fungicide (f1; KAB; MAB; X8824742); Gastroprotective (1; WHO); Glutathiogenic (1; JAC7:405); Gram(+)-icide (1; X16277395); Heme-Oxidase Inducer (1; X15356994); Hemostat (f; DAD); Hepatoprotective (12; AKT; APA; DAD; PH2; PNC; TRA); Hepatotoxic (1; MAB); Hypocholesterolemic (1; APA; MAB; TRA; WAM); Hypoglycemic (1; X16277395); Hypolipidemic (f1; MAB; VAD); Hypotriglyceridemic (1; TRA); Immunostimulant (1; BGB; TRA); Insectifuge (f1; PHR); I-Kappa-B-Kinase Inhibitor (1; X15356994); Laxative (f; BIB); Lice (f; HAD); Lipase Promoter (1; JEB50:167); Lipolytic (f; PH2); Litholytic (f1; HHB; MAB); Maltase Promoter (1; JEB50:167); Mucogenic (1; WHO); Mucolytic (f; AKT); Myorelaxant (f1; WHO); Nematocide (f1; MAB; X8221978); NF-Kappa-B Inhibitor (1; X15356994); NO-genic (1; PR14:443); NO Scavenger

(1; MAB); iNOS Inhibitor (1; X15356994); Oncogene Inhibitor (1; X15356994); Orexigenic (f12; BIB; PHR); Ornithine-Decarboxylase Inhibitor (1; MAB); Parasiticide (f; SUW); Phagocytotic (1; BGB: WHO); PKC Inhibitor (1; X15356994); Plasmodicide (1; X10865470); Protisticide (1; APA; MPI; PNC); Radioprotective (1; X10775394); Respiratonic (f; BOW); Secretogogue (1; TRA); Secretolytic (1; TRA); Sortase-A Inhibitor (1; X16277395); Stimulant (f; BIB; DEP; SUW); Stomachic (f; BIB); Sucrase Promoter (1; JEB50:167); TNF Inhibitor (1; MAB); Tonic (f1; SUW); Tyrosine Kinase Inhibitor (1; X15356994); Ubiquitin-Proteasome Inhibitor (1; X15356994); Ulcerogenic (1; APA; MAB; WHO); Uterotonic (1; AHP; LIB); Vulnerary (f1; AKT; KAB).

Indications (Turmeric):

Abscess (f1; FNF; TRA); Adenoma (1; X7954412); Adenopathy (1; DAD; JLH; X7954412); Allergy (f1; WAM); Alzheimer's (1; COX; FNF); Amenorrhea (f1; BGB; PH2; WHO); Anorexia (f12; BGB; BRU; PHR; PH2); Arthrosis (f1; COX; KAP; MAB; WAM; WHO); Asthma (f1; MAB; WHO); Atherosclerosis (1; MAB; SKY; VAD; JMF8:246); Athlete's Foot (1; FNF); Atony (f; DEP); Bacillus (1; X10552805); Bacteria (1; X10552805); Biliousness (f1; KAB; VAD); Bite (f; BIB; DEP; PH2); Bleeding (f; PH2); Boils (f1; DAD; WHO); Bowen's Disease (1; X11712783); Bronchosis (f; BIB; DEP; PH2); Bruise (f; DAV; DEP; PED; PH2; WHO); Bursitis (1; SKY); Cancer (f1; JLH; MAB); Cancer, abdomen (1; COX; FNF; JLH); Cancer, bladder (f1; X11712783); Cancer, breast (f1; COX; FNF; MAB); Cancer, cervix (f1; X11712783); Cancer, colon (f1; COX; FNF; JLH; JNU); Cancer, duodenum (f1; X7954412); Cancer, esophagus (f1; JAC7:405); Cancer, joint (f1; JLH; MAB); Cancer, liver (f1; JAC7:405); Cancer, mouth (f1; COX; FNF; JLH); Cancer, nose (f1; COX; FNF; JLH); Cancer, sinew (f1; COX; FNF; JLH); Cancer, skin (f1; X7954412); Cancer, stomach (f1; JAC7:405); Cancer, uterus (f1; X11712783); Cardiopathy (f1; AKT; MAB; X15622377); Cataracts (f1; MAB); Catarrh (f; DEP; UPW); Chest ache (f; PH2); Childbirth (f; DAD); Cholecocystosis (12; APA; PHR; VAD; JAF51:6802); Circulosis (f; BOW); Cold (f; DEP; KAP; NPM; PH2); Colic (f; APA; PED; PH2); Coma (f; DAD); Congestion (f; APA; BIB; DEP); Conjunctivosis (f; KAB; MAB; PH2; SUW), Constipation (f; PH2); Coryza (f; DEP; KAB); Cough (f; NPM); Cramp (f1; AKT; BIB; DAD); Cystosis (f; PH2); Dermatosis (f1; AKT; DEP; MAB; PH2; SUW; WHO; WOI); Diabetes (f1; BOW; JMF8:251); Diarrhea (f1; APA; DEP; WHO); Dropsy (f; DAD); Duodenosis (1; X7954412); Dysgeusia (f; KAB); Dyskinesia (f; VAD); Dysmenorrhea (f1; AKT; APA; DLZ; PED; WHO); Dyspepsia (f12; KOM; MAB; PH2; WHO); Dysuria (f; DAD); Eczema (f1; BGB; KAP; MAB); Edema (f1; KAP; PH2); Elephantiasis (f; DAD); Enterosis (f1; AKT; DAD; PH2; WHO); Epilepsy (f; WHO; X16028990); Epistaxis (f; DAD; PH2); Esophagosis (1; JAC7:405); Fever (f1; APA; BIB; DEP; COX); Fibrosis (1; BGB; MAB); Fit (f; DEP); Fungus (f; BIB; PH2); Gallstones (f1; APA; MAB); Gas (f1; APA; PH2); Gastrosis (f1; PH2; VAD); Gonorrhea (f; BIB; KAB); Gray Hair (f; HAD); Fungus (1; LIB); Headache (f; PH2); Hematemesis (f; DAD; PH2); Hematuria (f; DAD); Hemorrhage (f; PED); Hemorrhoids (f; MAB); Hepatosis (f12; DEP; MAB; PED; PHR; PH2; TRA); Herpes (f; EGG); High Blood Pressure (1; KAP); High Cholesterol (1; AKT; APA; MAB; TRA; VAD; JMF8:246); High Triglycerides (1; MAB; TRA); Hyperhomocysteinemia (1; X15622377); Hyperlipidemia (1; MAB; JMF8:256); Hysteria (f; DAD; DEP); IBS (1; PED); Infection (f12; MAB; MPI; PH2); Inflammation (f1; DEP; PHR; PH2; WAM; WHO); Itch (f; APA; KAP; PH2); Jaundice (f1; DEP; MAB; TRA); Laryngitis (f1; BIB; COX); Leprosy (f; PH2); Leishmania (1; X10865470); Leukemia (1; AKT); Leukoderma (f; DAD); Leukoplakia (1; X11712783); Lymphoma (1; BIB; COX; FNF); Malaria (f; KAB; KAP; PH2); Morning Sickness (f1; MAB); Mucososis (f; PH2); Mycosis (f1; DEP; PH2; X8824742); Nematode (1; X8221978); Nephrosis (f1; AKT; PH2); Ophthalmia (f1; AKT; DAD; DEP; PH2); Orbital Pseudotumor (1; PR14:443); Osteoarthrosis (1; MAB); Otorrhea (f; DEP); Ozoena (f; KAB); Pain (f1; BIB; DEP; COX; WHO; X16028990); Parasite (f; BIB; DAD; KAP LIB); Plasmodicide (1; X10865470); Polyp (f1; COX; JLH; JNU); Psoriasis (1; FNF; MAB); Puerperium (f; MAB); Radiation (1; AKT); Restenosis (1; MAB); Rheumatism (f1: BIB; COX; SKY); Rhinosis (f1; COX; JLH); Ringworm (f; APA; BIB; DEP; KAP; PH2); Scabies (f2; BGB;

DEP); Smallpox (f; DAD); Snakebite (1; JAF51:6802); Sore (f; PH2); Sore Throat (f; PH2); Sprain (f1; DEP; MAB; SUW); Staphylococcus (1; MPI; UPW); Sting (f; DEP); Stone (f1; HHB; MAB); Stroke (f; BOW; PH2); Swelling (f1; AKT; COX; NPM; PH2); Syphilis (f; DAD); Thrombosis (f1; VAD); Tonsilosis (f; NPM); Trauma (f; AKT; X16028990); Ulcer (f1; BIB; COX; PED; WHO); Uveosis (2; AKT); Venereal Disease (f; BIB; DAD); Vertigo (f; BIB; DEP; DAD); Virus (1; X10389986); Vomiting (f; PH2); Wart (f; JLH); Whitlow (f; JLH); Worm (f1; DEP; X8221978); Wound (f1; APA; BGB; PH2; SUW; WAM); Yeast (f1; PED).

Dosages (Turmeric):

FNFF = !!!
Rhizomes widely consumed as foods. 4.5–9 g/day, as tea (AHP); 0.5–1 g several ×/day, between meals; or 1.5–3 g/day, often with warm milk (APA); 300-mg capsules, to 3 ×/day (APA); 1 tsp/cup warm milk (APA); 300-mg capsule, 3 ×/day (APA); 1200 mg curcumin (APA); one 445-mg StX capsule 2–3 ×/day (JAD); 0.1 g up to 20 g day (HHB); 1.5–3 g rhizome (KOM); 4 g turmeric powder in water, 1–2 ×/day (MAB); 5–14 ml fluid extract (1:1), divided in 4 or 5 doses (MAB). 0.5 tsp turmeric powder boiled with 2 cups water for 5 minutes, cooled to lukewarm and gargled for colds, cough, and tonsillitis (NPM). 3–5 g fresh herb (PED); 0.3–0.5 g dry herb (PED); 0.4 g dry herb:2 ml alcohol/2 ml water (PED); 1.5–3 g/day crude drug (SHT); 400 mg curcumin, 3 ×/day (SKY); 3–9 g crude turmeric/day (WHO); 1.5–3 g powdered plant; 0.5–1 ml tincture (1:10) 3 ×/day; 0.5–1 g oral infusion 3 ×/day (WHO).

- Ayurvedics consider the rhizome alexiteric, anthelmintic, depurative, emollient, laxative, tonic, vulnerary, and useful for biliousness, boils, bruises, complexion, dysgeusia, dyspepsia, dysuria, elephantiasis, inflammation, leukoderma, ozoena, smallpox, sprains, and swelling (KAB).
- Bolivians, considering the rhizome antiscorbutic, aperitive, diuretic, stimulant, and tonic, use it in childbirth and dysmenorrhea (DLZ).
- Chinese use for amenorrhea, colic, congestion, and externally for dermatosis (KAB).
- Haitians use the cholagogue root for fever with jaundice (AHL).
- Madagascans use rhizome as aperient, astringent, carminative, diuretic, emmenagogue, stimulant, and tonic (KAB).
- Mohammedens, following doctrine of signature, use for hepatosis and jaundice (KAB).
- Nepali boil 1/2 tsp 5 minutes in 2 cups water and gargle with the lukewarm tea for colds, coughs, and tonsillitis (NPM).
- Peruvians apply grated root to herpes, rheumatism, and wounds, as a cicatrizant (EGG).
- Peruvians take a tablespoon of grated root for malaria and take the root decoction for infectious hepatitis, 1 or 2 (child or adult) spoonfuls a day for 10 to 15 days (DAV; EGG).
- Unani consider the rhizome useful for bruises, dysuria, hepatosis, jaundice, scabies, and strangury (KAB).

Downsides (Turmeric):

Class 2b. Emmenagogue; uterotonic. Counterindicated in patients with bile duct obstruction, gallstones, hyperacidity, stomach ulcers (AHP, 1997; AEH). While in moderate doses, turmeric is said to inhibit cancers, lymphomas, and ulcers, overdoses of curcuminoids may possibly be cytotoxic and ulcerogenic and may lead to dimunition of red and white corpuscles. Still, Commission E approves 1.5 to 3 g/day, not nearly enough to provide 1200 mg curcumin. Commission E also reports contraindications: biliary obstruction, adverse effects: GI-irritation from continued use; consult physicians before using with gallstones (BIS; KOM). At 10% of diet, turmeric caused some loss of hair in rats (MAB). Care should be taken in women who wish to conceive or patients complaining of alopecia (MAB). Women who are pregnant or children with gallbladder or liver disease or ulcers should avoid turmeric (WAM); limit internal use to 10 days (WAM). Rather frightening what one reads in UPW (2000): "Laboratory animals treated with it are reported to have been rendered entirely infertile."

Extracts (Turmeric):

Fond as I am of synergy and food farmacy, I like the following comments from Verma et al. (1997). Curcumin and genistein can inhibit estrogen-positive human breast cells induced by estradiol or pesticides individually or mixed. Curcumin and genistein were synergic, totally inhibiting induction *in vitro* (X9168916). Curcuminoids inhibit cancer at initiation, promotion, and progression *in vitro* and *in vivo* (MAB). Viva the curried bean soup I am having for lunch! Reportedly as effective as hydrocortisone acetate or indomethacin in experimental inflammation (WHO); both natural anti-inflammatory curcumin (1200 mg/day) and unnatural phenylbutazone (30 mg/day) improved joint swelling, morning stiffness, and walking time in rheumatoid arthritics, both better than placebo (WHO). Bruneton notes that the antiinflammatory ED50 or curcumin orally in rats is 48 mg/kg (= 4.8 g for me) and apparently devoid of side effects (BRU) while the ipr ED50 is only 2.1 mg/kg, suggesting that the ipr route is 20 times more effective. But I am not into injecting herbs. Enjoy your curried beans, counting on those synergies. Duke suggests that curcumin needs to be compared with Celebrex and Vioxx as a COX-2 inhibitor. Essential oil showed significant antihistaminic and anti-inflammatory activity, the latter at 0.1 ml/kg, which translates to 10 ml for me, a rather dangerous dose. At doses of 1.5 g/day for 30 days, turmeric reduced urinary excretion of mutagens in an uncontrolled trial of 16 chronic smokers. In six non-smoking controls, there was no change in urinary secretion. Turmeric had no effect on serum alanine aminotransferase, aspartate amino transferase, blood glucose, creatinine, and lipid profile (MAB). Turmeric extract (circa 20 mg curcumin/day) for 45 days dramatically decreased blood lipid peroxide levels in 18 male subjects (MAB). Curcumin is poorly absorbed (some 15 to 35% max in rats) orally; but if administered with piperine (from black and long pepper), absorption improves more than 150% in rats. However, in human volunteers, 20 mg piperine increases the bioavailability of curcumin 20-fold (MAB). One study indicated that curcumin and sodium curcuminate were more potent than phenylabutazone in acute and chronic arthritic models, while another found it only 1/10th as effective as ibuprofen. While ulcerogenic in large doses, curcumin is only about 1/3 as ulcerogenic as the phenylbutazone. In low doses, curcumin had antiulcer activity, protecting against the ulcerogenic activity of phenylbutazone (MAB). 1-Phenylhydroxy-N-pentane stimulates the secretion of secretin, gastrin, and bicarbonate, thus helping maintain the gastric pH, in dogs and humans (TRA). LD50 ether extracts 12,200 mg/kg orl rat (MAB); LDlo curcumin >2000 mg/kg orl mus (MAB); LDlo curcumin >5000 mg/kg orl rat (MAB); curcumin more potent against *Leishmania* than pentamidine (JAF51:6604).

PALMAROSA (*CYMBOPOGON MARTINI* (ROXB.) J.F. WATSON) ++ POACEAE

Synonyms:

Andropogon martini Roxb.; *Andropogon schoenanthus* var. *martini* Hook. f.

Notes (Palmarosa):

> **Take thou also unto thee principal spices, of pure myrrh five hundred shekels, and of sweet cinnamon half so much, even two hundred and fifty shekels, and of <u>sweet calamus</u> two hundred and fifty shekels.**
>
> **Exodus 30:23 (KJV)**

> **Take the finest spices: of liquid myrrh five hundred shekels, and of sweet-smelling cinnamon half as much, that is, two hundred and fifty, and of <u>aromatic cane</u> two hundred and fifty.**
>
> **Exodus 30:23 (RSV)**

As for you, take to yourself the choicest perfumes: myrrh in congealed drops five hundred units, and sweet cinnamon in half that amount, two hundred and fifty units, and sweet calamus two hundred and fifty units.

Exodus 30:23 (NWT)

To what purpose cometh there to me incense from Sheba, and the sweet cane from a far country? Your burnt offerings are not acceptable, nor your sacrifices sweet unto me.

Jeremiah 6:20 (KJV)

To what purpose does frankincense come to me from Sheba, or sweet cane from a distant land? Your burnt offerings are not acceptable, nor your sacrifices pleasing to me.

Jeremiah 6:20 (RSV)

What does this matter to me that you bring in even frankincense from Sheba, and the good cane from the land far away? The whole burnt offerings of you people serve for no pleasure, and your very sacrifices have not been gratifying to me.

Jeremiah 6:20 (NWT)

Zohary informs us that aromatic grasses were used daily in the biblical world, imported from the Near East or India, for cosmetics, flavorings, medicines, and perfumery. To wit, when they opened the tombs of the Pharaohs (20th and 21st dynasty) in 1881, circa 3000 years after burial, the aroma of *Cymbopogon* was still obvious. The Hebrew words *kanev hatov, knei bosem,* and sometimes *kaneh* alone were often used to convey the broad semi-taxonomic concept of aromatic grass, sweet cane, sweet grass. Admitting that it is hopeless to speculate about which of the possible species (*Cymbopogon citratus, Cymbopogon martinii,* or *Cymbopogon schoenanthus*) was intended by the biblical writers, Zohary led his discussion with ginger grass. One of them does grow wild in the Holy Land. Zohary quotes from the Revised Standard Version of the Bible (RSV© 1946, 1952, 1971, and 1973), which renders the *Cymbopogon* in the first quote above as aromatic grass (ZOH), whereas my KJV renders it as sweet calamus. In my first Bible book, I followed the Moldenke's suggestion that it could be *Andropogon schoenanthus* or *Andropogon muricatus* (which is apparently *Vetiveria*) and they leaned toward the vetiver. After reading Zohary, I am more inclined to side with him. No one seems to push *Acorus calamus,* which did not apparently occur in the Holy Land. It seems less likely to have been imported than the *Cymbopogon* or *Vetiveria*, to either of which the alternative translation "sweet cane" seems more appropriate. The aromatic grasses share many chemicals and activities. According to WOI, "Two varieties are known, *motia* and *sofia*, which are morphologically indistinguishable." Some of the activities and indications, even common names, might as well refer to *C. schoenanthus*, which EFS treated as synonymous with *A. martinii.* BOU and UPW entries below apply to North African *C. schoenanthus* (including *C. proximum*). Remains of *C. schoenanthus* have been found in Egyptian tombs. Hippocrates, the "Father of Medicine," is said to have used it circa 440 B.C. Further, it is said to have been used in the toilet and burial preparation of the Prophet Mohammed.

Common Names (Palmarosa):

Afar (Arab.; Mauritania; UPW); Agyaghas (Beng.; Hindi; NAD); Bhustrina (Sanskrit; EFS); Bhutrina (Sanskrit; NAD); Buluuje (Upper Volta; UPW); Camel Grass (Eng.; UPW); Camel's Hay (Arab.; BOU); Chiendent Pied de Poule (Fr.; UPW); Citronelle (Fr.; BOU); Gandhabena (Beng.; Hindi; India; EFS; NAD); Gandh Bel (Hindi; WOI); Geranium Grass (Eng.; BOU; FAC);

Geraniumgras (Ger.; HHB); Gingergras (Ger.; USN); Gingergrass (Eng.; Ocn.; AH2; ZOH); Halfet Hashma (Arab.; BOU); Idhkir (Arab.; BOU); Indian Geranium (Eng.; EFS); Jonc Aromatique (Fr.; BOU); Jonc Odorant (Fr.; BOU); Kaneh (Heb.; ZOH); Kanev Hatov (Heb.; ZOH); Kavatham Pillu (Tam.; WOI); Knei Bosem (Heb.; ZOH); Lemmad (Arab.; Mali; BOU; UPW); Mahareb (Arab.; Nig.; BOU; UPW); Makkah (Arab.; BOU); Mao Hsiang (China; EFS); Motiya (India; USN); Namar Grass (Eng.; EFS); Ñangulé (Bambara; Sen.; UPW); Nard (Fr.; EFS); Nemour Grass (Eng.; EFS); Nimar Grass (Eng.; EFS); Nobi (Hausa; (Niger; UPW); Oost Indische Geranium (Dutch; EFS); Ostindisches Geraniumgras (Ger.; EFS); Paille de la Mecue (Fr.; BOU); Palmarosa (Eng.; Ger.; Scn.; Sp.; AH2; HHB; USN); Palmarosa Indien (Fr.; EFS); Palmarosagras (Ger.; HHB); Palmaroza (Tur.; EFS); Rauuns (Guj.; WOI); Robisa (Ayu.; AH2); Rohisa (Sanskrit; WOI); Roosa Grass (Eng.; EFS); Rosha (India; USN); Rosha Grass (Eng.; Ocn.; AH2); Roshegavat (Mar.; WOI); Ruaghas (Hindi; India; EFS; NAD); Rusa (Eng.; EFS); Rusagras (Ger.; EFS); Rusha (India; USN); Russagras (Ger.; HHB); Scenanth (Eng.; BOU); Schoenanthe (Fr.; BOU); Schoenanthe Officinal (Fr.; BOU); Shakanarupillu (Tam.; NAD); Sha'ret et Trab (Arab.; BOU); Sofiya (India; USN); Sumpiga (Ghana; UPW); Sweet Calamus (Eng.; KJV; ZOH); Sweet Cane (Eng.; KJV); Tiberrimt (Ber.; BOU); Tibn Makkah (Arab.; BOU); Turkse Geranium (Dutch; EFS).

Activities (Palmarosa):

Abortifacient (f; UPW); Analgesic (f; UPW); Anthelmintic (1; X13680833); Antiseptic (1; X12809717); Antispasmodic (f; EFS); Aphrodisiac (f; UPW); Astringent (f; BOU); Carminative (f; BOU; EFS); Emmenagogue (f; BOU); Insectifuge (f1; WOI; X15119079); Nematicide (1; X13680833); Stimulant (f; EFS); Sudorific (f; BOU; EFS); Vulnerary (f; BOU).

Indications (Palmarosa):

Ache (f; UPW); Alopecia (f; WOI); Amenorrhea (f; BOU); Arthritis (f; WOI); Biliousness (f; WO2); Bleeding (f; BOU); Cancer, liver (f; UPW); Cancer, spleen (f; UPW); Cancer, stomach (f; UPW); Dermatosis (f; WOI); Enterosis (f; NAD); Fever (f; BOU; UPW); Gas (f; BOU; NAD); Guineaworm (f; UPW); Impotence (f; UPW); Infection (1; X12809717); Lumbago (f; WOI); Mania (f; UPW); Pain (f; UPW); Parasite (f; UPW); Rheumatism (f; BOU); Snakebite (f; UPW); Sore (f; UPW); Spasm (f; EFS); Worm (1; X13680833); Wound (f; BOU); Yeast (1; X12809717).

Dosages (Palmarosa):

FNFF = !
Essential oil used in baked goods, chewing gum, deserts, gelatin, and ice creams (FAC). Some Northeast Africans eat the inner core of the rhizome as an aphrodisiac (UPW).

- Ghanans apply leaves, pounded in a little water, to body aches and pains (UPW).
- Ghanans mash flowers to apply, or poultice ashes to guineaworm sores (UPW). Maybe I should try that on the next bot fly I get. The last one cost me more than a thousand dollars.
- Ghanans take a tea of the inflorescence for fever (UPW).
- Nigerians inhale the burning smoke to treat temporary mania (UPW).
- Togo and north Ghanan natives use the grass for snakebite (UPW).

Extracts (Palmarosa):

Containing up to 2250 ppm perillyl-alcohol, this species could well be important.

MALTESE MUSHROOM (*CYNOMORIUM COCCINEUM* L.) ++ BALANOPHORACEAE

NOTES (MALTESE MUSHROOM):

Who cut up mallows by the bushes, and juniper roots for their meat.

Job 30:4 (KJV)

They pick mallow and the leaves of bushes, and to warm themselves the roots of the broom.

Job 30:4 (RSV)

They were plucking the salt herb by the bushes, And the root of broom trees was their food.

Job 30:4 (NWT)

Because this mushroom-like parasite is edible, and has edible roots, while juniper (broom) roots are not very edible, I was influenced in my earlier version to conclude that *Cynomorium* constituted the juniper roots of Job. Or maybe it even grew as a root parasite among the stems and roots of the halophytes. Such "roots" are frequently eaten in times of scarcity, for example, on the Canary Islands. In Qatar, where it is given the Arabic name *tarthuth*, natives eat it. In Northern Africa, the roots are pulverized and used as a spice (BIB). But Zohary concludes that the juniper (broom) root of Job 3 is *Retama*. On Malta, where it was once considered endemic, *Cynomorium* was so highly prized for its supposed medicinal help in dysentery that military sentinel were posted around places where it occurred. In some cultures it is considered an aphrodisiac for males (suspected to increase the sperm count); in others, for females. Bedouins either peel the root and eat it, or grind it and make a sweetened tea for colic (BIB).

COMMON NAMES (MALTESE MUSHROOM):

Abushal (Arab.; BOU); Afdad (Ber.; BOU); Champignon de Malte (Fr.; BOU); Cynomoir Acarlate (Fr.; BOU); Hawkal (Arab.; BOU); Maltese Mushroom (Eng.; BOU); Marshoush (Arab.; BOU); Masrut (Arab.; BOU); Mazrour (Arab.; BOU); Mousowrar (Arab.; BOU); Raetem (?; TAN); Scarlet Cynomorium (Eng.; BOU); Tarthoorth (Arab.; GHA); Tarthuth (Arab.; Oman; Qatar; Saudi; BIB; GHA); Tartous (Ber.; BOU); Tartout (Arab.; BOU); Tartout el Beni Edem (Arab.; BOU); Terzous (Ber.; BOU); Zobb el Ard (Arab.; BOU); Zobb el Ghaba (Arab.; BOU); Zobb el Qa'a (Arab.; BOU); Zobb el Tourki (Arab.; BOU).

ACTIVITIES (MALTESE MUSHROOM):

Aphrodisiac (f; BOU; PR14:288); Astringent (f; BOU); Deobstruent (f; BOU); Gonadotrophic (1; PR14:288); Hypotensive (f; X683693); Laxative (f; GHA); Narcotic (f; PR14:288); Spermatogenic (f1; BOU; PR14:288; X11282435); Tonic (f; BOU; PR14:288).

INDICATIONS (MALTESE MUSHROOM):

Arthrosis (f; BIB); Back (f; BIB); Biliousness (f; BIB); Colic (f; BIB); Constipation (f; BIB; GHA); Dysentery (f; BIB); High Blood Pressure (f; X683693); Impotence (f1; BIB; PR14:288); Infertility (1; BIB; PR14:288); Nephrosis (f; BIB); Sterility (f; BIB; PR14:288).

Dosages (Maltese Mushroom):

FNFF = !
Tuareg used pulverized root as a spice; root apparently eaten by biblical Job. In Qatar, where it is given the Arabic name *tarthuth*, natives eat it (Batanouny, 1981). In Northern Africa, the roots are pulverized and used as a spice.

- Bedouins eat, or grind the peeled root to make a sweetened tea for colic (BIB).
- Chinese regard the herb for the back, kidney, and knee, using it for constipation, impotency, and sterility (BIB).
- Maltese prize the plant for dysentery (BIB).
- North Africans mix powdered plants with butter for biliary obstructions (BIB; TAN)

Extracts (Maltese Mushroom):

A Chinese species of this genus proved more estrogenic than kudzu, following after *Polygonum cuspidatum, Rheum palmatum, Cassia obtusifolia, Polygonum multiflorum, Epimedium brevicornum,* and *Psoralea corylifolia* (X15814262). On Malta, where it was once considered endemic, In some cultures it is considered an aphrodisiac for males (suspected to increase the sperm count); in others, for females. Bedouins either peel the root and eat it or grind it and make a sweetened tea for colic. North Africans mix powdered plants with butter for biliary obstructions (BIB).

PAPYRUS (*CYPERUS PAPYRUS* L.) ++ CYPERACEAE

Synonyms:

Cyperus olivaris Targioni-Tozzetti, *Cyperus tuberosus* Roxn., *Pycreus rotundus* (L.) Hayek

Notes (Papyrus):

> **That sendeth ambassadors by the sea, even in vessels of bulrushes upon the waters, saying, Go, ye swift messengers, to a nation scattered and peeled, to a people terrible from their beginning hitherto; a nation meted out and trodden down, whose land the rivers have spoiled!**
>
> **Isaiah 18:2 (KJV)**

First described by Theophrastus (circa 372–287 B.C.) from cultivated material on the Nile delta, the inflorescences were described as useful only for "garlands for the shrines of the Gods." But boats were made from the stalks, which were also important sources of parchment paper. Moses was laid in a cradle woven from the bulrushes of papyrus growing in the rivers of lower Egypt. Among these same bulrushes the ark was placed, to be discovered by Pharaoh's daughter who brought Moses up as her son. Galen, Dioscorides, and later Islamic pharmacologists (e.g., Ibn Gulgul and El Ghafiqi) included papyrus among medicinal plants. Common in the upper Jordan valley, the papyrus reaches its northern natural limits in Israel (BIB; UPW; ZOH).

Common Names (Papyrus):

Bardi (Arab.; BOU); Birdi (Arab.; Nig.; UPW); Bulrush (Eng.; UPW); Burdi (Arab.; Nig.; UPW); Castañuela (Sp.; POR); Coco-Grass (Eng.; POR); Coquito (Sp.; POR); Fole (Kanuri; Nig.; UPW); Gemi (Heb.; ZOH); Gomeh (Heb.; ZOH); Hamasuge (Japan; POR); Herbe-à-Oignon (Fr.; POR); Jonc du Nil (Fr.; BOU); Junça (Por.; POR); Kotolo (Kanuri; Nig.; UPW); Mothe (Nepal; POR); Nile Papyrus (Eng.; BOU); Paper Reed (Eng.; BOU); Papier du Nil (Fr.; BOU; USN); Papiro (Por.; USN); Papyrus (Eng.; Fr.; JLH; UPW; USN); Papyrusstaude (Ger.; USN); Souchet à Papier (Fr.; USN);

FIGURE 1.41 Papyrus (*Cyperus papyrus*).

Herbe-à-Oignon (Fr.; POR); Juncia (Sp.; POR); Qasab el Bardi (Arab.; BOU); Rundes Zypergras (Ger.; POR); Souchet à Papier (Fr.; BOU); Souchet à Tubercules (Fr.; POR); Souchet Rond (Fr.; POR); Suo Cao (China; POR); Umm Ganagan (Nig.; UPW); Xiang Fu Zi (China; POR); Ya Haeo Mu (Thai; POR); Ya Khon Mu (Thai; POR); Zigolo Infestante (It.; POR). Nscn.

Activities (Papyrus):

Vulnerary (f; BIB).

Indications (Papyrus):

Burn (f; JLH); Callus (f; JLH); Cancer (f; JLH); Dermatosis (f; JLH); Fistula (f; BIB); Induration (f; JLH); Ophthalmia (f; BIB); Sore (f; BIB); Stomatosis (f; BIB; BOU); Wound (f; BIB; BOU).

Dosages (Papyrus):

FNFF = !

Pith commonly eaten, raw or cooked. Starchy rhizomes and lowermost parts of the stem cut off and consumed raw, boiled, or roasted, or just chewed like sugarcane. Roasted rhizomes were once a fairly common food. The Roman poet Martial joked about the fiber left in the mouth after chewing it (BIB; IHB).

- Gabonese chew the dried rhizomes for use in warding off evil spirits (UPW).
- Old World inhabitants use ashes of burned papyrus like charcoal in ophthalmia (BIB).
- Old World inhabitants sometimes used the pith to widen fistulae (BIB).

- North Africans steeped the plant in vinegar, then dried and burned it, and used the ashes for preventing the spread of oral ulcers or to heal wounds (BIB; BOU).
- Tanganyikan women take root decoction with leaf sap from *Maytenus senegalensis* for sterility (UPW).

Extracts (Papyrus):

Octopamine and tyramine reported from the leaves.

CEYLON EBONY (*DIOSPYROS EBENUM* KOENIG.) + EBENACEAE

Synonyms:

Diospyros assimilis Bedd.; *Diospyros glaberrima* Rottb.; *Diospyros hebecarpa* A. Cunn.; *Diospyros sapota* Roxb.

Notes (Ceylon Ebony):

The men of Dedan were thy merchants; many isles were the merchandise of thine hand: they brought thee for a present horns of ivory and ebony.

Ezekiel 27:15 (KJV)

The men of Rhodes traded with you; many coastlands were your own special markets, they brought you in payment ivory tusks and ebony.

Ezekiel 27:15 (RSV)

The sons of De'dan were your traders; many islands were merchants in your employ; horns of ivory and ebony they have paid back as gift to you.

Ezekiel 27:15 (NWT)

All three versions in my trilogy call it ebony, a very important timber in the same genus with our eastern persimmon (*Diospyros virginianum*). Its puckery fruits are quite astringent, even eaten after frost. Many of the tropical persimmons are also astringent. More importantly, the heartwood of several tropical species is the source of ebony, a hard black wood used for piano keys. Ebony was used, of old as it is today, frequently inlaid with ivory. We read that 200 logs of ebony presented to the kings of Persia every year by the Ethiopians were originally from India or Sri Lanka. The royal throne of Pluto, king of the mythical underworld, was made of ebony, as were carvings of many Egyptian gods and goddesses, especially those of Darkness, Night, and Sorrow. *D. ebenum* is viewed as the best ebony timber, the only one that yields jet black heartwood without streaks or marking (BIB). Zohary identifies the biblical ebony, *hovenium* in Hebrew, *hbu* in Egyptian, as *D. ebenum*, admitting that many species of *Diospyros* yield this expensive wood (ZOH). The Bible is sketchy about commercial routes in biblical times. Although not sure that ebony and ivory reached Israel from India, Zohary seems sure that both Asian and African merchandise were shipped to the Phoenician commercial Dedan, on the Arabian coast (ZOH). Common names below can be viewed as more generic than specific, the EFS names applying to *Diospyros ebenum, D. embryopteris,* and other species. I have left out any I am sure apply to our eastern persimmon.

Common Names (Ceylon Ebony):

Abnes e Hindi (Arab.; EFS); Abnus (Arab.; Hindi; DEP; WOI); Abnush (Nepal; POR); Acha (Tam.; DEP); Avolio (It.; EFS); Bale (Kan.; KAB); Bois Noir (Fr.; KAB); Ceylon Ebony (Eng.; USN); Ceylon

FIGURE 1.42 Ceylon Ebony (*Diospyros ebenum*).

Persimmon (Eng.; POR); Chara (Sri.; KAB); Chërnoe Ebenovoe Derevo (Rus.; POR); Ebenuz (Sp.; POR); Ch'i Shih (China; EFS); Diosupirosu Ebenumu (Japan; POR); East Indian Ebony (Eng.; SKJ); Ebano (It.; Sp.; EFS); Ebans (Hindi; WOI); Ebbenhoutboom (Dutch; EFS); Ébène (Fr.; POR); Ébènier (Fr.; KAB; POR); Ébènier de Ceylan (Fr.; POR); Ébènier de Maurice (Fr.; POR); Ebenus (Latin; DEP); Ebenuz (Sp.; USN); Ebony (Eng.; POR; USN); Ebony Persimmon (Eng.; POR; USN); Echter Ebenholzbaum (Ger.; POR; USN); Gab (India; EFS); Hbu (Arab.; Egypt; ZOH); Hovenum (Heb.; ZOH); Indian Ebony (Eng.; IHB); Kaju Arang (Malaya; EFS); Kakkayttali (Tam.; KAB); Kalétja (Malaya; EFS); Kaluwara (Sing.; DEP); Kanka (Sanskrit; EFS); Karai (Tam.; WOI); Kare (Kan.; DEP); Karemara (Kan.; Mysore; SKJ; WOI); Karu (Mal.; KAB; WOI); Karunkali (Tam.; DEP; WOI); Kendhu (Oriya; WOI); Khenda (Oriya; DEP; KAB); Kinkini (Sanskrit; EFS); Legno Santo (It.; EFS); Malabar Ebony (Eng.; KAB); Mallali (Mancharabad; DEP); Mauritius Ebony (Eng.; POR); Mishatumpi (Mal.; KAB); Mushtumpi (Mal.; WOI); Nallavalludu (Tel.; WOI); Nalluti (Tel.; WOI); Pei Shih (China; EFS); Tai (Mar.; DEP); Temru (India; EFS); Tendu (Hindi; India; DEP; EFS); Tinduka (Sanskrit; EFS); Tseilonskoe Ebenovoe Derevo (Rus.; POR); Tumbi (Tam.; WOI); Tumbiri (Sanskrit; EFS); Tumiki (Tel.; KAB; WOI); Vayari (Kerala; Mal.; SKJ; WOI); Wu Mu (China; POR); Nscn.

Activities (Ceylon Ebony):

Astringent (f; EFS; SKJ; WOI); Attenuant (f; SKJ; WOI); Litholytic (f; SKJ; WOI); Piscicide (f; BIB; PCS).

Indications (Ceylon Ebony):

Cancer (f; JLH); Dermatosis (f; BIB); Excrescence (f; JLH); Itch (f; PCS); Leprosy (f; PCS); Infection (f; PCS); Mycosis (f; PCS); Ringworm (f; PCS); Stone (f; WOI).

Dosages (Ceylon Ebony):

FNFF = !
Fruits edible (BIB).

- Unani consider the plant astringent, attenuant, and litholytic (KAB).

COROMANDEL (*DIOSPYROS MELANOXYLON* ROXB.) + EBENACEAE

Synonyms:

Diospyros dubia Wall.; *Diospyros tupru* Buch.-Ham.; *Diospyros wightiana* Wall. *fide* DEP.

Notes (Coromandel):

They brought thee for a present horns of ivory and ebony.

Ezekiel 27:15 (KJV)

The heartwood of several tropical species is the source of ebony, a hard black wood used for piano keys. Ebony was used, of old as it is today, frequently inlaid with ivory. We read that 200 logs of ebony presented to the kings of Persia every year by the Ethiopians were originally from India or Sri Lanka. The royal throne of Pluto, king of the mythical underworld, was made of ebony, as were carvings of many Egyptian gods and goddesses, especially those of Darkness, Night, and Sorrow. *D. ebenum* is viewed as the best ebony timber, the only one that yields jet black heartwood without streaks or marking (BIB). Zohary identifies the biblical ebony, *hovenium* in Hebrew, *hbu* in Egyptian, as *D. ebenum*, admitting that many species of Diospyros yield this expensive wood (ZOH). The Bible is sketchy about commercial routes in biblical times. Although not sure that ebony and

FIGURE 1.43 Coromandel (*Diospyros melanoxylon*)).

ivory reached Israel from India, Zohary seems sure that both Asian and African merchandise were shipped to the Phoenician commercial Dedan, on the Arabian coast. (ZOH). Common names below can be viewed as more generic than specific, the EFS names applying to *Diospyros ebenum, D. embryopteris,* and other species. I have left out any I am sure apply to our eastern persimmon.

Common Names (Coromandel):

Abanasi (Kan.; WOI); Abnus (Arab.; Hindi; Iran; Urdu; DEP; KAB); Abuus (Arab.; DEP); Balai (Kan.; Kanari; DEP; KAB); Bale (Kan.; WOI); Bois de Coromandel (Fr.; KAB); Coromandel (Eng.; WOI); Dirghapatraka (Sanskrit; WOI); Ebony (Eng.; WOI); Ebony Persimmon (Eng.; WOI); Gatto lazo (It. KAB); Gora Tiril (Kol.; DEP); Jalaja (Karai; Tam.; WOI); Kakatembhurnia (Mar.; KAB); Kari (Mal.; WOI); Karundumbi (Tam.; KAB; WOI); Karunthumb (Tam.; DEP); Kend (Beng; NAD); Kendu (Hindi; Oriya; DEP; KAB); Kenduka (Sanskrit; DEP); Kiril (San.; KAB); Kiu (Beng.; DEP); Mancigata (Tel.; WOI); Manjigata (Tel.; KAB); Nallatumki (Tel.; WOI); Ouk Chin Ya (Burma; DEP; KAB); Schwartzholzbaum (Ger.; NAD); Tamrug (Baroda; Guj.; KAB; WOI); Tembhurni (Thana;

DEP; KAB); Temburni (Kolaba; KAB); Tendu (Baigas; Hindi; Mar.; DEP; KAB; WOI); Timberni (Bom.; KAB); Timburni (Hindi; Mar.; DEP; WOI); Timru (Raj.; SKJ); Timrug (Guj.; DEP); Tumari (Kan.; WOI); Tumbi (Tam.; WOI); Tumburnni (Bom.; DEP); Tumi (Tel.; DEP); Tumki (Tel.; WOI); Tummer (Gond.; DEP; KAB); Tumru (Mar.; WOI); Tumvuru (Sanskrit; NAD); Nscn.

Activities (Coromandel):

Antiseptic (f; SKJ); Astringent (f; DEP; EFS; WOI); Carminative (f; WOI); Depurative (f; WOI); Diuretic (f; WOI); Hemostat (f; WOI); Intoxicant (f; BIB); Laxative (f; WOI); Tonic (f; DEP; NAD).

Indications (Coromandel):

Adenopathy (f; BIB); Anemia (f; BIB); Biliousness (f; KAB); Bleeding (f; DEP); Blood (f; WOI); Burn (f; BIB); Cardiopathy (f; BIB); Cerebrosis (f; BIB); Constipation (f; WOI); Corneosis (f; BIB); Dermatosis (f; WOI); Diarrhea (f; NAD; WOI); Dysentery (f; BIB; DEP; NAD); Dyspepsia (f; NAD; WOI); Dysuria (f; WOI); Epistaxis (f; BIB); Infection (f; SKJ); Leukorrhea (f; BIB); Nyctalopia (f; BIB); Ophthalmia (f; BIB) Palpitations (f; BIB); Scabies (f; BIB) Smallpox (f; BIB); Sore (f; BIB); Splenosis (f; BIB) Trichiasis (f; BIB) Uterosis (f; BIB); Vaginosis (f; BIB); Wen (f; BIB); Intoxicant (f; BIB).

Dosages (Coromandel):

FNFF = !
Fruits edible (WOI).

- Asian Indians burn the bark to treat smallpox (BIB).
- Ayurvedics consider the fruits astringent, carminative, and good for biliousness (KAB).
- Hindus consider the seeds intoxicating, using them for heart palpitations, mental disorders, and nervous breakdowns (KAB).
- Indian Hakims apply powdered bark to corneal ulcers, using it internally with black pepper for diarrhea, dysentery, and dyspepsia (KAB).
- Unani use the leaves for burns, epistaxis, ophthalmia, scabies, trichiasis, tubercular glands, and wens; the flowers for anemia, leucorrhea, nightblindness, scabies, splenitis, and urinary discharges (KAB).

RUSSIAN OLIVE (*ELAEAGNUS ANGUSTIFOLIA* L.) ++ ELAEAGNACEAE

Synonyms:

Elaeagnus angustifolia var. *orientalis* (L.) Kuntze; *Elaeagnus hortensis* M. Bieb.; *Elaeagnus moorcroftii* Wall. *ex* Schltdl.; *Elaeagnus orientalis* L. *fide* (DEP; USN)

Notes (Russian Olive):

And that they should publish and proclaim in all their cities, and in Jerusalem, saying, Go forth unto the mount, and fetch olive branches, and pine branches, and myrtle branches, and palm branches, and branches of thick trees, to make booths, as it is written.

Nehemiah 8:15 (KJV)

And that they should publish and proclaim in all their towns and in Jerusalem, "Go out to the hills and bring branches of olive, wild olive, myrtle, palm, and other leafy trees to make booths, as it is written."

Nehemiah 8:15 (RSV)

FIGURE 1.44 Russian Olive (*Elaeagnus angustifolia*).

> **And that they should make proclamation and cause a call to pass throughout all their cities and throughout Jerusalem, saying "Go out to the mountainous region and bring in olive leaves, and the leaves of <u>oil trees</u> and myrtle leaves and palm leaves and the leaves of branchy trees to make booths, according to what is written."**
>
> **Nehemiah 8:15 (NWT)**

Some of the biblical references to oil or olive trees are believed to refer to the Russian olive, a common shrub in Palestine. I did not arrive at those conclusions; I just report them. The RSV makes a distinction between olive and wild olive, and the NWT distinguishes between the olive leaves and the oil trees. I like to think they are singling out leaves of *Olea*, fruits of which are one of the better sources of oleic acid, a heart-friendly monounsaturated fatty acid, and the *Elaeagnus*, the fruits of which are one of the better sources of prostate-friendly lycopene. The KHV does not suggest *Elaeagnus* to me at all. Speaking of Lebanon, my late friend, anthropologist Jane Philips, said,

"There is a sacralization of the olive and the oleaster in the area." The plant is said to bring some people back from the shadows of death. *Oleaster* was called "*umm-zayt*" by an illiterate villager, who responded no when asked if that was the name of the plant, "that was what it is" (BIB; HJP). I also confess to having trouble distinguishing the species. Kirtikar and Basu illustrate and key three species, as follows:

- Endocarp hard and bony:
  - —*E. angustifolia*
- Endocarp ribbed, coriaceous, closed inside with a dense felt of white hairs:
  - —Fruits 6 mm long: *E. umbellata*
  - —Fruit 2.5–3.8 cm long: *E. latifolia*

The oleaster yields an inferior oil, used as a medicine but not as a food. Spaniards use the flower juice for malignant fevers. The seed oil is used for bronchitis and catarrh. The leaves are astringent. Seeds have been used in homeopathy. Lebanese use all parts of the plant medicinally, including hot flowers compressed onto neuralgia and aching wounds. Persons near death are sometimes turned around by the flower infusion (BIB).

Common Names (Russian Olive):

Árbol del Paraíso (Sp.; USN); Árvore-do-Paraíso (Por.; USN); Bohemian Olive (Eng.; DEP); Bull (Arab.; Syria; HJP); Chalef (Fr.; USN); Chalef à Feuilles Étroites (Fr.; USN); Dar el Kalbah (Arab.; Syria; HJP); Eleagno (It.; HHB); Gewöhnliche Ölweide (Ger.; HHB); Jerusalem Willow (Eng.; DEP; KAB); Nuqd (Arab.; Syria; HJP); Oil Tree (Eng.; NWT); Oleaster (Eng.; Ger.; DEP; USN); Olivagno (It.; HHB); Olivier de Bohême (Fr.; DEP; USN); Olivier des Sables (Fr.; KAB); Olivier Sauvage (Fr.; KAB); Olivo de Bohemia (Sp.; USN); Panjino (Sp.; USN); Paradusbaum (Ger.; HHB); Russian-olive (Eng.; USN); Sanjata (Afg.; DEP; KAB); Sanjit (Afg.; Pushtu; DEP; KAB); Santil (Afg.; DEP; KAB); Sanzalai (Zhob; KAB); ShāZǎo (Pin.; DAA); Shiulik (Hindi; Nwp.; DEP; KAB; NAD); Silverberry (Eng.; FNF); Sinjid (Pishin; Toba; KAB); Sinjit (Kharan; Pushtu; KAB); Sinjli (Barkhan; Bori; Sanjawi; KAB); Sinzalae (Kohlu; Bori; Sharig; KAB); Sinzalai (Barkhan; Bori; Sanjawi; KAB); Sirshing (Tibet; DEP; WOI); Sirsing (Tibet; DEP); Sugarberry (Eng.; FNF); Trebizond-Date (Eng.; USN); Wild Olive (Eng.; BIB; JLH; RSV); Wilde Oelbaume (Ger.; DEP); Zaqqum? (Arab.; Syria; HJP); Zayzafûn (Arab.; Syria; HJP); Zineid (Iran; DEP). Nscn.

Activities (Russian Olive):

Allergenic (1; X15461599); Analgesic (f; BIB); Antiaggregant (1; FNF); Antibacterial (1; FNF); Anticancer (1; FNF); Anticarcinogenic (1; FNF); Antiglaucomic (1; JNU); Antihepatotoxic (1; FNF); Antihistaminic (1; FNF); AntiHIV (1; FNF); Antiinflammatory (1; FNF; X10967484); Antileukemic (1; FNF); Antileukotriene (1; FNF); Antilipoperoxidant (1; FNF); Antimutagenic (1; FNF); Antinitrosaminic (1; FNF); Antinociceptive (1; X10967484); Antioxidant (1; CRH (Feb.):13.1997; FNF); Antiperoxidant (1; FNF); Antipresbyopic (1; JNU); Antiprostatitic (1; CRH (Feb.):13.1997); Antiradicular (1; FNF; HAD); Antiseptic (f; BIB); Antispasmodic (1; FNF); Antitumor (1; FNF); Antitumor, bladder (1; NR56:35); Antitumor, brain (1; NR56:35); Antitumor, breast (1; NR56:35); Antitumor, cervical (1; NR56:35); Antitumor, colon (1; JNU); Antitumor, esophagus (1; JNU); Antitumor, lung (1; JNU); Antitumor, mouth (1; JNU); Antitumor, pancreas (1; JNU); Antitumor, prostate (1; NR56:35; JNU); Antitumor, rectum (1; JNU); Antitumor, stomach (1; JNU); Antiviral (1; FNF); Astringent (f; HHB); Cancer Preventive (1; 525); Cholagogue (1; FNF); Choleretic (1; FNF); COX-2 Inhibitor (1; FNF); Cyclooxygenase Inhibitor (1; FNF); Cytotoxic (1; FNF); Diuretic (1; FNF); Gastroprotective (f1; X12902057); Hepatoprotective (1; FNF); Hypocholesterolemic (1; FNF); Hypotensive (1; JBH); Immunostimulant (1; FNF); Lipoxygenase Inhibitor (1; FNF); Myorelaxant (1; X12648826); ODC Inhibitor (1; FNF); Prostaglandigenic (1; FNF); Sunscreen (1; FNF).

Indications (Russian Olive):

Bacteria (1; FNF); Bronchosis (f1; FNF; KAB; WOI); Burn (f; BIB; HJP); Cancer (f1; FNF; JLH; JNU); Cancer (1; FNF); Cancer, bladder (1; NR56:35); Cancer, brain (1; NR56:35); Cancer, breast (1; NR56:35); Cancer, cervical (1; NR56:35); Cancer, colon (1; JNU); Cancer, esophagus (1; JNU); Cancer, lung (1; JNU); Cancer, pancreas (1; JNU); Cancer, prostate (1; NR56:35; JNU); Cancer, rectum (1; JNU); Cancer, stomach (1; JNU); Cancer, mouth (f; JLH; JNU); Carcinoma (1; FNF); Catarrh (f; KAB; WOI); Cerebrosis (1; NR56:35); Cervicosis (1; FNF); Constipation (f; BIB; HJP); Cramp (1; FNF; X12648826); Cystosis (1; FNF); Dysuria (f; HAD); Enterosis (f1; BIB; JNU); Esophagosis (1; FNF); Fever (f; HHB; HJP); Gastrosis (f1; FNF; X12902057); Glaucoma (1; JNU); High Blood Pressure (1; FNF; JBH); High Cholesterol (1; FNF); HIV (1; FNF); Immunodepression (1; FNF); Infection (f; BIB; HJP); Inflammation (1; FNF; X10967484); Maculitis (1; FNF); Mastosis (1; FNF); Neuralgia (f; BIB; HJP); Pain (f1; BIB; HJP; X10967484); Pancreatosis (1; FNF); Presbyopia (1; JNU); Proctosis (1; FNF); Prostatosis (1; CRH (Feb.):13.1997; FNF); Pulmonosis (1; FNF); Stomatosis (f1; FNF; JLH); Ulcer (f1; X12902057); Water Retention (1; FNF); Wound (f; BIB).

Dosages (Russian Olive):

FNFF = !!

The fruit is small and insipid, or large and quite edible. I ate many of the astringent fruits as a boy in Carolina, not realizing I might be sharing a culinary experience with the children of Israel. The fruits, believed by some to be implied by some olive references in the Bible, are known as Trebizond dates, sometimes dried and powdered to make an Arabian breadstuff. An intoxicant is distilled from the fruits. Middle Easterners may boil the fruits (even spoiled fruits known as *afouna*) to express an oleaster oil, used rather like olive oil (BIB).

- Lebanese boil astringent leaves to treat enteric fevers (HJP).
- Lebanese consider the oil antiseptic, laxative, and apply it to burns (HJP).
- Lebanese apply heated flowers as a compress for neuralgia, pain, and wounds (HJP).

Natural History (Russian Olive):

Fruit is a favorite food of birds and mammals; leaves are eaten by sheep and goats (WOI).

Extracts (Russian Olive):

Eleagnine is a racemic form of tetrahydroharman (HHB). Many of the activities and indications above followed by FNF are scored 1 because of suspected high lycopene content.

ROCKET (*ERUCA SATIVA* MILL.) ++ BRASSICACEAE

Synonyms:

Brassica eruca L.; *Brassica erucoides* Roxb.

Notes (Rocket):

> **And one went out into the field to gather herbs, and found a wild vine, and gathered thereof wild gourds his lap full, and came and shred them into the pot of pottage: for they knew them not. So they poured out for the men to eat. And it came to pass, as they were eating of the pottage, that they cried out, and said, O thou man of God, there is death in the pot. And they could not eat thereof. But he said, Then bring meal. And he cast it into the pot; and he said, Pour out for the people, that they may eat. And there was no harm in the pot.**
>
> **2 Kings 4: 39–41 (KJV)**

FIGURE 1.45 Rocket (*Eruca sativa*).

One of them went out into the field to gather herbs, and found a wild vine and gathered from it his lap full of wild gourds, and came and cut them up into the pot of pottage, not knowing what they were. And they poured out for the men to eat. But while they were eating of the pottage, they cried out, "O man of God, there is death in the pot!" And they could not eat it. He said, "Then bring meal." And he threw it into the pot, and said, "Pour out for the men, that they may eat." And there was no harm in the pot.

2 Kings 4:39–41 (RSV)

> **Accordingly a certain one went out to the field to pick mallows, and he got to find a wild vine and went picking wild gourds from it, his garment full, and then came and sliced them into the stew-pot, for they were not acquainted with them. Later, they poured it out for the men to eat. And it came about, as soon as they ate from the stew, they themselves cried out and began saying, "There is death in the pot!, O man of the [true] God." So he said "FETCH then, flour. After he threw it into the pot, he went on to say, "Pour out for the people, that they may eat." And nothing injurious proved to be in the pot.**
>
> **2 Kings 4:39–41 (RSV)**

It seems that all my versions agree that edible herbs or mallows were gathered in the field. But something poisoned the pottage, perhaps corrected by the addition of flour. I have not seen any speculation on the wild gourd, but cucurbitacins in wild gourds could foul an edible green soup. Whether or not flour or meal would correct that problem, I do not know. Yes, some scientists agree that this is the garden vegetable mentioned in the Bible (Kings II 4:39–40) as "Oroth." Zohary notes that the word *oroth* is mentioned as a plant only once, in the quote above. Referring to the Gilgal area in the Jordan Valley, where the garden rocket (arabic *jarjir*) still occurs today, Bedouins collect it as potherb or salad. Since *oroth* also appears as *gargir* in the Talmud, it is plausible to identify it with the rocket. Oroth may not necessarily be a specific potherb, but the Aramaic translation as "vegetables" in the RSV may well be correct. This is supported by the biblical verb *aroh* meaning "to collect, pick, gather" (ZOH). Rabbi Yohanan tells us that Oroth clears the eyes (Talmud, 3rd century). "Both Dioscorides and Galen recommended eating seeds for increasing semen production." ZOH "…In the Talmud and in the Rabbinical literature of the tenth century of Irak al-Qazwine of the thirteenth century indicated that eating seeds with honey will stimulate sexual desire." ZOH "…The Jewish Mishnah mention(s) that Rocket was used as a pepper substitute. The seeds were crushed and the paste was used to flavor meat." ZOH

Common Names (Rocket):

Achnef (Ber.; BOU); Ackerrauke (Ger.; KAB); Ai'afein (Arab.; BOU); 'Aisha (Arab.; BOU); Arugula (Eng.; USN); Assu (Pun.; SKJ); Baglet (Arab.; BOU); Bhutaghna (Sanskrit; SKJ); Bimbata (Sanskrit; KAB); Bou Kahli (Arab.; BOU); Chara (Kum.; KAB); Cress (Eng.; HJP); Daradharsha (Sanskrit; WOI); Djedjir (Arab.; BOU); Dua (Kum.; DEP); Duan (Nwp.; KAB); Eihukan (Iran; NAD); Eruca (Malta; KAB); Fedorênte (Mad.; Por; PST); Garden Rocket (Eng.; USN); Gargir (Arab.; BOU; ZOH); Gery (Arab.; BOU); Horf (Arab.; BOU); Jamba (Pun.; WOI); Jambeh (Iran; DEP; KAB); Jambeho (Sin.; KAB); Jambho (Mah.; Sind; DEP; NAD); Jamnia (Pun.; DEP); Jarjir (Arab.; NAD); Jirjir (Arab.; Syria; HJP); Lalu (Nwp.; DEP); Mulai (Loralai; KAB); Oroth (Heb.; ZOH); Oruga Común (Sp.; USN); Rábano Silvestre (Por.; USN); Rocket-Salad (Eng.; USN); Gargir (Arab.; BOU); Jamba (India; USN); Kerkas (Arab.; BOU); Lalu (Nwp.; KAB); Mandao (Afg.; DEP; KAB); Ölrauke (Ger; USN); Oruga (Sp.; KAB); Rashad (Arab.; Syria; HJP); Rawq (Arab.; BOU); Roka (Arab.; Syria; HJP); Roqueta (Sp.; USN); Roquette (Eng.; Fr.; BOU; USN); Roquette des Jardins (Fr.; KAB); Roquette Vraie (Fr.; BOU); Rouka (Arab.; BOU); Ruca (Cat.; KAB); Rucheta (It.; KAB); Rucola (It.; KAB; USN); Rugula (Eng.; USN); Ruke (Ger.; USN); Safed Sarson (Hindi; KAB); Safed Sarsu (Bom.; NAD); Sahwan (Nwp.; KAB); Salad Rocket (Eng.; USN); Senfkohl (Ger.; Hindi; India; Nwp.; KAB; USN); Tanakfail (Ber.; BOU); Tara (Nwp.; Pun.; KAB; SKJ; KAB); Senfrauke (Ger.; USN); Seoha (Hindi; WOI); Shiltam (Arab.; BOU); Shwetsursha (Beng.; KAB; NAD; SKJ); Siddarthra (Sanskrit; SKJ); Suffed Shorshi (Beng.; DEP); Tamamira (Pun.; NAD); Tamarira (Hindi; India; Nwp.; KAB; USN); Taramira (Pun.; NAD); Taramiri (Pun.; NAD); Thorfel (Ber.; BOU); Tira (Nwp.; KAB); Usan (Pun.; KAB).

Activities (Rocket):

Antidiabetic (1; X11053894); Antioxidant (1; X11053894; X15796582); Antiscorbutic (f; BOU); Aphrodisiac (f; BOU; KAB); Bactericide (1; MPI); Deodorant (f; EB52:394); Depurative (f; HJP); Diuretic (f; HHB; KAB; SKJ; UPW); Epoxide Hydrolase Inducer (1; X15796582); Glutathionagenic

(1; X11053894); Glutathione-Transferase Inducer (1; X15796582); Phase-II-Detoxicant Inducer (1; X15796582); Quinone-Reductase Inducer (1; X15796582); Rubefacient (f; BOU; UPW); Stimulant (f; BOU; KAB; MPI); Spermagenic (f; EB52:394); Stomachic (f; KAB; MPI; UPW); Vesicant (f; KAB).

Indications (Rocket):

Acne (f; EB52:394); Adrenoleukodystrophy (1; FNF); Adrenomyeloneuropathy (1; FNF); Anemia (f; HJP); Bacteria (1; MPI); Bite (f; EB52:394); Cancer (f; JLH); Conjunctivosis (f; EB52:394); Dermatosis (f; KAB); Diabetes (1; X11053894); Epilepsy (f; KAB); Gastrosis (f; EB52:394); Hemorrhoid (f; KAB); Hepatosis (f; JLH); Hyperglycemic (1; X11053894); Impotence (f; BOU; EB52:394); Induration (f; JLH); Infection (1; MPI); Inflammation (f; KAB); Itch (f; KAB); Nausea (f; KAB); Nephrosis (f; EB52:394); Ophthalmia (f; EB52:394); Salmonella (1; MPI); Shigella (1; MPI); Toothache (f; KAB).

Dosages (Rocket):

FNFF = !!
Greens widely eaten, raw in salads or cooked. The Jewish Mishnah mentions that rocket was used as a pepper substitute. Crushed seeds were used to flavor meat. Rocket was used "in the Holy Land during the Hellenistic period," as a spice, a food, and a medicine. Mohammedens add the rocket juice to sour pomegranates to make them sweet (NAD; EB52:394).

- Ayurvedics view as cholagogue, stomachic, vermifuge, and use for dermatosis, epilepsy, hemorrhoid, inflammation, itch, leukoderma, nausea, and toothache (KAB).
- Egyptians eat the green salad as an aphrodisiac (BOU).
- Europeans consider the young leaves antiscorbutic, diuretic, stimulant, and stomachic (KAB).
- Israelis think that eating rocket on an empty stomach prevents sweat smells (EB52:394).
- Israelis apply ground seeds to the face for acne (EB52:394).
- Lebanese give a few drops of expressed juice to weak babies (HJP).
- Lebanese Gypsies use the herb for blood purification (BOU).
- Near Easterners around the Holy Land think that eating seeds or using ground powder under the arms functions as a deodorant (EB52:394).

Downsides (Rocket):

Eating too much may cause headache (EB52:394).

Extracts (Rocket):

Barillari et al. (2005) note that rocket is mentioned in traditional pharmacopoeia and ancient literature for several therapeutic properties, and contains several health-promoting agents (e.g., carotenoids, vitamin C, fibers, flavonoids, and glucosinolates). The latter gained attention as precursors of isothiocyanates, potent inducers of phase-II detoxication process, important in the detoxification of electrophiles, and protection against oxidative stress. The major glucosinolate in rocket seeds is glucoerucin, (circa 100–110 μM/ g ZMB) representing 95% of total glucosinolates. Glucoerucin is sometimes converted into sulforaphane, the most effective inducer of phase-II enzymes (X15796582). Eruca is the namesake of erucic acid, with some good and some bad activities. Erucic and oleic acids are constituents of the cinematic Lorenzo's oil, which led to complete normalization of plasma levels of saturated very-long-chain fatty acids. If given early enough to those genetically targeted, it may help; however, the oil has no substantial effect on childhood adrenoleukodystrophy once neurologic symptoms develop. The oil's erucic acid content varies from 33% to 47%, eicosenoic acid (C 20:1) from 7.3% to 9.8%. (EB52:394), and oleic acid circa 28% (HHB).

GALBANUM (*FERULA GUMMOSA* BOISS.) + APIACEAE

SYNONYMS:

Ferula galbaniflua Boiss. & Buhse; *Peucedanum galbaniflua* (Boiss. & Buhse) Baill.

NOTES (GALBANUM):

> **And the LORD said unto Moses, Take unto thee sweet spices, stacte, and onycha, and galbanum; these sweet spices with pure frankincense: of each shall there be a like weight.**
>
> **Exodus 30:34 (KJV)**

> **And the LORD said to Moses, "Take sweet spices, stacte, and onycha, and galbanum, sweet spices with pure frankincense (of each shall there be an equal part).**
>
> **Exodus 30:34 (RSV)**

> **And Jehovah went on to say to Moses, "Take to yourself perfumes; stacte drops and onycha, and perfumed galbanum, and pure frankincense. There should be the same portion of each."**
>
> **Exodus 30:34 (NWT)**

Galbanum was an ingredient in the incense burned at the golden altar in the Holy Place, consistently with stacte and onycha and frankincense. Recent authorities maintain that "incense" used in the Tabernacle services was a mixture, in definite proportions, of frankincense, galbanum (*Ferula gumosa*), onycha (*Styrax benzoin*), and stacte (*Styrax officinalis*). Use of any incense not composed of these four ingredients (in the proper proportions) was strictly forbidden. The galbanum is a fetid yellowish gum resin, containing a chemical substance called *umbelliferone*. The gum is collected by cutting the young stem a few inches above the ground. A milky juice flows out and soon hardens. Today it is used in the manufacture of varnish. Galbanum oils and resinoids are used as fragrance components in lotions, perfumes, and soaps. Galbanum's popularity has expanded because of the "herbaceous-green" odored personal care products on the market. Extracts of galbanum have preservative and antimicrobial properties. Aqueous, hydroalcoholic, and chloroform extracts are all antiseptic.

COMMON NAMES (GALBANUM):

Bariji (Arab.; Syria; HJP); Barzhad (Arab.; EFS); Galbanum (Eng.; Fr.; Scn.; Tur.; CR2; EFS); Galbanumbaum (Ger.; EFS); Galbensaft (Ger.; HH3); Gandhabiroza (India; EFS); Gaoshira (Sanskrit; EFS); Jawashir (India; EFS); Kinneha (Iran; EFS); Moederharsboom (Dutch; EFS); Muttergummi (Ger.; HH3); Mutterharzbaum (Ger.; EFS); Mutterharz (Ger.; HH3); Quanawashaq (Arab.; JLH); Qinnah (Arab.; Syria; HJP).

ACTIVITIES (GALBANUM):

Anticonvulsant (1; X12241984); Antiedemic (f; BIB; HJP); Antiepileptic (1; X12241984); Antiseptic (f1; BIB; PH2; X15567258); Antispasmodic (1; X11695880); Bactericide (1; HH3 X15567258); Emmenagogue (f; EFS); Emollient (f; BIB); Escherichia (1; X15567258); Expectorant (f; EFS; PH2); Gram(+)-icide (1; X15567258); Stimulant (f; EFS; PH2); Stomachic (f; BIB); Uterotonic (f; BIB); Vulnerary (f; HJP; PH2).

FIGURE 1.46 Galbanum (*Ferula gummosa*).

Indications (Galbanum):

Addiction (1; X11483380); Allergy (f; BIB; HJP); Amenorrhea (f; EFS); Asthma (f; SKJ); Bacteria (1; HH3; X15567258); Bronchosis (f; SKJ); Cancer (f; JLH); Cancer, abdomen (f; JLH); Cancer, breast (f; JLH); Cancer, colon (f; JLH); Cancer, gum (f; JLH); Cancer, liver (f; JLH); Cancer, parotid (f; JLH); Cancer, spleen (f; JLH); Cancer, stomach (f; JLH); Cancer, testicles (f; JLH); Cancer, uterus (f; JLH); Caries (f; BIB); Chilblain (f; BIB); Cold (f; BIB; HJP); Colic (f; BIB; HJP); Cramp (f; BIB); Diarrhea (f1; X11695880); Dyspepsia (f; BIB; PH2); Edema (f; BIB); Enterosis (f1; BIB; X11695880); Epilepsy (1; X12241984); Escherichia (1; X15567258); Gas (f; PH2); Gastrosis (f1; BIB; X11695880); Gingivosis (f; BIB); Hepatosis (f; JLH); Hysteria (f; BIB); Induration (f; JLH); Infection (f1; HH3; PH2; X15567258); Inflammation (f; JLH); Mastosis (f; BIB); Morphinism (1; X11483380); Neurosis (f; BIB); Orchosis (f; JLH); Otosis (f; BIB); Parotosis (f; JLH); Phymata (f; JLH); Polyp (f; BIB; JLH); Rheumatism (f; EFS); Scleroma (f; JLH); Spasm (1; X11695880); Splenosis (f; JLH); Staphylococcus (1; HH3); Swelling (f; BIB; HJP; JLH); Uterosis (f; JLH); Withdrawal (1; X11483380); Wound (f; HJP; PH2).

Dosages (Galbanum):

FNFF = !

Galbanum oils and resinoids are used as flavor components in many foods, including non-alcoholic beverages, baked goods, candies, condiments, gelatins, puddings, relishes; the oil is used in meats and gravies (BIB). 0.3–1 g resin (HHB).

- Lebanese use imported galbanum as stomachic tonic for colds and colic (HJP).
- Lebanese work resin into hot olive oil to dress wounds (HJP).

Extracts (Galbanum):

LD50 Sodium galbanate = 227 mg/kg ipr mus (HH3).

FIG (*FICUS CARICA* L.) ++ MORACEAE

Notes (Fig):

> **For Isaiah had said, Let them take a lump of figs, and lay it for a plaister upon the boil, and he shall recover.**
>
> **Isaiah 38:21 (KJV)**

> **Now Isaiah had said, "Let them take a cake of figs, and apply it to the boil, that he may recover."**
>
> **Isaiah 38:21 (RSV)**

> **And Isaiah proceeded to say, "Let them take a cake of pressed dried figs, and rub [it] upon the boil, that he may revive."**
>
> **Isaiah 38:21 (NWT)**

Do I poultice my boil with a lump of fresh figs, a cake of dried figs, or rub the boil with a cake of pressed dried figs? For myself, I would drip some of the milk from the stem on my boil if I had the plant handy, but I would try dried figs in a pinch. The leaves of the fig, first fruit recorded in the Bible, were used to cover Adam and Eve's nakedness. I do not know about Adam and Eve but, with me, the leaves cause violent itching in contact with my bare skin. Other members of the fig family, if not the fig itself, have been used to make bark cloth, which is much more comfortable. To sit under one's own vine and fig tree was the Jewish concept of peace and prosperity as indicated in I Kings 4:25. Fig leaves are still sewn together and used as wrappings for fresh fruit. To Egyptians, the fig represented the Tree of Life. Some suggest that the fig was the forbidden fruit of the Garden of Eden. They believe that eating the dried fruits facilitates conception.

Let me once again quote from one of hundreds of letters I received back when I was with the USDA, leading their Medicinal Plants Laboratory. "I have just read your article about searching for plants that contain anticancer chemicals. For a long time I have believed that figs would be used in the treatment of cancer. My reason — in II Kings Chap. 20 in the King James Version of the Bible beginning with the 1st verse through 7. Please read it and see what you think. I do pray to God that something will come through soon." I read the scripture he suggested. "And Isaiah said, Take a lump of figs. And they took and laid it on the boil, and he recovered." (II Kings 20:7). After reading that letter and scripture, I went to Jonathan Hartwell's *Plants Used against Cancer*; and there among more than three full pages of anticancer folklore, found that folklore suggested figs for many cancerous conditions (e.g., cancer of the gums and uterus; calluses; condylomata; corns; exacerbations; excrescences of the eyelids, vulva, or uterus; fibroids; impostumes; moles; myrmecia; neoplasms; polyps; scleroses of the cervix, kidney, limbs, liver, sinews, spleen, stomach, testicles, and uterus; thymi; tumors of the abdomen, bladder, fauces, feet, glands, liver, neck, parotid, uterus, and windpipe; warts; and wens (BIB).

Common Names (Fig):

A Tsang (China; EFS); Anjir (Afg.; Beng.; Guj.; Hindi; Iran; Kharan; Kon.; Mar.; Nepal; Urdu; KAB; NPM); Anjira (Bom.; Sanskrit; AH2; KAB); Anjra (Guj.; NAD); Anjur (Kon.; KAB); Anjura

FIGURE 1.47 Fig (*Ficus carica.*)

(Kan.; KAB); Anjuru (Tel.; KAB); Aviavimbazaha (Hova; KAB); Azart (Ber.; BOU); Bakhis (Ber.; BOU); Berbereira (Mad.; Por.; PST); Bilaitloa (Mun.; KAB); Bou (Provence; KAB); Breva (Sp.; AVP); Brevo (Sp.; JFM); Cabrahigo (Sp.; KAB); Caprifiguier (Fr.; AVP); Carique (Fr.; BOU); Chagar el tin (Arab.; AVP); Common Fig (Eng.; VOD); Doomoor (Beng.; NAD); Doomoot (India; EFS); Echte Feige (Ger.; USN); Emohi (Ber.; BOU); Fagari (Pun.; KAB); Fagu (Pun.; DEP; KAB); Faguri (Pun.; KAB); Feige (Ger.; AVP); Feigenbaum (Ger; EFS; KAB; USN); Fico (It.; KAB; USN); Fig (Creole; Eng.; Haiti; Scn.; AH2; CR2; NPM; VOD); Fig Frans (Creole; Haiti; VOD); Figener (Den.; EFS); Figo (It.; Por.; AVP; KAB); Figovoi Drava (Rus.; KAB); Figu (Ma.; JFM); Figue France (Haiti; AVP); Figueira (Mad.; Por.; KAB; PST); Figueira Brava (Por.; KAB); Figueira comun (Ma.; JFM); Figueira de Baco (Ma.; JFM); Figuera (Cat.; KAB); Figuera Borda (Cat.; KAB); Figuier (Fr.; AHL; BOU; KAB); Figuiera (Por.; AHL; USN); Figuier Blanc (Fr.; AHL); Figuier Commun (Fr.; USN); Fijge (Ma.; JFM); Fikontrae (Swe.; KAB); Fugefa (Hun.; KAB); Fugu (Pun.; KAB); Higo (Peru; Sp.; AHL; AVP; DAV; USN); Higo Extranjero (Dr.; Sp.; AHL); Higuera (Peru; Sp.; KAB; DAV); Higuera Comun (Sp.; USN); Hinjir (Sibi; KAB); Incir Agasi (Tur.; EFS); Inzar (Sibi; KAB); Kakodumbar (Sanskrit; KAB); Karm (Arab.; BOU); Karmus (Arab.; BOU); Kerma (Arab.; Tunisia; AVP; BOU); Kimri (Pun.; KAB); Kohianjir (Sarawan; KAB); Krade (Greek; JLH); Kuru Incir (Tur.; EB51:195); Lovea si Phle (Cam.; JLH); Manjimeda (Tel.; KAB); Manjula (Sanskrit; KAB); Medi (Tel.; KAB); Modipatu (Tel.; KAB); Moo Fah (China; EFS); Olynthoi (Greek; JLH); Pushposhunyo (Oriya; KAB); Ravi (Iran; EFS); Saphansi (Burma; NAD); Shimeatti (Tam.; NAD); Simaiyatta (Tam.; KAB); Simayatta (Mal.; KAB); Simayatti (Tel.; KAB); Simeyatti (Kan.; KAB); Smochin (Rom.; KAB); Smokovnitsa (Rus.; KAB); Sykas (Greek; JLH); Syki (Greek; KAB); Sykia (Greek; KAB); Taguerout (Ber.; BOU); Tamazate (Ber.; BOU); Tamehit (Ber.; BOU); Tanaglet (Ber.; BOU); Tazert (Ber.; BOU); Ten (Arab.; KAB); Tenach (Heb.; KAB); Tenatti (Tam.; KAB); Teneyatti (Tel.; KAB; NAD); Tiethie (Burma; KAB); Tin (Arab.; KAB); Tin Teen (Arab.; EFS); Tine (Arab.; BOU); Udeunbara (Sanskrit; EFS); Vijgeboom (Dutch; EFS); Vijgenboom (Dutch; KAB); Wu Hua Guo (Pin.; DAA; USN); Wu hua Kuo (China; EFS; KAB); Yemis (Tur., EB49:406); Ying Jeh Kuo (China; EFS); Yu T'an Po (China; EFS).

Activities (Fig):

Alexiteric (f; BIB); Allergenic (1; HH3); Analgesic (f; EB49:406); Anthelmintic (1; GHA; HHB; X15727070; X11473446); Anticancer (1; X11473446); Antidiabetic (1; X12682822); Antidote (f; BOU); AntiHSV-1 (1; X15613791); Antiinflammatory (f1; VAD); Antileukemic (1; X11473446); Antilymphomic (1; X114734460); Antimutagenic (1; X15131968); Antioxidant (1; X12682822); Antisarcomic (1; X11473446); Antiseptic (f; AHL; BIB); Antitumor (breast) (1; X11473446); Antitumor (prostate) (1; X11473446); Antitussive (f; DAV); Antiviral (1; X15613791); Aperient (f; BIB); Aphrodisiac (f; BIB; HH3); Ascaricide (1; WOI); Balsamic (f; VAD); Catabolic (1; X11473446); Demulcent (f; BIB; DEP; EFS; VOD); Deodorant (f; KAB); Digestive (f1; BIB; VAD); Diuretic (f1; BIB; GHA; HH3); Emollient (f; BIB; BOU; DEP; EFS; VOD); Expectorant (f; BIB; EFS); Hypocholesterolemic (1; X11032050); Hypoglycemic (1; X11473446); Lactagogue (f; DAA; NMH); Laxative (f; BIB; DEP; EFS; GHA); Litholytic (f; BIB; GHA; KAB); Mnemonic (f; RAR); Nematicide (1; X15727070); Pectoral (f; AHL; BIB); Phototoxic (1; HH3); Proteolytic (1; GHA; WOI); Purgative (f; BIB); Restorative (f; BIB); Stimulant (f; DAV; RAR); Stomachic (f; BIB); Suppurative (f; DEP); Tonic (f; BIB; BOU; GHA); Toxic (f; DAV); Vermifuge (f1; BIB; GHA).

Indications (Fig):

Abscess (f; BIB); Adenopathy (f1; HHB; JLH; SOU); Alopecia (f; BIB); Anemia (f; WOI); Ascaris (1; WOI); Asthma (f; AHL; BIB; JFM); Boil (f; BIB; VOD); Bronchosis (f; DEP); Burn (f; VAD); Callus (f; JLH); Cancer (f; BIB); Cancer, abdomen (1; FNF; JLH); Cancer, bladder (1; FNF; JLH); Cancer, breast (1; FNF; HHB; X11473446); Cancer, cervix (1; FNF; JLH); Cancer, colon (1; FNF;

JLH); Cancer, eye (1; FNF; JLH); Cancer, feet (1; FNF; JLH); Cancer, gum (1; FNF; JLH); Cancer, kidney (1; FNF; JLH); Cancer, liver (1; FNF; JLH); Cancer, mouth (1; FNF; JLH); Cancer, neck (1; FNF; JLH); Cancer, parotid (1; FNF; JLH); Cancer, prostate (1; X11473446); Cancer, spleen (1; FNF; JLH); Cancer, stomach (1; FNF; JLH); Cancer, testicle (1; FNF; JLH); Cancer, throat (1; FNF; JLH); Cancer, uterus (1; FNF; JLH); Cancer, vulva (1; FNF; JLH); Carbuncle (f; BIB); Catarrh (f; BIB; KAB); Cervicosis (f; JLH); Cheilosis (f; KAB); Childbirth (f; DAV); Cold (f; HH3; JFM; VOD); Colic (f; VAD); Condyloma (f; BIB; JLH); Conjunctivosis (f; BIB; BOU); Constipation (f; DEP; KAB; VAD); Corn (f; BIB; DAA); Cough (f; BIB; HH3); Cystosis (f; JLH); Dandruff (f; DAV; SOU); Depression (f; GHA); Diabetes (f1; JFM; X12682822); Diarrhea (f; EB51:195); Diphtheria (f; BIB); Dropsy (f; SOU); Dysentery (f; PH2); Emphysemic (f; VAD); Enterosis (f; PH2; VAD); Epistaxis (f; KAB); Fibroid (f; JLH); Flu (f; AHL; BIB); Fracture (f; DAV); Freckle (f; BOU; GHA); Furuncle (f; VAD); Gastrosis (f; DAA; JLH; VAD); Gingivosis (f; BIB; JLH); Glossosis (f; KAB); Gout (f; KAB); Hemorrhoid (f; BIB; HH3; NPM); Hepatosis (f; HH3; JLH); Hernia (f; DAV); Herpes (1; X15613791); High Triglycerides (1; X11473446); HSV-1 (1; X15613791); Impostume (f; JLH); Impotence (f; DEP); Induration (f; JLH); Infection (f1; FNF; KAB; X15613791); Inflammation (f; BIB); Leprosy (f; BOU; KAB); Leukemia (1; FNF; HHB; X11473446); Leukoderma (f1; DEP; FNF); Lymphoma (1; FNF; HHB; X11473446); Mastosis (f; EB51:195); Measles (f; BIB); Mole (f; JLH); Mucososis (f1; KAB; VOD); Mycosis (f1; FNF; KAB); Myrmecia (f; JLH); Nematode (1; X15727070); Nephrosis (f; GHA; JLH); Neurosis (f; GHA); Obesity (1; X11473446); Ophthalmia (f; BOU; JLH); Orchosis (f; JLH); Pain (f; BIB; EB49:50); Papillomatosis (1; X14720183); Paralysis (f; BIB); Parotosis (f; JLH); Pertussis (f; BIB; JFM); Pharyngosis (f; VAD); Phymata (f; JLH); Pimple (f; BIB); Polyp (f; BIB); Pulmonosis (f; JLH); Respirosis (f; VAD); Rhinosis (f; JLH); Ringworm (f1; FNF; KAB); Sarcoma (1; FNF; HHB; X11473446); Sclerosis (f; JLH); Scrofula (f; BIB; EFS); Sore (f; JLH); Sore Throat (f; BIB); Splenosis (f; HH3; JLH; NAD); Stomachache (f; DAA); Stomatosis (f; JLH; NAD; VAD); Stone (f; GHA; NAD); Stress (f; GHA); Swelling (f; JLH); Thirst (f; BIB); Thrush (f; BIB); Toothache (f; JFM); Tuberculosis (f; DEP); Tumor (f; BIB; VOD); Uterosis (f; JLH); Venereal Disease (f; BIB); Virus (1; X15613791); Wart (f1; BIB; NPM; VOD; X14720183); Wen (f; JLH); Worm (f1; BIB; X15727070); Wound (f; JFM); Yeast (f; BIB).

Dosages (Fig):

FNFF = !!!

Figs are eaten fresh or dried and threaded on long strings. "Cakes of Figs" are mentioned in I Samuel 25:18, and these were consumed for travel. North Africans make a tonic anise-flavored fig brandy (BIB; BOU); 30 g fig syrup (HH3).

- Africans drop fig latex in ant holes to drive them away (BIB).
- Africans use the fresh root in a lotion for thrush (BIB).
- Arabs deem the Smyrna fig a better aphrodisiac (DEP).
- Asian Indians apply leaf juice early in leukoderma (DEP) (furanocoumarins; JAD).
- Asian Indians suggest the fruit pulp with vinegar and sugar for pediatric bronchitis (DEP).
- Ayurvedics use the fruit for epistaxis, leprosy, and diseases of the blood and head (KAB).
- Chinese apply the leaves to hemorrhoids (BIB).
- Cubans drink strained leaf decoction for chest ailments (JFM).
- Haitians apply latex to warts, and roasted half figs to boils and tumors (VOD).
- Haitians eat raw, dry, or roasted fig, often with senna, as a laxative (VOD).
- Haitians take demulcent fig decoction for colds (VOD).
- Hispaniolans suggest aromatic leaf tea taken for asthma and flu (AHL).
- Latinos smoke the leaves for asthma (JFM).
- Lithuanians eat figs with dates, raisins, and wheat bread for cancer (JLH).

- North Africans suggest the leaf decoction to erase freckles (BOU).
- Latinos steep ripe fruits in booze overnite, then eat on empty stomach for pertussis (JFM).
- Latinos boil three sundried leaves 15 minutes in 300 g water for diabetes (JFM).
- Unani use the root for leucoderma and ringworm, the alexiteric, aphrodisiac, litholytic, purgative, tonic, fruit for alopecia, chest pains, hepatosis, fever, inflammations, paralysis, piles, splenosis, and thirst. They regard the milky juice as diuretic, expectorant, yet dangerous to the eyes (KAB).
- Yemeni eat mixed dates, figs, honey, and raisins for depression and nervous tension (GHA).

Downsides (Fig):

No health hazards or contraindications reported with proper administration of suggested therapeutic dosages (PH2) (but PH2 designates no specific quantified dosage! JAD).

Extracts (Fig):

Wang et al. (2004) found an antiviral (herpes simplex) effect in fig leaf extracts with relatively low toxicity (X15613791). Stepek et al. (2005) demonstrated an expected anthelmintic effect of natural plant cysteine proteinases against a GI nematode, *Heligmosomoides polygyrus, in vitro.* Cysteine proteinases from papaya, pineapple, fig, and Egyptian milkweed all damaged the cuticle of *H. polygyrus.* LD50 values indicated that the purified proteinases were more efficacious than the proteinases in the crude latex, with purified ficin, papain, chymopapain, Egyptian milkweed latex extract, and pineapple fruit extract containing fruit bromelain, having the most potent effect (X15727070). Agabeili et al. (2004) found antimutagenic and genoprotective activities with fig extracts (X15131968). Hemmatzadeh et al. (2003) successfully treated bovine papillomatosis with fig latex (comparable to salicylic acid) (X14720183). Perez et al. (2003) found that fig extracts are useful in diabetes; the extracts tend to normalize antioxidant status (X12682822). Poultice of dried figs in milk is said to deodorize malignant cancers (KAB).

SYCAMORE FIG (*FICUS SYCOMORUS* L.) + MORACEAE

Synonyms:

Ficus cocculifolia Baker; *Ficus gnaphalocarpa* (Miq.) A. Rich.; *Ficus sycomorus* subsp. *gnaphalocarpa* (Miq.) C.C. Berg; *Ficus trachyphylla* (Miq.) Miq.; *Sycomorus gnaphalocarpa* Miq.; *Sycomorus trachyphylla* Miq.

Notes (Sycamore Fig):

> **Then answered Amos, and said to Amaziah, I was no prophet, neither was I a prophet's son; but I was an herdman, and a gatherer of sycamore fruit.**
>
> **Amos 7:14 (KJV)**

> **Then Amos answered Amazi'ah, "I am no prophet, nor a prophet's son; but I am a herdsman, and a dresser of sycamore trees.**
>
> **Amos 7:14 (RSV)**

> **Then Amos answered and said to Amazi'ah, "I was not a prophet, neither was I the son of a prophet; but I was a herdsman, and a nipper of figs of sycamore trees.**
>
> **Amos 7:14 (RSV)**

FIGURE 1.48 Sycamore Fig (*Ficus sycomorus*).

The sycamore fig that Zacchaeus allegedly climbed to see Jesus pass is a curious tree combining the characteristics of fig and mulberry. Its porous but durable wood was used for temples and auditoria, as well as for fashioning mummy chests or coffins (sarcophagi) found in perfect condition after more than 3000 years (BIB; FP1). Others say it is only fit for fuel. African Masai use twigs in fire making. The milky latex, like many other fig species, contains rubber-like compounds. In the Holy Land, it is frequently planted as a shade tree, the shade reported to have embraced the Virgin Mary. "At Marhave is a large sycamore or Pharaoh's Fig, very old, but which bears fruit every year. They say that upon the Virgin passing that way with her son Jesus and being pursued by the people, this Fig tree opened to receive her and closed her in again, until the people had passed by and then opened again. The tree is still shown to travelers" (BIB). Zohary (FP1) notes that the plant is widely cultivated in the Holy Land (e.g., on the coastal plain and the Jordan Valley), but native to Ethiopia and elsewhere in tropical eastern Africa. Not setting viable seed, it is easily propagated by cuttings.

Common names (Sycamore Fig):

A Nak (Guinea; UPW); Djimez (Arab.; Niger; UPW); Figuier Sycomore (Fr.; USN); Ga (Mali; UPW); Gamiesa (Arab.; Nig.; UPW); Ganlu (Dahomey; UPW); Grande Sycomore (Fr.; UPW); Gummays (Arab.; Syria; HJP); Jiben Yadek (Gambia; UPW); Jummays (Arab.; Syria; HJP); Kankanga (Ivo.; UPW); Kilumpui (Ghana; UPW); Kobahi (Upper Volta; UPW); Madaka (Sen.; UPW); Mulberry Fig (Eng.; JLH; USN); Ndahi (Sierra Leone; UPW); Nouhe (Eng.; JLH); Pharaoh's Fig (Eng.; FAC); Shikmim (Heb.; ZOH); Shikmoth (Heb.; ZOH); Sicomoro (Sp.; USN); Sycamore Fig (Eng.; USN); Sycamore of Cyprus (Eng.; JLH); Sycomore (Eng.; USN); Sykomore (Ger.; USN); Tcheque (Guinea-Bissau; UPW); Tin el Jummays (Arab.; Syria; HJP); Nscn.

Activities (Sycamore Fig):

Antidote (f; HJP); Antiseptic (f1; HJP; X8170162); Cholagogue (f; UPW); Depurative (f; BIB; HJP); Lactagogue (f; BIB); Purgative (f; UPW).

Indications (Sycamore Fig):

Abrasion (f; BIB; HJP); Adenopathy (f; UPW); Bacteria (1; X8170162); Burn (f; JLH); Cancer (f; JLH); Cancer, fauces (f; JLH); Cancer, limbs (f; JLH); Cancer, spleen (f; JLH); Caries (f; UPW); Cerebrosis (f; UPW); Chest ache (f; UPW); Cirrhosis (f; BIB); Cough (f; BIB; UPW); Depression (f; BIB); Dermatosis (f1; X8170162); Diarrhea (f; BIB); Dropsy (f; UPW); Dysentery (f; UPW); Fever (f; BIB); Gastrosis (f; BIB); Hepatosis (f; UPW); Induration (f; JLH); Infection (f1; BIB; X8170162); Inflammation (f; BIB; JLH; UPW); Jaundice (f; UPW); Melancholy (f; BIB); Pertussis (f; UPW); Pulmonosis (f; BIB); Respirosis (f1; X8170162); Sarcoma (1; UPW); Scrofula (f; BIB); Snakebite (f; UPW); Sore (f; HJP); Sore Throat (f; BIB); Splenosis (f; JLH); Stomachache (f; UPW); Swelling (f; UPW); Tetanus (f; HJP); Tumor (f; JLH); Typhoid (f; BIB); Wart (f; JLH); Wound (f; BIB; HJP).

Dosages (Sycamore Fig):

FNFF = !!

Produced in several crops per year, the yellowish fruit smells like an ordinary fig but is inferior in taste and sugar content. In olden times, fruits were much consumed by the poor, raw or cooked, and even sold in the markets. Some Africans consume with millet or ferment a beverage. Leaves also eaten in soups or in peanut dishes. Latex serves as a vegetable rennet (BIB; FAC; ZOH).

- Egyptians apply the milk to burns, cancers, indurations, and warts (JLH).
- East Africans use the bark for sore throat, the Masai for diarrhea (BIB; UPW).
- Ethiopians use the root to prevent typhoid (BIB).
- Ghanans give bark decoction for cough and whooping cough (UPW).
- Ghanans, Senegalese, and Upper Voltans use for snakebite (UPW).
- Hausa Nigerians collect root sap in a cup to treat pediatric cough (UPW).
- Lebanese apply the latex to shallow abrasions and skin infections to ward off tetanus, using bark decoction for blood poisoning (HJP).
- Mali natives apply the latex to carious teeth (UPW).
- Senegalese use latex for dysentery, and the bark for chest ache, glandular inflammations, and stomach problems (UPW).
- Tenda women make leaf soup with millet to ensure adequate lactation (UPW).

Natural History (Sycamore Fig):

For complex reasons, the sycamore fig is completely dependent on man, who has saved it from extinction. Fertilization by wasps is necessary for ripening of the fruits, but no seeds are produced in the process, as the ovaries are turned into galls, which are inedible. Ancient Hebrews incised young fruits with a special knife, in a process called gashing (*balos*), mentioned in the RSV version of Amos 7:14, "I am a herdman, and a dresser of sycomore trees," but missed in the KJV. Cypriots and Egyptians use the same method. For some reason, the wasp-dependent variety in Israel was replaced by a parthenocarpic variety, which has no need for the wasp in ripening its seedless fruits. This species bears fruits several times a year in the Holy Land (ZOH). Zohary refutes some scholars' speculation that the sycamore fig was introduced from Africa, perhaps by Natufian Man circa 10,000 years ago. Zohary thinks it is more likely a tertiary relic of an earlier coastal tropical flora (with *Acacia albida, Ziziphus spina-christi*) (ZOH). Fruits and leaves are fed to cows to increase the flow of milk, especially in arid areas (BIB). Most animals and birds eat the fruits; cattle and sheep browse the leaves (UPW).

Extracts (Sycamore Fig):

Fruit extracts exhibited antitumor activity in potato disc bioassay, and had significant antibacterial activity but no antifungal activity (X8170162).

MANNA (FROM *FRAXINUS ORNUS* L.) ++ OLEACEAE

Notes (Manna):

> **Behold, we have sent you money to buy you burnt offerings, and sin offerings, and incense, and prepare ye manna.**
>
> **Baruch 1:10**

Manna, at least in this account of *Fraxinus*, refers to the exudate from the ash tree, not the ash tree itself, with many names and indications of its own. According to the Moldenkes, there are three distinct types of manna in the Bible, the more familiar first type secured by purchase and trade, consisting of the gummy exudates of *Fraxinus ornus, Alhagi maurorum*, or *Tamarix mannifera*. Danin (*Econ. Bot.* 26:373, 1972) adds to this list *Acacia raddiana, Anabasis setifera, Astragalus echinus, Capparis cartitaginea, Capparis spinosa, Gomphocarpus sinaicus, Hammada salicornica*, and *Pyrethrum santolinoides* as sources of manna (BIB). Some of these are treated elsewhere. Zohary explains that, etymologically, manna stems from *man* or *man ha* = "What is that?" And he too fails to answer the question authoritatively. He seems to favor the interpretation that manna was an exudate from scaly insects *Trabulina mannifera* or *Najacoccus serpentina* feeding on the tamarisk or, even more likely, the white hammada, *Hammada salicornia*, which is widespread in southern Sinai. They exude a sweet liquid that hardens and drops to the ground to be gathered by the Bedouins like honey or sugar. Yet another type grew up during the night when the ground was moist, but "withered away" and "stank" with the heat of the sun (Exodus 16). The Moldenkes suggest that this was *Nostoc*, a tiny blue-green algae that grows rapidly during the night. Soft and gelatinous, these algal growths "disappear as the sun evaporates the dew, only to reappear the next night if there is abundant dew." (Moldenke and Moldenke, 1952) A third type "fell from heaven" (Numbers 11). Botanists tend to suspect lichens of the genus *Lecanora*, which after periods of drought dry up, curl up, break loose from the ground, and are transported by the wind. Sheep relish these lichens and Bedouins make a bread therefrom. Circa 1889, a shower of such lichens fell into Iran during a great famine (BIB). Clearly, *Fraxinus ornus* does produce manna and there has been trade in that manna. Only *Fraxinus syriaca* is reported in the *Flora of Palestine* (FP3). So if the biblical manna was *Fraxinus*, it would have to have been from that Syrian species, or imported from outside. The *Fraxinus* manna can be secured either as flakes ("flake manna"), fragments ("common manna"), or a viscid mass ("fat manna"). A good ash tree can yield a pound or more per season. Annual production in Sicily, where manna was once produced commercially, was circa 750 tons. The first medicine mentioned in the Moldenke's book, manna is described as a gentle laxative, demulcent, and expec-

FIGURE 1.49 *Fraxinus ornus*. Source of manna.

torant. In Grieve's *A Modern Herbal,* we read that manna was chiefly used as a children's laxative or to disguise other medicines. In 1906, Dr. Steinberg is said to have recommended dulcinol, a mixture of manna and common salt as a sweetening agent in diabetes. Duke and Wain list the following as uses: aperient, debility, laxative, purgative, restorative, and tonic (DAW). The leaves of the manna ash contain, in addition to aesculetin, cichoriin, ornol, and sedoheptulose, two marginal antitumor compounds: ursolic acid and rutin. Aesuletin and aesculin are anti-inflammatory. According to Uphof, manna from *Fraxinus* contains glucose, levulose, manneotetrose, mannite, manninotriose, and resin. Ash was recommended by Lebanese for diarrhea and malaria and the bark flakes for fever. Algerians powdered the seeds in olive oil and honey for gonorrhea (BIB).

Common Names (Manna):

Manna (Eng.; CR2).

Activities (Manna):

Aperient (f; DAW); Astringent (f; MAD); Demulcent (f; BIB); Expectorant (f; BIB); Laxative (1; KOM; PH2); Purgative (f; MAD); Restorative (f; BIB); Tonic (f; DAW); Vermifuge (f; MAD).

Indications (Manna):

Constipation (f1; KOM; PH2); Debility (f; DAW); Diabetes (f; BIB); Hemorrhoid (1; KOM); Proctitis (1; KOM); Scrofula (f; MAD); Worm (f; MAD).

Dosages (Manna):

FNFF = !
10–50 g manna in milk (HHB); 20–30 g manna (adult) (KOM; PH2); 2–16 g manna (child) (KOM; PH2).

Downsides (Manna):

No health hazards or contraindications reported with proper administration of suggested therapeutic dosages (PH2). Can cause flatulence in sensitive patients (KOM). Contraindicated in cases of bowel obstruction (KOM). Do not take laxatives long term without consulting a physician (KOM).

Natural History (Manna):

Manna is extruded from scaly insects feeding on various trees. *Trabulina mannifera* or *Najacoccus serpentina* feed on the tamarisk. They exude a sweet liquid that hardens and drops to the ground and can be gathered as a sweet stuff. Named for manna, mannitol, a sugar, is found in leaves of *Fraxinus ornus* L., *F. angustifolia* Vahl., *Olea europaea* L., and *Phillyrea media* L., all members of the olive family. In the two ash species, mannitol content gradually increases in spring, peaking in summer, followed by a gradual decrease (260–720 μM/g ZMB). Rainfall seems negatively correlated with seasonal increase in mannitol content, reaching a maximum at the end of the dry season (X12197521).

Extracts (Ash):

As to the ash itself, Stefanova et al. found antidemic and antiinflammatory activity with bark extract injections, the inflammatory activity partially due to its coumarins (X7650947). Kostova reported hydroxycoumarins, secoiridoid glucosides, phenylethanoids, flavonoids, and significant antimicrobial, antioxidative, photodynamic damage prevention, wound healing, antiinflammatory, immunomodulatory, and antiviral activities support folkloric use of the bark (X11429238).

LEVANT COTTON (*GOSSYPIUM HERBACEUM* L.) + MALVACEAE

Synonyms:

Gossypium abyssinicum Watt.; *Gossypium africanum* Watt.; *Gossypium arboreum* var. *wrightianum* Tod.; *Gossypium cambayense* var. *wrightianum* Tod.; *Gossypium eglandulosum* Cav.; *Gossypium obtusifolium* Roxb. & G. Don; *Gossypium punctatum* var. *acerifolium* Tod.; *Gossypium wrightianum* Tod. *fide* HH2

Notes (Levant Cotton):

> **And when these days were expired, the king made a feast unto all the people that were present in Shushan the palace, both unto great and small, seven days, in the court of the garden of the king's palace; Where were white, green, and blue, hangings, fastened with cords of fine linen and purple to silver rings and pillars of marble: the beds were of gold and silver, upon a pavement of red, and blue, and white, and black, marble.**
>
> **Esther 1:5–6 (KJV)**

> **And when these days were completed, the king gave for all the people present in Susa the capital, both great and small, a banquet lasting for seven days, in the court of the garden of the king's palace. There were white <u>cotton</u> curtains and blue hangings caught up with cords of fine linen and purple to silver rings and marble pillars, and also couches of gold and silver on a mosaic pavement of porphyry, marble, mother-of-pearl and precious stones.**
>
> **Esther 1:5–6 (RSV)**

> **And when these days had come to the full, the king held a banquet for seven days for all the people that were found in Shu'shan the castle, both the great as well as the small, in the courtyard of the garden of the king's palace. There were linen, fine <u>cotton</u> and blue held fast in ropes of fine fabric and wool dyed reddish purple in silver rings and pillars of marble, couches of gold and silver upon a pavement of porphyry and marble and pearl and black marble.**
>
> **Esther 1:5–6 (NWT)**

Somehow the KJV left out the cotton and that is precisely why it was not covered in my first book of the Bible. Mentioned only once in the Holy Scriptures, cotton was used to wrap Egyptian mummies. Zohary notes that it was not grown early in the land of Israel, but was cultivated in the last centuries B.C., perhaps under the name *tzemer-gefen* (vine wool) because its leaves resembled the grape. Plants cultivated as an annual for the fiber among the seeds that furnish Asiatic or Levant cotton.

Common Names (Levant Cotton):

Species not necessarily distinguishable; activities and indications often lumped (see, e.g., JFM). CRC entries relate to *Gossypium hirsutum*. MAD entries relafe to *Gossypium herbaceum*. Many JLH entries assigned to *Gossypium* sp. Algodao (Ocn.; Por.; AH2; KAB); Algodoeiro (Mad.; JAD); Algodoeiro Asiático (Por.; USN); Algodon (Sp.; JLH); Algodoneiro (Brazil; Mad.; JLH); Algodonero Herbaceo (Sp.; EFS; USN); Ambara (Kan.; KAB); Anagnika (Sanskrit; EFS; KAB); Arabian Cotton (Eng.; USN); Aziatische Katoenplant (Dutch; HH2); Badara (Mal.; KAB); Badarika (Tel.; KAB); Balaccastilla (Pam.; KAB); Baumwollenstrauch (Ger.; EFS; HH2); Bombax (?; JLH); Bong Se (Annam; KAB); Bulac (Tag.; KAB); Bumbac (Rom.; KAB); Cadaba (Vis.; KAB); Common Cotton (Eng.; BUR); Cotone (It.; EFS); Cotone Asiatico (It.; HH2); Cotonnier (Fr.; EFS); Cotonnier d'Asie (Fr.; USN); Cotonnier de Malta (Fr.; KAB); Cotonnier Herbace (Fr.; NAD; USN); Cotonnier Serere (Fr.; Sen.; UPW); Cotton (Eng.; BUR); Edudi (Tel.; DEP); Fitan (Arab.; KAB); Hlopchatnik

FIGURE 1.50 Levant Cotton (*Gossypium herbaceum*).

(Rus.; KAB); Iladambarutti (Tam.; KAB); Indian Cotton (Eng.; KAB); Ingilma (Nig.; UPW); Kabsi (Uraon; KAB); Kadsom (Mun.; KAB); Kapas (Beng.; Bom.; Dec.; Guj.; Hindi; Malaya; EFS; KAB; NAD); Kapsini (Kon.; KAB); Karpas (Heb.; Sanskrit; EFS; ZOH); Karpasa (Ayu.; AH2); Karpasi (Sanskrit; DEP); Katoenstruik (Dutch; EFS; HH2); Korono-ni (Mali; UPW); Korpaso (Oriya; KAB); Koton Malti (Malta; KAB); Krabas (Cam.; KAB); Krautiger Baumwollstrauch (Ger.; USN); Ku Chung (China; EFS); Kupas (Hindi; DEP; KAB); Kurtam Ussul (Arab.; DEP); Kutn (Arab.; Syria; HJP); Kuttun (Arab.; EFS); Levant Cotton (Eng.; Scn.; AH2; CR2); Maltese Cotton (Eng.; USN); Mian Hua (Pin.; DAA); Mien Hua (China; EFS); Pambah (Iran; DEP; EFS); Pamuk (Tur.; EFS; HH2); Papas (India; EFS); Parti (Tulu; KAB); Paruthi (Tam.; NAD); Parutti (Tam.; DEP; NAD); Paththi (Tel.; NAD); Rimo (Sen.; UPW); Ru (Guj.; DEP); Rui (Hindi; Pun.; Urdu; DEP; KAB); Sea Island Cotton (Eng.; BUR); Short Staple American Cotton (Eng.; FAC); Short Staple Cotton (Eng.; USN); Syrian Cotton (Eng.; USN); Thao Mien (Ic.; KAB); Ts'ao Mien (China; EFS); Tsiahilika (Sakalaave; KAB); Tula (Beng.; DEP); Tzemer Gefen (Heb.; ZOH); Vaum (Sind,; DEP; KAB); Wa (Burma; DEP; KAB); Wah (Burma; DEP; KAB); Ya La Po Mien (China; HH2).

Activities (Levant Cotton):

Abortifacient (f1; BUR; CRC; FNF; PH2); Amebicide (1; X16076104); Antifeedant (f; X15074657); Antifertility (f1; HH2; JAC7:405); Antimutagenic (1; X3278214); Antioxidant (1; X15878283); Antiproliferant (1; X15878283); Antiseptic (f; BIB); Antispermatogenic (1; JAC7:405); Aphrodisiac (f; HJP; PH2); Apoptotic (1; X15949956; X15554914); Astringent (f; CRC); Calcineurin Inhibitor (1; X15621416); Contraceptive (f1; CRC; PH2); Cytostatic (1; PH2); Demulcent (f; NAD); Diuretic (f; CRC); Emmenagogue (f1; AHP; BUR; CRC; HJP; PH2); Emollient (f; BIB; CRC); Expectorant (f; NAD); Hemostat (f; CRC; MAD); Hypercholesterolemic (1; PH2); Hypertriglyceridemic (f; PH2); Hypocholesterolemic (f; PH2); Hypoglycemic (1; HH2); Lactagogue (f; CRC; HJP; KAB; NMH); Laxative (f; BIB); Nervine (f; BIB); Oxytocic (f; CRC; EFS; PH2); Protisticide (1; (X16076104); Uterotonic (f1; AHP; BUR); Vasoconstrictor (f; CRC); Vulnerary (f; BIB).

Indications (Levant Cotton):

Adenopathy (f; JLH); Ague (f; BUR); Ameba (1; (X16076104); Amenorrhea (f1; KAB; MAD; PH2); Anemia (f; MAD); Asthma (f; CRC; MAD); Bite (f; PH2); Bleeding (f; CRC; MAD; PH2); Bronchosis (f; CRC); Burn (f; JLH); Cancer (f1; CRC; FNF; X15878283); Cancer, abdomen (f1; JLH; X15812364); Cancer, breast (f1; JLH; X15812364); Cancer, colon (f1; JLH; X15812364); Cancer, larynx (1; X15812364); Cancer, lung (1; X15554914); Cancer, nose (f1; JLH; X15812364); Cancer, prostate (1; X15713891); Cancer, uterus (f1; JLH; X15812364); Carcinoma (1; X15570010); Cerebrosis (f; DAW); Childbirth (f; HHB; PH2); Climacteric (f; PH2); Cold (f; JFM); Constipation (f; PH2); Corn (f; JLH); Cough (f; MAD; PH2); Cramp (f; JFM); Cystosis (f; HH2; PH2); Debility (f; MAD); Dermatosis (f; KAB); Diarrhea (f; CRC; PH2); Dysentery (f; CRC; JFM; PH2); Dyslactea (f; JFM; PH2); Dysmenorrhea (f1; MAD; PH2); Dysuria (f; BUR; JFM); Earache (f; JFM); Enterosis (f; JFM); Epilepsy (f; PH2); Fever (f; BUR; HHB; JFM; MAD); Fibroid (1; CRC; JLH); Fibroma (f; JLH); Gonorrhea (f; HH2; PH2); Headache (f; CRC; PH2); Hemorrhoid (f; CRC; JFM); Herpes (f; KAB); High Blood Pressure (f; JFM); High Cholesterol (f; PH2); Hypochondria (f; CRC); Infertility (f; MAD); Inflammation (f; JFM; PH2); Laryngitis (f; JFM); Leprosy (f; BIB); Malaria (f1; BIB; KAB; X15978953); Menorrhagia (f; PH2); Metrorrhagia (f; PH2); Micromastia (f; BIB); Morning Sickness (f; MAD; PH2); Nausea (f; PH2); Neuralgia (f; DAW); Neuroblastoma (1; X15927359); Neurosis (f; PH2); Pain (f; JFM; PH2); Pharyngosis (f; BIB); Polyp (f; CRC; FNF; JLH); Psoriasis (1; X15878283); Pulmonosis (f; JFM; MAD); Rheumatism (f; JFM); Rhinosis (f; JLH); Scabies (f; BIB; KAB); Snakebite (f; PH2); Sore (f; BIB; HH2); Sting (f; NAD); Tumor (f1; CRC; FNF; HH2); Urethrosis (f; PH2); Uterosis (f; CRC; FNF); UTI (f; JFM); Virus (f; DEP); Wart (f; JLH).

Dosages (Levant Cotton):

FNFF = !
Seeds eaten in Egyptian cuisine; used for extraction of oil, used for cooking, and other purposes, roasted as coffee substitute. Oil a source of vitamin E, and used to substitute for sesame oil. Oil-cake used as fertilizer and fodder, and to make edible tempeh. Leaves eaten (BIB; FAC). 100 g root in 1 liter (l) water, reduced by boiling to 0.5 l; take 50 g liquid every half hour. Dangerous formula for abortion (CRC). 1–2 Tbsp fluid extract during childbirth (HHB). Three leaves in 1 liter water for hypertension (JFM). 6–8 g seed/150 g water or milk, 3 ×/day as lactagogue (f; JFM); 1.8–3.75 g root bark; 20–40 drops liquid extract (PH2); 2–4 ml liquid extract (PH2).

- Africans take the leaf decoction for dysentery and headache (BIB).
- Annamese use flowers for amenorrhea and dysmenorrhea, the seed oil for herpes, scabies, and wounds (KAB).
- Asian Indians use the roots for urinary disorders (BUR).
- Ayurvedics view flowers as antibilious, antihallucinogenic, lactogogue, refrigerant, and tonic, using leaves for anemia, oliguria, and otosis; and view seed as aphrodisiac and lactogogue. They use the plant for snakebite, scorpion stings, skin ailments, and uterine discharges (KAB).
- Bambas of northern Rhodesia apply the leaves to warts (JLH).
- Brazilians use the root bark for uterine fibromas (JLH).
- French Guianans use seeds in fumigations for indolent tumors (JLH).
- Lebanese use roots for malaria and Malta fever; impotent men use decoction cautiously; women use as emmenagogue and lactagogue (HJP).
- Lebanese consider cottonseed oil the more stimulating of the oils (HJP).
- Lebanese applied crushed seed (of *Gossypium barbadense*) to dermatosis such as poison ivy (HJP).
- Senegalese Soce give root macerate to newborn, sick, or rachitic children to make them grow big and strong (UPW).
- South American females take root decoction as a contraceptive, the seed as a lactagogue or breast enlarger (BIB).
- Unani regard the seeds as aphrodisiac, expectorant, and laxative, using them for orchitis; they poultice the flowers on burns, scabies, and scalds, and use them in syrup for hypochondria, and insanity; the leaves they take internally for dysentery, externally for gout; flowers are also used as analgesic for burning eyes and inflammation (KAB).

Downsides (Levant Cotton):

Class 2b; contraindicated in urogenital irritation or tendency to inflammation; may sterilize men (AHP). No health hazards or contraindications reported with proper administration of suggested therapeutic dosages (PH2). Drug toxic due to gossypol. Animals fed seed press cakes for 3 to 4 weeks showed enterosis, gastrosis, hematuria, jaundice, and nephritic and ophthalmic disorders (BUR; PH2).

Natural History (Levant Cotton):

Bezemer et al. (2004) found that in undamaged plants, terpenoid aldehydes were concentrated in the young immature main leaves. Aboveground feeding by *Spodopterta exigua* larvae on a mature leaf enhanced terpenoid (hemigossypolone, heliocides 1 and 4) concentrations in immature leaves but not in the damaged leaf. Root herbivory by wireworms (*Agriotes lineatus*) also resulted in increased terpenoid levels in foliage (X15074657).

Extracts (Levant Cotton):

The oilseed cake contains circa 4% glutamic acid. In its free state, L-glutamic acid is used to treat mental deficiencies in infants and adolescents. Gossypol shows antitumor activity in several NCI tumor systems. LD50 Gossypol = 20 mg/kg ipr mus (BIB). This species reportedly contains 3,500–18,100 ppm gossypol in the embryo, as in most cottons a mix of the (+) and (–) enantiomers. Some of this gets into cottonseed oils (X16076104). Talking about cotton in general, Stipanovic et al. (2005) state that in many ways the (+)-enantiomer is positive, from a health point of view and the (–)-enantiomer is negative from a food point of view, but positive for biological activities, including medicinal activities. For example, (–)-gossypol inhibits various cancer cells more effectively than the (+)-enantiomer. (–)-Gossypol is a more effective inhibitor of various enzymes than (+)-gossypol. (–)-Gossypol, but not (+)-gossypol, shows anti-HIV-1 activity in humans (9). (–)-Gossypol is effective against ameba. (–)-Gossypol, but not (+)-gossypol, has male antifertility activity and is more toxic to animals. Broiler chickens fed a diet containing 5% cottonseed [(+)- to (–)-gossypol ratio of 83:17] gained weight at the same rate as the 100% soybean control diet. Cumulative weight gains of the chickens decreased circa 126 g for each 100-mg increase in (–)-gossypol consumed, whereas the cumulative weight gains were not significantly altered with increased (+)-gossypol (X16076104).

GUNDELIA (*GUNDELIA TOURNEFORTII* L.) + ASTERACEAE

Notes (Tournefort's Gundelia):

O my God, make them like a wheel; as the stubble before the wind.

Psalms 83:13 (KJV)

In the RSV, wheel is replaced by whirling dust. In Isaiah 17:13, Zohary's and the RSV's whirling dust reads more like "a rolling thing before the whirlwind" in the KJV. All seem appropriate for *Gundelia*, known to travelers as the "steppe monster." The thistle-like heads, often joined into a group, may roll before the wind in the empty desert steppes, rather like a tumbleweed, scattering its edible seed. This is the only species in the genus *Gundelia*, largest of 30 Holy Land species called tumbleweed (ZOH).

Common Names (Gundelia):

Akov? (Heb..; ZOH); Akub (Arab.; FAC; HJP; ZOH); Akuvith (Heb.; ZOH); Cardi (?; FAC); Galgal (Heb.; ZOH); Gundelia (Eng.; HJP); Hakub (Arab.; FAC); K'aub (Arab.; HJP; ZOH); Kenger (?; FAC); Ku'ayb (Arab.; Syria; HJP); Steppe Monster (Eng.; ZOH); Tournefort's Gundelia (Eng.; ZOH); Tumbleweed (Eng.; ZOH); Nscn.

Activities (Gundelia):

Antiseptic (f; X11378279); Bactericide (f; X11378279); Emetic (f1; HJP; ZOH); MDR Inhibitor (f; X11378279); Vulnerary (f; HJP).

Indications (Gundelia):

Bacteria (f; X11378279); Infection (f; X11378279); Pseudomonas (f; X11378279); Snakebite (f; HJP); Sore (f; HJP); Wart (f; HJP).

Dosages (Gundelia):

FNFF = !

Middle Eastern peasants make a "delicious dish" from the young fleshy heads, rather like the related globe artichokes. The oil-rich ripe nuts are edible and tasty. Leaves, thick stems, undeveloped flower buds, and roots all eaten. Toasted seeds used as "kenger coffee." Sold in Lebanese and Syrian markets. Anatolians collect and dry the plants for winter fodder (FAC; ZOH).

- Lebanese suggest the latex for burning off warts, drying up sores, as an emetic, and snakebite cure (HJP).

Downsides (Gundelia):

Latex emetic.

Extracts (Gundelia):

Aburjai et al. (2001) screened methanolic extracts combined with seven different antibiotics to check the synergic activities against *Pseudomonas aeruginosa*, including a resistant strain. *Gundelia tournefortii* and *Lepidium sativum* inhibited the growth of the resistant strain. Chloramphenicol, gentamicin, and cephalosporin can be given with the plant material, to good advantage (X11378279).

ENGLISH IVY (*HEDERA HELIX* L.) + ARALIACEAE

Notes (English Ivy):

> **And in the day of the king's birth every month they were brought by bitter constraint to eat of the sacrifices; and when the fast of Bacchus was kept, the Jews were compelled to go in procession to Bacchus, carrying ivy.**
>
> **II Maccabees 6:7**

Zohary comments that ivy, now rare in upper Galilee and Samaria, may once have been more widely distributed. In earlier times, the leaves formed the poet's crown, as well as the wreath of Bacchus, to whom the plant was dedicated. Ivy was once bound around the brow to prevent intoxication. Hence, a garland of ivy was hung outside olden roadhouses to indicate that wine was sold therein. Greek priests presented a wreath of ivy to newly married persons, symbolizing fidelity. The leaf contains circa 10% saponin and has been used for washing wool. Leaves boiled with soda are said to be suitable for washing clothes. Young twigs are a source of yellow and brown dye. Hardwood can be used as a boxwood substitute in engraving. Extracts are found in French massage creams and soaps. Ivy leaves were once bruised, gently boiled in wine, and drunk to alleviate intoxication by wine. Flowers, decocted in wine, were used for dysentery. Plant said to have been used as an emetic and narcotic on at least three continents. Tender ivy twigs, boiled in butter, were a primitive approach to remove sunburn.

Common Names (English Ivy):

Afal (Ber.; BOU); Aise (Fr.; KAB); 'Amshak (Arab.; Syria; HJP); Arbambal (Hazara; KAB); Azemnoun (Ber.; BOU); Bach Euoc Ngo Cong (Ic.; KAB); Banda (Kum.; Barga; Arab.; BOU); Barren Black (Eng.; BUR); Barren Ivy (Eng.; KAB); Beglet el-berba (Arab.; BOU); Benewithtree (Eng.; KAB); Bentwood (Eng.; EFS); Bindwood (Eng.; KAB); Birdwood (Eng.; BUR); Black Ivy (Eng.; KAB); Bluszcz (Pol.; KAB); Borostyan (Hung.; KAB); Bourreau des Arbres (Fr.; KAB); Brumbrumdakari (Beas; KAB); Ch'ang Ch'ung T'eng (China; KAB); 'Cisus (Arab.; Syria; HJP); Common Ivy (Eng.; BOU); Creeping Ivy (Eng.; BUR); Dudela (Nepal; KAB); Duvar Sarmaşiğ (Tur.; EFS); Edera (It.; KAB); Eevy (Eng.; KAB); Efeu (Ger.; EFS); Eibhean (Ire.; KAB); Ellera (It.; EFS); English Ivy (Eng.; Scn.; AH2; BOU); Epheu (Ger.; EFS); Euné (Fr.; KAB); Eura (Cat.; KAB); Eurré (Fr.; KAB); Fai-Borostyan (Hung.; KAB); Ground Ivy (Eng.; KAB); Habl el-masakin (Arab.;

FIGURE 1.51 English Ivy (*Hedera helix*).

BOU; HJP); Halbambar (Jhelum; KAB); Hera (Por.; KAB); Hiedra Comun (Sp.; EFS); Hyven (Eng.; KAB); Iedere (Rom.; KAB); Immergroen (Ger.; KAB); Ivin (Eng.; KAB); Ivory (Eng.; KAB); Ivy (Eng.; CR2); Kadloli (Sutlej; KAB); Kaneri (Sutlej; KAB); Kaniuri (Sutlej; KAB); Karbaru (Sutlej; KAB); Karmora (Kas.; KAB); Karur (Ravi; KAB); Klimop (Dutch; EFS); Klyf (Dutch; KAB); Kossos (Ger.; KAB); Koubbar (Ber.; BOU); Kuri (Ravi; KAB); Kurol (Chenab; KAB); Lablab (Behar; KAB); Lablab Kibir (Arab.; Syria; HJP); Labiab Kebir (Arab.; BOU); Lablab (India; EFS); Leblab (Arab.; BOU); Leouno (Lan.; KAB); Liedna (Malta; KAB); Lierré (Fr.; BOU; KAB); Lierre Commun (Fr.; BOU; EFS); Lierre d'es Poëtes (Fr.; KAB); Lierre d'Europ (Fr.; KAB); Lierre Grimpant (Fr.; BOU; EFS); 'Maddada (Arab.; Syria; HJP); Mandia (Kas.; KAB); Maravala (Mal.; KAB); Maravalai (Mal.; KAB); Mithiari (Jaunsar; KAB); Murgroen (Swe.; KAB); Parwata (Pushtu; KAB); Pliusch (Rus.; KAB); Qessous (Arab.; BOU); Small Ivy (Eng.; BUR); Tassouflal (Ber.; BOU); Vedbende (Den.; KAB); Wintergroen (Ger.; KAB); Woodbind (Eng.; BUR); Yedra (Sp.; KAB); Yedra Comun (Sp.; EFS).

Activities (English Ivy):

Abortifacient (f; AAH); Allergenic (1; APA; CRC; PH2; X14513244); Amebicide (1; BGB; PNC); Anodyne (1; APA; BGB); Anthelmintic (f; BGB; PH2; PNC); Anticancer (1; BGB); Antiexudative (f; PH2); Antiinflammatory (f12; KOM; X12834000) Antileishmannic (1; BGB); Antimelanomic (1; BGB); Antioxidant (1; X15241892); Antiparasitic (1; BGB); Antitussive (f12; AAH; APA; X12006725); Antiseptic (f1; APA; BIB; CRC); Antispasmodic (f12; APA; KOM; PH2; PIP; PM10:213); Aphrodisiac (f; CRC; HJP); Astringent (f1; BUR; CRC); Bactericide (1; APA; PH2; PM10:213); Bronchodilator (1; PM10:213); Cathartic (f; CRC; BGB; PNC); Contraceptive (f; CRC); Cytotoxic (f; BGB; PH2; PNC); Diaphoretic (f; CRC); Emetic (f1; BOU; CRC); Diuretic (1; X12730261) Emmenagogue (f; BOU; CRC); Expectorant (f12; APA; KOM; PH2; PIP; X12006725); Febrifuge (f; BGB; DAA; PNC); Fungicide (1; APA; BGB; PH2; PNC); Hemolytic (1; CRC); Hemostat (f; AAH); Intoxicant (f; EFS); Irritant (2; KOM; PH2); Laxative (f; CRC); Litholytic (f; MAD); Molluscicide (1; BGB; PH2; PNC); Mucoirritant (2; KOM); Mucolytic (f; PM10:213); Narcotic (f; CRC); Natriuretic (1; X12730261); Pediculicide (f; CRC; KAB); Poison (f; BUR); Protisticide (1; APA); Purgative (f; CRC); Secretolytic (f; BIS; PM10:213); Sedative (1; APA; BGB); Stimulant (f; CRC); Sudorific (f; CRC; DAA); Trichomonicide (1; BGB); Vasoconstrictor (f; CRC); Vasodilator (f; CRC); Vermifuge (f; CRC).

Indications (English Ivy):

Abscess (f; KAB); Adenopathy (f; KAB); Amenorrhea (f; APA; BOU); Arthrosis (f; APA); Asthma (12; PM10:213; X12725580); Backache (f; AAH); Boil (f; BOU); Bronchosis (f12; APA; PHR; PIP; X12725580); Bunion (f; AAH); Burn (f; APA; MAD; PHR); Cacoethes (f; JLH); Callus (f; JLH; MAD; PHR; PH2); Cancer (f; BOU; CRC; JLH; KAB); Cancer, breast (f; JLH); Cancer, colon (1; X15740080; X15796588); Cancer, lymph (f; JLH); Cancer, nose (f; JLH); Cancer, uterus (f; JLH); Catarrh (2; KOM; MAD; PH2; PIP); Cellulitis (f; PHR; PH2); Chilblain (f; AAH); Cholecystosis (f; PH2); Conjunctivosis (f; AAH); Corn (f; JLH; MAD); Cough (f12; APA; CRC; PHR; X12006725); Cramp (f12; BOU; KOM; MAD); Dermatosis (f1; APA; BOU); Dysentery (f; CRC; GMH); Dysmenorrhea (f; APA; BOU); Dyspnea (12; X12006725); Eczema (f; AAH); Favus (1; MAD); Fever (f; BOU); Gout (f; MAD; PHR; PH2); Headache (f; JLH; MAD); Hemoptysis (f; CRC); Hepatosis (f; PHR; PH2; PNC); High Blood Pressure (1; CRC); Hydrocephaly (f; MAD); Impotence (f; HJP); Induration (f; JLH); Infection (1; APA); Inflammation (f12; BGB; KOM; MAD; PH2; X12834000); Intoxication (f; BGB; CRC; GMH); Jaundice (f; CRC; MAD); Leishmania (1; BGB; PR15:298); Leukorrhea (f; MAD); Lice (1; CRC); Liver Flukes (1; PNC); Malaria (f1; BIB; CRC; MAD); Measles (f; AAH); Menorrhagia (f; MAD); Mucososis (f; MAD); Mycosis (f; AAH); Nephrosis (f; AAH); Neuralgia (f; PHR; PH2); Pain (12; APA; X12006725); Parasites (1; APA; PHR; PH2); Parotitis (f; AAH); Pertussis (f1; AAH; APA; BIS); Phlebitis (f; PHR; PH2); Polyp (f; JLH; MAD); Rachitis (f; PH2); Respirosis (f12; APA; KOM; PH2; PIP; X12006725); Rheumatism (f; APA; BGB; PHR; PH2); Rhinosis (f; MAD); Ringworm (f; AAH); Scabies (1; APA; MAD); Sclerosis (f; CRC); Scrofula (f; BGB; CRC; PHR; PH2); Sore (f; BOU; PHR); Spasm (f; BOU); Splenosis (f; MAD; PHR; PH2); Sprain (f; AAH); Stomachache (f; MAD); Stone (f; MAD); Sunburn (f; CRC; GMH); Tetters (f; AAH); Toothache (f; APA; CRC; MAD); Tuberculosis (f; HJP; MAD); Wart (f; JLH; MAD); Wen (f; JLH); Worm (f; CRC); Wound (f; BOU; PHR).

Dosages (English Ivy):

FNFF = ?
According to Tanaka (TAN), "leaves are said to be added to beer to make it strong"; or bruised and boiled in wine to render it less intoxicating (GMH). 0.5 g/cup tea/1–3 ×/day (APA); 0.3 g/day leaf (KOM; PIP); 1/2 tsp (1 g) powdered leaf (MAD). 1 tsp herb/1/4 cup; steep 10 minutes 1–3 ×/day (PHR).

- British apply vinegar leaf macerate (or wear leaf in sock) for corns, bunions, and warts (AAH).
- British ingest berries for aches and pains, cold and cough (AAH).
- Devons take leaf/berry infusion for mumps (AAH).

- Lebanese (educated northerners) consider the plant an aphrodisiac (HJP).
- Lebanese believe crushed leaves and berries help tuberculosis (HJP).
- Lebanese suggest tendrils in yogurt as a skin lotion and bleach (HJP).
- Mediterraneans hint that ingesting 1 g powdered fruit induces sterility (BIB).
- Scots and Irish make a hat of ivy for children with eczema (AAH).
- South African whites apply the vinegar-steeped leaves to cancerous growths and corns (BIB).

Downsides (English Ivy):

No health hazards or contraindications reported with proper administration of suggested therapeutic dosages (PH2). None known or indicated (KOM; PIP). Can cause contact allergy (BIS). Palestinians regard the fruit as toxic to children (BIB).

Natural History (English Ivy):

Sheep and deer will eat the leaves in winter, although cows often will not (BIB).

Extracts (English Ivy):

Suleyman et al. (2003) demonstrated antiinflammatory effects of a crude saponin extract and saponin (IC77 = 100–200 mg/kg) (X12834000). Hofmann et al. (2003), in a review of randomized controlled trials, found that ivy drops were significantly superior to placebo in reducing airway resistance. Cough syrup and suppositories were modestly useful, and can improve respiratory functions of children with chronic bronchial asthma (X12725580). Hecker, Runkel, and Voelp (2002), studying a proprietary cough tablet (585–780 mg tablet corresponding to circa 100–130 mg of dried ivy leaf extract), reported relief in cough 92.2%; expectoration 94.2%; dyspnea 83.1%; and respiratory pain 86.9% (X12006725). Ridoux et al. (2001) reported antileishmanial activity of three ivy saponins (alpha-hederin, beta-hederin, and hederacolchiside A) in association with pentamidine and amphotericin B. The saponins exhibited a strong antiproliferative activity on all stages of development of the parasite but demonstrated a strong toxicity to human cells. Association of subtoxic concentrations of saponins with antileishmanial drugs such as pentamidine and amphotericin B demonstrated that saponins could enhance the efficiency of conventional drugs (X11406851). Polyacetylenes, especially falcarinol (also allergenic), may prevent colon cancer; and are also reportedly allergenic, analgesic, antileukemic, antimelanomic, antimycobacterial, antiseptic, antitubercular, antitumor, bactericide, chemopreventive, cytotoxic, and sedative (FNF; X15740080; X15796588). Gepdiremen et al. (2005) demonstrated antiinflammatory activity of four of the saponins: alpha-hederin, hederasaponin-C, hederacolchiside-E, and hederacolchiside-F in edema (0.02–20mg/kg). Indomethacin was the most potent drug in the first phase. For the second phase of acute inflammation, indomethacin and hederacolchiside-F were both very potent. Alpha-hederin was relatively ineffective. Hederasaponin-C, -E, and -F may block bradykinin or other inflammation mediators (X16008120).

BARLEY (*HORDEUM VULGARE* L.) ++ POACEAE

Synonyms:

Hordeum aegiceras Watt.; *Hordeum coeleste* Watt.; *Hordeum deficiens* Steud.; *Hordeum distichon* L.; *Hordeum gymnodistichum* Watt.; *Hordeum hexastichon* L.; *Hordeum intermedium* Kornicke; *Hordeum sativum* Jessen; *Hordeum zeocriton* L.

Notes (Barley):

Let thistles grow instead of wheat, and cockle instead of barley. The words of Job are ended.

Job 31:40 (KJV)

Let thorns grow instead of wheat, and foul weeds instead of barley. The words of Job are ended.

Job 31:40 (RSV)

FIGURE 1.52 Barley (*Hordeum vulgare*).

> **Instead of wheat, let the thorny weed go forth, and instead of barley stinking weeds. The words of Job have come to an end.**
>
> **Job 31:40 (RSV)**

Here among the weeds and thorns, we have two, barley and wheat, of the "seven species." Jewish folklore calls them the "seven species," the barley, wheat, vines, figs, pomegranate, oil (olive), and honey (date) of Deuteronomy. Mentioned more than 30 times in the Bible, at least 13 times with wheat, barley was considered inferior to wheat. Ripening a month or so earlier than wheat, it was taken for *omer* offerings at Passover, but the first grains of wheat were offered at the Feast of the Pentecost. Cultivation reportedly began circa 8000 B.C. in southwestern Asia (ZOH). Barley may be less important since the internal combustion engine replaced the horse, but stalks of the wild varieties still wave on the hills of Galilee and Judea, ancestors of the world's barley (BIB). Barley is the fourth most important cereal in the United States, but probably half of the production is used for livestock fodder, a quarter for the brewing industry (80% for beer, 14% distilled alcohol, and 6% malt syrup). Winter barley furnishes nutritive pasturage without seriously reducing yields (BIB; ZOH). Barley grain is demulcent and easily assimilable, and used in invalid and convalescent diets. Pearl barley is commonly used. Powdered parched grains used in the form of a gruel for painful and atonic dyspepsia. Barley water with honey is prescribed for bronchial coughs, and with gum arabic used for soothing irritations of the bladder and urinary passage. The seed meal is a folk remedy for cancer of the uterus, inflammatory tumors and gatherings, parotid gland tumors, and hard tumors. The seed flour is said to be a cure for anal condylomata, tumors behind the ears, scirrhus of the testicles, and spleen, whitlows, and tumors. Seed cataplasms are said to help breast cancers. Lebanese mix barley with olive oil for indolent ulcers (BIB).

Common Names (Barley):

Akshata (Sanskrit; KAB); Alcacer (Sp.; EFS); Arbus (Ashkobi; KAB); Arpa (Hun.; Tur.; DEP; KAB; EB49:406); Arpah (Tartary; KAB); Baarli (Kon.; NAD); Barley (Eng.; Scn.; AH2; CR2); Barliarisi (Tam.; WOI); Barlibiyam (Tel.; KAB; WOI); Barliyarisi (Tam.; KAB; WOI); Béchette (Fr.; KAB); Biugg (Swe., KAB); Brehnajau (Bagwana; KAB); Buza (Pun.; KAB); Byg (Den.; KAB); Cebada (Sp.; Peru; EGG; KAB); Cebada Común (Sp.; EFS); Cevad (Kon.; KAB); Cevada (Por.; KAB); Cevada das Cuatro Carreras (Por.; EFS); Cevada Sancta (Brazil; KAB); Cheno (Guj.; NAD); Da Mai (Pin.; DAA); Dhanya Bhedam (Tel.; DEP); Divya (Sanskrit; SKJ); Farro (It.; KAB); Garsten (Ger.; KAB); Gerst (Dutch; EFS; KAB); Gerste (Ger.; EFS); Gort (Dutch; EFS); Hadaka Mugi (Japan; TAN); Indarjawu (Nwp., KAB); Jab (Beng.; Iran; KAB; WOI); Jan (Hindi; NAD); Jao (Afg.; Iran; DEP; EFS; KAB); Jaon (Pun.; WOI); Jaoshirin (Afg.; DEP; KAB); Jaotursh (Afg.; KAB); Jau (Beng.; Guj.; Hindi; Nepal; Nwp.; Pun.; KAB; NPM; SKJ; WOI); Jav (Beng.; Bom.; Guj.; Hindi; Urdu; KAB; WOI); Java (Mar.; WOI); Jave (Hindi; NAD); Jave Godhi (Kan.; KAB; WOI); Jawa (Sin.; NAD); Jenczmien (Pol.; KAB); Jetschmen (Rus.; KAB); Jowakhar (Bihar; KAB; WOI); Kari (Arm.; KAB); Kordne (Lapland; KAB); Krithari (Greek; KAB); Kung Mai (China; EFS; KAB); Mai Ngai (Malaya; KAB); Mai Ya (Pin.; AH2); Mu Yao (Burma; DEP); Muyau (Burma; KAB); Nas (Bhutan; Tibet; DEP; TIB); Nigo omugi (Japan; TAN); No Mai (China; EFS); Ohva (Fin.; KAB); Omugi (Japan; TAN); Ordi (Cat.; KAB); Ordigal (Lan.; KAB); Orge (Fr.; EFS; TAN); Orge Anguleuse (Fr.; NAD); Orz (Rom.; KAB); Orzo di Spagna (It.; EFS); Orzo Nudo (It.; EFS); Pachcha Yava (Tel.; DEP); Pachcha Yavulu (Tel.; NAD); Paighambari (Tibet; DEP); Pearl Barley (Eng.; DEP; EFS); Perlgraupen (Ger.; EFS); Rasuli (Tibet; DEP); Rokuju omugi (Japan; TAN); Rollgerste (Ger.; EFS); Satu (Bom.; Dec.; Mar.; DEP; KAB); Sechszeilige Gerste (Ger.; NAD); Seorah (Heb.; ZOH); Seorim (Heb.; ZOH); Shaair (Arab.; DEP; KAB); Shijo Omugi (Japan; TAN); Situshaka (Sanskrit; WOI); Soah (Lassa; DEP; KAB); Suj (Hindi; KAB); Ta Mai (China; EFS; TAN); Tosa (Nepal; DEP; KAB); Urbusha (Harboi; Jhalawan; Sharig; KAB); Xghei (Malta; KAB); Yava (Ayu.; Sanskrit; EFS; KAB; NAD; WOI); Yavaka (Sanskrit; Tel.; DEP; WOI); Ymvah (Guj.; WOI); Ymwah (Guj.; DEP; KAB); Yurk (Nwp.; KAB).

Activities (Barley):

Abortifacient (f; BIB); Antiinflammatory (f; EGG); Antilactagogue (f; BIB; NMH); Antioxidant (1; X10888490); Aphrodisiac (f; EGG; KAB); Astringent (f; TIB); Demulcent (f; BIB; EFS); Digestive (f; BIB; TIB); Diuretic (f; BIB); Emollient (f; BIB; EFS); Expectorant (f; BIB); Febrifuge (f; BIB); Hypocholesterolemic (1; X12566485); Lactagogue (f; NMH); Refrigerant (f; TIB); Sedative (f; EGG); Stomachic (f; BIB); Tonic (f; SKJ; TIB).

Indications (Barley):

Abscess (f; EB51:195); Acrochordon (f; BIB); Adenopathy (f; BIB); Anemia (f; KAB); Anorexia (f; KAB); Asthma (f; KAB); Biliousness (f; KAB); Bladder (f; BIB); Bronchosis (f; BIB); Burn (f; BIB); Cancer (f; BIB); Cancer, bladder (f; BIB; JLH); Cancer, breast (f; BIB); Cancer, colon (f1; BIB; JLH; X834216); Cancer, feet (f; BIB; JLH); Cancer, parotid (f; BIB; JLH); Cancer, throat (f; BIB; JLH); Cancer, uterus (f; BIB; JLH); Carcinoma (f; BIB; JLH); Catarrh (f; BIB; DEP; EFS); Chest (f; BIB); Chilblain (f; BIB); Cholera (f; BIB); Cold (f; EB49:406); Colitis (f; PH2); Condyloma (f; BIB); Cough (f; BIB); Cystosis (f; BIB); Debility (f; BIB; DEP); Dermatosis (f; EB51:195); Diabetes (1; X8632219); Diarrhea (f; BIB); Dyspepsia (f; BIB; SKJ; WOI); Dysuria (f; KAB; EB51:195); Enterosis (f; JLH; PH2); Fever (f; BIB); Fig (f; BIB); Fistula (f; NAD); Gastrosis (f; PH2); Gingivosis (f; KAB); IBD (f; PH2); High Cholesterol (1; X12566485); Hyperlipidemia (1; X8632219); Impotence (f; EGG; KAB); Induration (f; BIB; JLH); Infection (f; BIB); Inflammation (f; BIB; EGG); Insomnia (f; EGG); Itch (f; EB51:195); Measles (f; BIB; EGG); Obesity (1; X8632219); Orchosis (f; BIB); Pain (f; KAB); Parotitis (f; BIB); Phthisis (f; BIB); Proctosis (f; BIB; NAD); Puerperium (f; BIB); Rheumatism (f; EB51:195); Scirrhus (f; BIB; JLH); Smallpox (f; EGG); Sore (f; BIB); Splenosis (f; JLH; BIB); Swelling (f; BIB); Syndrome X (1; X8632219); Thirst (f; KAB); Tuberculosis (f; BIB); Tumor (f; BIB); Urogenitosis (f; BIB); Uterosis (f; BIB); UTI (f; NAD); Wart (f; BIB); Whitlow (f; BIB).

Dosages (Barley):

FNFF = !!!

Cereal widely eaten, ground for breadstuffs, pearled for soups, brewed for beer; flaked for stout, fermented for Bavarian vinegars, or scorched for coffee; diastatic malt (malted barley flour) is added to feed the yeast in rising breads. Barley sprouts are often added to salads. Faddists use barley juice expressed from young seedlings as a health food (e.g., BarleyGreen or greenmagmax). It is essential in making miso. (Steamed barley is inoculated with koji fungus, added to steamed soybeans, and fermented into miso. Miso is supposed to be extra rich in estrogenic isoflavones.) In India, a cooling drink called *sattu* is made. Ashes of leaves used in Patna (India) to make sherbets (BIB; FAC; JAD; TAN; EB54:155). The biblical barley cake looks substantial, although not so good from the Atkins or Weston Price points of view: "Take thou unto thee wheat, and barley, and beans, and lentiles, and millet, and fitches, and put them in one vessel, and make thee bread thereof … And thou shalt eat it as barley cakes …" (Ezekiel 4:9 and 12). 450 mg (PH2).

- Algerians used barley after trepanning, and in poultices for fractures and swollen testicles and abscesses; and applied to the head for sunstroke (BIB).
- Asians use powdered seed for abdominal tumors (JLH).
- Ayurvedics use barley for anorexia, anemia, asthma, biliousness, bronchosis, burns, dysuria, impotence, and ulcers (KAB).
- Cereal boiled in yogurt is taken internally for rheumatism (EB51:195).
- Chinese use malt or germinated barley with the radicle attached as demulcent, expectorant, and stomachic (KAB).
- Iranians say, "What has disease to do with men who live upon barley-bread and buttermilk?" (BIB).
- Lebanese use cultivated barley for infections and tuberculosis (HJP).
- Peruvians apply barley flour to measles and smallpox (EGG).
- Peruvians suggest "mazamorra de cebada" for sexual debility (EGG).
- Peruvians view toasted barley decoction as antiinflammatory, depurative, and diuretic (EGG).
- Turks recommend the grain decoction for colds (EB49:406).
- Unani use barley for biliousness, chest pain, fever, headache, inflamed gums, and thirst (KAB).

Downsides (Barley):

Class 2b (AHP). No health hazards or side effects known with proper therapeutic dosages (PH2).

Natural History (Barley):

Many fungi attack barley and some cause serious damage in some areas. Agricultural agents should be consulted as to methods for control. Those reported on barley include the following species: *Alternaria tenuis, Ascochyta hordei, Aspergillus minutus, Botrytis cinerea, Calonectria graminicola, Camarosporium umbonatum, Candida variabilis, Cephalosporium curtipes, C. gramineum, Cephalothecium roseum, Cercosporella herpotrichoides, Cerebelia andropogonis, Cladosporium herbarum, Claviceps purpurea, Cochliobolus sativus, Corticium gramineum, C. solani, Cryptoascus graminis, Curvularia geniculata, Dendryphion laxum, Drechslera graminea, D. teres* (*Helminthosporium teres, Pyrenospora teres*), *Erysiphe graminis* f. sp. *hordei, Fusarium acuminatum, F. aquaeductum, F. avenaceum, F. concolor, F. culmorum, F. equiseti, F. graminearum, F. heterosporum, F. oxysporum, F poae, F. redolens, F. roseum* f. *cerealis, F. sambucinum* and var. *coeruleum, F. scirpi, F. solani, F. sporotrichioides, Gibberella saubinetti, G. zeae, Griphosphaeria nivalis, Helminthosporium oryzae, H. sativum, H. sorokinianum, H. teres, H. tetramera, H. zonatum, Heterosporium hordei, Lagena radicicola, Leptosphaeria herpotrichoides, Linocarpon cariceti, Macrophoma hennebergii, Marssonia graminicola* (*Rhynchosporium secalis*), *Monilia sitophila, Mucor* spp., *Mycosphaerella hordeicola, M. tassiana, M. tulasnei, Nigrospora sphaerica, Oidium monilioides, Olpidiaster radicis, Ophiobolus cariceti, O. graminis, O. herpotricus, Paecilomyces varioti, Papularia sphaerosperma, Penicillium* spp., *Phoma glomerata, Pleospora trichostoma, Puccinia coronata* and f. sp., *secalis, P. glumarum and* f. sp. *hordei, P. graminis* and several f. spp., *P. anomala, P. hirsutum, P. hordei, P. kapuscinski, P. purpurogenum, P. rubigovera (P. recondita), P. sanguineum, Pullularia pullulans, Pyrenophora grainea, P. japonica, P teres, Pythium aphanidermatum, P. arrhenomanes, P. debaryanum, P. iwayamai, P. volutum, Ramularia hordei, Rhizoctonia solani, Rhizophus arrizus, R. elegans, R. nigricans, Rhynchosporium graminicola, R. secalis* and f. sp. *hordei, Sclerophthora macrospora, Sclerotinia borealis, S. delphinii, S. sclerotium, S. rolfsii, Selenophoma donacis* var. *stomaticola, Selenophoma everhartii, Septoria avenae, S. hordei, S. nodorum, S. passerini, Sordaria finicola, Spongospora subterranea, Stemphyllium botryosum, Tilletia hordei, T. panicii, Torula antennata, T. graminicola, Trichoderma glaucum, T. kongingi, Ustilago avenae, U. hordei, U. nigra, U. segetum, U. tritici, U. zeae,* and *Wojnowicia graminis.* Virus diseases include the following: Barley stripe mosaic (False stripe), Oat pseudorosette, Rice streak, Rice black-streaked dwarf, Wheat green mosaic, Wheat rosette, Barley yellow dwarf, Barley yellow mosaic, Yellows, and False stripe. Bacterial diseases include those caused by the following species: *Bacillus hordei, Pseudomonas atrofaciens, P. hordei, P. striaefaciens* var. *japonica, P. translucens* and var. *undulosa,* and *Xanthomonas translucens* and f. sp. *hordei* and *hordei-avenae.* Plants may also be parasitized by *Cuscuta pentagona* and *Strigna lutea.* Nematodes isolated from barley include the following species: *Acrobeloides buetschlii, A. enoplus, Anguina tritici, Aphelenchoides parietinus, Aphelenchus avenae, Belonolaimus gracilis, Chiloplacus symmetricus, Criconemoides mutabile, Ditylenchus dipsaci, D. radicicola, Dorylaimus laetificans, D. nothus, D. obtusicaudatus, Eucephalobus striatus, Helicotylenchus dihystera, H. erythrinae, H. pseudorobustus, Heterodera avenae, H. hordecalis, H. latipons, H. zeae, Hoplolaimus galeatus, H. tylenchiformis, Meloidogyne artiellia, M. chitwoodi, M. incognita, M. incognita* var. *acrita, M. naasi, M. arenaria, Merlinius brevidens, Mesorhabditis monhystera, Mirolaimus mirus, Neocriconella mutabilis, Panagrolaimus rigidus, Pelodera lambdiensis, Plectus granulosus, Pratylenchus crenatus, P. neglectus, P. neocapitatus, P. penetrans, P. pinguicaudatus, P. pratensis, P. minyus, P. thornei, Punctodera punctata, Rhabditis gongyloides, Rotylenchtis erythrinae, Stibanguina radicicola, Trichodoras christiei, Tylenchus scandens, T. pratensis, T. spiralis, T. hordei, Tylenchorhynchus claytoni,* and *T. dubius* (HOE).

Extracts (Barley):

Behall et al. (2004) demonstrated that as with oats, increased soluble fiber via barley consumption in a healthy diet can reduce cardiovascular risk factors. Total cholesterol and LDL cholesterol after the high-soluble fiber diet were significantly lower (X14963054). Behall's research suggests that a biblical barley bread (beans, fitches, lentils, millet, and wheat) with exercise might be a nice herbal alternative to Zocor for high cholesterol. The following hierarchy of antioxidant activity was provided for 80% methanolic extracts originated from whole grain: buckwheat > barley > oat > wheat congruent with rye. (X10888490). Spent barley grain was very protective against colon cancer relative to wheat bran and commercial barley bran (X8234216).

HYACINTH (*HYACINTHUS ORIENTALIS NAME* L.) + LILIACEAE

Notes (Hyacinth):

I am the rose of Sharon, and the lily of the valleys.

Song of Solomon 2:1

Zohary does not index this species, so I may have to reconsider this as the biblical lily of the valley. It is a handsome ornamental, regarded as the biblical lily of the valley by some. It has long been a source of perfume, once extracted from the flowers in France and Holland. In classical mythology, Hyacinthus was a handsome young man, adored by Apollo as by Zephyrus. When the boy seemed to prefer Apollo, Zephyrus became jealous and deflected a quoit, which killed Hyacinthus. Apollo then transformed the body into the blue flowers whose beauty and fragrance were to be his eternal memorial. Grieving, Apollo sighed the universal eastern grieving wail "Ai, Ai," letters said to be inscribed to this day on the tepals. Because Ai was similar to the Greek word for eternal, the hyacinth is often carved into Greek tombs (BIB). Roots of the related *Hyacinthus nonscriptus* are used for leucorrhea; dried and powdered, they are used as a styptic (BIB; GMH).

Common Names (Hyacinth):

Bassal Sambal (Arab.; Syria; HJP); Eastern Hyacinth (Eng.; HJP); Khuzam (Arab.; Syria; HJP); Wild Hyacinth (Eng.; HJP); Nscn.

Indications (Hyacinth):

Dysuria (f; BIB; HHB); Jaundice (f; BIB; HHB); Venereal Disease (f;HJP).

Dosages (Hyacinth):

FNFF = X

- Lebanese treat venereal disease both topically and with dilute infusions (HJP).

Extracts (Hyacinth):

The essential oil contains the antitumor compound benzaldehyde.

FIGURE 1.53 Hyacinth (*Hyacinthus orientalis*).

GOLDEN HENBANE (*HYOSCYAMUS AUREUS* L.) X SOLANACEAE

NOTES (GOLDEN HENBANE):

> **And the border went out unto the side of Ekron northward: and the border was drawn to Shicron, and passed along to mount Baalah, and went out unto Jabneel; and the goings out of the border were at the sea.**
>
> **Joshua 15:11 (KJV)**

Zohary relates that Shicron can be identified as a place name in Judea, where this species of *Hysoscyamus*, the most common of five species in *Flora of Palestine*, is abundant, often growing in cracks in old ruins and walls, right there in Jerusalem. Zohary also mentions *H. muticus*, also a possibility, as the most poisonous species, widely used as a medicine and narcotic (ZOH). In a review of Israeli Solanaceae used medicinally, Dafni and Yaniv (1994) interviewed 106 local Israeli healers. Only four species (*Lycium europeaum, Solanum nigrum, Hyoscyamus aureus,* and *Hyoscyamus albus*) are still extensively used today; while the use of some species has been almost abandoned (*Datura* spp., *Mandragora autumnalis,* and *Withania somnifera*). Today, all the plants are applied externally, they are rarely used as narcotics, and most uses are variable and localized. Extensive distribution of safer modern anesthetics, narcotics, and sedatives reduced reliance on Solanaceae (X7990499). Apparently, all contain the dangerous but biologically active alkaloid, hyoscyamine (ZOH). If all contain hyoscyamine, and *Hyoscyamus muticus* is most poisonous, it may be best endowed with hyoscyamine, and probably exhibits many of the activities of hyoscyamine. Activities of hyoscyamine are listed below under "Extracts."

COMMON NAMES (GOLDEN HENBANE):

Banj (Arab.; Syria; HJP); Golden Flowered Henbane (Eng.; HJP); Golden Henbane (Eng.; JAD); Saccharus (Greek; ZOH); Sakaran es Shar (Arab.; Syria; HJP); Shicron (Heb.; KJV); Shikkeron (Heb.; RSV; ZOH); Shikrona (Heb.; ZOH); Nscn.

INDICATIONS (GOLDEN HENBANE):

Cancer, breast (f; HOC); Epilepsy (f; HJP).

DOSAGES (GOLDEN HENBANE):

FNFF = X

- Lebanese consider the Egyptian species is better for epilepsy (HJP).
- Lebanese midwives harvest and utilize this common wild weed (HJP).
- Lebanese smoke a pinch of leaf powder to relieve headache (HJP).

EXTRACTS (GOLDEN HENBANE):

Hyoscyamine: analgesic (WBB); anticholinergic 150–300 μg 4 ×/day/man (M29); antidote (anticholinesterase) (M29); antiemetic (JBH); antiherpetic (EMP5:193); antimeasles (EMP5:193); antimuscarinic (M29); antineuralgic; antiparkinsonian (JBH); antipolio (EMP5:193); antisialogogue (JBH); antispasmodic (JEB26:75); antiulcer (M29); antivertigo; antivinous; bronchodilator; bronchorelaxant; cardiotonic; CNS Depressant (KCH); CNS Stimulant (JBH); mydriatic (JBH); photophobigenic; psychoactive (WBB); sedative; toxic (PJB1(1):177); viricide (EMP5:193).

EGYPTIAN HENBANE (*HYOSCYAMUS MUTICUS* L.) X SOLANACEAE

Synonyms:

Hyoscyamus insanus fide EFS; *Hyoscyamus muticus* L. subsp. *falezlez* (Coss.) Maire *fide* USN; *Scopolia datora* Dunal; *Scopolia mutica* Dunal *fide* HHB

Notes (Egyptian Henbane):

> **And the border went out unto the side of Ekron northward: and the border was drawn to Shicron, and passed along to mount Baalah, and went out unto Jabneel; and the goings out of the border were at the sea.**
>
> **Joshua 15:11 (KJV)**

Zohary relates that Shicron can be identified as a place name in Judea, where this species of *Hyoscyamus*, the most poisonous and narcotic of five species in the *Flora of Palestine*, occurs. While favoring *H. aureus*, Zohary also mentions *H. muticus* as a possibility, as the most poisonous species, widely used as a medicine and narcotic (ZOH). In a review of Israeli Solanaceae used medicinally in Israel, Dafni and Yaniv (1994) interviewed 106 local Israeli healers in Israel. They suggest that today all such plants are applied externally, and rarely used as narcotics (X7990499). Apparently, all contain the dangerous but biologically active alkaloid, hyoscyamine (ZOH). If all contain hyoscyamine, and *Hyoscyamus muticus* is most poisonous (reportedly it contains the most alkaloids; WOI), it may be best endowed with hyoscyamine, and probably exhibits many of the activities of hyoscyamine as reported in my CRC *Handbook of Medicinal Herbs* (2nd edition) for *Hyoscyamus niger.* HHB suggests that its uses are similar to those of belladonna (HHB). EFS suggests that leaves of the toxic *H. muticus,* smoked by Faquirs, are used as *H. niger* (EFS). Sometimes used as a narcotic instead of opium (UPH). The plant is so poisonous as to have been used in Saharan warfare (BOU), and in cult rituals such as Datura (HJP).

Common Names (Egyptian Henbane):

Afalehlé (Ber.; BOU); Ägyptisches Bilsenkraut (Ger.; EFS); Banj (Arab.; Syria; HJP); Betina (Arab.; BOU); Egyptian Henbane (Eng.; JAD; WOI); Egyptisch Bilzenkruid (Dutch; EFS); Falezlez (Arab.; BOU); Folhas de Meimendro (Por.; HHB); Genegi (Ber.; BOU); Giusquiamo Egiziano (It.; HHB); Goungat (Arab.; BOU); Hojas de Beleño (Sp.; HHB); Indian Henbane (Eng.; KAB); Jusquiame d'Egypt (Fr.; EFS); Jusquiame du Désert (Fr.; BOU); Jusquiame Faleslez (Fr.; BOU); Koh Bana (Kharan; KAB); Koheebhang (Bal.; KAB); Kohibhang (Bal.; KAB); Kohibung (India; EFS); Saccharus (Greek; ZOH); Sakaran (Arab.; BOU; HJP); Sakaran Masrj (Arab.; Syria; HJP); Samm ul Far (Arab.; Syria; HJP); Shicron (Heb.; KJV); Shikkeron (Heb.; RSV; ZOH); Shikrona (Heb.; ZOH); Sikran (Arab.; Syria; HJP); Tataruh Sakaran (Arab.; Syria; HJP); Nscn.

Activities (Egyptian Henbane):

Analgesic (f; BOU); Anesthetic (f; BOU); Antidote (Lead) (f; BOU); Antispasmodic (f1; BOU; HJP); Aphrodisiac (f; BOU); Deliriant (1; KAB); Hallucinogenic (f; BOU); Hypnotic (1; EFS); Intoxicant (f; DAW); Mydriatic (1; EFS); Narcotic (1; EFS); Parasympathetic (f; HJP); Sedative (1; EFS); Toxic (1; EFS).

Indications (Egyptian Henbane):

Asthma (f; BOU; EB22:167); Bunion (f; HJP); Cancer (f; HJP); Cerebrosis (f; DAW); Colic (f; BOU); Corn (f; HJP); Cough (Lead) (f; BOU); Cramp (f; BOU); Cystosis (Lead) (f; BOU); Hemorrhoid (f; HJP); Hysteria (f; BOU); Insomnia (f; DAW); Nervousness (f; UPH); Neurosis (f; BOU); Pain (f; BOU); Spasm (f1; BOU; HJP); Spine (f; DAW); Toothache (f; DAW).

Dosages (Egyptian Henbane):

FNFF = X
200–400 mg leaf (HHB).

- Asians Indians use the plant as an intoxicant (KAB).
- North Africans poultice fresh leaves onto painful areas of the body (BOU).
- North Africans smoke the leaves for asthma (BOU).
- North African women use the seed to gain weight (BOU).

YELLOW FLAG (*IRIS PSEUDACORUS* L.) + IRIDACEAE

Synonyms:

Iris longifolia Lam. & DC; *Iris lutea* Lam.; *Iris palustris* Moench.; *Xiphium pseudoacorus* Schrank. *fide* HH3

Notes (Yellow Flag):

He shall grow as the lily, and cast forth his roots as Lebanon.

Hosea 14:5 (KJV)

One of the most handsome wild flowers of Europe, the yellow flag has been equated with the lily of Hosea by Moldenke and Moldenke, but not by Zohary. Who knows which scholars were correct? Remember, my reader, that when I drafted my first *Medicinal Plants of the Bible* (BIB; 1985), I had the Moldenke book but not the more recent Zohary book (ZOH). Ms. Duke's beautiful illustration of the yellow flag graced the cover of that book sold. I invested at least $4000 in that book, expecting 10, 12, or 15% of royalties after the first 1200 were published. However, the publisher never published that many, let alone sold that many. I did get ten copies of the book, nothing more. Now even I do not have a hard copy of this collector's item. But because this species was the cover picture on that collector's item, I retain the species in this new, faith-based revision (BIB; ZOH). The roots were once used like orris to scent linen closets. They are used also as a source of tannin and blue and black dyes. The flowers offer a yellow dye.

Common Names (Yellow Flag):

Acoro bastardo (Por.; EFS); Ácoro Falso (It.; Sp.; EFS); Akoron (?; JLH); Arab's Iris (Eng.; HJP); Burbit (Arab.; Syria; BOU; HJP); Butter and Eggs (Eng.; EFS); Daggers (Eng.; GMH); Dragon Flower (Eng.; GMH); Espadaña Fina (Sp.; EFS); Falscher Kalmus (Ger.; HH3); False Acorus (Eng.; JLH); Flaggon (Eng.; GMH); Flambe Bâtarde (Fr.; BOU); Flambe d'Eau (Fr.; BOU; EFS); Flambe des Marais (Fr.; BOU); Fleur-de-Lis (Fr.; USN); Fliggers (Eng.; GMH); Flower de Luce (Eng.; EFS); Gelbe Schwertlillie (Ger.; EFS; HH3); Gele Lis (Dutch; EFS); Giglio Gialla (It.; HH3); Giglio Gialla delle Paludi (It.; EFS); Gladon (JLH); Gladiolus (Eng.; JLH); Gladyne (Eng.; GMH); Glaieul des Marais (Fr.; BOU); Hashishet Yakub (Arab.; Syria; HJP); Iride Gialla (It.; EFS); Iris des Maurais (Fr.; EFS); Iris Faux-Acore (Fr.; BOU; EFS); Iris Jaune (Fr.; BOU; EFS; USN); Jacob's Sword (Eng.; BOU; GMH); Levers (Eng.; GMH); Lirio Espadañal (Sp.; EFS); Livers (Eng.; GMH); Meklin (Eng.; GMH); Myrtle Flower (Eng.; GMH); Pale-yellow Iris (Eng.; USN); Pakhana Bheda Lakri (Guj.; NAD); Sari Stisen (Tur.; EFS); Sawsan Asfar (Arab.; BOU); Segg (Eng.; GMH); Shalder (Eng.; GMH); Sheggs (Eng.; GMH); Siyaf (Arab.; Syria; BOU; HJP); Sowsan Asfar (Arab.; Syria; HJP); Spadella d'Acqua (It.; EFS); Sumpf Schwertlillie (Ger.; HH3); Susan (Arab.; Syria; HJP); Wasser Schwertlillie (Ger.; EFS; HH3); Water Flag (Eng.; EFS; USN); Yellow Flag (Eng.; EFS;

FIGURE 1.54 Yellow Flag (*Iris pseudacorus*).

GMH; HH3; TAN; USN); Yellow Flag Iris (Eng.; USN); Yellow Iris (Eng.; EFS; USN); Yellow Sedge (Eng.; EFS); Yellow Water Iris (Eng.; USN); Yreos (?; JLH); Nscn.

Activities (Yellow Flag):

Analgesic (f; GMH); Analeptic (f; BOU); Antidote (f; BOU); Astringent (f; EFS); Carminative (f; EFS); Cercaricidal (1; X15880993); Diuretic (f; BOU; EFS; NAD); Emetic (f; EFS; HJP); Febrifuge (f; EFS); Insecticide (1; X15880993); Larvicide (1; X15880993); Laxative (f; EFS); Miracide (1; X15880993); Mosquitocide (1; X15880993); Purgative (f; BOU; HJP); Schistosomicide (1; X15880993); Sternutatory (f; EFS); Stimulant (f; NAD); Stomachic (f; EFS); Tonic (f; EFS); Vermifuge (f; EFS); Vulnerary (f; BIB).

Indications (Yellow Flag):

Arthrosis (f; BOU); Bite (f; GMH); Bruise (f; BIB; GMH); Cacoethes (f; JLH); Cancer (f; JLH); Cancer, gum (f; JLH); Cancer, liver (f; JLH); Cancer, spleen (f; JLH); Childbirth (f; HJP); Cholera (f; BIB); Condyloma (f; JLH); Constipation (f; EFS); Convulsion (f; BIB); Cough (f; BIB); Cramp (f; BOU); Diarrhea (f; BIB; UPH); Dropsy (f; BOU); Dysmenorrhea (f; BIB; GMH); Epilepsy (f; HJP); Fever (f; EFS); Gas (f; EFS); Gingivosis (f; JLH); Halitosis (f; HJP); Headache (f; GMH); Hepatosis (f; BOU; JLH; NAD); Hypothermia (f; BOU); Infection (f; HJP); Jaundice (f; HJP); Leukorrhea (f; BIB; GMH; UPH); Myalgia (f; BIB); Ophthalmia (f; BIB); Pain (f; GMH); Polyp (f; JLH); Rabies (f; BOU); Rheumatism (f; BOU); Schirrus (f; JLH); Schistosoma (1; X15880993); Sciatica (f; BOU); Snakebite (f; BIB); Sore (f; HJP); Splenosis (f; JLH); Swelling (f; GMH); Toothache (f; AAH; BIB); Worm (f; EFS).

Dosages (Yellow Flag):

FNFF = ?
"Minced rhizomes mixed with couscous, a popular dish in North Africa, to serve as a condiment and to increase the girth of women" (BOU). Seeds have been used as a coffee substitute. Powdered root was once used as a snuff, and as a breath freshener (BIB; TAN).

- English use the root for toothache (AAH).
- Moroccans consider it a general antidote to poisons (BOU).
- Moroccans mix powdered rhizome with vinegar as diuretic and purgative (BOU).
- North Africans boil the fragmented rhizomes (one of Morocco's favorite alexiterics) for rheumatism and sciatica (BOU).

Natural History (Yellow Flag):

Flowers adapted to both honey bees and bumblebees, as well as long-tongued hover flies (*Rhingia rostrata*). Seeking honey, they push through the outer perianth segments and the style, the intermediate anther dusting pollen on the back (GMH).

ENGLISH WALNUT (*JUGLANS REGIA* L.) ++ JUGLANDACEAE

Synonyms:

Juglans duclouxiana Dode; *Juglans fallax* Dode; *Juglans kamaonica* (C. DC.) Dode; *Juglans orientis* Dode; *Juglans regia* subsp. *kamaonica* (C. DC.) Mansf.; *Juglans regia* var. *orientis* (Dode) Kitam. *Juglans regia* var. *sinensis* C. DC.; *Juglans sinensis* (C. DC.) Dode

Notes (English Walnut):

> **I went down into the garden of <u>nuts</u> to see the fruits of the valley, and to see whether the vine flourished, and the pomegranates budded.**
>
> **Song of Solomon 6:11 (KJV)**

FIGURE 1.55 English Walnut (*Juglans regia*).

I went down to the nut orchard, to look at the blossoms of the valley, to see whether the vines had budded, whether the pomegranates were in bloom.

Song of Solomon 6:11 (RSV)

To the garden of nut trees, I had gone down to see the buds in the torrent valley, to see whether the vine had sprouted, whether the pomegranate trees had blossomed.

Song of Solomon 6:11 (NWT)

Apparently this was *the* nut, a nut so common and important that it needed no denominator. One of King Solomon's most valuable fruit trees was the walnut, a handsome tree with smooth gray bark and fresh green leaves. Walnuts were widely cultivated in biblical times for the nuts and timber. Still, walnut is not mentioned by that name at all in the KJV, and the word *nut* only appears twice. In Jesus's time, walnut trees reportedly grew on the shores of the Sea of Galilee. There is a place called Valley of the Walnuts in eastern Jerusalem. Jesus's seamless coat was a rich brown, the dye allegedly of walnut leaves and nuts. The heavy green rind encasing the nut is steeped in boiling water to produce a brown dye. Walnuts are also used to tint gray hair black (BIB; ZOH). The Súbah of Kashmir use walnuts in divination; they throw a walnut into a pond; if it floats; it is a good omen; if it sinks, a bad omen (DEP).

Common Names (English Walnut):

Ak (Nwp.; KAB); Akhoda (Sanskrit; DEP); Akhor (Gurung; Hindi; KAB; NPM); Akhrot (Beng.; Guj.; Hindi; KAB; TAN); Akroda (Mar.; DEP; KAB); Akrodu (Kan.; DEP); Akrottu (Tam.; Tel.; DEP; KAB); Akschoda (Sanskrit; KAB); Akshota (Ayu.; AH2); Ankus (Arm.; KAB); Carpathian Walnut (Eng.; USN); Ceviz (Tur.; EB51:195; EB49:406); Ceviz Ağaci (Tur.; EFS); Chaharmaghz

(Iran; Pun.; KAB); Ch'iang T'ao (China; EFS); Circassian Walnut (Eng.; FAC); Dio (Hun.; KAB); Diólevél (Hun.; EFS); Djouz (Arab.; BOU); Dun (Kas.; KAB); Echter Walnußßaum (Ger.; USN); Egoz (Walnut; ZOH); English Walnut (Eng.; Scn.; AH2; CR2); Gemeiner Walnussbaum (Ger.; EFS); Gerga' (Arab.; BOU); Girdagan (Iran; DEP); Gognier (Fr.; KAB); Goz (Arab.; ZOH); Hardy Walnut (Eng.; FAC); Hup Thor (Malaya; KAB); Hu Tao (China; Pin.; AH2; TAN; USN); Hu Tao Ren (Pin.; AH2; DAA); Jaoz (Shingar; KAB); Jauz (Arab.; Quetta; KAB; ZOH); Jawz (Arab.; Syria; HJP); Jbuz (Arab.; KAB); Jouz (Arab.; EFS); Joz (Arab.; GHA); Kabsing (Assam; DEP; KAB); Kado (Tamang; NPM); Karaunt (Garhwal; KAB); Karydia (Greek; KAB); Kato (Tamang; NPM); Katu (Gurung; NPM); Khaisi (Rai; NPM); Kharat (Nwp.; DEP); Kharot (Kum.; DEP; KAB); Khayusing (Limbu; NPM); Khejik (Limbu; NPM); Khonsi (Newari; NPM); Knau (Lan.; KAB); Kolpot (Lepcha; NPM); Kotasi (Sherpa; NPM); Kowal (Lepcha; DEP; KAB); Krot (Kas.; DEP); Madeira Walnut (Eng.; USN); Naksh (Arab.; GHA); Noce (It.; KAB); Noce Comun (It.; EFS); Noeddetrae (Den.; KAB); Nogal (Sp.; KAB); Nogal Común (Sp.; USN); Nogal Europea (Sp.; USN); Nogal Inglés (Sp.; USN); Nogueira-comum (Por.; USN); Noguera (Cat.; Por.; KAB); Noix Commune (Fr.; EFS): Noyer (Fr.; BOU); Noyer Comun (Fr.; BOU; USN); Nuc (Rom.; KAB); Nussbaum (Ger.; KAB); Ockernootenboom (Dutch; KAB); Okhar (Bhojpuri; Chepang; Danuwar; Jaunsar; Majar; Nepal; Tharu; KAB; NPM; SUW); Orechovve Derevo (Rus.; KAB); Orzeszuma Wloska (Pol.; KAB); Persian Walnut (Eng.; Ocn.; AH2; USN); Phoro (Sunwar; NPM); Seiyo Gurumi (Japan; TAN); Siskhyasi (Burma; KAB); Sowak (Arab.; BOU); Starga (Lad.; MKK); Star-ga (Tibet; NPM); Tadjoudjte (Ber.; BOU); Tagashing (Bhutan; DEP; KAB); Than (Pangi; KAB); Thanka (Pun.; DEP); Thitkya (Burma; DEP); Tsouik (Ber.; BOU); Ughz (AFG); Vadnód (Den.; EFS); Valnoettraed (Swe.; KAB); Waghz (Afg.; DEP; KAB); Walnootboom (Dutch; EFS); Walnuss (Ger.; MAD); Yanggulk (Tur.; KAB).

Activities (English Walnut):

Alterative (f; PNC); Analgesic (f1; BIB; X14522443); Anthelmintic (f; DEP; NPM); Antiaflatoxin (1; X15053524); Antiaggregant (1; BIB; FNF); Anticancer (f1; BGB); Antidepressant (1; MAM); Antidote (f; BOU; DEP); Antifatigue (f1; MAM); Antiherpetic (1; BGB); Antiinflammatory (f1; APA; X14522443); Antioxidant (1: X10616967); Antiparasitic (f1; X15619562); Antiperspirant (1; APA; PH2); Antiradicular (1; X10616967); Antiseptic (f1; BIB; JAD; PNC); Antispasmodic (f; NAD); Antitumor (f1; APA; JAD); Antiulcer (f1; MAM); Antiviral (1; BGB; X14727919); Anxiolytic (1; MAM); Aphrodisiac (f; BOU; PH2); Astringent (f1; APA; BGB; HHB; NPM; PH2); Bactericide (1; BIB; X15068416; X14727919); Candidicide (1; BGB; BIB); Carcinogenic (1; BGB; PH2); Cardioprotective (f1; X15294495); Cardiotonic (f; KAB); Carminative (f; KAB); Cerebrotonic (f; EFS); Chemoprotective (1; X14580007); Cholagogue (f; NAD; NUT); COMT Inhibitor (1; MAM); Cytotoxic (1; X14727919); Dentifrice (f; BOU); Depurative (f; NUT; PHR; PH2); Digestive (f; NUT); Diuretic (f; NUT); Emollient (f; BGB); Euphoriant (1; MAM); Expectorant (f; KAB); Fungicide (1; APA; JAD; PH2); Hair Dye (1; PNC); Hemostat (f1; JAD; MAD; NUT); Herbicide (1; JAD); Hypocholesterolemic (1; X15294495); Immunostimulant (1; MAM); Insecticide (f; NUT); Lactifuge (f; NAD); Laxative (f; BIB; JAD; PNC); Leukoplakogenic (1; PH2); Lipolytic (1; X15294495); Litholytic (f; NUT); MAO (1; BIB; FNF); Molluscicide (1; WOI); Mutagenic (1; BGB; PH2); Myorelaxant (1; MAM); Myotonic (f; MAD); Neuroprotective (f1; MAM); Piscicide (1; WOI); Serotoninergic (1; MAM); Stimulant (f; NUT); Taenicide (f; NAD); Tonic (f; NPM; NUT); Vermifuge (f1; HHB; JAD; PH2); Vulnerary (f; JAD).

Indications (English Walnut):

Acne (f1; APA; X15974479); Adenopathy (f; MAD); Aegilops (f; JLH); Alopecia (f; BIB; BOU); Anemia (f; MAD); Anthrax (1; NUT; WOI); Anxiety (1; MAM); Aphtha (f; BIB); Arthrosis (f; MAD); Asthma (f; NUT; PH2); Atherosclerosis (1; BIB; FNF); Bacillus (1; WOI); Bacteria (1; BIB); Backache (f; NUT); Bacteria (1; X15068416); Beriberi (f; PH2); Bleeding (f; HJP); Bronchosis (f; KAB); Bruise (f; KAB); Burns (f; BGB; X15664457); Caligo (f; NAD; NUT); Cancer (f1; BIB; BGB; JLH);

Cancer, breast (f1; JLH); Cancer, colon (f1; JLH); Cancer, intestine (f1; JLH); Cancer, kidney (f1; JLH); Cancer, lip (f1; JLH); Cancer, liver (f1; JLH); Cancer, mouth (f1; JLH); Cancer, stomach (f1; JLH); Cancer, throat (f1; JLH); Cancer, uterus (f1; JLH); Candida (1; BGB); Carbuncle (f; BIB); Cardiopathy (f1; BIB; X15294495); Caries (f; BIB; MAD); Catarrh (f; MAD; PHR; PH2); Chancre (f; NUT); Cholera (1; WOI); Colic (f; NUT; PH2); Condyloma (f; JLH); Conjunctivosis (f; APA; BGB; NUT); Constipation (f; NAD; PH2); Corn (f; JLH); Coryza (f; BOU); Cough (f1; BIB; FNF; NUT); Dandruff (f; BGB; MAD); Depression (1; FNF); Dermatosis (f12; HJP; KOM; PHR; PH2; PNC); Diabetes (f; BOU; MAD); Diarrhea (f1; APA; HHB); Diptheria (1; WOI); Dogbite (f; BGB); Dysentery (f; NUT; PH2); Eczema (f1; APA; BGB; GHA; MAD; PNC); Enterosis (f; HHB; PHR; PH2); Epilepsy (f; PNC); Epithelioma (f; JLH); Escherichia (1; BIB; WOI); Exanthema (f; MAD); Favus (f; NUT); Fever (f; EB49:406); Fistula (f; MAD); Flu (f; BIB; BOU); Fungus (f; BIB; KAB); Gangrene (f; BIB; HJP); Gastrosis (f; HHB; MAD; PHR; PH2); Gingivosis (f1; APA; BIB; BOU); Gout (f; MAD); Gray Hair (1; PNC); Halitosis (f; BIB; BOU); Headache (f; BIB; NPM; PNC); Heartburn (f; KAB; NUT); Helicobacter (1; X15068416); Hemorrhoid (f; KAB); Herpes (f1; BGB; KAB; PNC); High Blood Pressure (1; BIB; FNF); High Cholesterol (1; X15294495); Hunger (f1; BIB); Hyperemesis (f; NAD); Impotence (f; MAD; NUT; PH2); Infection (f1; BGB; KAB); Infertility (f; BIB; EB51:195); Inflammation (f1; APA; BIB; PH2; X14522443); Insomnia (1; MAM); Laryngitis (1; BIB; FNF); Leprosy (f; KAB); Leukorrhea (f; NUT); Lupus (f; MAD); Lymphopathy (f; MAD); Malaria (f; EB49:406); Mercury Poisoning (f; MAD); Mycosis (f1; APA; KAB); Nephrosis (f; NUT); Nervousness (f; GHA; MAD); Neurosis (f1; MAD; MAM); Obesity (f1; BIB; FNF); Ophthalmia (f; HHB); Pain (f; NUT; EB49:406; X14522443); Parasite (f1; X15619562); Plaque (1; BIB); Pruritus (f; BGB); Pulmonosis (f1; BIB; FNF; MAD); Pyorrhea (f; BOU); Rabies (f; KAB); Rheumatism (f; APA; DEP; NUT; SUW); Rhinosis (1; BIB; FNF); Rickets (f; HHB; NAD); Ringworm (f1; APA; KAB); Salmonella (1; WOI); Scrofula (f; APA; BGB; BOU; HHB); Snakebite (f; BGB); Sore (f1; BGB; NUT; PNC; SUW); Sore Throat (f; KAB); Spasm (f; NAD); Spiderbite (f; BGB); Staphylococcus (1; BIB); Stomatosis (1; APA); Streptococcus (1; BIB); Sunburn (f; BGB); Sunstroke (f; EB49:406); Sweating (2; KOM; PHR); Swelling (f; BIB); Syndrome X (1; X15294495); Syphilis (f; BIB; MAD); Tapeworm (f1; BAD; WOI); Tuberculosis (f; MAD); Ulcer (f1; MAM; X15068416); Venereal Disease (f; BIB; MAD); Virus (1; X14727919); Wart (f; JLH); Whitlow (f; JLH); Worm (f1; APA; MAD); Wound (f1; APA); Yeast (1; BIB).

Dosages (English Walnut):

FNFF = !!!
Nuts consumed fresh, roasted, or salted; used in candies, pastries, and flavorings. Ground nut shells used as adulterant of spices. Fruits, when dry pressed, yield a valuable oil used in paints and soaps; when cold pressed, yield a light yellow edible oil, rich in omega-3s; used in foods as flavoring. Young fruits pickled. French make a liqueur from the green nut; green nuts also pickled; leaves rarely consumed as tea (FAC, TAN; EB54:155); 1/2 tsp chopped leaf/cup water (externally only; APA); 2–3 g leaf/100 ml water for compresses (KOM); 3–6 g (PHR); 4–8 ml liquid leaf extract (PNC).

- Algerians use leaf shampoo against alopecia (BOU), fresh leaves and bark as styptic (HJP), and burning sugar with the shells for headache (HJP).
- Arabians believed in the famed antidote of Mithridites: two walnuts, two figs, and twenty leaves of rice, rubbed together with a grain of salt (DEP).
- Arabians inhale fumes from burning nut shells for coryza and flu (BOU); or to repel insects (GHA).
- Asian Indians suggest spirits distilled from leaves or fruits as antispasmodic and in 1 to 2 drachms used to check morning sickness (NAD).
- Asian Indians suggest the edible cold pressed oil for dimness of vision. (NAD).
- Ayurvedics and/or Unani regard the therapeutic properties as analgesic, antibilious, aperitif, aphrodisiac, cardiotonic, carminative, constipating, expectorant used for blood disorders, bronchosis, bruises, cardiopathy, hemorrhoids, rabies, and ringworm (KAB).

- Chinese use leaves and powdered hull as astringent and depurative in syphilis (BIB).
- Greeks and Romans regarded walnuts as symbols of fecundity, and scattered walnuts about at weddings.
- Italians use the plant as an antiparasitic in cheeses (X15619562).
- Lebanese think the nut increases fertility, and use rancid old oil to treat old ulcers (HJP).
- Malayans say the kernels fatten the body and strengthen the muscles, suggesting them for colic, dysentery, and heartburn (KAB).
- Turks clothe nude patients in leaves for fever, rheumatic pain, and sunstroke (EB49:406).
- Orientals use the kernels for laryngeal and lung disorders, and mix them with almond and ginseng for chronic cough; they used the oil for skin ailments (BIB).
- Saudis used bark (bambar in Saudi Arabia, dandosa in Pakistan) soaked in water and rubbed on gums and teeth, both as cleanser and cosmetic, imparting reddish color to gums, considered attractive (GHA).
- Turks take the fruit decoction internally for malaria (EB49:406).
- Yemeni eat the nuts for nervous tension (GHA).

Downsides (English Walnut):

Class 2d. Juglone is mutagenic. External carcinogenic effects noted after chronic external use (AHP, 1997). None known for the leaf at proper dosage (KOM; PHR; PH2). Commission E reports fruit shell not permitted for therapeutic use; usefulness not documented adequately. Fresh shells contain the naphthoquinone constituent juglone, which is mutagenic and possibly carcinogenic. The juglone content of dried shells has not been studied adequately (AEH).

Natural History (English Walnut):

Pollination is often a problem, as English walnuts are monoecious, with separate staminate and pistillate flowers in different parts of the same tree. Staminate catkins are 10 to 15 cm long and produce 1 to 4 million pollen grains each. Sometimes, freshly picked catkins are put on paper in a room at 21°C and the shed pollen stored in a desiccator at 0°C. Then pollen is blown on trees by a fan mounted on a truck. Helicopters are sometimes used to blow pollen over an orchard. Seedlings are very susceptible to mushroom root rot, and Walnut girdle disease ("Blackline") is thought to occur when certain horticulture varieties of *Juglans regia* are grafted on rootstocks of *Juglans hindsii* and its hybrids, associated with graft incompatibility. Fungi known to attack Persian walnuts include *Alternaria nucis, Armillaria mellea, Ascochyta juglandis, Aspergillus flavus, Auricularia auricula-judae, A. mesenterica, Cerrena unicolor, Cladosporium herbarum, Coniophora cerebella, Coprinus micaceus, Coriolus tephroleucus, Cribaria violaceae, Cryptovalsa extorris, Cylindrosporium juglandis, C. juglandis, C. uljanishchevii, Cytospora juglandina, Cytosporina juglandina, C. juglandicola, Diplodia juglandis, Dothiorella gregaria, Erysiphe polygoni, Eutypa ludibunda, Exosporina fawcetti, Fomes fomentarius, F. igniarius, F. ulmarius, Fusarium avenaceum, F. lateritium, Ganoderma applanatum, Glomerella cingulata, Gnomonia ceratostyla, G. juglandis, G. leptostyla., Hemitricia leiotyichia, Hypoxylon mediterraneum, Inonotus hispidus, Laetiporus sulphureus, Lentinus cyathiformis, Licea tenera, Marsonia juglandis, Melanconis carthusiana, M. juglandis, Melanconium juglandis, M. oblongum, Melanopus squamosus, Microsphaera alni, M. juglandis, Microstroma juglandis, Mycosphaerella saccardoana, M. woronowi, Nectria applanata, N. cinnabarina, N. ditissima, Oxyporus populinus, Phellinus cryptarum, Phleospora multimaculans, Phoma juglandis, Phomopsis juglandis, Phyllactinia guttata, Phyllosticta juglandina, P. juglandis, Phymatotrichum omnivorum, Phytophthora cactorum, P. cinnamomi, P. citrophthora, Pleospora vulgaris, Pleurotus ostreatus, Polyporus hispidus, P. picipes, P. squamosus, Polystictus versicolor, Rhizopus nigricans, Stereum hirsutum, Trametes suaveolens, Tubercularia juglandis, T. vulgaris, Verticillium albo-atrum.* Bacteria attacking Persian walnut include *Agrobacterium tumefaciens, Bacillus mesentericus, Bacterium juglandis, Pseudomonas juglandis, Xanthomonas juglandis, Cuscuta pentagona,* also parasitized the tree. The following nematodes have been isolated from Persian walnut: *Cacopaurus pestis, Diplogaster striatus, Diplogaster coronata, Ditylenchus intermedius, Meloidogyne arenaria, M. javanica, M.* sp., *Pratylenchus coffeae, P. pratensis, P. vulnus, Rhabditis debilicauda, R. spiculigera,* and *Tylolaimophorus rotundicauda.* Among the insect pests of this walnut are the following: Walnut Blister mite (*Eriophytes tristriatus*), Walnut aphid (*Chromaphis juglandicola*), Italian pear scale (*Diaspis iricola*), Calico scale (*Eulecanium cerasorum*), Frosted scale (*Parthenolecanium Pruinosum*), Walnut scale (*Quadraspidiotus juglansregiae*), Codling moth (*Cydia pomonella*), Fruit tree leaf-roller (*Archips argyrospila*), Indian meal moth (*Plodia interpunctella*), Walnut caterpillar (*Datana integerrima*), Red-humped caterpillar (*Schizura concinna*), Walnut span worm (*Phigalia plumigeraria*), and Walnut husk fly (*Rhagolestis completa*).

Extracts (English Walnut):

Guarrera et al. (2005) mention walnut's antiparasitic use for cheeses (X15619562). Regular nut consumption can result in a 10% reduction in LDL cholesterol in a few weeks. Nuts often rich in arginine, vitamin E, folate, fiber, potassium, magnesium, tannins, and polyphenols. Although nuts contain approximately 80% fat, the nut feeding trials have not shown any associated weight gain in those ingesting nuts, suggesting that the addition of nuts in the diet may have a satiating effect. Daily ingestion of a small quantity is an acceptable lifestyle intervention to prevent coronary heart disease (X15294495). Gallic acid in the pellicle shows potential for inhibiting of aflatoxigenesis (X15053524). Colaric et al. (2005) list chlorogenic, caffeic, *p*-coumaric, ferulic, sinapic, ellagic, and syringic acid as well as syringaldehyde and juglone from ripe fruits of ten walnut cultivars. Not only in the kernel, but also in the pellicle, did syringic acid, juglone, and ellagic acid predominate (average values of 33.83, 11.75, and 5.90 mg/100 g of kernel; and 1003.24, 317.90, and 128.98 mg/100 g of pellicle, respectively), and the contents of ferulic and sinapic acid (average values of 0.06 and 0.05 mg/100 g of kernel and 2.93 and 2.17 mg/100 g of pellicle, respectively) were the lowest in all cultivars. It was found that the walnut pellicle is the most important source of walnut phenolics. The ratio between the contents in pellicle and kernel varied by at least 14.8-fold for caffeic acid (cv. Adams) and by up to 752.0-fold for *p*-coumaric acid (cv. Elit) (X16076123). One study indicates an absence of tocotrienol, while another found it consistently but at levels below 2 ppm (X15941326; X15969535). Good source of dietary serotonin, quickly broken down in the gut (where there are serotonin receptors). Serotonin is reportedly analgesic, antiaggregant, anticholinesterase, anticonvulsant, antiendotoxic, antigastrisecretogogic, antireserpinic, bronchoconstrictor, cardiovascular, coagulant, euphoriant, myorelaxant, myostimulant, neurotransmitter, oxytocic, teratogenic, ulcerogenic, and vasoactive [LD50 = 117 ipr rat; LD50 = 160 ivn mus; LD50 = 868 ipr mus]. Juglone reportedly is allelochemic, allergenic, anticariogenic, antidermatophytic, antiEBV, antifeedant, antiherpetic, antiparasitic, antiseptic, antitumor, antitumor-promoter, antiviral, bactericide, chemopreventive, dermatitigenic, fungicide, keratolytic, molluscicide, sedative, sternutatory, and viricide [LD50 = 2.5 ppm (orl mus)]. The combination of tannin with all its pesticidal activities and juglone may be pretty potent. Aqueous extracts of fresh walnut leaves, free of juglone, possess strong bactericidal activity against *Bacillus anthracis* and *Corynebacterium diptheriae,* weaker activity against *Bacillus subtilis, Escherichia coli, Micrococcus pyogens, Pneumococci, Proteus, Salmonella typhosa, S. typhimurium, S. dysenteriae, Streptcocci,* and *Vibrio* (WOI). Phillips et al. (2005) quantified the phytosterols in walnut: delta-5-avenasterol 73 ppm; delta-7-avenasterol; campestanol 24; campesterol 49; phytosterols 1060–1200; poriferasta-7,25-dienol 54; poriferasta-7,22,25-dienol; sitostanol <17; and beta-sitosterol 889 ppm (X16302759).

SOFT RUSH (*JUNCUS EFFUSUS* L.) ++ CYPERACEAE

Synonyms:

Juncus bogotensis Humb.; *Juncus communis* Mey.; *Juncus laevis* Wallr. *fide* HHB

Notes (Soft Rush):

> **And they shall turn the rivers far away; and the brooks of defence shall be emptied and dried up; the reeds and the flags shall wither.**
>
> **Isaiah 19:6 (KJV)**

> **And the parched ground shall become a pool, and the thirsty land springs of water: in the habitation of dragons, where each lay, shall be grass with reeds and rushes.**
>
> **Isaiah 35:7 (KJV)**

Moldenke and Moldenke list this first among five candidates for the flags and rushes of Isaiah and Job, doubting that any of them are the "nutritious river grass of Egypt" (BIB). Zohary does not even list this species, but rather the prickly sea rush (*Juncus acutus*) as a prominent member of aquatic vegetation along the edges of water bodies, with the papyrus (*Cyperus papyrus*), giant reed (*Phragmnites australis*), bramble bush (*Rubus sanguineus*), clammy inula (*Inula viscosa*), and even the oleander (*Nerium oleander*), but no mention of *Juncus effusus* (ZOH). In China, the pith is used for lamp wicks and mat making. Indians in California use the rush for domestic utensils and fodder.

Common names (soft rush):

Binse (Ger.; HHB); Bog Rush (Eng.; BUR); Bulrush (Eng.; Ocn.; AH2); Common Rush (Eng.; BUR); Deng Xin Cao (Pin.; AH2); Flatter Binse (Ger.; EFS); Flattersinse (Ger.; HHB); Giunco (It.; EFS); Hu Hsu Ts'ao (China; EFS); Jonc à Lier (Fr.; USN); Jonc à Mèche (Fr.; HHB); Jonc Commun (Fr.; HHB); Jonc Épars (Fr.; USN) Jonc Étenndu (Fr.; EFS); Junco de Esteiras (Sp.; EFS); Junquera (Sp.; EFS); Lamp Rush (Eng.; Ocn.; AH2); Pit Rush (Dutch; EFS); Rush (Eng.; EFS); Saz (Tur.; EFS); Soft Rush (Eng.; Scn.; AH2; BUR); Steinbinse (Ger.; HHB); Teng Hsin Ts'ao (China; EFS).

Activities (Soft Rush):

Antiinflammatory (f; DAW); Antiseptic (1; X12126307); Bactericide (1; X12126307); Candidicide (1; X12126307); Cathartic (f; BUR); Depurative (f; WOI); Discutient (f; DAW); Diuretic (f; BUR; DAW; HHB); Fungicide (1; X12126307); Lenitive (f; DAW); Litholytic (f; EFS); Nervine (f; DAW); Pectoral (f; DAW); Sedative (f; DAW).

Indications (Soft Rush):

Anuria (f; BIB); Candida (1; X12126307); Convulsion (f; DAW); Dropsy (f; BIB); Dysuria (f; DAW); Fear (f; DAW); Fistula (f; DAW); Fungus (1; X12126307); Hepatosis (f; DAW); Infection (1; X12126307); Insomnia (f; DAW); Jaundice (f; DAW); MDR (1; X12126307); Mycosis (1; X12126307); Sore (f; DAW); Sore Throat (f; BIB); Staphylococcus (1; X12126307); Stone (f; EFS; WOI); Strangury (f; DAW); Yeast (1; X12126307).

Dosages (Soft Rush):

FNFF = ?
Not indexed by FAC or TAN, but not listed as poisonous either.

- Chinese use the pith, depurative and diuretic, to keep fistulous sores open. The pith decoction is considered antilithic, discutient, and pectoral, and is prescribed for anuria, cough, dropsy, insomnia, micturition, and sore throat (BIB).

FIGURE 1.56 Soft Rush (*Juncus effusus*).

Downsides (Soft Rush):

In South Africa, the herb is suspected of causing "vlei poisoning" (WBB).

Extracts (Soft Rush):

Hanawa et al. (2002) found a phenanthrene (dehydroeffusol) and a dihydrophenanthrene (juncusol), both of which display enhanced antimicrobial activities in light. The antimicrobial activities against methicillin-resistant and -sensitive *Staphylococcus aureus* and *Candida albicans* were increased 16- and 2-fold, respectively, by irradiation with ultraviolet-A irradiation (X12126307).

GRECIAN JUNIPER (*JUNIPERUS EXCELSA* M. BIEB) + CUPRESSACEAE

Synonyms:

Juniperus excelsa Wall.; *Juniperus excelsa* var. *depressa* O. Schwarz; *Juniperus excelsa* var. *farreana* P.N. Mehra; *Juniperus excelsa* var. *polycarpos* (K. Koch) Silba; *Juniperus excelsa* subsp. *polycarpos* var. *pendula* (Mulk.) Imkhanitskaya; *Juniperus excelsa* subsp. *seravschanica* (Komarov) Imkhanitskaya; *Juniperus excelsa* subsp. *turcomanica* (B.A. Fedtsch.) Imkhanitskaya; *Juniperus foetida* var. *excelsa* (M.-Bieb.) Spach.; *Juniperus isophyllos* K. Koch; *Juniperus gossainthaneana* Loddig *fide* DEP; *Juniperus macropoda* Boiss.; *Juniperus olivieri* Carr.; *Juniperus polycarpos* K. Koch; *Juniperus polycarpos* var. *pendula* Mulk.; *Juniperus polycarpos* var. *seravschanica* (Komarov) Kitamura; *Juniperus recurva* Buch.-Ham; *Juniperus sabina* var. *excelsa* (M.-Bieb.) Georgi; *Juniperus sabina* var. *taurica* Pall.; *Juniperus seravschanica* Komarov; *Juniperus taurica* (Pall.) Lipsky; *Juniperus turcomanica* B.A. Fedtsch.; *Sabina excelsa* (M.-Bieb) Ant.; *Sabina isophyllos* (K. Koch) Ant.; *Sabina olivieri* Ant.; *Sabina polycarpos* Ant.; *Sabina religiosa* Ant.; *Sabina seravschanica* (Komarov) Nevski *fide* CJE.

Notes (Grecian Juniper):

> **They have made all thy ship boards of <u>fir</u> trees of Senir: they have taken cedars from Lebanon to make masts for thee.**
>
> **Ezekiel 27:5 (KJV)**

Zohary suggests that *Juniperus excelsa* (or *Juniperus foetidissima*), a stately Lebanese conifer, should fall under the collective term of conifers, *berosh*. Lebanese call it *brotha*, a "name surely identical with the *berothim* of the Song of Solomon" (ZOH). The Accadian word *burasu* could also refer to this species, as it was reportedly imported from the Zaigros mountains near Eilam where it still occurs. It also grows on Mt. Senir, mentioned in the quote above. Will we ever know what ancient non-botanical historians meant when they chose their words in recording and translating?

Common Names (Grecian Juniper):

Al'Allan (Arab.; Oman; GHA); Apurs (Bal.; DEP); Berosh (Heb.; ZOH); Berothim (Leb.; ZOH); Broth (Leb.; ZOH); Burasu (Accadian; ZOH); Cedrelate (?; JLH); Cedrus (?; JLH); Chandan (Nepal; WOI); Charai (Hazara; DEP); Chher Syukpa (Tibet; NPM); Chundun (Kum.; DEP); Dhang Ling (Tamang; NPM); Dhup (India; CJE); Dhupi (Nepal; DEP); Eastern Savin (Eng.; ZOH); Grecian Juniper (Eng.; HJP; USN); Greek Juniper (Eng.; CJE); Himalayan Pencil Cedar (Eng.; DEP); Indian Juniper (Eng.; WOI); Juniper (Eng.; JLH); Kūtrān (Arab.; Yemen; GHA); Lewar (Pun.; WOI); Lizzab (Arab.; Syria; HJP); Luir (Pun.; DEP); Padam (India; WOI); Paddam (Nwp.; DEP); Shukpa (Tibet; DEP); Syukpa (Sherpa; NPM); Weeping Blue Juniper (Eng.; NPM); Nscn.

Activities (Grecian Juniper):

Analgesic (f; GHA); Antiseptic (f1; HJP; X10234860); Bactericide (1; X10234860); Carminative (f; HJP); Diuretic (f; HJP; UPH); Emmenagogue (f; HJP); Hemostat (f; HJP); Hydragogue (f; HJP); Stimulant (f; HJP); Stomachic (f; HJP); Vermifuge (f; HJP); Vulnerary (f; HJP).

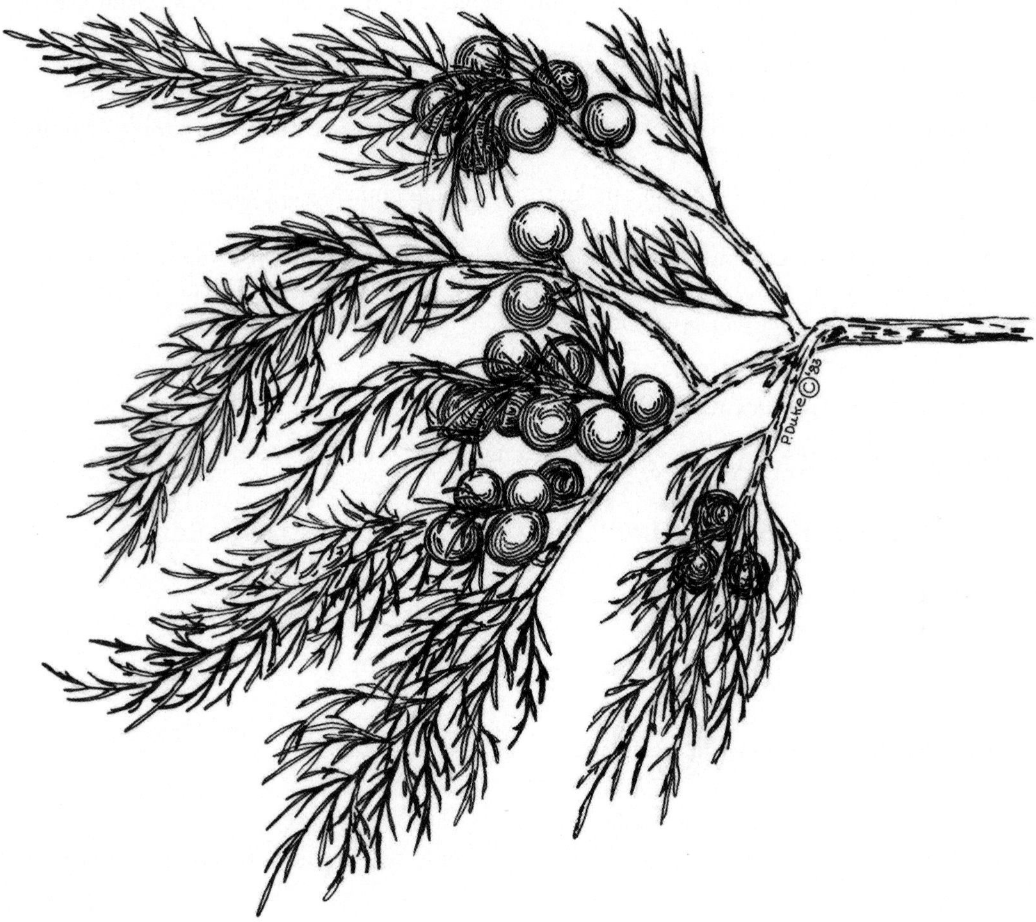

FIGURE 1.57 Grecian Juniper (*Juniperus excelsa*).

Indications (Grecian Juniper):

Amenorrhea (f; HJP); Arthrosis (f; HJP); Bacteria (1; X10234860); Cancer (f; JLH); Condyloma (f; JLH); Cough (f; HJP); Cystosis (f; HJP); Delirium (f; DEP; NAD; WOI); Dropsy (f; HJP); Dysmenorrhea (f; UPH); Dyspepsia (f; HJP; UPH); Enterosis (f; HJP); Epistaxis (f; HJP); Excrescence (f; JLH); Hepatosis (f; HJP); Infection (f1; HJP; X10234860); Jaundice (f; HJP); Myalgia (f; GHA); Mycobacterium (1; X10234860); Nephrosis (f; HJP); Pain (f; GHA); Paralysis (f; GHA); Polyp (f; JLH); Proctosis (f; JLH); Rheumatism (f; HJP); Rhinosis (f; JLH); Sore (f; HJP); Tuberculosis (1; X10234860); Venereal Disease (f; HJP); Wart (f; JLH); Worm (f; HJP); Wound (f; HJP).

Dosages (Grecian Juniper):

FNFF = ?

Although I find nothing on this as food in FAC or TAN, I suspect it parallels the edibility of other junipers: some relatively toxic, some relatively benign. WOI comments that the fruits and essential oil are very similar to the gin juniper, *J. communis*, but for gin flavoring, the pinene, smelling of turpentine, must be removed (WOI).

- Algerians snuff the powdered leaves for nosebleed (HJP), dressing wounds with leaves, olive oil, and saliva (HJP).
- Algerians sprinkle powdered leaves on circumcisions (HJP).
- Algerians steep boiled leaves and take with 1/2 cup Turkish coffee for cough (HJP).
- Arabians rub oil-soaked leaves on muscular pain and massage them onto paralyzed limbs or face (GHA).
- Asian Indians suggest inhaling smoke from burning branches in delirium of fever (DEP; NAD).
- Asian Indians suggest its uses are similar to the common juniper, *J. communis* (DEP).
- Lebanese think this is the best source of huile de cade, used in topical liniments and salves, and internally in cough and liver medicines (HJP).

Natural History (Grecian Juniper):

Wood rotting fungus (*Fomes juniperus*) may be a nusiance (WOI).

CADE JUNIPER *(JUNIPERUS OXYCEDRUS* L.) + CUPRESSACEAE

Synonyms:

J. macrocarpa fide CJE.

Notes (Cade Juniper):

For he shall be like the <u>heath</u> in the desert.

Jeremiah 17:6 (KJV)

In the *Flora of Palestine*, Zohary and associates list only two species of *Juniperus* native to the Holy Land:

- Plant monoecious shrub or small tree; mature cones with six to eight scales; mature leaves ares cale-like — *Juniperus phoenicia*
- Plant dioecious tree; mature cones with three to six scales; mature leaves are acicular, prickly — *Juniperus oxycedrus* (FP1)

Not even indexing *Juniperus excelsa*, which some books have identified as the biblical cedar, nor the common juniper, *Juniperus communis*, on which medicinal literature focuses, Zohary seems to have selected *J. phoenicia* as the most likely equivalent of the Aroer of the Arnon, but confesses that "[t]he identification of the biblical *arar* with *Juniperus* is based solely on the Arabic name given to this and other species of *Juniperus* in several Arabic-speaking countries" (ZOH). Other scholars have dwelt on *J. oxycedrus,* which led me to do likewise in my second biblical book (BI2) and in this newer compilation.

Common Names (Cade Juniper):

Algum (Bib.; Eng.; BI2); Appegggi (It.; EFS); 'Ar'ar (Arab.; Syria; HJP); Ardic Giligilisi (Tur.; EB49:406); Ardic Katrani (Tur.; EB49:406); 'Ar'or' (Heb.; BI2); Branket Enebaer (Den.; EFS); Cada (Sp.; VAD); Cade (Dutch; EFS); Cade Juniper (Eng.; USN); Ceder Wacholder (Ger.; CJE); Cedre Piquant (Fr.; CJE); Cedro de Espanha (Por.; EFS); Cirti (Tur.; EB51:195); Crvena Kleka (Serbia; CJE); Dikenli Ardic Evi (Tur.; EB49:406); Enebro de Bayas Rojas (Sp.; USN); Enebro de la Miera (Sp.; EFS); Genévrier Cade (Fr.; EFS; USN); Genévrier Epineux (Fr.; USN); Ginepro Pungente (It.; CJE); Kade (Ger.; USN); Kade (Ger.; EFS); Karaakatran (Tur.; EB49:406); Kedros

(Greek; JLH); Kyklan (Arab.; Syria; HJP); Oxicedro (Por.; EFS); Oxycèdre (Fr.; EFS); Prickly Cedar (Eng.; HOC); Prickly Juniper (Eng.; CJE; USN); Red-berry Juniper (Eng.; USN); Red Juniper (Eng.; EFS); Rotbeeriger Wacholder (Ger.; USN); Roter Wacholder Sariardic (Tur.; EB49:406); Sariardicevi (Tur.; EB49:406); Tikenardici (Tur.; EB49:406).

Activities (Cade Juniper):

Abortifacient (f1; VAD); Analgesic (f1; EFS; X9498241); Antiedemic (1; X9498241); Antiinflammatory (1; VAD; X9498241); Antipruritic (f; AYL; BIB); Antiseptic (f1; BRU; FNF; HOC); Bactericide (1; X12639746); Candidicide (1; X12639746); Carminative (f; BIB); Contraceptive (f; BIB); Diuretic (f; BIB; EFS); Emmenagogue (f1; VAD); Empyreumatic (f; HOC); Expectorant (f; VAD); Hypoglycemic (f; VAD); Insectiphile (1; X15279265); Keratolytic (f1; AYL; BIB; VAD); Parasiticide (1; BRU; FNF; HOC); Stimulant (f; BIB; EFS); Stomachic (f; BIB); Vermifuge (f; BIB; EFS); Vulnerary (f; BIB).

Indications (Cade Juniper):

Abscess (f; EB49:406); Alopecia (f; BIB); Arthrosis (f; VAD); Asthma (f; VAD); Bacillus (1; X12639746); Bacteria (1; X12639746); Bronchosis (f; VAD); Brucella (1; X12639746); Candida (1; X12639746); Cancer (f; BIB; JLH); Catarrh (f; VAD); Cold (f; EB49:406); Cough (f; EB49:406); Cystosis (f; BIB; VAD); Dermatosis (f1; BRU; VAD); Dislocation (f; EB51:195); Dysuria (f; EB49:406); Eczema (f; BRU; HOC); Edema (1; X9498241); Enteromonas (1; X12639746); Escherichia (1; X12639746); Favus (f; BIB); Fistula (f; EB49:406); Fracture (f; EB51:195); Fungus (1; X12639746); Gout (f; HOC; VAD); Hemorrhoid (f; HJP; EB49:406); Hepatosis (f; AYL; BIB); High Blood Pressure (f; VAD); Infection (1; BRU; FNF; X12639746); Inflammation (1; X9498241); Itch (f1; VAD); Jaundice (f; BIB); Keratosis (f; BRU; VAD); Leprosy (f; BIB); Mycosis (1; X12639746); Nephrosis (f; BIB); Neuralgia (f; VAD); Neurodermatosis (f; BRU; VAD); Ophthalmia (f; BIB); Pain (f1; X9498241); Parasite (f1; BRU; FNF; EB51:195); Pediculosis (f; BIB); Pharyngosis (f; VAD); Proctosis (f; EB49:406); Pseudomonas (1; X12639746); Psoriasis (f; BIB; HOC); Rheumatism (f; HOC; VAD; EB51:195); Rhinosis (f; VAD); Scabies (f1; BIB; VAD); Seborrhea (f; BIB); Sinusosis (f; VAD); Snakebite (f; BIB); Sore (f; BIB); Staphylococcus (1; X12639746); Stone (f; VAD); Strangury (f; VAD); Toothache (f1; BIB; EFS; VAD); Tumor (f; BIB); Urethrosis (f; VAD); Wound (f; HJP; EB49:406); Xanthomonas (1; X12639746); Yeast (1; X12639746).

Dosages (Cade Juniper):

FNFF = ?

3–5 drops cade oil in a little water, followed by a weak purgative, for worms (BIB). 3 cups decoction/day (20 g/l; boiled 3 minutes). 10–20 drops fluid extract 3 ×/day (VAD); 35–50 drops 1:10 tincture, 1–3 ×/day (VAD); 1–2 drops essential oil in oil or alcohol carrier or on a sugar cube (VAD).

- Algerians apply powdered rotten bark to facial cuts, leaf ashes in oil or water to piles (HJP).
- Lebanese crush and boil the berries, then soaking in oil for use in bladder and kidney ailments, or in alcohol as carminative and stomachic (HJP).
- Lebanese suggest berries crushed in sheep fat as an antiseptic vulnerary salve for wounds (HJP).
- Lebanese suggest the tea for jaundice and liver problems (HJP).
- Lebanese suggest crushed berries in tar or turpentine for sores and venereal disease (HJP).
- Palestinians use the dark brown tar (cade oil) for healing skin ailments (FP1).
- Turkish take fruits like a pill for bronchosis and colds (EB49:406).
- Turkish mix resin or tar mixed with honey for cough (EB49:406).
- Washingtonians apply the oil to external cancers (JLH).

Downsides (Cade Juniper):

Contraindicated in pregnancy (may be oxytocic), lactation, infancy, or neurologic patients or nephritis. Essential oil can be allergenic or irritant. Long-term use may lead to albuminuria and hematuria (VAD). The principal component of *Juniperus oxycedrus* tar is cadinene, a sesquiterpene, but cresol and guaiacol are also found. The oils derived from *Juniperus oxycedrus* tar were not skin irritants in animals. *Juniperus oxycedrus* tar was genotoxic in several assays. Clinical tests showed no evidence of irritation or sensitization with any of the tested oils, but some evidence of sensitization to the tar. A 2-year dermal carcinogenicity assay performed using National Toxicology Program (NTP) methods is needed. It was concluded that the available data are insufficient to support the safety of these ingredients in cosmetic formulations (X11558640).

Natural History (Cade Juniper):

Arceuthobium oxycedri (Loranthaceae) is a (semiparasitic) mistletoe found only on this juniper species.

PHOENICIAN JUNIPER (*JUNIPERUS PHOENICIA* L.) + CUPRESSACEAE

Synonyms:

Juniperus bacciformis Carr.; *Juniperus lycia* L.; *Juniperus oophora* Kunze; subsp. *eumediterreanea* P. Lebreton et S. Thivend; *Juniperus phoenicia* L. var. *lobelii* Guss.; *Juniperus phoenicia* L. var. *malacocarpa* Endl; *Juniperus phoenicia* L. var. *sclerocarpa* Endl.; *J. phoenicia* L. subsp. *turbinata* (Guss.) Nyman 1881; *Juniperus terminalis* Salisb. *Juniperus tetragona* Moench; *Juniperus turbinata* Guss.; *Sabina bacciformis* (Carr.) Antoine; *Sabina lycia* (L.) Antoine; *Sabina phoenicia* (L.) Antoine; *Sabina turbinata* (Guss.) Antoine; *Sabinella phoenicia* (L.) Nakai. *fide* GJE.

Notes (Phoenician Juniper):

> **From Aroer, which is by the brink of the river of Arnon.**
>
> **Deutoronomy 2:36 (KJV)**

In the *Flora of Palestine*, Zohary and associates list only two species of *Juniperus* native to the Holy Land:

- Plant monoecious shrub or small tree; mature cones with six to eight scales; mature leaves scale-like — *Juniperus phoenicia*
- Plant dioecious tree; mature cones with three to six scales; mature leaves acicular, prickly — *Juniperus oxycedrus* (FP1)

Not even indexing *Juniperus excelsa*, which some books have identified as the biblical cedar, nor the common juniper, *Juniperus communis*, on which medicinal literature focuses, Zohary seems to have selected *J. phoenicia* as the most likely equivalent of the Aroer of the Arnon, but confesses that "[t]he identification of the biblical *arar* with *Juniperus* is based solely on the Arabic name given to this and other species of *Juniperus* in several Arabic-speaking countries" (ZOH). Other scholars have dwelt on *J. oxycedrus.*

Common names (Phoenician Juniper):

Aifz (Ber.; BOU); Arar (Arab.; ZOH); 'Ar'ar (Arab.; Syria; BOU; HJP); Aroer (Heb.; ZOH); Dafran al 'Arr'arr (Arab.; Syria; HJP); 'Djineda (Arab.; BOU); Fausse Rouge (Fr.; BOU); French Cade

Tree (Eng.; HOC); Genévrier Phénicie (Fr.; CJE); Genévrier Rouge (Fr.; BOU); Phoenician Juniper (Eng.; ZOH); Shurbin (Arab.; Syria; HJP); Zimbreiro (Mad.; Por.; PST); Zimbro (Mad.; Por.; PST); Zimeba (Ber.; BOU); Nscn.

Activities (Phoenician Juniper):

Abortifacient (f; HJP); Analgesic (f; BOU); Anticancer (f1; JLH; X6253598); Antiinflammatory (f; BOU); Antiseptic (1; X12720394); Astringent (f; BOU); Bactericide (1; X12720394); Candidicide (1; X12720394); Diuretic (f; DAW); Emmenagogue (f; BOU); Fumigant (f; DAW); Fungicide (1; X12720394); Hemostat (f; HJP); Stomachic (f; DAW); Tonic (f; DAW).

Indications (Phoenician Juniper):

Bacteria (1; X12720394); Bleeding (f; HJP); Burn (f; JLH); Cancer (f1; JLH; X6253598); Candida (1; X12720394); Childbirth (f; BOU); Constipation (f; BOU); Cough (f; HJP); Dermatosis (f; BOU; HJP); Diarrhea (f; BOU); Dropsy (f; HJP); Dysuria (f; BOU); Enterosis (f; BOU); Epistaxis (f; HJP); Fungus (1; X12720394); Infection (1; X12720394); Inflammation (f; BOU); Mycosis (1; X12720394); Nephrosis (f; HJP); Pain (f; BOU); Polyp (f; JLH); Rhinosis (f; JLH); Staphylococcus (1; X12720394); Strangury (f; DAW); Wart (f; JLH).

Dosages (Phoenician Juniper):

FNFF = ?.

- Algerians boil berries and steep overnight, taking with a half cup of Turkish coffee, morning and evening, for cough (HJP).
- Algerians use powdered dry leaves, presumably in decoction or tea, to dilate the urinary tract to disinfect the intestines, and to treat mild dermal infections (BOU).
- Algerians snuff powdered leaves for nosebleed (HJP).
- Algerians sprinkle powdered leaves on circumcisions (HJP).
- Egyptians apply the berries to burns and cancers (JLH).
- Lebanese use boiled leaves as diuretic (HJP).
- Lebanese boil crushed berries and steep in oil as soothing vulnerary for skin conditions (HJP).
- North Africans consider the leaves emmenagogue, increasing uterine contractions during parturition.
- North Africans use hot leaf tea for children's diarrhea (BOU).
- Southern Europeans use the plant for nasal polyps and warts (JLH).

Downsides (Phoenician Juniper):

The International Journal of Toxicology (2001) published on the safety of *Juniperus phoenicea* extract, one of several juniper extracts used as biological additives in cosmetics. No genotoxicity data were available for extracts. Available data are insufficient to support the safety in cosmetic formulations (X11558640).

LETTUCE (*LACTUCA SATIVA* L.) ++ ASTERACEAE

Synonyms:

Lactuca sativa var. *angustana* Lam.; *Lactuca sativa* var. *capitata* L.; *Lactuca sativa* var. *crispa* L.; *Lactuca* L.H. *Bailey sativa* var. *longifolia*

Notes (Lettuce):

> **The fourteenth day of the second month at even they shall keep it, and eat it with unleavened bread and bitter herbs.**
>
> **Numbers 9:11 (KJV)**

> **In the second month on the fourteenth day in the evening they shall keep it; they shall eat it with unleavened bread and bitter herbs.**
>
> **Numbers 9:11 (RSV)**

> **In the second month on the fourteenth day between the two evenings, they should prepare it. Together with unfermented cakes and bitter greens they should eat it.**
>
> **Numbers 9:11 (NWT)**

Lettuce is the most popular of the salad vegetables, yet one of the bitter herbs of the Bible, at least according to some American writers, including myself. For example, Moldenke and Moldenke believed that *Cichorium endivia, Cichorium intybus, Lactuca sativa, Nasturtium officinale, Rumex acetosella,* and *Taraxacum officinale* were among the bitter green herbs of the Bible. More discriminating Israeli writers seem to favor chicory. (Such scholars regard endive as the bitter herb of Moses. Apparently, Zohary does not (ZOH), more convinced that *C. pumilum* Jacq. was the bitter herb. Not to worry. That is just a variety or subspecies of endive.) Zohary lists six species of *Lactuca* that could be called "wild lettuce," and are, I would wager, bitter, and many are consumed at times as bitter herbs. *Lactuca sativa* could be cultivated in Israel with irrigation. So I retain it as a remotely possible bitter herb of the Bible. Boulos tells us that in ancient Egypt, lettuce was a symbol of fertility (BOU). Primitive wild edible (bitter) lettuces *L. scariola* and *L. serriola* are reported in the *Flora of Palestine*, and even more probably bitter herbs of the Bible, at least in my book. Cultivated lettuce seems to cross readily with (or degenerate with reseeding into) *L. scariola* and *L. serriola*, which are themselves regarded as synonyms (USN; WOI); these being generally more bitter than derived lettuce and probably having more medicinal activities, especially compared to iceberg lettuce, which is mostly water.

Common Names (Lettuce):

Alface (Ma.; Por.; JFM; USN); Alface Comun (Ma.; Por.; JFM); Alface Cultivada (Ma.; Por.; JFM); Bazr ul Khas (Arab.; EFS); Cabbage Lettuce (Eng.; EFS); Ch'ien Chin Ts'ai (China; EFS); Chisha (Japan; TAN); Garden Lettuce (Eng.; USN); Garten Lattich (Ger.; EFS); Garten Salat (Ger.; EFS); Grüner Salat (Ger.; USN); Hakkarike (Kan.; NAD); Harouka (Arab.; BOU); Hovedsalat (Den.; EFS); Huvudsallat (Swe.; EFS); Insulata (It.; EFS); Kaha (Sin.; DEP); Kahu (Beng.; Hindi; Iran; DEP; EFS; WOI); Kavu (Tel.; DEP; WOI); Khas (Arab.; Hindi; DEP); Khass (Arab.; BOU); Khuss (Arab.; Syria; HJP); Kopfsalat (Ger.; EFS; USN); Kropsia (Dutch; EFS); Laitue (Fr.; USN); Laitue Cultive (Fr.; BOU; EFS); Laitue Pommee (Fr.; EFS); Lattich (Ger.; TAN); Lattuga (It.; TAN); Lattuga a Capuccio (It.; EFS); Lechuga (Cuba; Peru; Sp.; EGG; JFM; RyM; USN); Lechuga Acogollada (Sp.; EFS); Lechuga Arepollada (Sp.; EFS); Leti (Creole; Haiti; VOD); Lettuce (Eng.; USN); Marul (Tur.; EFS; EB54:155); Mesiouka (Arab.; BOU); Salad (India; WOI); Salada (Singh.; DEP); Salattu (Tam.; WOI); Saleet (Kan.; NAD); Shallatu (Tel.; NAD); Shatlatu Virai (Tam.; NAD); Sla (Dutch; EFS); Tuinsia (Dutch; EFS); Tukm-i-kahu (Iran; EFS); Wo Chu (China; TAN); Wo Ts'ai (China; EFS); Zaub (Hmong; EB57:365).

FIGURE 1.58 Lettuce (*Lactuca sativa*).

Activities (Lettuce):

Alexiteric (f; BIB); Anaphrodisiac (f; BIB; DAW); Anodyne (f1; DAW; JFM; FT67:215); Antidote (f; DAW); Antispasmodic (f; BIB); Anxiolytic (1; FT67:215); Aphrodisiac (f; BOU); Cardiac (f; BIB); Carminative (f; DAW); Demulcent (f; NAD); Diaphoretic (f; BIB); Digestive (f; EGG; HJP); Diuretic (f; EFS; SOU); Emollient (f; BOU; EFS; EGG); Expectorant (f; JFM; NAD); Febrifuge (f; BIB); Hypnotic (f; EFS); Lactagogue (f; BIB); Narcotic (f; EFS); Parasiticide (f; BIB); Poison (f; EFS); Psychdelic (f; BIB); Purgative (f; BIB); Refrigerant (f; DAW); Sedative (f; BOU; EFS; SOU; FT67:215); Stomachic (f; JFM); Tranquilizer (f; VOD).

Indications (Lettuce):

Asthma (f; DAW; WOI); Bronchosis (f; JFM; WOI); Bubo (f; BIB); Burn (f; WOI); Cancer (f1; JLH); Cancer, breast (f1; JLH; X15546249); Cancer, colorectal (f1; X15546249); Cancer, face (f; JLH); Cancer, gastric (f1; X15546249); Cancer, lung (f1; X15546249); Cancer, tongue (f; JLH); Cancer, uterus (f; JLH); Cardiopathy (f; DAW; JFM); Catarrh (f; JFM); Circulosis (f; DAW); Conjunctivosis (f; JFM); Constipation (f; JFM); Cough (f; JFM); Cystosis (f; JFM); Delirium (f; NAD); Dysmenorrhea (f; DAW; VOD); Dyspepsia (f; NAD); Edema (f; BIB); Fever (f; DAW); Hepatosis (f; NAD); High Blood Pressure (f; JFM); Hyperglycemia (f; DAW); Impotence (f; BOU); Infertility (f; BOU); Inflammation (f; JFM); Insanity (f; NAD); Insomnia (f1; DAW; SOU; VOD; FT67:215); Nephrosis (f; JFM); Nervousness (f; HJP; NAD; VOD); Neuralgia (f; DAW); Pain (f1; DAW; FT67:215); Palpitation (f; DAW; JFM; NAD); Pertussis (f; DAW); Sclerosis (f; JLH); Sore (f; WOI); Spasm (f; WOI); Spermatorrhea (f; NAD); Stress (1; FT67:215); Strangury (f; SOU); Swelling (f; BIB); Tuberculosis (f; BIB); Tumor (f; JLH); Typhoid (f; DAW); Urogenitosis (f; DAW).

Dosages (Lettuce):

FNFF = !!!

Leaves widely used as a salad, sometimes as vegetable. Leaves also eaten braised or wilted, or used in soups with broth, with bouillon cubes or spices. In stem lettuce varieties, young stems are peeled and cooked, but not the coarse unpalatable leaves (BIB; EGG; FAC; TAN).

- American hustlers sometimes promote the milky exudate as an opium substitute (BIB).
- Asian Indians suggest lettuce juice for nervousness and palpitations of the heart (NAD).
- Asian Indians suggest hot lettuce tea for dyspepsia and hepatoses (NAD).
- Asian Indians recommend one part lettuce seed and two parts poppy seed steeped to render a mucilage sweetened with sugar for insomnia (NAD).
- Dominicans consider the lettuce salad or tea as tranquilizer (VOD).
- Haitians take leaf tea for dysmenorrhea, insomnia, nervousness, and ophthalmia (VOD).
- Iranians suggest the seeds for typhoid (BIB).
- Lebanese occasionally apply wilted lettuce to abrasions, swellings, and wounds (HJP).
- Lebanese druggists keep powdered lettuce seed to calm feverish patients, and to deter boys from excessive masturbation, "but it is not strong enough to help mad people, those with fits, or excited women" (HJP).
- North Africans consider the plant a symbol of fertility, and the seed oil aphrodisiac (BOU).
- Panamanians claim that merely eating lettuce salad will promote sound sleep (JFM).
- Peruvians place lettuce with olive oil on the forehead for insomnia (VOD).
- Peruvians suggest the tea, sweetened with scorched sugar, for insomnia and strangury (SOU).
- Venezuelans suggest leaf decoction for cystosis, dysuria, and nephrosis (JFM).

Downsides (Lettuce):

Bolted lettuce eaten as a vegetable has been reported to cause coma (BIB).

Extracts (Lettuce):

Frequent intake of fruits, raw vegetables, carrots, pumpkin, cabbage, and lettuce, and frequent physical exercise, are associated with decreased risk for gastric, breast, lung, and colorectal cancer (X15546249).

CALABASH GOURD (*LAGENARIA SICERARIA* (MOLINA) STANDL.) ++ CUCURBITACEAE

Synonyms:

Cucurbita lagenaria L.; *Cucurbita leucantha* Duchesne; *Cucurbita longa* hort.; *Cucurbita siceraria* Molina; *Lagenaria lagenaria* (L.) Cockerell; *Lagenaria leucantha* Rusby; *Lagenaria vulgaris* Ser.

Notes (Calabash Gourd):

And Dilean, and Mizpeh, and Joktheel …

Joshua 15:38 (KJV)

After my first book was published, I was fascinated to read in Zohary that "the town-name of Dilean" was derived from *delaath*, a term occurring in postbiblical literature for the bottle-gourd, long and extensively cultivated. Specimens from Egyptian tombs date circa 3500 to 3,000 B.C., but in America date back to circa 7000 B.C. Zohary adds that scientists believe that these fruits floated from Africa to the other side of the Atlantic, the seeds remaining viable for 2 years. Be it an American or un-American species, Julia Morton's comments pushed it back even earlier, saying it was known and utilized in the Old World for 12,000 years, here in the New World for 15,000 years. It is now widely cultivated in the tropics and warm temperate zone as food, utensil, and medicine (JFM). AH2 has designated the standardized common name to be Calabash Gourd. So be it! I used Bottle Gourd in my CRC *Handbook of Medicinal Herbs* (Edition 2). Appropriately bowing to AHP, the American Herbal Products Association, for the betterment of the herbal industry, I will now use its standardized common name, Calabash Gourd, a vine, not to be confused with the Calabash tree (Crescentia). According to Dr. Max Beauvoir (VOD), rattles made from this gourd are very important in Haitian Voodoo. The rattle, known as *asson*, made from the shell of this gourd, is the symbol of the spiritual power of the priest (houngan) or priestess (mambo). The rattle is usually covered with a network of bead strings representing the world at large. There are also several snake vertebrae representing Damballah, and a bell to summon the attention of Iwa when rung. The gourd is reportedly the container of the Ossâim in the Afro-Brazilian Candomblé religion (VOD quoting Voeks, 1997).

Common Names (Calabash Gourd):

Abóbora Branca (Por.; AVP; KAB); Abóbora Carneira (Por.; POR); Abóbora d'Agua (Por.; AVP); Adanggu (Ewe; KAB); Akpaki (Ga; KAB); Alabu (Sanskrit; SKJ); Alaburu (Tel.; KAB); Alahko (Koasiti; AUS); Alava (Sanak; DEP; KAB); Anapa Kai (Mal.; NAD); Apakyi (Twi; KAB); Asmakabaği (Tur.; EFS); Bagaña (Dr.; AUS); Bairntua (Fanti; KAB); Bangaña (Dr.; AVP); Bau (Vn.; POR); Bau Nam (Annam; KAB); Bella Shora (Mal.; DEP; KAB); Benares Pumpkin (Eng.; NAD); Bhopala (Mar.; WOI); Bhopla (Mar.; KAB); Birsuku (Mun.; KAB); Boga Lao (Assam; DEP; SKJ; WOI); Bottiglia di Zucca (It.; AVP); Bottle Gourd (Eng.; Ocn.; BOU; CR2; FAC; USN; VOD); Bouteille (Fr.; BOU); Brujito (Pan.; IED); Bule (Ma.; JFM); Busin Swai (Burma; DEP;

KAB); Cabaceira (Brazil; AVP); Cabaceiro Amargóso (Brazil; AVP); Cabaço (Brazil; Por.; MPB; POR; USN); Cabaço Amargosa (Brazil; MPB); Cadungo Amargo (Sp.; Pr.; AVP); Cajombre (Sp.; POR; USN); Calabash (Eng.; FAC; USN); Calabash Cucumber (Eng.; BOU; IHB; NPM); Calabash Gourd (Eng.; Scn.; AH2; USN); Calabaza (Peru; Sp.; EGG; POR; RAR; USN); Calabaza de San Roque (Sp.; EFS; POR); Calabaza Vinatera (Sp.; AVP; KAB; POR); Calabazo (Col.; Dr.; AVP); Calbas Largu (Creole; JFM); Calebasse (Fr.; BOU; POR); Calebasse Courant (Fwi.; AUS); Calebasse Courge (Guad.; Mart. AVP); Calebasse d'Europe (Fr.; BOU); Calebasse d'Herbe (Fr.; KAB) Calebasse Douce (Guad.; Mart.; AVP); Calebasse Franc (Haiti; AUS; AVP); Calebasse Longe (Guad.; Mart.; AVP); Calebasse Longue (Haiti; AUS; AVP); Calebasse Musquée (Guad.; Haiti; Mart.; AUS; AVP); Calebasseterre (Fr.; Guiana; KAB); Calebassier (Fr.; Haiti; AVP; POR; USN); Calebassier Grimpant (Fr.; POR); Camasa (Sp.; Ven.; EFS; JLH); Camaza (Ma.; JFM); Caracho (Pr.; AVP); Carracho (Pr.; AVP); Chappu P'ege (Bol.; Callawaya; DLZ); Chiti Anab (Tel.; NAD); Chucña (Peru; EGG; RAR; SOU); Cocombro (Brazil; KAB); Cogorda (Sp.; POR); Cojombro (Ma.; JFM); Cojudito (Peru; EGG); Cojudo (Peru; EGG); Colombro (Por.; EFS); Colondro (Por.; EFS); Congourde (Fr.; Haiti; AHL; BOU); Cougourde (Fr.; POR); Courge Bouteille (Fr.; EFS; POR); Courge de Perelin (Fr.; AVP); Courge Massue (Fr.; POR); Courge Pélerine (Fr.; POR); Courge Siphon (Fr.; POR); Cucurbita (Sp.; RAR); Cucuzzi (It.; FAC); Cucuzzi Caravasi (It.; FAC); Cuia (Brazil; MPB); Cuieté (Brazil; MPB); Delaath (Isr.; ZOH); Diya Laba (Sin.; DEP); Diya Labu (Sin.; KAB; POR); Dodi (Hindi; POR); Douma (Sudan; AVP); Dubb'a (Arab.; BOU); Dudhi (Guj.; Hindi; India; POR; USN; WOI); Dudhiyun (Kathiawar; KAB); Dudhya (Mar.; DEP; WOI); Dudi (Hindi; POR); Dum (Chepang; NPM); Efepe (Creek; Muskogee; AUS); Fepe (Creek; Muskogee; AUS); Flaschenkürbis (Ger.; POR; USN); Flaskegræskar (Den.; POR); Flaskkurbits (Swe.; POR); Fleskalebas (Dutch; EFS; POR); Flessepompoen (Dutch; POR); Fran (Creole; Haiti; VOD); Geöhnlicher Flaschenkürbis (Ger.; USN); Ghiya (Pun.; WOI); Golkaddu (Bijnor; Hindi; DEP; KAB); Gourd (Eng.; JFM); Gourde (Fr.; BOU; POR); Gourde Bouteille (Fr.; POR; USN); Gourde Calebasse (Fr.; EFS); Gourde des Pèlerins (Creole; Haiti; AUS; VOD); Gourde Massue (Fr.; POR); Gourde Trompette (Haiti; AHL); Gourdo (Lan.; KAB); Gros Calebasse (Haiti; AHL); Gubba Kaya (Tel.; DEP); Güícharo (Pr.; AUS; AVP); Güiro (Cuba; Dr.; Haiti; Taino; AUS; AVP); Güiro Amargo (Sp.; POR; USN); Güiro Cimarrón (Cuba; AUS; AVP); Güiro Dulce de Nueva Guinea (Cuba; RyM); Güiro Guyaro (Cuba; AUS); Guyaro (Cuba; AUS); Halagumbala (Kan.; NAD; WOI); Healing Squash (Eng.; FAC); Hefepe (Creek; Muskogee; AUS); Hefepe-Nerkv (Creek; Muskogee; AUS); Hifipa (Mikasuki; AUS); Ho Lo (Ic.; KAB); Horoto (Arawak; Sur.; AUS); Hu Gua (China; PR); Hu Lu (China; PR); Hu Lu Gua (China; PR); Hu Zi (Pin.; DAA); Hurreakadu (Sin.; KAB); I'niizhe (Osage; AUS); Irao (Sin.; DEP); Iselwa (Zulu; ZUL); Italian Edible Gourd (Eng.; FAC); Jamaru (Brazil; MPB); Jomatapheng (Lepcha; NPM); Ka-Bed (Tibet; NPM); Kaddú (Hindi; Iran; DEP; NAD); Kadu (San.; DEP; KAB); Kadu Duddi (Kon.; NAD); Kaduasuki (Naguri; KAB); Kadubhopla (Bom.; SKJ); Kadugol (Urdu; KAB); Kadulau (Beng.; SKJ); Kadutalkha (Iran; KAB); Kadwitumbade (Guj.; NAD); Kahisore (Kan.; NAD); Kalabas (Den.; POR); Kalbas (Creole; Haiti; VOD); Kalbas di Core Abao (Ma.; JFM); Kalbas Kouran (Creole; Haiti; VOD); Kalbas Largoe (Ma.; JFM); Kalebass (Swe.; POR); Kalebassenkürbis (Ger.; POR); Kalubay (Vis.; POR); Karehulmar (Arab.; ?); Kashiphal (Hindu; Mah.; NAD; SKJ); Katutumbi (Sanskrit; DEP); Keedú (Pun.; DEP); Khaddu (Hindi; NAD); Khi Luu Saa (Thai; POR); Khlôôk (Khmer; POR); Ko (Biloxi; AUS); Kochi (Alabama; AUS); Kodu (Beng.; NAD); Kohla (Sin.; NAD); Koro (Carib; Sur.; AUS); Kukuk (Java; Sunda; IHB; POR); Kurlaru (Sanskrit; EFS); Labu (Malaya; IHB); Labu Air (Dei.; POR); Labu Ayĕr Bĕrleher (Malaya; IHB); Labu Ayĕr (Java; IHB); Labu Ayĕr Puteh (Malaya; IHB); Labu Botol (Malaya; POR); Labu Jantong (Malaya; IHB); Labuka (Sanskrit; KAB); Labu Kĕndi (Malaya; IHB); Labu Puteh (Malaya; EFS); Labu Putih (Dei.; POR); La Guasu (Chiriguano; DLZ); Lakttine (Ber.; BOU); Lau (Assam; Beng.; Nwp.; DEP; WOI); Lauka (Magar; Nepal; Newari; Sunwar; Tamang; Tharu; NPM); Laukaa (Nepal; POR); Lauki (Bhojpuri; Hindi; India; Nwp.; EFS; KAB; NPM; POR); Lek (Dwi.; JFM); Lokhi (Hindi; POR); Long White Gourd (Eng.; NAD); Lova (Chiriguano; DLZ); Lyiquisigua (Bol.; Chiriguano; DLZ); Mambiro (Cr.; AVP); Manamtao (Thai; POR); Mao Gua (China; POR); Mardudi

(Kon.; KAB); Marimbo (Pr.; AVP); Marimbo Almizcle (Sp.; AVP); Mate (Peru; EGG; RAR; SOU); Mathi (Bol.; Que.; DLZ); Mati (Bol.; Peru; Que.; DLZ; EGG; RAR; SOU); Matti (Peru; RAR); Mazon (Sp.; EFS); Mekuri (Naga; DEP; KAB); Mirango (Mandingo; KAB); Mo Kwa (Canton; POR); Naam Tao (Thai; POR); Nambiro (Cr.; Pr.; AUS; JFM); Namtao (Thai; IHB; POR); Namz Taux (Laos; POR); Ndopote (Manjia; KAB); New Guinea Bean (Eng.; FAC); New Guinea Butter Vine (Eng.; FAC); Oo Lo Kwa (Canton; POR); Opo (Tag.; FAC; KAB); Ouowi (Ber.; BOU); P'ao (China; EFS); Pehe (Omaha; Ponca; AUS); Peh Poh (Singapore; POR); Pepino Chino (Ma.; JFM); Phusi (Nepal; DEP; KAB); Pierna de Pobre (Sp.; EFS); Pilgrim Bottle (Eng.; AVP; IED); Poro (Bol.; Peru; DLZ; EGG; RAR; SOU); Porongo (Bol.; Brazil; DLZ; MPB); Poto (Peru; EGG; RAR); Poto Pate (Peru; EGG); Pullokurpitsa. (Fin.; POR); Pulu (Aym.; Bol.; DLZ); Púnu (Aguaruna; Peru; EGG); Pura (Peru; EGG; RAR); Puru (Peru; EGG; RAR; SOU); Purunga (Brazil; MPB; RAR); Qar'a Aslawiya (Arab.; BOU); Qar'a Dubba (Arab.; BOU); Qar'a Duruf (Arab.; BOU); Qar'a Tawil (Arab.; BOU); Qer'aa (Arab.; BOU); Qer'aa Beida (Arab.; BOU); Qer'aa el leben (Arab.; BOU); Qer'aa Gardousi (Arab.; BOU); Qer'aa Medwen (Arab.; BOU); Saffed Kadu (India; EFS); Shokshi (Chickasaw; AUS); Shora Kai (Tam.; DEP); Shorakkai (Tam.; KAB; SKJ; WOI); Shukshi Okpulo (Chickasaw; AUS); Shukshubok (Choctaw; AUS); Sicay (Vis.; KAB); Sinu (Sakai; IHB); Sísira (Garifuna; Nic.; IED); Sorakai (Tam.; NAD); Sorakaya (India; SKJ); Sorekayi (Kan.; WOI); Sorrakaya (Tel.; WOI); Su Kabagi (Tur.; EB54:155); Tafe-qeloujla (Ber.; BOU); Takhsait (Ber.; BOU); Tä 'Mbacti' (Yuchi; AUS); Taquera (Brazil; MPB); Tarro (Ma.; Sal.; AUS; JFM); Tecomate (Ma.; Sal.; AUS; JFM); Tecomatillo (Ma.; Sal.; AUS; JFM); Tembephutra (Limbu; NPM); Tikta Tumbi (Sanskrit; NAD); Tito Tumba (Nepal; POR); Tokal (Semang; IHB); Tol (Guat.; Ma.; Sal.; AUS; JFM); Totumo (Ma.; Sal.; Ven.; AUS; JFM); Trompetenkürbis (Ger.; POR); Tukal (Besisi; Pangan; IHB); Tula de Mate (Pan.; IED; JFM); Tumada (Guj.; DEP; KAB; WOI); Tumba (Pun.; KAB); Tumri (Hindi; Kum.; DEP; KAB; WOI); Upo (Tag.; POR); Vinotera (Peru; EGG; RAR); Voambahy (Sakalave; KAB); Voata Vomanta (Hova; KAB); Wachekalasi (Rai; NPM); Waluh Kĕnti (Java; IHB); Wamnuha (Dakota; AUS); White-flower Gourd (Eng.; USN); White Pumpkin (Eng.; NAD); Woo Lo Gwa (Canton; POR); Woo Lo Kwa (Canton; POR); Wu Lo Gwa (Canton; POR); Xataan (Amahuaca; Peru; EGG; RAR); Yumí (Aguaruna; Peru; EGG; SOU); Yuugao (Japan; FAC; POR); Zucca da Pescare (It.; HHB); Zucca da Tabacco (It.; POR); Zucca Lunga (It.; AVP).

Activities (Calabash Gourd):

Analgesic (f; AUS; DEM); Anthelmintic (f; BOU; WBB); Antibilious (f; WOI); Antidote (f; LMP); Antiperiodic (f; KAB); Antitussive (1; FNF; HAD); Antiulcer (1; FNF; HAD); Bitter (1; JFM); Carminative (f; JFM); Cardiotonic (f; KAB); Cerebrotonic (f; KAB); Cholinergic (1; FNF; HAD); Demulcent (1; FNF; HAD); Diuretic (f; AUS; BOU; EFS); Emetic (f; EFS); Febrifuge (f; EFS); Hemostat (1; FNF; HAD; KAB); Hepatoprotective (1; FNF); Laxative (f; AUS); Litholytic (f; EFS); Pectoral (f; WBB); Purgative (1; EFS; JFM); Ribosome Inactivator (1; X11104364); Taenicide (f; KAB); Tonic (f; KAB); Trypsin Inhibitor (1; X7889483); Vulnerary (f; DLZ; KAB).

Indications (Calabash Gourd):

Adenopathy (f; NAD); Alopecia (f; IHB; WOI); Alzheimer's (1; HAD); Asthma (f; AHL; AUS; KAB); Atheroma (1; HAD); Biliousness (f; EFS; KAB); Body ache (f; DEM); Boils (f; DEM); Bronchosis (f; KAB); Cancer (f1; HAD; JLH); Cancer, colon (f1; FNF; JLH; X7889483); Childbirth (f; MPB); Colic (f; LMP); Constipation (1; JFM); Corn (f; JLH); Cough (f1; AHL; AUS; FNF; HAD; LMP); Delirium (f; IHB; WOI); Dermatosis (f; DLZ; IHB; JFM); Diarrhea (f1; DEP; FNF; HAD); Dropsy (f; AUS; WOI); Dyskinesia (1; FNF; HAD); Dysuria (f; KAB); Earache (f; KAB); Favus (f; DLZ); Fever (f; BOU; IHB; LMP); Gas (f; JFM); Gastrosis (f; HHB); Gingivosis (f; LMP); Headache (f; DEM; JFM); Heart Problems (f; JFM); High Cholesterol (1; HAD); Hepatosis (1; FNF); Hoarseness (f; AHL); Hyperacidity (f; NPM); Inflammation (f; KAB); Insanity (f; DEM); Itch (f; AUS; DLZ); Jaundice (f;

NAD; SKJ); Leukorrhea (f; KAB); Malaria (f; KAB); Mange (f; JFM); Myalgia (f; KAB); Nephrosis (f; DAV; EGG; MPB; RAR); Nyctalopia (f; NAD); Ophthalmia (f; KAB); Pain (f; AUS; KAB); Pimple (f; IHB; LMP; WOI); Pregnancy (f; JFM); Rheumatism (f; WOI); Rhinosis (f; NAD); Ringworm (f; DLZ); Scabies (f; DLZ); Scrofula (f; NAD); Sore (f; MPB); Stomachache (f; HHB); Strangury (f; KAB); Thirst (f; IHB); Toothache (f; LMP); Tumors (f; JLH); Typhoid (f; HAD; IHB); Typhus (f; LMP); Ulcer (f; KAB; NPM); Uterosis (f; KAB); Vaginosis (f; KAB); Worm (f; BOU).

Dosages (Calabash Gourd):

FNFF = !!!
Young fruits widely eaten; seeds edible, yielding an oil that can be rendered into a tofu-like curd; leaves and young shoots cooked as potherbs (FAC, TAN; EB54:155). Young leaves and fruits, cooked, could (or maybe even should) be one of our five daily fruits and vegetables (100 g servings).

- Amazonians, Bolivians, and Brazilians all report the use of seeds for nephritis (MPB).
- Asian Indians apply the pulp around the seeds to the head in delirium (IHB).
- Asian Indians boil fruit juice with an equal amount of oil to massage scrofulous glands (NAD).
- Asian Indians suggest nose drops of the fruit juice for "atrophic rhinosis" (NAD).
- Asian Indians use fruit ashes with honey as collyrium for night blindness (NAD).
- Ayurvedic practitioners consider the leaves useful in biliousness, earache, leukorrhea, strangury, uterosis, and vaginosis; the fruit for asthma, bronchitis, cardiopathy, inflammation, pain, and ulcers (KAB).
- Bolivians apply the powdered seed in a lard pomade to favus, ringworm, scabies, and other dermatoses (DLZ).
- Brazilians apply heated leaves to aid labor and heal sores (MPB).
- Dominicans suggest the syrup of powdered fruits for asthma, cough, and hoarseness (AHL).
- East Asians apply the fruit juice with lime to pimples, and leaf juice to bald heads (IHB).
- Hindus prescribe the leaf decoction for jaundice (DEP).
- Indonesians use young fruit juice to quench thirst of typhoid patients (IHB).
- Malayans eat the fruits for colic and fever, also bathing their heads in the potlikker (IHB).
- Nepalese consider the fruit pulp emetic and purgative, using the juice for acid stomach, dyspepsia, and ulcers (NPM).
- Nicaraguan Garifuna use leaf and root decoction orally and topically as a digestive and laxative, and for skin rashes and sores (IED).
- Unani consider the fruits antibilious, diuretic, and febrifuge, and the seeds useful for ardor urinae, cough, earache, fever, inflammation, and strangury (KAB).
- Venezuelans poultice leaves with coconut oil onto tumors (JLH).
- Zulu use leaf and root infusion for stomachache (HHB).

Downsides (Calabash Gourd):

Processed fruits and leaves and flowers eaten on many continents, but still reported as potentially toxic, especially green fruits. Fruit flesh fed to rabbits, leads to restlessness and dyspnea, with paralysis and death from asphyxia (WBB).

Extracts (Calabash Gourd):

Leaves contain 1300 ppm steroidal steroids (ZMB; JFM). If data cited in WOI are correct, this remains my best source of choline in vegetables. For the many possible benefits of choline, see the USDA Phytochemcial database (http://www.ars-grin.gov/duke). Dietary fiber from the fruit suppressed colonic carcinogenesis in mice by lowering the bile acid concentration and pH in the colon.

The mechanisms of action differ from those of wheat bran (X7889483). Pectin is, of course, an anticancer compound. On a zero-moisture basis, the fruits can contain up to 21% pectin, cf. 35% for marshmallow roots; 19% for carrots; 14% for hops; and 11% for eggplant, rosehips, and sunflower; and 10% for flaxseed. Genetically targeted for colon cancer, I would enjoy more frequent indulgence, cooking these (minus the hops; instead, I might have a glass of beer) to make a pectin potpourri. At levels of 250 mg/kg, various extracts, including the ethanol extract of the fruits, exhibited liver-protecting properties.

BAY (*LAURUS NOBILIS* L.) ++ LAURACEAE

NOTES (BAY):

> **I have seen the wicked in great power, and spreading himself like a green bay tree.**
>
> **Psalms 37:35 (KJV)**

> **I have seen a wicked man overbearing, and towering like a cedar of Lebanon.**
>
> **Psalms 37:35 (RSV)**

> **I have seen the wicked a tyrant and spreading himself as a luxuriant [tree] in native soil.**
>
> **Psalms 37:35 (NWT)**

Zohary, noting that the laurel abounds on Mt. Carmel and in stony soils around Galilee, suggests that the word *orem* in Isaiah 44:14 is bay, although it is rendered as cedar in the RSV, and today orem means pine. Further complicating things, Zohary suggests that the cedar of Isaiah 44:14 (RSV) is laurel, as in the following quote: "He cuts down cedars; or he chooses a holm tree or an oak and lets it grow strong among the the trees of the forest; he plants a cedar and the rain nourishes it" (RSV). Zohary further comments that generally scholars think the laurel is not mentioned in the Bible, and then draws evidence to suggest that laurel in RSV is rendered as *cedar* (while in modern Hebrew oren = pine). In Aramaic, it is *aranye*; in Arabic, *ar*; and in Accadian, *eru*, according to Zohary (ZOH). Although not a huge tree, the evergreen bay can attain a height of more than 20 feet, maybe even 60 feet (GMH). Hence, I use the quote above as possibly referring to *Laurus nobilis* as the biblical green bay. Even in modern languages identifying a plant by nothing more than its common name is perilous at best. In biblical times, the bay was symbolic of wealth and wickedness. And that is why I use the Psalms quote above. The evergreen leaves, when broken, emit a sweet scent and furnish an extract used by the Orientals in making perfumed oil. In the ancient Olympic games, the victorious contestant was awarded a chaplet of bay leaves, placed on his brow. The Roman gold coin of 342 B.C. has a laurel wreath modeled on its surface. Oil replaces dried leaves to great advantage because it can be measured more precisely and provides more uniform results. The fat derived from the fruits has been used for soap making and veterinary medicine. Leaves twined into wreaths by ancient Greeks and Romans were used to crown their victors in sports and wars. The wood, resembling walnut, can be used for cabinetry.

COMMON NAMES (BAY):

Alloro (It.; EFS); Alloro Poetico (It.; HH3); Ar (Arab.; ZOH); Ar Atzil (Heb.; PAY); Aranye (Aramaic; ZOH); Asa musa (Arab.; BOU); Bahia (Sp.; EFS); Bakelaar (Dutch; EFS); Bay (Eng.; Scn.; AH2); Bay Laurel (Eng.; EFS); Bayleaf (Eng.; CR2); Bayleaf Laurel (Eng.; CR2; USN); Baz (India; EFS; NAD); Daphne (Greek; GMH; HJP); Defne (Tur.; EFS); Edler Lorbeerbaum (Ger.; EFS); Eru (Accadian; ZOH);

Habbula Ghara (Iran; EFS); Laurierboom (Dutch; EFS); Gekkeiju (Japan; HH3); Ghar (Arab.; BOU; HJP); Grecian Laurel (Eng.; BIB; EFS); Green Bay (Eng.; BIB; KJB); Hab el Ghar (Arab.; EFS); Habb Ghar (Arab.; BOU); Habul Ghar (India; DEP); Habula Ghara (Iran; NAD); Laurbaer (Den.; EFS); Laurel (Eng.; BOU); Laurél Común (Sp.; EFS); Laurier (Fr.; BOU); Laurier d'Apollon (Fr.; BOU; EFS); Laurier des Poetes (Fr.; BOU); Laurier Franc (Fr.; BOU; EFS); Laurier Sauce (Fr.; BOU; GMH; HH3); Laurierboom (Dutch; EFS); Lauro (Sp.; EFS); Lauro Franco (It.; EFS); Lauro Poetico (It.; EFS); Lauro Regale (It.; HH3); Lauro Regio (It.; EFS); Lorbeer (Ger.; HH3); Lorbeerbaum (Ger.; HH3; USN); Lorbeerstrauch (Ger.; USN); Loureiro (Mad.; Por.; EFS; USN); Louro (Por.; USN); Maraget Musa (Arab.; BOU); Noble Laurel (Eng.; HH3); Old World Bay (Eng.; HOC); Orem (Heb.; ZOH); Rand (Arab.; Syria; BOU; HJP); Roman Laurel (Eng.; BOU; GMH; HH3); Sweet Barg (Eng.; DEP); Sweet Bay (Eng.; BIB; BOU; EFS); Sweet Laurel Tree (Eng.; EFS); Taset (Ber.; BOU); True Laurel (Eng.; GMH); Victor's Laurels (Eng.; NAD); Yue Gui Zi (Pin.; DAA); Zafui (India; DEP).

Activities (Bay):

Abortifacient (f; HJP; SPI); Allergenic (1; CRC; PH2; TAD); Analgesic (f; PAY); Anodyne (f; CRC); Antioxidant (1; TAD); Antiperspirant (f; PAY); Antiseptic (f1; CRC; HHB; PAY; PH2); Antirheumatic (f; PHR); Antispasmodic (f; BOU; EFS); Antiulcer (1; PR14:581); Antiviral (1; APA); Aperitif (f; APA; CRC); Aphrodisiac (f; PAY); Bactericide (1; APA; CRC); Bitter (f; HHB); Candidicide (1; HH3); Cardiodepressant (1; PAY); Carminative (f1; APA; CRC; HHB; HJP; JFM); Cholagogue (f; PNC; VAD); Diaphoretic (f; APA; CRC; PNC; SPI); Digestive (f; BOW; JFM); Diuretic (f; CRC; HHB); Emetic (f; CRC); Emmenagogue (f; APA; BOU; CRC; HHB; JFM); Ethanol-Absorption Inhibitor (X11003152); Expectorant (f; GAZ; VAD); Febrifuge (f; APA); Fungicide (1; APA; CRC; PAY); Gastrotonic (f; BOU; CRC; JFM); Hepatotonic (f; CRC; HJP); Hypotensive (1; APA; PAY); Insectifuge (f1; HH3; PH2; TAD); Laxative (f; PAY); Molluscicide (1; HH3; PH2); Narcotic (f1; BIB; CRC); Nervine (f; CRC; EFS); NO Inhibitor (1; X10834299); Orexigenic (1; APA; CRC; VAD); Parasiticide (1; BOU; HHB); Pediculicide (1; VAD); Rubefacient (1; PHR; PH2); Sedative (f1; APA; CRC; JFM); Stimulant (f; CRC; PNC); Stomachic (f; BOU; CRC; PNC); Sudorific (f; BOU; CRC); Tonic (f; SPI); Vermifuge (f; PAY).

Indications (Bay):

Ague (f; GMH); Alzheimer's (1; COX; FNF); Amenorrhea (f; BOU; CRC; SPI); Anorexia (f1; APA; BOW; GMH; VAD); Arthrosis (f1; APA; COX; FNF; VAD); Asthma (f; VAD); Bacteria (1; CRC; HHB); BO (f; PAY); Bronchosis (f; VAD); Bruises (f; APA); Bug bite (f; APA); Cancer (1; CRC; JLH); Cancer, anus (1; JLH); Cancer, colon (1; COX; FNF); Cancer, eye (1; CRC; JLH); Cancer, face (1; CRC; JLH); Cancer, joint (1; CRC; JLH); Cancer, liver (1; CRC; JLH); Cancer, mouth (1; JLH); Cancer, parotid (1; CRC; JLH); Cancer, spleen (1; CRC; JLH); Cancer, stomach (1; CRC; JLH); Cancer, testicle (1; CRC; JLH); Cancer, uterus (f; CRC; JLH); Candida (1; HH3; SPI); Catarrh (f; BOU); Cheilosis (f; HH3); Cold (f; PAY); Colic (f; APA; CRC; SPI); Condyloma (f; CRC); Cough (f; CRC; HJP); Cramp (f; BOU); Cystosis (f; GAZ); Dandruff (f; APA); Deafness (f; JFM); Debility (f; JFM; PAY); Dermatosis (f; APA; HOC; SPI); Diabetes (1; FNF); Diarrhea (f; PAY); Dislocation (f; HJP); Dropsy (f; NAD); Dyspepsia (1; APA; JFM); Dysuria (f; NAD); Earache (f; CRC; GMH; PAY); Emphysema (f; VAD); Enterosis (f; VAD); Escherichia (1; X10438227); Fibroid (f; CRC; JLH); Fungus (1; BIB; CRC); Furuncle (f; HH3); Gas (f1; APA; BIB; SPI); Gastrosis (f; CRC; PAY); Halitosis (f; GAZ); Headache (1; FNF; HAD); Hemorrhoid (f; VAD); Hepatosis (f; CRC); High Blood Pressure (1; APA); Hysteria (f; CRC; GMH; SPI); Impostume (f; CRC; JLH); Impotence (f; PAY); Induration (f; JLH); Infection (1; CRC; SPI); Jaundice (f; PAY); Klebsiella (1; X10438227); Leukorrhea (f; GAZ; NAD); Lice (1; VAD); Mange (f; JFM); Migraine (1; FNF; HAD); Mycosis (f; CRC; SPI); Neuralgia (f; PAY); Ophthalmia (f; PAY); Orchosis (f; JLH); Pain (f; APA; NAD; PAY); Parasite (1; BOU; HHB; SPI); Pediculosis (f; HH3); Pharyngosis (f; VAD); Polyp (f; CRC); Proctosis (f; JLH); Rheumatism (f; BOU; CRC; HH3; HJP; NAD; PHR; PH2; SPI); Salmonella (1; HH3); Scabies (f; BOW); Scirrhus (f; JLH); Sclerosis (f; CRC); Sinusosis (f; VAD); Sore (f; APA; HH3; JFM); Spasm (f; CRC); Splenosis (f; HOC; JLH);

FIGURE 1.59 Bay (*Laurus nobilis*).

Sprain (f; APA; CRC; WOI); Staphylococcus (1; HH3; SPI); Sting (f; HOC; PAY); Stomachache (f; PAY); Stomatosis (f; HH3); Swelling (f; HOC); Ulcer (1; JFM; PR14:581); Urethritis (f; GAZ); Uterosis (f; JLH); Wen (f; CRC); Worm (f; PAY); Wound (1; APA); Yeast (1; X10438227).

Dosages (Bayleaf):

FNFF = !!!

Dried bay leaves are used to flavor meats, fish, poultry, vegetables, soups, and stews, and are especially popular in French dishes; also as an ingredient in pickling spices and vinegars. Leaves once used as a tea substitute. Priestesses of Apollo at Delphi in ancient Greece chewed laurel leaves to better prophecy (PAY). An essential oil, distilled from the leaves, is used for flavoring food products, such as baked goods, confectionary, meats, sausages, and canned soups, and in perfumery. Bay leaves GRAS (generally recognized as safe) at 1000 ppm, the oil at 200 ppm (BIB; TAD). 1–2 tsp leaf/cup water to 3 ×/day (APA); 1–2 drops essential oil added to brandy, honey, or tea (APA).

- Asian Indians suggest the oil expressed from boiling seeds is applied to rheumatic pain (NAD).
- British suggest bay oil for bruises, earache, and sprains (GMH).
- British suggest the berries as abortifacient, deobstruent, emmenagogue, and orexigenic (GMH).
- French use the leaves as carminative (GMH).
- Israelis drink fruit and leaf decoction to prevent diarrhea (PAY).
- Israelis massage joint and nerve pain with fruit oil (PAY).
- Israelis take laurel leaf tea to enhance potency (PAY).
- Israelis use vapor bath (laurel, rue, savory, three-lobed sage) for colds and debility (PAY).
- Israelis wash head with laurel decoction to strengthen the hair (PAY).
- Israelis wash with laurel decoction to expel perspiration odor (PAY).
- Lebanese mountaineers use raw berries to induce abortion (HJP).
- Lebanese tightly cork and steep leaves and berries in brandy in the sun for several days, then distill, and use as liniment for rheumatism and sprains, or as an emmenagogue (HJP).
- Spaniards suggest the fruit oil for arthritis and pediculosis (VAD).

Downsides (Bay):

Class 1 (AHP). None known at proper dosage (PHR). No health hazards or side effects known with proper therapeutic dosages (no dosage given, however) (PH2). Leaf and berry oil may cause severe lesions of the skin. Contact dermatosis from handling leaves or essential oil reported. Diarrhea, nausea, and vomiting from excessive doses of the essential may occur. Sesquiterpene lactones are aromatic compounds widely distributed in certain plant families, with highest concentrations–generally found in leaves and flowers. Sheep and cattle poisonings due to sesquiterpene lactone–containing species have been reported. Cases of allergic contact dermatosis in humans have also been reported (AEH1). There have been a few unfortunate fatalities to people perforating their intestines with fragmented laurel leaves. Always remove them from your spaghettis and stews (JAD, TAD). Artemorin, costunolide, costuslactone, deacetlylaurenobiolide laurenobiolide, reynosin, santamarin, and verlorin are eight alpha-methylene-gamma-butyrolactones documented to be the chief cause of allergy (contact dermatosis) in *Laurus* (TAD).

Extracts (Bay):

EO LD50 = 3950 mg/kg orl rat (HH3). Leaves bacteriostatic to *Bacillus subtilis*, *Escherichia coli*, *Hafnea alnei*, *Micrococcus luteus*, *Proteus vulgaris*, *Pseudomonas aeruginosa*, *Salmonella enteridis*, *Staphylococcus aureus*, and fungistatic or fungicidal toward *Aspergillus niger*, *A. terreus*, *Candida albicans*, *Fusarium moniliforme*, *Phytophthora capsici*, *Rhizoctonia solani*, and *Sclerotinia sclerotiorum*. Leaf extracts reduce aflatoxin production by *Aspergillus parasiticusm* and botulin toxin from *Clostridium botulinum* (HH3). Extracts active against *Acinetobacter baumanii*, *Aeromonas veronii* biogroup sobria, *Candida albicans*, *Enterococcus faecalis*, *Escherichia col*, *Klebsiella pneumoniae*, *Pseudomonas aeruginosa*, *Salmonella enterica subsp. enterica serotype typhimurium*, *Serratia marcescens*, and *Staphylococcus aureus* at concentrations at or below 2% (X10438227). With compounds like parthenolide and santamarin, this shares many of the antimigraine compounds of feverfew 3-alpha-acetoxyeudesma-1,4(15),11(13)-trien-12,6-alpha-+ ++olide (X11003152). Confirming my contention that it is a suite of phytochemicals rather than a single silver bullet, in this case for inhibiting absorption of ethanol, Yoshikawa et al. (2000) reported that sesquiterpenes having an alpha-methylene-gamma-butyrolactone moiety, costunolide, dehydrocostus lactone, zaluzanin D, reynosin, santamarine, and others, isolated from laurel leaves, selectively inhibit ethanol absorption (X11003152). Italian Scientists De Marino et al. (X15675799) reported several phytochemcials, some new, and their inhibitory effects on nitric oxide, a major component of inflammation. The leaves, widely used in Italian cooking and folk medicine, had five new megastigmane glucosides named laurosides A through E. The effects of isolated compounds on nitric oxide production in macrophages were examined (X15675799).

HENNA (*LAWSONIA INERMIS* L.) ++ LYTHRACEAE

Synonyms:

Lawsonia alba Lam.

Notes (Henna):

My beloved is unto me as a cluster of camphire in the vineyards.

Song of Solomon 1:14 (KJV)

Mentioned only once, in the very botanical Song of Solomon, henna (RSV; camphire in the KJV) was early used by the Egyptians as a cosmetic. Egyptian mummies were wrapped in henna-colored cloth. Mummies entombed for more than 3000 years still show the henna used on their nails Henna powder is on sale in all great Arab cities even today. Henna is valued, especially by women of Egypt, for it yields a powerful dye of a dark dusky red, rather like iron rust in color. The women use it to stain the palms of their hands and the soles of their feet. It is also an effective check to excessive perspiration. Distilled perfumes from the flowers lead to an essential oil, *mehendi*, important in religious feasts (BIB; ZOH). The dye itself was called *puker* (I am reminded of puca for red in Quechua) by ancient Eyptians, *kupr* or *kufer* by the *Copts*, and *kufra* in Aramaic and Accadian, a name that survived in postbiblical literature. The Hebrew word is *kopfer* (ZOH). Mohammedens claim that the prophet called it *syyadu riáhín* (= the best of herbs). The young leaves, dried, powdered, and then soaked in water with a little lime juice, constitute the dye. Leaves may be harvested in the second year, and the plants may live 15 years. In India and Pakistan, henna is widely used by both men and women for coloring nails, fingers, hands, and hair. Hair is dyed a brownish-chestnut shade, which turns black in conjunction with indigo. To dye the hair, an infusion of dried leaves to which has been added a little lime juice is used. Henna leaves dyed fingers, nails, hands, and feet a dull orange. A deep red color may be obtained when henna is mixed with catechu. Infusion of leaves also used for dyeing cotton fabrics is a light reddish-brown. Wool and silk may also be dyed by henna. Leaves also used in manufacture of perfumed oils and as a tanning agent. The rose-scented flowers, attractive to bees, give an essential oil (mehndi oil) long used in Indian perfumery. Plants are grown as hedge plants throughout India. Wood is used to make tool handles, tent pegs, and other articles (BIB).

Common Names (Henna):

Al Qatab (Arab.; Syria; HJP); Alcana (Sp.; USN) Alcana d'Oreint (Fr.; KAB); Alcanna (Eng.; Fr.; BOU); Alfeneiro (Por.; USN); Alheña (Sp.; USN); Alhenna (Arab.; KAB); Bapar (Heb.; KAP); Benjati (Oiya; WOI); Bhurara (Lambadi; KAB); Bri-mog (Tibet; NPM); Camphire (Eng.; BIB; BOU; KAB; KJV; NPM); Chi Giap Hoa (Ic.; KAB); Cinamomo (Pi.; KAB); Cypress Shrub (Eng.; KAB); Dambin (Burma; NAD) Dan (Burma; DEP); Danbin (Burma; KAB); Diabe (Ber.; BOU); Dvivantra (Sanskrit; KAB); Egyptian Privet (Eng.; Bhojpur; BOU; KAB; NPM); Faghia (Arab.; BOU);

FIGURE 1.60 Henna (*Lawsonia inermis*).

Flè Jalouzi (Creole; Haiti; VOD); Fleurs Jalousie (Haiti; AVP); Foudeoum (Ber.; BOU); Foundemm (Woloff; KAB); Góranta (Tel.; DEP); Górante (Kan.; DEP); Goranti (Kan.; MPI); Hamella (Ber.; BOU); Hana (Arab.; Yemen; X15890471); Henna (Arab.; Eng.; Pr.; Scn.; AH2; CR2; GHA; HJP; RSV; USN; ZOH); Hennastrauch (Ger.; USN); Henne (Bom.; Sp.; KAB); Henné (Fr.; Haiti; AVP; BOU; USN); Hina (Arab.; GHA); Hiná (Iran; DEP; NAD); Hinie (Mal.; NAD); Hinná (Arab.; Syria; DEP; GHA; HJP); Inai (Malaya; Sumatra; IHB; KAB); Inai Parasi (Sumatra; IHB); Iswan (Belgaum; KAB); Jamaica Mignonette (Eng.; KAB; USN); Kopfer (Heb.; ZOH); Korandam (Tam.; KAB); Korate (Kan.; KAB); Krapeu (Cam.; KAB); Krommi (Tel.; KAB); Kufer (Coptic; ZOH); Kufra (Accadian; Aramaic; ZOH); Kupf (Coptic; ZOH); Kuravaka (Sanskrit; DEP); Kuravamu (Tel.; NAD); Kuruvamum Goranti (Tel.; MPI); Kypros (Greek; KAB); Laincha (Newari; NPM); Lalle (Ber.; Hausa; BOU; KAB); Madarangi (Kan.; NAD); Madayanti (Ayu.; AH2); Madayantikaa (Sanskrit; MPI); Madirengi (Tulu; KAB); Mailanchi (Mal.; NAD); Manghati (Oriya; DEP); Marithondi (Tam.; DEP); Maritondi (Sin.; DEP); Maritondo (Sin.; KAB); Maruthani (Tam.; MPI; WOI); Mayilanchi (Mal.; DEP); Mayilangi (Mal.; KAB; MPI); Medi (Guj.; KAB); Médi (Guj.; DEP); Mehandi (Bhojpur; Danuwar; Mooshar; Nepal; Tharu; NPM); Mehendi (Raj.; Urdu; DEP; KAB); Mehndi (Hindi; Mar.; DEP); Mendí (Sin.; DEP); Mendika (Sanskrit; WOI); Methhi (Kon.; NAD); Meti (Kon.; KAB); Meritondi (Sin.; NAD); Mignonette (Eng.; St. Bart.; AVP; NPM); Mignonette Tree (Eng.; AVP); Mihndi (San.; DEP); Mindi (Mun.; KAB); Mindie (Eng.; KAB); Minyonèt (Creole; Haiti; VOD); Mohuz (Kas.; WOI); Monjuati (Oriya; KAB); Monz (Kas.; MPI); Muhanone (Swahili; KAB); Nakrize (Pun.; KAB); Pachar Kuku (Java; Sunda; IHB); Palphung (Limbu; NPM); Panwár (Dec.; Guj.; Mah.; Pun.; DEP); Parasi (Sumatra; IHB); Pontaletsche (Mal.; NAD); Pontlasi (Mal.; WOI); Puker (Egypt; ZOH); Raktagarba (Sanskrit; NAD); Reseda (Creole; Ger.; Haiti; AVP; VOD); Reséda (It.; AVP); Resedá (Cr.; Cuba; Dr.; Mex.; Por.; Pr.; Sp.; AVP); Réséda de Cayenne (Fr.; Guiana; KAB); Réséda de France (Guad.; Mart.; AVP); Réséda des Antilles (Fr.; AVP); Réséda du Pays (Fr.; AVP); Resedá Falso (Por.; AVP); Reseda Miñoneta (Sp.; AVP); Rhanni (Ber.; BOU); Shudi (Beng.; DEP); Simrú (Bhote; DEP); Syadu Riáhín (Arab.; BIB); Tanbalaksun (Rai; NPM); Tche Kia Hoa (China; KAB); Tien Děng (Thai; IHB); Tien Kāo (Thai; IHB); Tien King (Thai; IHB); Tihure (Magar; NPM); Tiure (Gurung; NPM); Tiuri (Tamang; NPM); Tree Mignonette (Eng.; KAB); Tyure (Nepal; Sunwar; NPM); Yoranná (Arab.; DEP); Zhi Jia Hua Ye (Pin.).

Activities (Henna):

Abortifacient (f; AHP; CRC; PH2); Allergenic (1; CRC; MPI; X10632222); Alterative (f; CRC); Analgesic (f1; X8966192); Anesthetic (f; GHA); Anthelmintic (f; BIB); Antiedemic (1; MPI); Antifertility (1; CRC; MPI); Antihyaluronidase (1; MPI); Antiinflammatory (f1; DEP; MPI; X8966192); Antileukemic (1; X7524165); Antioxidant (1; X15267142; X15813363); Antipapillomic (1; X12708740); Antiperspirant (f; CRC); Antiseptic (f1; BOU; X2125156; X15812867); Antispasmodic (1; CRC); Antitumor (1; X12708740); Antitussive (f; X15890471); Antiviral (1; X7524165); Astringent (f1; BOU; CRC; DEP; PHR; PH2); Bactericide (1; PHR; PH2; WOI; X2125156); Candidicide (1; MPI); Cardiodepressant (1; CRC); Cerebrotonic (f; CRC; KAB; WBB); Chemopreventive (1; X12708740); Collyrium (f; BIB); Deodorant (f; NAD); Diuretic (f1; IHB; PHR; PH2; KAB); Emetic (f; KAB); Emmenagogue (f; CRC; KAB); Emollient (f; KAB); Expectorant (f; KAB); Febrifuge (f1; KAB; X8966192); Fungicide (1; X620734); Gram(+)-icide (1; X11167035); Gram(−)-icide (1; X11167035); Hepatoprotective (1; X1620739; X15267142); Hypotensive (f1; CRC; HJP); Insecticide (f; BOU); Molluscicide (1; X15253044); Nematicide (1; MPI); Nitrate-Reductase Inhibitor (1; MPI); Peroxidase Inhibitor (1; MPI); Propecic (f; DEP); Protopectinase Inhibitor (1; MPI); Refrigerant (f; CRC; MPI); Reverse-Transcriptase Inhibitor (IC60 = 125 μg/ml) (1; X7524165); Sedative (f; CRC); Soporific (f; CRC); Sunscreen (f; CRC); Trypanocide (f; JEB79:279; JEB90:91); Trypsin Inhibitor (1; X12490230); Tuberculostatic (1; X2125156); Uterosedative (1; CRC); Vulnerary (f; BOU; KAB; VOD; X15890471).

Indications (Henna):

Abscess (f; HJP); Alopecia (f; DEP); Ameba (f; PHR; PH2); Amenorrhea (f; KAB; PH2); Anemia (f; PH2); Bacteria (1; CRC; PHR; X15812867); Baldness (f; KAB); Beriberi (f; CRC); Bleeding (f; MPI); Blenorrhea (f; BIB); Boil (f; WOI); Bronchosis (f; KAB; PH2); Bruise (f; DEP; WOI); Burn (f; BOU; DEP; KAB; NPM; WOI); Calculus (f; KAB); Cancer (f1; JLH; X12708740); Cancer, diaphragm (f; JLH); Cancer, joint (f; JLH); Cancer, liver (f; JLH); Cancer, mouth (f; JLH); Cancer, sinew (f; JLH); Cancer, spleen (f; JLH); Cancer, testicle (f; JLH); Cancer, uterus (f; JLH); Candida (f1; BIB; MPI; X15756356); Childbirth (f; CRC; IHB); Cholera (1; MPI); Circumcision (f; IHB); Condyloma (f; JLH); Cough (f; PH2); Dandruff (f; GHA; PHR; PH2); Delirium (f; DEP); Dermatosis (f; APA; CRC; GHA; MPI; NPM); Dropsy (f; HJP); Dysentery (f; PHR; PH2); Dysmenorrhea (f; PH2); Dyspepsia (f; CRC); Dysuria (f; PH2); Eczema (f; GAZ; PHR; PH2); Edema (1; MPI); Enterosis (f; GAZ; PHR; PH2); Epilepsy (f; VOD); Escherichia (1; WOI; X15756356); Fever (f1; GHA; PH2; X8966192); Fibroid (f; JLH); Foot (Hot Foot) (f; KAB); Fungus (1; PHR); Gastrosis (f; CRC; GAZ; PHR; PH2; VOD); Gingivosis (f; GHA; KAB); Glossosis (f; GHA); Gonorrhea (f; DEP; IHB; KAB); Gray Hair (f1; APA; NAD); Headache (f; DEP; GHA; KAB; MPI); Hemicrania (f; KAB; PH2); Hepatosis (1; X1620739); Herpes (f; CRC; IHB); High Blood Pressure (f1; BIB; CRC; HJP); Hoarseness (f; IHB); Hysteria (f; BIB); Induration (f; JLH); Infection (f1; CRC; PHR; PH2; X15812867); Inflammation (f1; DEP; GHA; JLH; MPI); Insanity (f; KAB); Itch (f; NPM); Jaundice (f; CRC; DEP; MPI); Leprosy (f; BOU; CRC; DEP; PH2; WBB); Leukemia (1; X7524165); Leukoderma (f; KAB); Leukorrhea (f; IHB; MPI); Lumbago (f; KAB); Malaria (f; PH2); Myalgia (f; MPI; SKJ); Mycosis (f1; GAZ; PHR; PH2; X620734; Neurosis (f; CRC); Numbness (f; GHA); Obesity (f; BIB; CRC); Onychyosis (f; IHB); Ophthalmia (f; BOU; IHB); Orchosis (f; JLH); Pain (f1; CRC; GHA; VOD; X8966192); Pediculosis (1; X12512805); Polyp (f; JLH); Pseudomonas (1; X15756356); Psychosis (f; PH2); Puerperium (f; IHB); Rheumatism (f; BOU; CRC; DEP; HJP; PH2; VOD); Sarcoma (1; CRC); Scabies (f; KAB; PHR; PH2); Scald (f; KAB); Scirrhus (f; JLH); Scurf (f; CRC); Shigella (1; X15756356); Smallpox (f; NAD); Sore (f1; APA; GHA; KAB; PH2; VOD); Sore Throat (f1; CRC; IHB; WOI); Spermatorrhea (f; DEP; KAB; MPI); Splenomegaly (f; CRC; DEP; KAB; MPI); Sprain (f; KAB; NAD); Staphylococcus (1; X15812867); Stomachache (f; VOD); Stomatosis (f1; APA; DEP; GHA; KAB); Streptococcus (1; CRC; MPI); Swelling (1; MPI); Syphilis (f; HJP); Tetanus (f; VOD); Toothache (f; NPM); Trypanosoma (f; JEB79:279; JEB90:91); Tuberculosis (1; X2125156); Typhoid (f; SKJ; MPI); Ulcer (f; GAZ; PHR); Uterosis (f; JLH); Vaginosis (f; NAD); Venereal Disease (f; CRC; DEP; KAB); Vibrio (1; MPI); Virus (1; X7524165); Whitlow (f; IHB; JLH); Wound (f; BOU; NPM; PH2); Yeast (1; MPI).

Dosages (Henna):

FNFF = ?
3 g powdered leaf, internally, for ameba and ulcers (HHB; PH2).

- Algerians poultice leaves for abscesses, bronchoses, and rheumatism, taking with juniper pitch for dropsy (HJP).
- Annamese use the leaves for herpes, jaundice, and leprosy (KAB).
- Arabs add leaves to food to lower fevers (GHA).
- Arabs chew leaves to treat inflammation and sores of gums, mouth, and tongue (GHA).
- Arabs mix fat and onions with powdered leaf paste, left on skin sores 5 days (GHA).
- Ayurvedics suggest seeds for bowel ailments, fever, and insanity, the leaves, considered emetic and expectorant, for leucoderma (KAB).
- Burmese apply the leaves to burning feet (DEP).
- Cambodians use diuretic, pectoral roots for bronchosis and gonorrhea (IHB; KAB).
- Haitians apply the leaf/flower tea topically to rheumatism sores, taking it orally for epilepsy, stomachache, and tetanus (VOD).

- Javans use the plant for leukorrhea (IHB).
- Konkanese mix the leaf juice with water for spermatorrhea (DEP).
- Lebanese suggest that the plant is febrifuge and hypotensive (HJP).
- Malayans use the root in abortifacient decoctions (KAB), the leaves in throat gargles (IHB).
- Malayans use the leaf decoction in childbirth, and for gastrosis and venereal disease. Mixed with the poisonous *Plumbago*, it is said to be abortifacient (BIB)
- Malayans apply the leaves to burning feet (IHB).
- Merjayouns use the leaf tea for fever and hypertension (BIB).
- Unani use the leaves for alopecia, amenorrhea, boils, bronchosis, headache, lumbago, ophthalmia, scabies, splenosis, stomatosis, and syphilis (KAB).
- Vytians (from Tamal, India) use leaves and tender shoots for leprosy (DEP).

Downsides (Henna):

Class 2a (AHP). Health risks or side effects following the proper administration of designated therapeutic dosages are not recorded (PH2). Tannin may cause stomach problems (PH2).

Extracts (Henna):

Extracts arrest, *in vitro*, Mycobacterium tuberculosis at 5 mg/ml (X2125156). Ethanol:water (1:1) extracts hepatoprotective with no effect on bile flow. MLD ≥ 2000 mg/kg orl mus (X1620739). LD50 (ethanolic extract) ≥ 1000 mg/kg ipr mus (MPI). Head lice were eradicated within a week treated by henna mixed with aqueous extract of sheah (100%) or mixed with helba (75%) or with karkada (50%) (X12512805). *In vitro* synergic interaction of crude extracts was detected with tetracycline (X15812867).

LENTIL (*LENS CULINARIS MEDIK*) +++ FABACEAE

Synonyms:

Cicer lens (L.) Willd.; *Ervum lens* L.; *Lens esculenta* Moench; *Lens lens* Huth.; *Lentilla lens* (L.) W. Wight ex D. Fairchild; *Vicia lens* (L.) Cosson & Germ.

Notes (Lentil):

Then Jacob gave Esau bread and pottage of lentiles; and he did eat and drink, and rose up, and went his way: thus Esau despised his birthright.

Genesis 25:34 (KJV)

Then Jacob gave Esau bread and pottage of lentils, and he ate and drank, and rose and went his way. Thus Esau despised his birthright.

Genesis 25:34 (RSV)

And Jacob gave Esau bread and lentil stew, and he went to eating and drinking. Then he got up and went his way. So Esau despised the birthright.

Genesis 25:34 (NWT)

Zohary points out that lentil was the first pulse mentioned in the Bible. I would like to add that it is the first one to be done when you are cooking a seven-pulse soup. Even dried lentils can be rendered edible with a half hour of simmering. Small wonder that lentils were important staples during biblical and postbiblical eras — maybe even earlier. Zohary reports carbonized lentil seeds dating to 6 or 7 millennia B.C. Since the Bronze Age, lentils have been found in association with barley and wheat (ZOH). My late friend, anthropologist Jane Philips, said that some "people think it was the food that made Daniel wise," or that it was the "mess of pottage for which Esau sold his birthright" (HJP). Lentils have been found in Syria, Iran, and Greece (with barley and wheat) prior to 5000 B.C. (Zohary, 1972). Takeoka et al. (2005), pegging domestication at circa 8000 B.C. in the Fertile Crescent, say they are rich in protein, dietary fiber, folate, iron, and phosphorus. Lentil flour today, as two millennia ago, can be added to "cereal flour" to make baby foods, breads, and cakes. In the United States, lentil production totaled circa 230 million pounds in 2002–2003, mostly in Washington, Idaho, Montana, and North Dakota (X15941338). Seeds are a source of commercial starch for the textile and printing industries. Green plants are used as green manure (BIB).

Common Names (Lentil):

Adah (Afg.; DEP; KAB); Adas (Afg.; Arab.; Iran; DEP; JNP; KAB; ZOH); Adasa (Arab.; NAD); Adashim (Heb.; ZOH); Aoi Mame. (Japan; POR); Arroufle (Fr.; KAB); Avangoule (Fr.; KAB); Buro Mussur (Beng.; DEP); Chanching (Pun.; DEP; POR); Channangi (Can.; NAD); Chiri Sanagalu (Tel.; DEP); Esse (Fr.; KAB); Gabholika (Sanskrit; KAB); Gram (Eng.; USN); Gurubija (Sanskrit; KAB); Halasaka (Sanskrit; KAB); Harashim (Heb.; KAB); Kacang Koro (Dei.; POR); Kacang Serinding (Malaya; POR); Kalyanabija (Sanskrit; KAB); Kerse (Pun.; DEP); Lânti (Khmer; POR); Lente (It.; POR); Lenteja (Peru; Sp.; EGG); Lenticchia (It.; POR); Lenticchie (It.; POR); Lentiho (Lan.; KAB); Lentil (Eng.; CR2; NAD); Lentilha (Mad.; Por.; POR); Lentilhas (Por.; POR); Lentille (Fr.; POR); Lentille Cultivée (Fr.; POR); Lentilles (Fr.; POR); Lins (Swe.; POR); Linse (Ger.; POR); Linsen (Ger.; POR); Linssi (Fin.; POR); Linte (Rou.; KAB); Linze (Dutch; POR); Linzen (Dutch; POR); Llantia (Cat.; KAB); Malka Masur (Pun.; NAD); Mangaliya (Sanskrit; KAB); Massur Chanangi (Kan.; DEP); Massur Moha (Assam; POR); Masur (Guj.; Hindi; Kon.; Mah.; Pun.; Sanskrit; Urdu; DEP; NAD; POR); Masura (Sanskrit; DEP; NAD); Masuraka (Sanskrit; KAB); Masuri (Beng.; Nwp.; NAD); Masuri Dal (Beng. Bom.; Guj.; DEP; NAD); Masurika (Sanskrit; KAB; NAD); Masurmoja (Assam; DEP; WOI); Mauri (Pun.; DEP); Mercimek (Tur.; POR; EB54:155); Miraju-Maka (Iran; NAD); Mirjumak (Iran; DEP); Missur Pappu (Tel.; NAD); Missur Purpur (Tam.; DEP; NAD); Mohi (Pun.; POR); Mohr (Pun.; DEP); Mohri (Pun.; DEP); Monhri (Pun.; DEP); Musri (Beng.; NAD); Musur (Nepal; POR); Nantille (Fr.; KAB); Nashik (Iran; KAB); Patani (Tag.; POR); Peni (Burma;); Pothundhambala (Sin.; POR); Prithubijaka (Sanskrit; KAB); Ragadali (Sanskrit; KAB); Shran Chung (Tibet; TIB); Shran Chung Leb Mo (Tibet; TIB); Sura (Sanskrit; KAB); Tambularaga (Sanskrit; KAB); Tchetchevitsa (Rus.; KAB); Thua Daeng (Thai; POR); Thua Raatcha Maat (Thai; POR); Tsentel (Lan.; KAB); Vrihikanchana (Sanskrit; KAB); Ye Bing Dou (China; POR).

Activities (Lentil):

Antioxidant (f; X14609082); Aperitif (f; KAB); Astringent (f; TIB); Diuretic (f; KAB); Hypocholesterolemic (1; X8672408); Laxative (f; KAB); Orexigenic (f; KAB); Refrigerant (f; TIB); Stimulant (f; NAD); Tonic (f; KAB).

Indications (Lentil):

Anemia (f; BIB; HJP); Anorexia (f; KAB); Aposteme (f; BIB); Backache (f; TIB); Biliousness (f; KAB); Bronchosis (f; KAB); Cancer (1; FNF; JLH); Cancer, abdomen (1; FNF; JLH); Cancer, anus (1; FNF; JLH); Cancer, ear (1; FNF; JLH); Cancer, eye (1; FNF; JLH); Cancer, gum (1; FNF; JLH); Cancer, parotid (f1; FNF; JLH); Cancer, throat (1; FNF; JLH); Cardiopathy (f; KAB); Childbirth (f; BIB); Condyloma (f; BIB); Constipation (f; BIB; DEP; LEG); Dermatosis (f; KAB; TIB); Dysentery (f; BIB); Enterosis (f; BIB; DEP; LEG); Fungus (f; TIB); Gingivosis (f; JLH); High Cholesterol (1; X8672408); Induration (f; BIB); Infection (f; TIB); Inflammation (f; BIB); Mastosis (f; KAB); Ophthalmia (f; KAB); Pain (f; TIB); Parotosis (f; JLH); Proctosis (f; JLH); Pulmonosis (f; KAB); Ringworm (f; TIB); Scirrhus (f; BIB); Scurvy (f; BIB); Smallpox (f; BIB; DEP; LEG); Snakebite (f; NAD); Sore (f; BIB; DEP; LEG); Stomatosis (f; KAB); Strangury (f; KAB); Ulcer (f; BIB); Urogenitosis (f; BIB); Wart (f; BIB).

Dosages (Lentil):

FNFF = !!!
Lentil is cultivated for its nutritious seed, considered one of the most nutritious of pulses. Split seeds used in soups; flour is used mixed with cereals, in cakes, and as a food for invalids and infants. In biblical times, as today, breads were made from lentils and barley. Along parts of the Nile, it is the only breadstuff. Ground seeds enter Ethiopian spice *mit'in shiro*; young pods cooked like haricot beans; sprouts and young pods eaten. Maybe even Esau's biblical red pottage was an early health food. Lentils are widely used as a meat substitute. That is a heart-friendly idea right there. Husks, dried leaves, stems, and bran are fed to livestock (BIB; FAC, TAN; EB54:155).

- Ayurvedics use seed for biliousness, dysentery, eye ailments, heart ailments, skin diseases, strangury, and tumors (KAB),
- Ethiopians use seeds for dysentery (BIB).
- Germans use lentil soup to facilitate eruption in smallpox, and as a poultice on ulcers after smallpox (BIB).
- Lebanese use lentils for anemia following dysentery; they poultice hot lentil soup, with or without onions, on sores (BIB).
- Iraqi use ground lentils to ease delivery (BIB). Seeds, often in cataplasms, are used for apostemes, cancers, condylomata, indurations, scirrhus, tumors, and warts (of the abdomen, anus, breast, ear, eye, face, fauces, feet, genitals, gums, intestines, parotids, rectum, etc.).
- Unani use seed for breast inflammation, blood ailment, bronchitis, chest ailments, eye disorders, and stomatitis (KAB)

WHITE LILY (*LILIUM CANDIDUM* L.) + LILIACEAE

Notes (White Lily):

… to feed in the gardens, and to gather <u>lillies</u>.

Song of Solomon 6:2 (KJV)

Zohary astutely observes that lily flowers show up in many ancient edifices, in the columns of Assyria, Egypt, the Minaons, and in Solomon's Temple in Jerusalem. Long a symbol of beauty, fertility, and fruitfulness, it became a church symbol of holiness, resurrection, and spiritual purity. Many paintings of Mary depict her with the "Madonna Lily." The lily has been found, doubtfully wild, all over the Mediterranean basin from France to the hills of Syria, suggestive of the old Roman Empire. Because few of these lily colonies set seed, it has been suggested that they were originally planted, perhaps, near Roman outposts as "Materia medica." Zohary also hints strongly that the crocus (*havatzeleth*) of Isaiah 35:1 and the rose in Song of Solomon 2:1 should both have been translated as "lily." In perfumery, the absolute of lily is used in high-grade compositions of floral and perfumes. It is an excellent fixative. MAD does not cover *L. candidum*, instead featuring tiger lily and, under that, the martagon. PH2 covers *Lilium candidum* and *L. martagon* but excludes *Lilium tigrinum* Ker Gawl, now better known as *Lilium lancifolium* Thunberg. The USDA and WOI view

FIGURE 1.61 White Lily (*Lilium candidum*).

all three as different, so I also will include all three in my herbal desk reference. EFS and HHB aggregated data for tiger lily and martagon. I treat them as separate.

Common Names (White Lily):

Annunciation Lily (Eng.; UPH); Azucena (Spain; VAD); Azucena Común (Sp.; EFS); Beyaz Zambak (Tur.; EFS); Bourbon Lily (Eng.; UPH); Brown's Lily (Eng.; TAN); Giglio Bianco (It.; EFS); Giglio della Madonna (It.; EFS); Hakata Yuri (Jap.; TAN); Havatzeleth (Heb.; ZOH); Hongkong Lily (Eng.; TAN); Lirio Blanco (Peru; Sp.; EGG); Lirio Común (Sp.; EFS); Lirio de San Antonio (Peru; Sp.; EGG); Lis Blanc (Fr.; EFS); Lis Candide (Fr.; EFS); Madonna Lily (Eng.; CR2; EFS; HJP; USN); Mar Yusef Zambak (Arab.; Syria; HJP); Marien Lilie (Ger.; EFS); Mary Lily (Eng.; EFS); Meadow Lily (Eng.; EFS); Pai Ho (China; TAN); Raziqi (Arab.; Syria; HJP); Sawsan Abyad (Arab.; Syria; HJP); Showshan (Heb.; ZOH); Shuwshan (Heb.; ZOH); Weisse Lilie (Ger.; EFS); White Lily (Eng.; CR2; EFS; USN); Witte Lelie (Dutch; EFS); Zanbaq (Arab.; Syria; HJP); Nscn.

Activities (White Lily):

Analgesic (f1; PH2; VAD); Antiinflammatory (f1; PH2; VAD); Antispasmodic (f; EFS; WOI); Antitumor (1; X11130250); Astringent (f1; BIB; FEL; HJP; PH2; WOI); Candidicide (1; X12501491); Cicatrizant (f1; VAD); Demulcent (f; BIB; FEL; HJP; WOI); Diuretic (1; BIB; PH2); Emollient (f; BIB; EGG); Expectorant (f1; BIB; PH2); Fungicide (1; HHB; 12501491); Phagocytotic (1; PNC); Stimulant (f; BIB); Sudorific (f; BIB); Tonic (f; EGG; FEL); Vulnerary (f1; HG57:40).

Indications (White Lily):

Abscess (f; VAD); Arthrosis (f; VAD); Bite (f; HHB); Bleeding (f; AAH); Blister (f; HG57:40); Boil (f; EGG; VAD); Bruise (f; EGG); Burn (f; HHB; PH2); Callus (f; BIB; VAD); Cancer (f; BIB; FEL; JLH); Cancer, breast (f; BIB; JLH); Cancer, colon (f; JLH); Cancer, ear (f; BIB; JLH); Cancer, intestine (f; JLH); Cancer, liver (f; BIB; JLH); Cancer, lung (f; BIB; JLH); Cancer, neck (f; BIB; JLH); Cancer, rib (f; BIB; JLH); Cancer, scrotum (f; BIB; JLH); Cancer, spleen (f; BIB; JLH); Cancer, stomach (f; JLH); Cancer, testicle (f; BIB; JLH); Cancer, throat (f; BIB; JLH); Cancer, uterus (f; BIB; JLH); Candidiasis (1; X12501491); Corn (f; BIB; HJP); Cramp (f; WOI); Dermatosis (f1; BIB; WOI; HG57:40); Dropsy (f; BIB; PNC; WOI); Dysmenorrhea (f; MNC); Earache (f; EGG); Eczema (f; VAD); Epilepsy (f; HJP; WOI); Erysipelas (f; EGG); Fibroid (f; JLH); Fungus (1; X12501491); Furuncle (f; PH2); Gynecosis (f; PH2); Hepatosis (f; JLH); Induration (f; JLH); Infection (1; HHB); Inflammation (f; BIB; FEL; PH2; PNC); Leukorrhea (f; FEL); Lumbago (f; HHB); Mastosis (f; VAD); Mycosis (1; HHB); Nausea (f; BIB); Orchosis (f; JLH); Otosis (f; FEL); Pain (f1; FEL; HG57:40); Polyp (f; JLH); Pregnancy (f; EGG); Prolapse (f; FEL); Skin (f; BIB); Sore (f1; BIB; FEL; HHB; PH2; PNC; HG57:40); Spasm (f; BIB); Splenosis (f; JLH); Swelling (f; HHB); Toothache (f; HHB); Tumor (f; BIB); Uterosis (f; FEL; JLH); Wart (f; JLH); Water Retention (f; HHB); Whitlow (f; VAD); Wound (f1; PH2; HG57:40; VAD); Yeast (1; X12501491).

Dosages (White Lily):

FNFF = ?

Tanaka hints that, like *L. brownii*, *Lilium candidum* Lour (non L.) is grown as a food plant (roots possibly edible) or ornamental. All TAN entries apply only to *Lilium brownii*. PH2 says to apply a thick paste, made from fresh or cooked onions = lily bulbs, I presume.

- Europeans staunch bleeding with wine or whiskey petal macerate (AAH).
- Europeans suggest the pollen for epilepsy (UPH).
- Lebanese use the cooked roots for everything from corns to epilepsy (HJP).

- Peruvians apply rum tincture of mashed petals as a poultice on bruises and contusions (EGG).
- Peruvians make a pomade of mashed petals, flax, matico leaves, and milk for erysipelas (EGG).
- Peruvians suggest boiled root as a tonic (EGG).
- Peruvians suggest mashed flowers in a drop of almond oil for earache and pains of pregnancy (EGG).
- Spaniards suggest for fissures in the nipples (VAD).
- Early Tibetan manuscripts recommended lily bulb baths to treat sores and wounds. Bulb infusions and decoctions promote healing in experimental rats. The tea eases pains, removes rashes and blisters, and promotes epithelization (Alakbarov, 2003).

Downsides (White Lily):

None reported at proper dosage, but no specific dosage indicated (PH2).

Natural History (White Lily):

The white flowers, smelling stronger at night than by day, attract pollinating hawk moths (ZOH).

Extracts (White Lily):

Antimycotic activity possibly due to gamma-methylenebutyrolactone (HHB).

FLAX (*LINUM USITATISSIMUM* L.) +++ LINACEAE

Notes (Flax):

And he took it down, and wrapped it in linen, and laid it in a sepulchre that was hewn in stone, wherein never man before was laid.

Luke 23:53 (KJV)

Then he took it down and wrapped it in a linen shroud, and laid him in a rock-hewn tomb, where no one had ever yet been laid.

Luke 23:53 (RSV)

And he took it down, and wrapped it up in fine linen, and he laid him in a tomb carved in the rock, in which no man had yet lain.

Luke 23:53 (NWT)

Linen is mentioned many, many times in the Bible, in the KJV in at least 23 books, as for example, Exodus, where the word *linen* shows up 36 times. Conversely, flax is mentioned only eleven times, in eight works, including two citations in Exodus. Ancient Egyptians depicted the growth of flax on papyri, the spinning of flax thread in murals, and the weaving of that thread into linen. Remains of the Pharaohs are bound in fine and delicate linen, woven with an expertise that is still today difficult to repeat 3000 to 4000 years later. The Gezer Calendar found at the ancient site, and dating back to the beginning of the Israelite civilization (circa 10th century B.C.), referred to flax cultivation. Zohary states that cultivation dates back to 5000 B.C. in the Middle East, including Israel, where flax probably originated and where the wild progenitor, *Linum bienne*, is still indigenous.

FIGURE 1.62 Flax (*Linum usitatissimum*).

Not only was the linen used in mummy cases, but the oil was used in embalming. The flax plant is commonly used to make linen, the most ancient of all textile fibers. Linen is the most important product made from the fiber of the flax plant. Flax fibers are soft, lustrous, and flexible, although not so flexible or elastic as those of cotton or wool. Seeds contain 24% protein, and also are the source of linseed oil. In some countries, it is also used as edible oil and in soap manufacture (BIB; ZOH). Because linen has a long anticancer folk history, it is interesting to see that *Linum* contains the anticancer agents 3′-demethylpodophyllotoxin, podophyllotoxin, and beta-sitosterol. L-glutamic acid is used to treat adolescent mental deficiencies. Seeds are considered emollient, demulcent, pectoral, diuretic, and astringent. Crushed seeds make a good poultice (for colds, pleurisy, etc.), either alone or with mustard; lobelia, or hollyhock seed, is added to the poultice for boils. Sometimes, seeds are roasted and used in a poultice. Hot seeds are applied to abscesses and rheumatism. Sometimes employed as an addition to cough medicines. Linseed oil mixed with an equal quantity of lime water, known as Carron oil, is an excellent application for burns and scalds. Oil mixed with honey is used as a cosmetic for removing spots from the face. In veterinary medicine, oil is used as a purgative for sheep and horses; and a jelly, formed by boiling seeds, is used as a purgative for sheep and horses, and is often given to calves (BIB).

Common Names (Flax):

AAkawa Yob (Rai; NPM); Aalas (Bhojpuri; Chepang; Danuwar; Magar; NPM); Agashi (Kan.; NAD); Agasi (Kan.; WOI); Alas (Nepal; SUW); Alashi (Bom.; Kon.; Ger.; KAB; NAD); Alasi (Urdu.; KAB); Ali (Madras; KAB); Alish (Kas.; Pun.; KAB; NAD); Alivari (Tam.; WOI); Alshi (Guj.; Tam.; KAB; NAD); Alshi Virai (Tam.; NAD); AlshiViral (Tel.; NAD); Alsi (Dec.; Guj.; Hindi; Kum.; Loralai; Nepal; KAB; NPM; WOI); Ana (Japan; TAN); Ankejil Guib (Sunwar; NPM); Thange (Limbu; NPM); Arasi (Tharu; NPM); Atasi (Sanskrit; Tel.; KAB; WOI); Auma (Sanskrit; KAB); Avisi (Tel.; WOI); Bad (Heb.; KAB); Bazarug (Iran; KAB); Bazarul Kattana (Arab.; EFS; KAB; NAD); Bazrukattan (Arab.; KAB); Bijri (Nwp.; DEP; KAB); Bizr el Oinab (Arab.; Syria; HJP); Blaebas (Eng.; KAB); Budrapatni (Sanskrit; KAB); Chanaka (Sanskrit; KAB); Cheruchanavittint-evilta (Kerala; Mal.; DEP; KAB; SKJ); Chikna (Behar; DEP; KAB); Delkmouch (Ber.; BOU); Devi (Sanskrit; KAB); Echter Lein (Ger.; EFS); Flachs (Ger.; KAB); Flax (Eng.; CR2; NAD); Flix (Eng.; KAB); Fluchs (Ger.; DEP); Gemeiner Flachs (Ger.; NAD); Gemeiner Lein (Ger.; NAD); Haarlinsen (Ger.; KAB); Hab e Kattan (Arab.; DEP; GHA); Haimwati (Sanskrit; KAB); Hoerfroe (Den.; KAB); Hoerre (Den.; KAB); Ho Ma (Annam; KAB); Hor (Den.; KAB); Horfro (Den.; EFS); Hou Ma Tse (China; KAB); Hu Ma Tzu (China; EFS); Icags Tig Dkar Po (Tibet; TIB); In (Rom.; KAB); Javas (Mar.; WOI); Javas (Bom.; Dec.; DEP); Javasa (Bom.; Marathri; DEP; KAB); Kattan (Arab.; KAB; HJP); Kettan (Arab.; BOU); Keten (Tur..; EFS; KAB); Kettan (Arab.; ZOH); Keun (Ger.; Kas.; KAB); Kittan (Arab.; Phoenician; Syria; HJP; ZOH); Kitu (Accadian; ZOH); Kshaumi (Sanskrit; KAB); Kshuma (Sanskrit; KAB); Kuman (Iran; KAB); Kutan (Iran; DEP); Lein (Ger.; KAB); Len (Hun.; Pol.; Rus.; EFS; KAB); Lijnzaad (Dutch; EFS); Lin (Celtic; Fr.; Swe.; BOU; DEP; EFS); Linaza (Sp.; EFS); Lin Chaud (Fr.; KAB); Lin Comun (Fr.; KAB); Lin Cultive (Fr.; KB; NAD); Linhaca (Por.; KAB); Linho (Por.; DEP; KAB); Lino (It.; Sp.; DEP; KAB); Linon (Greek; KAB); Linseed (Eng.; KAB; NAD); Lint Bells (Eng.; KAB); Lint Bennels (Eng.; KAB); Lint Bow (Eng.; KAB); Lli (Cat.; KAB); Lyne (Eng.; KAB); Lyon (Rus.; KAB); Madagandha (Sanskrit; KAB); Madan Ginjalu (Tel.; DEP; KAB); Madotkata (Sanskrit; KAB); Malina (Sanskrit; KAB); Masina (Beng.; Sanskrit; KAB); Masini (Beng.; NAD); Masrina (Sanskrit; KAB); Masruna (Sanskrit; KAB); Masuna (Sanskrit; KAB); Nilapushpi (Sanskrit; KAB); Nilpushpika (Sanskrit; KAB); Parvathi (Sanskrit; KAB); Pesheth (Heb.; ZOH); Pesu (Uriya.; KAB); Pichhila (Sanskrit; KAB); Pishtah (Heb.; ZOH); Roghani Zaghira (Iran; NAD); Rongonimbazaha (Kum.; KAB); Saatlein (Ger.; KAB); Saghir (Iran; DEP); Sagrek (Tur.; EB49:406); San (Sanskrit; KAB); Sib Muma (Arab.; Yemen; GHA); Sonnbiam (Kon.; KAB); Sunila (Sanskrit; KAB); Suverchala (Sanskrit; KAB); Tailottama (Sanskrit; KAB); Thange (Limbu; NPM); Tifert (Ber.; BOU); Tisi (Beng.; Hindi; Kum.; Mooshar; Newari; KAB;

NPM; WOI); Tisinam (Lepcha; NPM); Tukhhmehkatan (Iran; KAB); Tukhmizaghira (Iran; NAD); Tukeme Katan (Iran; DEP); Ulla Sulu (Tel.; DEP); Uma (Sanskrit; EFS; NAD); Vlas (Dutch; DEP; EFS; KAB); Vlix (Eng.; KAB); Yabani Keten (Tur.; EB54:155); Ya Ma Zi (Pin.; DAA); Zaghir (Iran; KAB); Zaghu (Iran; EFS; KAB; NAD); Zar Ma (Tibet; TIB); Zegrek (Tur.; EB54:155); Zeyrek (Tur.; EB54:155); Ziggar (Tur; KAB); Zighir (Kashghar; DEP; KAB);.

Activities (Flax):

Allergenic (1; APA; X1257320); Analgesic (f; BOU); Anodyne (f; MAD; NAD); Antiaggregant (1; APA; PH2); Antiallergenic (1; WAM; X12418783); Antiatherogenic (12; APA; BGB; X15777541); Antidiaphoretic (f; NAD); Antiestrogenic (1; APA; PH2); Antiinflammatory (1; APA; BGB; WAM; X12418783); Antioxidant (1; PH2; X15853420); Antiseptic (1; VAD); Antitumor (1; APA; BGB; PH2); Antitussive (f; PNC); Aphrodisiac (f; DEP; MAD); Astringent (f; BIB; KAB); Cardiotonic (f; DEP); Cerebrotonic (f; KAB); Cyanogenic (1; BIB); Demulcent (f12; APA; BGB; FEL; GHA; WAM); Deobstruent (f; NAD); Digestive (1; WAM); Diuretic (f; BIB; NAD); Emetic (f; FAD); Emmenagogue (f; BIB); Emollient (1; APA; BGB; FEL; PNC); Estrogenic (1; APA); Expectorant (f; BIB; NAD); Fungicide (f; PH2); Hypocholesterolemic (12; PH2; X15777541); Hypoglycemic (1; PH2; JAC7:405); Hypolipemic (1; VAD); Lactagogue (f; BIB); Laxative (f12; BGB; PH2; SHT; WAM); Lipolytic (2; APA; BGB; PHR); Nervine (f; SUW); Pectoral (f; PNC); Peristaltic (2; PIP); Purgative (2; FAD); Refrigerant (f; TIB); Suppurative (f; BIB); Vulnerary (f; BIB).

Indications (Flax):

Abscess (f; BGB; EB49:406); Acid Stomach (1; WAM); Acne (1; WAM); Adenopathy (f; JLH); Allergy (1; X12418783); Ameba (f; BIB); Aposteme (f; JLH); Arthrosis (f1; APA; BGB; GHA); Ascarides (f; FEL); Atherosclerosis (1; X15777541); Backache (f; BIB); Biliousness (f; BIB; KAB); Bleeding (f; KAB; MAD); Boil (f; BIB; BOU; FAD; NPM); BPH (f; MAD); Bronchosis (f; MAD; PH2; PNC); Bruise (f1; VAD); Burn (f; DEP; FAD); Cancer (1; FAD); Cancer, abdomen (1; FNF; JLH); Cancer, anus (1; FNF; JLH); Cancer, breast (1; APA; FNF; JLH); Cancer, cervix (1; FNF; JLH); Cancer, colon (1; APA; FNF; JLH); Cancer, gland (1; FNF; JLH); Cancer, liver (1; FNF; JLH); Cancer, mouth (1; FNF; JLH); Cancer, neck (1; FNF; JLH); Cancer, parotid (1; FNF; JLH); Cancer, skin (1; FNF; JLH); Cancer, spleen (1; FNF; JLH); Cancer, stomach (1; FNF; JLH); Cancer, testis (1; FNF; JLH); Cancer, throat (1; FNF; JLH); Cancer, uvula (1; FNF; JLH); Cancer, uterus (1; FNF; JLH); Carbuncle (f; NAD); Cardiopathy (f; DEP); Catarrh (f; PHR; PH2); Cellulitis (f; NAD); Cervicosis (f; JLH); Childbirth (f; BIB); Cholecystosis (f; PHR); Cold (f; FAD; FEL); Colic (f; FEL; HHB); Colitis (2; APA; FEL; KOM; SHT); Condyloma (f; JLH); Conjunctivosis (f; BIB; NAD); Constipation (f12; APA; BGB; GHA; PH2; WAM); Corn (f; JLH); Cough (f; APA; FAD; FEL; MAD; PH2); Cramp (f; BIB); Cystosis (f1; HJP; MAD; PH2; VAD); Dermatosis (12; APA; BGB; PH2; WAM; X12418783); Diabetes (1; MAD; PH2; X14995053); Diarrhea (f; BOU; PH2); Diverticulitis (2; APA; PH2; SHT); Dysentery (f; FEL); Dysuria (f; FEL); Eczema (1; WAM; VAD); Enterosis (f2; BOU; KOM; PIP; PH2); Fever (f; FAD); Furuncle (f; MAD; VAD); Gallstone (f; HHB; MAD); Gastrosis (f12; APA; BOU; KOM; PH2; PIP; VAD); Glossosis (f; MAD); Gonorrhea (f; DEP; GHA; MAD; PH2); Gout (f; FAD; MAD); Hayfever (f; HHB); Head cold (f; KAB); Hemorrhoid (1; FEL); Hepatosis (f; HJP; JLH); High Cholesterol (12; APA; PH2; X15777541); Hyperactivity (1; WAM); Hyerglycemia (1; X14995053); Hysteria (f; NAD); IBD (f; PH2); IBS (f; PH2); Impetigo (f; MAD); Induration (f; JLH); Infection (1; PH2; VAD); Inflammation (f12; APA; BOU; FAD; FEL; PH2; WAM; X14995053); Itch (f; HHB); Kidney stone (f; MAD); Leprosy (f; BIB); Lumbricus (f; FEL); Lupus (2; APA; BGB); Lymphangitis (f; MAD); Malaria (f; APA; BGB); Measles (f; MAD); Mucososis (1; APA); Mycosis (1; PH2); Nausea (f; MAD); Nephrosis (f12; APA; BGB; FEL; GHA; HHB; HJP); Obesity (1; VAD); Ophthalmia (f; PH2); Orchosis (f; JLH); Otitis (1; WAM); Pain (f; BOU; FEL; GHA; NAD); Parotitis (f; JLH); Pericardosis (f; NAD); Peritonosis (f; NAD); Pleurisy (f; BIB; FAD; HJP); Pneumonia (f; FAD); Proteinuria (1; APA); Proctosis (f; FEL; JLH); Puerperium (f; NAD); Pulmonosis (f; FAD; HHB); Pyelitis (f; HHB); Quinsy (f; NAD); Rheumatism (1; BGB; FAD; MAD); Sapraemia (f; NAD); SLE (1; BGB); Sore (f; NAD); Sore Throat (f; APA); Splenosis (f; JLH); Stone (f; HHB; HJP); Strangury (f; NAD); Stroke (1; X15777541; Swelling (f; JLH; MAD); Sycosis (f; JLH); Syndrome-X (1; X14995053); Urethrosis (f; BOU; PH2); Urogenitosis (f; DEP; KAB); UTI (f; APA; NAD); Vaginosis (f; NAD); Venereal Disease (f; DEP; PH2); Wart (f; JLH); Whitlow (f; JLH); Wound (f; GHA; NPM).

Dosages (Flax):

FNFF = !!!

Seeds widely eaten, even substituted for eggs in baking; The oil is used in Egyptian cuisine (e.g., ful medames, a stewed (faba) bean dish served with garlic, onions, and cumins). Talk about food "farmacy," I would recommend that as food farmacy (replacing a fatty meat dish), for all the major "civilized" diseases of civilization; cancer, coronary, diabetes, even impotence and Parkinson's disease! Ethiopians substitute for butter, roasting the seeds and pasting them on breads, and make a health food mixing seeds with honey and water. Parched seeds serve as coffee substitute. Sprouted seeds added to salads. Unripe seed pods cooked in chutneys. Transylvanians eat cooked leaves with lemon and sour cream. Linseed is often employed with other seeds as food for small birds (BIB; FAC; TAN; EB54:155; Cunnane and Thompson, 1995; Judd, 1995). One 1300 mg StX for 740 mg ALA softgel (APA); 1 Tbsp whole or crushed (not ground) seed/150 ml liquid 2–3 ×/day (APA; PIP). 1 oz seed/1 qt water at boiling point (FEL). 1 Tbsp bruised or whole (not powdered) seed, with at least 150 ml water 2–3 ×/day (PH2). For hypercholesterolemia, 35–50 g crushed seed/day (PH2). As antiaggregant, 1–2 Tbsp flaxseed oil/day (PH2).

- Arabians use ground seed with butter or decocted in water for arthritic pain and wounds (GHA).
- Arabians use seed decoction with cumin and sugarcane for kidney pain (GHA).
- Arabians use seed infusion for digestive and urinary tract inflammation (GHA).
- Arabians use 1 to 2 tsp seed with water for constipation or gonorrhea (GHA).
- Ayurvedics use leaves for asthma and cough; and seeds for backache, biliousness, consumption, inflammation, leprosy, ulcers, and urinary discharges (KAB).
- Ethiopians suggest the seeds for amebic dysentery (BIB).
- Lebanese apply hot flax seed for abscess, pain, and rheumatism; the oil for burns, constipation, and fever; linseed tea, with lemon, for cystitis, gallstones, gravel, hepatitis, and kidney stones (HJP).
- Turks pound seeds in milk and apply to abscess (EB49:406).
- Unani use the seeds, considered aphrodisiac, diuretic, emmenagogue, and lactagogue, for cough and kidney ailments; its oil for "bad blood," internal wounds, and ringworm; and the burnt bark for bleeding and wounds. They use the bark and leaves for gonorrhea (KAB).

Downsides (Flax):

Class 2d. Contraindicated in bowel obstruction (AHP, 1997). No health hazards or side effects known with proper therapeutic dosages (PH2); (PH2 designates no dosage!, JAD). Commission E reports contraindications: ileus; interactions; reduced absorption of other drugs possible. No adverse effects when used with a sufficient amount of liquid. Other sources report contraindications: intestinal obstruction; patients with inflammatory intestinal diseases should only use the seed in swollen state. Abuse of high doses may result in electrolyte losses (AEH). Ayurvedics hint that seed may harm eyes and virility (KAB). Under pessimal conditions, 100 g linseed can liberate up to 50 mg HCN, enough to bring about symptoms of poisoning. But these conditions rarely, if ever, maintain.

Most HCN is metabolized via the enzyme rhodanase, which can convert 30 to 60 mg HCN per hour into the relatively toxic thiocyanate. Even doses of 300 g ground linseed evoked no symptoms of poisoning in volunteers. Single doses of 100 g do not cause a significant rise in blood hydrocyanic acid and thiocyanate levels. Taking 15 g linseed thrice daily for 3 to 4 weeks raised thiocyanate levels in blood and urine (BIS). May adversely affect absorption of drugs (as with any mucilaginous preparation). Bulking agents should not be taken by those with stenotic lesions of the GI tract. May lead to bowel obstruction if fluid intake is inadequate. Do not take bulking agents when lying down or at bedtime. Do not use with antiperistaltics (such as e.g., loperamide) (SHT). Do not use seed in diverticulitis patients (WAM). Do not confuse flaxseed oil with more poisonous denatured linseed oil (WAM).

Natural History (Flax):

When grown for seed, flax is self-pollinating. Cross-pollination does not seem to increase seed yield consistently. Advisable to isolate different varieties by at least 100 to 300 m when producing seed for propagation. Many fungi have been found on flax, but the most serious diseases are flax wilt, several rusts, seedling blights, and Pasmo. Causative agents are *Alternaria brassiceae, A. lini, A. linicola, A. solani, A. tenuis*; *Ascochyta linicola, Botrytis cinerea, Colletotrichum lini, C. linicola, Corticium solani, Diplodia lini, Erysiphe cichoracearum, E. polygoni, E. polyphaga, Fusarium acuminatum, F. avenaceum, F. lini, Kabatiella lini, Leveillula linacearum, Melampsora lini, Mycosphaerella linorum, M. tulasnei, Oidiopsis lini, Oidium lini, Phoma lini, P. linicola, Polyspora lini, Pythium* spp., *Rhizoctonia solani, Sclerotinia fuckeliana, S. libertiana, S. minor, S. sclerotiorum, Sclerotium rolfsii, Septoria linicola, Sphaerella linorum, Thielaviopsis basicola*, and *Trichothecium roseum*. Important viruses causing disease in flax are Aster yellows (*Chlorogenus callistephi*), Beet curly top, and Yellows. Flax may be parasitized by *Cuscuta epilinum, C. epithymum, C. indecora, C. pentagona*, and *Striga lutea*. The bacteria *Agrobacterium tumefaciens* and *Pseudomonas atrofaciens* also cause diseases. Insects are not a serious problem in flax production. Nematodes isolated from flax include the following species: *Ditylenchus dipsaci*, Helicotylenchus erythrinae, H. spp., *Heterodera schachtii, Meloidogyne arenaria, M. hapla, M. incognita, M. incognita acrita, M. thamesi, M. javanica, Paratylenchus* sp., *Pratylenchus coffeae, P. penetrans, Tylenochorhynchus* spp., and *Xiphinema* spp. (HOE).

Extracts (Flax):

Seeds, like those of Perilla, are one of the best sources of the omega-3 fatty acid called ALA (alpha-linolenic acid), which reportedly has antiaggregant, antiinflammatory (IC50 = 42 μM), antileukotriene-D4, antimenorrhagic, antimetastatic, antiprostatitic, cancer-preventive, hypotensive, immunostimulant (0.12 ml/man/day) activities; lignans antiestrogenic, antimycotic, antioxidant, and antitumor (PH2). See Rudin's *The Omega-3 Phenomenon* for an incredible list of supposed indications for flaxseed oil, an entire book full. Brooks et al. (2005) showed that flaxseed supplementation alters estrogen metabolism in postmenopausal women more than soy supplementation [soy (25 g soy flour), or flaxseed (25 g ground flaxseed) muffin for 16 wk]. Urinary concentrations of 2-hydroxyestrone (little biological activity), but not of 16-alpha-hydroxyestrone (estrogenic agonistic causing proliferation), increased significantly in the flaxseed group. Improving this ratio (as also happens in women consuming Brassicaceae) improves the odds of patients against breast cancer (HG67:34) (X14749240). All this suggests consumption of more biblical beans, crucifers, and flaxseed, in lieu of animal fat, improves one's odds against certain hormone-dependent cancers (JAD). In a GMO approach, Lorenc-Kukula et al. (2005) generated flax plants with increased antioxidant properties. The simultaneous expression of genes encoding chalcone synthase, chalcone isomerase (CHI), and dihydroflavonol reductase (DFR) resulted in a significant increase in flax antioxidant capacity. The increase in yield of transgenic plants and their higher resistance to *Fusarium culmorum* and *Fusarium oxysporum* correlated with a significant increase in monounsaturated fatty acids and a slight increase in lignans (X15853420). Prasad (2005) demonstrated the hypocholesterolemic and antiatherosclerotic effect of the lignan complex containing secoisolariciresinol diglucoside antioxidant, 3-hydroxy-3-methylglutaric acid (hypocholesterolemic) and cinnamic acids (antioxidants) (X15777541). Bloedon and Szapary (2004) state that flaxseed, rightly or wrongly, is the richest known source of both alpha-linolenic acid (ALA) and the phytoestrogeric lignans, as well as a good source of soluble fiber. Human studies have shown that flaxseed can modestly reduce serum total and low-density lipoprotein cholesterol concentrations, reduce postprandial glucose absorption, decrease some markers of inflammation, and raise serum levels of omega-3 fatty acids, viz. ALA and eicosapentaenoic acid. Data on the antiplatelet, antioxidant, and hypotensive effects of flaxseed, however, are inconclusive (X14995053). Phillips et al. (2005) presented phytosterol data on flaxseed, averaging the ground and unground accessions values. Such phytosterols are medicinally important, for example, in BPH and in high cholesterol (X16302759).

STORAX (*LIQUIDAMBAR ORIENTALIS* MILL. AND *L. STYRACIFLUA* L.) + + HAMAMELIDACEAE

NOTES (STORAX):

Is there no balm in Gilead; is there no physician there? why then is not the health of the daughter of my people recovered?

Jeremiah 8:22 (KJV)

Is there no balm in Gilead? Is there no physician there? Why then has the health of the daughter of my people not been restored?

Jeremiah 8:22 (RSV)

Is there no balsam in Gilead? Or is there no healer there? Why is it, then, that the recuperation of the daughter of my people has not not come up?

Jeremiah 8:22 (NWT)

Like many responsible authors (e.g., HHB; LAF), when covering species of *Liquidambar*, I aggregated American Storax and Oriental Storax for the second edition of my *Handbook of Medicinal Herbs*. Distinguishing liquidambar and styrax resins seems rather difficult. Hence I believed that reporting separately would be rather artificial. Few people alive can guarantee the taxonomic source of the resin itself. And I suspect there are few if any chemical analyses based on vouchered specimens, and many more based on aggregates of miscellaneous resins. Nonetheless, AH2 maintained them apart, *Liquidambar orientalis* representing the oriental species, the resin of which is commonly known as Levant Storax. This clearly would be the biblical storax as well. AH2 maintains as distinct the American storax or sweet gum, *Liquidambar styraciflua* L. Even Zohary in Israel states that the two species are the same. This oriental species seems to be extinct in Israel, if it was ever there, and is limited to southwest Turkey and possibly Lebanon, even historically in Gilead, northeast of Israel. But it is long gone in Gilead. Still it is too often referred to as balm of Gilead.

COMMON NAMES (STORAX):

American Storax (Eng.; LAF); American Styrax (Eng.; FAC); Anbar Sa'el (Arab.; Syria; HJP); Asht Loban (India; EFS); Asian Styrax (Eng.; FAC); Asiatic Storax (Eng.; LAF); Asle Luban

FIGURE 1.63 Storax (*Liquidambar* spp.).

(Iran; DEP); Asli (Iran; EFS; NAD); Balm (Eng.; BIB; RSV); Estoraque (Sp.; EFS); Liquidambar (Sp.; EFS); Levant Storax (Eng.; AH2; LAF); Liquidámbar (Sp.; EFS); Liquid Storax (Eng.; NAD); Liquid Styrax (Eng.; JLH); Lubhani (Iran; EFS; NAD); Maiaa (JLH); Meah (Arab.; DEP); Meihsila (Hindi; Iran; DEP; NAD); Meri Arishappal (Tam.; NAD); Miah Sayelaha (Arab.; EFS; NAD); Nantayu (Burma; DEP); Nataf (Heb.; ZOH); Neri Arishappal (Tam.; DEP; NAD); Official Storax (Eng.; HJP); Orientalischer Amberbaum (Ger.; USN); Oriental Sweet Gum (Eng.; Scn.; AH2); Rasamalla (Mal.; Malaya; Tam.; DEP; NAD); Rose Malloes (Eng.; NAD); Sagia (?; JLH); Selaras (Guj.; DEP); Shilarasam (Tel.; DEP; WOI); Shilbaka (Ayu.; AH2); Sigala Agaci (Tur.; EFS); Silajit (Sanskrit; EFS); Silaras (Beng.; Guj.; Hindi; Kan.; Mah.; Tel.; EFS; NAD; WOI); Silarasa (Mar.; DEP); Silha (Hindi; DEP); Silhaka (Sanskrit; EFS; NAD); Stacte (Heb.; ZOH); Storace (It.; EFS); Storax (Eng.; Fr.; Ger.; CR2; EFS; FAC); Storax Calamite (?; JLH); Storaxbaum (Ger.; USN); Storaxboom (Dutch; EFS); Storiaux Liquida (?; JLH); Styrax (Eng.; Fr.; Ger.; EFS; JLH); Styrax Linquide (Fr.; NAD); Su He Ziang (Pin.; DAA); Tzori (Heb.; ZOH); Usturak (Sanskrit; EFS); Usturuk (Bom.; DEP).

Activities (Storax):

Allergenic (f; GAZ); Antidote (f; DAA); Antiinflammatory (1; PH2); Antiscabies (f; EFS); Antiseptic (f1; DEP; HJP; LAF; NAD; SKJ); Astringent (f; DEP; NAD); Bactericide (f; GAZ); Diaphoretic (1; PH2); Diuretic (f; NAD); Expectorant (f1; DEP; EFS; FEL; HJP; LAF; SKJ); Fungicide (f; GAZ); Parasiticide (f; DAA); Pectoral (f; DEP); Stimulant (f; EFS; HJP; PH2; SKJ); Suppurative (f; DEP); Tonic (f; DEP; NAD); Vulnerary (f; EFS).

Indications (Storax):

Adenopathy (f; NAD); Asthma (f; FEL); Bacteria (f; GAZ); Bronchosis (f; FEL; NAD; PH2); Burn (f; JLH); Cancer (f; DAA; JLH; LAF); Cancer, liver (f; JLH); Cancer, spleen (f; JLH); Cancer, uterus (f; JLH); Catarrh (f; FEL; NAD); Childbirth (f; HOC); Cold (f; DAA); Colic (f; NAD); Convulsion (f; DAA); Cough (f; FEL; PH2); Cystosis (f; GAZ; NAD); Dermatosis (f; DEP; HHB; PH2; WOI); Diarrhea (f; HOC); Dysentery (f; HOC); Dysuria (f; NAD); Enterosis (f; HOC); Epilepsy (f; DAA; PH2); Fever (f; PH2); Fistula (f; HOC); Frostbite (f; DAA); Fungus (f; GAZ); Gastrosis (f; JLH); Gleet (f; FEL); Gonorrhea (f; FEL); Hepatosis (f; JLH); Induration (f; JLH); Infection (f1; GAZ; HJP); Inflammation (f; GAZ); Itch (f; GAZ; PH2); Lactose-Intolerance (f; PH2); Leprosy (f; PH2); Leukorrhea (f; FEL); Malaria (f; DAA); Mycosis (f; FEL; GAZ); Orchosis (f; DEP; NAD); Pain (f; NAD); Parasite (f; DAA); Phthisis (f; DEP); Phymata (f; JLH); Pityriasis (f; NAD); Polyp (f; JLH); Pulmonosis (f; FEL; NAD); Pyelitis (f; NAD); Respirosis (f; GAZ); Ringworm (f; FEL); Scabies (f; DEP; FEL; HHB; HJP); Scleroma (f; JLH); Sclerosis (f; JLH); Sore (f; FEL; PH2); Sore Throat (f; DEP; NAD); Splenosis (f; JLH); Sting (f; NAD); Swelling (f; DEP); Syncope (f; PH2); Tinea (f; FEL); Urethrosis (f; HJP); Uterosis (f; JLH); Venereal Disease (f; FEL); Worm (f; HOC); Wound (f; DAA; PH2).

Dosages (Storax):

FNFF = !
GRAS; "approved for food use" but at low levels, for example, circa 15 ppm in candy and 25 ppm in baked goods (LAF). American styrax resin is chewed like gum to freshen breath and clean teeth. Oriental styrax is used to flavor baked goods, candy, chewing gum, and tobacco (FAC; TAN). 10 to 20 grains resin (FEL).

- American styrax mixed with honey and consumed in childbirth (HOC).
- British once applied the resin to indurations of the liver and spleen (JLH).
- Egyptians applied powdered wood to burns and cancers (JLH).
- Germans applied the resin to indurations of the uterus (JLH).
- Hondurans mix American styrax with garlic, onions, and hot water for worms (HOC).
- Lebanese suggest the resin as expectorant, scabicide, stimulant, and urinary antiseptic (HJP).

Downsides (Storax):

No health hazards or contraindications reported with proper administration of suggested therapeutic dosages (PH2) (dosage not indicated; PH2). Ingestion may lead to diarrhea. Possibly allergenic (GAZ; PH2). Broad topical application may lead to transdermal absorption followed by kidney damage (albuminuria; hemorrhagic nepritis) (PH2).

Extracts (Storax):

Purified storax contains circa 33 to 50% storesin, an alcoholic resin, both free and as cinnamic esters. Contains 5 to 15% cinnamic acid, 5 to 15% cinnamyl cinnamate or styracine, and circa 10% phenylpropyl-cinnamate — plus small amounts of ethyl cinnamate, benzyl cinnamate, and styrene (phenylethylene, even traces of vanillin (WOI). Also a volatile oil (styrol, styracin, etc.) (NAD). LAF report also a resin (storesin) containing triterpenic acids (oleanolic and 3-epioleanolic acids) (LAF).

DARNEL (*LOLIUM TEMULENTUM* L.) X POACEAE

SYNONYMS:

Lolium arvense With.; *Lolium robustum* Reich.; *Lolium temulentum* var. *leptochaeton* A. Braun; *Lolium temulentum* var. *muticum* Noulet *fide* DEP, etc.

NOTES (DARNEL):

But while men slept, his enemy came and sowed tares among the wheat.

Matthew 13:25 (KJV)

The grass called darnel or the biblical *tares* is reasonably safe; but once an ergot fungus gets into the seeds, it can be quite dangerous. But generally speaking, the things that make it poisonous also make it strong medicine, as so often happens in nature. Growing throughout the Middle East, its grains are reported from Egyptian tombs at least 4000 years old. Affording nutritive feed for livestock, it should not be grazed after the seeds set. Human deaths are attributed to eating the infected seed, ground up with wheat. It serves as chicken and pigeon feed, but even this is discouraged. According to homeopath Hernandez Mesa, it is used for St. Vitus' dance and idiocy, having been dropped by the allopaths (BIB).

COMMON NAMES (DARNEL):

Aera (f; JLH); Airon (JLH); Aqoullab (Ber.; BOU); Bearded Darnel (Eng.; BUR; EFS); Borrachuela (Sp.; EFS); Cheat (Eng.; BOU); Cizana (Sp.; EFS); Cockle (Eng.; JLH); Cockul (?; JLH); Cockyll (?; JLH); Cominillo (Sp.; EFS); Crop (JLH); Danaqah (Arab.; BOU); Darnell (Eng.; BOU; CR2); Darnell Grass (Eng.; BUR); Delice (Tur.; EFS); Dolik (Dutch; EFS); Dudackorn (JLH); Gioglio (It.; EFS); Git (JLH); Italian Ryegrass (Eng.; JLH); Ivary (Eng.; BOU); Ivraie (Fr.; BOU); Ivraie Enivrante (Fr.; EFS); Joio (Mad.; Por.; EFS); Joyo (Sp.; EFS); Kockul (JLH); KoKkul (JLH); Laichour (Ber.; BOU); Loglio (It.; EFS; JLH); Lolium (Fr.; BOU); Lover's-Steps (Eng.; BUR); Machni (Hindi; India; DEP; EFS; SKJ); Mostaki (Pun.; WOI); Nigella (JLH); Pane Vino (It.; EFS); Poison Rye Grass (Eng.; BUR); Raaigras (Dutch; EFS); Rabillo (Sp.; EFS); Raygrass (Eng.; BUR); Ryegrass (Eng.; BOU); Samma (Arab.; BOU); Schlafweisen (Ger.; EFS); Shaylam (Arab.; BOU); Sikra (Arab.; BOU); Suwal (Arab.; BOU); Tare (Eng.; ZOH); Tares (Eng.; BUR); Taumellolch (Ger.; EFS; MAD); Vallico (Sp.; EFS); Wenwort (Eng.; JLH); Wray (JLH); Zizanie (Fr.; BOU); Ziwan (Heb.; ZOH); Zizzania (It.; EFS); Zun (Heb.; ZOH); Zuwan (Arab.; BOU).

ACTIVITIES (DARNEL):

Anodyne (f; BUR; CRC); Deliriant (f; BUR); Depurative (f; CRC; MAD); Hemostat (f; BOU); Narcotic (1; CRC; EFS); Nephrotoxic (1; PHR); Poison (1; CRC); Sedative (f; SKJ).

INDICATIONS (DARNEL):

Arthrosis (f; BOU; JLH; MAD); Bleeding (f; BOU; CRC); Cancer (f; JLH); Cancer, ear (f; JLH); Cancer, joint (f; JLH); Cancer, knee (f; JLH); Cancer, liver (f; JLH); Cancer, spleen (f; JLH); Cancer, uterus (f; JLH); Colic (f; CRC); Condylomata (f; JLH); Cramp (f; BOU; PHR; PH2); Dermatosis (f; BIB; CRC; HHB); Dysuria (f; CRC; PH2); Eczema (f; CRC); Enterosis (f; BOU; PHR; PH2); Epistaxis (f; BOU; CRC; PHR; PH2); Fracture (f; BIB; CRC); Freckle (f; SKJ); Gangrene (f; BIB; CRC); Gastrosis (f; PHR; PH2); Headache (f; CRC); Hepatosis (f; CRC); Incontinence (f; BIB; BOU; CRC; PH2); Induration (f; CRC; JLH); Insomnia (f; PHR; PH2; SKJ); Kernel (f; CRC); Leprosy (f; BIB; CRC); Lichen (f; MAD); Meningosis (f; BIB; CRC); Menopause (f; BIB; CRC); Migraine (f; BIB; CRC); Nausea (f; BOU); Neuralgia (f; BOU; CRC; MAD; PH2); Pain (f; CRC; HHB; PHR); Paralysis (f; MAD); Polyuria (f; HHB); Rheumatism (f; BOU; CRC; HHB); Sciatica (f; CRC); Scirrhus (f; CRC); Splenosis (f; CRC); Splinter (f; BIB); Tenesmus (f; MAD); Toothache (f; BIB; CRC); Tremor (f; BOU; MAD); Tumor (f; CRC); Uterosis (f; JLH); Vertigo (f; CRC; MAD; PH2); Wen (f; CRC).

Dosages (Darnel):

FNFF = X
Although Indians make food of it (e.g., pinole from the seed in California), it could be poisonous (BUR; TAN). Do not take. Many of the indications and dosages are homeopathic only.

- Asian Indians suggest pounded seed as a sedative poultice and to erase freckles (SKJ).
- British use the wenwort in salves on wens (JLH).
- Lebanese hint of a mystic mountain cult that infuses the grass or soaks the seeds to extract the ergot, which is then used to induce religious ecstasy.
- Lebanese women made a tea of the whole grass for children with colic; adults used the ground seed for blood poisoning, leprosy, migraine, rheumatism, and toothache (HJP).
- Moroccans use the plant decoction for hemorrhage and urinary incontinence (BIB).

Downsides (Darnel):

Can be toxic. No cases of poisoning are known in recent times (PHR). Symptoms include apathy, colic, confusion, cramping, delirium, diarrhea, gastroenterosis, giddiness, headache, mydriasis, paralysis, somnolence, speech problems, staggering, tinnitus, vertigo, visual disorders, and vomiting, rarely death through respiratory failure (BUR; DEP; MAD; PH2). Only mentioned as food for Native Americans (DEM).

BOXTHORN (*LYCIUM EUROPAEUM AUTH.*) + SOLANACEAE

Synonyms:

Lycium barbarum L. var. *brevilobum*; *Lycium mediterraneuum* Dunal; *Lycium orientale* Miers. *fide* ZOH

Notes (Boxthorn):

> **Now the Philistines gathered together their armies to battle, and were gathered together at Shochoh, which belongeth to Judah, and pitched between Shochoh and *Azekah*, in Ephes-dammim.**
>
> **I Samuel 17:1 (KJV)**

Zohary argues that Azekah is a place name, named for the abundant boxthorns there, noting that such is frequent in the Bible; names for acacia, olive, oak, pistachio, pomegranate, and willow also used as place names in the Holy Land (ZOH). EFS entries below may apply as well to *L. barbarum* or *halimiflorum*.

Common Names (Boxthorn):

Alquitira (Sp.; EFS); Artos (Sp.; EFS); Ausseg (Arab.; ZOH); Awsadsch (JLH); Azekah (Heb.; ZOH); Bastard Jasmine (Eng.; EFS); Bocksdorn (Ger.; EFS); Boksdorn (Dutch; EFS); Boxthorn

(Eng.; EFS; USN); Cambronera (Sp.; EFS); Caprifolium (?; JLH); Espinheiro (Mad.; PST); Ganger (India; EFS); Granévano (Sp.; EFS); Jasmin Bâtard (Fr.; EFS); Kanu (India; EFS); Licium (JLH); Mater Silvi (?; JLH); Matrisilvi (?; JLH); Prickly Box (Eng.; EFS); Spina Cristi (It.; EFS); Tea Tree (Eng.; EFS); Teufelsdorn (Ger.; EFS); EFS names may also apply to *Lycium barbarum* and or *Lycium halimifolium*; Nscn.

Activities (Boxthorn):

Antispasmodic (f; EFS); Diuretic (f; EFS); Laxative (f; EFS); Toxic (f; EFS).

Indications (Boxthorn):

Cancer (f; JLH); Cramp (f; EFS); Gingivosis (f; JLH); Hepatosis (f; JLH); Induration (f; JLH); Spasm (f; EFS); Splenosis (f; JLH).

Dosages (Boxthorn):

FNFF = ?
Most species, including *L. arabicum*, to which TAN refers this species, have edible fruits and many have leaves used in teas. I would not be afraid to sample *L. europaeum* as a fruit or leaf tea, but I would be reluctant to recommend it carelessly (FAC, TAN).

Natural History (Boxthorn):

The small edible fruits are dispersed by birds (ZOH).

APPLE (*MALUS DOMESTICA* BORKH.) +++ ROSACEAE

Synonyms:

Malus communis DC.; *Malus pumila* Mill.; *Malus sylvestris* Hort, non Mill.; *Pyrus malus* L.; *Pyrus pumila* L. *fide* WOI

Notes (Apple):

Stay me with flagons, comfort me with apples: for I am sick of love.

Song of Solomon 2:5 (KJV)

Sustain me with raisins, refresh me with apples; for I am sick with love.

Song of Solomon 2:5 (RSV)

Do you people refresh me with cakes of raisins, sustain me with apples; for I am lovesick.

Song of Solomon 2:5 (NWT)

The word *apple* shows up twelve times in my King James version, scattered through eight works, most frequently (four citations) in the Song of Solomon. Although I concluded that apricot or

FIGURE 1.64 Apple (*Malus domestica*).

pomegranate might better be the apple of the Garden of Eden (BIB), Israeli botanist Michael Zohary seems to favor the apple over the apricot, although apples were not mentioned in narratives regarding the Garden of Eden. The Hebrew *tappuah* occurs five times in the Bible as an apple tree, six times as a place name, and once as a proper name. Other biblical scholars have sometimes referred to the *tappuah* as apricot or bitter orange, although these plants were reportedly introduced much later than the apple, which dates in Europe from the Neolithic. Still, although no remains of apple are found among the prehistoric remains of the Middle East, ancient Egyptian papyri circa Ramses II (1298–1235 B.C.) show Nile delta fields with pomegranate, apple, olive, and fig. Zohary adds that the Arabic *tuffah* refers exclusively to apple trees, justifying its identification with the Hebrew *tappuah*. Apples may have been introduced to the Holy Land from Armenia, Iran, Syria, or Turkey circa 4000 B.C. (ZOH). AH2 and USN maintain *Malus domestica* (USN) or *Malus pumila* (AH2) for the apple, and both retain for the crab apple name, *Malus sylvestris*. I would wager that the crab apple is better medicine, ounce for ounce. As a matter of fact, I would speculate that one crabapple (proportionately more phytochemicals) would almost equate (dosage-wise) to one big juicy McIntosh (proportionately more water).

Common Names (Apple):

Abbel (Eng.; JLH); Abreletrae (Den.; KAB); Acuk (Tur.; EB49:406); Acuk Pekmezi (Tur.; EB49:406); Alema (Kalmuk; KAB); Almafa (Hun.; KAB); Amra (Tibet; NPM); Apfel (Ger.; TAN;

USN); Apfelbaum (Ger.; KAB; USN); Appeltraed (Swe.; KAB); Apple (Eng.; Scn.; AH2; CR2; TAN); Appleboom (Dutch; EFS; KAB); Badara (Sanskrit; KAB); Boquettier (Fr.; AVP); Borsdorf Pippin (Austria; JLH); Chui (Pun.; KAB); Chur (Pangi; KAB); Crabapple (Eng.; NAD); Croisson (Fr.; AVP); Eima (Tur.; EB49:406); Eima Azmasi (Tur.; EB49:406); Elmai (Tur.; EB54:155); Jablon (Pol.; KAB); Jabuka (Illyria; KAB); Kashu (Pun.; DEP); Kushu (Ladak; DEP; KAB); Lin Ch'in (China; EFS; KAB); Lipot (Lpcha; NPM); Macanceira (Por.; EFS); Maceira (Por.; KAB; USN); Macieira (Por.; AVP; EFS); Mana (Pushtu; KAB); Manra (Pushtu; DEP); Manzana (Peru; Sp.; EGG; USN); Manzano (Peru; Sp.; EGG; KAB; USN); Mar (Rom.; KAB); Maschanzkerapfel (Austria; JLH); Melo (It.; KAB); Meto (It.; AVP); Milia (Greek; KAB); Nagnse (Limbu; NPM); Omena (Fin.; KAB); Palu (Kunawar; KAB); Paoma (Hova; KAB); Perote (Ca.; Sp.; ROE); P'ing Kuo (China; TAN); Pome Dols (Cat.; KAB); Pomme (Fr.; TAN); Pomme d'Arbre (Haiti; AVP); Pomme de France (Haiti; AVP); Pommier (Fr.; Haiti; AVP0; KAB); Pommier Commun (Fr.; USN); Pomo (Malta; KAB); Ringo (Japan; TAN; USN); Safarchand (Mah.; NAD); Sapukawasi (Rai; NPM); Sco (Nwp.; KAB); Seb (Beng.; Bhojpuri; Hindi; Mooshar; Tharu; DEP; NPM); Seba (Sanskrit; DEP); Sebhaphala (Sanskrit; EFS; NAD); Seb Safarang (Hindi; NAD); Sebu (Kan.; KAB; WOI); Seo (Hindi; KAB; WOI); Servu Kittalay (Can.; NAD); Sev (Urdu; NAD); Sevu (Kan.; DEP); Sheo (Nwp.; DEP); Shewa (Afg.; DEP; KAB); Shivinthaka (Sanskrit; NAD); Sib (Iran; DEP); Sir (Iran; KAB); Soof (Sin.; NAD); Suf (Sin.; DEP; KAB); Sufferjang (Bom.; Guj.; EFS; NAD); Syau (Danuwar; Magar; Nepal; Newari; Sunwar; Tamang; NPM); Taph (Arab.; ZOH); Tappuah (Heb.; ZOH); Tiffah (Arab.; Syria; HJP); Tuffah (Arab.; DEP; HJP; ZOH); Tyffah (Arab.; KAB); Uril (Mongolia; EB54:528); Urilin Chai (Mongolia; EB54:528); Yablon (Rus.; KAB).

Activities (Apple):

Anthelmintic (f; CRC; WOI); Antiallergic (1; JAF51:3806); Antialopecic (1; PR15:311); Antiandrogenic (1; JNU); Antiinflammatory (f1; JNU); Antimelanogenic (1; X16029003); Antioxidant (1; JNU; PR15:311; X15769178); Antiproliferant (1; JAF51:609); Antiseptic (f1; NAD; X15713000); Antitoxic (1; JAF51:3806); Antiviral (1; X15713000); Aperitif (f; CRC); Apoptotic (1; JNU); Bactericide (1; CRC); Cardioprotective (1; JNU; WOI); Carminative (f; CRC; ROE); Chemopreventive (f1; X15769178); Cicatrizant (f; VAD); Cyanogenic (f1; CRC); Demulcent (f; VAD); Depurative (f; CRC; NAD); Digestive (f; CRC; EFS); Diuretic (f; CRC; EFS); Emollient (f; CRC); Gastroprotective (1; X15647180); Hypnotic (f; CRC; NAD); Hypoglycemic (f1; JNU; EB49:406); Hypotensive (1; JAF51:3806); Hypouricemic (f; NAD); Laxative (f; CRC); Lipolytic (f1; JAF51:3806); Mucoprotective (f; VAD); Noogenic (f; NAD); Orexigenic (f; EFS); PKC Inhibitor (1; PR15:311); Poison (f; CRC); Propecic (1; PR15:311); Refrigerant (f; CRC); Sedative (f; CRC); Tonic (f; CRC; LMP); Tyrosinase Inhibitor (1; X16029003); Vulnerary (f; VAD).

Indications (Apple):

Alcoholism (f; NAD); Allergy (1; JAF51:3806; X15630302); Alopecia (1; PR15:311; JAF51:3806); Anemia (f; HJP; LMP); Aphonia (f; DEM); Arthritis (f; TOM); Asthma (f1; JNU; NAD); Bacteria (1; CRC; WOI); Biliousness (f; CRC; KAB; LMP; NAD); Blindness (f; DEM); BPH (f; JNU); Bruise (f; DEM); Cacoethes (f; CRC; JLH); Calculus (f; NAD); Callus (f; JLH); Cancer (f1; CRC; JLH; HJP); Cancer, breast (f1; JLH; X15769178); Cancer, colon (f1; FNF; JLH); Cancer, eye (f; JLH); Cancer, liver (f1; JNU); Cancer, lung (f1; JNU); Cancer, prostate (f1; JNU); Cardiopathy (f1; JNU; WOI); Caries (f1; JAF51:3806); Catarrh (f; CRC; NAD; ROE); Chilblain (f; AAH); Childbirth (f; ROE); Cholera (1; JAF51:3806); Colic (f; ROE); Condyloma (f; JLH); Conjunctivosis (f; FEL; KAB); Constipation (f1; CRC; KAB; NAD); COPD (1; Associated Press, Jan. 19, 2000); Cough (f; HJP); Dermatosis (f1; HHB; X15630302); Diabetes (f; CRC; ROE); Diaper Rash (f; TOM); Diarrhea (f1; PH2);

Diverticulosis (1; JNU); Dropsy (f; FEL; HJP); Duodenosis (f1; HJP); Dysentery (1; CRC; PH2); Dyspepsia (f1; CRC; HHB; HJP; NAD; PH2; EB49:406); Dyspnea (f; DEM); Earache (f; AAH; DEM); Eczema (f; NAD); Enterosis (f; CRC); Epilepsy (f; HJP); Erysipelas (f; FEL); Fever (f; CRC; KAB); Flu (1; X15713000); Flux (f; CRC); Gallstone (f; DEM); Gastrosis (f1; TOM; X15647180); Gout (f1; JNU; NAD); Gravel (f; BUR; FEL); Halitosis (f; TOM); Hay Fever (1; JNU); Headache (f; HJP); Heart (f; CRC); Hemorrhoid (f; DEM; HJP); Hepatosis (f; NAD); Herpes (1; JNU); High Blood Pressure (1; JAF51:3806); Hive (1; JNU); Hoarseness (f; DEM); Hyperemesis (f; ROE); IBS (1; JNU); Infection (f1; NAD; X15713000); Inflammation (f1; JNU; CRC; FEL); Insomnia (f; CR2; ROE); Leukemia (f; JLH); Lumbago (f; NAD); Malaria (f1; CR2; KAB); Neuralgia (f; NAD); Neuritis (f; NAD); Obesity (f1; WOI; JAF51:3806); Ophthalmia (f; JLH; KAB); Pancreatosis (1; JNU); Paralysis (f; HJP); Pertussis (f; CRC); Phthisis (f; HJP); Pneumonia (f; ROE); Poison Ivy (f; TOM); Pregnancy (f; ROE); Prostatosis (1; JNU); Rash (f; TOM); Rheumatism (f; HJP; NAD; ROE); Scarlatina (f; FEL); Sciatica (f; NAD); Scrofula (f; HJP); Scurvy (f; CRC); Sore (f; AAH; JLH); Sore Throat (f; FEL); Spasm (f; CRC); Sprain (f; AAH); Stomachache (f; TOM); Stone (f; NAD); Stress (f; WOI); Stroke (f1; HJP; JNU); Sunstroke (f; EB49:406); Swelling (f; HJP); Thirst (f; CRC); Tumor (f1; JLH); Ulcer (f1; HJP); Venereal Disease (f; JLH); Virus (1; JNU; X15713000); Wart (f; CRC; KAB); Wound (f; VAD).

Dosages (Apple):

FNFF = !!!
Fruit widely eaten raw, dried into chips, or cooked; in apple brandies, apple butters, apple ciders, apple jacks, apple sauces. Fruit puree sometimes used in making ketchups. Fruit is a commercial source of pectin (TAN). Flowers are eaten, often fried in batter. Peels are used in Turkish teas (FAC). Leaves are used to make tea (Mongolia; EB54:528). One British cheese called *applewood* is smoked over applewood fires (FAC).

- Alabamans suggest apple vinegar for diaper rash, halitosis, poison ivy, and rheumatism (TOM).
- Americans suggest boiling a tart apple in one pint of water for fever (FEL).
- Americans take bark tea for bilious, intermittent and remittent fevers (KAB).
- Asian Indians suggest that alcoholics might try apples or apple juice instead of whiskey (NAD).
- Ayurvedics suggest the fruit for biliousness, constipation, fever, and impotence (KAB).
- British suggest apple peel for cancer (JLH).
- Cornell scientists showed whole apple extracts prevent mammary cancer in rats dose-dependently (comparable to 1, 3, and 6 apples a day in humans; X15769178).
- English apply rotten apple on sore spots, earache, and rheumatic or weak eyes (AAH; KAB).
- French apply roasted apple poultice for inflamed eyes (KAB).
- Lebanese suggest a spoonful of cider vinegar in a glass of water for stroke (HJP).
- Lebanese suggest compresses of sweetened apple sauce for cancer and swellings (HJP).
- Peruvians suggest tea of apple, orange peel, and flax for nausea of pregnancy, tea of apple or peach leaves with matico to facilitate delivery (ROE).
- Turks eat apples to reduce libido (EB49:406).
- Turks take boiled fruits as hypoglycemic, tonic; for sunstroke (EB49:406).
- Yugoslavs suggest eating raw apples for anemia, cough, dyspepsia, headache, and phthisis (HJP).

Natural History (Apple):

Apples are not dependably self-fertile, so provisions for cross-pollination should be made. Bees and other insects are agents. Honeybees bring about 90% of pollen transfer in the orchard. One colony of bees of medium size (15,000 to 20,000 bees) per acre is sufficient. Mice are very destructive to apple trees. Keep a 1-m diameter space around tree free of weeds or mulched. Rabbits also eat bark. Protect with wire screen. Apple trees are attacked by a great many fungi, perhaps 150 worldwide, and many are very serious, but can be controlled with various sprays. Among them are the following: *Actinopetle dryina, Alternaria mali, A. tenuis, Armillaria mellea, Ascochyta mali, Aspergillus terreus, Botryodiplodia theobromae, Botryosphaeria ribis, Botrytis cinerea, B. mali, Cephalosporium carpogenus, C. roseum, Cephaothecium roseum, Cepholeuros virescens, Cercospora mali, C. pyri, Cicinnobolus cesatii, Cladosporium herbarum, Clasterosporium carpophilum, Clitocybe tabescens, Colletotrichum fructus, Coniocybe nivea, Coniophora cerebella, Coniothecium chomatosporum, Coniothyrum fuckelii, C. olivaceum, C. pirinum, Corticium centrifugum, C. galactinum, C. laetum, C. litschaueri, C. salmonicolor, Coryneum foliicola, C. longistipitatum, Creonectria purpurea, Cylindrocarpon angustum, C. mali, Cylindrocladium scoparium, Cyphella albo-violascens, C. marginata, Cytoplea cinerea, Cytospora carphosperma, C. leucostoma, C. mali, Cytosporina ludibunda, Daedalea confragosa, Daldinia concentrica, Dematophora necatrix, Dermatea corticola, Diaporthe perniciosa, Didymella voglinoi, Didymosphaeria microstictica, Diplodia griffoni, D. mutila, D. natalensis, Elsinoe pyri, Endomyces mali, Entomosporium maculatum, Epicoccum granulatum, Eutypella stellulata, Fabraea maculata, Fomes applanatus, F. annosus, F. fomentarius, F. pinicola, F. pomaceus, Fracciaea heterogenea, Fumago vagans, Fusarium acuminatum, F. avenaceum, F. culmorum, F. dimerum, F. equiseti, F. fructigenum, F. herbarum, F. lateritium, F. moniliforme, F. poae, F. scirpi, F. solani, F. vasinifectum, Fusicladium dendriticum, Ganoderma curtisii, Gibberella baccata, G. fujikuroi, Gliocladium viride, Gloeosporium album, G. frustigenum, G. perennans, Gleodes pomigena, Gymnosporangtum juniperinum, G. sabinae, G. tremelloides, G. clavipes, G. globosum, G. juniperi-virginianae, G. nidus-avis, Glomerella cingulata, G. rubicola, Glutinium macrosporum, Haplosporella mali, Helminthosproium papulosum, Hendersonia cydoniae, Hendersonula toruloidea, Hormodendron cladosporioides, Hydnum setosum, Hymenochaete agglutinans, Hypholoma sublateritium, Illosporium malifoliorum, Lambertelia corni-maris, Lentinus tigrinus, Leptosphaeria coniothyrium, Leptothyrium pomi, Leucostoma persooni, Liinospora ochracea, Marasmius pyrinus, Merulinus corium, M. papyrinus, Microdiplodia pirina, Monilia cinerea, M. fructigena, M. laxa, Monochaetia mali, Mucor piriformis, Mycosphaerella pomi, M. sentina, M. tulasnei, Myriangium asterinosporum, Myxosporium corticola, M. mali, M. microsporum, Nectria cinnabarina, N. coccinea, N. ditissima, N. galligena, Neofabraea malicorticis, N. perennans, Nummularia discreta, Oidium farinosum, Oothecium indicum, Pellicularia koleroga, Penicillium expansum, Peniophora cinerea, Peyronellaea veronensis, Pezicula corticola, Pezizella oenotherae, Phacidiella discolor, Phacidiopycnis malorum, Phellinus friesianus, Phialophora goidanichii, Ph. malorum, Pholiota adiposa, Phoma glomerata, Ph. pomi, Ph. prunicola, Phomopsis mali, Ph. perniciosa, Phyllactinis suffulta, Phyllosticta mali, Ph. persicae, Ph. pirina, Ph. prunicola, Ph. solitaria, Phymatotrichum omnivorum, Physalospora cydoniae, Ph. mutila, Ph. Obtusa, Ph. malorum, Ph. rhodina, Phytophthora cactorum, Ph. citricola, Ph. palmivora, Ph. parasitica, Ph. primulae, Ph. syringae, Ph. vignae, Plenodomus fuscomaculans, Pleospora fructicola, P. herbarum, P. mali, Pleurotus ostreatus, P. ulmarius, Podosphaera clandestina, P. leucotricha, P. oxycanthae, Polyporus purpureus, P. ignarius, P. versicolor, Puccinia heterospora, Pythium aphanidermatum, P. debaryanum, P. intermedium, P. oligandrum, P. ultimum, P. vexans, Radulum aterriumum, Rhabdospora rhodina, Rhizoctonia aderholdii, Rh. solani, Rhizopus arrhizus, Rh. nigricans, Roesleria hypogaea, Rosellinia necatrix, R. radiciperda, Schizophyllum commune, Sclerotinia cydonia, S. fruticola, S. fructigena, S. laxa, S. mali, S. nipponica, S. sclerotiorum, Sclerotium rolfsii, Scolecosporium pedicellatum, Septobasidium mariana, S. pedicellatum, S. pseudopedicellatum, Septoria piricola, S. pyri, Sphaeropsis malorum, S. pomorum, Sphaerotheca gestum, Stereum hirsutum, S. purpureum, Stromatinia fructigena, Trametes hispida, Trichoderma viride, Trichoseptoria fructigena, Trichothecium roseum, Tympanis conspersa, Ulocladium consortiale, Ustulina zonata, Valsa albiens, V. americana, V. leucostoma, Valsella melastoma, V. papyriferae, Venturina inaequalis, Verticillium dahliae, Xylaria longeana.* Apple trees may be parasitized by the following flowering plants: *Cuscuta monogyna, Psittacanthus cuneifolius, Comandra pallida, Phoradendron flavescens, Viscum album, V. cruciatum,*

Loranthus oleifolius, and *L.* virescens. The main bacterial diseases of apples are caused by the following: *Agrobacterium rhizogenes*, *A. tumefaciens*, *Erwinia amylovora*, *Pseudomonas melophthora*, *Ps. papulans*, and *Ps. syringae*. Some of the viruses causing diseases in apples are Apple chat fruit, Dwarf, Epinasty, Flat limb, Green crinkle, Green mottle, Leafspot, Line pattern, Apple mosaic, Pyrus virus #2, Proliferation, Ringspot mosaic of pear, Apple rosette, Rough skin, Rubbery wood, Scaly bark, Striped mosaic, Stunt, Tulare apple mosaic, Variegation, Yellow dragon, and Yellow mosaic (*Marmor rosae*). Apples have many physiological ailments, linked to weather conditions, mineral deficiencies, water supply, storage, and genetics. The following nematode list includes species known to cause problems in apples: *Criconemella curvata*, *Crossonema multisquamatum*, *Doryllium minor*, *Globodera mali* (a cyst nematode in the areas of the old USSR), *Longidorus macromucronatus*, *L. maximus*, *Meloidogyne hapla*, *M. incognita acrita*, *M. mali*, *Merlinius brevidens*, *Neotylenchus* sp., *Paratylenchus amblycephalus*, *P. hamatus*, *Pratylenchus coffeae*, *P. penetrans*, *P. pratensis*, *P. thornei*, *P. vulnus*, *Trichodorus viruliferus*, *Tylenchorhynchus dubius*, *T. maximus*, and *Xiphinema americanum* (Golden, 1984). The worst apple pests in the United States include fruit-tree leafroller (*Archips argyrospilus*), Redbanded leafroller (*Argyrotaenia velutinans*), Plum curculio (*Conotrachelus nenuphar*), Wooly apple aphid (*Eriosoma lanigerum*), Codling moth (*Laspeyresia pomonella*), Tarnished plant bug (*Lygus lineolaris*), Oyster shell scale (*Lepidosaphes ulmi*), European red mite (*Panonychus ulmi*), Spider mite (*Tetranychus modanieli*), and White apple leafhopper (*Typhlocyba pomaria*).

Downsides (Apple):

No health hazards or side effects known with proper therapeutic dosages (PH2) (but PH2 designates no dosage; JAD).

Extracts (Apple):

Liu and Chen (2005) state, rightly or wrongly, that "Apples are … the major contributors of phytochemicals in human diets." Extracts exhibit strong antioxidant and antiproliferative activities. Whole apple extracts prevent mammary cancer in rats (comparable to a human eating one, three, and six apples a day (X15769178). Hamauzuta et al. (2005) evaluated phenolics in apple and quince; the latter, with more 3-caffeoylquinic acid and 5-caffeoylquinic acid and polymeric procyanidins, had more antiflu activity (X15713000). Apple antioxidants may help prevent gastric diseases (X15647180). Several procyanidin fractions inhibit tyrosinase and melanogenesis (X16029003). Epicatechin and procyanidin B2 were the major contributors to the antioxidant activity of whole apple. Hydroxycinnamic acids may have a significant role in the flesh (X15941346).

HIGH MALLOW (*MALVA SYLVESTRIS* L.) +++ MALVACEAE

Synonyms:

Malva ambigua Guss.; *Malva elata* Salisb.; *Malva erecta* C. Presl; *Malva glabra* Desr.; *Malva mauritiana* L.; *Malva obtusa* Moench.; *Malva ruderalis* Salisb.; *Malva vulgaris* S. F. Gray *fide* HH2

Notes (High Mallow):

Who cut up mallows by the bushes, and juniper roots for their meat.

Job 30:4 (KJV)

They pick mallow and the leaves of bushes, and to warm themselves the roots of the broom.

Job 30:4 (RSV)

FIGURE 1.65 High Mallow (*Malva sylvestris*).

They were plucking the salt herb by the bushes, And the root of broom trees was their food.

Job 30:4 (NWT)

Some authors argue for *Malva*, others for *Atriplex*, for the mallow of the Bible. Zohary and I think that the Hebrew word *halamuth* is probably malvaceous, most likely Malva or Alcea. Both provide edible potherbs. Modern Hebrew renders *Malva* as *halamith*. Because *Malva nicaeensis* (petals less than 3 times as long as sepals) is more common in Israel than *M. sylvestris* (petals more than 3 times as long as sepals), either could have been the mallow of Job (ZOH). Flowers are collected and exported from Iran as medicinal. Once, flowers were strewn before peasants' doors and woven into garlands. The flower tincture is a delicate test for alkali (BIB).

Common Names (High Mallow):

Ad Dahza (Arab.; Syria; HJP); Ad-dama (Arab.; BOU); Amedjir (Ber.; BOU); Baerwinde (Ger.; KAB); Bereut (Fr.; EFS); Blue Mallow (Eng.; FAC); Bread and Cheese (Eng.; KAB); Cam Quy (Ic.; KAB); Cheese Flower (Eng.; BUR); Chin K'uei (China; EFS); Common Mallow (Eng.; BOU; BUR; HJP; TAN); Country Mallow (Eng.; BUR); Djir (Ber.; BOU); Ebegümesi (Tur.; EFS; KAB); Ebemgumesi (Tur.; EB54:155); Fausse Guimauve (Fr.; HH2); Fouassier (Fr.; BOU); Fromageon (Fr.; BOU); Grande Mauve (Fr.; BOU; HH2); Grosse Käsepappel (Ger.; HH2); Ground Dock (Eng.; BUR); Gulikhadmi (Afg.; DEP; KAB); Gulkehr (Sanskrit; EFS); Gulkhair (Hindi; KAB); Gul Kheir (Hindi; NAD); Halamith (Heb.; ZOH); Halamuth (Heb.; ZOH); High Mallow (Eng.; Scn.; AH2; SKJ; TAN); Hobbeiza (Malta; KAB); Hubeize (Arab.; ZOH); Ibeqquola (Ber.; BOU); Imejjir (Ber.; BOU); Kanji (Nwp.; DEP); Käsekraut (Ger.; HH2); Käsepappel (Ger.; EFS); Katost (Den.; Swe.; EFS; KAB);

Khabaji (Sin.; KAB); Khabajni (Sin.; DEP); Khalazi (Kas.; WOI); Khatmi (Beng.; Orissa; Patna; DEP; KAB; SKJ); Khitami i Kuchaka (Iran; EFS); Khitmi (Arab.; KAB); Khobbeiza (Arab.; BOU); Khokorteen (Egypt; DEP); Khubaji (Urdu; KAB); Khubasi (Bom.; DEP; KAB); Khubasi (Arab.; Bom.; DEP; NAD); Khubas Towdrie (Iran; DEP); Khubbazi (Arab.; Syria; EFS; HJP); Klein Kaajeskruid (Dutch; EFS); Kunzi (Hindi; WOI); Lyesnaia Malva (Rus.; KAB); Malachi (Greek; KAB); Mallow (Eng.; CR2); Maller (Sussex; KAB); Maluwe (Dutch; KAB); Malva (Peru; Por.; Sp.; EFS; EGG; HH2); Malva Domestica (It.; EFS); Malva Major (Cat.; KAB); Malva Ordinaria (Por.; KAB); Malva Riondela (It.; HH2); Malva Silvestre (Peru; Por.; Sp.; EFS; EGG); Malve (Ger.; EFS); Malvilla (Dr.; AHL); Malvone (It.; KAB); Malyva Kerek (Hun.; EFS); Mamejjirt (Ber.; BOU); Marsh Mallow (Eng.; AAH; BOU); Mauretanische Malve (Ger.; HH2); Mauritian Mallow (Eng.; FAC); Mauve (Fr.; Reunion; EFS; KAB; TAN); Mauve Sauvage (Fr.; BOU; HH2); Mejjir (Ber.; BOU); Meule (Fr.; BOU); Mohrenmalve (Ger.; HH2); Nalba (Rom.; KAB); Nanakillagah (Iran; KAB); Ouabejjir (Ber.; BOU); Pancake Plant (Eng.; BUR); Papsajt (Hun.; KAB); Patari (Kon.; NAD); Raqma (Arab.; BOU); Raqmiya (Arab.; BOU); Rosspappel (Ger.; HH2); Sannabindigegida (Kan.; Mysore; KAB; SKJ); Shaz Ziele (Pol.; KAB); Suchel (Kas.; MKK); Tilchuni (Nwp.; DEP; KAB); Vilayati Kangai (Hindi; DEP); Vilayatikangoie (Dec.; KAB); Zeni-aoi (?; FAC); Zhanba (Mongolia; X12795226).

Activities (High Mallow):

Analgesic (1; BRU); Anticomplement (1; HH2; X2379192); Antiinflammatory (f1; EGG; FAD; GAZ; WAM); Antioxidant (1; X14630594); Antiperoxidant (1; X14630594); Antiseptic (f1; WOI; X14698521); Antispermatogenic (1; HH2); Antitussive (f1; VAD); Antiulcer (1; PR14:581); Astringent (f1; BOU; FAD; HHB); Demulcent (12; DEP; KOM; PIP; WAM); Diuretic (f; DEP; FAD); Emollient (f1; BOU; DEP; EFS; GAZ; WOI); Enterotonic (f; BOU; WOI); Expectorant (f; EFS; EGG; MAD); Febrifuge (f; EFS; KAB); Iron Chelator (1; X14630594); Laxative (f; BOU; EFS); Pectoral (f; BOU; BUR; EFS; PNC); Refrigerant (f; EFS); Stimulant (f; BOU); Sudorific (f; EGG); Uterotonic (f; WOI).

Indications (High Mallow):

Abscess (f; VAD); Adenopathy (f; AAH); Anemia (f; MAD); Angina (1; HHB); Aphtha (f; VAD); Asthma (f; AAH); Backache (f; AAH); Blepharosis (f; KAB); Boil (f; AAH); Bronchosis (f12; EGG; FAD; PHR; PH2); Bruise (f; AAH); Bug bite (f; PNC); Cancer (f; JLH); Cancer, breast (f; JLH); Cancer, neck (f; JLH); Cancer, penis (f; JLH); Cancer, stomach (f; JLH); Cancer, uterus (f; JLH); Catarrh (f; AHL; EFS; HHB; PH2); Childbirth (f; MAD); Cholecystosis (f; DEP; MAD; PHR; PH2); Cold (f1; AHL; EFS; PNC; VAD); Colitis (f1; GAZ); Consumption (f; MAD); Corn (f; JLH); Cough (f12; BOU; DEP; KOM; MAD; PHR; PH2; PIP; VAD); Colosis (1; BRU); Cough (1; BRU); Cramp (1; BRU); Cystosis (f; DEP; KAB); Dandruff (f; MAD); Dermatosis (f1; BOU; BRU; DEP; EGG; GAZ; WAM); Diarrhea (f1; BOU; WAM); Duodenosis (f; VAD); Dysuria (f1; BOU; WAM; WOI); Earache (f; MAD); Eczema (f1; GAZ); Emphysema (f; MAD; VAD); Enterosis (f1; EFS; HHB; PH2); Epilepsy (f; MAD); Favus (f; MAD); Fever (f; DEP; MAD); Flu (f; VAD); Fracture (f; AAH; HJP); Furuncle (f; VAD); Gastrosis (f1; GAZ; HHB; PHR; TOM; WAM); Glossosis (f; VAD); Gonorrhea (f; DEP; NAD); Hoarseness (f1; CAZ); Induration (f; JLH); Infection (f1; WOI; X14698521); Inflammation (f; EGG; KAB); Itch (1; BRU); Jaundice (f; KAB); Laryngitis (f; GAZ; MAD); Mastosis (f; JLH); Mucososis (f2; BOU; KOM; MAD); Nephrosis (f; AAH; MAD); Obesity (f; VAD); Oligolactea (f; MAD); Oliguria (f; VAD); Ophthalmia (f; BOU; MAD); Pain (1; BRU); Parotitis (f; AAH); Pertussis (f; WOI); Pharyngosis (f12; BRU; GAZ; PH2; PIP); Pulmonosis (f; AAH; HHB; NAD; WOI); Sore (f; HJP); Sore Throat (f12; GAZ; KOM; MAD; PIP); Splenosis (f; FAD; KAB); Sprain (f; AAH); Sting (f; VAD); Stomachache (f; FAD); Stomatosis (f12; BRU; KOM; PH2; PIP; VAD); Strangury (f; DEP; KAB); Swelling (f; JLH); Tenesmus (f; KAB); Toothache (f; AAH; MAD); Tuberculosis (f; MAD); Ulcer (f1; VAD; PR14:581); Urethrosis (f1; GAZ); Uterosis (f; JLH); Vaginosis (f; EGG; VAD); Venereal Disease (f; DEP); Wound (f; HJP; PHR; PH2).

Dosages (High Mallow):

FNFF = !!!

Widely eaten; a wholesome vegetable when boiled (even by Romans; DEP); leaves used as tea substitute; flowers eaten as garnish or veggie, also used to color wines red; unripe fruits (cheeses) eaten as a nibble; seeds also eaten (BIB; FAC; TAN; WOI; EB54:155). 1.5 g/cup (HHB); up to 5 g/day (HHB); 2–3 tsp (3.2–4.8 g) leaf in cold tea (MAD). 3–5 g/cup 2 to 3 ×/day (PH2); 5 g/day flower or leaf (PIP); 2–8 ml liquid extract (PNC).

- Algerians use infusions as demulcent and pectoral (HJP); applying powdered leaves in milk to jaw fractures (HJP).
- Dutch suggest flowers of this species (leaves of *M. neglecta*) for cough and cold (EFS).
- Egyptians poultice pounded leaves (of *Malva parviflora*) on scorpion stings (BIB).
- Europeans use the infusion for colds, coughs, gravel, and strangury (BIB).
- Gypsies use mallow in a treatment for tuberculosis, the root for dysentery (HJP).
- Iranians regard the mallows as antitussive, cooling, and mucilaginous, and prescribe mallow, with *Alhagi*, *Cordia*, *Nymphaea*, *Viola*, and *Ziziphus*, as a purgative (BIB).
- Irish bathe sprains or fractures with the root decoction (AAH).
- Lebanese compress the root onto infections, taking plant internally for cough (HJP).
- Palestinians use leaves and/or flowers (known as foliae malva or flos malvae) as a gargle and mouthwash (BIB).
- Peruvians suggest decocting 10g/leaf + flower/liter H_2O as expectorant, sudorific (EGG).
- Peruvians suggest floral salve or decoction for bronchitis, sitz bath in floral/foliar decoction for urinary inflammation, or bathing dermal infections with decoction (EGG).
- Spanish suggest fresh plant juice for bites and stings (VAD).
- Unani regard the plant as cooling, febrifuge, and mucilaginous, suggesting it for blepharitis, inflammation, jaundice, scorpion sting, sore throat, splenomegaly, strangury, and urinary discharge, and as an emollient cluster in tenesmus (KAB).

Downsides (High Mallow):

Class 1 (AHP). No health hazards or contraindications reported with proper administration of suggested therapeutic dosages (PH2). None reported (PIP). None known (WAM).

Extracts (High Mallow):

Demulcent and mucilaginous, the plant (5 g leaf/day) has been approved by Germany's Commission E for irritations of the mucosa of the mouth and throat and associated dry cough (KOM). *M. neglecta*, *parviflora*, and *sylvestris* used interchangeably (FAH).

Proestos et al. (2005) checked the species for flavonoids and phenolics and their antioxidant and antimicrobial activity, finding circa 11 ppm gallic acid, 43 ppm ferulic acid, and 14 ppm hydroxybenzoic acid in high mallow (X15713039).

MANDRAKE (*MANDRAGORA OFFICINARUM* L.) X SOLANACEAE

Synonyms:

Atropa mandragora L.; *Mandragora acaulis* Gaertn.; *Mandragora mas* Gersault.; *Mandragora microcarpa* Bertol.; *Mandragora officinalis* Mill.; *Mandragora praecox* Sweet; *Mandragora vernalis* Bertol.

Notes (Mandrake):

The mandrakes give a smell, and at our gates are all manner of pleasant fruits, new and old, which I have laid up for thee, O my beloved.

Song of Solomon 7:13 (KJV)

The mandrakes give forth fragrance, and over our doors are all choice fruits, new as well as old, which I have laid up for you, O my beloved.

Song of Solomon 7:13 (RSV)

The mandrakes themselves have given [their] fragrance, and by our entranceways there are all sorts of the choicest fruits. The new ones as well as the old. O my dear one, I have treasured up for you.

Song of Solomon 7:13 (NWT)

And Jacob came out of the field in the evening, and Leah went out to meet him, and said, Thou must come in unto me; for surely I have hired thee with my son's mandrakes. And he lay with her that night.

Genesis 30:16 (KJV)

When Jacob came from the field in the evening, Leah went out to meet him, and said, "You must come in to me; for I have hired you with my son's mandrakes." So he lay with her that night.

Genesis 30:16 (RSV)

When Jacob was coming from the field in the evening, Léah went on out to meet him and then said: "It is with me you are going to have relations, because I have hired you outright with my son's mandrakes." Accordingly he lay down with her that night.

Genesis 30:16 (NWT)

As suggested in the story of Leah and Rachel, mandrake may increase fertility and/or lust. Arabs called it "devil's apples" because of its supposed powers to excite to sex. If my earlier readings were correct (BIB), the yellow plum-like fruit is found ripe in Israel during wheat harvest. If so, this would suggest the spring flowering *M. officinarum* (*M. vernalis*) rather than the autumn flowering *M. autumnalis* (roots not macroscopically distinguishable [HH2]). Both are loaded with potent poisonous alkaloids. The flavor is sickeningly sweet, although rather insipid. Eaten in quantity, it produces dizziness, and may even stimulate men and women to insanity. It is also thought to stimulate conception and has a folk history of being used in love potions and incantations. As late as 1630 in Hamburg, Germany, three women were executed for possession of mandrake root, supposed "evidence" that they were involved in witchcraft. The "witches of Salem," on the other hand, had the American mandrake, *Datura stramonium*, which contains the same transdermally active, "high-flying" alkaloids. Mandrake has a large root, dark brown and rugged, sometimes shaped like the human body. It has long been an object of superstition. Jews considered the mandrake a charm against evil spirits. Others believed that mischief-making elves would find its strange odor unbearable. Old folk tales say that Jews tied a dog to the plant to pull it from the ground, as it would

FIGURE 1.66 Mandrake (*Mandragora officinarum*).

kill a man to touch it so fresh without certain precautions; "the mandrake shrieked, the dog died, rendering the root harmless thereafter" (BIB). Once esteemed for its medicinal and narcotic properties, mandrake still may have orgiastic and magical applications among cults involving the sexes and has been used as an aphrodisiac (BIB).

Common Names (Mandrake):

Adamotu (Tur.; EFS); Addüsselâmotu (Tur.; EFS); Alraune (Ger.; HH2); Alruin (Dutch; EFS); Astrang (Arab.; EFS; NAD); Bayd ul Jinn (Arab.; Syria; HJP); Berenjilla (Sp.; EFS); Bhagener (Hindi; NAD); Dastam Haryah (Arab.; EFS; NAD); Devil's Apple (Eng.; EFS); Dudaim (Heb.; ZOH); Erdmännlein (Ger.; HH2); Giatya Bruz (Iran; EFS; NAD); Hexenkraut (Ger.; EFS); Hundsapfel (Ger.; EFS); Kaatjuti (Tam.; DEP; WOI); Katavjati (Tam.; NAD); Kattai Jati (Madras; Tel.; NAD; SKJ); Lakmani (Hindi; India; EFS; NAD); Lakshamana (Sanskrit; EFS); Lakshmana (Hindi; Sanskrit; SKJ); Lang Tu (China; EFS); Lebruj (Baz.; India; EFS; NAD); Loofahat (Mal.; DEP); Lufahat (Malaya; EFS; NAD); Luckmuna (Hindi; DEP; WOI); Luckmunie (Hindi; DEP); Lufah (Arab.; Hindi; DEP; WOI); Mandragora (It.; EFS); Mandrágora (Por.; EFS); Mandrágora Macho (Sp.; EFS); Mandragore (Fr.; EFS); Mandrake (Eng.; Scn.; AH2; CR2); Mano di Gloria (It.; EFS); Mardami (Iran; EFS; NAD); Putrada (Sanskrit; EFS; NAD); Rabuhi (Arab.; Syria; HJP); Raktavindu (Sanskrit; EFS; NAD); Satan's Apple (Eng.; EFS); Serag al Coshrob (Arab.; DEP); Sirag el Qutrub (Arab.; Syria; HJP); Toverwortel (Dutch; EFS); Tufah ul Shitan (Arab.; DEP); Tuffahhel Maganin (Arab.; Syria; HJP); Ussul ul Lufah (Arab.; DEP); Ustrung (Arab.; DEP); Yabruhh (Arab.; Syria; HJP); Yavruhim (Aramaic; Mishnaic; ZOH); Yebruz (Beng.; WOI); Zauberwurzel (Ger.; HH2).

Activities (Mandrake):

Anesthetic (f1; BIB; CRC; SKJ; WOI); Anticholinergic (1; CRC; PH2); Antisialogogue (1; CRC); Aphrodisiac (f1; BIB; CRC; MAD); Cardiotonic (1; CRC); Cathartic (f; CRC); Cholagogue (f; CRC; EFS); Emetic (f; CRC); Emmenagogue (f; MAD); Hallucinogenic (f1; BIB); Hepatotonic (f; EFS); Hypnotic (f; CRC); Mydriatic (1; CRC; WOI); Narcotic (f1; CRC; EFS; MAD; SKJ); Nervine (f; CRC); Poison (1; CRC; SKJ); Purgative (f; CRC); Radioprotective (f; MAD); Refrigerant (f; CRC); Sedative (f1; BIB; CRC; HHB; MAD); Stimulant (f; CRC; EFS).

Indications (Mandrake):

Adenopathy (f; JLH); Arthrosis (f; CRC; MAD); Asthma (f; CRC; HHB; PH2); Bronchosis (f; HHB); Cancer, uterus (f; JLH); Colic (f; CRC; PH2); Condyloma (f; JLH); Convulsion (f; BIB; CRC); Cough (f; CRC; HH2); Depression (f; HHB); Duodenosis (f; HHB); Dysmenorrhea (f; HHB; HH2); Dystony (f; HHB); Frigidity (f; NAD); Gas (f; HHB); Gastrosis (f; PH2); Gout (f; MAD); Hayfever (f; CRC; PH2); Headache (f; HHB); Hemorrhoid (f; HHB); Hepatosis (f; CRC; HHB); Induration (f; JLH); Infertility (f; BIB); Inflammation (f; JLH); Kernel (f; JLH); Pain (f1; BIB; CRC); Pertussis (f; HHB; HH2; PH2); Rheumatism (f; CRC; HHB); Schizophrenia (f; BIB); Sciatica (f; HHB); Scirrhus (f; JLH); Sclerosis (f; CRC); Scrofula (f; BIB); Seborrhea (f; MAD); Sore (f; CRC); Swelling (f; NAD); Tenesmus (f; HHB); Tumor (f; CRC; JLH); Ulcer (f; CRC; HHB; PH2); Uterosis (f; JLH); Wound (f; MAD).

Dosages (Mandrake):

FNFF = X
15–30 drops tincture (HHB). Mostly homeopathic today (PH2).

- Asian Indians take root bark and leaves as anesthetics for painful swellings (NAD).
- Ancient Greeks used it as an anesthetic as Nazis used Datura, containing the same twilight-inducing alkaloids (HJP; JAD).
- Greeks soak the "love apple" in wine as an aphrodisiac (ZOH), believing it helps barren wives conceive (ZOH).
- Lebanese consider the root a powerful cathartic and emetic, using it for arthritis and exorcism (schizophrenia?) (HJP).
- Syrians (in Syria or Brooklyn) still regard this as the famous biblical aphrodisiac (HJP).

Downsides (Mandrake):

Class 3 (AHP). Not allowed in food products in Canada (AHP).

Extracts (Mandrake):

Atropine is anticholinergic, both central and peripheral. It tends to reduce secretions (gastric, intestinal, nasal, saliva, sweat, teats), decrease gastric and intestinal motility, and increase heart rate. It also causes pupil dilation, increases intraocular pressure, and photophobia; l-hyoscyamine and l-scopolamine have essentially the same activities except that scopolamine is a powerful hypnotic and usually slows the heart rate. Scopolamine-containing plants have been used as anesthetics for centuries in traditional Chinese medicine (BIB).

WILD MINT (BIBLICAL MINT) (*MENTHA LONGIFOLIA* (L.) L.) +++ LAMIACEAE

Synonyms:

Mentha asiatica Boriss.; *Mentha calliantha* Stapf.; *Mentha candican*s Crantz.; *Mentha capensis* Thunb.; *Mentha concolor* Stapf; *Mentha hamadanensis* Stapf.; *Mentha incana* Willd.; *Mentha longifolia* subsp. capensis; *Mentha longifolia* subsp. *hymalaiensis*; *Mentha longifolia* subsp. *longifolia*; *Mentha longifolia* subsp. *polyadenia*; *Mentha longifolia* subsp. *typhoides*; *Mentha royleana* Benth.; *Mentha spicata* var. *longifolia* L.; *Mentha sylvestris* L.; *Mentha viridis* var. *canascens* Fries *fide* DEP; HH2, etc.

Notes (Biblical Mint):

> **But woe unto you, Pharisees! for ye tithe mint and rue and all manner of herbs, and pass over judgment and the love of God: these ought ye to have done, and not to leave the other undone.**
>
> **Luke 11:42 (KJV)**

> **But woe to you Pharisees! for you tithe mint and rue and every herb, and neglect justice and the love of God; these you ought to have done, without neglecting the others.**
>
> **Luke 11:42 (RSV)**

> **But woe to you Pharisees, because you give the tenth of the mint and the rue and of every [other] vegetable, but you pass by the justice and the love of God! These things you were under obligation to do, but those other things not to omit.**
>
> **Luke 11:42 (NWT)**

I was really taking a leap of faith when I used the name "Biblical Mint" instead of "Wild Mint" as in CR2. Zohary makes me feel better when he says this species is the most common and medicinally used of the three species in Israel. It grows there in moist ditches, stream sides, and swamps (ZOH).

FIGURE 1.67 Wild Mint (Biblical Mint) (*Mentha longifolia*).

This species is highly variable, including at least 21 subspecies and circa 150 types. Jews, very scrupulous in paying the tithe, served mint with their meat dishes, especially at the Spring Feast of the Paschal Lamb. The ancient Hebrews, Greeks, and Romans used it. Jews strewed synagogue floors with mint so that its perfume scented the place. Pliny gave 41 remedies in which mint was considered efficacious (BIB; WOI; ZOH).

The three Israeli species are keyed as follows:

- Throat of calyx naked; floral verticels crowded:
  - — Calyx 3.5–4 mm long; verticels circa 2 cm in diameter — *Mentha aquatica*
  - — Calyx 1.2–1.5 mm long, verticels circa 1 cm in diameter — *Mentha longifolia*
- Throat of calyx hairy; floral verticels remote — *Mentha pulegium*

Common Names (Biblical Mint):

Baburi (Pun.; DEP); Belanne (Pun.; WOI); Biblical Mint (Eng.; JAD); Boo Dee Na (Burma; DEP); Byi Rug (Tibet; TIB); English Horsemint (Eng.; PH2); Fan-ho (Chi.; EFS); Gha-Gha (Arab.; GHA); Habak (Arab.; GHA); Hertsmint (Dutch; EFS), Horsemint (Eng.; EFS; FAC; USN; WOI); Hortelã (Mad.; JAD); Hortela silvestre (Por.; EFS), Jangli pudina (Lad.; MKK); Jungli Pudina (Hindi; WOI); Koshu (Pun.; WOI); Mastranzo nevado (Sp.; EFS), Menta salvatica (It.; EFS), Mentastio (It.;

EFS), Mentastro (It.; EFS), Mentastro longa (Sp.; EFS), Menthe sauvage (Fr.; EFS), Na'ana (Arab.; GHA); Nagbo (Iran; EFS), Padina (Nwp.; DEP); Pahadi-pudina (India; EFS), Pa-ho (Chi.; EFS), Podina (Hindi; DEP); Po-ho (Chi.; EFS), Pudang (Iran; EFS), Pudina (Bom.; WOI); Pudnakushma (Pun.; WOI); Ross-Minze (Ger.; EFS), Shamshahai (Pushtu; DEP); Sudanaj (Arab.; EFS), Ufuthane lomhlange (Zulu; ZUL); Vartalau (Bom.; DEP; WOI); Wild mint (Eng.; EFS; USN), Wilde Minze (Ger.; EFS), Wu-pa-ho (Chi.; EFS), Yabani nane (Tur.; EFS).

Activities (Biblical Mint):

Anodyne (f; BIB; ZOH); Antiacetylcholinesterase (1; TIB); Antiaggregant (1; MAM); Antiatherosclerotic (1; MAM); AntiHIV (1; X15058498); Antiinflammatory (1; TIB); Antileukotriene (1; TIB); Antioxidant (f1; TIB; X12802721); Antiprostaglandin (1; MAM); Antiradicular (1; X12802721); Antiseptic (f1; EFS; VAG; VVG); Antispasmodic (f1; VAG; VVG); Antithrombic (1; MAM); Antiviral (1; TIB; X15058498); Aphrodisiac (f; BIB); Astringent (f1; BIB; TIB); Bactericide (1; TIB); Candidicide (1; X12802721); Carminative (f1; BIB; DEP; EFS; FNF; PH2; ZOH); CNS Depressant (1; TIB); CNS Stimulant (1; TIB); Cyclooxygenase Inhibitor (1; MAM); Decongestant (f1; VAG; VVG); Diaphoretic (f; VVG); Diuretic (f; WBB; ZUL); Emmenagogue (f; ZUL); Fungicide (1; TIB; X12009988); Reverse-Transcriptase Inhibitor (1; X15058498); Sedative (1; TIB); Spasmogenic (1; TIB); Stimulant (1; EFS; PH2; ZOH); Stomachic (f; BIB).

Indications (Biblical Mint):

Adenopathy (f; VVG); Alzheimer's (1; MAM); Apoplexy (f; BIB; DEP; SKJ); Arthrosis (1; MAM); Asthma (f; ZUL); Backache (f1; FNF; WBB); Bacteria (1; TIB); Bronchosis (f; WBB; ZUL); Cancer (f; FNF); Candida (1; X12802721); Chest Colds (f; BIB); Childbirth (f; ZUL); Chill (f; GHA); Cold (f; BIB; WBB; ZUL); Cough (f; GHA; WBB; ZUL); Cramps (f; VVG); CVI (1; VVG); Dermatosis (f; BIB); Dysentery (f; MKK); Dysgeuzia (f; DEP); Dysmenorrhea (f1; FNF; ZUL); Dyspepsia (f; PH2; SKJ; ZUL); Dyspnea (f1; GHA; VVG); Dysuria (f; VAG); Epilepsy (f; VAG); Fever (f; BIB; HHB); Fungus (1; TIB; X12802721); Gas (f1; BIB; PH2; ZUL); Gastrosis (f; GHA); Headache (f; BIB; PH2; UPW; VAG; WBB; ZUL); HIV (1; X15058498); Hysteria (f; VVG; ZUL); Impotence (f; BIB); Induration (f; JLH); Infection (f1; TIB; X15058498); Inflammation (f1; TIB); Insomnia (f; VAG); Labor (f; ZUL); Mycosis (1; PH2; X12802721); Pain (f; DEP; ZOH); Pharyngosis (f; TIB); Respirosis (1; VAG; ZUL); Rheumatism (f; BIB; DEP; HHB; TIB); Sclerosis (f; JLH); Sinusitis (f; WBB); Sore Throat (f; TIB); Stomachache (f1; BIB; ZUL); Swelling (f; ZUL); UTI (f; VAG; VVG); Virus (1; TIB; X15058498); Wound (f1; FNF; TIB; UPW; WBB; ZUL).

Dosages (Biblical Mint):

FNFF = !!
As early as A.D. 37, mint was often mentioned in cooking recipes. The plant is eaten in chutnies, on cottage cheeses and curds, candied, and made into teas. Oil used as a substitute for peppermint oil for flavoring confectionery (BIB; FAC; WOI).

- Arabians take tea with honey for chill, dry cough, dyspnea, fever, headache, or gastrosis (GHA).
- Asian Indians of Trans-Indus use the decoction for fever and heat apoplexy (DEP).
- British considered mint juice with vinegar an aphrodisiac (stirred up venery and bodily lust) (BIB).
- Egyptian farmers regard the flowering tops and leaves as carminative (BIB).
- Himalayans use the herb as anodyne, astringent, and carminative for rheumatism (DEP).
- Nepalese use mint juice as antiseptic on wounds. The decoction is used for fever, and with *Glycyrrhiza* and *Bombax* for sore throat (TIB).

Extracts (Biblical Mint):

Amzazi et al. (2003) observed that nontoxic concentrations (10 μg/ml) of, in particular, methanol (Ext-1) and ethyl acetate (Ext-3) extracts significantly inhibited HIV infection by circa 40% and 55%, respectively. One extract showed significant inhibition of HIV-1 reverse transcriptase (X15058498). Iranian scientists (X15597306) showed that 1 μM piperitone potentiated nitrofurantoin 3- to 20-fold in different resistant strains of enterobacteriaceae (X15597306). Flavonoids, mainly flavones, may be the major inhibitors of HIV infection (X15058498). Fractions containing phenolics may exhibit CNS-stimulant and spasmogenic activities. But extracts or populations rich in essential oils may be CNS depressant and somnifacient (TIB). If antiacetylcholinesterase activity is useful in Alzheimer's disease, this biblical mint (like lemon balm, rosemary, sage, and watermint) is well endowed. (–)-Borneol, IC38 = 164 μg/ml; (–)-carvone, IC43 = 164 μg/ml; 1,8-cineole, IC50 = 41 μg/ml; elemol, IC50 = 34 μg/ml; (–)-limonene, IC27 = 164 μg/ml; (–)-linalol, IC37 = 164 μg/ml; (–)-linalyl-acetate, IC38 = 82 μg/ml; (+)-menthofuran, IC33 = 82 μg/ml; (–)-menthol, IC38 = 164 μg/ml; (–)-menthone, IC39 = 164 μg/ml; (–)-menthyl acetate, IC35 = 41 μg/ml; (+)-piperitenone –oxide, IC50 = 64 μg/ml; piperitenone, IC50 = 110 μg/ml; (–)-piperitonee, IC50 = 136 μg/ml; (+)-pulegone, IC50 = 136 μg/ml; and most potent, viridiflorol, IC50 = 25 μg/ml (JAF46:3434).

BLACK MULBERRY (*MORUS NIGRA* L.) +++ MORACEAE

Notes (Black Mulberry):

> **And to the end they might provoke the elephants to fight, they shewed them the blood of grapes and mulberries.**
>
> **I Maccabees 6:34 (KJV)**

> **They showed the elephants the juice of grapes and mulberries, to arouse them for battle.**
>
> **I Maccabees 6:34 (RSV)**

Not finding Maccabees in my Jehovah's Witness Bible, I do see that ba'ca bushes is used rather than mulberry in 2 Samuel 5:23, and again in 24, and in I Chronicle 14:14, and again in 15. Thus far, I have not figured out where ba'ca came from. And Zohary does not mention the ba'ca or mulberry of Samuel or Chronicles. According to some traditionalists, a palm tree is customarily planted in the courts of the southern Holy Land, while to the north, it is the mulberry, the pleasant juice of whose fruit, mingled with water, in which the sweet-scented violet has been infused, forms one of the most grateful kinds of sherbert (BIB). The red juice was used to incite the elephants of Antioch into battle. Burmans worship the mulberry, while some Europeans believe the devil stains his shoes black therewith. In old Palestine, both the black and white mulberry were cultivated to feed silkworms. In parts of China, they make a thick preserve on the 15th day of their first month. Trees are often planted as ornamentals. Berries are used to fatten sheep, after which their meat is believed to be more digestible (BIB). I suspect that most birds and few foresters, herbalists, publishing phytochemists and phytotherapists, and even plant taxonomists do not know which species, white, black, or red mulberry, respectively, *M. alba*, *M. nigra*, or *M. rubra*, they last ingested for edible or medicinal reasons. Further, I doubt that most chemical and overseas clinical trials have been vouchered. Still, I kept the species separate in my Herbal Desk Reference, knowing that I have white and black fruited trees volunteering, almost weeds here in the Green Farmacy Garden. I would myself use them generically for food and medicine, although as a botanist, I cannot prescribe such. See other species for other indications. I will bet that in 10 years, someone bolder than I will treat them all generically as mulberry, *Morus* spp.

FIGURE 1.68 Black Mulberry (*Morus nigra*).

Common Names (Black Mulberry):

Agud (Rom.; KAB); Amoreira (Mad.; Por.; EFS; KAB); Amoreira Negra (Por.; KAB); Amourié (Lan.; KAB); Black Mulberry (Eng.; FAC; KAB); Cel Tree (Eng.; BUR); Common Mulberry (Eng.; KAB); Dud (Rom.; KAB); Dut Agaci (Tur.; EFS); Dut Pekmezi (Tur.; EB49:406); European Mulberry (Eng.; TAN); Gelso (It.; EFS); Hei Sang (China; TAN); Karadut (Tur.; EB49:406); Maulbeerbaum (Ger.; EFS; KAB); ?Messikanu (Sumerian; ZOH); ?Mesucan (Heb.; ZOH); Meurier (Fr.; KAB); Moerbeiboom (Dutch; EFS); Moerbezieboom (Dutch; KAB); Mon (Cam.; KAB); Moral (Sp.; VAD); More Celse (?; JLH); More Matura (?; JLH); Morera negra (Cuba; Cat.; Sp.; EFS; KAB; RyM); Moro (It.; EFS); Mûrier (Fr.; EFS); Mûrier Noir (Fr.; KAB); Persian Mulberry (Eng.; FAC); Sang (China; EFS); Schwarzer Maulbeer (Ger.; MAD); Shahtut (Bal.; KAB); Shetura (Bom.; Guj.; NAD); Shetuta (Hindi; NAD); Sicaminum (?; JLH); Siccomorum (?; JLH); ?Sukannu (Heb.; ZOH); ?Sycamine (Heb.; ZOH); Tchelkovitsa (Rus.; KAB); Toola (Sanskrit; EFS); Tut (Arab.; India; EFS; GHA); Tût Aswad (Arab.; Syria; HJP); Tût Shami (Arab.; Syria; HJP); Tuta (Arab.; Iran; EFS; NAD); Tutovoi Drava (Rus.; KAB); Ud al Tut (Arab.; GHA).

Activities (Black Mulberry):

Analgesic (f1; VAD; X10817216); Anthelmintic (f; EFS); Antidiabetic (f1; VAD); Antidote (Aconite) (f; MAD); Antihemolytic (1; X15173994); Antiinflammatory (f1; VAD); Antinociceptive

(1; X10817216); Antioxidant (1; X15173994); Antiperoxidant (1; X15173994); Astringent (f; EFS); Capillariprotective (f1; VAD); Cathartic (f; DEM); Cicatrizant (f1; VAD); Depurative (f; BIB); Diuretic (f1; EFS; PNC; WOI); Emetic (f; DEM); Expectorant (f; MAD; PNC); Febrifuge (f; EFS); Hypoglycemic (1; VAD; WOI); Hypotensive (1; PNC; WOI); Lactagogue (f; DEP); Laxative (f1; EFS; MAD; PH2; PNC; VAD); Nervine (f; EFS); Purgative (f; BIB); Refrigerant (f; BIB; PNC); Taenicide (1; BUR); Tonic (f; EFS; GHA); Vermifuge (f; BIB; FEL).

Indications (Black Mulberry):

Alopecia (f; EB49:406); Baldness (f; EB49:406); Bile Problems (f; DEM); Bleeding (f; MAD); Bronchosis (f; PHR); Burn (f; MAD); Cancer (f; JLH); Cancer, spleen (f; JLH); Cancer, throat (f; JLH); Cancer, uterus (f; JLH); Catarrh (f; PHR); Constipation (f1; FEL; PNC; VAD); Cough (f; BIB); Depression (f; DEP); Diabetes (f1; GHA; MAD; WOI); Diarrhea (f; EFS; MAD); Dysentery (f; MAD); Dysmenorrhea (f; BIB; GHA); Dyspepsia (f; DEP); Enterosis (f; DEM); Fever (f; BIB; EFS; FEL); Gastrosis (f; JLH; EB49:406); GI Problem (f; DEM); Hemorrhoid (f1; VAD); Hepatosis (f; MAD); Herpes (f; EB49:406); High Blood Pressure (f1; GHA; WOI); Hysteria (f; MAD); Induration (f; JLH); Infection (f; EB49:406); Inflammation (f1; PH2; VAD); Insomnia (f; EFS); Menorrhagia (f; MAD); Mucososis (f; PH2); Pain (f1; VAD; X10817216); Pharyngosis (f; VAD); Phlebitis (f; VAD); Quinsy (f; FEL); Respirosis (f; PH2); Sore (f; MAD); Sore Throat (f; BIB; DEP; FEL; JLH; MAD; VAD); Splenosis (f; JLH); Stomachache (f; MAD; EB49:406); Stomatosis (f; VAD); Swelling (f; BIB); Tapeworm (f; FEL); Thirst (f; BIB); Tonsilosis (f; MAD); Toothache (f; BIB; MAD; VAD); Tumor (f; BIB); Uterosis (f; JLH); Varix (f1; VAD); Water Retention (f; EFS); Worm (f; EFS; FEL; MAD).

Dosages (Black Mulberry):

FNFF = !!

Fruits eaten raw, dried, juiced, or cooked or fermented. Widely cultivated for its edible fruit, which may be sun-dried and stored as winter food. Fruit should be picked very ripe, when they are sweet and better flavored. Easily squashed, they stain the skin. To remove stain, juice from unripe fruit is rubbed over the skin (BIB; FAC; TAN). 1.7–3.5 g fruit syrup (MAD); 2–4 ml fruit syrup (PHR; PH2). Steep 1 Tbsp in 1 cup of water for 10 minutes; take 3 cups/day (VAD). Use leaf tea (30–50 g/l) as gargle. Use root decoction (30–50 g/l) as gargle (VAD).

- Cambodians use the leaves for conjunctivitis (KAB).
- Iranians use root bark for dysmenorrhea (BIB).
- Lebanese may mix cow manure, crushed bilberry leaves, and olive oil, wrapped in fresh mulberry leaves, for earache (HJP).
- Pakistanis use leaf tea to lower blood pressure and blood sugar (GHA).
- Turks use concentrated fruit juice (1 tsp) as gargle or wash for herpes (Turkey; EB49:406).
- Yemeni use fruits infused with Salvadora as tonic to regulate menstruation (GHA).

Downsides (Black Mulberry):

Morus alba was classed Class 1. None known (PHR). No health hazards or contraindications reported with proper administration of suggested therapeutic dosages (PH2) (but PH2 designates no specific quantified dosage!; JAD). VAD says counterindicated in gastrosis, ulcers; tannins in leaf teas and root decoctions may irritate the mucosa (VAD).

Natural history (Black Mulberry):

Wild birds, poultry, and hogs are fond of fruits (BIB). Leaves of the species *Morus nigra* said to be inferior to those of *M. alba* for feeding silkworms (WOI).

Extracts (Black Mulberry):

Naderi et al. (2004) reported antioxidant effects of the fruit juice, hydroalcoholic and polyphenolic extracts, suggesting that the fruit protects against peroxidative damage to biomembranes and biomolecules (X15173994).

MYRTLE (*MYRTUS COMMUNIS* L.) ++ MYRTACEAE

Notes (Myrtle):

> **I will plant in the wilderness the cedar, the shittah tree, and the *myrtle*, and the oil tree; I will set in the desert the fir tree, and the pine, and the box tree together.**
>
> **Isaiah 41:19 (KJV)**

> **I will put in the wilderness the cedar, the acacia, the *myrtle*, and the olive; I will set in the desert the cypress, the plane and the pine together.**
>
> **Isaiah 41:19 (RSV)**

> **Instead of the thorn shall come up the fir tree, and instead of the brier shall come up the *myrtle* tree: and it shall be to the LORD for a name, for an everlasting sign that shall not be cut off.**
>
> **Isaiah 55:13 (KJV)**

> **Instead of the thorn shall come up the cypress; instead of the brier shall come up the *myrtle*; and it shall be to the LORD for a memorial, for an everlasting sign which shall not be cut off.**
>
> **Isaiah 55:13 (RSV)**

> **Instead of the thicket of thorns the juniper tree will come up; instead of the stinging nettle the *myrtle* tree will come up; and it must become for JEHOVAH something famous, a sign to time indefinite that will not be cut off.**
>
> **Isaiah 55:13 (NWT)**

At least the King James and the Revised Standard Versions are consistent as to translating myrtle. But the acuminate reader, looking at the quotes above, will see differences in the botanical common names in the different versions. That is why my third biblical account has more species than the first edition. The first quote rather suggests what might be a good approach, for example, in a deforested Haiti. The second shows some of the differences in the translations, RSV using cypress where KJV uses fir. Then the New World translation renders it as juniper. So the compiler is left with the quandary: which is it, cypress, fir, or juniper?

The useful myrtle has been grown for millennia for its fragrant, aromatic flowers, leaves, and bark. Jews collect myrtle to adorn their sheds and booths at the Feast of Tabernacles. Purplish-black berries known as mursins have medicinal value, and are also eaten. All parts of the plant are dried for perfume. Around Rabat, leaves are mixed with shampoos, believed to darken the hair. The shrub is still used today by Jews at the Feast of the Tabernacles. Sprigs with three leaves in a whorl (which are not common) are especially esteemed. Referred to chiefly as a symbol of divine generosity. Greeks consider it a symbol of love and immortality, and used it for crowning their priests, heros, and outstanding men. Emblematic of peace and joy in the Bible. To ancient Jews it was symbolic of peace and justice. In Jerusalem and Damascus bazaars, the flowers, leaves, and fruit are sold

FIGURE 1.69 Myrtle (*Myrtus communis*).

for making perfume. Arabs say that myrtle is one of three plants taken from the Garden of Eden because of its fragrance. Turkish and Russian leather is tanned aromatically with the bark and roots. It is said to have been a symbol of sensual love and passion to Venus and is placed on Bohemian caskets as a symbol of immortality. The leaves are used for massage to work up a glowing skin. The fragrant leaf oil is used in perfumery. The oil is also used in toilet waters, especially eaux de cologne. The wood is very hard and of interesting texture and grain. Plants are often ground for ornament, as it makes a nice evergreen hedge in appropriate Mediterranean climates (BIB).

COMMON NAMES (MYRTLE):

Aas (Arab.; BOU); Abhulas (Sin.; DEP); Adess (Eritrea.; KAB); Arrayán (Sp.; KAB; USN); As (Arab.; Iran; DEP; KAB; ZOH); Asbiri (Iran; DEP); Asu (Accadian; ZOH); Braut Myrte (Ger.; HHB); Echte Myrte (Ger.; HHB; HH2); Habhul (Pun.; DEP); Habulas (Arab.; Beng.; Iran; Pun.; Urdu; DEP; KAB); Hadas (Aramaic; Heb.; Yemen; GHA; ZOH); Halmoush (Arab.; BOU); Herbe du Lagui (Fr.; KAB); Hodem (Heb.; KAB); Isferem (Iran; DEP); Ismar (Iran; DEP); Kulinaval (Tam.; KAB); Mara (Bori; Sanjawi; KAB); Maurid (Iran; DEP); Mersin (Arab.; Tur.; BOU; EFS); Mirt (Rom.; KAB); Mirto (It.; Sp.; HH2; KAB; USN; VAD); Mirtus (Hun.; KAB); Mortella (It.; HH2); Motellina (It.; HH2); Murad (Hindi; Pun.; DEP; EFS; KAB); Murta (Cat.; Por.; EFS; KAB); Myron (Greek; ZOH); Myrsini (Greek; KAB); Myrt (Rus.; KAB); Myrt (Dutch; EFS); Myrte (Fr.; Ger.; BOU; KAB; MAD); Myrtem (Swe.; EFS); Myrtle (Eng.; Malta; Scn.; AH2; CR2; KAB); Rihan (Arab.; BOU; ZOH); Shalmun (Arab.; BOU); Sutre Sowa (Arab.; EFS); Sutrsowa (Arab.; Beng.; DEP; KAB); Tarihant (Ber.; BOU); Vilayati Mehndi (Hindi; Pun.; DEP); Wax Myrtle (Eng.; Ocn.; AH2); Yas (Oman; GHA).

ACTIVITIES (MYRTLE):

Alpha-glucosidase Inhibitor (1; X15704495); Analgesic (f; FEL; KAB); Antidiabetic (1; X15704495); Antiedemic (f1; MPI; PH2); Antigenotoxic (1; X15474415); Antiinflammatory (1; X15061659); Antimutagenic (1; X12889542); Antioxidant (1; X15061659); Antiradicular (1; X15474415); Antiseptic (f1; BIB; BOU; FEL; GHA; KAB; PH2; VAD); Antispasmodic (f; BIB); Astringent (f1; BIB; BOU; GHA; VAD); Bactericide (1; BIB; PH2); Bronchosecretolytic (1; PH2); Candidicide (1; FT75:74); Carminative (f; BIB; DEP); Cerebrotonic (f; KAB); Chemopreventive (1; X15474415); Cicatrizant (1; VAD); Collyrium (f; BIB); Culicide (1; X11997977); Decongestant (f; BOW); Deodorant (1; FEL; PH2); Depressant (1; PH2); Digestive (f1; VAD); Diuretic (f; KAB); Emmenagogue (f; KAB); Expectorant (f; VAD); Fungicide (1; PH2); Gram(+)–icide (1; VAD); Hemostat (f1; GHA; KAB; VAD); Hypoglycemic (1; PH2; X15234770); Insecticide (f1; GHA; X11997977); Larvicide (1; X11997977); Laxative (f; KAB); Parasiticide (1; BIB; FT68:276); Pectoral (f; BIB); Propecic (f; KAB); Rubefacient (f; BIB); Sedative (f1; BIB; VAD); Stimulant (f; BIB; BOU; GHA); Stomachic (f; BIB; HHB); Tonic (f; BIB); Vermifuge (f; BIB).

INDICATIONS (MYRTLE):

Abscess (f; BOU); Acne (f; BOW); Adenopathy (f; JLH); Alopecia (f; DEP; KAB); Aphthae (f; BIB; DEP; KAB); Aposteme (f; JLH); Arthrosis (f1; MPI); Asthma (f; BOU); Bacteria (1; BIB; WOI); Bleeding (f; BIB; DEP; GHA; KAB); Blister (f; GHA); Boil (f; BOU); BPH (f; PH2); Bronchosis (f1; BIB; FEL; HHB; PH2); Cacoethes (f; BIB); Cancer (f; JLH); Cancer, breast (f; JLH); Cancer, colon (f; JLH); Cancer, gum (f; JLH); Cancer, liver (f; JLH); Cancer, spleen (f; JLH); Cancer, throat (f; JLH); Cancer, uterus (f; JLH); Candida (1; FT75:74); Catarrh (f; FEL); Cerebrosis (f; BIB; DEP); Chest (f; BIB); Childbirth (f; GHA); Cold (1; PH2); Colic (f; GHA); Condylomata (f; BIB); Conjunctivosis (f; BOU; FEL); Cough (f; GHA; MAD); Cystosis (f1; BIB; FEL; KAB; PH2); Diabetes (f1; X15704495; X15234770; X6482478); Diarrhea (f1; BIB; BOU; GHA; MAD; PH2; VAD); Dropsy (f; MAD); Dysentery (f; BIB); Dyspepsia (f; BIB; DEP; KAB); Eczema (f; BIB); Edema (f1; MPI); Emphysema (f; VAD); Enterosis (f; GHA; JLH); Epilepsy (f; BIB; DEP; WOI); Epistaxis (f; GHA); Fatigue (f; PH2); Fever (f; BIB; GHA); Fibroid (f; JLH); Fungus (1; FT75:74); Gangrene (f; FEL); Gas (f; DEP); Gastrosis (f; BIB; DEP; KAB; MAD); Gingivosis (f; BOW; JLH); Gonorrhea (f; MAD); Gray Hair (f; BIB; BOU); Headache (f; BIB; GHA); Hemorrhoid (f; FEL; PH2; VAD); Hepatosis (f; BIB; DEP; JLH; KAB; WOI); Induration (f; JLH); Infection (1; BOU; PH2; VAD); Inflammation (1; MPI); Insomnia (f1; VAD); Intertrigo (f; FEL); Laryngitis (f; VAD); Leishmania (1; FT68:276); Leukorrhea (f; BIB; FEL; PH2); Mastosis (f; JLH); Menorrhagia (f; FEL); Nephrosis (f; FEL; VAD); Night sweats (f; BIB); Otosis (f; PH2; VAD); Pain (f; Bou; FEL; MAD); Parotosis (f; JLH); Pertussis (1; PH2); Pharyngosis (f; FEL; VAD); Phthisis (f; BIB); Pleurodynia (f; MAD); Polyp (f; BIB; JLH);

Proctosis (f; JLH); Prolapse (f; BIB); Prostatitis (f; VAD); Pulmonosis (f1; BIB; BOU; MAD; VAD); Pyelitis (f; BIB; VAD); Respirosis (f; BOU; KAB); Rheumatism (f; BIB; DEP; KAB); Sinusitis (f; PH2); Smallpox (f1; BIB; BOU); Sore (f; BIB; GHA; KAB); Sore Throat (f1; VAD); Splenosis (f; JLH); Sting (f; GHA); Stomatosis (f; KAB); Tonsilosis (f; JLH); Tuberculosis (1; MAD; PH2; WOI); Ulcer (f; DEP; KAB); Urogenitosis (f; BIB); Uterosis (f; BIB; JLH); UTI (f; BOW); Vaginosis (f; BOW; VAD); Wart (f; JLH); Whitlow (f; BIB); Worm (f; PH2); Wound (f1; BIB; GHA).

Dosages (Myrtle):

FNFF = !!
Dried or fresh fruits a spice; buds eaten in Italy and used in liqueurs. Leaves, made into tea, considered an alternative to *buchu*. The sprigs were formerly added to wine to increase its potency. Leaf oil used to replace dried leaves in various culinary compositions, especially table sauces. Green and dried fruits sometimes used as a condiment (FAC; HOS; TAN). Fifteen to 40 grains powdered myrtle for catarrh, nephrosis, and phthisis (FEL). one to 2 drops, several times a day (MAD); 15 to 30 g leaf/liter water for tea, but take only 0.2 g leaf per day (PH2).

- Arabs take ground leaves orally for nosebleeds (GHA).
- Algerians recommend the leafy infusion for asthma (BIB).
- Arabians take leaf tea taken with lemon and salt for colic (GHA).
- Iranians make a hot poultice for boils from the plant (BIB).
- Lebanese consider the plant binding and diuretic, believing it holds loose things in place, the bowels, the emotions, or the teeth (HJP).
- North Africans use the dry flower buds for smallpox (BOU).
- Saudis use leaves as astringent, antiseptic, and stimulant; mixed with other herbs and taken 40 days after childbirth (GHA).
- Unani direct smoke from the leaves onto hemorrhoids, using the fruit for bronchitis, headache, and menorrhagia (KAB).
- Yemeni use dry leaf decoction for cough, diarrhea, fever, and headache (GHA).

Downsides (Myrtle):

No health hazards or contraindications reported with proper administration of suggested therapeutic dosages (PH2). Large doses may cause diarrhea, nausea, and vomiting. More than 10 g myrtle oil can threaten life, due to high cineole content (myrtle contains 135–2250 ppm cineole according to my calculations, meaning 10 g myrtle would contain a maximum 22.5 mg cineole). Several herbs may attain higher levels of cineole, including bay, bee balm, betel pepper, biblical mint, boldo, cajeput, cardamom, eucalyptus, ginger, greater galangal, horse balm, hyssop, lavender, nutmeg, rosemary, sage, spearmint, star anise, sweet annie, thyme, and turmeric. So reductionisticall assuming no synergies or antagonisms or additivities, a ridiculous assumption, one would assume that any good (and evils) accruing to the cineole in myrtle should apply even more so to those listed above, which theoretically at least, may attain higher levels of cineole, some (e.g., cardamom) attaining levels more than 20-fold higher. Symptoms of this alleged cineole intoxication may include circulatory disorders, collapse, lowered blood pressure, and respiratory failure. Do not apply oil to face of children as it may trigger asthma-like attacks, bronchial spasm, glottal spasms, or even respiratory failure (PH2).

Extracts (Myrtle):

Of five herb studies, Traboulsi et al. (2002) (X11997977) reported that myrtle extracts (flowers and leaves) were most active against fourth-instar larvae of the mosquito *Culex pipiens molestus*. LC50 = 16 mg/l (X11997977). Hayder et al. (2004) (X15474415) reported on antigenotoxic and free-radical scavenging activities of myrtle-leaf extracts. Several different extracts significantly decreased the SOS response, suggesting their potential in chemoprevention (X15474415). Onal et al. (2005) (X15704495) showed that alpha-glucosidase inhibition by aqueous extracts might help with diabetes mellitus (IC50 = 38 μg/ml) (X15704495). Aqueous and ethanolic extracts of leaves, roots, and stems are active against Gram-negative and Gram-positive bacteria. The plant contains antibacterial phenols. One thermolabile principle was highly active against *Micrococcus pyogenes* var. *aureus*. The principle resembled streptomycin in its action on *Mycobacterium tuberculosis* (WOI). Aqueous berry extract active against carrageenan-induced edema in the rats paw. comparable to oxyphenylbutazone] Mono- and sesquiterpenes antibacterial, antifungal, and antiseptic (PH2). EO or extract larvicidal (LC50 = 16 mg/l) (X11997977). Methanol seed extracts active against clotrimazole-resistant *Candida albicans* (MinIC = 1.25 mg/kg) (FT75:74). One fraction strongly inhibited alpha-glucosidase (IC50 = 38 μg/ml) (X15704495).

DAFFODIL (*NARCISSUS TAZETTA* L.) X AMARYLLIDACEAE

NOTES (DAFFODIL):

The wilderness and the solitary place shall be glad for them; and the desert shall rejoice, and blossom as the rose.

Isaiah 35:1 (KJV)

Widely cultivated as a beautiful ornamental, this narcissus also has very aromatic flowers, in the Holy Land white outside, golden yellow inside, and to me more aromatic than our American roses. Zohary is skeptical about this representing either the biblical lily or the biblical rose, as some non-Israeli writers had suggested. Of the two species native to Israel, this one seems common in damp alluvial soils and also on rocky hillsides, even in the northern Negev, where it flowers in November, going dormant in February. It grows wild in the desert from the Mediterranean Sea to the center of Palestine, near Joppa. Palestinians, like me, delighting in the fragrance, carry blossoms onto the streets and into their homes. Flowers yield an essential oil, the absolute of which is a valuable adjunct in high-grade French perfume. It blends well with jasmine perfume. The plant holds a deserved place among oriental medicinals. Bulbs are imported into Bombay, dried, sliced, and sold. The root extract in olive oil is a skin lotion, in vinegar a shampoo rinse, and in brandy an aphrodisiac to be used only with great caution. The anticancer folklore is interesting because narcissus has shown genuine antitumor activity, perhaps due to the presence of antitumor compounds such as lycorine and benzaldehyde. *Hager's Handbook* suggests that this species and *N. pseudonarcissus* share the same activities and indications, and possibly many common names as well (BIB, HHB; HJP; ZOH).

COMMON NAMES (DAFFODIL):

Ain Serdouk (Arab.; BOU); Behar (Arab.; BOU); Berengat (Arab.; BOU); Bunchflower Daffodil (Eng.; USN); Bunchflower Narcissus (Eng.; USN); Chinese Sacred Lily (Eng.; USN); Daffodil (Eng.; CR2; USN); Dohn en Nirdjis (Arab.; JLH); French Daffodil (Eng.; BOU); Ghafu (Arab.; Syria; HJP); Irisa (Pun.; DEP; WOI); Jirundus (Arab.; Syria; HJP); Khenounat en-Nebi (Arab.; BOU); Nardjes (Arab.; BOU); Narciso (Sp.; EGG); Narcisse (Fr.; BOU); Narcisse Tazette (Fr.; BOU); Narcissus (Eng.; HJP; ZOH); Nargis (Arab.; Pun.; BOU; DEP; HJP); Nowar Zouawa (Arab.; BOU); Polyanthus Narcissus (Eng.; USN; WOI); Ranjis (Arab.; Syria; HJP); Shoshan (Heb.; ZOH); Shui Hsien (China; JLH); Shui Xian Hua (Pin.; DAA); Tazetta (Eng.; Australia); Teif Eddib (Arab.; BOU); Tikheloulin en-Nebi (Ber.; BOU).

ACTIVITIES (DAFFODIL):

Abortifacient (f; CRC; HHB); Analeptic (1; FNF); Analgesic (f; BIB; CRC); Antiacetylcholinesterase (1; X10869205); Anticancer (1; CRC; FNF; X3802955); Antidote (f; BIB); AntiHIV (1; X10934347); Antiinflammatory (f; CRC); Antileukemic (1; X3368194); Antimalarial (1; X14669261); Antimitotic (1; X5340258); Antiproliferative (1; X15522215); Antisyncytial (1; X11012085); Antiviral (1; CRC; HHB; X176907); Aphrodisiac (f; BIB); Cholinergic (1; FNF); Convulsant (1; DAA); Cytotoxic

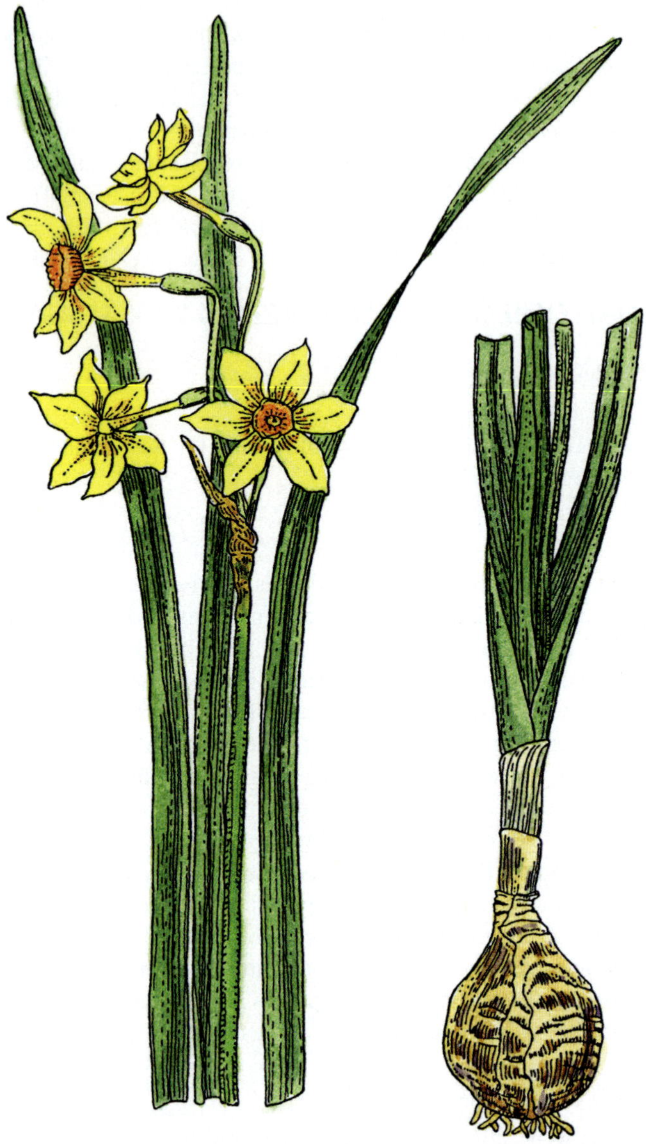

FIGURE 1.70 Daffodil (*Narcissus tazetta*).

(1; CRC; X11683132); Demulcent (f; BIB); Diuretic (f; CRC; WOI); Emetic (f1; CRC; DEP; FNF; NAD); Febrifuge (f; CRC); Hemagglutinant (1; X10945441); Immunomodulatory (1; X15522215); Memorigenic (1; BRU; FNF); Neurotonic (1; FNF); Paralytic (1; DAA); Peristaltic (1; FNF); Poison (1; CRC; DAA); Purgative (f; CRC; WOI); Sedative (1; FNF); Sialogogue (1; FNF).

Indications (Daffodil):

Abscess (f; CRC; DAA); Alzheimer's (1; FNF; X10869205); Boil (f; CRC); Bronchosis (f; HHB); Cancer (f1; CRC; FNF; HHB); Cancer, breast (f1; BIB; FNF); Cancer, ear (f1; BIB; FNF); Cancer, lung (1; X3802955); Cancer, uterus (f; JLH); Constipation (f; CRC); Cramp (f; HHB); Dermatosis (f; CRC; DAA); Diarrhea (f; HHB); Dysentery (f; HHB); Epilepsy (f; BIB); Esophagosis (f; CRC); Fever (f; CRC);

Fit (f; BIB); Gynecopathy (f; DAW); Headache (f; DEP); HIV (1; X10934347); Induration (f; JLH); Infection (1; X176907); Inflammation (f; CRC; DAA); Itch (f; CRC; DAA); Leukemia (1; X3368194); Malaria (1; X14669261); Mastosis (f; BIB; CRC); Myasthenia (1; FNF); Neurosis (1; FNF); Ophthalmia (f; CRC; DAA); Pain (f; CRC; DAA); Pertussis (f; HHB); Polio (1; FNF); Rhinosis (f1; HHB; X176907); Rhinotracheitis (1; X176907); Sclerosis (f; BIB); Sinew (f; JLH); Sore (f; CRC; HHB; HJP); Swelling (f; CRC); Tumor (f; JLH); Ulcer (f; CRC); Uterosis (f; JLH); Virus (1; CRC X10934347).

Dosages (Daffodil):

FNFF = X
Facciola mentions only *Narcissus jonquili* as having edible flowers (FAC).

- Chinese apply the bulb as a liniment on tumors (JLH).
- Chinese used bulb as a demulcent bolus to carry bones out of the esophagus (DAA).
- Chinese use juice of the bulb for eye ailments (DAA).
- Lebanese pour boiling water over flowers, and steep 2 or 3 minutes as a stomach tonic (HJP).
- Lebanese use roots for epilepsy and fits (petit mal, sometimes worsened by bulbs) (HJP).
- North Africans ingest ground dried bulbs for poisoning and ill health (BOU).
- North Africans apply crushed bulbs mixed with honey to burns (BOU).
- Orientals use dried flowers for female fevers (LMP).

Downsides (Daffodil):

Narcissus has been suspected of giving off an evil emanation, producing dullness of the intellect, insanity, and even death (BIB). In animals, it may induce enterosis, fever, gastrosis, and rapid pulse; larger doses can cause convulsions, paralysis, even death (DAA).

Natural History (Daffodil):

Flowers are visited by long-tongued insects, attracted by the intense contrasting yellow and white, and the pleasant aroma, stronger by night (ZOH).

Extracts (Daffodil):

One study more than 30 years ago hints that narcissus might be useful in some sorts of bird flus (Papas et al., 1973), at least inhibiting DNA polymerase of avian myeloblastosis virus (X4123212). Galanthamine, an alkaloid found in this and other species of *Narcissus*, and in *Galanthus* and *Leucojum*, is now marketed in the United States as a prescription drug for Alzheimer's. Chu and Ng (2004) reported immunomodulatory and antiproliferative activities for a glutamine-rich antifungal peptide from bulbs of the Chinese daffodil (X15522215). Sener et al. (2003) found antimalarial activity in Amaryllidaceae. 6-Hydroxyhaemanthamine, haemanthamine, and lycorine were most potent against *P. falciparum*, and galanthamine and tazettine were least potent (X14669261). The last four are reported for this species.

JATAMANSI (*NARDOSTACHYS GRANDIFLORA* DC) ++ VALERIANACEAE

Synonyms:

Fedia grandiflora Wall. ex DC; *Nardostachys gracilis* Kitamura; *Nardostachys jatamansi* auct.; *Patrinia jatamansi* auct.; *Valeriana jatamansi* Wall. *fide* DEP; EFS

Notes (Jatamansi):

Thy plants are an orchard of pomegranates, with pleasant fruits: camphire with spikenard.

Song of Solomon 4:13 (KJV)

FIGURE 1.71 Jatamansi (*Nardostachys grandiflora*).

With names such as spikenard (*Nardostachys*) and nard (*Cymbopogon*) long confused, it is difficult to know which was meant in biblical texts, but most biblical scholars, including Zohary, seem to favor *Nardostachys* as the biblical spikenard. Zohary states that in biblical times, spikenard came from India with such spices as cassia and cinnamon. Spikenard was then used in incenses and perfumes, but is today all but obsolete (ZOH). For more than a century, the alpine plant was so rare that Bhutan prohibited export, according to DEP, which also mentions *Nardostachys* as the spikenard of the Song of Solomon, St. John, and St. Mark. Because it was reportedly poured, it must have been an oil rather than a balm. Scholars assume it may then have been a mix of oils, much as used by modern Indian women to wash their hair (DEP). Occurring in alpine India and vicinity, elevation 11,000 to 17,000 feet, the plant is endangered in the wild. But spikenard is still valued in India, though endangered. The drug Jatamansi, or Nardus Root, consists of short, thick, dark grey rhizomes crowned with reddish brown tufted fibrous remains of the petioles of radical leaves. The rhizome is used in medicinal oils; reported to promote hair growth, imparting blackness (BIB; DEP; ZOH). Spikenard oil possesses antiarrhythmic activity of possible therapeutical use in auricular flutter; it is less effective than quinidine, but less toxic. In moderate doses it has a distinct depressant action on the central nervous system; and relaxes the skeletal and smooth muscles.

Common Names (Jatamansi):

Achte Narde (Ger.; NAD); Akashamansi (Sanskrit; KAB); Amritajata (Sanskrit; KAB); Balachhada (Urdu; KAB); Balacharea (Bom.; DEP; NAD); Balcchar (Hindi; DEP); Balchir (Hindi; Pun.; DEP; NAD); Balchkar (Ic.; KAB); Balu Char (Hindi; DEP); Bekh Kurphus (Behar; DEP; KAB); Bhultya (Nepal; NPM); Bhutajata (Sanskrit; KAB); Bhutakeshi (Sanskrit; KAB); Bhutijatt (Kas.; KAB; WOI); Bhut Jatt (Kas.; DEP; NAD); Bhytajata (Sanskrit; NAD); Billi Lotan (Dec.; DEP); Chakravartini (Sanskrit; KAB); Epi de Nard (Fr.; KAB); Espica Nardo (Sp.; HH2; KAB); Espiga Nard (Cat.; KAB); Gandhamansi (Sanskrit; KAB); Gan Song (Pin.; DAA; HH2); Gauri (Sanskrit; KAB); Haswa (Nepal; DEP; WOI); Hinsra (Sanskrit; KAB); Hint Sümbülü (Tur.; EFS); Indian Nard (Eng.; CR2; WOI); Indian Spikenard (Eng.; NAD); Indian Valerian (Eng.; NAD); Indische Narde (Ger.; KAP); Jadamansi (Sanskrit; KAB); Janani (Sanskrit; KAB); Jaramanshi (Sin.; NAD); Jaramansi (Sin.; DEP; KAB); Jatala (Sanskrit; KAB); Jatalasi (Hindi; KAB); Jatamamshi (Kan.; Mal.; Tel.; DEP; KAB; NAD); Jatamanchi (Mal.; NAD); Jatamangsi (Nepal; DEP; WOI); Jatamansi (Ayu.; Beng.; Bhutan; Eng.; Gurung; Hindi; Nepal; Sanskrit; Scn.; AH2; CR2; DEP; KAB; NPM; WOI); Jatamashi (Hindi; Pun.; Tam.; DEP; NAD; WOI); Jatamasi (Guj.; DEP; NAD; WOI); Jatamavashi (Kan.; NAD); Jatamavshi (Mar.; NAD; WOI); Jatamensis (Ic.; KAB); Jatamsi (Tel.; NAD); Jatavali (Sanskrit; KAB); Jati (Sanskrit; KAB); Jatila (Sanskrit; KAB); Jeta Manchi (Mal.; DEP; NAD); Jetamanshi (Mal.; EFS; KAB); Jetamavashi (Kan.; DEP; KAB); Jhatamansi (Dec.; DEP); Kalicchad (Guj.; DEP; KAB; WOI); Kan Sung (China; HH2); Kan Sung Hsiang (China; EFS; KAB); Kanuchari (Hindi; KAB); Keshi (Sanskrit; KAB); Khasambhava (Sanskrit; KAB); Kiratini (Sanskrit; KAB); Kravyadi (Sanskrit; KAB); Krishnajata (Sanskrit; KAB); Kukilipot (Kas.; DEP; KAB; NAD; WOI); Laghumansi (Sanskrit; KAB); Limasha (Sanskrit; KAB); Mansi (Sanskrit; KAB); Mansini (Sanskrit; KAB); Masi (Garhwal; DEP; KAB; WOI); Mata (Sanskrit; KAB); Mishika (Sanskrit; KAB); Misi (Sanskrit; KAB); Mountain Nard (Eng.; EFS); Mrigabhaksha (Sanskrit; KAB); Muskroot (Eng.; EFS; NAD); Naird (Heb.; ZOH); Nalada (Sanskrit; KAB; ZOH); Naorochi (Khaling; NPM); Narada (Sanskrit; ZOH); Nard (Eng.; Heb.; BIB; ZOH); Nardenähre (Ger.; USN); Nard du Gange (Fr.; KAB); Nard Indien (Fr.; CR2); Narde Indike (Greek; NAD); Nardin (Arab.; Iran; Syria; ZOJ); Nard Indique (Fr.; KAB); Nardo Indico (Sp.; KAB); Nardos (Greek; ZOH); Nard Syriaque (Fr.; KAB); Naswa (Nepal; DEP; WOI); Naswan (Newari; NPM); Nerd (Heb.; ZOH); Niralambi (Sanskrit; KAB); Pampe (Bhutan; DEP); Pampi (India; DEP); Pangbu (Sherpa; NPM); Parvatavasini (Sanskrit; KAB); Paumpe (Bhutan; DEP); Peshi (Sanskrit; KAB); Peshini (Sanskrit; KAB); Pishachi (Sanskrit; KAB); Pishita (Sanskrit; KAB); Poi (Tamang; NPM); Putena (Sanskrit; KAB); Sambul (Arab.; HH2); Sevali (Sanskrit; KAB); Shvetakeshi (Sanskrit; KAB); Spang-Spos (Tibet; NPM); Span Pos (Tibet; KAP); Spicanard (Fr.; HH2); Spignard (It.; EFS); Spikenard (Eng.; KAB; NPM); Sukshmajatamansi (Sanskrit; KAB); Sukshmapatri (Sanskrit; KAB); Sumbul (Bom.; DEP); Sumbula theeb (Iran; NAD); Sumbulul-Assaffir (Arab.; DEP; KAB); Sumbul -i'l hind (Arab.; DEP; NAD); Sunbuluttibe-Hindi (Arab.; DEP); Sunbuluttib (Iran; DEP; KAB; NAD); Tamasi (Sanskrit; KAB); Tapasvini (Sanskrit; KAB); Tapaswini (Sanskrit; NAD); True Spikenard (Eng.; EFS); Vahini (Sanskrit; KAB); Valériane d'Inde (Fr.; EFS).

Activities (Jatamansi):

Alexiteric (f; KAB); Anthelmintic (f1; MPI); Antiarrhythmic (1; MPI; WOI; JAC7:405; X13331599); Anticonvulsant (1; PH2; MPI; WOI; X580202); Antidote (f; PH2); Antidyspneic (1; MPI); Antiemetic (1; KAP; MPI); Antiestrogenic (1; MPI); Antihistaminic (1; MPI); Antiischemic (1; X12479970); Antioxidant (1; X9102390); Antiperoxidant (1; X9102390); Antiseptic (1; MPI; SKJ; SUW); Antiserotonin (1; MPI); Antispasmodic (f1; DEP; KAB; MPI); Antiulcer (1; HH2; PH2; X580202); Bactericide (1; MPI); Bitter (f; KAB; SUW); Bronchodilator (1; MPI); Candidicide (1; MPI); Cardioprotective (f1; WOI); Carminative (f; KAB; SUW); CNS Depressant (f1; MPI; WOI; X11143748); Deobstruent (f; DEP; MPI); Deodorant (f1; DAA); Depurative (f; DEP; KAB); Diuretic (f; DEP; KAB; MPI; SUW); Emmenagogue (f1; AHP; DEP; KAB; NAD; SUW); Febrifuge (f; KAB); Fungicide (1; KAP; MPI; JAC7:405); GABA-nergic (1; X8202559); Hepatoprotective (1; X10940571); Hypocholesterolemic (f; JAC7:405); Hypolipidemic (1; X3215683); Hypotensive (1; KAP; MPI; X13522275); Laxative (f; KAB; SKJ; SUW); Lipogenic (f; KAB); Narcotic (1; WOI); Nervine (f; BIB); Neuritogenic (1; X14604758); Neurosedative (f; EFS); Neurotonic (f; EFS); Orexigenic (f; KAB; NAD); Pectoral (f; KAB); Propecic (f; DEP; KAB; MPI); Sedative (f12; MPI; NAD; SKJ; WOI; X580202); Serotoninergic (1; X8202559); Stimulant (f; KAB; MPI; SUW); Stomachic (f; KAB; SUW); Taenicide (1; MPI); Tonic (f; DEP; KAB; SUW); Tranquilizer (f1; KAP; MPI; X580202); Uterotonic (1; AHP).

Indications (Jatamansi):

ADD (2; MPI); Adenitis (f; NAD); Aggressiveness (1; KAP); Alopecia (f; KAB; MPI); Amenorrhea (f; DEP); Anorexia (f; KAB; NAD); Arrhythmia (1; MPI; X13331599); Asthma (1; MPI); Bacteria (1; MPI); Biliousness (f; KAB); Bite (f; KAB); Bronchosis (f1; KAB; MPI; SKJ); Cancer, abdomen (f; JLH); Cancer, bladder (f; BIB); Cancer, eye (f; JLH); Cancer, kidney (f; JLH); Cancer, larynx (f; BIB; JLH); Cancer, liver (f; JLH); Cancer, mouth (f; JLH); Cancer, rectum (f; JLH); Cancer, spleen (f; JLH); Cancer, stomach (f; JLH); Cancer, uterus (f; JLH); Cancer, vagina (f; JLH); Candida (1; MPI); Cardiopathy (1; KAB; MPI; NAD); Cerebrosis (f; NAD); Childbirth (f; EB24:259); Cholera (f1; DEP; MPI; SUW; EB24:259); Chorea (f; BIB); Circulosis (f; MPI); Colic (f; KAB; MPI; SUW); Complexion (f; KAB); Consumption (f; SKJ); Convulsion (f1; DEP; KAB; MPI; SKJ; SUW; X580202); Cough (f; KAB; MPI); Cramp (f; DEP; NAD; PH2); Debility (f; NAD); Delirium (f; EB24:259); Depression (f; BIB); Dermatosis (f; KAB); Dysentery (f; BIB); Dysmenorrhea (f; SKJ); Dyspnea (1; MPI); Dysuria (f; SKJ); Enterosis (f; JLH; KAB; PH2; SUW); Epilepsy (f; DEP; KAB; PH2; SUW); Erysipelas (f; KAB); Escherichia (1; MPI); Fever (f; KAB); Fungus (1; KAP; X749414); Gas (f; DEP; KAB; KAP; PH2); Gastrosis (f; DEP; JLH; KAP; PH2); Gleet (f; KAB); Gray Hair (f; DEP); Headache (f; KAP; PH2); Hemorrhoid (f; NPM); Hepatosis (f1; JLH; NAD; PH2; X10940571); High Blood Pressure (12; KAP; MPI; X13522275); High Cholesterol (f; JAC7:405); High Triglycerides (1; JAC7:405); Hyperkinesis (2; MPI); Hysteria (f; DEP; KAB; NAD; SUW); Infection (1; X749414); Inflammation (f; KAB; PH2); Insanity (f; BIB); Insomnia (f1; HH2; MPI; PH2; X580202); Ischemia (1; X12479970); Jaundice (f; DEP; NAD; PH2); Kidney stone (f; NAD); Leprosy (f; KAB; MPI; NAD); Lumbago (f; KAB); Madness (f; MPI); Malaria (f; DAA); Menopause (f; NAD; PH2); Mycosis (1; KAP; MPI; X749414); Nausea (f1; MPI; PH2); Nephrosis (f; JLH; KAB; NAD; PH2); Neurasthenia (1; MPI); Neurosis (f; NAD; PH2; ZOH); Obesity (1; X3215683); Ophthalmia (f; KAB; PH2); Pain (f; KAB; PH2); Palpitation (f1; DEP; KAB; MPI; NAB; WOI); Phymata (f; JLH); Proctosis (f; JLH); Respirosis (f; DEP); Rhinosis (f; SKJ); Salmonella (1; MPI); Scirrhus (f; JLH); Sclerosis (f; JLH); Smallpox (f; MPI; SKJ; EB24:259); Snakebite (f; KAB); Sore (f; KAB; MPI; EB24:259); Sore Throat (f; KAB); Spasm (f; BIB); Spermatorrhea (f; NAD); Splenosis (f; JLH); Staphylococcus (1; MPI); Sting (f; KAP; NAD); Stomachache (f; NAD; PH2); Stone (f; NAD); Streptococcus (1; MPI); Stress (f1; KAP; MPI; X580202); Syncope (f; EB24:259); Tachycardia (1; MPI; WOI); Tuberculosis (f; NAD; EB24:259); Tumor (f; BIB; JLH); Ulcer (1; HH2; PH2); Uterosis (f; PH2); Vaginosis (f; JLH); Worm (f1; MPI); Wound (f; KAB); Yeast (1; MPI).

Dosages (Jatamansi):

FNFF = X

No food references uncovered. 10–20 grains powdered root (DEP; HH2); 14–56 ml fluid extract (KAP); 28–56 ml infusion (KAP); 28–56 ml decoction (KAP); 1–1.5 g powdered herb (KAP). 0.6–1.3 g drug as single dose (PH2). 5 g root 3 ×/day (PH2). Wine glass full 3 x/daily (1:10 tincture or 1:40 infusion with 2 g max per single dose) (PH2).

- Asian Indians suggest 10 to 20 grains or 1 to 2 oz root in tea for chorea, flatulence, heart palpitations, and hysteria (DEP; NAD).
- Ayurvedics, viewing the roots alexipharmic, antipyretic, depurative, fattening, and tonic; suggest it for the complexion, and biliousness, blood disorders, dermatosis, erysipelas, leprosy, sores, and throat problems (KAB).
- Chinese suggest the plant for chorea, colic, coronary palpitations, hysteria, and malaria (DAA).
- Indochinese use the rhizome for epilepsy and hysteria (KAB).
- Iranians use the root tea for cardiac and nervous disorders (BIB).
- Nepalese suggest pasting the rhizome onto hemorrhoids (NPM).
- Santals use the plant for bronchitis, cholera, convulsions, cough, epilepsy, madness, smallpox, ulcers, and unconsciousness after childbirth (MPI).
- Unani, viewing the root as carminative, diuretic, emmenagogue, orexigenic, pectoral, stimulant, stomachic, and tonic, recommend it for baldness, chest pain, cough, enteritis, gleet, gray hair, lumbago, nephrosis, and wounds (KAB).

Downsides (Jatamansi):

Class 2b. Emmenagogue/uterine stimulant (AHP). No health hazards or side effects known with proper therapeutic dosages (PH2). Contraindicated during pregnancy (PH2). Lethal doses cause convulsions and death within a few hours (BIB; DAA).

Extracts (Jatamansi):

LD50 of jatamansone (ipr mus) 350 mg/kg, vs. 900 mg/kg for the whole essential oil. (MPI). EO hypotensive in dogs, dosages for hypotensive humans, less than lab animal dosage (MPI). Preliminary clinical trials of jatamansone exhibited reduced aggressiveness, restlessness, stubbornness, as well as less insomnia (MPI). With 28 hyperkinetic children, jatamansone and placebo were compared for 11 months with D-amphetamine and chlorpromazine. Jatamansone and especially amphetamine significantly improved behavior, amphetamine better helping aggressiveness and restlessness. Mentally retarded children showed little response to any of the drugs. Jatamansone had fewer side effects than the pharmaceuticals. Amphetamine exacerbated symptoms early on and caused insomnia in one patient (MPI).

Salim et al. (2003) demonstrated neuroprotective effects in rodent models of acute cerebral ischemia, probably via antioxidant activity (X12479970).

WATERCRESS (*NASTURTIUM OFFICINALE* R. BROWN) ++ BRASSICACEAE

Synonyms:

Baeumeria nasturtium Gaertn., *Cardamine fontana* Lam., *Cardamine nasturtium* Moench, *Nasturtium fontanum* (Lam.) Asch., *Nasturtium nasturtium-aquaticum* (L.) H. Karst. nom inval., *Radicula nasturtium* Cav. nom. illeg., *Radicula nasturtium-aquaticum* (L.) Rendle & Britten, *Rorippa nasturtium* Beck. nom. illeg., *Rorippa nasturtium-aquaticum* (L.) Hayek, *Sisymbrium nasturtium* Thunb. nom. illeg., *Sisymbrium nasturtium-aquaticum* L. *fide* HH2 and POR.

FIGURE 1.72 Watercress (*Nasturtium officinale*). Source: Regina Hughes in Reed (1970).

Notes (Watercress):

The fourteenth day of the second month at even they shall keep it, and eat it with unleavened bread and bitter herbs.

Numbers 9:11 (KJV)

In the second month on the fourteenth day in the evening they shall keep it; they shall eat it with unleavened bread and bitter herbs.

Numbers 9:11 (RSV)

In the second month on the fourteenth day between the two evenings, they should prepare it. Together with unfermented cakes and bitter greens they should eat it.

Numbers 9:11 (NWT)

Watercress is considered by some non-Israeli writers to be one of the bitter herbs of the Passover (with chicory, endive, and lettuce); but frankly, I think horseradish is more important in modern times. Israeli botanist Michael Zohary does not even mention watercress in his *Plants of the Bible*, but he does list it in the *Flora of Palestine* (FP1), as pluriregional, tropical, and temperate, in brooks and streams. Even I think of it as more pleasingly pungent than bitter, so I will not argue that it is the best candidate for the long list of bitter herbs. It is, however, my favorite among the many candidates, and possibly one of the healthiest (although often harboring human parasites when harvested wild). As a salad, it is said to promote the appetite. As a cosmetic, its bruised leaves are said to clear up facial blemishes, blotches, and spots. The juice, mixed with egg whites is said to help carcinoma. Made into snuff, it is a "cure" for polyps (JLH). Cress in vinegar is one remedy for anthrax. It is believed by some to interfere with implantation of the ovum or gestation. In small quantities, it is thought to act as an oral contraceptive and produce temporary sterility.

Common Names (Watercress):

Agretto (It.; AVP; KAB); Agrião (Mad.; Por.; EFS; POR); Agrião da Fonte (Por.; AVP); Agrião das Horias (Por.; AVP); Anandrano (Hova; KAB); Bachkresse (Ger.; HH2); Berra (Sp.; EFS); Berro (Peru; Sp.; EFS; EGG; HH2; VAD; VOD); Berro Cruz (Ma.; JFM); Berro de Agua (Sp.; EFS); Biolar (Irish.; KAB); Brancuta (Rom.; KAB); Braunkersch (Ger.; KAB); Brown Cress (Eng.; EFS); Brønnkarse (Nor.; POR); Brunnenkresse (Ger.; HH2; KAB; MAD; POR); Cailli (Fr.; KAB); Cardam de Izfor (Rom.; KAB); Chijchi (Peru; EGG); Creiso (Provenzal; KAB); Creixans (Cat.; KAB); Crescione (It.; KAB; POR); Crescione Acquatico (It.; POR); Crescione de Fonte (It.; EFS); Cresciune (Malta; KAB); Creson (Ma.; JFM); Cressione (It. HH2); Cressione Aquatico (It. HH2); Cresson (Fr.; Haiti; AVP; KAB); Cresson aquatique (Fr.; KAB); Cresson D'eau (Fr.; AVP; HH2; POR); Cresson de Fontaine (Fr.; EFS; HH2; POR); Cresson de Ruisseau (Fr.; KAB); Cresson Officinal (Fr.; KAB); Creixns (Cat.; KAB); Dou Ban Cai (China; POR); Gemeine Brunnerkresse (Ger.; HH2); Grabenkresse (Ger.; HH2); Gredas (Cat.; KAB); Grundkresse (Ger.; HH2); Horf el Ma (Arab.; AVP); Hrenita (Rom.; KAB); Kiaelkrasse (Swe.; KAB); Kioelkarssa (Swe.; AVP); Kersch (Ger.; KAB); Kowhiti-Whiti (Nz.; KAB); Kreson (Creole; Haiti; Rus.; POR; VOD); Kressa (Rus.; KAB); Kzezucha (Pol.; AVP); Lutpuriah (Dec.; KAB); Mastuerzo Acuatico (Sp.; HH2; KAB); Mizu Garashi (Japan; POR); Mul Naeng Ee (Korea; POR); Nasitord (Fr.; KAB); Nasturel (Rom.; KAB); Nasturzio (It.; EFS; HH2); Nasturzio Acquatico (It.; EFS; POR); Occoruro (Peru; EGG); Oranda Garashi (Japan; POR; TAN); Phàk Kàat Náam (Thai; POR); Piriya-Halim (Him.; Kum.; KAB; NAD); Quell Ranke (Ger.; KAB); Rechad (Tunis; AVP); Rukiew Wodna (Pol.; POR); Salat-Brunnenkresse (Ger.; EFS); Sante du Corps (Fr.; KAB); Selada Air (Indonesia; POR); Shahat (Iran; POR); Shamrock (Eng.; KAB); Shui Han Cai (China; POR); Shui Tian Jie (China; POR); Sija (Malta; KAB); Simsaag (Nepal; POR); Sisimbrio (It.; KAB); Sisimbrio acquatico (It.; POR); Sisymbrion (Greek; KAB); Sisimbro (It.; KAB); Stertion (Eng.; KAB); Suteresi (Tur.; EFS); Ting-Li (China; EFS); Ugas-Potocarka (Yug.; AVP); Wasserhanf (Ger.; EFS); Wasserkraute (Ger.; KAB); Wasserkresse (Ger.; EFS; HH2); Watercress (Eng.; CR2; EFS; VOD); Water Crishies (Ma.; JFM); Water Kers (Dutch; AVP); Water Cushie (Ma.; JFM); Weind Kresse (Den.; AVP); Wiesenkresse (Ger.; KAB); Witte Waterkers (Dutch; EFS); Xi Yang Cai Gan (Pin.; DAA); Zorret el Ma (Arab.; AVP); Nscn.

Activities (Watercress):

Abortifacient (f; BIB); Antidote (Nicotine) (f; BIB); Antihistaminic (1; X10746164); Antiimplantation (f; BIB); Antiinflammatory (1; X15917216); Antimitotic (1; HH2); Antiproliferant (1; X15016658); Antisarcomic (1; HH2); Antiseptic (1; HH2; PHR; PH2); Antitumor (1; BGB; PH2; X15016658); Antitussive (f;

VAD); Aperitif (f; PHR); Aphrodisiac (f; AHL); Apoptotic (1; X15016658; X15668997); Bactericide (1; BGB; WOI); Bitter (1; PHR); Chemopreventive (1; BGB; X12716290; X8634661); Cholagogue (1; APA; BGB); Cicatrizant (f; AHL; VOD); Contraceptive (f; BIB); COX-2 Inhibitor (1; X15917216); Cyanogenic (f; BIB); Cytoprotective (1; X15016658); Decongestant (1; APA); Depurative (f; APA; BGB; FAD; JFM; MAD); Detoxicant (1; JNU); Digestive (1; APA; PH2); Diuretic (f1; APA; FAD; MAD; PH2; VOD); Emmenagogue (f; MAD); Expectorant (f; AHL; BGB; EFS; MAD); Goitrogenic (1; WO2); Hypoglycemic (f; VAD); iNOS Inhibitor (1; X15917216); Laxative (f; BIB); NF-kappa-B Inhibitor (1; X15917216); Orexigenic (f1; APA; BIB; DEP; EFS; NAD; PH2); Propecic (f; AHL); Rubefacient (f; VAD); Sterilant (f; BIB); Stimulant (f; EFS; KAB); Tonic (1; APA); Vermifuge (f; EFS; MAD); Vulnerary (f; MAD; VOD).

Indications (Watercress):

Acne (f; BIB; MAD); Adenopathy (f; JLH); Alopecia (f; JNU; VAD); Anemia (f; JFM; VAD); Anorexia (f; DEP; PHR; VAD); Anthrax (f; BIB); Arthrosis (f; HH2; PHR; PH2); Ascites (f; MAD); Asthenia (f; BGB; VOD); Asthma (f; BIB; WOI); Bacteria (1; WOI); Blemish (f; BGB); Boil (f; JLH); Bronchosis (f12; AHL; FAD; PHR; VOD); Cancer (f1; APA; JLH); Cancer, breast (f1; APA; JLH); Cancer, colorectal (f1; X12716290); Cancer, esophagus (f1; JNU); Cancer, face (f1; APA; JLH); Cancer, gland (f1; APA; JLH); Cancer, lung (f1; APA; JNU; X8634661); Cancer, nose (f1; APA; JLH); Cancer, prostate (f1; X15016658); Cardiopathy (f; BIB; FAD); Catarrh (f12; BGB; KOM; PH2; VOD); Cold (f1; BIB; FNF); Colic (f; X15878246); Congestion (2; APA; KOM); Cough (2; APA; DAA; MAD; PHR); Cystosis (f; MAD); Debility (f; VOD); Dermatosis (f; BIB; EGG; JFM; JNU); Diabetes (f; MAD; VAD); Dropsy (f; MAD); Dyscrasia (f; MAD); Dyskinesia (f; VAD); Dysmenorrhea (f; BIB); Dyspepsia (f; APA; MAD); Earache (f; MAD); Eczema (f; MAD); Enterosis (f; EGG); Exanthema (f; MAD); Fever (f; MAD); Flu (f; BIB); Freckle (f; BIB); Gallstone (f; HNI); Gastrosis (f; DAA); Gingivosis (f; MAD; VOD); Glossosis (f; MAD); Goiter (f; BIB; FAD; WOI); Gout (f; MAD; VAD); Gravel (f; MAD); Head cold (f; BIB; MAD); Hepatoma (1; X15668997); Hepatosis (f; EGG; JFM; MAD; X15878246); Herpes (f; MAD); High Blood Pressure (f; VAD); Hyperazotemia (f; VAD); Impetigo (f; BIB); Impotence (f; AHL); Infection (f; MAD); Inflammation (2; APA); Insomnia (f; BIB; MAD; NAD); Ischia (f; MAD); Jaundice (f; MAD); Kidney stone (f; MAD); Laryngitis (f; VAD); Leprosy (f; BIB); Lethargy (f; FAD); Liver Fluke (f; BGB); Liver Spot (f; BIB); Mucosis (KOM); Nephrosis (f; BIB; X15878246); Nervousness (f; APA); Obesity (f; VAD); Odontosis (f; VAD); Pertussis (f; BIB); Pharyngitis (f; VAD); Phthisis (f; AHL); Polyp (f; BIB; JLH; WOI); Pulmonosis (f; BIB; KAB; MAD; VOD); Respirosis (f12; APA; BGB; KOM; PH2; VAD); Rheumatism (f; FAD; MAD; PHR); Rhinosis (f; BIB; JLH; WOI); Scabies (f; BIB); Sciatica (f; BIB); Scrofula (f; MAD); Scurvy (f123; FAD); Seborrhea (f; VAD); Sore (f; MAD); Sore Throat (f; WOI); Splenosis (f; MAD); Staphylococcus (1; HH2); Stomatosis (f; DAA; MAD); Stone (f; MAD; VAD); Strangury (f; WOI); Swelling (f; HH2); Tonsilosis (f; EGG); Toothache (f; JNU; MAD); Tuberculosis (f; BIB; JFM; MAD); Tumor (f1; BIB); Urethrosis (f; VAD); UTIs (f1; BGB; BIB; PH2); Wart (f; JLH); Wen (f; JLH); Worm (f; BIB; MAD); Wound (f; AHL; VOD).

Dosage (Watercress):

FNFF = !!!

Watercress is grown for the pungent leaves and young stems, and is used widely for garnishing meats, salads, and other dishes, even biscuits. The pungent flavor is due to gluconasturtin. Whole plant consumed as salad or vegetable. Seeds used like mustard (BIB; FAC; TAN); 2 g fresh herb/cup tea (APA); 20–30 g fresh herb/day (APA; KOM); 60–150 g freshly pressed juice (APA; KOM); 20–30 g fresh herb 3x to 150 g/day (HH2); 4–6 g dry herb/day (KOM); 2–3 cups preprandial tea (2 g (1–2 tsp) steeped 10–15 minutes in 150 ml boiled water) (PH2).

- Africans use chopped watercress, covered with honey overnight, as cough syrup (BIB).
- Chinese use the plant for asthma (BIB).

- Haitians apply chopped or crushed leaves as cicatrizant or vulnerary (VOD).
- Haitians consume leafy stems as depurative, diuretic, for bronchitis and pulmonary catarrh (VOD).
- Haitians take the leaf broth or decoction for asthenia or weakness (VOD).
- Haitians use chopped leafy stems to treat gum disease (VOD).
- Lebanese use seeds as alterative and depurative, applying the herb in yogurt to acne (BIB; HJP).
- Native Americans consumed watercress to alleviate gallstones (HNI).
- Panamanians believe this a sure cure for anemia (JFM).
- Peruvians eat the plant as depurative, diuretic for bronchitis and worms (EGG).
- Peruvians gargle the macerated plant for tonsilitis (EGG).
- Salvadorans steep the flowering plant in wine as an antiscorbutic, depurative, and tonic (JFM).
- Venezuelans use cress juice for tuberculosis, poulticing the leaves on dermatosis (JFM).
- Venezuelans use 4 Tbsp fresh plant juice with cold Guazuma bark tea (like slippery elm bark) with 4 Tbsp red wine, 2 to 3 ×/day for hepatosis (JFM).
- Western Europeans consider the plant ecbolic in large doses, emmenagogue in smaller doses (BIB).

Downsides (Watercress):

Class 2b, 2d (AHP, 1997). No health hazards or contraindications reported with proper administration of suggested therapeutic dosages (PH2). KOM and PH2 report contraindications: peptic ulcer, nephrosis; not to be used by children younger than 4 years; adverse effects: GI complaints (rarely). Since viewed as emmenagogue and uterocontractant, contraindicated in pregnancy. Irritation of gastric mucosa may develop following chronic ingestion of large quantities. (AEH). I suspect that much of what is said here could be said about horseradish, wasabi, and even the milder crucifers such as broccoli, etc. (JAD). VAD cautions against use with gastritis, ulcers, UTIs (although also listed as an indication), and hypothyroidism. Be careful to clean watercress thoroughly as it may harbor dangerous aquatic microbes (VAD).

Natural History (Watercress):

Bees and flies are the chief pollinators. According to Eastman, the only pest is the caddisfly, *Limnephilus lunatus*, an aquatic case-making larva, like the cress native to Europe. Eating above water are such species as bean aphids (*Aphis rumicis*; blackish sapsuckers clustered under leaves) and watercress leaf beetles (*Pheodon aeruginosa* bronze-black oval chrysomelids). Pierid butterfly caterpillars, called whites, feed on many mustard species. The diamondback moth caterpillar (*Plutella xylostella*) eats holes in the lower leaf surface, pupating in mesh-like cocoons. Underwater, the plant may provide food all year. There may be amphipods (*Gammarus*), which feed on yellowed leaves, the fresh leaves toxic to them, and water sowbugs (*Asellus*). Trout consume young leaves and attached animals. The plants are fairly high in vitamins, minerals, and protein. The leaves are eaten by ducks, muskrats, and deer, and serve as shelter for small aquatic animal life. Regrettably, the plant accumulates heavy metals. There goes another healthy human food, spoiled by humans, like the Maine fish, polluted by the lumber industry, canaries in the aquatic microcosm.

Extracts (Watercress):

Rose et al. (2005) (X15917216) showed that beta-phenylethyl- and 8-methylsulphinyloctyl isothiocyanates suppress nitric oxide and prostaglandin E2. Both have known chemopreventive properties. Overproduction of nitric oxide (NO) and prostaglandins (PGE) are associated with pathological conditions (e.g., inflammation and cancer). The reduction in both iNOS and COX-2 expression were associated with inactivation of nuclear factor-kappaB and stabilization of IkappaBalpha (X15917216). Plants become bitter in flowering, better for medicine, worse for food (JFM). Watercress is the richest source of phenethylisothiocyanate, which also occurs in bok choy, turnips, and turnip greens (JNU). Singling out watercress, Chiao ct al. (2003) (X15016658) note that consumption of an isothiocyanate metabolite inhibits growth of human prostate cancer cell xenografts by apoptosis and cell cycle arrest. Epidemiology indicates that crucifer consumption is inversely related to prostate cancer incidence. The N-acetylcysteine (NAC) conjugate of phenethylisothiocyanate (PEITC-NAC), abundant in watercress, inhibited proliferation and tumorigenesis. The supplemented diet group (8 μM/g) showed significant reductions in tumor size in 100% of the mice (X15016658). Linking this study to an earlier Maori study seems to suggest this bitter herb be included in your diet if you fear colon or prostate, or any, cancer. Thompson and Shaw (X12716290) noted that New Zealanders have a high rate of colorectal cancer, but the Maori have a lower incidence than non-Maori New Zealanders (22.2 and 43.7 per 100,000, respectively). "Maori eat more in total, eat more red meat, drink more alcohol, consume more saturated fat, have a higher prevalence of obesity, and have a lower proportion of individuals consuming a given level of fruit and vegetables per day." *Sonchus* and *Nasturtium officinale* are foods with plausible cancer protective properties and are components of the Maori, but not the non-Maori diet (X12716290). All this tells me that if you must eat, drink, and be merry, include some bitter herb — especially watercress — in your fare.

OLEANDER (*NERIUM OLEANDER* L.) X APOCYNACEAE

Synonyms:

Nerium indicum Mill.; *Nerium latifolium* Mill.; *Nerium odorum* Soland. (Treated separately in HH2; aggregated here and in AH2.)

Notes (Oleander):

Hearken unto me, ye holy children, and bud forth as a rose growing by the brook of the field.

Ecclesiasticus 39:13 (KJV)

Back in 1985, I postulated, along with other writers, that the rose of Ecclesiaticus was *Nerium*. But not Zohary, who renders the rose of Ecclesiasticus as the genus *Rosa*. Zohary does include oleander in his *Plants of the Bible*, equating it with the place name Ardat (II Esdras 9:26), which he deems cognate with *ardaf*, the Hebrew name for oleander, and close kin to Armenian *ardab*. According to other students of the Scriptures, the oleander is the "rose of the waterbrooks" — the "rhododendron" or "rose tree" of the Greeks. Either way, oleander is clearly a medicinal and poisonous ornamental common in Israel. To the Spanish, oleander is known as *laurel*, and is their favorite shrub for parks and gardens. As an evergreen summer favorite, it is termed "tough and attractive," and does well in almost any soil. In Greece, India, and Italy, it is a funeral plant. It is used to decorate Hindu temples. Zohary notes that oleander is common on riverbanks in Israel. Palestinians secure from it a very active cardiac glucoside used in pharmacy. It is used as a rat poison in Europe. Honey from the flowers may even be poisonous. It is widely planted as an ornamental in tropical and subtropical countries. Some suggest that it is the "willow of the brook" of Leviticus used for constructing booths for the Feast of Tabernacles. Some view it as the Jericho rose because on the eastern side of Jordan, it becomes a tree 25 feet tall. With a copious folk repertoire of anticancer activity, oleander will probably be found to contain more proven anticancer agents than just the rutin and ursolic acid. Leaves are dangerously applied to cutaneous eruptions; the decoction is used to destroy maggots in wounds. In Lebanon, as perhaps elsewhere, informants contradict, consider it calming yet irritating, a cause yet a cure for sore eyes, a medicine yet a poison. Such contradictions fan the flames of homeopathy (BIB).

FIGURE 1.73 Oleander (*Nerium oleander*).

Common names (Oleander):

Adelfa (Sp.; EGG; HH3; RAR; USN); Adelfa Laurel (Peru; EGG); Alari (Mal.; DEP); Alelí (Pr.; LWW); Alelí Extranjero (Pr.; LWW); Alhelí (Pr.; LWW); Alili (Ber.; BOU); Amancayo (Peru; RAR); Anini (Ber.; BOU); Arali (Tam.; SKJ); Aralivayr (Tam.; DEP); Ardab (Arm.; ZOH); Ardaf (Heb.; ZOH); Ariri (Ber.; BOU); Azuceno de la Habana (Col.; LWW); Baladre (Sp.; EFS); Balandre (Sp.; USN); Barhamase (Nepal; SUW); Bau Swan (Newari; NPM); Beijos (Por.; AVP); Belladonna (Fr.; AVP); Ceylon Rose (Eng.; EFS); Chu Au (China; EFS); Delfa (Arab.; AVP; BOU); Dephali (Iran; EFS; NAD); Dhavekanera (Kon.; NAD); Difla (Arab.; ZOH); Diflah (Arab.; Syria; HJP); Duftender Oleander (Ger.; HH3); Espirradeira (Por.; AVP); Flor del Perú (Dr.; Peru; AHL; AVP); Franse

Bloem (Dwi.; LWW); Gandeli (BAL; KAB); Ganderai (Pushtu; DEP); Ganhira (Pun.; DEP); Ganira (Bom.; DEP); Gheneru (Tel.; DEP); Haban (Arab. GHA); Hamancay (Que.; RAR); Harduf (Heb.; ZOH); Indian Oleander (Eng.; SKJ); Jaur (Bal.; DEP); Jiia Zhu Tao (Pin.; AH2); Jowari (Bal.; KAB); Kagaer (Guj.; SKJ); Kanagale (Kan.; DEP); Kanagilu (Kan.; NAD); Kanaveeram (Tam.; NAD); Kaner (Hindi; Nepal; EFS; NPM); Kanera (Guj.; Mah.; DEP; NAD); Kaneri (Mar.; DEP); Kanero (Orissa; SKJ); Kanher (Guj.; NAD); Kaniha-Swaa (Nepal); Kanirkejur (Dec.; DEP); Kanyur (Kum.; DEP); Karabi (Beng.; India; DEP; EFS; NAD); Karavira (Ayu.; Sanskrit; AH2; DEP; EFS); Karber (Hindi; NAD); Karbir (Nepal; NPM); Kasturipatte (Tel.; NAD); Kharazahra (Arab.; EFS; NAD); Khar Sahrah (Iran; DEP); Kiah (China; EFS); Kuruvira (Hindi; DEP); Kyochiku-to (Japan; USN); Laurel (Peru; RAR); Laurel Blanco (Sp.; LWW); Laurel Colorado (Sp.; LWW); Laurelles (Fr.; AVP); Laurel Rosa (Bol.; Peru; Sp.; DLZ; EGG; USN); Laurel Rosado (Ecu.; LWW); Laurier (Fr.; Haiti; AHL); Laurier Blanc (Haiti; AHL); Laurier des Jardins (Haiti; AVP); Laurier Rose (Fr.; Haiti; BOU; EFS; HH3; LWW; USN); Laurier Tropical (Haiti; AHL); Laurose (Fr.; Gaud.; AVP); Lawrin (Bol.; Callawaya; DLZ); Leandro (It.; EFS); Loandro (Por.; AVP); Loendro (Por.; EFS; HH3); Lorbeer-rose (Ger.; EFS); Loryé Rose (Creole; Haiti; VOD); Loryé Wos (Creole; Haiti; VOD); Loureiro Rosa (Por.; AVP; HH3); Loyee (Haiti; TRA); Martinica (Dr.; His.; Peru; AHL; AVP; LWW); Massa di S. Giovanni (It.; EFS); Massa di S. Giuseppe (It.; EFS); Naranjillo (Peru; EGG; RAR); Narciso (Cr.; Cuba; Hon.; LWW; TRA); Narciso Rosado (Sal.; LWW); Nérier à Fles (Fr.; AVP); Nérion (Fr.; AVP); Oleander (Eng.; Dwi.; Scn.; AH2; BOU; VOD); Oléandre (Fr.; BOU; EFS; USN); Oleandro (It.; EFS; HH3); Pascua (Sp.; USN); Pirulí (Dr.; His.; AHL); Rajabaka (San.; SKJ); Rajbaka (San.; DEP); Rhododendron (Eng.; AVP); Rhododendron de Pline (Fr.; AVP); Rosa de Berberiá (Sp.; Ven.; EFS; LWW); Rosa del Perú (Dr.; AHL); Rosa Francesa (Cuba; Peru; RAR; RyM; TRA); Rosage (Fr.; AVP; EFS); Rose (Eng.; BIB); Rose Bay (Eng.; CR2; SKJ; RAR); Roseberry Spurge (Eng.; EFS; NAD); Rose Laurel (Eng.; CR2; HH3; VOD); Rosen Lorbeer (Ger.; EFS; HH3); Sevadilha (Por.; EFS); South Sea Rose (Eng.; EFS; HH3); Sum el Himar (Arab.; DEP); Sumala Himara (Arab.; EFS); Svetapushpa (Sanskrit; EFS); Sweet-scented Oleander (Eng.; EFS; NAD); Ward el Hhimar (Arab.; Syria; HJP); Ward el-homar (Arab.; BOU); Welriekende Oleander (Dutch; EFS); Wohlriechende (Ger.; NAD); Wohlriechende Oleander (Ger.; EFS; HH3); Zabb el Hhimar (Arab.; Syria; HJP); Zakkum (Tur.; EFS).

Activities (Oleander):

Abortifacient (f1; BOU; HH2; WWB); Adaptogenic (1; KAP); Analgesic (f1; KAP; MPI); Anticancer (f1; CRC; MPI; X14609747; X15730243); Antiedemic (f; MPI; X14522443); Antiinflammatory (f1; KAP; MPI; TRA; X14609747; X15730243); Antinociceptive (1; X14522443); Antiseptic (f1; JFM; WBB); Antistress (1; MPI); Antiulcer (1; MPI); Antiviral (1; TRA); Aphrodisiac (f1; KAB); Apoptotic (1; X14609747); Bactericide (1; WBB; WOI); Bradycardic (1; MPI); Cardiac (1; BIB; CRC); Cardioactive (1; X3086679); Cardiotonic (1; AHL; BIB; CRC; KAP; LWW; WOI); CNS Depressant (1; KAP); Cyanogenic (1; BIB; CRC); Diuretic (f1; BIB; CRC; EFS; KAP; WOI); Emetic (f1; CRC; MPI); Emmenagogue (f1; BIB; CRC; HH2; MAD); Febrifuge (f1; BOU; KAP; MPI); Fungicide (f1; MPI); Insecticide (f1; BOU; CRC; NPM); Insectifuge (f; BIB; CRC); Lachrymatory (f; KAP; WOI); Larvicide (f; BIB); Negative Chronotropic (1; PH2); Neuroprotective (1; X15492866); Paralytic (f; BOU); Parasiticide (f; BIB; CRC); Poison (f1; CRC; DEP; SUW); Positive Inotropic (1; PH2); Larvicide (f; WOI); Neurogenic (1; X12542335); Pediculicide (f; FEL); Piscicide (f1; WBB; X14505708); Purgative (f; BIB; CRC; JFM); Rodenticide (f1; CRC; LWW); Spasmogenic (f; KAP; MPI); Sternutatory (f; BIB; CRC; JFM); Stimulant (f; CRC); Tonic (f; CRC; EFS); Toxic (f; EFS); Uterotonic (1; TRA); Vermifuge (f; JFM); Vulnerary (f; TRA).

Indications (Oleander):

Alopecia (f; BOU); Angina (f; MAD); Apoplexy (f; MAD); Aposteme (f; CRC); Arrhythmia (f1; MAD; TRA); Arthrosis (f; KAB; MAD); Asthma (f; CRC); Atheroma (f; CRC); Bacteria (1; TRA; WBB); Bronchosis (f; GHA); Cancer (f; BOU); Carcinoma (f; CRC); Cardiopathy (f1; BIB; BOU;

PH2; TRA); Chancre (f; KAB; MPI; NAD; SUW; WOI); Circulosis (1; TRA); Conjunctivosis (f; CRC); Corn (f; CRC; JLH); Cough (f; GHA); Dermatosis (f; CRC; DEP; NPM; PH2; SKJ; SUW; WOI); Dysmenorrhea (f; CRC; WBB); Dyspnea (f; FEL); Eczema (f; CRC; MAD); Edema (f; FEL); Epilepsy (f; CRC; FEL; WBB); Epithelioma (f; CRC; JLH); Fungus (f; NPM); Furuncle (f; HH2); Gingivosis (f; BIB; BOU); Headache (f; KAP; MAD); Hemorrhoid (f; KAP; MPI; PH2); Herpes (f; CRC; NPM); High Blood Pressure (f; BIB); Impetigo (f; BIB); Impotence (f; KAB); Induration (f; JLH); Infection (f1; JFM; WBB); Inflammation (f1; KAB; MPI; TRA); Insomnia (f; MAD); Itch (f; BOU; KAB); Leprosy (f; BIB; DEP; HJP; SUW); Leukoderma (f; KAB); Lice (f; FEL); Lumbago (f; KAB); Maggot (f; CRC); Malaria (f; CRC; MAD; WBB); Mange (f; DLZ; VOD); Myalgia (f; KAB); Mycosis (f1; MPI; NPM); Myocardosis (f; MAD); Neurosis (1; X15492866); Odontosis (f; BIB; BOU); Ophthalmia (f; CRC; KAP; NAD; PH2); Pain (f1; KAB; X14522443); Paralysis (f; MAD); Parasite (f; AHL; JFM); Pediculosis (f; VOD); Psoriasis (f; CRC; SUW; X15613134); Respirosis (f; TRA); Rhinosis (f; BIB); Ringworm (f; BIB; CRC; NPM); Salmonella (1; TRA); Scabies (f; CRC; MAD; PH2; WBB); Sinusosis (f; CRC; GHA; JFM); Snakebite (f; CRC; HH2; NAD; NPM; WBB); Sore (f; CRC; SUW; VOD; WOI); Staphylococcus (1; WBB); Sting (f; NAD); Stomatosis (f; BIB); Stress (f1; KAP); Swelling (f1; DEP; JFM; KAP; NAD; SUW; X14522443); Syphilis (f; BOU; NAD); Tachycardia (f; MAD); Tinea (f; DLZ); Tumor (f; CRC; JLH); Urethrosis (f; KAB); Venereal Disease (f; BIB; BOU; JFM); Vertigo (f; MAD); Virus (1; TRA); Wart (f; BIB); Worm (f; KAB); Wound (f; KAB).

Dosages (Oleander):

FNFF = X
Equivalent to 1–3 grains fresh bark or dried fruit (FEL); 50 mg leaf (MAD).

- Algerians gargle leaf decoction to strengthen gums and teeth, use it as nose drops (BOU), and use leaves as a fumigant for fever (HJP).
- Asian Indians apply a paste of ground root in water to chancres and sores on the penis (NAD).
- Ayurvedics use the plant for bronchosis, conjunctivosis, dermatosis, fever, hemorrhoids, itch, leukoderma, urethrosis, worms, and wounds (KAB).
- Bolivians suggest the powdered dried leaves as a powerful sternutatory, or in lard, or decocted in oil for mange and tinea (DLZ).
- Curacaons boil the flowers and gargle the decoction, and bathe dermatitic children in the leaf decoction (JFM).
- Dominicans infuse the leaves in oil to alleviate dermatosis and parasites, rubbing the leaves on lice (AHL; TRA).
- Ethiopians dress skin diseases on the head with the leaves (BIB).
- Haitians use the vinegar leaf macerate for head lice and sores (VOD).
- Lebanese suggest the plant as a dangerous antiedemic, antiitch, cardiotonic, hypotensive, the decoction useful for impetigo and ringworm, root extracts for amenorrhea, leprosy. and venereal disease (HJP).
- Mohammedans suggest the leaf decoction to reduce swelling, and oil with root bark (external only) for dermatosis and leprosy (KAB).
- Munda use the seeds to poison jackals and pigs (KAB).
- Nepalese use a root bark paste for dermatoses, herpes, and ringworm (NPM).
- Ukranians use the plant as an abortifacient (HJP).
- Unani suggest the flowers for arthralgia, headache, impotence, inflammation, lumbago, myalgia, scabies, and the dangerous root for pain in the abdomen and joints (KAB).
- Venezuelans suggest inhaling the steam from boiled shoots for sinusitis (JFM).
- Venezuelans use the latex to repel flies (AHL)

Downsides (Oleander):

No health hazards or contraindications reported with proper administration of suggested therapeutic dosages (PH2) (but PH2 designates no dosage!; JAD). Commission E reports accidental and therapeutic use of leaf has resulted in partially fatal poisonings (AEH). Overdoses may cause arrhythmia, bradycardia, cardiodepression, confusion, cyanosis, diarrhea, headache, hyperkalemia, nausea, neurodepression, stupor, and vomiting (BRU; DEP; MAD; PH2). More than 100 years ago we read that "Several cases are on record of fatal results from the internal administration" (DEP). Over a 6-year period (1989–1995), 142 bovines died in California (BRU). During the Persian campaign, Alexander's army lost horses that had fed on the shrub, and some soldiers died who had grilled their meat on skewers made from the wood (WBB). 15–20 g fresh leaf can kill a horse, 10–20 a cow, 1–5 a sheep. Children have died after ingesting a handful of flowers (JFM).

Extracts (Oleander):

Leaves, flowers, and stem bark possess cardiotonic properties, especially the leaves. Cornerine has proved effective against cardiac ailments in clinical trials, particularly improving the heart muscle functions. Oleandrin is diuretic and stimulates the heart. The flavonal glycosides influence vascular permeability and possess diuretic properties (BIB).

BLACK CUMIN (*NIGELLA SATIVA* L.) ++ RANUNCULACEAE

Synonyms:

Nigella indica Roxb.; *Nigella truncata* Viv.

Notes (Black Cumin):

> **For the <u>fitches</u> are not threshed with a threshing instrument, neither is a cart wheel turned about upon the cummin; but the fitches are beaten out with a staff, and the cummin with a rod.**
>
> **Isaiah 28:27 (KJV)**

> **<u>Dill</u> is not threshed with a threshing sledge, nor is a cart wheel rolled over cummin; but dill is beaten out with a stick, and cummin with a rod.**
>
> **Isaiah 28:27 (RSV)**

> **For it is not with a threshing instrument that *black cumin* is given a treading; and upon cumin no wheel of a wagon is turned. For it is with a rod that black cumin is generally beaten out, and cumin with a staff.**
>
> **Isaiah 28:27 (NWT)**

Here the three versions differ in the translation of the Hebrew *ketzah*, and I frankly like the NWT version best. Of the three, only the NWT renders it black cumin (*Nigella* of the Ranunculaceae), which is not taxonomically related to cumin or cummin (*Cuminum* of the Apiaceae). The KJV renders it fitches, which is *Nigella*. Zohary adds that the identity of the Hebrew *ketzah* with "black cummin" or "nutmeg flower" is not only linguistically supported but attested by the postbiblical custom of sprinkling the seeds over bread and cake and of flavoring dishes with them. The Arabic and Aramaic name is *kazha* (ZOH). This herb may be even more important to the Muslims than to the Christians and Jews. According to an Arab proverb, "in the black seed is the medicine for every disease except death." Nearly a millennium ago, Rhineland abbess Hildegard von Bingen (1098–1179), described by translator Throop as artist, Benedictine abbess, healer, linguist, musician,

FIGURE 1.74 Black Cumin (*Nigella sativa*).

mystic, playwright, poet, preacher, spiritual counselor, theologian, and visionary, published her *Physica*. She suggested pounding the seed with honey as fly paper, streaking it where you wish to attract the flies, which, on tasting, will sicken and die (HIL).

COMMON NAMES (BLACK CUMIN):

Agenuz (Sp.; POR); Agenuz Comun (Sp.; POR); Ajenuz (Sp.; USN); Araignee (Fr.; BOU); Arañuel (Sp.; USN); Aranyajeeraka (Sanskrit; NAD); Bashpika (Sanskrit; KAB); Black Caraway (Eng.; USN); Black Cumin (Eng.; Ocn.; AH2; CR2; KAB; USN); Chernushka Posevnaia (Rus.; POR); Cinnamonea (It.; KAB); Cominho-Negro (Por.; USN); Çöreg Otu (Tur.; POR); Çörek Otu (Tur.; POR); Corekotu (Tur.; EFS); Cumin Noir (Fr.; EFS; KAB; NAD; POR); Cuminella (It.; KAB; POR); Czarnuszka Siewna (Pol.; POR); Dill (Eng.; ZOH); Erbe Spezie (It.; KAB; POR); Faux Cumin

(Fr.; KAB); Fennel Flower (Eng.; HJP); Fitch (Eng.; BIB, HJP); Gandana (Iran; KAP); Gemeiner SchwartzKuemmel (Ger.; KAB); Gith (Fr.; EFS; KAB); Gitono (It.; EFS); Gittaione (It.; POR); Haba Saoda (Arab.; Yemen; X15890471); Habba Sooda (Arab.; GHA); Habba Suda (Arab.; BOU); Habbatoussouda (Arab.; DEP); Habbatussuda (Arab.; KAB); Habbet el-barakah (Arab.; BOU); Hubsindee (Egypt; KAB); Jintan Hitam (Malaya; EFS; POR); Kadujeeru (Guj.; NAD); Kalajaji (Sanskrit; KAB); Kala Jira (Hindi; India; DEP; EFS; KAB); Kalanji (Hindi); Kalenjira (Bom.; KAB); Kalenjire (Bom.; DEP; NAD); Kalijeeri (Gwalior; KAB); Kalijira (Beng.; KAB; NAD); Kalonji (Bom.; Hindi; KAB); Kalonjijirum (Guj.; KAB); Kaluduru (Sin.; KAB; NAD); Kalzira (Beng.; KAB); Kammun Aswad (Arab.; BOU); Kammun el Akhal (Arab.; BOU); Kamûn Awad (Arab.; Syria; HJP; POR); Kamune Asvad (Arabic; EFS; NAD); Kamuneavad (Arab.; KAB); Karamuk (Tur.; POR); Karava (Sanskrit; KAB); Karavi (Sanskrit; KAB); Karejirage (Kan.; KAB); Karijirigay (Kan.; KAB; POR); Karijirige (Kan.; KAB); Karimsiragam (Kan.; DEP; KAB); Karinchirakam (Mal.; NAD); Karunchirakam (Mal.; DEP; KAB); Karunjirajam (Tam.; KAB); Karunjirakam (Mal.; POR); Karunshiragam (Mal.; Tam.; DEP; KAB); Karunshirogam (Tam.; NAD); Kazha (Arab.; Aramaic; ZOH); Kelanji (Bom.; NAD); Kemun (Arab.; Libya; POR); Ketzah (Heb.; ZOH); Kommen (Den.; EFS); Krishnajira (Mah.; KAB); Krishna-jiraka (Ayu.; Sanskit; AH2; DEP; KAB); Kulanjan (Dec.; Urdu; KAB; POR); Kunchi (Sanskrit; KAB); Kunchika (Sanskrit; KAB); Kunjika (Sanskrit; KAB); Kurras (Arab.; KAP); Kusheng (Chinese; Malaya; EFS; KAB); Love in a Mist (Eng.; RFS); Melanthion (Greek; NAD); Melanzio Domestico (It.; POR); Mugrela (Beng.; Hindi; DEP; KAB; NAD); Mugrelo (Nepal; KAP; SUW); Mungrela (Beng.; KAB); Musavi (Sanskrit; KAB); Nallajilakara (Tel.; DEP; NAD); Nallajirakara (Tel.; POR); Neguilla (Sp. EFS; POR); Nellajeelakaira (Tel.; KAB); Nielle de Crete (Fr.; KAB); Nielle Romaine (Fr.; KAB); Nigela (Mad.; Por.; POR); Nigella (Eng.; Scn.; AH2); Nigelle cultivee (Fr.; KAB); Nigelle de Crète (Fr.; USN); Nigelle Romaine (Fr.; KAB); Nigera (Japan; KAP); Nullajilakara (Tel.; KAB); Nutmeg Flower (Eng.; KAB; ZOH); Pei Hei Zhong Cao (China; POR); Prathvika (Sanskrit; KAB); Prithvi (Sanskrit; KAB); Prithu (Sanskrit; KAB); Prithuka (Sanskrit; KAB); Qahta Sauda (Yemen; GHA); Quatre Epice (Fr.; BOU); Quesah (Heb.; KAB); Roman Coriander (Eng.; HOS; USN); Romische Schwartz Kuemmel (Ger.; KAB); Samonne (Burma; KAB); Samon-ne (Burma; DEP; NAD); Sanoudj (Arab.; BOU); Satmung (Burma; NAD); Schwartz Kummel (Ger.; EFS; KAB); Sehniz (Tur.; POR); Shanaza (Syria; HJP); Shewadaru (Afg.; DEP; KAB); Shouniz (Arab.; NAD); Shûnîz (Arab.; Iran; KAB; POR); Siyahbiranj (Iran; KAB); Siyahdanah (Iran; EFS; KAB; NAD); Siyahdaru (Afg.; KAB; NAD); Small Fennel (Eng.; DEP; KAB; NAD); Sthulajiraka (Sanskrit; KAB); Sushavi (Sanskrit; KAB); Svartkummin (Swe.; POR); Tchernushka (Mal.; KAB); Tikammin (Ber.; BOU); Toute Épice (Fr.; BOU; USN); Tukhmigandha (Kas.; DEP; KAB; NAD); Upakunchika (Ayu.; Sanskrit; AH2; KAP); Upakunchiraka (Sanskrit; KAB); Zerara (Ber.; BOU); Zwarte Komijn (Dutch; EFS; POR); Zwarte Kummel (Dutch; EFS).

Activities (Black Cumin):

Abortifacient (f; BIB; DEP; KAP); Amebicide (1; MPI); Analgesic (f1; X10552840); Anesthetic (f1; NAD); Anorectic (1; X15325727); Anthelmintic (f; BIB; DEP; KAP); Antiamphetamine (f; MPI); Antibilious (f; BIB; EFS); Anticancer (f1; FNF); Anticonvulsant (f1; X15795687); Antidiabetic (f1; X15589481); Antidote (Hg) (f; SKJ); Antiedemic (f1; X10552840); Antiepileptic (1; X15795687); Antifibrinolytic (1; X15693715); Antifibrosarcomic (1; X11531013); Antihistaminic (1; HAD; HHB; WOI); Antihypertensive (1; FNF); Antiinflammatory (f1; BOW; X10552840); Antileukemic (1; X1270717); Antileukotriene (1; X15648658); Antimetastatic (1; X15693715); Antinephrotic (1; X10755708); Antinociceptive (f1; HAD); Antioxidant (1; X10925395); Antioxytocic (1; FNF); Antiperoxidant (1; X15785320); Antiproteinuric (1; X10755708); Antiradicular (1; PR14:323); Antiseptic (f1; WOI; X10548758); Antispasmodic (f1; BOU; HAD; HHB; WOI); Antitumor (colon) (f1; X15375533); Antitussive (f; X15890471); Antiulcer (1; HOS); Antiviral (1; HAD); Apoptotic (1; X15375533); Bactericide (1; HAD; HHB; WOI); Bronchodilator (f1; HAD); Calcium Antagonist (1; X11381824); Candidicide (1; FNF); Cardiodepressant (1; MPI); Cardioprotective (1; X9700580);

Carminative (f1; BOU; DEP; HHB; MAD; SKJ; SUW; WOI); Chemopreventive (1; X11531013; Choleretic (1; HHB); Cholecomimetic (1; MPI); CNS Depressant (1; MPI); Contraceptive (f; HAD); Cyclooxygenase Inhibitor (1; FNF); Cytotoxic (1; X1270717); Detoxicant (1; X11531013); Diaphoretic (f; BIB); Digestive (f; BIB; DEP); Diuretic (f1; BIB; BOU; EFS; HHB; MAD; SUW; WOI; X10967716); Emmenagogue (f; BIB; BOU; DEP; EFS; MAD; SUW; WOI); Fungicide (1; MPI); Gamma-Glutamyl Transpeptidase Inhibitor (1; X15785320); Glutathiogenic (1; X11531013; X15785320); Gram(+)-icide (1; YAB); Gram(–)-icide (1; YAB); Hepatoprotective (f1; X10883736); Histaminergic (1; X8270170); Hyperthermic (1; DEP); Hypoglycemic (1; X11744291); Hypolipidemic (1; X10755708); Hypotensive (1; MPI); Hypotriglyceridemic (1; X5326549); Hypouricemic (1; HHB); Immunostimulant (1; HAD); iNOS Inhibitor (1; X15589481); Insecticide (f1; HIL; MPI); Insectifuge (1; WOI); Insulinogenic (1; X15482373); Interferogenic (1; FNF); Lactagogue (f1; BIB; BOU; DEP; EFS; HAD; HHB; KAB; MAD; SUW; WOI); Laxative (f; BOW); 5-Lipoxygenase Inhibitor (1; FNF); Lipolytic (f1; X15325727); Nephroprotective (1; X5326549); NO Inhibitor (1; X15589481); Orexigenic (f; BIB; HAD); Pediculicide (f; DEP); Phagocytotic (1; HAD); Plasminogen Activator (1; X15830841); Protein Kinase Inhibitor (1; X15589481); Protisticide (1; MPI); Purgative (f; BIB; EFS); Quinone Reductase Inducer (1; X11531013); Respirostimulant (f1; X8270170); Secretagogue (f; DEP); Stimulant (f; EFS); Stomachic (f; DEP; EFS); Taenicide (f1; BOU; MPI); Tissue-type Plasminogen Activator (t-PA) Inhibitor (1; X15693715); Tonic (f; EFS); Toxic (1; HHB); Uterocontractant (f; KAP); Uterotonic (f; BOW); Vermifuge (f; BIB; BOU; DEP; HAD; MAD); Xanthine-Oxidase Inhibitor (1; X15785320).

Indications (Black Cumin):

Achylia (f; MAD); Allergy (f; HAD); Ameba (1; MPI); Amenorrhea (f; BOU; KAP); Anorexia (f1; BOU; HAD); Arthrosis (f1; HAD); Ascites (f; BIB); Asthma (f1; BOU; GHA; HAD; HHB; MAD; SKJ; WOI); Bacillus (1; X10548758); Bacteria (1; HAD); Biliousness (f; KAP); Bite (f; HAD); Bronchosis (f1; BOU; HAD; HHB; WOI); Bronchospasms (f1; WOI); Cachexia (f; SKJ); Callus (f; BIB; JLH); Cancer (f1; FNF; BIB; HAD); Cancer, abdomen (f1; FNF; JLH); Cancer, colon (f1; FNF; JLH); Cancer, eye (f1; FNF; JLH); Cancer, kidney (f1; JLH; X15785320); Cancer, liver (f1; FNF; JLH); Cancer, nose (f1; FNF; JLH); Cancer, uterus (f1; FNF; JLH); Candidiasis (1; FNF); Cardiopathy (1; X8270171); Catarrh (f; DEP; HHB); Chest (f; GHA); Childbirth (f; GHA; SUW); Cholera (1; MPI); Cold (f; DEP); Colic (f; BIB); Congestion (f; GHA); Conjunctivosis (f; GHA); Constipation (f; GHA; SKJ); Convulsion (1; X15795687); Corn (f; BIB; JLH); Cough (f1; BOU; SKJ; WOI; X15890471); Cramp (f1; HHB; MAD); Dermatophyte (1; X15908151); Dermatosis (f1; HAD; SUW; WOI; X15908151); Diabetes (f1; HAD; X11744291; X5326549); Diarrhea (f1; MAD; X11381824); Dysentery (f; HHB; SKJ); Dysmenorrhea (f; BOU; DEP; KAP); Dyspepsia (f; BIB); Dyspnea (f1; GHA; NAD); Eczema (f; DEP; NAD); Emaciation (f; SKJ); Enterosis (f; BIB; BOU; MAD); Epilepsy (1; X15795687); Eruptions (f; BIB); Escherichia (1; KAP; MPI); Fever (f1; BIB; MAD; SUW; WOI); Fibrosarcoma (1; X11531013); Flu (f; BIB; BOU; HJP); Fungus (1; X10548758; X15908151); Gas (f; BOU; MAD); Gout (1; HHB); Halitosis (f; NAD); Headache (f; BIB; BOU; HJP); Hemorrhoid (f; BIB; BOU; GHA; KAB); Hepatosis (f1; BIB; JLH; MAD; X10883736); High Blood Pressure (1; MPI; X8270171); High Cholesterol (1; HAD); High Triglycerides (1; X5326549); HIV (1; HAD); Hydrophobia (f; BIB); Hyperlipidemia (1; X10755708); Induration (f; JLH; MAD); Infection (f1; HAD; X15908151); Inflammation (f1; X10552840); Jaundice (f; BIB; HHB; MAD); Leprosy (f; BOU; SKJ); Leukemia (1; X1270717); Leukorrhea (f; MAD); Lice (f; DEP); Malaria (f; KAP; NAD); Migraine (f; BOU); Mycosis (1; X10548758); Myrmecia (f; BIB); Nephrosis (1; X10755708; X5326549); Obesity (f1; FNF); Ophthalmia (f; HAD; KAB); Orchosis (f; BIB); Pain (f1; HAD; NAD); Paralysis (f; BIB); Parasite (f1; BOU; HAD); Pityriasis (f; DEP; NAD); PMS (f1; HAD; KAP); Polio (f; GHA); Proctosis (f; SKJ); Prolapse (f; SKJ); Proteinuria (1; X10755708); Pityriasis (f; DEP); Puerperium (f1; KAB; WOI); Pulmonosis (f; HAD; HHB; MAD); Respirosis (f; BOU); Rheumatism (f; PR14:323; X15890471); Rhinosis (f; BIB; JLH; NAD); Salmonella (1; HAD);

Sclerosis (f; BIB); Sinusosis (f; BOU); Smallpox (f; SKJ); Snakebite (f; BIB); Sniffles (f; MAD); Splenosis (f; MAD); Staphylococcus (1; HAD; MPI); Sting (f; HAD; SUW); Stomachache (f; BIB; MAD); Stomatosis (f; HAD); Swelling (f; BIB; NAD); Syphilis (f; SKJ); Taenia (f1; BOU; MPI); Thrombosis (1; X15830841); Toothache (f; BOU; MAD); Tumor (f; BIB; HAD); Ulcer (f1; HOS); Uterosis (f; JLH); Vibrio (1; MPI); Virus (1; HAD); Worm (f; BOU; MAD); Wound (f; HAD); Yeast (1; X10548758).

Dosages (Black Cumin):

FNFF = !!

Seeds eaten as spice; mixed in bread in Europe; used whole or ground as a flavoring, especially in oriental cookery (BIB; TAN). Whole seeds used in Russian rye and Turkish breads; sprinkled on food or mixed with pepper. Seed oil also used culinarily. Arabs mix the seed with honey as a confectionary. Ethiopians add them to *Capsicum* pepper sauces, or use with *Aframomum*, *Piper*, and *Zingiber* in local alcoholic beverages (BIB). 1–5 g seed (KAP); 10–20 g for amenorrhea, dysmenorrhea (KAP); 0.6–1.2 g seed (HHB; MAD); 1 tsp seed in hot tea (MAD).

- Algerians take the roasted seeds with butter for cough, and with honey for colic (BIB, HJP).
- Arabians rub crushed seed with *Rhazya* leaf, *Teucrium*, *Zingiber*, and oil onto polio (GHA).
- Arabians use seed as lactagogue, soaking in rose oil as eyedrops for eye infections (BIB; GHA).
- Asian Indians use bruised and fried seeds, in muslim (via inhalation), for catarrh and cold (NAD),
- Asian Indians use the seed tincture for anorexia, diarrhea, dyspepsia, and fever (KAP).;
- Asian Indians use seed oil or powdered seed in sesame oil for dermatosis (KAB; KAP).
- Ayurvedics consider the herb anthelmintic, carminative, emmenagogue, and stimulant (KAB).
- Ethiopians mix seed with melted butter, wrap in cloth, and sniff for headache (BIB).
- Indonesians add seeds to astringent medicines for abdominal disorders (BIB).
- Lebanese take seed extract for liver ailments (BIB).
- Malayans poultice seeds onto abscesses, headache, nasal ulcers, orchitis, and rheumatism (BIB).
- North Africans mix seed with honey for morning aperitif (BOU).
- One Pakistani B'hai was very impressed with how the plant lowered recalcitrant high triglycerides (JAD).
- Unani, considering the plant abortifacient, anthelmintic, diuretic, and emmenagogue, use it for cough, fever, hydrophobia, jaundice, paralysis, pulmonosis, and sore eyes (KAB).
- Yemeni use the seeds for hemorrhoids (GHA).

Downsides (Black Cumin):

Seeds poisonous to man in high doses (BOU).

Extracts (Black Cumin):

The Prophet Mohammed reportedly recommended black cumin as a cure-all to his associates in Arabia 1400 years ago. It remains one of the most famous medicinal herbs in the Moslem world. Nigellone protects guinea pigs from histamine-induced bronchospasms (WOI), suggesting one rationale behind its use in asthma, bronchitis, and cough. The lipid portion of the ether extract of the seeds has shown lactagogue activity in rats, verifying its folk usage as a lactagogue. In large quantities, the seeds are also used to induce abortion. LD50 alcoholic extract is 540–580 mg/kg ipr mus MPI. The seed oil produces a concentration-dependent inhibition of tissue-type plasminogen activator (t-PA), urokinase-type plasminogen activator (u-PA), and plasminogen activator inhibitor type 1 (PAI-1). The seed oil decreases the fibrinolytic potential of human fibrosarcoma cells *in vitro*, possibly slowing local tumor invasion and metastasis (X15693715). Khan and Sultana (2005) show inhibition of renal carcinogenesis, oxidative damage, and hyperproliferation by *Nigella sativa*. Rats treated orally with 50 and 100 mg/kg body weight showed decreased gamma-glutamyl transpeptidase, lipid peroxidation, xanthine oxidase, H_2O_2 generation, blood urea nitrogen, serum creatinine, renal ODC activity, DNA synthesis ($P < 0.001$), and the incidences of tumors, and glutathione and antioxidant enzymes were increased significantly. In rats, *Nigella* could be a potent chemopreventive agent suppressing oxidative stress, hyperproliferative responses, and renal carcinogenesis in rats (X15785320). Hosseinzadeh et al. (2005) showed that intracerebroventricular administration of thymoquinone suppresses epileptic seizures in rats. Thymoquinone is anticonvulsant, hypnotic, and muscle relaxant and alters motor coordination and locomotor activity (X15795687). Alpha-spinasterol significantly (circa 1000 x simvastin, a coenzyme-A inhibitor) modulates development and/or progression of diabetic nephropathy. It reduced significantly attendant increases of serum triglycerides, renal weight, and urinary protein excretion in diabetic mice (X15326549).

WHITE WATERLILY (*NYMPHAEA ALBA* L.) +++ NYMPHAEACEAE

Synonyms:

Castalia alba (L.) Woodv. & Wood; *Castalia minoriflora* Simonk; *Castalia speciosa* Salisb.; *Leuconymphaea alba* (L.) Kuntze; *Nymphaea alba* f. *csepelensis* Soó; *Nymphaea alba* f. *limosa* Soó; *Nymphaea alba* var. *melocarpa* Casp.; *Nymphaea alba* subsp. *occidentalis* (Ostenf.) Hyl.; *Nymphaea erythrocarpa* Hentze; *Nymphaea exumbonata* Rupr.; *Nymphaea melocarpa* (Casp.) Asch. & Graebn.; *Nymphaea minoriflora* (Simonk.) E. D. Wissjul.; *Nymphaea occidentalis* (Ostenf.) Moss; *Nymphaea officinalis* Gaterau; *Nymphaea parviflora* Hentze; *Nymphaea polystigma* E. H. L. Krause; *Nymphaea* rotundifolia Hentze; *Nymphaea splendens* Hentze; *Nymphaea suaveolens* Dumort.; *Nymphaea urceolata* Hentze; *Nymphaea venusta* Hentze

Notes (White Waterlily):

And the chapiters that were upon the top of the pillars were of lily work in the porch, four cubits.

I Kings 7:19 (KJV)

Numerous sculptured representations in ancient Egyptian tombs show their concern with waterlilies, probably *N. alba* or *N. lotus*. Zohary reports *N. alba* as the only white-flowered waterlily in Israel. UPW lists *Nelumbo lotus* as the Egyptian waterlily. As UPW notes, the common names used in Africa (see UPW entries below, ascribed to the closely related *Nymphaea lotus*) could apply to any white-flowered waterlily. The flowers are still admired by Egyptians, whose belles often wear them in their headdresses. The flower was sacred to Egyptians more than 4000 years ago. Waterlilies prevail in such distant cultures as Mexico, Japan, India, and China. Even ancient Greeks had a legend that a beautiful nymph, deserted by Hercules, flung herself into the Nile to be transformed into a white lotus. It is the national flower of Thailand. Some suspect waterlilies of being narcotic. Rhizomes are used for tanning. There was, in Homer's *Odyssey*, a mythical nation where the people subsisted wholly on lotus, and lived in dreamy indolence induced by the diet. In later stories, these people became the North African lotophagi or "Homeric lotus eaters." Osinis, the great Egyptian God of the Underworld, and Judge of the Dead, is represented as wearing a crown of sacred lotus flowers. Horus, the God of Silence, sits, like Buddha, on a lotus with a finger to his lip, commanding silence (BIB). Regarded as an aphrodisiac, astringent, diaphoretic, hemostat, narcotic, and sedative, this species has reportedly served as a folk remedy for cancer, diarrhea, spasms, and tumors (e.g., of the testes).

FIGURE 1.75 White Waterlily (*Nymphaea alba*).

Common Names (White Waterlily):

Alga Palustris (?; JLH); Bach Lieu Tu (Ic.; KAB); Bado (Hausa; KAB); Beyaz Nilüfer (Tur.; EFS); Bhash es Samak (Arab.; Syria; HJP); Bishine (Nig.; UPW); Bololol (Upper Volta; UPW); Bondai (Niger; UPW); Brimposh (Kas.; DEP; NAD); Cacabus Veneris (?; JLH); Carfano (It.; EFS; KAB); Cobetenas (Sp.; EFS); Duilleoghuidhebhaitighe (Irish; KAB); European White Waterlily (Eng.; Ocn.; AH2); Flatterdock (Eng.; TAN); Golfão Branco (Por.; EFS); Grzybienie (Pol.; KAB); Hung Pai (China; KAB); Jaberri Faro (Gambia; UPW); Kamud (Kas.; DEP); Kien Thiet (Vn.; KAB); Kubishka (Rus.; KAB); Kumuda (Ayu.; Sanskrit; AH2; NAD); Lis Blanc des Étangs (Fr.; EFS); Lis d'Eau (Fr.; KAB); Lotus Lily (Eng.; KAB); Ndairi (Arab.; Sen.; UPW); Nénuphar Blanc (Fr.; EFS); Nénuphar Blanco (Sp.; EFS); Nilofar (Kas.; DEP); Nilufar (Arab.; KAB); Nimphea Branca (Por.; KAB); Ninfea Bianca (It.; EFS); Ninfea Blanca (Sp.; KAB); Ntanowa (Ghana; UPW); Nttanowa (Akwapim; KAB); Nungu (Guinea; UPW); Nymphéa Blanc (Fr.; USN); Pandharenkamal (Bom.; KAB); Papaver Palustre (?; JLH); Papeo (It.; EFS); Platos (Sp.; EFS); Pluta (Rom.; KAB); Rosa de Amor (Sp.; EFS); Sasa (Sierra Leone; UPW); Sebluomen (?; JLH); Sidi (Greek; KAB); Sjoeblad (Swe.; KAB); Soeblomster (Den.; KAB); Sórose (Den.; EFS); Tchingalir (Guinea-Bissau; UPW); Tertermantrer (Ga.; KAB); Ul Haudhan (Arab.; Syria; HJP); Water Rose (Eng.; KAB); Weisse Seeblume (Ger.; KAB); Weisse Seerose (Ger.; EFS; USN); Weisse Teichrose (Ger.; EFS); White Pondlily (Eng.; EFS); White Waterlily (Eng.; Scn.; AH2; EFS; USN); Witte Plomp (Dutch; EFS); Witte Waterlelie (Dutch; EFS); Yi a Kwe (Liberia; UPW).

Activities (White Waterlily):

Anaphrodisiac (f; DEP; EFS); Anodyne (f; HJP); Antiinflammatory (f; GAZ); Antioxidant (1; X15881650); Antiperoxidant (1; X15881650); Antiproliferant (1; X15881650); Antispasmodic (f1; EFS; WOI); Astringent (f; DEP; EFS); Chemopreventive (1; X15881650); Demulcent (f; NAD); Digitalic (1; WOI); Diaphoretic (f; DEP); Diuretic (1; MPI); Glutathiogenic (1; X15881650); Hemostat (f; EFS; HJP); Narcotic (f; DEP; EFS); Nephroprotective (1; X15881650); Paralytic (f1; WOI); Sedative (f1; WOI); Vulnerary (f; HJP); Xanthine-Oxidase Inhibitor (1; X15881650).

Indications (White Waterlily):

Bleeding (f; HJP); Boil (f; HJP); Cancer (f1; HJP; JLH; X15881650); Cancer, kidney (1; X15881650); Cancer, testes (f; JLH); Cardiopathy (f; HJP); Cramp (f; EFS); Diarrhea (f; DEP; EFS; KAB); Dysentery (f; DEP; KAB; HJP); Gastrosis (f; HJP); Inflammation (f; GAZ); Insomnia (f; GAZ); Leukorrhea (f; GAZ); Nephrosis (1; X15881650); Nymphomania (f; GAZ); Orchosis (f; JLH); Pain (f; HJP); Pharyngosis (f; GAZ); Parasite (f; HJP); Stomatosis (f; GAZ); Ulcer (f; HJP); Vaginosis (f; GAZ); Venereal Disease (f; HJP).

Dosages (White Waterlily):

FNFF = !!
Seeds, roots, and stalks are common foodstuffs in Egypt. The seeds are ground into flour for bread, or are roasted and eaten like a nut. Rhizomes are boiled before consumption. Rootstocks and seeds also eaten elswehere; starch-rich roots are a source of starch; fermented to make a French beer (BIB; DEP).

- Gabonese eat stewed leaves in pregnancy, applying leaves to itch (*N. lotus*) (UPW).
- Hausa use seed (*N. lotus*) for eruptive fevers (UPW).
- Iranians take flowers, often mixed with sugar, *Adiantum, Viola, Ziziphus,* and "oxtongue," for chest ailments and fevers (BIB).
- Ivory Coastals use decoction for bronchitis and cough (*N. lotus*) (UPW).
- Lebanese slice and boil dried or powdered roots for boils, dysentery, parasites, ulcers, and venereal disease (HJP).

- Nigerians take leaf decoction (*N. lotus*) as anodyne and to prevent abortion (UPW).
- Nigerians use seed (*N. lotus*) for skin diseases (UPW).
- Pennsylvanians apply the root to tumors (JLH).
- Senegalese attribute (*N. lotus*) antiemetic, sedative, and vermifugal properties (UPW).
- Tanganyikans sprinkle fruit ashes on burns, using root decoction for insanity (*N. lotus*) (UPW).

Downsides (White Waterlily):

None reported (GAZ).

Extracts (White Waterlily):

Roots contain gallic acids. The alkaloid nymphaeine is present in all parts of the plant, except the seeds. Toxic to frogs, it produces tetanus-like symptoms. Alcoholic extracts of the rhizome (containing the alkaloid) are mildly sedative and spasmolytic; they do not significantly depress the heart; in large doses, they paralyze the medulla. A glycoside nymphalin with digitalis-like action occurs in the flowers (BIB). Khan and Sultana (2005) showed anticarcinogenic effects against oxidative damage, hyperproliferative response, and renal carcinogenesis in Wistar rats. Treatment with 100 and 200 mg/kg body weight significantly decreased gamma-glutamyl transpeptidase, lipid peroxidation, xanthine oxidase, H_2O_2 generation, blood urea nitrogen, serum creatinine, renal ODC activity, DNA synthesis, and incidence of tumors. Renal glutathione also recovered (X15881650).

OLIVE (*OLEA EUROPEA* L.) ++ OLEACEAE

Notes (Olive):

His branches shall spread, and his beauty shall be as the olive tree, and his smell as Lebanon.

Hosea 14:6 (KJV)

His shoots shall spread out; his beauty shall be like the olive, and his fragrance like Lebanon.

Hosea 14:6 (RSV)

His twigs will go forth, and his dignity will become like that of the olive tree, and his fragrance will be like that of Lebanon.

Hosea 14:6 (NWT)

Olive is mentioned in 33 works of the KJV, including, for example, 5 citations in Deuteronomy.

As one of the most valuable of Holy Land trees, small wonder that it is also one of the most frequently mentioned. Even my best African resource book (UPW), naming many antimalarial plants in Africa, notes that it is not only oft mentioned in the Old Testament, but it also features prominently in Greek mythology. "It has become symbolic of peace, plenty, prosperity, and achievment" (UPW), and let me add health and the Mediterranean diet. Romans thought it the idyllic status symbol to have juice of the grape, alias wine, inside, and oil of the olive, alias olive oil, outside (UPW). And that oil was olive oil, one of our better sources of monounsaturated fatty acids (MUFAs). Biblical passages, where the word *anoint* appears, usually mean anoint with olive oil, often used as the carrier for perfumes. One scholar has said that no tree is more closely associated with the history of man and development of civilization than the olive. An Arabic proverb has it that gardens are folly while olives are kings. Olive oil was the base of the perfumed ointments sold in classic Athens and Rome. It is also used in the textile industry in wool combing. Olive pomace, the residue after milling, is used in animal feeds. The stones (seeds) are used in the manufacture of molded products and plastics. The bitter glucoside, oleuropein, of green olives is usually neutralized with lye or caustic soda before pickling. Wood is hard, beautifully grained, used in turnery and cabinet making. In Italy, an olive branch is hung over the door to keep out evil spirits (BIB).

Common Names (Olive):

Aceituno (Peru; Sp.; KAB); Amil (Ber.; BOU); Amourgha (Arab.; BOU); Arbre Eternel (Fr.; EFS); Azeboudj (Ber.; BOU); Azemmour (Ber.; BOU); Bouttaillaou (Lan.; KAB); Elaia (Greek; KAB); Itm (Arab.; GHA); Jaituni (Fula; Nig.; UPW); Jirjir (Arab.; Malaya; IHB); Julipe (Kan.; KAB); Karazeytin (Tur.; EB51:195); Man Zaitun (Hausa; UPW); Maslin (Rom.; KAB); Mitan (Dho.; Oman; GHA); Ölbaum (Ger.; HH2); Oleastro (It.; EFS); Olijfboom (Dutch; EFS); Oliva (Rus.; KAB); Olive (Eng.; Scn.; AH2; CR2); Oliveira (Por.; EFS); Oliveira Brava (Por.; UPW); Oliven Ölbaum (Ger.; EFS); Olivenbaum (Ger.; HH2); Olivera (Cat.; KAB); Olivier (Fr.; BOU); Olivier Sauvage (Fr.; BOU); Olivio (Sp.; EFS); Olivo (It.; Peru; Sp.; EFS; HH2; KAB; ROE); Olivo Selvatico (It.; EFS); Oliwa Drzewo (Pol.; KAB); Oljetroee (Den.; KAB); Oljetroeed (Swe.; KAB); Saisun (Tam.; KAB); Tahatimt (Tamachek; MALI); Tazbboujt (Ber.; BOU); Thatimt (Ber.; BOU); Tsi T'un (China; TAN; UPW); Tzetta (Ber.; BOU); Ulivo (It.; KAB; HH2; UPW); Zait (Arab.; Malaya; IHB); Zaitun (Arab.; Malaya; IHB); Zayit (Heb.; KAB); Zaytoun (Arab.; BOU); Zebboudj (Arab.; BOU); Zebbour (Arab.; BOU); Zebbug (Malta; KAB); Zeytin (Tur.; EB49:406); Zeytin Agaci (Tur.; EFS); Zeytun (Arab.; Malaya; Mali; IHB; UPW); Zzit (Ber.; BOU).

Activities (Olive):

ACE Inhibitor (1; VVG); Antiadrenergic (1. × 15070161); Antiarrhythmic (1; PHR; PH2; X15070161); Antiatherosclerotic (1; X12648829); Antidysrhythmic (1; X15070161); Antihypertensive (1; APA; X12648829); Antiischemic (1; X15070161); Antioxidant (1; APA); Antipyretic (f; VVG); Antisclerotic (f; ZUL); Antiseptic (f; EFS); Antispasmodic (1; PHR; PH2; ZUL); Astringent (f1; EFS; KAB; ZUL); Bactericide (f1; BIB; BOU); Beta-Adrenergic Antagonist (1; X15070161); Beta-Blocker (1; X15070161); Cardioprotective (1; X15070161); Cardiotonic (1; X15070161); Cholagogue (f; BIB; BOU; EFS); Collyrium (f; ZUL); Depurative (f; ZUL); Diuretic (f1; APA; BOU; PHR; VVG; X12648829); Dromotropic (1; X15070161); Febrifuge (f; EFS); Hypocholesterolemic (1; HH2); Hypoglycemic (1; APA; BIB; BOU; HH2; VVG); Hypotensive (f1; APA; BIB; BOU; PH2; SKJ; VVG; ZUL; X15070161); Hypouricemic (1; ZUL); Lactogogue (f; NMH); Lipoxygenase Inhibitor (1; X15086818); Molluscicide (1; HH2); Myorelaxant (1; ZUL); Nephrotonic (f; VVG); Pectoral (f; BIB); Positive Inotropic (1; X15070161); Tonic (f; VVG); Vasodepressor (1; X15070161); Vasodilator (1; ZUL).

Indications (Olive Leaf):

Aphtha (f; BOU); Arrhythmia (1; PH2); Atherosclerosis (1; HH2; X12648829); Bacteria (f1; BIB; BOU; ROE); Boil (f; GHA); Cancer (f; JLH); Cancer, breast (f; JLH); Cancer, gum (f; JLH); Cardiopathy (1; APA; PH2; X15070161); Cataract (f; GHA); Colic (f; VVG; ZUL); Condyloma (f; BIB; JLH); Constipation (f; GHA); Corns (f; ZUL); Cough (f; BOU); Cystosis (f; ZUL); Dermatosis (f; GHA); Diabetes (f1; APA; BIB; BOU; HH2; ROE); Diarrhea (f; UPW; VVG); Diptheria (f; ZUL); Dysrhythmia (1; X15070161); Fever (f; APA; HH2; UPW; VVG); Fracture (f; GHA); Gingivosis (f; GHA; JLH); Gout (f; HH2; ROE); Headache (f; ZUL); Hemorrhoid (f; EB51:195); Hepatosis (f; BIB); High Blood Pressure (1; APA; PH2; ROE; X15070161); High Cholesterol (1; HH2); Hypertonia (1; PHR); Malaria (f; FEL; KAB; ZUL); Mastosis (f; JLH); Ophthalmia (f; UPW; VVG; ZUL); Pain (f; EB51:195); Rheumatism (f; HH2; ZUL); Schistosomiasis (1; HH2); Scrofula (f; EFS; ZUL); Sore (f; UPW); Sore Throat (f; UPW; VVG; ZUL); Sprain (f; EB51:195); Stenocardia (1; X15070161); Stomatosis (f; BOU); Swelling (f; EB51:195); UTIs (f; ZUL); Venereal Disease (f; JLH); Virus (1; X15869811); Wart (f; BIB; JLH); Wounds (f; APA).

FIGURE 1.76 Olive Leaf (*Olea europea*).

Dosages (Olive):

FNFF = !!!

Although processed olives are a mainstay in American culture, it takes a lot of processing to make them edible. The relatively inedible fruits are pickled or cured with brine, lye, oil, salt and/or water. They can be sun-cured and eaten as relish. Leaves of the African variety are used as a condiment. (FAC). Leaves are capable of exuding manna (UPW): 7–8 g dry leaf in 150 ml water, 3–4 ×/day (that totals to an ounce of leaf (APA); 2 tsp leaf in hot water and steep 30 minutes (PHR).

- Algerians chew the leaves for toothache and oral sores caused by excess tobacco (HJP).
- Dhofari apply pounded leaves to boils, dermatoses, and itch, and apply burnt leaf ashes to blisters and sores (GHA).
- Dhofari take bark decoction or macerate for constipation (GHA).
- Eclectics suggest a "wineglass" of leaf decoction every 3 hours for malaria (FEL).
- Levant citizenry boil a handful in a quart of water until reduced to a pint, and drink for obstinate fevers (GMH).
- North Africans use leaves or their extracts as an antibacterial, antidiabetic, cholagogue, diuretic, hypoglycemic, hypotensive, and pectoral, the wood decoction for aphtha and stomatitis (BOU).
- Peruvians suggest the diuretic leaf decoction for high blood pressure (EGG).
- Saudis use the stems as toothbrushes to keep the gums healthy (GHA).

Downsides (Olive):

None known at proper dosage (PHR).

Natural History(Olive):

Beta-sitosterol, more so than beta-sitosteryl-D-glucoside, stimulated feeding of the olive weevil (*Dyscerus perforatus*) (X12872942). (–)-Olivil and (+)-1-acetoxypinoresinol, minor lignans, were significantly higher feeding attractants for the female than for the male weevil (X12729011).

Extracts (Olive):

From extra-virgin olive oil, Beauchamp et al. (2005) describe a new antiinflammatory COX-2-inhibiting phytochemical, oleocanthal, comparable, they say, to ibuprofen, even in pungency (X16136122). Micol et al. (2005) demonstrated antiviral activity against viral haemorrhagic septicaemia rhabdovirus (VHSV), a salmonid rhabdovirus, for leaf extracts and oleuropein, reducing viral infectivity 10 to 30%, respectively (X15869811). Somova et al. (2004) demonstrated cardiotonic and antidysrhythmic activities of oleanolic acid, ursolic acid, methyl maslinate, and uvaol, all of which showed low toxicity on brine shrimp. Oleanolic acid and methyl maslinate displayed a significant, dose-related vasodepressor effect on sinus bradycardia, acting as beta-adrenergic antagonists, blocking the effect of adrenaline and isoprenaline. They also exhibited positive inotropic and dromotropic effects (X15070161). Here is a suite of compounds acting on cardiopathy at many levels: antiarrhythmic, antidysrhythmic; anti-ischemic; beta-blocker; cardioprotective; cardiotonic; dromotropic, hypotensive, positive-inotropic; and vasopressor. LD50 = 1300 mg/lg ipr mus; >3000 mg/kg orl mus.

OLIVE OIL (*OLEA EUROPEA* L.) +++

Notes (Olive Oil):

The Mediterranean diet (high in olive oil, vegetables, fruit, fish, and lean meat) significantly reduces recurrence of adenoma, and may reduce the recurrence of colorectal adenomas, at least in women (X15677892).

Activities (Olive Oil):

Adaptogenic (1; X15566625); Antiaggregant (1; APA); Antiatherosclerotic (1; X15585759); Antidote (f; BOU); Antidote (lead) (f; FEL); Antiinflammatory (1; X15665734); Apoptotic (1; X15642702); Cardioprotective (1; APA); Chemopreventive (1; X15677892; X15642702); Cholagogue (f; BOU); Demulcent (f; KAB; WOI); Emollient (f1; APA; KAB; WOI); Enteroprotective (1; APA; X15677892);

Gastroprotective (f1; APA); Hypercholesterolemic (1; APA; X15572303); Hypoglycemic (1; APA); Laxative (f1; APA; KAB; WOI); Orexigenic (f; BOU); Purgative (f; BOU); Vulnerary (f; KAB).

Indications (Olive Oil):

Abscess (f; BOU); Adenoma (1; X15677892); Anorexia (f; BOU); Antidote (poison) (f; GMH); Arthrosis (f; GMH; ROE); Asthenia (f; BOU); Atherosclerosis (1; X15585759); Biliousness (f; ROE); Bite (f; FEL); Blister (f; GHA); Boil (f; BOU); Bruise (f; FEL); Burn (f; FEL); Calculus (f; ROE); Callus (f; JLH; X15677892); Cancer (f1; JLH; X15677892); Cancer, breast (f1; JLH; X15677892; X15642702); Cancer, colon (f1; JLH; X15677892); Cancer, liver (f1; JLH; X15677892); Cancer, neck (f1; JLH; X15677892); Cancer, spleen (f1; JLH; X15677892); Cancer, stomach (f1; JLH; X15677892); Cardiopathy (1; APA); Cholecocystosis (1; PHR); Colic (f; ROE); Constipation (f; BOU; ROE); Corns (f; ZUL); Dermatosis (f; GHA; GMH); Diabetes (1; APA); Dropsy (f; GMH); Earache (f; FEL); Earwax (f1; APA); Endotoxic Shock (1; X15665734); Enterosis (f1; APA; GMH); Exanthema (f; FEL); Gallstone (f; GMH); Gastrosis (f1; APA; JLH); Hepatosis (f; BOU; JLH; ROE); High Cholesterol (1; APA); Induration (f; JLH; X15677892); Inflammation (1; X15665734); Itch (f; GMH); Mastosis (f; FEL); Myalgia (f; GMH); Ophthalmia (f; GHA; JLH); Phthisis (f; FEL); Plague (f; FEL; GMH); Proctosis (f; FEL); Pulmonosis (f; GMH); Rheumatism (f; ROE); Scarlatina (f; FEL; GMH); Shock (1; X15665734); Sore (f; GHA); Splenosis (f; JLH); Sting (f; ROE); Stone (f; FEL; ROE); Stress (1; X15566625); Typhoid (f; GMH); Worm (f; FEL; GMH); Wound (f; FEL).

Dosages (Olive Oil):

FNFF = !!!
Romans thought it an idyllic status symbol to have juice of the grape, alias wine, inside, and oil of the olive, alias olive oil, outside. Olive oil has been deemed one of the best MUFA oils. As early as 1931 we read that it is best for cooking, and a valuable article of diet for both sick and healthy of all ages. Delicate babies absorb its nourishing properties through the skin (GMH). 1–2 oz olive oil as a laxative (APA); 1 Tbsp olive oil in the morning to protect the GI tract linings (APA); 1 oz as purge (FEL); 15–60 ml olive oil (PNC).

- Algerians use for baldness, cough, earache, fractures, gonorrhea, hemorrhage, hernia, impotence, liver congestion, skin diseases, sprains and stones (HJP).
- Arabs apply fruit juice around the eyes to soothe (GHA).
- Arabs mix powdered fruits with dates and salt to paste on fractures (GHA).
- Dutch East Indians apply olive oil or fruits to cancers (JLH).
- Germans and North Americans apply olive oil poultice to breast cancer (JLH).
- Latinos rub arthritic areas with olive oil (JAD).
- Peruvians treat cancer with olive oil/*Plumbago* salve (JLH).

Downsides (Olive Oil):

Class 1 (JAD). Commission E reports the oil should not be used in patients with gall- or bile stones because of the risk that a biliary colic is induced. Topical application rarely results in allergic skin reactions (AEH).

Extracts (Olive Oil):

Olive oil contains 0.05–1% phenolics (3–11 ppm p-hydroxyphenylethanol; 1.4–5.5 ppm 3,4-dihydroxy-phenylethanol; 0.8–3.2 ppm protocatechuic acid; 0.9–3.5 ppm p-hydroxybenzoic acid, 0.5–2.2 ppm vanilic acid; 0.4–1.8 ppm syringic acid, 0.3–1.1 ppm cinnamic acid, 0.3–1.2 ppm p-coumaric acid, 0.3–1.2 ppm o-coumarinic acid, 0.4–11.7 ppm caffeic-acid); 0 0.0125–0.75% carbohydrates, 0.125–0.25%

sterols, 500 ppm triterpenealcohol and hydroxytriterpene acids, 175–200 ppm tocopherol, 40–135 phospholipids, 3–13 ppm carotenoids, 1–10 ppm chlorophyll, and 0.2–20 ppm phaeophytine. The oil contains 1310 ppm beta-sitosterol, 58 ppm delta-7-stigmasterol, 29 ppm delta-5-avenasterol, 28 ppm campesterol, and 14 ppm stigmasterol. Obied et al. (2005) reported the following chemicals and activities in olive mill wastes, estimating recovery of 98% of the biophenols; that suggests to me that one would be 49 times better off eating the residues as the olive, at least as far as biophenols are concerned (e.g., caffeic acid (antiatherogenic, antidepressant, antiinflammatory, antioxidant, antiseptic, chemopreventive); catechol (antioxidant, antiseptic, antitumor, carcinogenic, herbicide); p-coumaric acid (antioxidant, antiseptic, chemopreventive); elenoic acid (antiseptic, antiviral); hydroxytyrosol (antiatherogenic, antiinflammatory, antioxidant, antiseptic, atheroprotective, cardioprotective, chemopreventive, whitener); oleuropein (antiatherogenic, antiinflammatory, antioxidant, antiseptic, antiviral, cardioactive, hypoglycemic); rutin (antiatherogenic, antiinflammaory, antioxidant, cardioactive, cytostatic, thyrostimulant); tyrosol (antiatherogenic, antiinflammatory, antioxidant, cardioactive, cytostatic, thyrostimulant); vanillic acid (antibacterial, antifungal, antioxidant, antiseptic); and verbascoside (antiatherogenic, antiinflammatory, antioxidant, chemopreventive) (FNF; X15712986). It is thus another recitation of the same old story — the unprocessed food is an order of magnitude better than the processed. Our paleolithic foods were better than that brought to us by our USDA and food processing industry.

LEBANESE OREGANO (*ORIGANUM SYRIACUM* L.) +++ LAMIACEAE

SYNONYMS:

Amaracus syriacus (L.) Stokes; *Marjorana crassa* Moench.; *Marjorana crassifolia* Benth Raf.; *Marjorana maru* (L.) Brig.; *Marjorana nervosa* Benth.; *Marjorana scutellifolia* Stokes; *Marjorana syriacum* (L.) Raf.; *Origanum crassa* (Moench.) Chev.; *Origanum maru* L.; *Origanum pseudo-onites* Lindberg *fide* HH2

NOTES (LEBANESE OREGANO):

Purge me with hyssop, and I shall be clean: wash me, and I shall be whiter than snow.

Psalms 51:7 (KJV)

Purge me with hyssop, and I shall be clean; wash me, and I shall be whiter than snow.

Psalms 51:7 (RSV)

May you purify me from sin with hyssop, that I may be clean; may you wash me, that I may become whiter even than the snow.

Psalms 51:7 (NWT)

This is the first time I remember both the KJV and the RSV reading exactly alike, in these two in the imperative. And the meaning is pretty much the same in the NWT. As almost always, at least in my limited purview, the NWT takes more space to say the same thing. However, it is difficult to simplify a Psalm such as this one. Recent scholars suggest that the hyssop of the Old Testament is most probably *Origanum syriacum*, not *Origanum maru*, as I concluded in my 1985 book. The herb more usually called hyssop, *Hyssopus officinalis*, does not even grow in Israel or Sinai (ZOH). Tucker and DeBaggio, like Zohary, refer bibliophiles to *Origanum syriacum* as the plant name for hyssop in the Bible. This species is abundant in the Holy Land, usually among dwarf shrubs on stony grounds. Alternatively, many scholars tend to agree that the hyssop of the Crucifixion is a sorghum. The common oregano (*O. vulgare*), herb of the year in 2005, so well known in gardens, grows to the north of the biblical settings, while *O. syriacum* abounds throughout the central hills. An aromatic substance is obtained from the crushed and dried leaves. The "hyssop" of the Scriptures was used to sprinkle the doorposts of the Israelites in Egypt with the blood of the Paschal Lamb so that the angel of death would pass by that house. It was employed in the purification of lepers and leprous houses, suggesting the Psalmists purge.

Common Names (Lebanese Oregano):

Biblical Hyssop (Eng.; TAD); Echter Staudenmajoran (Ger.; HH2); Egyptian Marjoram (Eng.; BIB); Ezov (Heb.; TAD; ZOH); Hyssop (Eng.; BIB; TAD); Lebanese Oregano (Eng.; TAD; USN); Syrian Hyssop (Eng.; TAD; ZOH); Syrian Majoram (Eng.; HH2); White Oregano (Eng.; TAD); Wild Marjoram (Eng.; X12009988); Ysop (Eng.; HH2); Za'atar (Arab.; TAD; ZOH). One author mentioned more than fifty species in six plant families going under the confusing common name "oregano" (EB42:232). Nscn.

Activities (Lebanese Oregano):

Analgesic (f; HH2); Antiacetylcholinesterase (1; X15652288); Antioxidant (1; X14969528); Antiradicular (1; X14969528); Antispasmodic (f; BIB); Diuretic (f; BIB); Emmenagogue (f; BIB); Fungicide (1; TAD); Insecticide (1; HH2); Iron Chelator (1; X14969528); Laxative (f; BIB); Purgative (f; BIB); Stimulant (f; BIB); Sudorific (f; BIB).

Indications (Lebanese Oregano):

Cancer (f; JLH); Cardialgia (f; HH2); Cardiopathy (f; HH2); Cold (f; BIB); Colic (f; BIB); Constipation (f; BIB); Cough (f; HH2); Debility (f; HH2); Dermatosis (f; BIB); Dysmenorrhea (f; HH2); Fibroid (f; BIB); Fungus (1; TAD); Infection (1; TAD); Leprosy (f; BIB); Mycosis (1; TAD); Pain (f; HH2); Paralysis (f; BIB); Polyp (f; BIB); Rheumatism (f; BIB); Sprain (f; BIB); Swelling (f; BIB); Toothache (f; HH2); Tumor (f; JLH); Uterosis (f; JLH).

Dosages (Lebanese Oregano):

FNFF = !!

Fruits of Sicilian sumac are crushed with *Origanum syriacum* to constitute main ingredients of the Middle Eastern spice mixture called za'atar (FAC; TAD). Arabs use it in teas and cook it in baked foods. Sold in the markets, it is a popular Arab spice (ZOH). The uses of the Syrian marjoram, if in fact it is specifically distinct, are not expected to differ from those of the true marjoram (BIB).

- Lebanese apply leaves to rheumatic sprains and swellings (BIB).
- Lebanese take tea of *Origanum maru* for childrens colds and colic (HPP)

Extracts (Lebanese Oregano):

As theoretically one of the richest sources of carvacrol in my USDA database (up to 5% essential oil; up to 80% of which can be carvacrol), this plant probably shares many of the biological activities of carvacrol.

STAR OF BETHLEHEM (*ORNITHOGALLUM UMBELLATUM* L.) + LILIACEAE

Notes (Star of Bethlehem):

> **And there was a great famine in Samaria: and, behold, they besieged it, until an ass's head was sold for fourscore pieces of silver, and the fourth part of a cab of <u>dove's dung</u> for five pieces of silver.**
>
> **2 Kings 6:25 (KJV)**

FIGURE 1.77 Star of Bethlehem (*Ornithogallum umbellatum*).

And there was a great famine in Sama'ria, as they besieged it, until an ass's head was sold for eighty shekels of silver, and the fourth part of a kab of dove's dung for five shekels of silver.

2 Kings 6:25 (RSV)

In time a great famine arose in Sama'ria, and look! they were besieging it until an ass's head got to be worth eighty silver pieces, and the fourth of a cab measure of dove's dung was worth five silver pieces.

2 Kings 6:25 (NWT)

Whether in the KJV, the RSV, or the NWT, in 2 Kings 6:25 it is consistently "dove's dung," apparently a famine food that commanded good money during a famine in Samaria. To this day, I still think of it only as famine food. And yet Zohary neither indexes *Ornithogalum* nor "dove's dung," nor the more euphonious "Star of Bethlehem," alluding to the six white points of the flower. So be it. Bulbs were used for food in Syria. In Dioscorides' day, the bulbs were commonly gathered, ground into meal, and mixed with flour to make bread. Modern Italians in time of scarcity eat the bulb. As Moldenke and Moldenke (1952) note, "These apparently authentic reports are remarkable since chemical analysis shows that the entire plant is intensely poisonous. Grazing animals avoid it, or, if they do eat of it are poisoned … the bulbs are edible only after being thoroughly roasted or boiled." Having read such alarming notes, I ventured out one March morning and boiled a few of the bulbs from the dark green weedy patches in my lawn. I boiled the tubers vigorously, without salt, and then cautiously consumed one. It had a saponaceous quality, suggesting to me that I might be consuming a hemolytic saponin from a dangerous family. There was a bitter aftertaste. Then I salted the bulb, which was a bit more palatable. I would need to be near starvation to consume more of these. I experienced a shortness-of-breath following the ingestion of only two bulbs. Philips (HJP) said that Egyptians and Syrians stored the bulbs for their pilgrimages to Mecca. My favorite foraging book for use in the field, *Edible Wild Plants of Eastern North America*, also treats them as edible, while my favorite poisonous plants text reports that the bulbs have caused death in cattle in the United States. Because the pretty stars open rather late in the day, they have been called "Sleepy Dick" or "Eleven O'Clock Lady." They marked 11:00 A.M. in Linnaeus's floral clock. I find that even as a floral clock species, it is not very timely. Specimens pulled during the day and left under fluorescent bulbs until 11:00 P.M. never close; hence, they do not open the next day. However, similar batches of specimens placed on a table in the unlit gazebo of the Green Farmacy Garden do close, early on dark days, later on bright days, to reopen with the morning light (BIB).

Common Names (Star of Bethlehem):

Aglio Florido (It.; HHB); Bath Asparagus (Eng.; GMH); Bella d'Undici Ore (Fr.; EFS); Belle d'Onze Heures (Fr.; EFS); Bogelmelk (Dutch; EFS); Cipollone Bianco (It.; HHB); Dame d'Onze Heures (Fr.; EFS); Dolden Milchstern (Ger.; USN); Doldiger Milchstern (Ger.; HH2); Dove's Dung (Eng.; GMH); Ebenstraussige (Ger.; HHB); Great Arabic Star Flower (Eng.; GMH); Leche de Gallina (Sp.; EFS); Leite de Galinha (Por.; EFS); Lesser Spanish Star (Eng.; GMH); Milchstern (Ger.; EFS); Nap at Noon (Eng.; HOC; USN); Ornitagalo (Sp.; EFS); Ornithogalum (Greek; GMH); Sleepy Dick (Eng.; USN); Snowdrop (Eng.; HHB; HOC); Star of Bethlehem (Eng.; CR2; EFS); Star of Hungary (Eng.; GMH); Stern von Bethlehem (Ger.; USN); Tükrükotu (Tur.; EFS); Vogelmilch (Ger.; EFS; HHB); White Field Onion (Eng.; GMH); Nscn.

Activities (Star of Bethlehem):

Cardiotonic (1; HOC); Digitalic (1; HOC); Emollient (f; HHB; HOC); Poison (f; CRC).

Indications (Star of Bethlehem):

Adenopathy (f; CRC; HJP); Cancer (f; GMH); Cardiopathy (1; HOC); Debility (f; CRC; HJP); Lymphosis (f; CRC). Bulbs of other eastern species were used for cachexia, infections, parotitis, scabs, ulcers, and wasting disease.

Dosages (Star of Bethlehem):

FNFF = ?
Although EFS defines it as nutritive, and FAC, GMH, HOW, and TAN, and even Dioscorides, say that bulbs are edible, raw or cooked, I no longer feel safe with these bulbs as food. Facciola says the flowers are eaten baked in bread (FAC).

- Lebanese used the bulb for lymphatic ailments and recommended them in diets for debility (HJP).

Downsides (Star of Bethlehem):

For a change, I should like to mention Peggy Duke, who illustrated Steve Hill's and Peggy Duke's 1985 *100 Poisonous Plants of Maryland.* "The bulbs contain toxic alkaloids that have killed sheep and cattle. There have been heavy losses in Maryland, where more than 1000 sheep were lost in a single year after eating bulbs that were brought to the surface by frost heaves. Apparently the leaves are not poisonous" (Hill and Duke, 1985).

Extracts (Star of Bethlehem):

Although early reports of the gout medicine colchicine have been extricated from the credible literature, there are still reports of other toxins (e.g., convallotoxin, convalloside, and strophanthidin) in *Ornithogalum.*

CHRIST'S THORN (*PALIURIS SPINA-CHRISTI* MILL.) + RHAMNACEAE

Synonyms:

Paliurus aculeatus Lam.; *Paliurus australis* Gaertn.; *Rhamnus paliurus* L.

Notes (Christ's Thorn):

> **And when they had platted a crown of thorns, they put it upon his head, and a reed in his right hand: and they bowed the knee before him, and mocked him, saying, Hail, King of the Jews!**
>
> **Matthew 27:29 (KJV)**

> **And plaiting a crown of thorns they put it on his head, and put a reed in his right hand. And kneeling before him they mocked him, saying, "Hail, King of the Jews!"**
>
> **Matthew 27:29 (RSV)**

> **And they braided a crown out of thorns and put it on his head, and a reed in his right hand. And kneeling before him they made fun of him, saying, "Good day, you King of the Jews!"**
>
> **Matthew 27:29 (NWT)**

To my surprise, there are two Rhamnaceous crown of thorns: one in the genus *Paliurus*, and the other is *Ziziphus*. *Paliurus* has a dry flattened, probably inedible fruit with a wing-like margin; *Ziziphus* has a fleshy globular edible fruit. Pliny the elder reported it useful for inflamed tumors (JLH).

Common Names (Christ's Thorn):

Christ's-thorn (Eng.; USN); Farah Joli (Tur.; GEP); Paliure (Sp.; VAD); Samur (Arab.; Syria; GEP); Nscn.

Activities (Christ's Thorn):

Anticathartic (f; FP2); Antiinflammatory (f; JLH); Astringent (f; FP2); Diuretic (f; FP2; VAD); Hypocholesterolemic (f; VAD); Hypotensive (f; VAD); Tonic (f; FP2).

Indications (Christ's Thorn):

Arteriosclerosis (f; VAD); Cancer (f; JLH); Cardiopathy (f; VAD); High Blood Pressure (f; VAD); High Cholesterol (f; VAD); Inflammation (f; JLH); Oliguria (f; VAD); Stone (f; VAD); Tumor (f; JLH); Urolithiasis (f; VAD).

Dosages (Christ's Thorn):

FNFF= ?
Facciola erroneously equates this species with the edible *Ziziphus spinus-christi*. Otherwise, I find no reference to this species being edible. 30g/l in tea, 3 to 4 cups a day (VAD); 30 drops fluid extract (1:1) 3 ×/day (VAD); 50–100 drops tincture (1:5) 1–3 ×/day (VAD).

SEA DAFFODIL (*PANCRATIUM MARITIMUM* L.) + AMARYLLIDACEAE

Notes (Sea Daffodil):

> **The wilderness and the solitary place shall be glad for them; and the desert shall rejoice, and blossom as the rose.**
>
> **Isaiah 35:1 (KJV)**

> **The wilderness and the dry land shall be glad, the desert shall rejoice and blossom; like the crocus.**
>
> **Isaiah 35:1 (RSV)**

> **The wilderness and the waterless region will exult, and the desert plain will be joyful and blossom as the saffron.**
>
> **Isaiah 35:1 (RSV)**

Like *Panax*, the name *Pancratium* implies all powerful, alluding to its healing potential (Greek *pan* = all; *krotion* = power). Widely cultivated as an ornamental, the plant also volunteers along tropical seashores, sometimes reaching higher latitudes. Zohary is skeptical about this representing either the biblical lily or the biblical rose, as some non-Israeli writers had suggested. Looks like RSV and NWT identified it with the saffron crocus.

Common Names (Sea Daffodil):

Busayl (Arab.; Syria; HJP); Qa'bul (Arab.; Syria; HJP); Qu'bul (Arab.; Syria; HJP); Sanbak Bahari (Arab.; Syria; HJP); Sand Lily (Eng.; HJP); Sea Daffodil (Eng.; FAC; TAN; ZOH); Shoshan (Heb.;

ZOH); Soosan (Arab.; Egypt; X9617056, 1998); Susan (Arab.; Syria; HJP); Susan al Bahr (Arab,; ZOH); Nscn.

Activities (Sea Daffodil):

Acaricide (1; X9617056); Anticancer (1; X15909123); Antimalarial (f; X14669261); Antinociceptive (1; X9379365); Aphrodisiac (f; HJP); Apoptotic (1; X15909123); Emetic (f; DAW); Larvicide (1; X9617056); Mosquitocide (1; X9617056); Poison (f; HJP); Purgative (f; DAW).

Indications (Sea Daffodil):

Impotence (f; HJP); Malaria (f; X14669261); Pain (1; X9379365); Sore (f; HJP); Splenosis (f; DAW); Venereal Disease (f; HJP); Wound (f; HJP).

Dosages (Sea Daffodil):

FNFF = !
"Bulbs may be edible" (TAN). Edible bulbs exhibited with other foods at the International Exhibition of 1862 (FAC; TAN). Seeds apparently eaten in ancient Greece (GAC).

Downsides (Sea Daffodil):

As of July 2004, the FDA Poisonous Plant Database listed titles alluding to toxicity of this species.

Natural History (Sea Daffodil):

The sea daffodil is regarded as a herald of rain in Israel. Flowering late in summer in the Holy Land, the leaves develop later. Flowers, like the evening primrose, open late in the afternoon, and are pollinated by nocturnal hawk moths — during their "one night stand." The plant, growing on unstable beaches and shorelines, has contractive roots that pull exposed roots deeper into the ground (ZOH).

MILLET (*PANICUM MILIACEUM* L.) ++ POACEAE

Synonyms:

Panicum asperimum L.; *Panicum effusum* R. Br.; *Panicum miliaceum* L. convar. *effusum* Alef.; *Panicum miliaceum* L. var. *effusum* Alef.; *Panicum miliaceum* L. var. *ruderale* Kitagawa; *Panicum milium* Pers.; *Panicum ruderale fide* DEP and POR (Kitag.) Chang; *Panicum spontaneum* Lysov ex Zuk

Notes (Millet):

> **Take thou also unto thee wheat, and barley, and beans, and lentiles, and millet and fitches, and put them in one vessel, and make thee bread thereof, according to the number of the days that thou shalt lie upon thy side, three hundred and ninety days shalt thou eat thereof.**
>
> **Ezekiel 4:9 (KJV)**

> **And you, take wheat and barley, beans and lentils, millet and spelt, and put them into a single vessel, and make bread of them. During the number of days that you lie upon your side, three hundred and ninety days, you shall eat it.**
>
> **Ezekiel 4:9 (RSV)**

> **And as for you, take for yourself wheat and barley, and broad beans and lentils, and millet and spelt, and you must put them in one utensil, and make them into bread for you, for the number of days that you are lying upon your side, three hundred and ninety days you shall eat it.**
>
> **Ezekiel 4:9 (NWT)**

FIGURE 1.78 Millet (*Panicum miliaceum*).

Ezekiel is said to have received an order from God to make bread with wheat, barley, beans, lentils, and pannag (millet), (and spelt or fitches, depending on the version). The mixture was moistened with camel's milk, oil, or butter. It was the main food that the common people ate. And as I stated previously, it certainly sounds healthier than some of today's breads — even the fortified breads. Zohary notes that millet or *dohan* was mentioned only once in the scriptures, suggesting that it may have become more popular after biblical times. He suggests that it derived from Ethiopian *Panicum callosum*. Relics are found in Mesopotamia as early as 3000 B.C., but no traces have been found in Israel, where it requires irrigation. Some writers suggest that *pannag* may be etymologically related to the Greek *panexia*, meaning a universal medicine or panacea, considered by Greek physicians as the cure for many ailments. It is eaten, often cooked unground like rice, during the religious fasts of Hindus. Proso millet is grown mainly in the United States as a grain crop, but may occasionally be grown for forage, but as forage the stems are coarse, hairy, and unpalatable. The seeds are chewed and the juice is applied to children's sores. Decoction is used as an antidote to *Momordica* poisoning (BIB).

Common Names (Millet Panic):

Acte hirse (Ger.; NAD); Ægte Hirse (Den.; POR); Akdari (Tur.; EFS); Anne (Pun.; DEP); Anu (Sanskrit; DEP); Arzan (Iran; DEP); Azhaum (Ashkobi; KAB); Azhdan (Kila Saifulla; KAB); Azhdun (Tobu; KAB); Bansi (Bundel.; KAB); Barag (Mar.; KAB); Baragu (Kan.; NAD; WOI); Bili Baragu (Kan.; DEP); Borona de Filipinas (Sp.; EFS); Bread Millet (Eng.; HHB); Broomcorn Millet (Eng.; Ocn.; AH2; NPM); Cavers (Tur.; POR); Chabor (Shoran; KAB); Cheena (Beng.; WOI); Chenaa (Hindi; Pun.; POR); Cheno (Mah.; NAD); Chi (China; EFS); Chin (Hindi; DEP; KAB); China (Beng.; Hindi; Sanskrit; EFS; NAD); Chinh (Bihar; DEP); Chino (Dec.; Mah.; Sindh; NAD); Chinu (Sin.; DEP; KAB); Chinwa (Kas.; DEP; KAB); Ciinaa (Guj.; POR); Cino (Nepal; POR); Cinu (Nepal; POR); Common Millet (Eng.; Ocn.; AH2; DEP; NPM); Dhengali (Mah.; NAD); Dhengli (Mah.; NAD); Dhurah Hhamrâ' (Arab.; POR); Dick Hirse (Ger.; EFS; HHB); Dohan (Heb.; ZOH); Dohna (Arab.; ZOH); Dokhu (Arab.; DEP); Dudha Vari (Mah.; NAD); Duhn (Arab.; ZOH); Dukhn (Arab.; POR); Echte Hirse (Ger.; EFS; HHB); Flatter Hirse (Ger.; EFS; HHB); Gadio (Guj.; NAD); Gamh (Quetta; KAB); Gemeiner Hirse (Ger.; EFS); Gewöhnliche Rispenhirse (Ger.; POR); Ghoti Savi (Mah.; NAD); Hairy Millet (Eng.; POR); Harilik Hirss (Estonia; POR); Hirs (Swe.; POR); Hirse (Den.; POR); Hirssi (Fin.; POR); Hog Millet (Eng.; NPM); Indian Buffalo Grass (S. Afr.; KAB); Indian Millet (Eng.; POR); Ji (China; POR); Kadukanni (Tam.; WOI); Katakanai (Tam.; DEP); Khra ma (Tibet; NPM); Kibi (Japan; POR); Klumpe Hirse (Ger.; EFS; HHB); Köles (Hun.; EFS); Kuri (Guj.; Nwp.; DEP; NAD); Mainairi (Sin.; DEP; POR); Miglio (It.; EFS); Miglio Nostrale (It.; EFS); Miglio Nostrano (It.; Swiss; POR); Mijo (Sp,; USN); Mijo Común (Sp.; EFS); Mijo Mayor (Sp.; EFS); Mil (Fr.; POR); Mil en Branches (Fr.; KAB); Milho Miudo (Por.; EFS); Milho Painço (Por.; POR); Millet (Eng.; Scn.; AH2); Millet Commun (Fr.; EFS); Millet d'Inde (Fr.; EFS); Millet Panic (Eng.; USN); Millet Paniculé (Fr.; EFS); Millet Rond (Fr.; NAD); Milocorn (Eng.; HHB); Panico Coltivato (It.; Swiss; POR); Panico Miglio (It.; POR); Panivaragu (Tam.; POR); Phikai (Bundel.; DEP); Pliumgierst (Dutch; EFS); Pluimgierst (Dutch; POR); Proso Millet (Eng.; Ocn.; AH2; NPM); Proso Obyknovennoe (Rus.; POR); Proso Posevnoe (Rus.; POR); Proso Sornoe (Rus.; POR); Proso Zwyczajne (Pol.; POR); Rad (Sanskrit; DEP); KRalle (Mah.; NAD); Rispenhirse (Ger.; POR); Russian Millet (Eng.; HHB); Salar (Pun.; KAB); Sama (Bom.; KAB); Samli (Guj.; DEP; KAB); San Zhi Ji (China; POR); Sava (India; EFS); Save (Kan.; KAB; NAD); Sawan Chaitwa (Nwp.; DEP); Sawanjethwa (Nwp.; KAB); Shamakh (Dec.; DEP); Shu (China; DEP; POR); Small Millet (Eng.; NAD); Thulo Kaguno (Nepal; POR); Trosgierst (Dutch; POR); True Millet (Eng.; POR); Tzedze (Ladak; DEP; KAB); Vara (Mah.; NAD); Varagu (Tam.; KAB; POR); Varaka (Sanskrit; KAB); Vari (Bom.; Dec.; Mah.; Sindh; DEP; NAD); Variga (Tel.; WOI); Varo (Mar.; WOI); Viljahirssi (Fin.; POR); Vogelgierst (Dutch; POR); Waaraagaalu (Tel.; POR); Wadi (Bom.; DEP); Wari (Dec.; KAB); Wild Millet (Eng.; POR); Wild Proso Millet (Eng.; POR); Wilde Rispenhirse (Ger.; POR); Worga (Tel.; DEP; KAB); Worglo (Arab.; KAB); Ye Sheng Ji (China; POR); Ye Sheng Ji Cao (China; POR).

Activities (Millet Panic):

Antidote (Cinnabar) (f; DAW); Antidote (*Momordica*) (f; DAW); Demulcent (f; DAW; EFS; NAD); Diuretic (f; DAW; EFS); Intoxicant (f; DAW); Pectoral (f; DAW; EFS); Refrigerant (f; BIB; DAW).

Indications (Millet Panic):

Abscess (f; DAA; DAW); Bleeding (f; DAW); Boil (f; DAW); Cancer (f; DAA); Cancer, breast (f; DAW); Childbirth (f; DAW); Cough (f; DAW); Dermatosis (f; DAA); Diarrhea (f; NAD); Fever (f; DAW); Gonorrhea (f; KAB); Hematuria (f; DAW); Mastosis (f; BIB); Sore (f; DAA; DAW); Venereal Disease (f; DAW).

Dosages (Millet Panic):

FNFF = !!!
As human food, millet is used as meal for making bread and other baked foods, as a paste from pounded wet seeds or as a boiled gruel. Prepared with milk and sugar, it is frequent at Indian marriage ceremonies. In Bihar, it is boiled and parched to make *markha.* In eastern Europe, the Balkans, Caucasus, and Asia, it is used to make an alcoholic beverage. The grain is eaten readily by livestock (mainly hogs, cattle, and poultry), but is not suited for horses. It is also grown for commercial birdfeed. It should be ground for livestock feed, equal to or superior in food value to oats (BIB; DEP; NPM).

- Baluchistanis use the plant to treat gonorrhea (KAB).
- Germans paste powdered seeds onto mammary cancers, (it was even tried at Memorial Hospital in New York) (JLH).

OPIUM POPPY (*PAPAVER SOMNIFERUM* L.) (+++ SEEDS) (XXX OPIUM) PAPAVERACEAE

Notes (Opium Poppy):

They gave him vinegar to drink mingled with gall: and when he had tasted thereof he would not drink.

Matthew 27:34 (KJV)

They offered him wine to drink, mingled with gall; but when he tasted it, he would not drink it.

Matthew 27:34 (RSV)

They gave him wine mixed with gall to drink; but after tasting it, he refused to drink.

Matthew 27:34 (RSV)

Unlike Zohary (ZOH), I am still inclined to believe that the biblical gall was opium. First I quote from my 1985 book, now out of print: "Walker equates this gall with *Papaver somniferum* while Moldenke and Moldenke equate it with *Citrullus colocynthis*, not even considering the opium poppy." The gall added to the vinegar and offered to Jesus was the juice of the opium poppy, a flower thriving in the Holy Land. The plant provides a narcotic that induces a heavy sleep. When the Roman soldiers at Golgatha took pity on their prisoner on the cross, they added poppy juice to the sour wine. Opium is the air-dried milky exudation obtained from excised unripe fruits. Egyptians claim to become more cheerful, talkative, and industrious following the eating of opium. When falling asleep, they have visions of orchards and pleasure gardens embellished with many trees, herbs, and various flowers (BIB). Jewish authorities maintain that the plant and its stupefacience were well known among the Hebrews more than 2000 years ago. The Jerushalmi warns against opium eating (BIB). Perhaps the following from Associated Press will strengthen my case: Researchers uncovered evidence of a thriving Bronze Age drug trade which supplied narcotics to ancient Mediterranean

FIGURE 1.79 Opium Poppy (*Papaver somniferum*).

cultures to ease pain of childbirth and disease. Ancient ceramic pots, most nearly identical in shape and about five inches long, found in settlements throughout the Middle East, date as far back as 1400 BC, according to Joe Zias, anthropologist, Hebrew University, Jerusalem. If turned upside down, the thin-necked vessels with rounded bases each resemble a poppy pod. The Mycenaean ceramics, analyzed with gas chromatography, turned up traces of opium. (Associated Press, August 8, 2002). And now there is new evidence as to why this plant has been a balm (and bane) to mankind for at least 5 millennia.

Common Names (Opium Poppy):

Abin (Sin.; DEP; NAD); Abini (Tam.; Tel.; DEP; NAD); Abkini (Tel.; DEP); Abou en Noum (Arab.; BOU); Abunom (Arab.; DEP); Adormidero (Sp.; EFS); Afim (Dec.; Hindi; Kasahmir; Nepal; Pun.; DEP; KAB; NAD); Afioun (Arab.; BOU); Afu (Mah.; NAD); Afyun (Arab.; Hindi; GHA; KAB); Agria (Greek; KAB); Ahiphena (Sanskrit; DEP; NAD; WOI); Amapola (Peru; Sp.; EGG); Anfião (Por.; POR); A Phien (Ic.; KAB); Aphim (Bom.; Guj.; Mah.; Nepal; NAD; POR); Aphina (Guj.; DEP); Aphioni (Greek; POR); Aphu (Mar.; DEP; KAB); Aphukam (Sanskrit; POR); Apkim (Nepal; DEP; KAB); Appo (Bom.; DEP; KAB); Balewort (Eng.; KAB); Bhain (Burma; DEP); Bhainzi (Burma; KAB); Bhinbin (Burma; NAD); Bilgasgase (Kan.; KAB; WOI); Birkes (Den.; POR); Bizrulkhashkhash (Arab.; KAB); Blauwmaanzaad (Dutch; POR); Bou en Noum (Arab.; BOU); Boudi (Arab.; BOU); Boundi (Arab.; BOU); Bungapion (Malaya; KAB); Cascall (Cat.; KAB); Chosa (Sanskrit; KAB); Doda (Kachhi; Pun.; DEP; KAB); Dormideira (Por.; EFS); Dormidera (Sp.; KAB); Gartenmohn (Ger.; EFS); Gasagasala (Tel.; KAB); Gasagase (Kan.; DEP); Gasalu (Tel.; WOI); Gashagasha (Tam.; NAD); Harir Igran (Arab.; BOU); Hashas (Tur.; EFS); Hashash (Tur.; KAB); Heul (Dutch; KAB); Hishas (Arab.; GHA); Kasakase (Kan.; Kon.; NAD); Kaskakasha (Mal.; KAB); Keshi (Japan; POR); Khas Khas (Sanskrit; EFS); Khashkhash Aswad (Arab.; Iran; BOU; DEP; KAB); Khashkhashsufaid (Urdu; KAB); Kheskkhash (Arab.; BOU); Khuskhus (Guj.; Mar.; WOI); Koknar (Iran; KAB); Kuru (Mal.; DEP); Maankop (Dutch; EFS; POR); Maanzaad (Dutch; POR); Mák (Hun.; EFS); Mak Lekarski (Pol.; POR); Mak Opiinyi (Rus; POR); Mak Opijnyj (Rus.; POR); Mák Sety (Czech; POR); Mak Snotvornyi (Rus.; KAB; POR); Oeillette (Fr.; POR); Oopiumiunikko (Fin.; POR); Opievallmo (Swe.; POR); Opium Poppy (Eng.; Scn.; AH2; CR2); Opiumvallmo (Swe.; POR); Opiumvalmue (Den.; Nor.; POR); Papavero (It.; EFS); Papavero da Oppio (It.; POR); Papoula Branca (Por.; KAB); Parag Tarbuti (Heb.; POR); Pasto (Beng.; KAB); Pavot (Fr.; BOU); Pavot Somnifére (Fr.; EFS); Peony Poppy (Eng.; KAB); Pest (Hindi; KAB); Pianta da Oppio (Malta; KAB); Pikincha (Sa.; ROE); Pioniunikko (Fin.; POR); Pionvallmo (Swe.; POR); Posht (Kum.; DEP); Post (Beng.; Hindi; DEP); Posta (Oudh; DEP); Postaka (Tam.; KAB; WOI); Posta Katol (Tel.; NAD); Posto Dheri (Beng.; NAD); Saphenaka (Sanskrit; NAD); Schlafmohn (Ger.; EFS); Slaapbol (Dutch; EFS; POR); Slaappapaver (Dutch; POR); Somnisor (Rom.; KAB); Sufeed Srah (Hindi; NAD); Tilidout (Ber.; BOU); Uniko (Fin.; POR); Vallmo (Swe.; KAB); Valmúafræ (Iceland; EFS); Valmue (Den.; Nor.; EFS; POR); Valmue Frø (Den.; POR); Vrtni Mak (Croatia; POR); Yang Gwi Bi (Korea; POR); Yanko Maiwa (Sa.; ROE); Ya Pin (China; NAD); Ying Su (Pin.; DAA); Ying Tzu Shu (China; EFS); Za Zang (Laos; POR).

Activities (Opium Poppy):

Abortifacient (f; SKJ); Analgesic (f1; APA; CRC; PHR); Anaphrodisiac (f1; FEL); Anodyne (f1; CRC; KAP); Anorectic (1; PR14:401); Antidiarrheal (f; PNC); Antidote (Atropine) (f; FEL); Antidote (Physostigmine) (f; FEL); Antidote (Strychnine) (f; FEL); Antiinflammatory (f; DEP); Antinociceptive (f1; PR14:401); Antispasmodic (f1; APA; DEM; DEP; FEL; PNC); Antitussive (f1; APA; PHR; PNC); Aphrodisiac (f; CRC; KAB); Astringent (f1; CRC; DAA; KAB); Bactericide (1; BIB); Bradycardic (1; PR14:401); Calmative (f1; CRC); Carminative (f; BIB; CRC); Cerebrostimulant (1; KAP; FEL); Chemopreventive (1; JAC7:405); Constipative (f1; PR14:401); Decongestant (1; CRC);

Deliriant (f; KAB); Demulcent (f; BIB; CRC); Diaphoretic (f; FEL; PNC); Diuretic (f; KAB); Emmenaogue (f; BOU); Emollient (f; CRC); Euphoric (f1; APA); Expectorant (f; CRC; ROE); Febrifuge (f1; FEL); Glutathiogenic (1; JAC7:405); Hemostat (f; CRC; KAB); Hypotensive (f; BIB; CRC); Hypnotic (1; APA); Intoxicant (f1; CRC); Lipogenic (f; KAB); Myorelaxant (f1; APA; FEL); Narcotic (f1; APA; CRC; SUW); Nervine (f; BIB; CRC; EFS); Refrigerant (f; KAB); Sedative (f1; APA; CRC; KAP); Spinostimulant (f; FEL); Stimulant (1; APA); Sudorific (f; CRC); Tonic (f; BIB; CRC); Tranquilizer (f; DEM); Vasodilator (1; CRC).

Indications (Opium Poppy):

Abscess (f; NAD); Amenorrhea (f; BOU); Anemia (f; KAB); Angina (1; DAA); Anxiety (f1; APA); Asthma (1; APA; CRC; FEL); Bleeding (f; KAB); Boil (f; BIB; CRC); Bronchosis (f; KAP; PHR); Bruise (f; CRC); Calculus (f1; FEL; NAD); Cancer (f1; CRC; FNF; JAC7:405); Cancer, bladder (f1; JLH); Cancer, breast (f1; JLH); Cancer, colon (f1; JLH); Cancer, ear (f1; JLH); Cancer, esophagus (f1; JAC7:405); Cancer, eye (f1; JLH); Cancer, liver (f1; JLH); Cancer, nose (f1; JLH); Cancer, skin (f1; JLH); Cancer, spleen (f1; JLH); Cancer, stomach (f1; JLH); Cancer, throat (f1; JLH); Cancer, tongue (f1; JLH); Cancer, uterus (f1; JLH); Cancer, uvula (f1; JLH); Cancer, vagina (f1; JLH); Carbuncle (f; NAD); Cardiopathy (f; NAD; WOI); Catarrh (f; CRC; FEL; ROE); Childbirth (f1; FEL); Cholecocystosis (f; PHR); Cholera (f; DEP; FEL; NAD); Cold (f; CRC); Colic (f; DEP; PHR; PH2); Condyloma (f; JLH); Conjunctivosis (f; CRC; FEL; NAD; PH2); Convulsion (f; KAP); Cough (f1; APA; FEL; PHR; PNC; ROE); Cramp (f1; APA; BOU; DEM; PH2); Cystosis (f1; BIB; CRC; DEP; PH2); Delirium (f; DEP; FEL); Depression (f; PH2); Dermatosis (f; FEL; JLH; ROE); Diabetes (f; NAD); Diarrhea (f1; APA; BOU; CRC; FEL; PH2); Duodenitis (f; WOI); Dysentery (f1; CRC; DEP; FEL; PH2); Dysmenorrhea (f; CRC; DEP); Dyspepsia (f; FEL; NAD); Earache (f; NAD); Eclampsia (1; FEL); Embolism (1; WOI); Enterosis (f1; APA; BOU; CRC; FEL; PH2); Epistaxis (f; BIB); Erysipelas (f; FEL); Fever (f; CRC; DEP; PH2); Flu (f; ROE; WOI); Flux (f; CRC); Gallstone (f; PH2); Gangrene (f; DEP); Gastrosis (f1; APA; DEP); Gonorrhea (f; FEL); Gout (1; FEL); Headache (f; CRC; DAA); Hemicrania (f; BIB; CRC; NAD); Hemorrhoid (f; FEL; NAD); Hepatosis (f; DEP; JLH); Hernia (f; NAD); High Blood Pressure (f; CRC); Hyperacidity (f; WOI); Hypochondria (f; CRC); Hysteria (f; CRC; FEL); Induration (f; JLH); Inflammation (f; CRC; EGG; PH2); Inhibition (1; APA); Insomnia (f1; APA; CRC; GHA); Itch (f; BIB); Kidney stone (f; PH2); Labor (f; NAD); Leprosy (f; NAD); Leukorrhea (f; CRC; DAA); Lumbago (f; NAD); Malaria (f; CRC; NAD); Mania (f; BIB; CRC); Mastosis (f; JLH); Melancholy (f; CRC); Menorrhagia (f; DEP); Metritis (f; NAD); Mucososis (f; FEL); Myocardosis (f; WOI); Nausea (f; CRC; FEL); Nephrosis (f; DEP; FEL); Neuralgia (f; CRC; FEL); Neurosis (f; DEP; GHA); Ophthalmia (f; DEP; PH2); Otosis (f; CRC); Pain (f1; APA; BOU; FEL); Peritonosis (f; DEP; FEL); Pertussis (f; CRC; WOI); Phthisis (f; DEP; FEL); Polyp (f; JLH); Proctosis (f; CRC; FEL; PH2); Prolapse (f; CRC; PH2); Pulmonosis (f; FEL; ROE); Raynaud's (1; WOI); Respirosis (f1; APA; FEL); Rheumatism (f; CRC; DEP); Scirrhus (f; JLH); Scrofula (f; NAD); Smallpox (f; NAD); Snakebite (f; CRC; NAD); Sore Throat (f; EGG; JLH); Spasm (f1; PHR); Spermatorrhea (f; CRC; DAA); Splenosis (f; JLH); Sprain (f; BIB; CRC); Stomachache (f1; APA; BIB; CRC); Stomatosis (f; EGG); Sunstroke (f; NAD); Swelling (f; CRC); Tenesmus (f; FEL; NAD); Tetanus (f; DEP; FEL; NAD); Toothache (f; CRC; DAA); Tuberculosis (f; PH2; WOI); Tumor (f; CRC); Typhoid (f; FEL); Typhus (f; NAD; PH2); Ulcer (f; CRC; PH2; WOI); Urethrosis (f; NAD); Urogenitosis (f1; BOU; PHR); Uterosis (f; DEP; FEL; JLH; NAD); Uvulosis (f; JLH); Vaginosis (f; JLH); Vomiting (f; DAA); Wart (f; CRC; JLH); Wound (f1; PHR).

Dosages (Opium Poppy):

FNFF = !!!
Seeds widely eaten or used as oil seed. Seeds contain no opium, and are used extensively in baking and sprinkling on rolls and bread. Although the seeds contain no narcotic alkaloids, urinalysis

following their ingestion may suggest morphine or heroin use. Leaves not so widely eaten as potherb or salad (BIB; DEP). Prescription only (for opiates). I do not believe I would recommend the dosage in KAP that is, 30–125 mg).

- Algerians tamp opium into tooth cavities (BIB).
- Ayurvedics, consider seeds aphrodisiac, constipating, and tonic, the fruit antitussive, binding, cooling, deliriant, excitant, and intoxicant, yet anaphrodisiac if freely indulged, the plant aphrodisiac, astringent, fattening, stimulant, tonic, and good for the complexion (KAB).
- Chinese use poppy heads for diarrhea, dysentery, and fluxes (KAB).
- Iranians use the seeds for epistaxis; applying a paste made from *Linum*, *Malva*, and *Papaver* to boils (BIB).
- Lebanese use opium wisely, to quiet excitable people, to relieve toothache, headache, incurable pain and for boils, coughs, dysentery, and itches (HJP).
- Peruvians suggest decoction of white flowers for flu, with milk for cough (ROE).
- Peruvians suggest floral or capsular tea for oral inflammation (EGG).
- Unani medicine suggests the fruit for anemia, chest pains, dysentery, fever. Deemed hypnotic, narcotic, and perhaps harmful to the brain (KAB).

Downsides (Opium Poppy):

Seeds Class 1 (AHP, 1997). Not indexed in Commission E. Opium overdoses can cause cold clammy skin, fast weak pulse, fluid in the lungs, cyanosis, pupil constriction, and possible death from circulatory and respiratory failure. Opium addicts can reportedly tolerate 2000 mg over 4 hours, but 300 mg will kill many naive subjects. Opiates have been detected in urine of poppy seed eaters as much as 48 hours after ingestion.

Natural History (Opium Poppy):

Although some self-pollination occurs before the flowers open, cross-pollination by insects also occurs. Some of the fungi attacking opium poppy include the following species: *Alternaria brassicae* var. *somniferi*, *Cladosporium herbarum*, *Erysiphe polygoni*, *Fusarium scirpi* var. *caudatum*, *Heterosporium echinulatum*, *Macrosporium papaveris*, *M. bresdolae*, *Mucor mucedo*, *Ophiobolus sativus*, *Oidium erysiphoides*, *Peronospora arborescens*, *P. papaveracea*, *Rhizoctonia solani*, *Sclerotinia sclerotiorum*, *Trichothecium roseum*. Plants are also attacked by the bacteria *Bacillus* (Erwinia) *papaveri*, causing bacterial blight, and *Xanthomonas papavericola*. The following nematodes have been isolated from the opium poppy: *Ditylenchus dipsaci*, *Longidorus maximus*, *Meloidogyne sp.*, *Pratylenchus crenatus*, *P. penetrans*, and *P. pratensis*. Insect pests include *Aphis papaveris*, *Ceutorhynchus abbreviatus*, *C. albovittatus*, *C. maculaalba*, *Cynips minor*, *Dasynevra papaveris*, *C. callida*, *Lestodiplosis callida*, *Mamestra brassicae*, *Phytomiza albiceps*, *Sciophila wahlbomiana*, and *Stenocarus fuliginosus* (HOE).

Extracts (Opium Poppy):

Like ginkgo, it "increases blood flow to the brain" (APA). But there is more. Poeaknapo (2005) reports de novo formation of morphine in human cells. Morphine, the major alkaloid of opium of *Papaver somniferum*, is one of the strongest analgesics known. "Endogenous morphine" has been long isolated and authenticated by mass spectrometry in trace amounts from specific animal and human tissue or fluids. The most widely accepted explanation presently is that morphine detected in human and animal tissues is of exogenous sources (e.g., dietary origin). Poeaknapo concludes that morphine, reticuline, and norlaudanosoline are unequivocally biosynthesized by cultured human cells, the precursors conclusively shown to be oxygen, tyramine, reticuline, and thebaine (X15874902). Phillips et al. (2005) quantified the phytosterols in poppy seed: delta-5-avenasterol, 177 ppm; delta-7-avenasterol; campestanol, 26 ppm; campesterol, 290 ppm; phytosterols, 1850 ppm; poriferasta-7,25-dienol, 89 ppm; poriferasta-7,22,25-dienol; sitostanol, <13 ppm seed; beta-sitosterol, 1093 ppm; spinasterol; stigmastanol; delta-7-stigmastenol and stigmasterol, 68 ppm (X16302759).

DATE PALM (*PHOENIX DACTYLIFERA* L.) +++ ARECACEAE

Notes (Date Palm):

> **And the greater house he cieled with fir tree, which he overlaid with fine gold, and set thereon palm trees and chains.**
>
> **2 Chronicles 3:5 (KJV)**

> **The nave he lined with cypress, and covered it with fine gold, and made palms and chains on it.**
>
> **2 Chronicles 3:5 (RSV)**

> **And the great house he covered with juniper wood, after which he covered it with good gold, and then he brought up upon it palm tree figures and chains.**
>
> **2 Chronicles 3:5 (NWT)**

The versions are consistent with the palm, but the ceiling was fir in KJV, cypress in RSV, and juniper in NWT. Zohary says the date palm is one of the Holy Land's most ancient fruit trees (cultivated remains found in Chalcolithic [circa 3700 B.C.] and Ubaidian [circa 4000 B.C.] strata at several sites in the Near East). In Judges 4:5, Deborah sat under the palm tree, poetic symbol of justice, righteousness, and upright stature. It continues to symbolize holiness and resurrection in Christian worship. Jericho was described as the "city of palm trees" (Deuteronomy 34:3). Date palm has long been associated with Palestine, even being the symbol on its coinage. Arabs say that there are as many uses for dates as there are days in the year. There is an Arab adage: "its head should be in fire (sunshine) and its feet in water." And Psalmists say "the righteous shall flourish like the palm tree." There is sap in the palm tree that, after fermentation, is used as a liquor. This may be some of the strong drink or wine of the Bible (BIB; ZOH).

Common Names (Date Palm):

Abdandan (Kej.; KAB); Agjjuf (Ber.; BOU); Arabian Date (Eng.; KAB); Balah (Arab.; Nig.; Syria; AVP; BOU; HJP; UPW); Begamjangi (Panjgur; KAB); Blah (Arab.; BOU); Chhomer (Heb.; KAB); Chohoraa (Nepal; POR); Chhuharra (Bom.; NAD); Chuara (Bom.; DEP); Chuhara (India; EFS); Curmal (Rom.; KAB); Dabino (Gambia; UPW); Dabinos (Sudan; AVP); Daddel (Den.; Nor.; POR); Daddelpalme (Den.; Nor.; POR); Dadelpalm (Dutch; EFS); Dadels (Dutch; AVP); Daktyle (Pol.; AVP); Daktylowiec (Pol.; POR); Date (Eng.; USN); Date Palm (Eng.; Scn.; AH2; CR2); Dátil (Cuba; Peru; Sp.; AVP; EGG); Datilero (Col.; Peru; Sp.; AVP; EGG); Datte (Fr.; Haiti; AVP); Dattel (Ger.; AVP); Dattelpalm (Ger.; EFS); Dattelpalme (Ger.; POR); Datteri (It.; AVP); Dattero (It.; EFS); Dattier (Fr.; Haiti; AVP; BOU; EFS); Dattier Commun (Fr.; POR); Dipya (Sanskrit; KAB); Dvash (Heb.; ZOH); Dwane (Ivo.; UPW); Echte Dadelpalm (Dutch; POR); Echte El-nakheil (Arab.; BOU); Edible Date

FIGURE 1.80 Date Palm (*Phoenix dactylifera*).

(Eng.; NAD); Finikovaia Pal'ma (Rus.; POR); Gajjira (Badaga; KAB); Gewone Dadelpalm (Dutch; POR); Gharar Khejur (Beng.; NAD); Gijjira Hannu (Kan.; NAD); Hai Zao (China; POR); Hazdacht (Ber.; BOU); Hurma Ag̮aci (Tur.; EFS); Ichu (Tam.; KAB); Indi (Singh.; DEP; KAB); Inthaphalam (Thai; POR); Isgaren (Ber.; BOU); Ita (Tel.; KAB); Itta (Mal.; KAB); Ittappazham (Mal.; WOI);

Kajura (Kan.; Nwp.; Pushtu; DEP; KAB); Karchuram (Tam.; POR); Karek (Guj.; DEP; KAB); Kerjura Kaya (Tel.; NAD); Karmah (Tur.; DEP); Kasser (Bhutan; DEP); Khaji (Hindi; Pun.; KAB; POR); Khajur (Beng.; Guj.; Hindi; Mar.; Kon.; Pun.; Sharig; KAB; NAD; POR; WOI); Kharakia (Guj.; NAD); Kharchuram (Tam.; WOI); Kharik (Mah.; NAD); Kharjjuraha (Sanskrit; DEP); Kharjur (Mar.; KAB); Kharjura (Ayu.; Kan.; AH2; DEP; WOI); Kharjuramu (Tel.; WOI); Khorjjuri (Oriya; KAB; WOI); Khourma (Tur.; AVP); Khurma (India; Nasiribad; Urdu; EFS; KAB); Khurmae Yabis (Arab.; DEP); Khurmal Kshusk (Iran; EFS; NAD); Khurmal Yabis (Arab.; EFS; NAD); Kurma (Sin.; DEP); Mach (Kohhaja; KAB); Mtende (Swahili; POR); Nakhel (Arab.; GHA); Nakhl (Arab.; Iran; Syria; BOU; DEP; GHA; HJP); Nakhleh (Arab.; KAB); Natchla (Arab.; Mali; UPW); Natsuma yashi (Japan; TAN); Natsume Yashi (Japan; POR); Nekhla (Arab.; BOU); Ntamaro (West Cameroons; UPW); Palma (It.; Malta; KAB); Palma Datil (Sp.; EFS); Palma Datilifera (Sp.; EFS); Pal'ma Finikovaia (Rus.; POR); Palmeira (Por.; KAB); Palmera (Sp.; AVP); Palmera Datilera (Sp.; POR); Palmier Dattier (Fr.; BOU; USN); Palmera de Dátiles (Sp.; POR); Palmizio (It.; POR); Palmtrae (Swe.; KAB); Perichchankay (Tam.; DEP; NAD; POR); Perita (Tel.; DEP); Phinikovoe Dyerevo (Rus.; KAB); Phoinix (Greek; KAB); Pinda Kharjura (Sanskrit; NAD); Pindakhejur (Hindi; India; EFS; NAD); Salma (Hindi; POR); Sendhi (Hindi; POR); Sunbalun (Burma; KAB); Swonpalwon (Burma; DEP); Taatelipalmu (Fin.; POR); Tafinaout (Ber.; BOU); Tamalo (Sierra Leone; UPW); Tamar (Arab.; Heb.; POR; ZOH); Tamara (Por.; AVP); Tamareira (Por.; Mad.; AVP); Tamaruy (Sen.; UPW); Tambaroohi (Upper Volta; UPW); Tammar (Arab.; GHA); Tammr (Arab.; BOU; GHA); Tanekht (Ber.; BOU); Tar (Sin.; KAB); Tayniyut (Ber.; BOU); Tazdait (Ber.; BOU); Teeney (Niger; UPW); Temer (Arab.; POR); Tenitta (Mal.; WOI); Tiyni (Ber.; BOU); Tomer (Heb.; KAB); Ton Inthaphalam (Thai; POR); Ttmer (Arab.; BOU); Uttatti (Kan.; NAD); Vrai Dattier (Fr.; UPW); Wu Lou Zi (Pin.; DAA; EFS; KAB); Ye Zao (China; POR); Zao Ye (China; POR); Zao Ye Zi (China; POR).

Activities (Date):

Allergenic (1; X1485659); Antihistaminic (1; X15814265); Antiinflammatory (f; KAB); Antimutagenic (1; X11804538); Antioxidant (1; X15814265); Antiradicular (1; X11804538); Aphrodisiac (f; BIB; BOU; DEP; EFS; GHA); Candidicide (1; FNF); Contraceptive (f; BIB); Demulcent (f; BIB; DEP); Deobstruent (f; HJP); Depurative (f; KAB); Diuretic (f; BIB; EFS); Emollient (f; BIB); Estrogenic (1; BIB; FNF); Expectorant (f; BIB); Fungicide (1; FNF); Gastroprotective (1; X15814265); Hepatotonic (f; KAB); Immunostimulant (1; X10904150); Laxative (f; BIB); Nephrotonic (f; KAB); Pectoral (f; BIB); Purgative (f; BIB); Refrigerant (f; BIB); Tonic (f; BOU; EFS; GHA).

Indications (Date):

Adenopathy (f; JLH); Ague (f; BIB); Anemia (f; BIB); Asthma (f; BIB; KAB); Bleeding (f; BOU); Blepharosis (f; BOU); Bronchosis (f; BIB; PH2); Bruise (f; GHA); Cancer (f; BIB); Cancer, abdomen (f; JLH); Cancer, colon (f; JLH); Cancer, liver (f; JLH); Cancer, mouth (f; JLH); Cancer, parotid (f; JLH); Cancer, spleen (f; JLH); Cancer, stomach (f; JLH); Cancer, testes (f; JLH); Cancer, throat (f; JLH); Cancer, uterus (f; JLH); Cancer, vagina (f; JLH); Candida (1; FNF); Catarrh (f; BIB); Chest (f; BIB); Coma (f; KAB); Condylomata (f; BIB); Cornea (f; NAD); Cough (f; BIB; BOU; KAB); Diarrhea (f; BIB; BOU; DEP); Enterosis (f; KAB); Fatigue (f; BIB); Fever (f; BIB); Flu (f; BIB); Gastrosis (f1; JLH; PH2; X15814265); Gonorrhea (f; BIB; KAB); Halitosis (f; DEP); Hangover (f; NAD); Headache (f; GHA; PH2); Hemorrhoid (f; BIB); Hepatosis (f; JLH); Impotence (f; BIB); Induration (f; BIB; JLH); Infertility (f; BIB; BOU; UPW); Inflamation (f; PH2); Intoxication (f; NAD); Jaundice (f; BOU); Keratitis (f; DEP); Leprosy (f; KAB); Longevity (f; BIB); Malaria (f; NAD); Mastosis (f; HHB); Mycosis (1; FNF); Nausea (f; KAB); Nephrosis (f; PH2); Opacity (f; NAD); Ophthalmia (f; BIB; NAD; PH2); Orchosis (f; JLH); Paralysis (f; KAB); Parotosis (f; JLH); Pterygia (f; BIB); Pulmonosis (f; KAB); Smallpox (f; NAD); Sore (f; BOU); Sore Throat (f; JLH); Splenosis (f; BIB;

JLH); Sterility (f; BIB); Stomachache (f; BIB); Stomatosis (f; JLH); Thirst (f; BIB; BOU); Toothache (f; BIB); Tuberculosis (f; BIB); Ulcer (f1; X15814265); Unconsciousness (f; KAB); Urogenitosis (f; BIB); Uterosis (f; JLH); Vaginosis (f; BIB); Wart (f; BIB); Whitlow (f; BIB); Wound (f; PH2); Yeast (1; FNF).

Dosages (Date):

FNFF = !!!
Fruits widely eaten; green fruits pickled and eaten; spathes soaked in water and chewed; male inflorescence (with estrone-containing pollen) eaten; pollen eaten; seeds occasionally eaten; pressed for edible oil; sap tapped for sugar or fermented. In some areas, 95% of the people survive on dates 9 months of the year. Fruits often preserved by drying or pressing them together into large cakes. Other products include date "honey" (bees are mentioned only four times in the Bible, while "honey" is mentioned 49 times), made from the juice of fresh fruit; date sugar; date sap, often made into a fermented beverage; date palm flour, made from pith of tree; oil from seeds; the kernels are ground up or soaked in water for days and used for animal food; both wine and honey are derived from the date; Nigerians feed dates with bran and *Sterculia* to immature young heifers to make them more prolific (BIB; FAC; TAN).

- Algerians smoke the seed powder for fever (HJP).
- Arabians consider the estrogen-containing pollen aphrodisiac and tonic (GHA).
- Arabians paste fruit on head and eyes for headache, salted fruits on bruises (GHA).
- Arabians use dates folklorically for gastric ulcers, and they work (X15814265).
- Arabians use green fruits as an astringent for hemorrhoids, applying powdered seeds or directing their smoke onto any affliction (BIB).
- Ayurvedics, viewing fruits as alexiteric, aphrodisiac, and tonic, use them for asthma, bronchosis, enterosis, fatigue, fever, leprosy, thirst, tuberculosis, and unconsciousness (KAB).
- Hausa add dates with hot peppers to native beer to make it less intoxicating (BIB).
- Lebanese believe the sugar from the fruits helps hepatitis (HJP).
- North Africans use fruit in vaginal pessary with other herbs to improve fertility (BOU).
- North Africans plaster powdered seeds on genital sores (BOU).
- North Africans ingest terminal bud for diarrhea, internal bleeding, and jaundice (BOU).
- North Africans use seed ashes in ophthalmic collyria (BIB).
- Unani consider the leaves aphrodisiac, hepatotonic; the flowers depurative, expectorant, febrifuge; the fruits aphrodisiac, nephrotonic, used for paralysis and pulmonosis; they apply the antiinflammatory seed to wounds (KAB).

Downsides (Date):

No health hazards or contraindications reported with proper administration of suggested therapeutic dosages (PH2) (but PH2 designates no dosage!; JAD).

Natural History (Date):

The Agriculture Handbook No. 165 lists the following diseases affecting the date palm: *Alternaria* sp. (leaf spot), *Alternaria citri* (brown spot of fruit), *Alternaria stemphylioides* (fruit spoilage), *Aspergillus niger* (calyx-end rot), *Auerswaldia palmicola* (on leaves), *Catenularia fuliginea* (fruit rot), *Ceratostomella radicicola* (root rot), *Colletotrichum gloeosporioides* (on leaves), *Diplodia phoenicum* (leafstalk rot, shoot blight, fruit rot), *Endoconidiophora paradoxa* (black scorch, heart bud rot), *Fusarium* spp. (inflorescence blight, fruit rot), *Graphiola phoenicis* (leaf spot, false smut), *Meliola furcata* (black mildew), *Meloidogyne* spp. (root knot nematodes), *Omphalia pigmentata* (decline disease), *Omphalia tralucida* (decline disease), *Penicillum roseum* (fruit rot), *Pestalotia* sp. (leaf spot), *Phomopsis phoenicola* (fruit rot), *Phymatotrichum omnivorum* (on roots), *Pleospora herbarum* (fruit rot, mold), and *Poria* spp. (wood rot). Popenoe (1920) assesses the percent damage caused by some of the major insect pests. Tackholm and Drar (1969–1973) give a good account of Egyptian diseases and pests.

Extracts (Date):

A 5% date extract showed better growth inhibition on *C. albicans* as compared to amphotericin B. The date extract caused a leakage of cytoplasmic contents from the yeast cells (Sallal, El-Teen, and Abderrahman, 1998). Al-Qarawi et al. (2005) demonstrated ameliorative effects of dates on ethanol-induced gastric ulcer in rats. Aqueous extracts of the fruit demonstrated potent antioxidant and antimutagenic properties (X11804538). The ethanolic undialyzed extract was more effective than other extracts tried (X15814265). Vayalil (2002) demonstrated potent antioxidant and antimutagenic properties of the aqueous extracts of the fruits (X11804538).

COMMON REED (*PHRAGMITES AUSTRALIS* (CAV.) TRIN. EX STEAD.) ++ POACEAE

Synonyms:

Arundo phragmites L.; *Arundo vulgaris* Lam.; *Phragmites communis* Trin.; *Phragmites communis* var. *longivalvis* (Steud.) Miq.; *Phragmites longivalvis* Steud.; *Phragmites vulgaris* (Lam.) Crép.; *Phragmites vulgaris* var. *longivalvis* (Steud.) W. Wight

Notes (Common Reed):

> **For the LORD shall smite Israel, as a reed is shaken in the water, and he shall root up Israel out of this good land, which he gave to their fathers, and shall scatter them beyond the river, because they have made their groves, provoking the LORD to anger.**
>
> **1 Kings 14:15 (KJV)**

Zohary adduces evidence for identifying *kaneh* with "reed," in analogies with the reed pen that I mentioned in my first Bible book (III John 13), the broken reed of a staff (II Kings 18:21; measures of length, Ezekiel 40:5), and the shaft of the lamp stand (Exodus 25:31). During the biblical period, reeds were extensively grown and used for field hedges, flutes, housing, mats, pens, scales, and walking scales (ZOH). Extensively used in Mediterranean regions and elsewhere for building dwellings, lattices, fences, arrows by Indians, and for weaving mats and carrying nets. A variegated form is used as an ornamental grass. The reed is useful in the manufacture of pulps for rayon and paper. It contains more than 50% cellulose and has a fiber 0.8 to 3.0 mm long and 5.0 to 30.5 μm in diameter. It is also useful in the production of homogeneous boards of good strength. It can be processed into a fine fibrous material suitable as a filling material in upholstery. Flowering stalks yield a fiber suitable for rope making. It is also used for thatching and for making partitions, fences, coarse mats, baskets, sandals, etc. Panicles are used for making brooms and for decoration. Pens for writing on parchment were cut and fashioned from this reed, and the stems were used as a linear measuring device. Bedouins use the stem to make flutes. Chinese commonly use the stem for fuel. Common reed provides high-quality, warm-season forage and is readily eaten by cattle and horses. However, it becomes tough and unpalatable after maturity. Animals grazing this grass during the winter should be fed a protein concentrate (BIB).

Common Names (Common Reed):

Aqraban (Arab.; GHA); Aranim (Ber.; BOU); Ashi (Japan; TAN); Bog Reed (Eng.; EFS); Bous (Arab.; BOU); Bus (Arab.; Syria; HJP); Cana (Sp.; TAN); Cañeta (Sp.; EFS); Caniço (Por.; EFS);

FIGURE 1.81 Common Reed (*Phragmites australis*).

Canna da Spazzole (It.; EFS); Canna di Palude (It.; EFS); Cannuchia (It.; EFS); Carrizo (Sp.; EFS); Common Reed (Eng.; Ocn.; AH2; BOU; CR2; WOI); Common Reed Grass (Eng.; TAN); Dila (Pun.; DEP); Ditch Reed (Eng.; BOU); Djaboub (Ber.; BOU); Ghab (Arab.; BOU; HJP); Hagna (Arab.; BOU); Hajna (Bahrain; Sausi; GHA); Heish Moddeid (Arab.; BOU); Ilili (Ber.; BOU); Jonc à

Balais (Fr.; BOU); Kamis (Tur.; EB54:155); Kaneh (Heb.; ZOH); Kasab (Arab.; Syria; HJP); Kita Yoshi (Japan; TAN); Lang Ku Ten (China; EFS); Lu Gen (Pin.; AH2; DAA); Lu Jing (China; AH2); Lu Ti Ken (China; EFS); Lu Wei (China; AH2; EFS); Phragmites (Scn.; AH2); Qasba (Arab.; BOU); Reed (Eng.; BUR); Riet (Dutch; EFS); Roseau (Fr.; BOU); Roseau Balais (Fr.; EFS; BOU); Roseau Commun (Fr.; BOU); Roseau de Maris (Fr.; BOU); Sazkamisi (Tur.; EFS); Schilfrohr (Ger.; EFS); Sokarik Otu (Tur.; EB54:155); Tagasiba (Ber.; BOU); Tebu Salah (Malaya; EFS); Tiouli (Ber.; BOU); Tiranimine (Ber.; BOU); Tissendjelt (Ber.; BOU); Tra'a (Arab.; BOU); Wild Broomcorn (Eng.; FAC).

Activities (Common Reed):

Alexiteric (f; BIB); Antiemetic (f; BOU; GHA; WOI); Depurative (f; BUR); Diaphoretic (f; BOU; HHB; JFM; PH2; WOI); Diuretic (f; BOU; BUR; EFS; HHB; JFM; PH2); Emetic (f; DEM); Expectorant (f; DEM); Febrifuge (f; BIB; BOU); Litholytic (f; JFM); Sialogogue (f; BIB); Stomachic (f; BIB; BOU); Sudorific (f; EFS).

Indications (Common Reed):

Abscess (f; BOU); Arthrosis (f; BOU); Bite (f; PH2); Bleeding (f; BIB); Boil (f; DEM); Bronchosis (f; BIB); Cancer (f; JLH); Cancer, breast (f; JLH; PH2); Carbuncle (f; DEM); Cholera (f; BIB); Condyloma (f; BIB); Cough (f; AAH); Dermatosis (f; DEM); Diabetes (f; HHB; PH2; WOI); Diarrhea (f; DEM); Dropsy (f; EFS); Dysuria (f; BIB); Earache (f; BIB); Fever (f; BIB; BOU); Food Poisoning (f; BOU); Fracture (f; DEM; HJP; WOI); Gastrosis (f; BIB; DEM; HJP); Gout (f; EFS); Hematuria (f; BIB); Hiccup (f; BIB); Induration (f; JLH); Jaundice (f; BOU); Kidney stones (f; JFM); Leukemia (f; HHB; JLH; PH2); Mastosis (f; BIB; JLH); Measles (f; BIB); Nausea (f; BIB); Pain (f; BIB); Pneumonia (f; DEM); Pulmonosis (f; BOU; DEM); Rheumatism (f; EFS; WOI); Sore (f; BIB); Sore Throat (f; BIB); Sunstroke (f; BIB); Thirst (f; BIB); Toothache (f; BIB); Typhoid (f; BIB).

Dosages (Common Reed):

FNFF = !

Young shoots sometimes eaten like bamboo shoots; grain edible; partly unfolded leaves eaten as a vegetable; young leaves of var. *longivalvis* are dried, ground, and made into dumplings with cereal flour; rhizomes sometimes cooked and eaten like potatoes; sugar extracted from rhizome; scorched plant used as coffee substitute. In Russia, they are harvested and processed into starch. Stalks exude a manna-like gum, which is eaten. (BIB; EFS; FAC; HHB; TAN; EB54:155).

- Arabians make a cooling and antiemetic beverage from the plant (GHA).
- Chinese use the plant for leukemia (JLH).
- Africans use the sugary exudate for chest pain and pneumonia (BIB).
- Cape Africans apply powdered seed to burns (BIB).
- Chinese used as a remedy for hiccups and poisoning from eating stale seafood (BIB).
- East Asians use the plant for rheumatic ailments (WOI).
- Hebrides inhabitants make a cough medicine from reed and stinging nettle (AH2).
- Lebanese pack fractures with broken reeds (HJP).
- Orientals make a packing of reeds as a splint for fractures (BIB).

Downsides (Common Reed):

No health hazards or contraindications reported with proper administration of suggested therapeutic dosages (PH2) (but PH2 designates no dosage!; JAD).

CALABRIAN PINE (*PINUS BRUTIA* TEN.) +++ PINACEAE

SYNONYMS:

Pinus brutia subsp. *eldarica* (Medw.); *Pinus brutia* var. *pendulifolia* Frankis.; *Pinus brutia* var. *pityusa* (Steven) Silba; *Pinus brutia* var. *stankewiczii* (Sukaczev) Frankis.; *Pinus halepensis* subsp. *brutia* (Ten.) Holmboe; *Pinus halepensis* var. *brutia* (Ten.) A. Henry; *Pinus halepensis* var. *brutia* (Ten.) Elwes et Henry; *Pinus halepensis* subsp. *brutia* (Tenore) Holmboe; *Pinus persica* Strangw.; *Pinus pityusa* Steven *fide* CJE and USN

NOTES (CALABRIAN PINE):

And that they should publish and proclaim in all their cities, and in Jerusalem, saying, Go forth unto the mount, and fetch olive branches, and pine branches, and myrtle branches, and palm branches, and branches of thick trees, to make booths, as it is written.

Nehemiah 8:15 (KJV)

And that they should publish and proclaim in all their towns and in Jerusalem, "Go out to the hills and bring branches of olive, wild olive, myrtle, palm, and other leafy trees to make booths, as it is written."

Nehemiah 8:15 (RSV)

And that they should make proclamation and cause a call to pass throughout all their cities and throughout Jerusalem, saying "Go out to the mountainous region and bring in olive leaves, and the leaves of oil trees and myrtle leaves and palm leaves and the leaves of branchy trees to make booths, according to what is written."

Nehemiah 8:15 (NWT)

Zohary, working with the RSV, comments that pine forests were once copious in the Holy Land. The Hebrew word for pine today (*oren*) meant laurel in biblical times. But *etz shemen* occurs five times side by side with the olive and should be rendered "pine tree" and not "olive wood" or "wild olive" as in the RSV. Jewish villages in North Kurdistan, where *Pinus brutia* forms natural forests, preserve the name etz shemen, probably since the Babylonian exile. This is very closely related to the Aleppo pine (*Pinus halepensis*, which see), and probably shares many phytochemcials and medicinal indications. Zohary prefers the Aleppo pine to the Calabrian pine. I include accounts for both, not knowing who is correct. Quite candidly, I would be as happy to consider this "wild olive" following olive in the RSV to be the *Elaeagnus*, which see. After Nehemiah, the word "pine" is said not to reoccur in the Bible, but about 500 years later it is mentioned by Josephus, who says Solomon had pine wood brought in ships from Ophir, which was used for pillars and support to the King's temple and palace, and partly for musical instruments (e.g., cymbals, harps, and psalteries), for the glorification of God by the Levites. Some writers equate the pine branch of Nehemiah with *Elaeagnus* and the *thick trees* with *Pinus brutia*, whose boughs might have been used to make booths for the Feast of the Tabernacles (Moldenke and Moldenke, 1952). Whether or not *Pinus brutia* is properly identified as the biblical pine branch, the turpentine was doubtless used for medicine, like other turpentines. This one certainly is close to *Pinus halepensis*. I have not assigned all the medicinal virtues of generic turpentine or other pine extracts, but I frankly suspect they can be used interchangeably. The substance *burasu*, common in the herbals of ancient Assyria, was prepared by soaking some pine wood in water. It was applied externally to muscles and ligaments as an embrocation. Internally, it was taken for kidney or liver ailments (BIB).

FIGURE 1.82 Calabrian Pine (*Pinus brutia*).

Common Names (Calabrian Pine):

Calabrian Pine (Eng.; CJE; USN; X14987727); Etz Shemen (Heb.; KJV; ZOH); Kizilççam (= Red pine) (Tur.; CJE); Oren (Heb.; KJV; ZOH); Pitys (Greek; CJE); Turkish Pine (Eng.; CJE; USN); Türkische Kiefer (Ger.; USN); Wild Olive (Eng.; KJV; ZOH); Nscn.

Activities (Calabrian Pine):

Antiseptic (1; X10548751); Bactericide (1; X10548751).

Indications (Calabrian Pine):

Bacillus (1; X10548751); Bacteria (1; X10548751); Enterobacter (f; X10548751); Escherichia (f; X10548751); Infection (f; X10548751); Klebsiella (f; X10548751); Listeria (f; X10548751); Mycobacterium (f; X10548751); Pneumonia (f; X10548751); Proteus (f; X10548751); Pseudomonas (f; X10548751); Staphylococcus (f; X10548751).

Natural History (Calabrian Pine):

This is reportedly the most important forest tree in the northeastern Mediterranean area. A sap-sucking insect, *Marchalina hellenica*, produces large amounts of honey-dew, harvested by honeybees and sold as "pine honey" (CJE).

ALEPPO PINE (*PINUS HALEPENSIS* MILL.) ++ ABIETACEAE

Synonyms:

Pinus abasica Carr.; *P. alepensis* Poir. in Lamarck; *Pinus arabica* Sieber ex Spreng.; *Pinus carica* D. Don in Fellows; *Pinus ceciliae* Llorens et L. Llorens; *Pinus genuensis* S.E. Cook; *Pinus halepensis* Mill. var. *abasica* (Carr.) Carr.; *Pinus halepensis* Mill. var. *carica* (D. Don) Carr.; *Pinus halepensis* Mill. var. *ceciliae* (Llorens et L. Llorens) Rosello et al.; *Pinus halepensis* Mill. var. *genuensis* (S.E. Cook) Antoine; *Pinus halepensis* Mill. var. *minor* Antoine; *Pinus hispanica* J. Cook; *Pinus loiseleuriana* Carr.; *Pinus maritima* Mill.; *Pinus maritima* Aiton non Mill.; *Pinus parolinii* Vis.; *Pinus penicillus* Lepeyr.; *Pinus pseudohalepensis* Denhardt ex Carr.; *Pinus x saportae* Rouy; *Pinus sylvestris* L. var. *maritima* Aiton *fide* CJE

Notes (Aleppo Pine):

As for the stork, the fir trees are her house.

Psalms 104 (KJV)

Source of Greek turpentine, the Aleppo "pine" is a handsome tree. Most of the "fir tree" references in the Bible are now believed to refer to the Aleppo pine, *Pinus halepensis*. The Bible tells us the timber is quite useful; it is used for construction (doors, homes, rafters, ships) and musical instruments. The bark contains up to 20% tannin, and is used for tanning. Zohary tells us that pine forests were once widespread in the Holy Land, but Aleppo pine is much scarcer now, due to abuse throughout the years. There are still a few stands in Mediterranean parts of Israel, some attaining 20 m tall and living 150 years. The Hebrew *etz shemen*, which occurs in the Bible five times, should be rendered pine, not olive wood. The modern Hebrew word for pine, *oren*, was in biblical times used for laurel (ZOH). The myriad uses of "turpentine" from any species of pine might as well accrue to *Pinus halepensis* turpentine as well. "Turpentine" is loosely defined as the oleoresin obtained from longleaf and slash pines and other pines that yield exclusively terpene oils, or the essential oil obtained from the oleoresin. Turpentine from one species or another has been used for catarrh, cough, dysuria, dyschezia, gonorrhea, leucorrhea, and rheumatism. Unquestionably, turpentine has antiseptic, counterirritant, and rubefacient properties; and it is apparently also allergenic and tumorigenic, causing albuminuria, coma, cough, erythema, hematuria, headache, insomnia, nausea, and urticaria. The rosin from various pine species has been used for abscesses, boils, and cancers. Pine tar has been used in expectorant cough syrups for bronchitis, catarrh, colds, etc. Rosin has also been used for skin diseases such as psoriasis, ringworm, and toothache. In Russia, steroids have been extracted from pine pulp. In Dioscorides' day, the seeds were used for cough, and the cones for stomach ailments (BIB). And I suspect that the pine bark will also be well endowed with OPCs (oligomeric procyanidins), like the French Maritime Pine, whose bark and OPCs and pycnogenol gets so much praise.

Common Names (Aleppo Pine):

Aleppo Pine (Eng.; USN); Amelzi (Ber.; BOU); Etz Shemen (Heb.; ZOH); Hab Krash (Arab.; Syria; HJP); Igengen (Ber.; BOU); Jerusalem pine (Eng.; USN); Oren (Heb.; ZOH); Ouazouri (Ber.; BOU); Pin Blanc (Fr.; BOU); Pin d'Alep (Fr.; BOU); Pin de Jérusalem (Fr.; BOU); Pinheiro Francês (Mad.; JAD); Pinheiro do Alepo (Mad.; JAD); Pino Carrasco (Sp.; USN); Sanawbar (Arab.; BOU; HJP); Sanawbar Hhlab (Arab.; Syria; HJP); Senouber (Arab.; BOU); Seekiefer (Ger,; USN); Snawbar Barri (Arab.; Syria; HJP); Snouber (Arab.; BOU); Taida (Ber.; BOU); Zgougou (Arab.; BOU); Nscn.

Activities (Aleppo Pine):

Analgesic (f; HJP); Antioxidant (1; X10694042); Antiseptic (f; BOU; HJP); Astringent (f; BOU); Diaphoretic (f; HJP); Diuretic (f; HJP); Spermagenic (f; BOU); Vulnerary (f; BOU).

Indications (Aleppo Pine):

Chafing (f; HJP); Cold (f; HJP); Cough (f; HJP); Dermatosis (f; BOU; HJP); Gastrosis (f; HJP); Hemorrhoid (f; HJP); Impetigo (f; HJP); Infection (f; BOU; HJP); Infertility (f; BOU); Pain (f; HJP); Sore (f; HJP); Toothache (f; HJP); Venereal Disease (f; HJP); Wound (f; BOU).

Dosages (Aleppo Pine):

FNFF = !!
Thin-shelled seeds are prized by animals and humans. North Africans add them to festive cakes, and mix with sorghum to make a Ramadan beverage. Ground seeds are sprinkled over Tunisian pastries. Resin is used in fermenting wines (FAC).

- Lebanese scrape resinous exudates to make pills and suppositories (HJP).
- Lebanese use crude sap, called pitch or tar, internally for colds and coughs, externally for sores and venereal ailments (HJP).
- North Africans apply astringent powdered bark to wounds and use the tar as antiseptic (BOU).
- North Africans eat the seeds in honey first thing in the morning to augment sperm (BOU).

Downsides (Aleppo Pine):

As of July 2004, the FDA Poisonous Plant Database listed titles alluding to toxicity of this species.

Natural History (Aleppo Pine):

Common crossbills (*Loxia curvirostra*) feed on seeds of Aleppo pine, as do European red squirrels (*Sciurus vulgaris*). On the Iberian Peninsula, *Sciurus* exerted directional selection favoring larger cones with larger scales, which has caused cones there to be larger than in the Balearic Islands where *Sciurus* are absent (X15715841).

STONE PINE (*PINUS PINEA* L.) + ABIETACEAE

Notes (Stone Pine):

> **He heweth him down cedars, and taketh the cypress and the oak, which he strengtheneth for himself among the trees of the forest: he planteth an ash, and the rain doth nourish it.**
>
> **Isaiah 44:14 (KJV)**

> **He cuts down cedars; or he chooses a holm tree or an oak and lets it grow strong among the trees of the forest; he plants a cedar and the rain nourishes it.**
>
> **Isaiah 44:14 (RSV)**

> **There is one whose business is to cut down cedars; and he takes a certain species of tree, even a massive tree, and he lets it become strong for himself among the trees of the forest. He planted the laurel tree, and the pouring rain itself keeps making it get big.**
>
> **Isaiah 44:14 (NWT)**

There is much ambiguity between the versions above, and you will not find "pine" in any of them, although you will find cedar in all three passage versions. Working with the RSV, which translated "holm" rather than the underlined "cypress" above, Zohary notes that this is the only place where the Hebrew word *tirzah* occurs in the scriptures. He notes that this may have led Saadia Gaon, translator of the Bible into its first Arabic version (10th century), to render *tirzah* as "stone pine." Indeed, in Arabic, as in many languages, the names of several conifers include the radical rz or arz. Jewish villages in north Kurdistan, where closely related *Pinus brutia* forms natural stands, have probably preserved the name *etz shemen* for that pine, perhaps since the Babylonian exile. That was partly why I included *Pinus brutia* in my first biblical book. Meanwhile, stone pine was reportedly common in the coastal plain of Palestine, forming extensive forests during the last century. Groves of stone pine at Yarka on the coastal plain of Galilee and on Mt. Carmel may represent remnants of extensive old groves, planted or spontaneous. At one time, the nuts were exported as pignolia nuts (ZOH). More than just edible, the nuts are considered aphrodisiac. The Roman poet Ovid (born in the 1st century B.C.), in his *The Art of Love*, lists aphrodisiacs including pine nuts. The Greek physician Galen (2nd century A.D.) suggests pine seeds, honey, and almonds, taken before bedtime three nights in a row, to increase potency. Apicius, a Roman celebrity, recommended pine nuts, cooked onions, white mustard, and pepper as an aphrodisiac. Some Arabian sources suggest popularly 20 almonds and 100 pine nuts with a glassful of thick honey three nights before bedtime (CJE).

Common Names (Stone Pine):

Cypress (Eng.; ZOH); Holm (Eng.; ZOH); Italian Stone Pine (Eng.; FAC; USN); Nuces de Pino (JLH); Parasol Pine (Eng.; USN); Pignolia-nut Pine (Eng.; USN); Pin Parasol (Fr.; USN); Pin Pignon (Fr.; USN); Pinheiro Manso (Mad.; Por.; JAD); Pinie (Ger.; USN); Schirmkiefer (Ger.; USN); Stone Pine (Eng.; HJP; USN; ZOH); Tirzah (Heb.; ZOH); Umbrella Pine (Eng.; USN); Nscn.

Activities (Stone Pine):

Acaricide (1; X12137480); Allergenic (1; X12911512); Antiseptic (f; HJP); Aphrodisiac (f; CJE); Culicide (1; X15662650); Fungicide (f; HJP); Hemostat (f; HJP); Insecticide (1; X15662650); Insectifuge (1; X15662650); Larvicide (1; X15662650); Purgative (f; HJP).

Indications (Stone Pine):

Arteriosclerosis (f1; HOC); Arthrosis (f; JLH); Bleeding (f; HJP); Burn (f; HJP); Callus (f; JLH); Cancer (f; JLH); Cancer, breast (f; JLH); Cancer, bladder (f; JLH); Cancer, kidney (f; JLH); Cancer, liver (f; JLH); Cancer, throat (f; JLH); Cancer, uterus (f; JLH); Condyloma (f; JLH); Constipation (f; HJP); Cystosis (f; JLH); Fungus (f; HJP); Hepatosis (f; JLH); Impotence (f; CJE); Induration (f; JLH); Infection (f; HJP); Mastosis (f; JLH); Mycosis (f; HJP); Nephrosis (f; JLH); Phymata (f; JLH); Ringworm (f; HJP); Tumor (f; JLH); Uterosis (f; JLH); Worm (f; HJP).

Dosages (Stone Pine):

FNFF = !!!
One of the best of edible pine seeds, eaten raw or roasted or used in cakes, cookies, dolmas, pesto, picada, pilaf, sauces, soups. Romanians grind up whole green pine cones as a spice for game dishes. Gourmet oils produced from seeds in France (BIB; FAC).

- Druse apply the oil and resin to circumcision wounds (HJP).
- Early Greeks and Romans suggested almonds, honey, and pine nuts for aphrodisia (CJE).
- Lebanese use the sap or oil (called "priest's oil") as a purgative and an emollient for burns (HJP).

Downsides (Stone Pine):

Several abstracts refer to anaphylactic reactions to these nuts.

Extracts (Stone Pine):

Macchioni et al. (2002), "studying essential oils of four pine species", found that *P. pinea* oil and its two constituents (1,8-cineole and limonene) were most effective, showing 100% acaricidal activity at 6 and 8 μl, respectively (X12137480). Traboulsi et al. (2005) found extracts against fourth-instar larvae of the mosquito *Culex pipiens molestus* and some repelled mosquito bites. Terpineol and 1,8-cineole were the most effective at preventing bites, offering complete protection for 1.6 and 2 h, respectively (X15662650). Fortunately for nut-lovers, Phillips et al. (2005) quantified the phytosterols in five accesions of nuts. Such phytosterols are medicinally important, for example, in BPH and in high cholesterol (X16302759). The phytosterols in pine nuts (not necessarily *P. pinea*) based on five accessions were as follows: delta-5-avenasterol, 139–403 ppm; campestanol, 26—38 ppm; campesterol, 137–198 ppm; phytosterols 1470–2370 ppm; poriferasta-7,25-dienol, 66–177 ppm; sitostanol, <17–59 ppm; beta-sitosterol, 1044–1320 ppm; and stigmasterol, <17 ppm (X16302759).

MOUNT ATLAS MASTIC (*PISTACIA ATLANTICA* DESF.) ++ ANACARDIACEAE

Synonyms:

Pistacia atlantica var. *latifolia* DC.; *Pistacia mutica* Fisch & Mey

Notes (Mount Atlas Mastic):

Although less common than the other *Pistacias* in Palestine, this handsome long-lived species was probably adored and even deified by the ancients (FP2). Like oaks, terebinths were deified and revered by early Hebrews and other peoples. The Hebrew name *elah*, like that of the oak, stems from the Hebrew *el* (God), associated with strength and sturdiness. Zohary notes that "many translators and exegetes, unacquainted with the local flora, and embarrassed by the frequent occurrence in the Bible of *elah, elon*, *el*, *alah*, and *allon*, have seriously misunderstood these names. There are too many variations in the translations of different authors, and (as in the RSV) much inconsistency even within any given translation." (ZOH) Zohary concludes that, in general, allon and elon should be rendered oak, and *elah* and *alah* should be rendered terebinth. Terebinth stands often served as places of worship and incense burning, and even as revered burial sites. Terebinths feature in many biblical quotations. Saul and sons were buried under a terebinth (I Chronicles 10:12). Jacob buried idols under a terebinth (Genesis 35:4). An angel appeared to

Gideon under a terebinth (Judges 6:11). David slew Goliath in the Valley of Elah (which is Hebrew for terebinth) (I Samuel 17:2). David's son perished when his hair was caught in terebinth branches (II Samuel 18:9) (ZOH). Of the five species of *Pistacia* native to Israel and/or Sinai and Edom, the terebinth could have been either of the deciduous species, according to Zohary, who suspects the Atlantic terebinth more likely. While not mentioning it as a biblical species, Zohary does note that *Pistacia lentiscus* is co-dominant with carob in an evergreen scrub forest that grows in the foothills west of the mountain range, from Judea to the Lebanese border, as well as some of the eastern slopes of the mountains of Galilee and Samaria. So, *P. lentiscus* must be considered a potential biblical species as well. Hence, I have included the following species as biblical, with some redundancy, while Zohary may not have:

- Deciduous trees: (*Pistacia vera* is also deciduous):
  - —Leaf rhachis winged; leaflets obtuse, muticous: *P. atlantica*
  - — Leaf rhachis not winged; leaflets acute to acuminate: *P. palaestina*
- Evergreen shrub or tree:
  - — *P. lentiscus* (FP2)

Common Names (Mount Atlas Mastic):

Alah (Heb.; ZOH); Alk el Anbat (Arab.; BOU); Atlantic Pistacio (Eng.; BOU); Atlantic Terebinth (Eng.; ZOH); Betoum (Fr.; BOU); Botoum (Arab.; BOU); Elah (Heb.; ZOH); Gatouf (Arab.; BOU); Gueddain (Ber.; BOU); Hwadja (Arab.; BOU); Idj (Ber.; BOU); Iqq (Ber.; BOU); Khathiri (Arab.; BOU); Lggt (Ber.; BOU); Liez ou Illeg (Ber.; BOU); Mt. Atlas Mastictree (Eng.; USN); Pistachier de l'Atlas (Fr.; USN); Tecemlall (Ber.; BOU); Terebinth (Eng.; ZOH); Nscn.

Activities (Mount Atlas Mastic):

Allergenic (1; X3608141); Alpha-Amylase Inhibitor (1; X15182916); Hypoglycemic (f; X15182916).

Indications (Mount Atlas Mastic):

Diabetes (f; X15182916); Scrofula (f; BOU).

Dosages (Mount Atlas Mastic):

FNFF = !!
Acid fruits edible; sold in markets; used to season dates. Kernel used in pastries. Ripe fruits (*P. palaestina*) used in mideastern Zaatar, a mix of aromatic and food plants (BOU; FP2; X14759150).

- Jordanian herbalists recommend the species for hypoglycemic activity, which did not prove out in laboratory tests (X15182916).
- North Africans plaster leaves for scrofula (BOU).

Natural History (Mount Atlas Mastic):

In planted groves in Lahav Forest, Israel, scientists measured bird microhabitat selection in fruit-manipulated trees, trapping a total of 2357 birds. Sylviids exhibited a higher frugivory level than turdids. Sylviids selected densely foliated trees, while turdids were randomly distributed. Both species groups selected fruit-rich stopover habitats before further migration. Predation avoidance explains the sylviids' microhabitat selection; the migrants used foliage cover to reduce bird detectability by raptors (X15455207). Leaf galls produced by *Baizongia pistaciae*, on *Pistacia palaestina* (X14759150).

MASTIC (*PISTACIA LENTISCUS* L.) ++ ANACARDIACEAE

NOTES (MASTIC):

As the turpentine tree I stretched out my branches, and my branches are the branches of honour and grace.

Sirach (Apocrypha) 24:15 (KJV)

Like a terebinth I spread out my branches, and my branches are glorious and graceful.

Sirach (Apocrypha) 24:15 (RSV)

The mastic is a tree of spreading habit, with a thick trunk. The wood is hard and white. Its foliage is dense enough to cast a heavy shade on the deserts heated in the sun. When the bark is cut, Chian turpentine flows out; this has an agreeable perfume, not unlike jessamine, and is mild to the taste. Exposure to the air solidifies it to a transparent gum. This Gilead turpentine probably formed part of the spicery carried into Egypt from Gilead by the Ishmaelites as mentioned in Genesis 37:25. Bark and leaves are a source of tannin. The astringent leaves are also used for dyeing. The gum is used to sweeten the breath (e.g., in Tehran). Few resins have a greater "repertoire" in anticancer folklore than this plant, used for ascites, calluses, cancers (of the breast, face, lip, liver, medullary, pylorus, rectum, spleen, testicle, tongue, uterus, vagina), carcinoma, corn, cysts, epithelioma, excrescences, fungoids, inflammation, melanosis, polyps, sclerosis (breast, liver), skin ailments, and tumors (especially of the spleen) (JLH). With tannins, sitosterol, and shikimic acid reported, perhaps this cancer "repertoire" is justified. The gum is similar to "Chian turpentine which was recommended about fifty years ago as a remedy for cancer." (BIB) Leaves are used as an emmenagogue and for albuminuria and diarrhea. Lebanese infuse the leaves for diarrhea and fever. Dioscorides suggested that terebinth or turpentine was antidotal, aphrodisiac, diuretic, and expectorant (BIB).

COMMON NAMES (MASTIC):

Almecegueira (Por.; USN); Arbre au Mastic (Fr.; USN); Battoum (Arab.; BOU); Chios Mastictree (Eng.; USN); Darw (Arab.; Dhou (Ber.; BOU); Dirw (Arab.; BOU); Dro (Arab.; BOU); Drw (Arab.; BOU); Fadhiss (Ber.; BOU); Fethies (Ber.; BOU); Fustuq Sharqi (Arab.; BOU; HJP); Goudhim (Ber.; BOU); Goudhoum (Ber.; BOU); Gueddain (Ber.; BOU); Imidek (Ber.; BOU); Kinnah (Iran; EFS); Kinneh (Iran; DEP); Kinnoli (Iran; DEP); Kundari (Iran; NAD); Kundurumi (Beng.; Hindi; DEP; TAN); Lentisco (It.; Sp.; EFS; USN); Lentisk (Eng.; BOU; EFS; FAC); Lentisk Pistache (Eng.; FAC; UPH); Lentisque (Fr.; BOU; EFS; USN); Mastaka i Rumi (Iran; DEP); Mastic (Eng.; Fr.; CR2; EFS; USN); Mastiche (Eng.; NAD); Mastic tree (Eng.; BOU); Mastik (Arab.; Syria; HJP); Mastikboom (Dutch; EFS); Mastiki (Hindi; DEP); Mastixbaum (Ger.; EFS; USN); Mistaka (Arab.; GHA); Moesstakim (Malaya; EFS); Mustagi Rumi (India; EFS); Mustaka Sultani (Arab.; GHA); Mustiva (Arab.; GHA); Mustoka (Arab.; DEP); Rumi Mastaki (India; EFS); Rumi Mastungi (Beng.; DEP; SKJ); Rumi Mustiki (Hindi; SKJ); Sakir Rumi (Iran; NAD); Sakiz Agaci (Tur.; EFS); Saris (Arab.; Syria; HJP); Shagar el Mastika (Arab.; BOU; HJP); Sondro (It.; EFS); Tadist (Ber.; BOU); Tantarik (Pun.; DEP); Terebinth (Eng.; BIB; RSV); Tidekst (Ber.; BOU); Tidekt (Ber.; BOU); Turpentine (Eng.; BIB; KJV); Uluk Bagh Dame (Arab.; EFS); Uluk Baghdani (Arab.; DEP).

ACTIVITIES (MASTIC):

Allergenic (f; CRC); Analgesic (f; BIB; BOU; CRC); Antiatherogenic (1; X15136059); Antioxidant (1; X15848018); Antiperoxidant (1; X15848018); Antisarcomic (1; HH3); Antisecretory (1; X3724207); Antiseptic (1; HH3); Antitumor (1; HH3); Antitussive (f; BIB; BOU; CRC); Antiulcer (f1; GAZ; HH3; PH2; X3724207); Aphrodisiac (f; CRC; DEP; EFS; IHB); Apopotic (1; X15796160); Astringent (f1; EFS; PHR; PH2); Bactericide (1; HH3); Candidicide (1; HH3); Cardioprotective

(1; X15136059); Carminative (f; CRC; UPH); Cathartic (f; CRC; FDA); Corroborant (f; DEP); Culicide (1; X11997977); Depilatory (f; BIB; BOU); Diuretic (f; CRC; DEP; EFS; HH3); Emmenagogue (f; BOU); Expectorant (f; BOU; CRC; EFS); Fungicide (f; HH3); Hemostat (f; CRC; EFS); Hepatoprotective (f1; X12413719); Hypotensive (1; HH3; X1409845); Insecticide (1; X11997977); Larvicide (1; X11997977); Masticatory (1; CRC; BIB); Orexigenic (f; CRC); Sedative (f; BOU); Stimulant (f; DEP; EFS; HH3; UPH); Stomachic (f; CRC; EFS); Sudorific (f; CRC; EFS).

Indications (Mastic):

Adenopathy (f; JLH); Amenorrhea (f; BOU); Aphthae (f; NAD); Aposteme (f; CRC; JLH); Atherosclerosis (1; X15136059); Bacteria (1; X8808717); Bleeding (f; CRC; ERS); Blenorrhea (f; CRC); Boil (f; BIB; BOU; CRC); Bronchosis (f; FEL; NAD); Cancer (f; CRC; JLH); Cancer, anus (f; CRC; JLH); Cancer, breast (f; CRC; JLH); Cancer, colon (1; X15796160); Cancer, intestine (f1; BIB; JLH; X15796160); Cancer, liver (f; CRC; JLH); Cancer, parotid (f; CRC; JLH); Cancer, spleen (f; CRC; JLH); Cancer, stomach) (f; CRC; JLH); Cancer, testicle (f; CRC; JLH); Cancer, throat (f; CRC; JLH); Cancer, uterus (f; CRC; JLH); Candida (1; HH3; X8808717); Canker (f; BIB; CRC); Carbuncle (f; BOU; CRC); Cardiopathy (f; BOU); Caries (f; CRC; FEL); Catarrh (f; CRC; FEL; HH3; NAD); Childbirth (f; BOU); Cholecocystosis (f; BIB; CRC; HJP); Cirrhosis (f; CRC; HH3); Condyloma (f; CRC; JLH); Cough (f; BIB; BOU; GHA); Cystosis (f; GAZ); Debility (f; CRC; NAD); Dermatosis (f; GHA); Diarrhea (f; CRC; HH3; HJP); Dysentery (f; CRC; HH3); Enterosis (f; GAZ); Escherichia (1; HH3); Fever (1; GHA); Fracture (f; HJP); Fungus (1; HH3; X8808717; X126288418); Gastrosis (f; BIB; BOU; CRC); Gingivosis (f1; FEL; PHR; PH2); Glossosis (f; NAD); Gonorrhea (f; CRC; HH3); Gout (f; HH3); Halitosis (f; BIB; BOU; CRC; DEP; FEL; PHR); Heart (f; CRC); Hepatosis (f1; BIB; CRC; HH3; X12413719); High Blood Pressure (1; HH3; X1409845); Impotence (f; DEP); Induration (f; CRC; JLH); Infection (1; X8808717); Inflammation (f; JLH); Itch (f; BIB; BOU); Jaundice (f1; X12413719); Leukorrhea (f; CRC; HH3); Mastosis (f; BOU; CRC); Mucososis (f; CRC; UPH); Mycosis (1; HH3; X8808717; X126288418); Myosis (f; BOW); Nephrosis (f; FEL); Pain (f; BOU; CRC; GHA); Phymata (f; CRC); Proctosis (f; JLH); Pulmonosis (f; GHA); Respirosis (f; NAD); Rheumatism (f; BIB; BOU; HH3); Ringworm (f; BOW); Scirrhus (f; CRC; JLH); Sclerosis (f; CRC); Sore (f; HH3); Spermatorrhea (f; NAD); Splenosis (f; JLH); Staphylococcus (1; HH3); Stomatosis (f; GAZ; NAD); Throat (f; BOU); Toothache (f; CRC); Tumor (f; CRC); Ulcer (f1; BOU; PH2; X3724207); Venereal Disease (f; CRC; HH3); Urethrosis (f; GAZ); Wound (1; GHA); Yeast (1; HH3; X8808717).

Dosages (Mastic):

FNFF = !!

Mastic widely chewed; Romans used fruits as spice; seed kernels yield the edible shina oil of Cyprus; bark used as spice; wood used to smoke meat. Turks use in preparing the liqueur *raki* (DEP; FAC; TAN). 150 ml 10% aqueous resin decoction per day (HH3).

- Arabians chew the resin to enhance appetite and improve breath (GHA).
- Asian Indians suggest the tincture to stop leech bites (NAD).
- Jordanians treat jaundice with aqueous extract (both boiled and non-boiled), which showed antihepatotoxic activity (X12413719).
- Lebanese take the resin tincture with lemon for cholecocystosis, diarrhea, and hepatosis (HJP).
- Mohammeden physicians consider it aphrodisiac, diuretic, and stimulant (DEP).
- North Africans use oil from peeled nuts for itch and rheumatism (BOU).
- North Africans boil resin in milk for throat troubles (BOU).
- North Africans take root decoction for cough (BOU).
- North Africans take 1 tsp mastic pounded with 1 tsp honey each morning for 3 weeks for ulcer (BOU).

Downsides (Mastic):

No health hazards or side effects known with proper therapeutic dosages (PH2) (PH2 designates no dosage!; JAD).

Extracts (Mastic):

Ethanolic extracts antiseptic, bactericidal at 1 g/l, candidicidal at >l g/l. Decoction kills *Candida parapsilopsis*, *Escherichia*, *Sarcina*, and *Staphylococcus* at 312 mg/l; *Candida albicans* and *Cryptococcus* at 625 mg/l (HH3). Lyophilized aqueous extracts hypotensive in normotensive rats as 25 mg/kg orally. Tannins, especially ellagitannins, arrested tumor growth (5–10 mg/kg ipr mus (HH3); LD50 of the lyophilized aqueous extract is 680–1120 mg/kg ipr mus (HH3). Extracts or essential oil effective against *Culex larvae* (LC50 = 70 mg/l) (X11997977); Dedoussis et al. (2004) demonstrated cardioprotectrive antiatherogenic effects of the resinous exudate resin (used culinarily in some Mediterranean diets (X15136059). Balan et al. (2005) note that Chios mastic extracts induce apoptosis in human colon cancer (X15796160).

TEREBINTH (*PISTACIA TEREBINTHUS* L.) ++ ANACARDIACEAE

Notes (Terebinth):

I like the DEP account better than most because it seems to lump all the potential biblical species under this catchall species. My accounts of the mastic and the terebinth are very similar. The species and their folklore have been confused for centuries. Over a hundred years ago, DEP fingered var. *mutica* as the biblical *alah* of the Old Testament. DEP notes that it is the true primitive turpentine celebrated as the finest, superior to pine resins and mastic. Yet, DEP states that "The resin of var. *mutica* sesembles that of *P. lentiscus* and is used in the East as a substitute for mastic." Variety *mutica* also carries the same vernacular names and is considered identical with the resin of *Pistacia vera*. NAD followed DEP in treating *P. cabulica*, *P. khinjuk*, and *P. mutica* as synonyms. USN keeps them all separate. So, although starting to view them as all separate as I started working on *P. atlantica*, I evolved to think of them as one great taxon, with a great overlap in common names and uses. But I keep them distinct for now, more for the readers' convenience than for science. These turpentines, like pine turpentines, share many chemicals, activities, and indications. KAB treated only one species, not this one. When the bark is cut, Chian turpentine flows out; this has an agreeable perfume, not unlike jessamine, and is mild to the taste. Exposure to the air solidifies it to a transparent gum. This Gilead turpentine probably formed part of the spicery carried into Egypt from Gilead by the Ishmaelites as mentioned in Genesis 37:25. Few resins have a greater "repertoire" in anticancer folklore than this plant, used for ascites, calluses, cancers (of the breast, face, lip, liver, medullary, pylorus, rectum, spleen, testicle, tongue, uterus, vagina), carcinoma, corn, cysts, epithelioma, excrescences, fungoids, inflammation, melanosis, polyps, sclerosis (breast, liver), skin ailments, and tumors (especially of the spleen) (JLH). With tannins, sitosterol, and shikimic acid reported, perhaps this cancer "repertoire" is justified. According to Hooper, the gum is similar to "Chian turpentine, which was recommended about 50 years ago as a remedy for cancer." Leaves are used as an emmenagogue and for albuminuria and diarrhea. Dioscorides suggested that terebinth or turpentine was antidotal, aphrodisiac, diuretic, and expectorant (BIB).

Common Names (Terebinth):

Ban (Bal.; DEP); Bargabana (Iran; DEP); Baume de Cypres (Fr.; EFS); Bombay Mastiche (Eng.; NAD); Butz (Arab.; Syria; HJP); Butz Saqis (Arab.; Syria; HJP); Buzaganja (Bom.; NAD); Chian Turpentine (Eng.; BIB; EFS); Chios Terpentijn Boom (Dutch; EFS); Chios Terpentinbaum (Ger.; EFS); Cornalheiro (Por.; EFS); Cornicabra (Sp.; EFS); Cyprian Turpentine (Eng.; EFS); Cyprischer Chio (It.; EFS); Cyprus Turpentine (Eng.; SKJ; USN); East Indian Mastiche (Eng.; NAD); Guli

Pistah (Hindi; Iran; NAD); Gwan (Bal.; DEP); Habba Khadhra (Algiers; Arab.; JLH); Kabuli Mustaki (Bom.; Hindi; NAD; SKJ); Kanjak (Afg.; DEP); Khinjak (Pun.; DEP); Kinjad (Iran; DEP); Kunjad (Iran; DEP); Mastaki (Bom.; Hindi; DEP); Menengiç (Tur.; EFS); Pistachier Térébinthe (Fr.; USN); Qalafournis (Egypt; JLH); Scornobecco (It.; EFS); Terpentinbaum (Ger.; EFS); Terebinth (Eng.; HJP; NAD; USN); Terebinthe de Chio (Fr.; EFS); Terebinto (Por.; Sp.; EFS); Terebinto di Chio (It.; EFS); Wan (Bal.; DEP); Nscn.

Activities (Terebinth):

Antidote (f; HJP); Antiinflammatory (f; X11988853); Antiseptic (1; X126288418); Antitussive (f; HJP); Aphrodisiac (f; DEP; HJP); Astringent (f; HJP; SKJ); Deodorant (f; HJP); Digestive (f; HJP); Diuretic (f; HJP); Emmenagogue (f; BIB); Expectorant (f; HJP); Febrifuge (f; HJP); Fungicide (1; X126288418); Hemostat (f; DEP); Sedative (f; DEP); Stimulant (f; DEP; HJP); Stomachic (f; DEP); Vulnerary (f; DEP).

Indications (Terebinth):

Adenopathy (f; JLH); Albuminuria (f; BIB; HOC); Amenorrhea (f; BIB); Arthrosis (f; JLH); Ascites (f; DAW); Bite (f; HJP); Bleeding (f; DEP); Callus (f; JLH); Cancer (f; DEP; HJP); Cancer, brain (f; JLH); Cancer, breast (f; HJP); Cancer, diaphragm (f; HJP); Cancer, face (f; HJP); Cancer, lip (f; HJP); Cancer, liver (f; HJP); Cancer, medullary (f; HJP); Cancer, pylorus (f; HJP); Cancer, rectum (f; HJP); Cancer, spleen (f; HJP); Cancer, stomach (f; JLH); Cancer, testicle (f; HJP); Cancer, tongue (f; HJP); Cancer, uterus (f; HJP); Cancer, vagina (f; HJP); Carcinoma (f; JLH); Caries (f; EFS); Cheilosis (f; JLH); Colic (f; DEP); Corn (f; JLH); Cough (f; HJP); Cyst (f; JLH); Dermatosis (f; HOC; JLH); Diarrhea (f; BIB); Diaphragmosis (f; HJP); Dyspepsia (f; DEP; HJP); Encephalosis (f; JLH); Epithelioma (f; JLH); Excrescences (f; JLH); Fever (f; HJP); Fungus (f; X126288418); Gastrosis (f; JLH); Glossosis (f; JLH); Gout (f; HOC); Halitosis (f; HJP); Hepatosis (f; JLH); Impotence (f; HJP); Induration (f; JLH); Infection (1; X126288418); Inflammation (f1; HJP; X11988853); Mastosis (f; JLH); Melanosis (f; JLH); Mycosis (1; X126288418); Nausea (f; DEP); Orchosis (f; JLH); Parotosis (f; JLH); Polyp (f; JLH); Proctosis (f; JLH); Rhinosis (f; JLH); Scirrhus (f; JLH); Scleroma (f; JLH); Sore (f; JLH); Splenosis (f; JLH); Swelling (f; JLH); Tophus (f; JLH); Uterosis (f; DEP; HOC); Vaginosis (f; JLH); Vomiting (f; DEP); Wound (f; DEP).

Dosages (Terebinth):

FNFF = !
Iranians use the resin as a chewing gum (HJP); Southern Afghans and Baluchistani eat the fruits (*shiné*); kernel oil eaten with bread and relish (DEP).

- Algerians used the seeds in liniments for tumors (JLH).
- Americans, Australians, and Europeans used the Chian turpentine for scores of types of cancer (JLH)
- Asian Indians suggest 5 to 10 grains for cancer (NAD).
- Egyptians use the plant to treat excrescences (JLH).
- Iranians chew the gum to sweeten the breath (BIB).
- Iranians use the turpentine with pomade of cinnabar for cancers (JLH).
- Lebanese chew the resin to sweeten breath and improve digestion (HJP).
- Lebanese infuse the leaves for diarrhea and fever (BIB).
- Syrians use the "Cyprus turpentine" for cancer "cures and indurations of the liver" (HJP).

Natural History (Terebinth):

Pinkish galls (*khinjak* in Punjab) on the leaves, with aroma of turpentine, appear to be caused by an *Aphis*. Leaves eaten by camels, goats, and sheep (DEP).

PISTACHIO *(PISTACIA VERA* L.) +++ ANACARDIACEAE

SYNONYMS:

Pistacia narbonensis L.; *Pistacia nigricans* Crantz.; *Pistacia officinarum* Ait.; *Pistacia reticulata* Willd.; *Pistacia terebinthis* Mill. non L.; *Pistacia trifolia* L.; *Pistacia variifolia* Salisb. *fide* HH3

NOTES (PISTACHIO):

And their father Israel said unto them, If it must be so now, do this; take of the best fruits in the land in your vessels, and carry down the man a present, a little balm, and a little honey, spices, and myrrh, nuts, and almonds.

Genesis 43:11 (KJV)

Then their father Israel said to them, "If it must be so, then do this: take some of the choice fruits of the land in your bags, and carry down to the man a present, a little balm and a little honey, gum, myrrh, pistachio nuts, and almonds."

Genesis 43:11 (RSV)

Most commentators agree that the "nuts" of Jacob were pistachio nuts. And the RSV edition and NWT available to me on computer even specify pistachio nuts. Over a hundred years ago, DEP noted that the tree was extensively cultivated in Palestine, Persia, and Syria, and introduced in Italy and Spain. Zohary notes that it is mentioned only once in the Bible. The tree has long been cultivated in Israel. Nuts are found in the late Neolithic strata in Greece. They are imported to India as dyeing and tanning agents. Pistacio kernels yield circa 50% of a low-melting fatty oil used to a small extent in confectionery as spice oil and in medicine (BIB).

COMMON NAMES (PISTACHIO):

Alfóncigo (Sp.; EFS; USN); Alhócigo (Sp.; EFS); Botnim (Arab.; Heb.; ZOH); Butm (Arab.; ZOH); Buzaganja (Bom.; Hindi; NAD); Echte Pistazie (Ger.; HH3); Elah (Heb.; ZOH); Fistuk Baladi (Arab.; Syria; HJP); Fistuk Karmidi (Arab.; Syria; HJP); Fustuk (Arab.; Syria; HH3; HJP); Green Almond (Eng.; HH3); Grüner Mandelbaum (Ger.; EFS); Guli Pistah (Bom.; Hindi; NAD); Hu Chên Tzu (China; EFS); It Dirsegi (Tur.; EB54:155); O Yüeh Chün Tzu (China; EFS); Pimpernuss (Ger.; HH3); Pipernuss (Ger.; HH3); Pista (Afg.; Beng.; Bom.; Hindi; India; Iran; EFS); Pistaccio Vero (It.; EFS); Pistache Noot (Dutch; Fr.; EFS); Pistachero (Sp.; USN); Pistachier (Fr.; EFS); Pistachier Cultivé (Fr.; USN); Pistachio (Eng.; Scn.; AH2; EFS); Pistacho (Sp.; EFS); Pistakinuss (Ger.; HH3); Pistakion (Greek; JLH); Pistazie (Ger.; EFS; USN); Pistazienbaum (Ger.; USN); Pistazier (Den.; EFS); Piste (Iran; HH3); Pisteh (Iran; NAD); Pisutachio (Japan; USN); samfistigi Agaci (Tur.; EFS); Wu Ming Tzu (China; EFS).

ACTIVITIES (PISTACHIO):

Allergenic (1; X9140524); Anodyne (f; DAW); Antiseptic (1; X15881833); Antiviral (1; X15881833); Aphrodisiac (f; DEP); Astringent (f1; NAD); Decoagulant (f; DAW); Demulcent (f; DEP); Digestive (f; BIB; DEP); Fungicide (1; X15881833); Phytoalexin (1; X15941348); Sedative (f; BIB; DEP; NAD; SKJ); Tonic (f; BIB; DEP; NAD; SKJ).

INDICATIONS (PISTACHIO):

Abscess (f; DAW); Amenorrhea (f; DAW); Anodyne (f; DAW); Bacteria (1; X15186116); Bruise (f; DAW); Cancer, breast (1; X15941348); Cancer, liver (f; JLH); Chest (f; DAW); Circulation (f; DAW); Cough (f; BIB); Debility (f; NAD); Dermatosis (f; DAW); Dysentery (f; BIB; DAW); Dysmenorrhea (f; DAW); Dyspepsia (f; DEP); Enterosis (f; DAW); Fungus (1; X15881833; X126288418); Gastrosis (f; NAD); Gynecopathy (f; DAW); Hepatosis (f; JLH); Impotence (f; HJP); Infection (1; X15881833);

FIGURE 1.83 Pistachio (*Pistacia vera*).

Infertility (f; HJP); Mycosis (1; X15881833; X126288418); Nausea (f; DEP); Pain (f; DAW); Pruritis (f; DAW); Rheumatism (f; DAW); Sclerosis (f; JLH); Sore (f; DAW); Trauma (f; DAW); Virus (1; X15881833); Vomiting (f; DEP); Wound (f; HJP); Yeast (1; X15186116).

Dosages (Pistachio):

FNFF = !!!
Nuts widely eaten. Pista kernels have a delicious nutty flavor and are much used as ingredients of sweetmeats, confectionery, and ice creams. Pista is also eaten as a dessert; salted and roasted, it is much relished. Fruit husks are reported to be made into marmalade in Iran (BIB; FAC; TAN; EB54:155).

- Algerians used the powdered root in oil for children's cough (BIB).
- Asians use the expressed oil for stomach problems (NAD).
- Iranians infused the fruit's outer husks for dysentery (BIB).
- Middle Easterners consider the nut a "hot food" (GHA).
- Lebanese apply the resin to cuts (HJP).
- Lebanese compress the leaves (BIB; HJP).
- Lebanese think the nuts enhance fertility and virility (BIB; HJP).

Downsides (Pistachio):

Pistacia pollen is a major source of allergy (X9140524).

Natural History (Pistachio):

Nuts favored by squirrels, blue jays, and red-headed woodpeckers (NUT). Camels, goats, and sheep feed on the leaves, hence the name of the galls *boz ghanj.* One common name for the galls translates as "goat's sore" (DEP). The leaves of *P. vera* bear small, irregularly spheroid galls (Bokhara galls), which have been reported to be imported into India for dyeing and tanning purposes; galls contain 50% tannins (WOI).

Extracts (Pistachio):

Fortunately for nut lovers, Phillips et al. (2005) quantified the phytosterols in four accessions of pistachio nuts. Pistachio was the richest of the biblical nuts in total phytosterols (cf. sunflower in the American nuts). Such phytosterols could be medicinally important (X16302759). Phytosterols in pistachio nuts (four accessions): delta-5-avenasterol (262 ppm); delta-7-avenasterol; campestanol (50 ppm); campesterol (101 ppm); phytosterols (2740–2870 ppm); poriferasta-7,25-dienol (126 ppm); poriferasta-7,22,25-dienol; sitostanol (12 ppm); beta-sitosterol (2098 ppm); spinasterol; stigmastanol; and stigmasterol (23 ppm) (X16302759). Ozcelik et al. (2005) demonstrated antibacterial, antifungal, and antiviral activities for lipophilic extracts of various parts of the plant (leaf, branch, stem, kernel, shell skins, seeds) e.g., *Escherichia coli*, *Pseudomonas aeruginosa*, *Enterococcus faecalis*, *Staphylococcus aureus*, *Candida albicans* and *C. parapsilosis*, *Herpes simplex* (DNA), and Parainfluenza viruses (RNA). The extracts showed antifungal but little antibacterial activity at 128 to 256 μg/ml. Kernel and seed extracts showed significant antiviral activity (X15881833). Alma et al. (2004) found that the essential oil contained alpha-pinene (75.6%), beta-pinene (9.5%), trans-verbenol (3.0%), camphene (1.4%), trans-pinocarveol (circa 1.20%), and limonene (1.0%). The antimicrobial results showed that the oil inhibited nine of thirteen bacteria and all three yeasts studied. The essential oil of the gum was better than Nystatin against yeast, but weaker than ampicillin sodium, and streptomycine sulfate against bacteria (X15186116). Tokusoglu et al. found traces of resveratrol, comparable to levels in peanuts, in Turkish pistachios, 0.09–1.67 μg/g (av = 1.15 μg); cf. 0.03–7.17 μg/g in grapes and wines. There was more cis-transveratrol in pistachios than peanuts. In peanut, if not pistachio, resveratrol increases after biotic or abiotic stress. "Trans-resveratrol is a chemopreventive agent against human breast cancer" (X15941348).

ORIENTAL PLANE TREE (*PLATANUS ORIENTALIS* L.) ++ PLATANACEAE

Synonyms:

Platanus vulgaris Sah. *fide* DEP

Notes (Oriental Plane Tree):

> **And Jacob took him rods of green poplar, and of the hazel and chestnut tree; and pilled white strakes in them, and made the white appear which was in the rods.**
>
> **Genesis 30:37 (KJV)**

FIGURE 1.84 Oriental Plane Tree (*Platanus orientalis*).

Then Jacob took fresh rods of poplar and almond and plane, and peeled white streaks in them, exposing the white of the rods.

Genesis 30:37 (RSV)

Then Jacob took for his use staffs still moist of the storax tree and of the almond tree and of the plane tree and peeled in them white peeled spots by laying bare white places which were upon the staffs.

Genesis 30:37 (NWT)

Here the KJV version renders the Hebrew *armon* to chestnut while the RSV more properly renders it to plane tree. (I will puzzle over the hazel in the KJV as opposed to almond elsewhere, but I do not remember the hazel being biblical. And then I will puzzle over the storax in the NWT as opposed to the poplar in the KJV and RSV). Life gets complicated. The plane tree is oft cultivated and highly valued as an ornamental tree. It is said to be the tree under which Socrates enthralled his students. It has a short trunk, a roundish spreading crown, and is mostly grown for shade in parks and on roadsides. Seldom felled, it is allowed to grow to large dimensions. The wood is white, tinged with yellow or red; heartwood not distinct, fine-grained, moderately hard and heavy (weight circa 657 kg/cu. m), but strong. It warps during seasoning and is durable only under cover. It is easy to saw and presents a decorative figure when quarter sawn. It can be finished to a smooth surface that takes a beautiful polish. In Kashmir, the wood is mostly used for small boxes, trays, and similar articles that are lacquered and painted. In western Asia and Europe, it is used for cabinet making, furniture, veneers, carving, coach building, general turnery, and for wood pulp. It has been reported to be suitable for boot lasts (BIB).

Common Names (Oriental Plane Tree):

Aramon (f; KAB); Armon (Heb.; ZOH); Asiatische Platane (Ger.; EFS); Bhunj (Kas.; KAB); Boin (Kas.; DEP; KAB); Bonin (Kas.; DEP; KAB); Buin (India; Kas.; EFS; KAB; NAD); Buna (Kas.; DEP; KAB); Chanar (Pun.; DEP; KAB); Chinar (Chaman; Eng.; Iran; Pishin; Pun.; Quetta; Rus.; Urdu; EFS; DEP; KAB); Chintar (Pushtu; DEP; KAB); Çinar (Tur.; EFS); Dilba (Arab.; Aramaic; ZOH); Dulah (Iran; KAB); Jing Tu Shu (China; USN); Morgenländische Platane (Ger.; EFS); Oosterse Plataan (Dutch; EFS); Oriental Plane Tree (Eng.; EFS; USN; ZOH); Plataan (Dutch; KAB); Platan (Rom.; KAB); Platane (Fr.; Ger.; KAB); Platane d'Oreint (Fr.; EFS); Platano (It.; Por.; EFS; KAB); Plátano de Levante (Sp.; EFS); Plátano de Sombre (Sp.; EFS); Plátano do Oriente (Por.; EFS); Nscn.

Activities (Oriental Plane Tree):

Antirheumatic (f; EFS); Antiscorbutic (f; EFS); Antiseptic (1; X8302950); Bactericide (1; X8302950); Cyanogenic (1; EB30:402); Cytotoxic (1; X10712831); Gram(+)-icide (1; X8302950); Gram(–)-icide (1; X8302950); Tonic (f; EFS); Vulnerary (f; BIB).

Indications (Oriental Plane Tree):

Bacteria (1; X8302950); Bite (f; KAB); Bronchosis (f; KAB); Cancer (f1; JLH; X10712831); Carcinoma (f; BIB); Diarrhea (f; DEP; NAD; WOI); Dysentery (f; WOI); Hernia (f; BIB); Infection (1; X8302950); Inflammation (f; BIB); Leukemia (1; X10712831); Leukorrhea (f; KAB); Nephrosis (f; KAB); Ophthalmia (f; DEP; EFS; NAD); Pharyngosis (f; KAB); Pulmonosis (f; KAB); Rheumatism (f; EFS); Scurvy (f; EFS); Sore Throat (f; KAB); Toothache (f; BIB); Tumor (f; BIB); Voice (f; KAB); Wound (f; KAB).

Dosages (Oriental Plane Tree):

FNFF = ?

- Asian Indians suggest applying bruised leaves to the eye in ophthalmia (DEP; KAB).
- Asian Indians suggest bark, boiled in vinegar, for diarrhea, dysentery, hernia, and toothache (DEP; KAB).
- Unani use the bark for animal bites and leucoderma, the fruits and leaves for lacrymation, leucoderma, ophthalmia, toothache, and wounds, and disorders of the kidney, lungs, throat, and voice (KAB).

Natural History (Oriental Plane Tree):

Affected by the leaf spot disease caused by *Stigmina platani* and *S. visianica*. Dead wood attacked by beetles *Aeolesthes sarta*, *Batocera rufomaculata*, and *Capnoidis miliaris* (WOI).

Extracts (Oriental Plane Tree):

Contains active allantoin and asparagine (EFS).

Demetzos et al. (2000) isolated antileukemic cytotoxic compounds (platanoside and tilirioside) from methanolic bud extracts (X10712831). Mitrokotsa et al. (1993) studied antimicrobial activity against Gram-positive and Gram-negative organisms (X8302950).

WHITE POPLAR (*POPULUS ALBA* L.) ++ SALICACEAE

Notes (White Poplar):

And Jacob took him rods of green poplar, and of the hazel and chestnut tree; and pilled white strakes in them, and made the white appear which was in the rods.

Genesis 30:37 (KJV)

Then Jacob took fresh rods of poplar and almond and plane, and peeled white streaks in them, exposing the white of the rods.

Genesis 30:37 (RSV)

Then Jacob took for his use staffs still moist of the storax tree and of the almond tree and of the plane tree and peeled in them white peeled spots by laying bare white places which were upon the staffs.

Genesis 30:37 (NWT)

Zohary, like Moldenke and Moldenke, concluded that water sprouts of *Populus alba* are the most likely interpretation of Jacob's poplar rods. The NWT translates it as storax but Zohary rules that out on a phytogeographic basis. Native to Syria and Lebanon; for example, the white poplar thrives along water courses. Whether it is native to northern Israel is questionable. The young buds are covered with a resinous varnish with a balsamic aroma in the spring. Bruised buds produce a fragrant resin that may have been the incense burned by Ephraim in the groves of poplars. Because it casts a dense shade, white poplar has been extensively cultivated in the Holy Land. Bitten by a poisonous snake, Hercules found a mythical antidote in poplar leaves. The tonic bark is used for strangury and blood and skin diseases (BIB).

FIGURE 1.85 White Poplar (*Populus alba*).

Common Names (White Poplar):

Abbey (Eng.; EFS); Abele (Eng.; EFS; USN); Able (Eng.; DEP); Ak Kavak (Tur.; EFS); Álamo Bianco (Por.; EFS); Álamo Blanco (Sp.; EFS; USN); Alberaccio (Malta; KAB); Alberbaum (Ger.; KAB); Alberello (It.; EFS); Aouba (Lan.; KAB); Aspen (Eng.; BUR); Baid (Pun.; DEP); Blanc de Hollande (Fr.; KAB); Chanuni (Pun.; DEP); Chita Bagnu (Pun.; DEP; KAB); Chittabagun (Himalaya; KAB); Chopo (Sp.; EFS); Choupo Branco (Por.; EFS); European White Poplar (Eng.; BUR); Fras (Kas.; Pun.; DEP; KAB); Gattice (It.; EFS); Gin Doro (Japan; USN); Great Aspen (Eng.; HOC); Hakuyo (Japan; USN); Ilba (Cat.; KAB); Ispedar (Jhalawan; KAB); Jangli Frast (Pun.; DEP); Livneh (Heb.; ZOH); Mal (Pun.; DEP); Nyárfarügy (Hun.; EFS); Pai Yang (China; EFS); Peuplier Blanc (Fr.; EFS; USN); Pobo (Sp.; EFS); Popolo Bianco (It.; EFS); Prasti (Pun.; DEP); Prist (Pun.; DEP); Rikkan (Pun.; DEP); Safeda (Pun.; DEP); Sannan (Pun.; DEP); Silber Pappel (Ger.; EFS; USN); Silver-leaf Poplar (Eng.; UPH; USN); Speda (Afg.; DEP); Spedar (Chaman; Quetta; KAB); Spelda (Afg.; DEP); Sperdor (Afg.; DEP); Sufeda (Nasiribad; Sharig; Sibi; KAB); Topol

(Rus.; KAB); Urajiro Hako Yanagi (Japan; USN); Weiss Pappel (Ger.; EFS); White Asp (Eng.; EFS); White Poplar (Eng.; USN); Witte Abeel (Dutch; EFS); Witte Populier (Dutch; EFS); Xin Bai Yang (China; USN); Nscn.

Activities (White Poplar):

Antiperiodic (1; WOI); Antiseptic (f; DAW); Astringent (f; DAW); Bitter (f; DAW); Depurative (f; DEP); Diaphoretic (f; AAH); Diuretic (f; DAW; EFS); Febrifuge (f1; DAW; EFS; ZOH); Insectifuge (1; WOI); Revulsive (f; DAW); Stimulant (f; DAW; EFS); Tonic (f; DEP; EFS; ZOH); Uricosuric (1; WOI).

Indications (White Poplar):

Bone (f; DAW); Caries (f; DAW); Cold (f; DAW); Colic (f; DAW); Complexion (f; DAW); Coryza (1; WOI); Cystosis (f; HOC); Dermatosis (f; DAW; DEP); Dyspepsia (f; AAH); Fever (f; DAW); Flux (f; DAW); Goiter (f; DAW); Hematochezia (f; DAW); Hemorrhage (f; DAW); Herpes (f; DAW); Malaria (1; DAW; WOI); Necrosis (f; HOC); Neuraglia (1; WOI); Night Sweats (f; AAH); Osteosis (f; HOC); Rheumatism (f1; DAW); Sciatica (f; HOC); Splenomegaly (f; DAW); Strangury (f; DAW; DEP).

Dosages (White Poplar):

FNFF = !
Inner bark used as a flour substitute in survival breadstuffs (TAN).

- Chinese use leaf decoction for bone necrosis and tooth decay (HOC).
- Somerset British use bark infusion for dyspepsia, fever, and night sweats (AAH).

Natural History (White Poplar):

A rust, *Melampsora rostrupii*, and a powdery mildew, *Uncinula salicis*, affect the leaves. Defoliators and borers are reported from the tree (WOI).

Extracts (White Poplar):

Bark contains populnin and salicin, both of which can help eliminate uric acid (WOI).

EUPHRATES POPLAR (*POPULUS EUPHRATICA* OLIV.) ++ SALICACEAE

Notes (Euphrates Poplar):

> **He took also of the seed of the land, and planted it in a fruitful field; he placed it by great waters, and set it as a willow tree. And it grew, and became a spreading vine of low stature, whose branches turned toward him, and the roots thereof were under him: so it became a vine, and brought forth branches, and shot forth sprigs.**
>
> **Ezekiel 17:5–6 (KJV)**

Who would confuse a willow with a poplar? This Euphrates poplar has narrower, more willow-like leaves on younger shoots, but broader, poplar-like leaves on older shoots. Zohary notes that the Euphrates poplar grows characteristically on river banks along the Jordan, often coexisting with the wild date along brackish water courses. And like the Moldenkes, Zohary believes that the "willows" (Psalms 137:1–3) on which the Jews hung their harps were, in fact, the Euphrates poplar. Some versions of the Bible even say poplars instead of willow. Early Christian legendry, assuming that Jesus's cross was made of aspen (= poplar), has it that aspen trees everywhere started shuddering

FIGURE 1.86 Euphrates Poplar (*Populus euphratica*). Source: KAB

when nails were driven into the wood and have trembled ever since. Judas was said to have hanged himself on *Populus*, but some suggest instead *Cercis*, or *Ficus*, or *Pistacia* (BIB; ZOH).

Common Names (Euphrates Poplar):

Aspen (Eng.; BIB); Bahan (Bom.; Pun.; Pushtu; Sin.; KAB); Bahanr (Las Bela; KAB); Bahun (Shah Bilawul; Sibi; KAB); Ban (Bom.; KAB); Benti (Pun.; KAB); Bhakaim (Heb.; KAB); Bhan (Pun.; Sin. KAB); Bhani (Pun.; KAB); Euphrates Aspen (Eng.; BIB); Euphrates Poplar (Eng.; BIB); Gharab (Arab.; Iraq; ZOH); Hodung (Ladak; KAB); Hotung (Ladak; KAB); Hu Yang (China; USN); Junglibenti (Pun.; KAB); Labhan (Pun.; KAB); Pada (Afg.; Pishin; Quetta; KAB); Padak (Afg.; KAB); Padar (Bal..; KAB); Patk (Jhalawan; Kharan; Sarawan; KAB); Patki (Brahui; KAB);

Poplar (Eng.; BIB; ZOH); Putki Gundava; KAB); Safeda (Sin.; KAB); Safedar (Pun.; KAB); Safsaf (Arab.; Egypt; ZOH); Safsaf el Abiad (Arab.; Egypt; ZOH); Spana (Kohlu; KAB); Sperawan (Pun.; KAB); Spina (Kila Saifulla; Nasirabad; Sibi; KAB); Tzaftzafah (Heb.; ZOH); Nscn.

Activities (Euphrates Poplar):

Dentifrice (f; WOI); Vermifuge (f; KAB).

Indications (Euphrates Poplar):

Odontosis (f; WOI); Worm (f; KAB).

Dosages (Euphrates Poplar):

FNFF = ?
I have no reports on this one being used for food. The inner bark of other poplar species serves as a flour substitute; a few have sprouts that are eaten as greens and salads; the sap can serve as a source of sugar, possibly fermentable (FAC; TAN).

APRICOT (*PRUNUS ARMENIACA* L.) ++ ROSACEAE

Synonyms:

Armeniaca vulgaris Lamarck *fide* AH2

Notes (Apricot):

A word fitly spoken is like apples of gold in pictures of silver.

Proverbs 25:11 (KJV)

A word fitly spoken is like apples of gold in a setting of silver.

Proverbs 25:11 (RSV)

As apples of gold in silver carving is a word spoken at the right time for it.

Proverbs 25:11 (NWT)

An apple a day keeps the doctor away. An apricot (without sulfites) a day keeps the undertaker away. In biblical days, Solomon said, "comfort me with apples for I am sick." Could he have meant apricots? Some maintain that Abraham used dried apricot pulp on his journey from Ur (HJP). I do not know whether the apple or apricot, or even possibly the pomegranate, was intended in the above passages. Apple, apricot, and pomegranate, all possible, are good health foods, and all have been proposed as the apple of the Bible. Perhaps all are! Hunzas reportedly live well into old age, free of cancer and cardiopathy. Was it fresh air and glacial pure water, was it apricot, or was it the real apple? One visitor described the Hunza four meals a day as flatbread with fresh or boiled apricots for breakfast, ditto plus vegetables for lunch, apricot soup for dinner, and vegetables and fruit for late dinner (no mention of yogurt) (JNU). Could they have meant apricot pits in the Garden of Eden? Zohary does not even mention *Prunus* in his discussion of the biblical apple, believing that apples (*Malus domestica*) had been introduced into the Holy Land by 4000 B.C. As Milton says, "The fruit of that forbidden tree whose mortal taste brought death into the world, and all our woe." The seeds of both (and many other rose relatives) do contain laetrile-like compounds that can cure or kill, depending on dosage. Were it my Garden of Eden, or even my Garden of Solomon, I would have wanted both, and all other edible members of the *Prunus* and *Malus* genera. Bitter apricot kernel is highly toxic because of the potential cyanide levels. Expressed oil, known as Persic oil or apricot oil, is used as a pharmaceutical vehicle; it is obtained by the same process as bitter almond oil. Pit shells have been used to prepare activated charcoal, via destructive distillation. Perhaps the tylenol-HCN connection accomplished more to discourage laetrile than did NCI (BIB; ZOH).

Common Names (Apricot):

Abricoquer (Cat.; KAB); Abricotier (Fr.; EFS; KAB); Abrikoosboom (Dutch; EFS; KAB); Abrikos (Den.; EFS); Abrikosovoi Dyerevo (Rus.; KAB); Albaricoque (Mex.; KAB); Albaricoquero (Sp.; KAB; VAD); Albercocco (It.; KAB); Albercooque (Sp.; EFS); Alberge (Sp.; EFS); Albergero (Sp.; EFS); Albicocco (Malta; KAB); Albricoquiero (Por.; EFS); Alperciero (Por.; EFS); Alperchiero (Por.; EFS); Alukashmiri (Pun.; KAB); Anzu (Japan; TAN); Apricot (Eng.; Scn.; AH2; CR2; KAB); Aprikose (Ger.; HHB); Aprikosenbaum (Ger.; KAB); Binkook Tuffa Armina (Arab.; India; EFS; NAD); Binkuk (Arab.; KAB); Cais (Rom.; KAB); Cherkish (Kas.; KAB); Chhappuh (Heb.; KAB); Chilu (Hindi; KAB); Chinaru (India; EFS; NAD); Chinaru (India; NAD); Chinese Bitter Almond (Eng.; Ocn.; AH2); Chola (Kum.; KAB); Chuari (Hindi; DEP; WOI); Chuaru (Kum.; NAD); Chuli (Bhoti; Ladak; Pun.; DEP; KAB; WOI); Chulu (India; NAD); Ciruela (Sp.; EFS); Damasco (Por.; Sp.; KAB; USN); Damasquiero (Mad.; Por.; EFS; JAD); Damasquino (Sp.; USN); Galdam (Tibet; DEP; KAB); Gardali (Pun.; NAD); Gurdalu (Kas.; DEP); Gurdlu (Pun.; MPI; NAD; SKJ); Hanh (Ic.; KAB); Hari (Hazara; Pun.; DEP; KAB; WOI); Hing (China; TAN); Hsing (China; EFS; KAB); Iser (Kas.; NAD); Jaldaru (Sutlej; NAD); Jardal (Kon.; KAB); Jardalu (Hindi; Pushtu; NAD); Kayisi (Tur.; EB54:155); Kayisi Agaci (Tur.; EFS); Kham Bu (Tibet; NPM); Khista (Pushtu; DEP); Khorpani (Newari; NPM); Khubani (Hindi; Pushtu; Urdu; KAB; NAD; WOI); Khurpani (Nepal; NPM); Ku Xing Ren (Pin.; AH2); Kushm Aru (Kum.; DEP); Kushmiaru (Hindi; WOI); Mandata (Pun.; Pushtu; DEP; KAB); Marille (Ger.; HHB; USN); Meliaco (It.; EFS); Mishmis (Iran; NAD); Mishmish (Arab.; Iran; Syria; DEP; HJP); Moon of the Faithful (Eng.; DEP); Mushmush (Arab.; Syria; HJP); Pak Hang (Malaya; KAB); Pating (Bhote; DEP); Salkunamu (Korea; TAN); Sargabarackfa (Hun.; KAB); Sari (Pun.; WOI); Shiran (Pun.; NAD); Siberian Apricot (Eng.; USN); T'ien Mei (China; EFS); Tuffa Urmena (Arab.; DEP); Urumaana (Sanskrit; MPI); Xing (Pin.; AH2); Xing Ren (Pin.; DAA); Zardalu (Afg.; Iran; Kohlu; Quetta; Sharig; KAB).

Activities (Apricot):

Analgesic (1; X15744067); Anthelmintic (f; CRC; KAB); Antidote (f; CRC; DAA); Antiinflammatory (1; X15744067); Antioxidant (1; X15723750); Antiprostaglandin (1; X15744067); Antispasmodic (f; CRC; EFS); Antithyroid (1; AKT); Antitumor (f; APA), Antitussive (f; CRC); Aphrodisiac (f; CRC; EFS); COX-2 Inhibitor (1; X15744067); Cyclooxygenase Inhibitor (1; X15744067); Cyanogenic (f; CRC); Demulcent (f1; CRC; VAD); Emetic (f; KAB); Emollient (f; CRC; VAD); Expectorant (f; BIB; CRC); Hemostat (f; BIB); iNOS Inhibitor (1; X15744067); Laxative (f1; HJP; NAD; VAD); NO Inhibitor (1; X15744067); Pectoral (f; CRC; EFS); Peristaltic (1; VAD); Poison (f1; CRC); Refrigerant (f; NAD); Sedative (f; CRC); Tonic (f; CRC); Vermifuge (f; CRC); Vulnerary (f; CRC).

Indications (Apricot):

Aging (f1; VAD); Anemia (f; CRC; EFS); Asthma (f; APA; CRC; DAA); Bleeding (f; BIB; CRC); Bronchosis (f12; APA; CRC; DAA); Cancer (f1; APA; CAN; JLH); Catarrh (f; CRC; DAA); Childbirth (f; CRC); Cold (f; CRC; DAA); Congestion (f; APA), Conjunctivosis (f; CRC); Constipation (f; APA; CRC); Cough (f; APA; CRC); Deafness (f; KAB); Dermatosis (1; VAD); Diarrhea (f; KAB); Earache (f; KAB); Fever (f; CRC; KAB); Heart (f; CRC); Hemorrhoid (f; KAB); Hepatosis (f; KAB); Ichthyosis (f1; VAD); Infertility (f; BIB; CRC); Inflammation (f1; CRC; X15744067); Itch (f; APA); Laryngitis (f; CRC); Meningitis (f1; HHB); Mucosis (f1; VAD); Ophthalmia (f; BIB); Otosis (f; KAB); Pachymeningitis (f1; HHB); Pain (1; X15744067); Puerperium (f; BIB); Rheumatism (f; CRC); Snakebites (f; APA), Sore (f; APA; JLH); Sore Throat (f; BIB; CRC); Soroche (f; BIB; NAD); Spasm (f; CRC); Swelling (f; CRC; JLH); Thirst (f; CRC); Toothache (f; APA); Trichomoniasis (2; APA); Tumor (f; APA; CRC; JLH); Ulcer (f; JLH); Vaginosis (2; APA); Vulvosis (2; APA); Worm (f; DAA); Wound (f; DAA); Xeroderma (f1; VAD).

Dosages (Apricot):

FNFF = !!!

Ripe fruits widely eaten, fresh, preserved, or converted to juice, brandy, liqueurs; unripe fruits made into preserves and syrups; seeds, especially the bitter seeds, more medicinal, but eaten, like those of the almond and peach; bitter kernels used in macaroons called Ameretti di Saronno and liqueurs called Amaretto di Saronno; seed oil sometimes extracted. Kernels produce a sweet edible oil sometimes used as substitute for almond oil. Chinese almonds are the seed kernels of several sweet varieties of apricot, used for almond cookies, eaten salted and blanched, or made into gruel or flour. Afghans also use the seeds as almonds (FAC; TAN; EB54:155).

- Afghans use dried fruits as laxative and refrigerant in fevers (DEP; KAB).
- Chinese use fruits for asthma, bronchosis, cancer, cardiopathy, cold, cough, insomnia, worms, wounds, seeds for asthma, bronchosis, catarrh, cough, considering roots antidotal to seeds (DAA).
- Koreans use the expectorant kernel to treat dry throat (BIB).
- Malayans used dried fruits to quench thirst and allay fever (KAB).
- Tibetans apply fruits, after chewing them, in ophthalmia (DEP).
- Unani use the tonic seed for deafness, earache, hepatitis, piles, and worms, the fruit for diarrhea, fever, and thirst (KAB).

Downsides (Apricot):

Class 3 (AHP). Also known as bitter almond. One of the products apricot seed components break down into in the human body is highly toxic prussic acid. Fifty to sixty kernels of apricot seed can kill adults; 7 to ten kernels can be fatal in children. My CRC handbook says that a "double kernel is said to be enough to kill a man" (CRC). I think that is overkill. Headache and nausea occurred in some patients being treated for chronic bronchosis with a paste of apricot seed and sugar. Possible adverse interaction when taken with astragalus, skullcap, or kudzu root (TMA, 1996). Do not use, says APA, probably referring to the cyanogenic seed. Newall, Anderson, and Phillipson (1996) warn that the cyanogenetic glycosides in the seed can cause cyanide poisoning. Because of cyanide toxicity, apricot use in pregnancy and lactation should be avoided. Kernels may cause contact dermatosis. Avoid in pregnancy; ingestion of cyanogenic substances may be teratogenic. More than 20 deaths have been reported from laetrile and apricot kernel ingestion (CAN). Symptoms of acute intoxication include convulsions, dizziness, drowsiness, dyspnea, headache, hypotension, nausea, paralysis, coma, and then death. Death may occur from 1 to 15 minutes after ingestion. Antidotes for cyanide poisoning include aminophenol, cobalt edetate, hydroxocobalamin, nitrite, and thiosulphate. Symptoms of chronic intoxication (from HCN, cyanogenic foods, or drugs such as laetrile) include ataxia, blindness, cretinism, goiter, hypertonia, increased blood thiocyanate, lesions of the optic nerve, mental retardation, and thyroid cancer. Demyelinating lesions and other neuromyopathies may occur secondary to chronic cyanide exposure, including long-term laetrile therapy. Agranulocytosis has also been attributed to long-term laetrile therapy. Laetrile spelled with a capital "L" signifies a synthetic patented in 1961 but never sold in the United States. Spelled with a small "l," laetrile is, for lay purposes, synonymous with amygdalin. After making this distinction, APA spells it with a small "l" except, of course, at the beginning of a sentence where they say, "Laetrile consists of 6% cyanide (prussic or hydrocyanic acid), a highly poisonous substance that can kill by depriving the brain of oxygen." Then they unleash a hyperbolic error I presume, "the laetrile content of apricot pits varies from as much as 8% in some apricot varieties to twenty times that amount in wild varieties." Krebs marketed laetrile with a small "l" as vitamin B-17 (AHA).

Extracts (Apricot):

Among 37 varieties, total carotenoid content ranged from 15 to 165 mg kg (edible portion); beta-carotene led, followed by beta-cryptoxanthin and gamma-carotene, with less phytoene, phytofluene, gamma-carotene, lycopene, beta-cryptoxanthin, and lutein (X16076120). I am not surprised by the antioxidant capacities reported by Scalzo et al. (2005) for wild strawberry, six varieties of cultivated strawberry, and five varieties of apple and with apricot and peach grafts. Wild strawberries >> cultivated strawberries >> kiwifruit = apples = apricots = peaches. This tells me that the American wild strawberry, as I would have predicted, was better than the cultivated strawberry, apple, or apricot (X15723750). A retrospective analysis of laetrile in cancer patients showed slight activity. A subsequent clinical trial concluded that laetrile was ineffective in cancer treatment. Claims for laetrile were based on three different theories. Theory (1) claimed that cancerous cells contained copious beta-glucosidases, which release HCN from laetrile via hydrolysis. Normal cells were reportedly unaffected because they contained low concentrations of beta-glucosidases and high concentrations of rhodanese, which converts HCN to the less toxic thiocyanate. Later, however, it was shown that both cancerous and normal cells contain only trace amounts of beta-glucosidases, and similar amounts of rhodanese. Also, it was thought that amygdalin was not absorbed intact from the gastrointestinal tract (CAN). Theory (2) proposed that after ingestion, amygdalin was hydrolyzed to mandelonitrile, transported intact to the liver, and converted to a beta-glucuronide complex, which was then carried to the cancerous cells, hydrolyzed by beta-glucuronidases to release mandelonitrile and then HCN. This was believed an untenable theory. Theory (3), calling laetrile vitamin B-17, proposed that cancer results from B-17 deficiency. It postulated that chronic administration of laetrile would prevent cancer. No evidence was adduced to substantiate this hypothesis. Furthermore, it was even claimed that patients taking laetrile reduced their life expectancy, both through a lack of proper medical care and chronic cyanide poisoning. To reduce potential risks to the general public, amygdalin was made a prescription-only medicine in 1984 (CAN).

ALMOND (*PRUNUS DULCIS* (MILL.) D.A. WEBB) ++ ROSACEAE

Synonyms:

Amygdalus communis L.; *Amygdalus dulcis* Mill.; *Prunus amygdalus* Stock

Notes (Almond):

> **And their father Israel said unto them, If it must be so now, do this; take of the best fruits in the land in your vessels, and carry down the man a present, a little balm, and a little honey, spices, and myrrh, nuts, and almonds.**
>
> **Genesis 43:11 (KJV)**

> **Then their father Israel said to them, "If it must be so, then do this: take some of the choice fruits of the land in your bags, and carry down to the man a present, a little balm and a little honey, gum, myrrh, pistachio nuts, and almonds.**
>
> **Genesis 43:11 (RSV)**

So Israel their father said to them, "If, then, that is the case, do this: Take the finest products of the land in your receptacles, and carry them down to the man as a gift, a little balsam, and a little honey, labdanum and resinous bark, pistachio nuts, and almonds.

Genesis 43:11 (NWT)

Historically, some have argued that almond did not grow naturally in Egypt, because Jacob's sons took almonds to Joseph. (We often take almonds on trips although they are available at either end of our trip.) Zohary states, "Flowering almonds are not found in the Sinai today," (ZOH) but infers that they may have been there historically, as they do occur in the Negev Hills. Possibly related trees such as the Sinai Hawthorn were used as a substitute for almond in the floral candelabrum. Nowadays, the almond is widespread in the Holy Land, one of the earliest trees to flower. In Tuscany, almond branches were reportedly used as divining rods to locate hidden treasure. There is the legendary story of Charlemagne's troops' spears (almond) sprouting in the ground overnight and shading the tents the next day. Almonds are also valued for their ornamental flowers, one of the first trees to flower in the Palestinian spring. Because of their association with spring, the flower is associated with life after death or immortality. Modern English Jews carry flowering branches into the synagogue on spring festival days, as a herald of spring in Israel (BIB; ZOH).

Common Names (Almond):

Acibadem Agaci (Tur.; EFS); Allozo (Sp.; EFS); Almendro (Peru; Sp.; Spain; EFS; EGG; USN; VAD); Almond (Eng.; Scn.; AH2; CR2; USN); Amendi (Kon.; KAB); Amygdalia (Greek; KAB); Anande (Fr.; EFS); Amandelboom (Dutch; EFS); Amandier (Fr.; EFS; USN); Amandier Commun (Fr.; USN); Amendo (Japan; USN); Amendoeira (Por.; EFS; KAB); Amendosu (Japan; TAN); Amendoeira (Por.; USN); Ametlle (Cat.; KAB); Archin (Pab; KAB); Badam (Bom.; Dec.; Guj.; Hindi; Iran; Mal.; Mar.; Nepal; Pun.; Surab; Tur.; DEP; EFS; KAB; SUW); Badama (Sanskrit; EFS; NAD); Badamamu (Tel.; KAB); Badami (Kan.; DEP; KAB); Badamitte (Sanskrit; DEP); Badamo (Oriya; KAB); Badamshirin (Urdu; KAB); Badamu (Kan.; NAD); Badamvittilu (Tel.; DEP); Badan (Burma; DEP; KAB); Bademi (Tur.; EB54:155); Bilatibadam (Beng.; DEP; KAB; NAD); Bitter Almond (Eng.; USN); Bittermandelbaum (Ger.; USN); Emmellié (Fr.; KAB); Hadankyo (Japan; TAN); Kahero (Wad; KAB); Karamomo (Japan; TAN); Lauz (Arab.; GHA); Lawz (Arab.; Syria; HJP); Lawz Myrr (Arab.; Syria; HJP); Louz (Arab.; DEP); Louza (Arab.; EFS; NAD); Loz (Arab.; GHA); Lujaalhulu (Arab.; KAB); Luz (Arab.; Aramaic; Heb.; ZOH); Mandelbaum (Ger.; EFS; USN); Mandorlo (It.; EFS; KAB); Mandulafa (Hun.; KAB); Migdal (Pol.; Rom.; KAB); Mindalnoi Dyerevo (Rus.; KAB); Pa Tan Hsing (China; KAB); Parsivadumai (Tam.; KAB); Rattokotamba (Sih.; KAB); Shaged (Heb.; KAB); Shaked (Heb.; ZOH); Sweet Almond (Eng.; USN); Tatlibadem Agaci (Tur.; EFS); Vadamkottai (Tam.; DEP); Vadumai (Tam.; SKJ); Vatamkotta (Mal.; DEP); Vatavairi (Sanskrit; KAB). Many countries recognize a sweet (low cyanide) var. *dulcis* and a bitter (high cyanide) var. *amara*, appending their adjective for bitter or sweet to their word for almond. AH2 used bitter almond and sweet almond as other common names for the standardized common name "almond."

Activities (Almond):

Allergenic (1; JAF49:2131); Alterative (f; BIB); Antifeedant (1; X11902971); Antiinflammatory, Antispasmodic (f1; BOW; FNF; VAD); Antitussive (f1; FNF; VAD); Aphrodisiac (f; BIB; GHA); Astringent (f; BIB; DEP); Bactericide (1; APA; MPI); Carminative (f; BIB); Cerebrotonic (f; NAD); Cyanogenic (f; BIB); Demulcent (f1; APA; BIB; EFS; PH2); Deobstruent (f; DEP); Discutient (f; BIB; WOI); Diuretic (f; BIB; DEP); Emollient (f1; APA; BIB; EFS); Expectorant (f1; FNF; VAD); Hepatoprotective (f; VAD); Hypocholesterolemic (1; X15746835); Hypotensive (f; VAD); Laxative (f1; APA; BIB; VAD); Insectifuge (1; X11902971); Lipolytic (1; X15746835); Litholytic (f; BIB; DEP; WOI); Memorigenic (f; GHA); Nervine (f; BIB; SUW; WOI); Pectoral (f; VAD); Pediculicide

FIGURE 1.87 Almond (*Prunus dulcis*). Source: KAB

(f; DEP); Sedative (f; BIB); Spermatogenic (f; NAD); Stimulant (f; BIB); Sudorific (f; VAD); Tonic (f; BIB); Vermifuge (f; GHA); Vulnerary (f; KAB).

Indications (Almond):

Acne (f; BIB); Adenopathy (1; JLH); Ascites (f; BIB); Asthma (f; BIB); Biliousness (f; BIB; KAB); Bronchosis (f; BIB; NAD); Burn (f; VAD); Callus (f; BIB; JLH); Cancer (f; BIB; JLH); Cancer, bladder (f1; APA); Cancer, breast (f1; APA; JLH); Cancer, colon (f1; FNF); Cancer, gland (f1; FNF; JLH); Cancer, liver (f1; FNF; JLH); Cancer, mouth (f1; APA); Cancer, spleen (f1; FNF; JLH); Cancer, stomach (f1; FNF; JLH); Cancer, uterus (f1; FNF; JLH); Cardiopathy (1; APA; FNF; X15469659); Chafing (f; GAZ); Cold (f; BIB; FNF); Colic (f; BIB); Condyloma (f; BIB; JLH); Constipation (f1; APA); Corn (f; BIB; JLH); Cough (f1; BIB; DEP; FNF; GHA; PH2); Cramp (f; BIB); Cystosis (f; BIB; JLH); Delirium (f; BIB); Dermatosis (f; BIB; PH2; WOI); Diabetes (f; DAA; NAD); Dysmenorrhea (f; DEP); Dyspnea (f; BIB; GHA); Dysuria (f; NAD); Earache (f; BIB); Enterosis (f; DEP); Furuncle (f; BIB); Gallstone (f; BOW); Gastrosis (f; JLH); Gingivosis (f; BIB; DEP); Gleet (f; BIB; KAB); Gravel (f; BIB); Headache (f; BIB; DEP); Heartburn (f; BIB); Hepatosis (f; BIB; DEP; JLH); Herpes (f; GHA); High Blood Pressure (f; VAD); High Cholesterol (1; APA; X15746835); Hoarseness (f; NAD); Hydrophobia (f; BIB); Ichthyosis (f; VAD); Impotence (f; BIB; GHA); Induration (f; BIB; JLH); Infection (f; NAD); Inflammation (f; BIB; JLH); Itch (f; BIB; WOI); Kidney stone (f; BOW); Leukoderma (f; BIB); Mastosis (f; JLH); Nausea (f; PH2); Nephrosis (f; BIB; NAD); Neuralgia (f; DEP; KAB); Obesity (1; X15746835); Ophthalmia (f; DEP); Pain (f; DEP; KAB); Pediculosis (f; KAB); Polyuria (f; NAD); Psoriasis (f; VAD); Pulmonosis (f; BIB); Respirosis (f; EFS); Sclerosis (f; JLH); Sore (f; BIB; JLH); Sore Throat (f; BIB; KAB); Splenosis (f; BIB; DEP; JLH); Staphylococcus (1; MPI); Stomatosis (f; BIB; DEP; JLH); Stone (f; BOW); Streptococcus (1; MPI); Swelling (f; JLH); Ulcer (f; BIB); Uterosis (f; JLH); Venereal Disease (f; BIB); Vomiting (f; PH2); Worm (f; GHA).

Dosages (Almond):

FNFF = !!!

Nuts widely eaten, raw or roasted; seeds can be blended into almond milk or almond butter; used in baked goods and candies; seed oil quite delectable, used in flavoring baked goods, the bitter almond oils ending up in confections such as Maraschino cherries and liqueurs such as amaretto. Benzaldehyde may be used for almond flavoring, usually being cheaper than almond oil (BIB; FAC; TAN; EB54:155). 2–4 tsp oil as laxative (VAD).

- Arabians believe the seed kernels will improve sexual potency (GHA).
- Asian Indians suggest "almond nut cream" for "brain workers" (three almonds, two walnuts, two ounces of pine kernels crushed and steeped overnight in orange or lemon juice) (NAD).
- Ayurvedics consider the fruit, the seed, and its oil aphrodisiac, using the oil for biliousness, headache, and the seed as a laxative (KAB).
- Icelanders take bitter almonds (imported I suppose) for hepatic and splenic indurations (JLH).
- Iranians make an ointment from bitter almonds for furuncles (BIB).
- Lebanese believe that almonds and/or almond oil restore virility (HJP).
- Lebanese use the oil for skin trouble, including white leukoderma-like patches (HJP).
- Middle Easterners use the oil as an emollient to alleviate itching. Raw oil from the bitter variety is used for acne. Almond and honey was given for cough. Thin almond paste was added to wheat porridge to pass gravel or stone (BIB; DEP; HJP).
- Mohammedens recommend a plaster of bitter almonds with vinegar for neuralgia, with starch and peppermint for cough (DEP).
- Pakistanis eat five, seven, or eleven almonds before breakfast to improve the memory (GHA).
- Unani use the seed for ascites, bronchitis, colic, cough, delirium, earache, gleet, hepatitis, headache, hydrophobia, inflammation, renitis, skin ailments, sore throat, and weak eyes (KAB).

Downsides (Almond):

No health hazards or side effects known with proper therapeutic dosages (PH2) (PH2 designates no dosage! JAD). Ten bitter almonds said to be fatal to children, 60 to an adult (PH2). 0.5% of United States citizens show sensitivity to tree nuts, 0.6% to peanuts, and an additional 0.3% allergic to nuts but not specifying or differentiating between tree nut and peanut allergy.

Natural History (Almond):

Prominent diseases in India include "shot hole" caused by *Clasterosporium carpophilum* (Lev.) Aderh., "white spongy rot" due to *Fomes lividus* K1, "brown patchy leaf rot" due to *Phyllosticta prunicola* (Spiz) Sacc., "brown rot" due to *Sphaerotheca pannosa* (Wallr.) Lev., and a mosaic disease due to virus plague almond. The chrysomelid *Mimastra cyanura* Hope and the almond weevil *Myllocerus laetivirens* Marshall feed on the leaves. The San Jose scale *Quadraspidiotus perniciosus* Comstock is a minor problem. The almond moth *Ephestia cautella* Wlk. infests shelled almonds and dried apricot, currant, date, fig, peach, and plum (HOE). Resistance to the buprestid beetle, *Capnoidis tenebrionis*, may be proportional to the prunasin content of the roots (X11902971).

Extracts (Almond):

Amandin (almond major protein) is a complex protein with at least 28 peptides and accounts for circa 65% of total aqueous extractable almond protein (JAF49:2131). De Pascual et al. (1998) note that green almond extracts contain two monomers — (+)-catechin and (–)-epicatechin — and 15 oligomeric procyanidins (six dimers, seven trimers, and two tetramers) (J. Am. Diet. Assoc., 105(3):449–454, 2005). Almonds in the diet simultaneously improve plasma alpha-tocopherol concentrations and reduce plasma lipids (X15746835). The objective of this study was to assess the dose-response effect of almond intake on plasma and red blood cell tocopherol concentrations in healthy adults enrolled in a randomized, crossover feeding trial. Participants were 16 healthy men and women, aged 41 ± 13 years. After a 2-week run-in period, participants were fed three diets for 4 weeks each: a control diet, a low-almond diet, and a high-almond diet, in which almonds contributed 0, 10, and 20% of total energy, respectively. Changes in blood tocopherol levels were assayed by high-pressure liquid chromatography. Incorporating almonds into the diet helped meet the revised Recommended Dietary Allowance (RDA) of 15 mg/day alpha-tocopherol and increased lipid-adjusted plasma and red blood cell alpha-tocopherol concentrations. A significant dose-response effect was observed between percent energy in the diet from almonds and plasma ratio of alpha-tocopherol to total cholesterol (X15746835). Those who know me have probably heard me urging a Gatesian computerized approach, analyzing 1000 of the important food farmacy plants, for 1000 important phytochemicals, so we could then, via computer, determine which foods were best for which malady of mankind. Phillips et al. (2005) certainly dampen my enthusiasm for such an approach, showing that just the nuts present huge analytical challenges. My decades of compiling have shown me what megavariation there is in the quantitation of phytochemicals within a single species. I fear a megagatesian megacybernetic quandary. "The results of this study also illustrate the complexity that can be involved in evaluating food phytochemical data. Determination of the phytosterol composition of nuts and seeds is not amenable to 'production-scale' analysis of predetermined components using existing standard values" (X16302759). What is the most efficacious mix of the subinfinite combinations of the various phytosterols in a given nut, the phytosterols alone or with all the other phytochemicals, or in a biblical seven-nut/grain mixture, some possibly synergic, some probably additive, some possibly even antagonistic? I still optimistically suspect that all these sterols have been known to our genes for millions of years of co-evolution, and hence the body homeostatically grabs those it needs from the sterol mix, if they are needed, excluding them if unneeded. Fortunately for nut lovers, Phillips et al. (2005) quantified the phytosterols in nuts and seeds commonly consumed in the United States. Such phytosterols are medicinally important, for example, in BPH and in high cholesterol (X16302759). Here is what they found in almonds, based on four accessions: delta5-avenasterol (197 ppm), campestanol (33 ppm), campesterol (49 ppm) phytosterols (1930–2080 ppm), poriferasta-7,25-dienol (101 ppm), sitostanol (32 ppm), beta-sitosterol (1434 ppm), and stigmastanol (50 ppm) (X16302759).

RED SANDALWOOD (*PTEROCARPUS SANTALINUS* L. F.) ++ FABACEAE

NOTES (RED SANDALWOOD):

And the king made of the almug trees pillars for the house of the LORD, and for the king's house, harps also and psalteries for singers: there came no such almug trees, nor were seen unto this day.

I Kings 10:12 (KJV)

Many biblical scholars, including Zohary, believe the "almug" of Kings was the red sandalwood or red saunders Pterocarpus santalinus L. Anyhow, it is consistently called almug in KJV, NWT, RSV; and its hard and heavy wood, red to garnet colored, takes a good polish, well suited for Solomon's purposes. It was used in construction of the House of the Lord, and is still used for lyres and other musical instruments (ZOH). The wood is extremely hard and resistant to termites. Recently, the wood has been more important as a dye source, used for imprinting a red or pink color to calico, cotton, or silk. Red sandalwood is well known in Europe as an ingredient of "French polish" (BIB). JLH and IHB combined medicinal activities of *P. indicus* and *P santalinus.*

COMMON NAMES (RED SANDALWOOD):

Agaru Gandhamu (Ap.; SKJ); Algum (Eng.; Heb.; ZOH); Almug (Eng.; Heb.; BIB; ZOH); Atti (Tamilnadu; SKJ); Buckum (Iran; DEP); Caliatur Wood Tree (Eng.; EFS); Chan Chandanam (Mal.; NAD); Chandana (India; Sanskrit; EFS; JLH); Chandan Lal (Pun.; DEP); Chendana Dangi (Malaya; IHB); Chendana Mera (Malaya; EFS); Dul Surkh (Iran; DEP); Dunkelroche (Ger.; NAD); Erra Chandanam (Tel.; DEP); Erra Gandhapu-chekka (Tel.; DEP); Erra Gandhamu (Tel.; NAD); Flugal Frucht (Ger.; NAD); Gerra Chandan (Tel.; DEP); Honne (Karnataka; SKJ); Kaliaturholzbaum (Ger.; EFS); Kirmizi santal (Tur.; EFS); Kuchandana (Sanskrit; DEP); Kuchunduna (Beng.; DEP); Kuchandanam (Tel.; DEP); Lal Chandan (Tel.; DEP); Lala Chandan (Hindi; SKJ); Lalachandana (Bom.; DEP); Lalchandana (Hindi; NAD); Lenyo Caliatur (Sp.; EFS); Patrangan (Kerala; SKJ); Nasa-Ni (Burma; DEP); Rachandana (Kon.; NAD); Rakta Chandan (Nepal; Sanskrit; SKJ; SUW); Rakta Chandana (India; Sanskrit; DEP; EFS); Raktachandau (Nepal; SUW) Ratanjali (Guj.; SKJ); Sanders Tree (Eng.; EFS); Ragat Chandan (Hindi; DEP); Rakta Chandan (Nepal; SUW); Rakta Chandana (Beng.; SUW); Rakta Gandhamu (Tel.; NAD); Rakta Sandana (Beng.; Hindi; Kan.; Mah.; Sanskrit; NAD); Rangana (Beng.; DEP); Ratanili (Bom.; Guj.; NAD); Ratanjli (Bom.; Guj.; DEP); Red Sandalwood (Eng.; CR2; EFS); Red Sanders (Eng.; SKJ); Red Sanders Tree (Eng.; EFS); Red Sanders Wood (Eng.; SUW); Red Saunders (Eng.; USN; ZOH); Rod Sandel (Den.; EFS); Rood Sandelhoutboom (Dutch; EFS); Rotes Sandelholz (Ger.; EFS); Sandaku (Burma; DEP); Sandale Ahmar (Arab.; DEP); Sandale Surkh (Iran; EFS; NAD); Sandalia (Sp.; EFS); Sandalo (It.; EFS); Sandalo Rose (It.; DEP); Sandalo Vermelho (Por.; JLH); Sandel Hout (Den.; DEP); Santal Rouge (Fr.; DEP; EFS; NAD); Seyapu Chandanum (Tam.; DEP); Shen Chandanam (Tam.; NAD); Sun (Iran; DEP); Tilaparni (Beng.; DEP); Tambada Chandana (Mar.; DEP); Tambada Gand-hacha-chekka (Mar.; DEP); Tilapari (Sanskrit; DEP); Tjendana Djenggi (Malaya; EFS); Tzu T'an (China; EFS); Undum (Arab.; Hindi; Iran; DEP); Uruttah Chundanum (Mal.; DEP).

ACTIVITIES (RED SANDALWOOD):

Alexiteric (f; KAB); Allergenic (1; X8789238); Analgesic (f; HHB); Anthelmintic (f; KAB); Anticonvulsant (1; HH3); Antidiabetic (1; PH2; X11137350); Antiexudative (1; PH2); Antihyperglycemic

(1; X12033810); Antiinflammatory (f1; SKJ; X12413723); Antiproliferant (1; X11217086); Antispasmodic (1; PH2); Aphrodisiac (f; KAB); Astringent (f; SUW; WOI); Chemopreventive (1; X12413723); COX-2 Inhibitor (1; X12413723); CNS Depressant (1; HH3; PH2); Collagenic (1; X15866805); Depurative (f; EFS; KAB); Diaphoretic (f; SUW; WOI); Diuretic (f; BIB); Emetic (f; BIB); Expectorant (f; KAB); Febrifuge (f; KAB); Fungicide (1; WOI); Hemostat (1; NAD); Hypoglycemic (1; HHB); Insecticide (1; PH2); Nematicide (1; HH3); Refrigerant (f; SUW); TNF-alpha Inhibitor (1; X11217086); Tonic (f; NAD; SUW; WOI); Tranquilizer (1; HH3); Vulnerary (f1; X15866805; X15866819).

Indications (Red Sandalwood):

Biliousness (f; SUW; WOI); Bleeding (f; KAB); Blepharosis (f; BIB); Boil (f; BIB; DEP; IHB; SUW); Burn (f1; X15866805); Cancer (f1; JLH; X12033810); Cancer, abdomen (f; JLH); Cancer, breast (f1; X12033810); Cancer, colon (f; JLH); Cancer, mouth (f; JLH); Cerebrosis (f; KAB); Cholecocystosis (f; HH3; PH2); Conjunctivosis (f; NAD); Debility (f; HH3); Dermatosis (f; SUW; WOI); Diabetes (1; PNC; X11137350); Diarrhea (f; PH2); Dysentery (f; DEP; KAB; SKJ; WOI); Enterosis (f; JLH); Fever (f; DEP; HH3; PH2; SUW); Fungus (1; WOI); Gastrosis (f; PH2); Genitosis (f; NAD); Headache (f; DEP; HHB; PH2; SUW; WOI); Hemicrania (f; KAB); Hemorrhoid (f; NAD); Impotence (f; KAB); Inflammation (f1; SKJ; SUW; WOI; X12413723); Mastosis (f1; X12033810); Metrorrhagia (f; KAB); Mycosis (1; WOI); Neck ache (f; KAB); Ophthalmia (f; DEP; PH2; SUW); Prickly Heat (f; IHB); Snakebite (f; PH2); Sore (f; BIB; IHB); Sting (f; BIB); Stomatosis (f; IHB JLH); Swelling (f; JLH; SUW); Syphilis (f; IHB); Thrush (1; IHB); Toothache (f; KAB; PH2); Ulcer (f; PH2); Vaginosis (f; BIB); Venereal Disease (f; IHB); Vomiting (f; PH2); Wound (f1; X15866805; X15866819).

Dosages (Red Sandalwood):

FNFF = !

Powdered redwood used as a red food dye in commercial spice mixes and sauces; in Old England, it is used to dye gingerbread, jelly, pottages, etc. (FAC); 5 g tincture (200 parts powdered wood/1000 parts ethanol, PH2).

- Asians use the plant in bolmes, enemas, ghees, or powders for abdominal tumors (JLH).
- Asian Indians use the wood, lathered up in water, to wash blepharitis and superficial excoriations of the genital organs (BIB).
- Ayurvedics, regarding the wood as alexiteric, anthelmintic, aphrodisiac, and refrigerant, use it for biliousness, blood disorders, eye ailments, fever, mental aberrations, and ulcers (KAB).
- Indonesians consider it a secret remedy for poisoning (BIB).
- Iranians sells chips of the wood for use against dysentery (BIB).
- Unani use seeds for dysentery and urethral hemorrhage; applying the wood externally for fever, headache, hemicrania, inflammation, neckache, and toothache (KAB).

Downsides (Red Sandalwood):

Class 1. In the United States, allowable as flavor in alcoholic beverages only (AHP). No health hazards or contraindications with proper administration of suggested therapeutic dosages (PH2).

Extracts (Red Sandalwood):

Extracts potently inhibited COX-2 (IC >80% = 10 μg/ml) (X12413723). Compared with diabetic rats treated with glibenclamide, the antihyperglycemic activity of ethanolic bark extract at 20 mg/kg was more effective (X11137350). Biswas et al. (2004) report a vulnerary ointment from the bark, effective and with no toxic effects (X15866819).

POMEGRANATE (*PUNICA GRANATUM* L.) +++ PUNICACEAE

SYNONYMS:

Granatum punicum St.-Lag., *Punica florida* Salisb., *Punica multiflora* Hort. ex Siebold & Voss, *Punica nana* L., *Punica spinosa* Lam. *fide* POR

NOTES (POMEGRANATE):

I would cause thee to drink of spiced wine of the juice of my pomegranate

Song of Solomon 8:2 (KSV)

I would give you spiced wine to drink, the juice of my pomegranates.

Song of Solomon 8:2 (RSV)

I would give you a drink of spiced wine, the fresh juice of pomegranates.

Song of Solomon 8:2 (NWT)

It is nice to see near unanimity in the three versions of this rather sexy story in Song of Solomon. "Pomegranate" literally means "apple with grains," the reference being to the many clear, ruby-colored seeds, covered with a thin skin and full of juice, found in each fruit. Jewish legends suggest that the pomegranate has about 613 seeds, the same number of laws God gave to Israel. I have come to believe, unlike other scholars, including Zohary, that this estrogenic fertility-symbolizing fruit could well be the tree of knowledge. If you wish to read some of the sexiest passages in the Bible, look up the verses mentioning the pomegranate in the Song of Solomon. Zohary clearly thinks it important, if not the tree of knowledge. Commenting on some Solomon references, he says "the woman's beauty is likened to its beautiful shape, its many seeds symbolize fertility," (ZOH) the red juice is viewed as lover's nectar, and the aromatic flowers stand for the beautiful awakening of spring. I have one plant of this "yin" species that is hardy on my south-facing "yang" slope here in the Green Farmacy Garden. Rinds are used for tanning Morocco leather, giving a yellow color. Flowers give a red dye. Plants make a good ornamental hedge, especially in dry climates. Cut flowers are long lasting in arrangements. Pomegranate is the national flower emblem of Spain. Wood, although scanty, is hard and can be used for small objects and for walking sticks. Flowers are used by some women to give a red color to the teeth, and rind is used in Polynesia to give shining black color to teeth. In some areas, nonfading ink is made from the rind. Dried rind, called Malicorium, is sold in curved brittle fragments. In China, the pomegranate symbolizes fertility; women offer pomegranates to the Goddess of Mercy in the hope of being blessed with children. Boulos reports that the seed oil is estrogenic, perhaps providing a rationale for the Chinese beliefs (BIB). Other scientists report human-identical estrone, some at levels such that one fruit would provide a 2-day dosage of ERT (estrogen replacement therapy), but such quantities need to be verified. Most important is the specificity of the root bark for tapeworm.

COMMON NAMES (POMEGRANATE):

Aboda (Ewe; KAB); Al Lufân (Arab.; Syria; HJP); An Shih Liu (China; EFS; KAB); An Thatch Luu (Ic.; KAB); Anangani (Sinjawi; KAB); Anar (Bhojpuri; Dec.; Hindi; Kas.; Kotra; Lepcha; Mooshar; Nepal; Nwp.; Tharu; KAB; MKK; NAD; NPM; SUW); Anar Dakum (Sin.; KAB); Anar

FIGURE 1.88 Pomegranate (*Punica granatum*).

ke per (Hindi; NAD); Anara (Bom.; KAB; NAD); Anarbedama (Quetta; KAB); Anardaru (Mun.; KAB); Anarthamitha (Urdu; KAB); Apencoya (Peru; EGG); Apinhoya (Andes; ROE); Apongabe-andanitra (Hova; KAB); Armoun (Ber.; BOU); Aroumane (Ber.; BOU); Balaaustier (Fr.; KAB); Balustier (Fr.; EFS); Bijapura (Sanskrit; KAB); Carthagian Apple (Eng.; EFS); Cay Luu (Annam; KAB); Daariim (Nepal; POR); Dadam (Guj.; KAB; WOI); Dadima (Ayu.; Sanskrit; Tel.; AH2; JLH; NAD); Dadima Phalima (Sanskrit; EFS; NAD); Dadiman (Mal.; KAB); Dahrun (Sibi; KAB); Dalim (Assam; Beng.; Dec.; KAB; NAD); Dalima (Java; Jolo; Malaya; IHB; KAB); Dalimba (Bom..; Tel.; Kon.; Mah.; KAB); Dalimbay (Kan.; KAB); Dalimbu Hannu (Kan.; KAB); Dalimgachh (Beng.; KAB); Dallimbini (Kon.; KAB); Danimma (Tel.; WOI); Danoi (Jaunsar; KAB); Darakhtenar (Iran; KAB); Daraknar (Iran; KAB); Darim (Beng.; Danuwar; Dec.; Gurung; Magar; Nwp.; Sunwar; Swe.; Tamang; AVP; KAB; NPM); Darimba (Sanskrit; NAD); Daru (Pun.; KAB); Datema (Rus.; AVP); Delima (Bali; Malaya; Tag.; IHB; POR); Delumgaha (Sin.; NAD); Delungaha (Sin.; KAB); Dhale (Nepal; Newari; NPM); Dhalim (Hindi; NAD); Dhaun (Kas.; NAD); Djolanar (Arab.; BOU); Drzewo Granatowe (Pol.; KAB); Dulim (Iran; IHB); Dulima (Iran; IHB); Gemeiner Granatbaum (Ger.; TAN); Gharnangoi (Pushtu; KAB); Graanatapfel (Ma.; Sur.; AVP; JFM); Granaatappel (Dutch; POR); Granaatboom (Dutch; AVP); Granada (Sp.; Peru; Pi.; Pr.; KAB; DAV; LWW; ROE); Granada Agria (Ma.; Sp.; JFM; ROE); Granada de China (Sp.; ROE); Granadero (Ma.; Pr.; JFM; LWW); Granado de China (Mex.; KAB); Granado Enano (Ma.; JFM); Granat (Den.; Rus.; EFS;

POR); Granatæle (Den.; POR); Granatappel (Dwi.; Ma.; JFM; LWW); Granatäpple (Swe.; POR); Granatbaum (Ger.; AVP; NAD; MAD); Granatboom (Dutch; KAB); Granate (Ger.; AVP); Granatnik (Rus.; AVP); Granátovník (Cze.; POR); Granatowiec (Pol.; AVP); Granatrad (Swe.; KAB); Granattraee (Den.; KAB); Grenad (Creole; Haiti; VOD); Grenade (Fr.; Ma.; JFM; LWW); Grenadier (Haiti; Ma.; AVP; JFM); Grenadier Comun (Fr.; TAN); Grenadier Cultive (Fr.; NAD); Grenadyé (Creole; Haiti; VOD); Gronuto (Potenza; KAB); Gulnar (Iran; EFS; KAB); Hanor (Kharan; KAB); Jaman (Pun.; KAB); Kanthakasi (Rai; NPM); Karakamu (Tel.; KAB); Kok Mak Phi La (Laos; POR); Komamanga (Swahili; POR); Kuchaphala (Sanskrit; EFS; KAB); Kudhumani (Swahili; POR); Lalimse (Limbu; NPM); Liépou Pi (China; AVP); Ma Ko (Thai; POR); Madala (Michi; KAB); Madalai (Tam.; KAB); Madalam (Tam.; KAB); Madalangkai (Tam.; KAB); Madhubiija (Sanskrit; POR); Madulai (Tam.; WOI); Madulam (Tam.; KAB); Magragnar (Verona; KAB); Magraner (Cat.; KAB); Mangano (Sp.; USN); Mangrano (Sp.; EFS); Matalam (Mal.; KAB; POR; WOI); Melagranato (It.; AVP; KAB); Melgarne (Romagna; KAB); Melograno (It.; EFS; KAB); Migraine (Fr.; KAB); Miouganier (Fr.; EFS); Nar (Tur.; AVP); Nara Aci (Tur.; EFS; KAB); Nargosa (Shahrig; KAB); Nârumschk (Arab.; JLH); Naspal (Hindi; Mah.; KAB); Oschnoe Derewoe (Rus.; AVP); Pitligean (Rom.; KAB); Pomanzeira (Por.; EFS); Pomegranate (Eng.; Scn.; AH2; CR2; JFM; NPM); Pomeira (Por.; EFS); Pyé Grenad (Creole; Haiti; VOD); Rannua (Arab.; AVP); Rimani (Hausa; KAB); Rimaus (Heb.; KAB); Rimmon (Heb.; Isr.; ZOH); Roia (Greek; KAB); Roma (Brazil; Ma.; JFM; KAB; POR); Roman (Por.; AVP); Romanzeira (Ma.; Por.; AVP; JFM; KAB); Romãzeira (Mad.; JAD); Romeira (Ma.; Por.; AVP; JFM); Rommana (Tun.; AVP); Rouman (Arab.; AVP; BOU); Rum N (Arab.; GHA); Rummân (Arab.; Syria; HJP); Rummân Hamid (Arab.; Syria; HJP); Rummien (Malta; KAB); Sale Bin (Burma; KAB); Se-Bru (Tibet; NPM); Seok Ryu (Korea; POR); Shajratur Rumman (Arab.; EFS; KAB); Shak Liu (Malaya; KAB); Shih liu (China; AH2; POR; TAN); Shi Liu Hua (Pin.; AH2); Shi Liu Pi (Pin.; AH2; DAA); Shi Liu Ye (Pin.; AH2); Shi Liu Zi (Pin.; AH2); Shukadana (Sanskrit; NAD); Sor (Jhalawan; KAB); Tab Tin (Thai; IHB); Talibin (Burma; KAB); Tarmint (Ber.; BOU); Taroumant (Ber.; BOU); Thap Thim (Thai; POR); Yanuko (Ma.; JFM); Zakuro (Japan; TAN; USN).

Activities (Pomegranate):

Abortifacient (f; PH2; WBB); ACE Inhibitor (2; X11500191); Alpha-amylase inhibitor (1; X11223231); Alpha-Glucosidase Inhibitor (1; X15894133); Amebicide (1; HH2; PH2); Anodyne (f; BIB); Anthelmintic (1; HH2; KAP; PH2; SUW; VVG; WBB); Antiaging (1; X12570329); Antiatherogenic (12; X10799367; X12224378; X11500191); Antibiotic (1; VVG; WBB); Anticancer (1; X12002340); Antieicosanoid (1; X14585180); Antienteric (1; X15476301); Antifertility (1; MPI); Antiherpetic (1; X8679095; X15478204; JAF50:81); Antileukemic (1; X14585180); AntiMDR (1; X15882206); Antimutagenic (1; X12570329); Antimycobacterial (1; PR14:303); Antioxidant (1; X10799367); Antiprostaglandin (1; X14585180); Antiseptic (1; X10548758); Antispasmodic (f1; WOI); Antitubercular (1; WOI); Antiulcer (1; PR14:581); Antiviral (1; VVG; WOI; X11788838); Apopotic (1; X14585180); Aphrodisiac (f; BIB); Astringent (2; BIB; NPM; PH2; SUW; WBB); Bactericide (1; BIB; VGG; WBB; X10548758); Bechic (f; BOU); Cardiotonic (f; BIB; SUW; VOD; WOI); Catalase-genic (1; X15752628); Chemopreventive (1; X12002340; X14585180); CNS Stimulant (1; PHR); Collyrium (f; GHA; VOD); Contraceptive (f; JAF50:81); COX-2 Inhibitor (1; X14585180); Cytotoxic (1; HH2); Diuretic (1; VVG); Fungicide (1; MPI; WOI); Emmenagogue (f; WBB); Estrogenic (1; FNF; VOD); Febrifuge (1; HH2; NPM; SUW; VVG); Fungicide (1; MPI; WOI); Glutathione-peroxidase-genic (1; X15752628); Hemolytic (1; WOI); Hemostat (1; BIB; GHA); Hepatotoxic (1; VOD); Hypocholesterolemic (2; JNU); Hypoglycemic (1; HH2; VVG; X10837992; X15894133); Lipogenic (f; KAB); Lipoxygenase Inhibitor (1; X14585180); Molluscicide (1; X11050667); Nematicide (f; PH2); Orexigenic (f; KAB); Ornithine-Decarboxylase Inhibitor (1; X14585180); Paraoxonasigenic (1; JNU); Parastiticide (f; BIB); Pectoral (f; BOU); Phospholipase-A2 Inhibitor (1; X14585180); Radioprotective (1; X12570329; X15493960); Refrigerant (f; BIB; EFS; NPM; SUW); SOD-genic

(1; X15752628); Stimulant (f; BIB); Stomachic (1; DEP; MPI; SUW; WOI); Taenicide (f1; BIB; HH2; KAP; NAD; PH2; SUW; VOD); Uterorelaxant (f; BOU); Uterotonic (1; HH2; MPI); Vermifuge (f; BIB; NPM; SUW; VOD).

Indications (Pomegranate):

Abortion (f; SOU); Acne (1; JAF50:81); Ameba (1; X2131771); Amygdalosis (f; BIB); Anorexia (f; KAB); Asthma (f; BIB; VOD); Atherosclerosis (12; JNU; X10799367; X12224378; X11500191); Bacillus (1; X10548758); Bacteria (1; VVG; X2636992); Biliousness (f; BIB; KAB; NAD; ROE); Bleeding (f1; BIB; DEP; FNF); Bronchosis (1; BIB; KAB; KAP; MPI; NPM; WOI); Burn (f; GHA; NPM); Cancer (1; FNF; JLH); Cancer, abdomen (1; FNF; JLH); Cancer, anus (1; FNF; JLH); Cancer, breast (1; X12002340); Cancer, colon (f1; JLH; X16448212); Cancer, ear (1; FNF; JLH); Cancer, genital (1; FNF; JLH); Cancer, gum (1; FNF; JLH); Cancer, mouth (1; FNF; JLH); Cancer, neck (1; FNF; JLH); Cancer, prostate (1; X15744587); Cancer, skin (1; X14585180); Cancer, stomach (1; FNF; JLH); Cancer, throat (1; FNF; JLH); Cancer, uterus (1; FNF; JLH); Cancer, uvula (1; FNF; JLH); Candida (1; X12801361); Cardiopathy (f12; BIB; KAB; VOD; WOI; X11500191); Cerebrosis (f; BIB; KAB); Childbirth (f; JFM); Cholera (1; MPI; X8018898); Colic (f; BIB; KAB); Colitis (1; KAB; WBB); Condyloma (f; JLH); Conjunctivosis (f; BIB; IHB; KAP; MPI); Consumption (f; NAD); Corn (f; JLH); Cough (f; BIB; ROE); Cramp (f1; WOI); Dermatosis (f; GHA; IHB; ROE); Diabetes (1; VVG; X10837992; X15894133); Diarrhea (f1; BIB; EGG; HH2; KAP; NPM; PHR; PH2; SUW; VVG); Dysentery (f1; BIB; DEP; HH2; HJP; KAP; NPM; PHR; PH2; SUW; VVG); Dysmenorrhea (f; BIB); Dyspepsia (1; DEP; KAB; MPI; WOI); Earache (f; BIB); Endometriosis (f; MAD); Enterosis (f1; KAB; PH2; VOD; X15476301); Epistaxis (1; BIB; DEP; NAD; ROE); Escherichia (1; X10548758); Fever (1; BIB; HH2; VVG); Folliculosis (1; JAF50:81); Fungus (1; KAP; MPI; X2801361); Gall (f; JLH); Gastrosis (f; JLH; MPI; PH2); Gingivosis (f; JLH; KAB; VOD); Gonorrhea (f; WBB); Heartburn (f; MPI); Hemophilia (f; DEP); Hematuria (f; KAP; MPI; NAD); Hemoptysis (f; MPI; NAD); Hemorrhoid (f1; BIB; JLH; KAP; NAD; NPM; PHR; VOD); Hepatosis (f; GHA; KAB); Herpes (1; JAF50:81); High Cholesterol (2; JNU); HIV (1; Herpes (1; X8679095); Impotence (f; ROE); Infection (f1; MPI; VOD; X12636992); Infertility (f; JNU); Inflammation (f; BIB; NPM; VOD; WBB); Jaundice (f; GHA); Keratosis (f; BIB); Leukemia (1; X14585180); Leukorrhea (f; BIB; KAB; KAP; WBB); Malaria (f; BIB; KAB); Mastosis (f; BIB; JAF50:81); Melanoma (1; HH2); Menopause (1; FNF); Menorrhagia (1; BIB); Metrorrhagia (1; BIB); Mycosis (1; MPI; X12801361); Nausea (1; BIB); Nephrosis (f; KAB); Neurosis (f; MAD); Night sweats (f; BIB); Ophthalmia (f; BIB; KAB); Oxyuriasis (f; BIB); Pain (f; BIB; JFM); Paralysis (f; BIB); Pimple (f; BIB); Plague (f; BOU); Pneumonia (1; MPI); Proctosis (f; JLH; KAP); Prolapse (f; BIB; KAP); Pterygia (f; JLH); Pulmonosis (f; JFM; KAB); Rhinosis (f; JLH); Ringworm (1; PH2); Salmonella (1; MPI; X15476301); Scabies (f; BIB; KAB); Shigella (1; MPI); Snakebite (f; BIB); Sore (f; GHA; JFM; WBB); Sore Throat (f1; BIB; DEP; PHR; PH2; VOD); Splenosis (f; BIB; KAB); Staphylococcus (1; X2636992; X15882206); Stomachache (1; KAB; WBB; VVG); Stomatosis (1; BIB; JFM; KAB; X12801361); Swelling (f; ROE); Tapeworm (1; BIB; PHR; VVG); Thirst (f; NPM); Throat (f; ROE); Tonsilosis (f; VOD); Tuberculosis (f1; NAD; WOI; PR14:303); Tympanosis (f; JAF50:81); Ulcer (f1; BOU; PR14:581); Urogenitosis (f; BIB); Uterosis (f; DEP; JLH; KAP); Uvulosis (f; JLH); Vaginosis (f; BOU; JAF50:81); Venereal Disease (1; JAF50:81); Vertigo (f; ROE); Virus (1; VVG; WOI; JAF50:81); Vomiting (f; KAB; PH2); Wart (f; JLH); Water Retention (1; VVG); Whitlow (f; JLH); Worm (f1; BIB; KAP; PH2; SUW; VOD); Wound (f; EGG); Yeast (1; X10548758).

Dosages (Pomegranate):

FNFF = !!!

Fruits and cooked leaves food farmacy (FAC; JAD; TAN). The first sherbet may well have been a preparation of pomegranate juice mixed with snow. The acid pulp surrounding the seeds is the edible portion of the fruit, used as a salad or table fruit, or made into beverages or jellies. In Syria and Iran, fruit is cut open, seeded, strewn with sugar, and sprinkled with rose water. Wine is made from fruits, and seeds are used in syrups, preserves, gelatin desserts, icings, puddings, and sauces. As fruits ferment easily, they are used in Egypt to make a wine. Grenadine is a soft drink based on pomegranate, and grenadine syrup is used to flavor drinks (BIB). 1–2 g day (HHB). 7 g fl/300 cc water for inflamed mouth and throat (JFM). 4–5 g powdered flower (KAP). 1–3 g powdered root (KAP). 1–3 g powdered stem bark (KAP). 4–8 g powdered fruit (KAP). 1 part pericarp, root, or stem bark to 5 parts water (PH2). 250 parts powdered bark in 1500 parts water and boiled for 30 minutes (PH2).

- Asian Indians chew fruit rind with belleric for bronchosis and bronchorrhea (WOI).
- Ayurvedics use the fruit rind, appropriately enough, for diarrhea, dysentery, and worms; the root for worms; the flowers for epistaxis; the bark and seeds for bronchitis; and the ripe fruit, considered astringent, aphrodisiac and tonic, for biliousness, burning sensations, fever, heart disease, sore throat, and stomatitis (KAB).
- Cubans apply grated fruit rind to ulcers (JFM).
- Germans take 5–20 g bark as taenifuge (MAD).
- Haitians take the flower tea for asthma, the root and stem decoction for intestinal worms, and the rind infusion for diarrhea and dysentery (VOD).
- Iranians use powdered flowers with *Nummulites* sp. and *Rhus coriari*a for painful gums (BIB).
- Latinos take 5–12 g bark (RB or SB) in 240 cc water boiled until 1/3 gone; in 3 hourly doses on empty stomach 2 hours after taking 40 cc castor oil for tapeworms (JFM).
- Nepalese take 5 tsp rind juice 2 ×/day for diarrhea and dysentery (NPM).
- Peruvians take the bark tea as a genital tonic (EGG), using the fruit and bark for cancer and nasal polyps (JLH).
- Filipinos gargle with the leaf decoction for mouth problems (BIB).
- Puerto Ricans take juice sacs with mashed seeds to expel worms (JFM).
- North Africans (Cairo; Rabat) use fruit rind for bleeding, dentifrice, diarrhea, ulcers, with the decoction a specific vaginal plug for treating leukorrhea (BOU).
- Unani use the astringent bark for anal prolapse, colic, and piles; the flowers for biliousness, hydrocele, nausea, sore eyes, and sore throat; the green fruit for inflammation and keratitis; the ripe fruit for brain disorders, bronchitis, chest ailments, earache, scabies, sore eyes, sore throat, splenitis, and thirst; the seeds for biliousness, bowel ailments, hepatitis, liver ailments, nausea, scabies, and sore eyes (KAB).

Downsides (Pomegranate):

Bark Class 3. Pericarp Class 2d. Contraindicated with diarrhea; not to be taken with fats or oils when taken to kill parasites (AHP). Health hazards not known with proper administration of designated therapeutic dosages (PH2). Strong doses emetic, nauseant, and vertigogenic (JFM). Stronger doses (>80 g) may cause chills, collapse, dizziness, hematemesis, and visual disturbances, possibly even amaurosis and death (MAD; PH2). Even Mauritians believe the bark should be contraindicated in geriatrics, pediatrics, and pregnancy.

Extracts (Pomegranate):

Bark extract (with casuarinin, ellagitannin, and punicortein C cytotoxic to melanoma (ED50 = 2-– µg/ml, cf. <0.01 µg/ml for actinomycin) HH2. Extracts of abortive male flowers are hypoglycemic in diabetic rats (X10837992). Pomegranate juice has antiatherogenic effects in mice that may be attributable to its antioxidative properties (X10799367). Antiaging ellagitannin mixtures slowed aging-type mutations (X12570329). Huang et al. (2005) demonstrated that pomegranate flower extracts improve cardiac lipid metabolism in diabetic rat models (X15880139). Voravuthikunchai and Kitpipit (2005) (X15882206) found that ethanolic extracts inhibited all of 35 hospital isolates of methicillin-resistant *Staphylococcus aureus*. MIC = 200–400 mg/ml (X15882206). Sudheesh and Vijayalakshmi (2005) demonstrated that flavanoid-rich fruit fractions (10 mg/kg/day orl rat) had antiperoxidative potential, decreasing liver levels of malondialdehyde, hydroperoxides, and conjugated dienes, increasing activities of catalase, superoxide dismutase (SOD), glutathione peroxidase, and glutathione reductase. Tissue concentrations of glutathione also increased (X15752628). Azadzoi et al. (2005), studying oxidative stress in arteriogenic erectile dysfunction (ED), found pomegranate was the better free radical scavenger. The rabbit model of arteriogenic ED demonstrated decreased intracavernous blood flow, erectile dysfunction, loss of smooth muscle relaxation, decreased endothelial NOS and neuronal NOS, increased inducible NOS expression, and diffused cavernous fibrosis. Long-term pomegranate juice intake increased intracavernous blood flow, and improved erectile response and smooth muscle relaxation. Antioxidant therapy could help prevent smooth muscle dysfunction and fibrosis in ED, Erectile Dysfunction (X15947695).

TABOR OAK (*QUERCUS ITHABURENSIS* DECNE) ++ FAGACEAE

Synonyms:

Quercus aegilops var. *ithaburensis* Decne

Notes (Tabor Oak):

> **They sacrifice upon the tops of the mountains, and burn incense upon the hills, under oaks and poplars and elms, because the shadow thereof is good: therefore your daughters shall commit whoredom, and your spouses shall commit adultery.**
>
> **Hosea 4:13 (KJV)**

> **They sacrifice on the tops of the mountains, and make offerings upon the hills, under oak, poplar, and terebinth, because their shade is good. Therefore your daughters play the harlot, and your brides commit adultery.**
>
> **Hosea 4:13 (RSV)**

> **On the tops of the mountain they sacrifice and on the hills they make sacrificial smoke, under massive tree and storax tree and big tree because its shade is good. That is why your daughters commit fornication and your own daughters in law commit adultery.**
>
> **Hosea 4:13 (NWT)**

Going with the flow in my first crack at *Medicinal Plants of the Bible*, I selected the three oak species that other non-Israeli writers had selected as probably representing the oak of the Bible. Now armed with Zohary's *Flora of Palestine* (FP1, 1966) and Zohary's *Plants of the Bible* (ZOH, 1992), I will go with the Israeli flow. Zohary notes that of some 500 oak species worldwide, there are only three species that occur in Israel. They can be keyed as follows:

- Leaves evergreen, 2–4 cm long, sometimes prickly: *Q. calliprinos*
- Leaves deciduous, 4–10 cm long, margin not prickly, although dentate:
  - — Adult leaves glabrous on both sides; cupule circa 1 cm diameter: *Q. boissieri*
  - — Adult leaves tomentose below; cupules broader than 1 cm diameter: *Q. ithaburensis*

Zohary favors the deciduous Tabor Oak (*Q. ithaburensis*) and the evergreen oak (*Q. calliprinos*) as the rendition of the Hebrew *allon* and *elon*. The evergreen oak was discussed in my *Medicinal*

Plants of the Bible, under the name *Quercus coccifera*. Under good environmental conditions, both can grow to huge trees, symbolic of longevity, power, pride, and splendor. They were often involved in burials, offerings, and religious, reverent, and ritual customs and worship. And the wood had many uses. Zohary notes that "many translators and exegetes, unacquainted with the flora of the Holy Land, and embarrassed by the frequent occurrence in the Bible of *elah*, *elon*, *el, alah*, and allon, have seriously misapplied these names…. There are too many variations in the translations of different authors, and (as in the RSV) much inconsistency even within any given translation." (ZOH) Zohary concludes that in general *allon* and *elon* should be rendered as *oak*, and *elah* and *alah* should be rendered as terebinth.

Common Names (Tabor Oak):

Allon (Heb.; ZOH); Elon (Heb.; ZOH); Tabor Oak (Eng.; ZOH).

Activities (Other Oaks):

Anthelmintic (1; PH2); Antiinflammatory (2; KOM; SHT); Antiperspirant (1; APA; MAD); Antiseptic (1; APA; PNC); Antitumor (1; FAD); Antiviral (1; SHT); Astringent (f1; APA; MAD; SHT; VAD); Bactericide (1; BGB); Carcinogenic (1; FAD); Cicatrizant (f1; VAD); Diuretic (f1; VAD); Emetic (f; PED); Expectorant (1; BGB); Hemostatic (f1; APA; PNC; VAD); Immunostimulant (1; PHR); Litholytic (1; BGB); Vasoconstrictor (f1; VAD); Vulnerary (1; APA).

Indications (Other Oaks):

Adenopathy (f; JLH); Albuminuria (f; MAD); Anemia (f; MAD); Angina (f; MAD); Apoplexy (f; MAD); Asthma (f; MAD); Bacteria (1; VAD); Bleeding (f1; PH2; MAD; VAD); Blennorrhagia (f; MAD); Blepharosis (f; VAD); Bronchosis (2; MAD; PHR; PH2); Bruise (1; APA); Burns (f; FAD; HJP); Cacoethes (f; JLH); Callus (f; JLH); Cancer (f; FAD; JLH); Cancer, anus (f; JLH); Cancer, brain (f; JLH); Cancer, breast (f; JLH); Cancer, colon (f; JLH); Cancer, ear (f; JLH); Cancer, gum (f; JLH); Cancer, intestine (f; JLH); Cancer, lip (f; JLH); Cancer, liver (f; JLH); Cancer, mouth (f; JLH); Cancer, neck (f; JLH); Cancer, nose (f; JLH); Cancer, stomach (f; JLH); Cancer, tonsil (f; JLH); Cancer, uterus (f; JLH); Cancer, uvula (f; JLH); Carbuncle (f; MAD); Chilblain (f; VAD); Chilblains (1; APA); Chlorosis (f; MAD); Cirrhosis (f; MAD); Cold (2; PHR); Colitis (f; VAD); Consyloma (f; JLH); Conjunctivosis (1; APA; PH2; VAD); Cough (2; PHR; PH2); Cystosis (f; VAD); Debility (f; MAD); Dermatosis (f12; BGB; KOM; PH2; SHT); Diarrhea (f12; APA; KOM; MAD; PED; PH2; SHT; VAD); Dysentery (1; BBG; BIS; FAD); Dysmenorrhea (f; VAD); Dysuria (f; VAD); Eczema (1; APA; MAD; PH2; VAD); Encephalosis (f; JLH); Enterocolitis (1; APA; BIS); Enterosis (1; APA; MAD; VAD); Enuresis (f; MAD); Epistaxis (f; VAD); Fever (f12; HJP; PHR; VAD); Fibroma (f; JLH; MAD); Gastroenterosis (1; BIS); Gastrosis (f1; MAD; VAD); Genitalitis (2; APA; KOM); Gingivosis (f1; APA; JLH); Gout (f; MAD); Hematuria (1; MAD); Hemoptysis (1; MAD); Hemorrhage (1; BGB); Hemorrhoid (f1; APA; PED; PH2; PNC; VAD); Hepatosis (f; JLH; MAD); Hyperhidrosis (f; PH2); Incontinence (f; VAD); Induration (f; JLH); Infection (12; APA; PHR); Inflammation (12; BGB; VAD); Intertrigo (f; MAD; PH2); Itch (1; APA); Kidney stone (f; BGB); Leukorrhea (1; BGB; MAD); Malaria (f; BGB); Marasmus (f; MAD); Mastosis (f; GAZ); Metrorrhagia (f; VAD); Mucososis (1; APA); Mumps (f; VAD); Nephrosis (f; MAD); Nipple (f; GAZ); Pain (f; BGB; JLH); Periodontosis (f1; VAD); Pharyngosis (f12; KOM; PH2; VAD); Phthisis (f; MAD); Poison Ivy (f; FAD); Polyp (f; JLH); Prolapse (f; MAD); Proctitis (f12; APA; JLH; KOM; MAD; PH2); Psoriasis (f; MAD); Rheumatism (f; MAD); Rhinosis (f; JLH); Scrofula (f; MAD); Sore (f1; APA; HJP); Sore Throat (f1; APA; BGB; PNC; VAD); Splenosis (f; MAD); Stomatosis (f12; APA; KOM; MAD; PH2; VAD); Stone (f; VAD); Swelling (f; JLH); Tonsilosis (f; JLH; MAD); Ulcer (f; MAD); Urethrosis (f; VAD); Uterorrhagia (1; PH2); Uterosis (f; MAD; PH2); Uvulosis (f; JLH);

Vaginosis (1; BGB; VAD); Varicosities (1; APA; PH2); Virus (1; PH2); Wart (f; JLH); Whitlow (f; JLH); Worm (f; PH2); Wound (f1; HJP).

Dosages (Other Oaks):

FNFF = !
Most people disdain acorns as food although they have been mainstays in some cultures. Acorn-fed pork from Spain is world renowned. 1 tsp (circa 3 g) bark /cup water/day (APA); gargle with 2 tsp bark/pt water (APA); 3 g dry bark (KOM); 1–1.5 tsp (≈4-6 g) bark/day in hot or cold tea (MAD). 5 g bark/l water for baths; 20 g bark/l water for compresses; 2–4 Tbsp fresh bark (PED); 3–6 g dry bark (PED); 4.5 g dry bark:22 ml alcohol in 23 ml water (PED); 0.5–5 ml liquid bark extract (PNC).

Downsides (Other Oaks):

Class 2d. Contraindicated even for external use when skin is extensively damaged, especially with weeping eczema, febrile and infectious disorders, and cardiac insufficiency states III and IV; hypertonia stage IV (AHP, 1997). Uterosis (f; PH2); Commisson E reports for bark, interactions: reduced absorption of alkaloids, and other basic substances (AEH). Should not be taken for more than 3 or 4 days (SHT).

Natural History (Other Oaks):

The oak from which the "scarlet" of the Old Testament derived was probably the evergreen shrub, now known as *Quercus calliprinos*, or Kermes oak, attaining 10 to 20 feet. Its young shoots are covered with white, soft down, the breeding grounds of the kermes insect, *Chermes ilicis* (*Coccus ilicis*). These creatures yield a beautiful, rich, long-lasting dye. I would bet, but do not know, that galls would contain more tannins and medicinally active anthocyanins. The scarlet was known commercially as "grain" and "scarlet grain." When the bark is steeped in boiling water, it can yield a black dye, once used to dye hair. The Dyers Company of England selected three sprigs of this plant for their heraldic crest, granted by charter in 1420 and perhaps still used by the company (BIB).

ASIAN BUTTERCUP (*RANUNCULUS ASIATICUS* L.) + RANUNCULACEAE

Notes (Asian Buttercup):

> **But go into a field of flowers, where no house is builded, and eat only the <u>flowers of the field</u>; taste no flesh, drink no wine, but eat flowers only.**
>
> **2 Esdras 9:24 (KJV)**

This is one of 18 species covered in the *Flora of Palestine*. Zohary says this is the most showy crowfoot in all of Israel, and the one most likely to penetrate into arid areas. It is one of few with crimson flowers, most being yellow, a few being white with yellow. He says it can be regarded as one of the "flowers of the field." Although I have seen reference to eating some flowers in this family, I consider that somewhat risky because of the almost universal presence of caustic or vesicant anemonin or protoanemonin. I will not go with the edict in Esdras. Still, Tanaka mentions several "edible species" in the genus *Ranunculus*, nowhere specifying flowers. Facciola mentions four species, adding that flowers of *Ranunculus bulbosus* are pickled; flower buds of *Ranunculus ficaria* (also in Israel) make a good substitute for capers. I think of this more as an ornamental than a medicine, and certainly not a food plant, even if the Bible recommends eating the "flowers of the field." There are many other safer flowers in my fields. Finding no medicinal reports regarding this lovely species, I append some generic information, derived from other species of the same genus, *Ranunculus*, after the few common names I find reported for the Asian buttercup.

Common Names (Asian Buttercup):

Asian buttercup (Eng.; USN); Persian Buttercup (Eng.; USN); Persian Crowfoot (Eng.; USN).

Activities (Other Buttercups):

Allergenic (1; FNF); Anodyne (f; WOI); Anticancer (f; DAA); Antiscorbutic (f1; EFS; FNF); Antiseptic (f1; PNC; WOI); Antispasmodic (f; WOI); Antiviral (1; WOI); Astringent (f; EFS; PNC); Bactericide (1; WOI); Candidicide (1; WOI); Cyanogenic (f; EB30:403); Diuretic (f; WOI); Emmenagogue (f; EFS; SKJ); Fungicide (f1; PNC; WOI); Herbicide (f; GMH); Irritant (1; FNF); Lachrymatory (1; WOI); Lactafuge (1; WOI); Lactagogue (f; EFS; SKJ); Poison (1; DEP; WOI); Rubefacient (1; WOI); Sialogogue (1; PH2); Stimulant (f; DAW); Stomachic (f; WOI); Tonic (f; KAB; WOI); Toxic (f; EFS); Trypanocide (1; FT73:569); Vasoconstrictor (1; WOI); Vermifuge (1; WOI); Vesicant (1; WOI); Viricide (1; WOI).

Indications (Other Buttercups):

Abscess (f; DEM; GMH); Arthrosis (f; FAD; PH2); Asthma (f; WOI); Bacillus (1; DAA); Bacteria (1; WOI); Bleeding (f; PH2); Blister (f; PH2); Boil (f; DEM; FAD); Bronchosis (f; HHB; PH2); Cancer (f; JLH; MIC); Cancer, breast (f; AAH; JLH); Cancer, mouth (f; JLH); Cancer, spleen (f; JLH); Candida (1; WOI); Cold (f; DAA; DEM); Condyloma (f; DAW); Conjunctivitis (f; AAH; HHB; NPM); Consumption (f; GMH); Corn (f; AAH; JLH); Cramp (f; WOI); CVI (f; BRU); Debility (f; DAA); Dermatosis (f; PH2); Diptheria (1; WOI); Dropsy (f; MAD); Dysuria (f; WOI); Escherichia (1; WOI); Flu (f; WOI); Grippe (f; WOI); Fungus (f1; PNC; WOI); Gastrosis (f; NPM); Glossosis (f; MAD); Gingivosis (f; PH2); Gonorrhea (f; HHB); Gout (f; HHB; MAD); Halitosis (f; WOI); Headache (f; MIC); Hemorrhoid (f; AAH; BRU; HAD; MAD; PNC); Hemiplegia (f; MAD); Hepatosis (f; MAD); Induration (f; JLH); Infection (f1; PNC; WOI); Insomnia (1; WOI); Leukoderma (f; HHB; PH2); Mange (f; MAD); Mastosis (f; JLH); Measles (f; AAH); Mucososis (f; HHB); Mycosis (f1; PNC; WOI); Myosis (f; HHB; PH2); Nephrosis (f; KAB; MAD; WOI); Neuralgia (f; FAD; HHB; PH2); Ophthalmia (f; AAH); Pain (f; WOI); Pancreatosis (f; MAD); Phobia (f; MIC); Pleurisy (f; KAB); Pleuritis (f; HHB); Pneumonia (f; WOI); Rheumatism (f; FAD; HHB; PH2); Scabies (f; HHB; PH2); Sciatica (f; WOI); Scurvy (f1; FNF; PH2); Sniffles (f; MAD); Sore (f; JLH); Sore Throat (f; GMH); Spasm (f; WOI); Spermatorrhea (f; DAA); Staphylococcus (1; WOI); Stitch (f; KAB); Stomatosis (f; JLH; MAD); Streptococcus (1; DAA); Swelling (f; PH2); Thrush (f1; DEM; WOI); Tumor (1; DAA); Venereal Disease (f; HHB); Virus (1; WOI); Wart (f; AAH; JLH); Wen (f; JLH); Worm (1; WOI); Wound (f; PH2); Yeast (1; WOI).

Dosages (Other Buttercups):

FNFF = ?!

Swedes sometimes ate the leaves as a boiled potherb (GMH). Young leaves of some are eaten raw in salads and sandwiches or cooked as a potherb. Bleached stems are cooked and eaten. Bulbils, which form in the leaf axils and roots, can be cooked. Flower buds even said to substitute for capers. I do not recommend any member of the buttercup family for food, although others may do so. 2–5 g herb, or in tea, 3 ×/day (CAN); 2–5 ml liquid herbal extract (1:1 in 25% ethanol) 3 ×/day (CAN); 3% topical ointment (CAN); 30% fresh pilewort in benzoinated lard (CAN). 2.5–5 ml liquid extract (PNC).

- British apply root decoction as hot compress or mild ointment to hemorrhoids (AAH).
- British from Kent use the leaves for cleaning teeth (AAH).
- British from Norfolk use the floral tea to treat sore eyes and measles (AAH).
- British Highlanders put the roots under the arms to treat breast cancer or nodules (AAH).
- British make oil, ointment, or salve of leaf or root decoction for hard wens or tumors (JLH).
- Europeans use the root for cancers of the mouth (JLH).

Downsides (Other Buttercups):

In view of the data, excessive use, especially during lactation and pregnancy, should be avoided. Not recommended for internal consumption (CAN). Very acrid, causing burning of mouth and mucous membranes; blistering skin. Avoid (FAD).

Natural History (Other Buttercups):

Bright burnished petals have honey sacs at their base, but flowers may come out earlier in spring than most bees; hence, few flowers get pollinated, at least in Britain. Where upper leaves (of *R. bulbosus*) meet the stem, one may find minute bulbular propagules the size of a grain of wheat, which may float away with rains (GMH). Some 250 years ago, Linnaeus advised farmers to eradicate the weed because it was unattractive to cattle and because of its herbicidal activities. (Treating with coal-ash or wood ash was said to destroy the plant.) Underground parts reportedly work against the fungus *Venturia inaequalis*. A fungal plant pathogen, *Sclerotinia sclerotiorum* occurs at least in Norwegian populations (X11742545).

Extracts (Other Buttercups):

Protoanemonin antibacterial (against Gram-positive and Gram-negative bacteria, Candida, diptheria toxin, *Escherichia*, *Staphylococcus*), antiviral, cytopathogenic, vesicant, and vermicide (WOI). Closely related *R. sceleratus* contains the vasoconstrictor serotonin, with six other tryptamine derivatives and "two unidentified anti-5-hydroxytryptamine derivatives." (ZOH)

WHITE BROOM (*RETAMA RAETAM* FORSSK.) WEBB & BERTHEL.) + FABACEAE

Notes (White Broom):

> **But he himself went a day's journey into the wilderness, and came and sat down under a *juniper* tree: and he requested for himself that he might die; and said, It is enough; now, O LORD, take away my life; for I am not better than my fathers.**
>
> **I Kings 19:4 (KJV)**

> **But he himself went a day's journey into the wilderness, and came and sat down under a broom tree; and he asked that he might die, saying, "It is enough; now, O LORD, take away my life; for I am no better than my fathers."**
>
> **I Kings 19:4 (KJV)**

> **And he himself went into the wilderness a day's journey into the wilderness, and at length came and sat down under a certain broom tree. And he began to ask that his soul might die, and to say, "It is enough; now, O Jehovah, take my soul away; for I am no better than my fore fathers."**
>
> **I Kings 19:4 (NWT)**

It is called juniper in the KJV, broom in the RSV, a certain broom in the NWT, and white broom by Zohary, who adds that it is a common tall shrub in the Arabian, Israeli, and Saharan deserts. Bedouins indicate their respect for plants "by the life of the plant and our worshiped Lord." Small wonder they prohibit the felling of desert shrubs like *Acacia*, *Pistacia*, and *Retama*. In some places, *Retama* is the only shade-casting tree on the desert. And it makes the finest charcoal, which burns with intense heat.

Arabs claim it holds its heat for a year. In the Cairo market, it fetches a much higher price than any other fuel species. The expressions "coals of Juniper" used in Psalms 120, "burning coals," "live brown coals," "coals of broom," and "coals that lay waste" in biblical books indicate the popularity of the wood for charcoal. One legend suggests that when Jesus was praying in Gethsemane, he was disturbed by the cracking of the broom in the breeze. When finally led off by the soldiers, he said to the broom: "May you always burn with as much noise as you are making now." Another legend has it that the crackling of broom plants among which they hid almost revealed Mary and baby Jesus to Herod's soldiers. The branches are used in desert homes as coarse cords. At weddings and other ceremonies, Bedouins fasten sprigs of green plants like white broom to the tent entrance. Green, the color of live plants, is a symbol of life and vitality. The roots are used to insulate the handles of Bedouin coffee pots. Bedouins use the plant to make pins that fasten their curtains and as pins for their camel saddles (BIB).

Common names (White Broom):

Aligo (Ber.; BOU); Alouga (Ber.; BOU); Besliga (Arab.; BOU); Retem (Arab.; BOU); Retem Behan (Arab.; BOU); Rothem (Heb.; ZOH); Telit (Ber.; BOU); Tselgoust (Ber.; BOU); White Broom (Eng.; ZOH).

Activities (White Broom):

Abortifacient (f; UPH); Anticarcinomic (1; X15305322); Antidiabetic (1; X15852497); Antioxidant (1; X15305322); Cytotoxic (1; X15305322); Diuretic (1; X15848016); Ecbolic (f; BIB); Hypoglycemic (1; X15852497); Lipolytic (1; X15013197); Poison (f; BIB); Purgative (f; BIB; UPH); Toxic (f; BOU); Vermifuge (f; BIB; UPH); Vulnerary (f; BOU).

Indications (White Broom):

Amenorrhea (f; BIB; UPH); Carcinoma (1; X15305322); Constipation (f; BIB); Diabetes (1; X15852497); Diarrhea (f; BOU); Obesity (1; X15013197); Ophthalmia (f; BOU); Worm (f; BOU); Wound (f; BOU).

Dosages (White Broom):

FNFF = ?

- Bedouins grind and heat branches and green leaves over live coals until hot, place them in a thin cloth, and apply to arthritic pain (BIB).
- Bedouins grind, powder, and apply the plant to wounds (BIB).
- North Africans use the root against diarrhea, the branches for fever and wounds (BOU).
- Palestinians make a collyrium from the branches to wash their eyes (BIB).

Natural History (White Broom):

The pea-like flowers are pollinated by bees (ZOH). The exceedingly long roots enable the plant to tap deep water sources, so it thrives in the dry season; the twigs are photosynthetic (ZOH).

Extracts (White Broom):

Moroccan scientists (Maghrani et al., 2005) demonstrated an acute diuretic effect of the aqueous extract in rats. (5mg/kg/h ivn rat) Furosemide at 0.1mg/kg/h had similar effects (X15848016). Aqueous extracts showed lipid-and body-weight-lowering activities in rats after repeated oral administration at 20 mg/kg (X15013197).

PALESTINE BUCKTHORN (*RHAMNUS PALAESTINA* BOISS) + RHAMNACEAE

NOTES (PALESTINE BUCKTHORN):

He that diggeth a pit shall fall into it; and whoso breaketh an hedge, a serpent shall bite him.

Ecclesiastes 10:8 (KJV)

He who digs a pit will fall into it; and a serpent will bite him who breaks through a wall.

Ecclesiastes 10:8 (RSV)

He that is digging a pit will himself fall right into it; and he that is breaking through a stone wall, a serpent will bite him.

Ecclesiastes 10:8 (NWT)

There is no plant mentioned as such in the RSV and NWT, but that "hedge" mentioned in the KJV version has been hinted to represent *Rhamnus palaestina*, a little-known shrub akin to the well-known buckthorns and cascara sagradas, and probably sharing many of their chemicals and activities. That has prompted me to do another generic compilation showing the activities and indications attributed to other species of the genus *Rhamnus*. Like many other thorny shrubs of the Bible, this one has been used as a fence to exclude grazing animals from gardens and orchards (BIB).

COMMON NAMES (PALESTINE BUCKTHORN):

Palestine Buckthorn (Eng.; BIB); Hedge (Eng.; BIB). Ncsn.

ACTIVITIES (OTHER RHAMNUS):

Antiherpetic (1; APA; HH2); Antileukemic (1; APA); Antiseptic (1; APA; DEM); Antispasmodic (f; PED); Antitumor (1; FNF; HOX; PNC); Antiviral (1; APA; HH2); Bactericide (1; HH2); Bitter (1; PED); Cathartic (f12; APA; DEM; KOM; PED); Collyrium (f; DEM); Depurative (f; DEM; HHB; MAD; PHR; PH2); Diuretic (f; EFS; HHB; MAD; PHR; PH2); Emetic (f1; APA; DEM; EFS); Fungicide (1; FNF); Hepatotonic (f; PED); Hydrogogue (1; PH2); Laxative (12; APA; EFS; HH2; KOM; PNC; PH2; SKY); Peristaltic (f12; PHR; PNC); Poison (f; DEM); Prostaglandigenic (1; PH2); Purgative (2; FNF); Sunscreen (f; APA); Tonic (f; DEM; PNC).

INDICATIONS (OTHER RHAMNUS):

Anemia (f; MAD); Appendicitis (f; MAD); Arthritis (f; DEM); Asthma (f; MAD); Bacteria (1; HH2); Biliousness (f; DEM; FEL); Cachexia (f; MAD); Cancer (f1; APA; FNF; HOX; JLH; PNC); Catarrh (f; FEL); Chickenpox (1; APA); Chlorosis (f; MAD); Colic (f; MAD); Constipation (2; FEL; KOM; PHR; SKY); Diarrhea (f; MAD); Dropsy (f; MAD); Duodenosis (f; FEL); Dysentery (f; DEM); Dyspepsia (1; FEL; PNC); Exanthema (f; MAD); Flu (1; APA); Gas (f; PED); Gastrosis (f; DEM; FEL); Gonorrhea (f; DEM); Gout (f; MAD); Headache (f; FEL); Hemorrhoids (12; KOM; PHR; PH2; PNC); Hepatosis (f; DEM); Herpes (f1; APA; HH2; MAD); Infection (f; DEM); Itch (f; DEM); Jaundice (f; FEL); Nausea (f; MAD); Obesity (f; MAD); Ophthalmia (f; DEM); Proctosis (f; PH2); Rheumatism (f; DEM; MAD); Ringworm (1; FNF); Sciatica (f; DEM); Shingles (1; APA); Sore (f; DEM; MAD); Stomatosis (f; MAD); Swelling (f; DEM); Uremia (f; MAD); Venereal Disease (f; DEM); Virus (1; HH2); Water Retention (f; HH2); Worm (f; DEM); Wound (f; DEM).

Dosages (Other Rhamnus):

FNFF = ?
One-half tsp powdered bark/cup water, morning and/or night, for short periods (APA); 1/2 tsp tincture (APA); 2–6 ml fluid extract (APA); 2–5 ml liquid bark extract (CAN, PNC); 1 g bark (HHB). Two 450-mg capsules at bedtime (NH); 1–3 g dry bark (PED); 2 g dry bark:10 mg alcohol/10 mg water (PED); 1–2.5 g powdered bark (PNC); 100–300 mg dry bark extract (PNC); 1 ml (circa 10 drops) StX (fluid); 1–5 ml tincture (SKY).

Downsides (Other Rhamnus):

Class 2b, 2c, 2d. Contraindicated in hemorrohoids and nephropathy (CAN), intestinal obstruction, abdominal pain of unknown causes, any enteritis (appendicitis, colitis, Crohn's disease, irritable bowel syndrome), nephropathy, and menstruation (AHP, 1997). Not for children under 12 years old. Do not use bark that has not been heat-treated or aged 1 year (AHP, 1997). Do not use if you have abdominal pain or diarrhea. Consult a health care provider prior to use if pregnant or nursing. Discontinue use in the event of diarrhea or watery stools. Do not exceed recommended dose. Not for long-term use (AHP). While widely used, anthranoid-containing laxatives can be habit-forming; some contain compounds suspected of being cytotoxic, genotoxic, mutagenic, and even tumorigenic; epidemiological studies in Germany reveal that abusers of anthranoid laxatives have 3 times higher rate of colon carcinoma (AEH, 115). Newall, Anderson, and Phillipson (1996) caution that anthraquinones are purgative and an irritant to the GI tract. Because of the anthraquinones, nonstandardized preparations should be avoided in pregnancy and lactation (CAN). "Some herbal laxative preparations such as cascara and senna, for example, can cause an increase in the potency of digoxin" (D'epiro, 1997). Chronic use may lead to hypokalemia, increasing efficacy, perhaps dangerously, of cardiac glycosides, perhaps antiarrhythmics. Hypokalemia can be increased with corticosteroids, diuretics, and licorice roots (KOM).

Extracts (Other Rhamnus):

Cathartic cascarosides induce the large intestine to increased peristalsis, inducing bowel movement. Clinical comparison of patients preparing for colonoscopy showed that GoLytely alone and *Senna* alone with enema did better than a mix of GoLytely with Cascara. Cascara was the last choice as far as cleanliness and quality of the exam (PH2.)

CASTOR (*RICINUS COMMUNIS* L.) + EUPHORBIACEAE

Notes (Castor):

And the LORD God prepared a gourd, and made it to come up over Jonah, that it might be a shadow over his head, to deliver him from his grief. So Jonah was exceeding glad of the gourd.

Jonah 4:6 (KJV)

And the LORD God appointed a plant, and made it come up over Jonah, that it might be a shade over his head, to save him from his discomfort. So Jonah was exceedingly glad because of the plant.

Jonah 4:6 (RSV)

Accordingly Jehovah God appointed a bottle-gourd plant, that it should come up over Jonah, in order to become a shade over his head, to deliver him from his calamitous state. And Jonah began to rejoice greatly over the bottle gourd plant.

Jonah 4:6 (NWT)

FIGURE 1.89 Castor (*Ricinus communis*).

We may never know which version of Jonah 4:6 is botanically more accurate. Castor can become a big shade tree; bottle gourds trained on a trellis can cast shade. Zohary thinks that castor is most likely. The huge leaves of this plant are excellently adapted for producing ample shade when growing alongside a bower, booth, or hut or overhanging a bench. Castor bean is cultivated for the seeds, which yield a fast-drying, non-yellowing oil, used mainly in industry and medicines. Consistent with the classical writers, including Strabo, Pliny, and Theophrastrus, Copley et al. (2005) found castor oil among the lighting oils used in archaeological lamps from Egypt, along with animal, flaxseed, radish (or some crucifer), and sesame oils (X15912234). The oil was extensively used also by Hebrews, as one of the five oils sanctioned by rabbinical tradition. Oil used in coating fabrics and other protective coverings, in the manufacture of high-grade lubricants, transparent typewriter and printing inks, in textile dying (when converted into sulfonated Castor Oil or TurRed Oil, for dying cotton fabrics with alizarine), in leather preservation, and in the production of Rilson, a polyamide nylon-type fiber. Dehydrated oil is an excellent drying agent that compares favorably with tung oil and is used in paints and varnishes. Hydrogenated oil is utilized in the manufacture of waxes, polishes, carbon paper, candles, and crayons. Blown oil is used for grinding lacquer paste colors; and when hydrogenated and sulfonated, it is used for the preparation of ointments. South Africans mix castor oil with kerosene as a culicide; the oil prevents tabanid flies from attacking camels. Castor oil pomace, the residue after crushing, is used as a high-nitrogen fertilizer. The pomace is said to induce asthma among individuals who inhale it. Although it is highly toxic due to the ricin, a method of detoxicating the meal has now been found, so that it can safely be fed to livestock. Stems are made into paper and wallboard. Moldenke and Moldenke remarked that neither the ancient Hebrews nor modern inhabitants of Palestine and Syria use it for medicine (BIB). Egyptians speak of *kaka* as the plant source of the *kiki* oil, mentioned by Herodotus as used for lighting. Castor has been found in 6000-year old Egyptian tombs. Zohary states that "The Talmud sometimes refers to *kikayon* as a plant yielding the castor oil long known in medicine" (ZOH). It has even worked its way into Vodou, associated with the deities Yemanjá and Omolu (Abaluaiê) in the Afro-Brazilian Candomblé religion, wherein the seed oil is used as a purgative (VOD). Wish my mother had read the warning — "not to be administered to children under 12 years" — and I might still like orange juice; too early in life, my mother assumed that castor oil, cut with orange juice, was a panacea. So to this day, I often imagine the taste of castor oil with my orange juice. Judi duCellier, my secretary for three decades, once took some castor bean seeds home to poison some moles. Judi's grandson chewed on one of those seeds, but was given ipecac immediately after Judi called me for advice that frightening Sunday morning.

Common Names (Castor):

Aa Ma (Newari; NPM); Aaril (Nepal; NPM); Akhilwane (Ber.; BOU); Alama (Nepal; SUW); Amanakkam-chedi (Tam.; NAD); Amanakku (Tam.; NAD; WOI); Amidamu (Tel.; NAD); Amudam (Tel.; NAD); Amudamuchettu (Tel.; WOI); Ander (Nepal; SUW); Andela (Nepal; SUW); Andi (Danuwar; Mooshar; NPM); Arand (Pun.; NAD); Aranda (Beng.; NAD); Arash (Arab.; GHA); Arend (Nepal; NPM); Arer (Nepal; NPM); Areth (Chepang; NPM); Ater (Chepang; NPM); Audla (Kan.; NAD); Avanakku (Mal.; WOI); Avend (Nepal; SUW); Awrioun (Arab.; BOU); Ayrunkukri (Sin.; NAD); Bazanjir (Afg; KAB); Bedanjir (Iran; NAD); Bheranda (Beng.; WOI); Bherenda (Beng.; NAD); Bi Ma (Pin.; AH2); Bi Ma Gen (Pin.; AH2); Bi Ma Ye (Pin.; AH2); Bi Ma You (Pin.; AH2); Bi Ma Zi (Pin.; AH2; DAA); Bois de Carapat (Fwi.; AVP); Buzanjir (Afg; KAB); Carapate (Guad.; AVP; TRA); Carrapa (Sp.; AVP); Carrapateira (Mad.; Por.; AVP); Castor (Eng.; Scn.; AH2; AVP; VOD); Castor Bean (Eng.; CR2; VOD); Castor Oil Plant (Eng.; AVP); Catapuzia Maggiore (It.; AVP); Causirro (Bol.; Chiriguano; DLZ); Cawapat (Dom.; TRA); Chittamanakku (Tam.; NAD); Chittmani (Tam.; NAD); Coch (Ma.; JFM); Daldo (Rai; NPM); Dandarobi (Tamang; NPM); Dan-khra (Tibet; NPM); Dar-ta (Tibet; NPM); Diveli (Bom.; Guj.; NAD); Diveligo (Guj.; WOI); Djarak Malkarone; (Arab.; AVP); Endaru (Sin.; NAD); Endi (Hindi; NAD); E-ra (Tibet; NPM); Eramudapu (Tel.; NAD); Eranda (Ayu.; Sanskrit; AH2; NAD); Erandi (Hindi; Mar.; WOI); Erandthailam (Tel.; NAD); Erendi (Guj.; Mah.; NAD); Eri

(Assam; NAD); Feni (Ber.; BOU); Feuille Grain (Haiti; AVP); Feuille Mascriti (Haiti; AVP); Figo do Inferno (Por.; AVP); Gab (Uriya; NAD); Gandharva Hastah (Sanskrit; NAD); Girgilla (Peru; SOU); Gourd (Eng.; BIB); Haralu (Kan.; WOI); Hedera (Heb.; ZOH); Herani (Sin.; NAD); Higuera del Diablo (Ma.; Sp.; JFM); Higuera Infernal (Mex.; Pan.; AVP; MPG); Higuerilla (Sp.; AVP; EGG); Higuerilla de la Tierra (Ma.; JFM); Higuerilla Mexicana (Ma.; JFM); Higuerillo (Ecu.; Sp.; AVP; BEJ); Higuerita (Sp.; AVP); Higuero (Sal.; AVP); Huile Mascriti (Haiti; AVP); Huiso Mero (Peru; Shipibo/Conibo; EGG); Iguerilla (Ma.; JFM); Indeyo (Raute; NPM); Jambalin (Nic.; IED); Kaka (Arab.; Egypt; ZOH); Kesusi (Burma; NAD); Kharvah (Tur.; AVP); Kharwah (Arab.; GHA); Kherwa' (Arab.; BOU); Khirva (Arab.; NAD); Khirwah (Arab.; GHA); Khurwa'a (Arab.; GHA); Kikayon (Heb.; ZOH); Koch (Ma.; JFM); Kolukanti (Heb.; ZOH); Kottai Muthu (Tam.; WOI); Krank (Ber.; BOU); Krapata (Ma.; JFM); Kukat (Chepang; NPM); Lidis (Chepang; NPM); Lirraiq (Ber.; BOU); Macoroco (Bol.; DLZ); Madhishe Aril (Majhi; NPM); Mamona (Por.; AVP; RAR); Mamoneira (Por.; AVP); Mascriti (Haiti; AVP); Maskèti (Haiti; TRA); Maskriti (Creole; Haiti; VOD); Masquiti (Ma.; JFM); Mbaicibo (Chiriguano; DLZ); Miniakjarah (Mal.; NAD); Oil Nut Tree (Jam.; AVP); Ourioura (Ber.; BOU); Palma Christi (Col.; Eng.; Fr.; Haiti; Mart.; Ocn.; AH2; AVP; BOU; MPG; TRA); Panchangulam (Sanskrit; NAD); Piojo del Diablo (Peru; EGG); Purgeer-Konr (Den.; AVP); Racznik (Pol.; AVP); Relajar (Col.; IED; MPG); Reyar (Tharu; NPM); Ricin (Fr.; AVP; BOU); Ricino (It.; Pan.; Por.; Sp.; AVP; EGG; MPG); Risen (Creole; Haiti; VOD); Ri'zinus (Ger.; AVP); Sabadillo (Sal.; AVP); Sadabherenda (Beng.; NAD); Shemouga (Arab.; BOU); Soubagabanan (Sudan; AVP); Ta-Ma-Tze (China; AVP); Tapaniquich (Chiquitano; DLZ); Tartago (Ven. AVP); Tartaku (Aym.; Bol.; DLZ); Thaturi (Tamang; NPM); Udu Kaju (Akha; EB40:38); Unapalan (Ulwa; ULW); Undertroed (Swe.; AVP); Vatari (Sanskrit; NAD); Verenda (Beng.; NAD); Wanderbaum (Dutch; AVP); Wunderbaum (Ger.; AVP); Yanyan (Gurung; NPM); Zait (Arab.; GHA); Zourma (Sudan; AVP).

Activities (Castor):

Allergenic (1; BOU); Analgesic (f; BOU); Anodyne (f1; APA; BOU; CRC); Antiabsorptive (f; PH2); Antiamebic (f; MPG); Anticholestatic (1; HH2); Anticoagulant (1; RCP7(3)); Anticonvulsant (f; MPG); Antidiabetic (1; HH2); Antidote (f; CRC); Antiedemic (f; BOU); Antifertility (1; X12748988); Antifilarial (1; MPG); AntiHIV (f; APA); Antiinflammaory (f; BOU); Antilactagogue (f; DEP); Antileukemic (1; APA); Antiseptic (1; HH2; PH2; TRA); Antispermagenic (1; X12748988); Antitumor (1; TRA); Antitussive (f; DAA); Antiviral (1; AAB; PHR); Aperient (f; CRC); Atticide (1; X15382509); Bactericide (1; CRC; HH2; TRA); Candidicide (1; HH2); Cathartic (f1; APA; CRC); Collyrium (f; GHA); Contraceptive (1; HH2; PH2); Culicide (f; BIB); Cyanogenic (f; CRC); Diaphoretic (f; JFM); Discutient (f; CRC; DAA); Diuretic (f; TRA); Embryotoxic (1; MPG); Emetic (f1; BOU; CRC; FAD); Emmenagogue (f; BOU; KAB; KAP); Emollient (f1; AAB; APA; BOU; PNC); Expectorant (f; CRC; DAA); Febrifuge (f; ULW); Fungicide (1; HH2; X15382509); Hepatoprotective (1; HH2); Hypoglycemic (f; MPG); Hypotensive (1; HH2); Immunostimulant (f; HH2); Insecticide (1; CRC; X14667057); Lactagogue (f12; AAB; BIB; CRC; FAD; NMH; NPM); Larvicide (f; CRC); Laxative (f12; CRC; FAD; PH2); Lipolytic (1; X11535138); Lymphocytogenic (f; HH2); Ovicide (1; X14667057); Parturient (f; FAD); Peristaltic (1; VVG); Piscicide (f; SUW); Poison (1; CRC); Purgative (f12; BOU; CRC; EGG; FAD; SUW; VVG); Pyrogenic (1; HH2); Ribosome Inactivator (1; X12447536); Secretagogue (1; PH2); Secretomotor (1; TRA); Spermicide (f; TRA); Tonic (f; CRC; JFM); Vermifuge (f; BIB).

Indications (Castor):

Abscess (f; APA; BOU; CRC; PHR; PH2); Adenopathy (f; DAA; JLH); Amenorrhea (f; BOU; DEP; KAP); Anasarca (f; BIB; CRC); Arthrosis (f; BOU; CRC; HH2; JFM; PH2); Asthma (f; AHL; BOU; CRC; TRA); Bacteria (1; HH2); Bite (f; CRC); Bleeding (f; DAA; HH2); Blenorrhea (f; EGG); Blister (f; GHA); Boil (f; BOU; JFM; PHR; SUW; VVG); Bruise (f; DLZ; EGG); Bunion (f; BIB); Burn (f; CRC; NPM); Cancer (f1; BOU; DAD; HH2; MPG); Cancer, breast (f; JLH); Cancer, colon (f; JLH); Cancer, stomach (f; JLH); Candida (1; HH2); Carbuncle (f; CRC; PH2); Caries (f; NPM; WOI);

Catarrh (f; AHL; BIB; CRC); Chancre (f; BIB; CRC); Childbirth (f; AAB; CRC; DAA; JFM; VOD); Cholera (f; CRC); Cold (f; CRC; JFM); Colic (f; BOU; CRC; JFM; NAD; VOD); Congestion (f; AAB); Conjunctivosis (f; GHA; NAD); Constipation (f1; APA; PH2; ULW); Convulsion (f; CRC); Corn (f; CRC); Cramps (f; JFM); Craw-craw (f; CRC); Cyst (f; APA); Cystosis (f; BOU); Dandruff (f; FAD); Deafness (f; CRC; DAA); Delirium (f; BIB; CRC); Dermatosis (f; BOU; CRC; EGG; FAD; JFM; PH2; VOD; EB40:38); Diabetes (f; HH2); Diarrhea (f; BIB); Dislocation (f; VOD); Dropsy (f; CRC); Dyslactea (f; BIB; EGG); Dyspepsia (f; PH2); Dystonia (1; DAD); Dysuria (f; EGG; NAD); Eczema (f; MPG); Edema (f; JFM); Elephantiasis (f; BIB); Enterosis (f; BOU; CRC; HH2; JFM; PH2); Epilepsy (f; BIB; BOU; CRC; UPW); Erysipelas (f; BIB; CRC; JFM); Escherichia (1; HH2); Fever (f; AAB; CRC; DAV; EGG; HH2; NPM; ULW); Flu (f; CRC); Fracture (f; BEJ); Fungus (1; HH2); Furuncle (f; PH2); Ganglion (f; TRA) Gastrosis (f; BOU); Gingivosis (f; JFM); Gonorrhea (f; HH2; VOD); Gout (f; CRC; DAA; NPM); Guinea worm (f; CRC); Halitosis (f; GHA); Headache (f; BOU; CRC; EGG; JFM; PH2; SUW; VOD); Hematoma (f; VOD); Hemorrhoid (f; EGG; FAD; JFM); Hepatosis (f; NPM); Hernia (f; DLZ); HIV (1; DAD); Hypothermia (f; BOU); Induration (f; CRC); Infection (1; TRA); Inflammation (f; BOU; CRC; EGG; PH2); Ischia (f; HH2); Itch (f; AAB; FAD; VOD); Jaundice (f; BOU; NPM); Leprosy (f; BIB; BOU; IED); Leukemia (1; MPG); Lichen (f; MAD); Lumbago (f; BOU; HH2; KAP); Mastosis (f; BOU); Measles (f; AAB); Migraine (f; PHR); Mole (f; CRC); Myalgia (f; CRC; DAV); Mycosis (1; HH2); Nephrosis (f; BOU); Neuralgia (f; EGG); Neurosis (f; BIB; BOU; CRC; NPM); Ophthalmia (f; BOU; PNC); Osteomyelitis (f; BIB; CRC); Otitis (f; HH2; PHR; PH2); Pain (f; GHA; JFM; NPM; PH2; VOD); Palsy (f; CRC; DAA); Paralysis (f; PH2); Parasite (f; BOU; EGG); Parotitis (f; MPG); Peritonitis (f; DLZ); Pharyngosis (f; BOU); Pleurodynia (f; DEP; NAD); Pneumonia (f; SKJ); Proctosis (f; DAA); Prolapse (f; CRC; DAA); Prostatosis (f; JFM); Rash (f; BIB; CRC); Rheumatism (f; BIB; CRC; JFM; KAP; NPM; VOD); Ringworm (f; BIB; FAD; NPM); Salmonella (1; TRA); Scabies (f; NPM); Scald (f; CRC); Sciatica (f; BOU; DEP; NAD); Scrofula (f; CRC); Seborrhea (f; BIB; CRC); Schistosomiasis (f; UPW); Sciatica (f; KAP); Shigella (1; TRA); Sinusitis (f; AAB); Sore (f; APA; FAD; VVG); Sore Throat (f; PH2); Splenosis (f; EGG; VOD); Sprain (f; BEJ); Staphylococcus (1; HH2; TRA); Sting (f; CRC; SUW); Stomachache (f; CRC; IED; VVG); Stomatosis (f;? Strabismus (f; CRC; DAA); Sunstroke (f; KAB); Swelling (f; BIB; BOU; CRC; DAA; JFM; KAP; VOD); Tapeworm (1; JFM); Thrombosis (1; RCP7(3)); Toothache (f; CRC; DAA; JFM; KAP); Tuberculosis (f; BIB; CRC); Tumor (f; CRC; JFM); Tympanitis (f; DEP); Typhus (f; MAD); Ulcer (f; BIB); Urethrosis (f; CRC; DAA; NAD); Uterosis (f; CRC; DAA); Vaginosis (f; AAB); Venereal Disease (f; BOU; CRC; DAA; JFM; VOD); Vertigo (f; BOU); Virus (1; MPG); Wart (f; APA; CRC); Whitlow (f; CRC); Worm (f1; BOU; CRC; PH2; TRA; VOD); Wound (f; BOU; CRC; DAA; NPM; VVG); Yeast (1; HH2).

Dosages (Castor):

FNFF = X
15–60 ml oil (APA); 5–28 ml castor oil (KAP). 9–15 g leaf paste (KAP). 3–6 g root paste (KAP). Five 2-g or ten 1-g capsules (PHR); 5–20 ml oil (PNC).

- Algerians use castor oil with rabbit blood as a contraceptive (BIB).
- Ayurvedics use the root for ascites, asthma, bronchitis, eructation, fever, inflammation, leprosy, and diseases of the head, glands, and rectum; the leaves for burns, dyslactea, earache, nightblindness, strangury, and worms; flowers for anal troubles, glandular tumors, and vaginalgia; fruit for hepatosis, pain, splenosis, and tumors; seed and/or oil for amenorrhea, asthma, ascites, backache, boils, convulsions, dropsy, elephantiasis, fever, hepatosis, inflammation, leprosy, lumbago, pain, paralysis, piles, rheumatism, ringworm, tumors, and typhoid; the root bark for skin ailments (KAB).
- Bahamans crush and boil the seed to get the oil, which is taken for colds and as a child's and new mother's tonic (JFM).
- Brazilians bathe hemorrhoids with the leaf decoction (JFM).

- Caribbeans use castor oil in a purgative tea after childbirth (VOD).
- Cubans suggest the root decoction as diuretic in prostatitis (JFM).
- Dominicans apply crushed seed to dislocation and hematomas (VOD).
- Dominicans use heated leaves for massaging the spleen, and for internal pain (VOD).
- Egyptian farmers poultice fresh leaves onto boils (BIB).
- Ghanans grind a cleaned root into a paste inserted in the nose for headache (UPW).
- Gurungs sometimes apply leaf juice to burns, taking it for diarrhea and dysentery (NPM).
- Haitians apply boiled leaves to swellings and wounds (VOD).
- Haitians use the seed oil as hair tonic, purgative and vermifuge, rubbing on burns, dermatosis, itch, and rheumatism, taking orally for respiratory ailments (VOD).
- Hausa in Africa use the root extract as a mouthwash for toothache (UPW).
- Lebanese use leaves and crushed beans as a topical dressing, not internally as a purgative (HJP).
- Mexicans place scalded leaves on the breasts of nursing mothers to increase milk (JFM).
- Nepalese smash cotyledons and paste onto gout and scabies (NPM).
- Nigerians burn the stem with *Calotropis* for chancre (BIB).
- Peruvians suggest the crushed leaves on the face or forehead for neuralgia (EGG).
- Peruvians use leaves heated in olive oil to relieve hemorrhoids (SOU).
- South Africans use the root for toothache (BIB).
- Terai of Nepal take one cotyledon a day following menstruation for birth control (NPM).
- TRAMIL (Traditional Medicine in the Islands) Caribbeans use the oil (topically or orally) for asthma, bronchoses, burns, earache, gangliosis, rheumatism, toothache, etc. (TRA).
- Zulus administer the leaves for stomachache, orally or rectally (BIB).
- In Guiana, the leaves are applied to the breast to augment the secretion of milk (BIB).

Downsides (Castor):

Class 2b, 2d. Contraindicated in intestinal obstruction and abdominal pain of unknown origin, appendicitis, inflammatory bowel disease; do not use for more than 8 to 10 days (AHP, PHR). No health hazards or side effects are known with proper administration of designated therapeutic dosages of castor oil (PH2). Overdoses can cause colic, drastic diarrhea, gastralgia and gastrosis, queasiness, and vomiting (PHR). The seeds contain 2.8 to 3% toxic substances, requiring 2.5 to 20 seeds to kill a man (chewing a single seed may be fatal to a child), 4 to kill a rabbit, 5 a sheep, 6 an ox, 6 a horse, 7 a pig, 11 a dog, but 80 for a cock or duck. The principal toxin is the albumin, ricin (DAD). Although some recommend in pregnancy, others say no. Midwives sometimes use the oil to induce labor (AHP, 1997). Refined oil contraindications: intestinal obstruction, unexplained stomachache; adverse effects include the following: frequent use produces electrolyte losses (interaction with cardiac glycosides), also gastric irritation, and allergic skin reactions. Should not be used for prolonged periods of time (AEH). Not for children under 12 years old (PHR). May induce dermatosis as well as cure it (FAD).

Natural History (Castor):

Castor bean is both self- and cross-pollinated by wind, varying from 5 to 36%, depending on the weather conditions. Pollen sheds readily between 26 and 29°C, with a relative humidity of 60%. For single cross-hybrid seed production, strains giving a 1:1 ratio or pistillate and heterozygous monoecious plants are used, the latter being rogued 1 to 5 days before flowering begins. Three-way cross-hybrids can also be used. For open pollinated types, roguing of all off-types is done after the last cultivation, and for pure seed production isolation necessity depends on the wind velocity. For hybrid and open pollinated types in the United States, stands are isolated 300 to 720 m; but in areas of less wind velocity, less distance may be sufficient. Fungi known to attack Castor bean plants include *Alternaria compacta, A. ricini, A. tenuis, A. tenuissima, Aspergillus itaconicus, A. niger, A. quercinus, Botrydiplodia manilensis, B. ricinicola, B. theobromae, Botryotinia ricini, Botrysphaeria ribis, Botrytis cinerea* (Gary mold), *Cephalosporium curtipes, Cercospora canescens, C. coffeae, C. ricinella, Cercosporella ricinella* (Leaf spot), *Cladosporium herbarum, Clitocybe tabescens, Colletotrichum bakeri, C. erumpens, C. ricini, Corticium solani, Didymella ricini, Diplodia natalensis, D. organicola, D. ricinella, D. ricini, Discosporella phaeochlorina, Epicoccum nigrum, Erysiphe cichoracearum, Fusarium moniliforme, F. orthoceras, F. oxysporum, F. sambucinum, F. semitectum, Gibberella pulicarus, Glomerella cingulata, G. ricini, Haplosporella manilensis, Lecanidion atratum, Leveillula lanata, L. taurica, Macrophomina phaseoli, Macrophoma phaseoli, Ph. ricini, Macrosporium cavarae, M. ricini, Melampsora euphorbiae, M. ricini, Melampsorella ricini, Mecrostroma minimum, Mucor fragilis, Mycosporella ricinicola, M. tulasnei, Myrothecium roridum, Oidiopsis taurica, Peniophora cinerea, Phoma macropyrena, Ph. ricini, Phomopsis ricini, Ph. ricinella, Phyllosticta bosensis, Ph. ricini, Phymatotrichum omnivorum (Root rot), Physalospora abdita, Ph. propinqua, Ph. rhodina, Ph. ricini, Ph. obtusa, Phytophthora cactorum, Ph. cinnamomi, Ph. palmivora, Ph. parasitica, Pleospora herbarum, Pythium aphanidermatum, P. debaryanum, P. gracile, P. intermedium, P. proliferum, P. ultimum, P. vexans, Rhabdospora ricini, Rhizoctonia solani, Schiffnerula ricini, Schizophyllum commune, Sclerotinia fuckeliana, S. minor, S. ricini, S. sclerotiorum, Scierotium rolfsii*, and *sphaceloma ricini.* The following bacteria also cause diseases: *Agrobacterium tumefaciens, Bacterium lathyri, B. ricini, Pseudomonas solanacearum, Xanthomonas ricini*, and *X. ricinicola. Striga lutea* parasitizes the plants. Nematodes isolated from Castor bean include *Aphelenchoides asterocaudatus, A. bicaudatus, A. subtenuis, Helicotylenchus cavenssi, H. pseudorobustus, H. schachtii, Meloidogyne arenaria* and var. *thamesi, M. hapla, M. incognita, M. incognita acrita, M. javanica, M. thamesi, Merlinius brevidens, Pratylenchus brachyurus, P. neglectus, P. pratensis, P. scribner, P. vulnus, P. zeae, Radopholus similes, Scutellonema clathricaudatum, Tricephalobus longicaudatus*, and *Tylenchorhychus mashhoodi* (Golden, 1984). Several insects are pests. In India, the Capsule borer (*Dichocrocis punctiferalis*) bores into young and ripening capsules; and Castor semiloopers (*Achoea janata*) are the worst pests. In Tanganyika damage by capsid and myrid bugs are a limiting factor causing immature fruit to drop. Green stinkbugs, leaf-hoppers, leaf-miners, and grasshoppers are pests that feed on the leaves. Most insects can be controlled by insecticides. Because some of the varieties are quite tall, wind storms are a potential hazard to a crop (HOE).

Extracts (Castor):

Ricin, the deadly poison, can be attached to monoclonal antibodies that only attack cancer cells, a technique reportedly tried in 1000 cancer patients (DAD). The AIDS virus can infect an immune cell by locking onto its cell receptor protein CD4. By genetically attaching the ricin to genetically engineered CD4 proteins, one obtains CD4-ricin, which will lock onto the external viruses of infected cells, 1000 times more often than onto healthy cells, possibly killing enough infected cells to prevent the spread of the disease symptoms. Like the botulism toxin, ricin can be used to kill overstimulated nerve endings in patients with dystonias (DAD). Ricinoleic acid has served in contraceptive jellies (DAD). Bigi et al. (2004) reported activity of extracts (their fatty acids in particular) and ricinine against the leaf-cutting ant *Atta sexdens rubropilosa* and the symbiotic fungus *Leucoagaricus gongylophorus* (X15382509).

DOG ROSE (*ROSA CANINA* L.) +++ ROSACEAE

Synonyms:

Rosa armata Stev. ex Besser; *Rosa caucasica* Pall.; *Rosa frondosa* Stev. ex Spreng; *Rosa glauca* Schot. ex Besser; *Rosa lutetiana* Lem.; *Rosa taurica* M. Bieb. *fide* HH3

FIGURE 1.90 Dog Rose (*Rosa canina*).

NOTES (DOG ROSE):

And as many fountains flowing with milk and honey, and seven mighty mountains, whereupon there grow roses and lilies, whereby I will fill thy children with joy.

11 Esdras 2:19 (KJV)

And the same number of springs flowing with milk and honey, and seven mighty mountains on which roses and lilies grow; by these I will fill your children with joy.

11 Esdras 2:19 (RSV)

A search of just the KJV turned up 42 sources, in which roses were cited (some as the past tense of rise). Most were in Genesis (21 matches), Judges (16), 1 Maccabees (Apocrypha) (10), Exodus (10), and Acts (9). However, there are almost as many interpretations of the word "rose." In my earlier Bible book (BIB), I cited Moldenke and Moldenke noting that crowning oneself with rosebud at a feast is a purely Greek custom borrowed by the Romans. But clearly, rose gardens were and are important in Israeli culture. A century or so later, Egyptians were growing roses under glass to send to Rome for banquets. Zohary suggests that there were cultivated roses already in biblical times in the Holy Land. But the Hebrew word *vered* is mentioned only in postbiblical times — for example, several times in the Talmud. "No gardens and orchards should be established in Jerusalem, with the exception of rose gardens that have existed there since the Early Prophets" While rose in the Bible may mean many species (e.g., Cistus, Hibiscus, Nerium, Rosa), it is concluded that they meant *Rosa* in several biblical quotations. It is nice that our national capital has its "rose garden." Would that powerful faith-based individuals insist on a second garden, even closer to the White House, of our wholesome biblical medicinal plants, often competitive with the unwholesome pharmaceuticals that fewer and fewer of us can afford. Zohary lists only two species of *Rosa* in the *Flora of Palestine* (FP2), but later mentions that four species are native to Israel (ZOH).

- — Inflorescences few-flowered; styles free; flowers mostly pink; fruit 1 to 2 cm long, scarlet, finally turning blackish crimson — *Rosa canina*
- — Inflorescences many-flowered; styles united into a column; flowers white; fruit circa 1cm long, brick red — *Rosa phoenicia*

Zohary seems to favor *Rosa phoenicia* (Phoenician Rose (Eng.; ZOH); Vered (Hebrew; ZOH)) as most likely in 1982 (ZOH). So did I in 1983 (BIB). But, in the intervening years, I encountered no new information on that species, so I will discuss the much better *Rosa canina* as a very similar species also present in the Holy Land and with a much bigger medicinal repertoire. Much of the German literature refers to *cynosbati*, alluding to the fruits and seeds of *Rosa canina*, or a fungus thereon (see EFS). While AH2 selected dog rose as the standardized common name, they also boldface **rose hips,** saying it is an acceptable, and even preferable, Standardized Common Name (AH2).

Common Names (Dog Rose):

Achdirt (Ber.; BOU); Agabanzo (Sp.; EFS); Azenzou (Ber.; BOU); Bédégar (Fr.; EFS); Bou Soufa (Arab.; BOU); Brier Rose (Eng.; Ocn.; AH2); Canker Flower (Eng.; BOU); Csipkebogyó Cynorrhodon (Fr.; EFS); Dog Brier (Eng.; Ocn.; AH2); Dog Rose (Eng.; Scn.; AH2; CR2); Eglantier (Fr.; BOU); Eglantine (Fr.; BOU); Escaramujo (Sp.; Chile; EFS); Galabardera (Sp.; EFS); Hagdorn (Ger.; HH3); Hagebuttenstrauch (Ger.; EFS); Hagrose (Ger.; HH3); Hekenrose (Ger.; EFS); Heprose (Eng.; BOU); Hip Rose (Eng.; EFS); Hondsros (Dutch; EFS); Hundrose (Ger.; MAD); Hybener (Den.; EFS); Kusbumu (Tur.; EFS); Monholinos (Sp.; EFS); Nab el Kalb (Arab.; BOU); Nesri (Arab.; BOU); Nisrin (Arab.; BOU); Pepins de Rosier Sauvage (Fr.; HH3); Rosa Brava (Por.; GEP); Rosa Canina (It.; HH3); Rosa de Cão (Por.; EFS); Rosa di Macchia (It.; EFS); Rosa Selvatica (It.; EFS; HH3); Rosa Silvestre (Sp.; EFS); Rosal Silvestre (Spain; VAD); Rose des Haies (Fr.; HH3); Rose Hips (Eng.; Ocn.; AH2; CR2); Rose Sauvage (Fr.; BOU); Rosenschwamm (Ger.; EFS); Rozenbottelstruik (Dutch; EFS); Rosier des Chiens (Fr.; BOU); Rosier Sauvage (Fr.; BOU); Schlafapfel (Ger.; EFS); Semance de Cynorrhodon (Fr.; HH3); Silva Macha (Por.; EFS); Tafrha (Ber.; BOU); Tigourma (Ber.; BOU); Ward Barri (Arab.; Syria; BOU; HJP); Ward es Seni (Arab.; BOU; HJP);

Ward es Siyag (Arab.; BOU; HJP); Ward ez Zeroub (Arab.; BOU); Wild Brier (Eng.; EFS); Wildrose (Ger.; HH3); Zarza Rosa (Sp.; EFS); Zaunrose (Ger.; HH3).

Activities (Dog Rose):

Analgesic (12; X15330493); Antiarthritic (12; X15330493); Antidiarrheic (f; PNC); Antiinflammatory (12; X12880322; X15330493); Anti-MDR (1; X14734860); Antimycobacterial (1; PR14:303); Antioxidant (1; HH3); Antiradicular (1; HH3); Antiscorbutic (f1; BOU); Antiseptic (f; PED); Antispasmodic (f; PED); Antitubercular (1; PR14:303); Antiulcer (f1; X12902057); Astringent (f12; APA; BOU; KOM; WAM); Bactericide (1; PR14:303); Beta-Lactamase Inhibitor (1; X14734860); Cicatrizant (f1; VAD); Demulcent (f1; WAM); Diuretic (f1; APA; BOU; HH3; PED; PHR; PH2); Gastroprotective (f1; X12902057); Hypoglycemic (1; HH3); Laxative (f1; APA; PHR; PH2); Lipoxygenase Inhibitor (1; HH3); Nervine (f1; WAM); Tonic (f; VAD); Vermifuge (f; HH3; MAD).

Indications (Dog Rose):

Albuminuria (f; MAD); Arteriopathy (f; VAD); Arthrosis (f12; PHR; X15330493); Asthenia (f; VAD); Bacteria (1; PR14:303; X14734860); Bleeding (f; HH3; PH2); Blepharosis (f; VAD); Burn (f; MAD); Cancer (f1; FNF; JLH); Cancer, genital (f1; FNF; JLH); Cancer, kidney (f1; FNF; JLH); Cancer, mouth (f1; FNF; JLH); Cancer, throat (f1; FNF; JLH); Cancer, uterus (f1; FNF; JLH); Capillary Fragility (f1; PED; VAD); Carcinoma (f; JLH); Catarrh (f; HJP; MAD); Chills (f; PHR); Cold (f; APA; PHR; PH2; VAD; JMF5:137); Conjunctivosis (f; VAD); Consumption (f; JEB79:57); Cough (f1; HJP; WAM); Cystosis (f; VAD); Dermatosis (f; VAD); Diarrhea (f1; BOU; PED; WAM); Dropsy (f; PHR); Dyspepsia (f; PH2; VAD); Dysuria (f; EFS; MAD; PHR; VAD); Edema (f; PH2; VAD); Enterosis (f; MAD; PH2); Exanthema (f; MAD); Flu (f; APA; PHR; PH2; VAD); Gallstone (f; MAD; PH2); Gastrosis (f; PED); Gonorrhea (f; MAD); Gout (f; PHR; PH2; VAD); Headache (f; APA; MAD); Hematoptysis (f; JEB79:57); Hemorrhoid (f; VAD); Hepatosis (f; JLH); High Blood Pressure (f; VAD); Hyperacidity (f; PH2); Hyperuricemia (f; VAD); Induration (f; JLH); Infection (1; PED; PHR; PH2; PR14:303; X14734860); Inflammation (f12; JLH; X15330493); Ischia (f; HH3); Kidney stone (f; MAD; PH2); Leukorrhea (f; PH2); Nausea (f1; WAM); Nephrosis (f; HJP; JLH; PH2); Nervousness (1; WAM); Obesity (f; VAD); Oliguria (f; VAD); Ophthalmia (f; JLH; VAD); Osteoarthrosis (12; X15330493); Pain (12; X15330493); Periodontosis (f; VAD); Rheumatism (f12; HH3; PHR; PH2; X15330493); Scar (1; HH3); Sciatica (f; PHR); Sore (f; APA; VAD); Sore Throat (f1; APA; WAM); Splenosis (f; JLH); Staphylococcus (1; X14734860); Stomatosis (f; APA; JLH); Stone (f1; PH2; VAD); Stress (f; PED); Swelling (f; VAD); Thirst (f; APA); Tuberculosis (1; JEB79:57; PR14:303); Tumor (f; JLH); Ulcer (f1; X12902057); Urethrosis (f; PH2; VAD); Uterosis (f; JLH); Varicosity (f; VAD); Wart (f; JLH); Worm (1; HH3); Wrinkle (1; HH3); Wound (f; VAD).

Dosages (Dog Rose):

FNFF = !!

Fruits are edible and vitamin rich. Seeds roasted as coffee substitute. Leaves used as tea substitute. Petals used to make candies, sandwiches, teas, added to honey, liqueurs, vinegars (EFS; FAC). 0.75–3 tsp chopped fruit/cup water/1–3 ×/day (APA, JAD, WIC); 2–5 g in infusion (HH3). 1–2 g dry fruit/cup water (PHR); 1/4–1/2 cup fresh fruit (PED); 6–12 g dry fruit (PED); 9 g dry fruit/cup boiling water (PED). 2 g (PH2). 1 tsp leaf or flower per cup water; 3–4 cups/day (VAD).

- Chileans use the plant for kidney cancers (JLH).
- Lebanese Gypsies use fruits for catarrh, nephritis, and reproductive organs (HJP).
- Lebanese use young leaves in healthful teas (HJP).
- Turks use the fruits for ulcers, and they proved out in rats (X12902057).
- Ukranians use floral/foliar tea for cough (HJP).

Downsides (Dog Rose):

Class 1 (AHP, 1997). No health hazards or contraindications reported with proper administration of suggested therapeutic dosages (PH2). None known for the fruits (PHR; WAM).

Extracts (Dog Rose):

Danish scientists (Rein et al., 2004) showed that powdered fruits (Hyben Vital) reduced joint pain and improved well-being in a double-blind, placebo-controlled, randomized trial with osteoarthritis. No major side effects occurred. Hyben Vital reduces symptoms of osteoarthritis (X15330493). Shiota et al. (2004) showed that tellimagrandin potentiated the activity of beta-lactams against methicillin-resistant *Staphylococcus aureus* (X14734860). Gurbuz et al. (2003) found 100% antiulcerogenic activity for the fruits in rats, comparable to misoprostol at 0.4 mg/kg (X12902057).

MADDER (*RUBIA TINCTORIUM* L.) X RUBIACEAE

Notes (Madder):

> **And after Abimelech there arose to defend Israel Tola the son of Puah, the son of Dodo, a man of Issachar; and he dwelt in Shamir in mount Ephraim.**
>
> **Judges 10:1 (KJV)**

Strange that if madder was important in the Holy Land in biblical times that it would not have gotten honorable mention as a plant in the Good Book. But it is mentioned only three times in the Bible, and then only as a proper name, as above. Zohary says it was important as a dye, and cultivated in all the Near East countries, in separate plots or intercropped with olive. Although early on, more used as a dye, it later assumed some medicinal and symbolic virtues as well, mentioned by such early greats as Pliny and Dioscorides (ZOH). Although important to early Greeks and Romans, it is not mentioned by the Sanskrit.

Common Names (Madder):

Alizari (Arab.; Ger.; BOU; KAB); Aroubian (Ber., BOU); Bacho (India; Pun.; DEP; EFS; KAB); Boyacikökü (Tur.; EFS); Dyer's Madder (Eng.; BOU; CR2); Erythrodanon (Greek; DEP; KAB); European Madder (Eng.; WOI); Färberröte (Ger.; EFS); Farberwete (Ger.; KAB); Fauwa (Arab.; Yemen; GHA); Fowwa (Arab.; BOU); Fuah (Heb.; ZOH); Fuwwa (Arab.; ZOH); Fuwwah (Arab.; Syria; HJP); Fuwwah es Sabbaghin (Arab.; Syria; HJP); Fuwwat as Sabbaghin (Arab.; BOU); Garance (Fr.; BOU; EFS; KAB); Garance des Teinturiers (Fr.; BOU; KAB); Garence (Fr.; KAB); Granza (Cat.; Sp.; EFS; KAB); Grapp (Ger.; KAB); Jên Ku Tan (China; EFS); Krap (Rus.; KAB); Krapp (Dutch; Ger.; EFS; KAB); Krapprod (Den.; EFS); Madder (Eng.; Scn.; AH2; BOU; CR2); Manyounth (India; EFS); Manyunth (Bom.; Sind; DEP; KAB; NAD); Marena (Rus.; KAB); Mariona (Rus.; DEP; KAB); Mee (Dutch; DEP); Meekrap (Dutch; EFS; KAB); Oroug Sabbaghin (Arab.; BOU); Ourouq Homor (Arab.; BOU); Parson (Syria?; JLH); Patachina (Rom.; KAB); Potha (Syria?; JLH); Puah (Heb.; ZOH); Puvah (Heb.; ZOH); Red Purgative (Eng.; Leb.; HJP); Robbia (It.; DEP; EFS); Rodan (Afg.; Iran; DEP; KAB); Rodang (Afg.; Iran; DEP; KAB); Rodea (Rom.; KAB); Roiba (Rom.; KAB); Rubia (It.; Sp.; EFS; KAB); Tarioubia (Ber., BOU); Taroubent (Ber., BOU); Taroubia (Ber., BOU); Tefthrion (Greek; KAB); Yin Khoot Tan (China; EFS).

Activities (Madder):

Abortifacient (f; BOU; HJP); Antigenotoxic (1; JAF51:3334); Antimutagenic (1; JAF51:3334; X10792014); Antiseptic (1; X15752641); Antispasmodic (f; GAZ); Aphrodisiac (f; BOU); Astringent (f; WOI); Carcinogenic (1; FNF; GAZ; X1370725); Chemopreventive (1; JAF51:3334); Contraceptive (f; HJP); Depurative (f; BOU); Digestive (f; WOI); Diuretic (f; BOU; EFS; GAZ; HHB; WOI); Emmenagogue (f; BOU; EFS; HJP; KAB); Expectorant (f; BOU); Fungicide (1; X15752641); Genotoxic (1; X1370725); Hydragogue (f; HJP); Hypotensive (f; BOU); Litholytic (f; GAZ; PH2; WOI); Mutagenic (1; GAZ; PH2; X11301857); Orexigenic (f; BOU); Poison (1; PH2); Purgative (f; HJP); Tonic (f; BOU; EFS; HHB); Vermifuge (f; BOU); Vulnerary (f; BOU).

Indications (Madder):

Amenorrhea (f; HHB; MAD); Anemia (f; BOU; HHB; MAD); Anorexia (f; BOU); Arthrosis (f; HHB; WOI); Bladderstone (f; HHB); Bruise (f; BOU; MAD); Calculus (f; WOI); Cancer (f; JLH); Cancer, bladder (f; JLH); Cancer, kidney (f; JLH); Cancer, liver (f; JLH); Cancer, skin (f; JLH); Cancer, stomach (f; JLH); Cancer, uterus (f; JLH); Childbirth (f; GHA); Chlorosis (f; MAD); Cholecystosis (f; KAB; WOI); Constipation (f; HJP); Cystosis (f; HHB); Decubitis (f; WOI); Dermatosis (f; JLH); Diarrhea (f; BOU; HHB); Dropsy (f; JLH); Dysentery (f; MAD); Dysmenorrhea (f; GHA; KAB; MAD; PNC); Dysuria (f; PNC); Enterosis (f; MAD); Enuresis (f; MAD); Fungus (1; X15752641); Gastrosis (f; JLH; MAD); Gout (f; MAD); Gravel (f; HJP); Hepatosis (f; HHB; MAD; PNC); High Blood Pressure (f; BOU); Induration (f; JLH); Infection (1; X15752641); Jaundice (f; HJP; MAD; ZOH); Kidney stone (f; HJP; PH2); Lichen (f; MAD); Malaria (f; MAD); Mycosis (1; X15752641); Nephrosis (f; JLH; HHB); Neurosis (f; NAD); Puerperium (f; GHA); Pyelonephrosis (f; MAD); Scab (f; MAD); Sciatica (f; BOU); Scrofula (f; MAD); Sore (f; HHB); Splenosis (f; HHB; JLH; KAB; MAD); Stone (f; HHB; WOI); Synovia (f; MAD); Tuberculosis (f; HHB; MAD); Uterosis (f; JLH); Water Retention (f; WOI); Worm (f; BOU); Wound (f; BOU; HHB).

Dosages (Madder):

FNFF = X

Do not take (JAD; PH2). 30 grains root, 3–4 ×/day (FEL). 1 g powdered root/3 ×/day (MAD).

- Algerians use as emmenagogue, hydragogue, litholytic, as a poultice, and as an abortifacient and contraceptive (HJP).
- Asian Indians report its use for cholecocystosis, hepatosis, jaundice, splenosis (KAB).
- Asian Indians say it works on the nervous and uterine systems (NAD; SKJ).
- Lebanese immigrants in New York refer to it as the red purgative (HJP).
- Russians consider this an important litholytic herb (WOI).
- Syrians use the plant for indurations of the liver and spleen (JLH).
- Yemeni women use crushed roots in a tonic tea after childbirth (GHA).
- Yemeni women use roots with berries of *Morus* and *Salvadora* for irregular menstruation (GHA).

Downsides (Madder):

The court is still out on whether or not alizarin, lucidin, and purpurin from *Rubia tinctorum* exert a disintegrating effect on the surface of bladder and kidney stones containing calcium. Because extracts of the root are mutagenic and contain genotoxic and tumorigenic compounds, it is not to be recommended (De Smet, 1993). Drug should not be administered (PH2).

Natural History (Madder):

Plant, perhaps dangerously, used as fodder. Camels are said to be fond of it. Bones of animals ingesting the plant may turn red, as do claws and beaks of birds.

BIBLICAL BRAMBLE (*RUBUS SANCTUS* SCHREB.) +++ ROSACEAE

SYNONYMS:

Rubus sanguineus Friv.

NOTES (BIBLICAL BRAMBLE):

> **For every tree is known by his own fruit. For of thorns men do not gather figs, nor of a bramble bush gather they grapes.**
>
> **Luke 6:44 (KJV)**

Finding almost nothing on *Rubus sanctus* or its synonym *R. sanguineus*, including common names, I, poetically licentious at times, have now denominated it the *biblical bramble*. Zohary called it the *true bramble*, but with lower case, perhaps to distinguish it from many other thorny plants, common in Israel, as in other arid areas. PubMed yielded one useful title on chemistry. Hussein, Ayoub, and Nawwar (2003) isolated two new natural caffeoyl esters (3,6-di-O-caffeoyl-(alpha/beta)-glucose and 1-O-caffeoyl-beta-xylose) and a new natural tannin (2,3-O-hexahydroxydiphenoyl-4,6-O-sanguisorboyl-(alpha/beta)-glucose) (X12895538). And there was one useful title on medicine. Turkish scientists (Erdemoglu, Kupeli, and Yesilada, 2003) demonstrated antinociceptive activities for aerial parts and roots (X14522443). With no more than that to offer, I instead aggregate my CRC (Edition 2) accounts of activities of other species of the *Rubus* genus, namely blackberries and raspberries below (Other *Rubus*).

COMMON NAMES (BIBLICAL BRAMBLE):

Biblical Bramble (Eng.; JAD); Bramble (Eng.; ZOH); Bramble Bush (Eng.; KJV); Sina (Heb.; ZOH); Sinaia (Heb.; ZOH); Sinim (Heb.; ZOH); Thorn (Eng.; BIB); True Bramble (Eng.; ZOH); Tzinim (Heb.; ZOH); Tzininim (Heb.; ZOH).

ACTIVITIES (OTHER *RUBUS*):

Anticancer (1; JNU); Anticholinesterase (1; CAN); Antidote (f; DEM); Antiinflammatory (1; APA; FAD); Antioxidant (1; JNU; JAF50:2926); Antiproliferative (1; JAF50:2926); Antiradicular (1; X1332092); Antiseptic (f1; PED); Antispasmodic (1; APA; CAN; PED); Antitumor (f; APA); Antiviral (1; CAN); Apoptotic (f; JNU); Astringent (f1; CAN; CEB; FAD; PH2); Bactericide (1; MAD); Chemopreventive (1; FNF; X11799774; X11181460); Choleretic (f; KOM); Depurative (f; APA; KOM; PH2); Detoxicant (1; JNU); Diaphoretic (f; KOM; MAD); Diuretic (f; DEM; EFS; KOM); Febrifuge (f; EFS); Fungicide (f; MAD); Hemostat (1; APA); Hypocholesterolemic (1; JNU); Myostimulant (1; CAN); NO Inhibitor (1; JAF50:850); Postparturient (f; CAN); Purgative (f; PH2); Stimulant (f; DEM; PED); Tonic (f; APA; DEM; EFS); Uterorelaxant (1; APA; FAD; PNC); Uterotonic (1; APA; FAD); Xanthine-oxidase Inhibitor (1; X1332092).

INDICATIONS (OTHER *RUBUS*):

Angina (f; MAD); Appendicitis (f; MAD); Bacteria (1; MAD); Biliousness (f; DEM); Bleeding (f1; APA; DEM; FEL; MAD); Boil (f; DEM); Bronchosis (f; CEB; MAD); Cancer (1; JLH; JNU; X11799774); Cancer, colon (1; JLH); Cancer, mouth (1; JLH); Cancer, stomach (1; JLH); Cancer, throat (1; JLH); Canker (f; APA; MIC); Cardiopathy (f; PHR; PH2); Cataract (f; DEM); Catarrh (f; DEM); Childbirth (f; CEB; DEM; FEL; PH2); Cholera (f; CEB; FEL); Cold (f1; DEM; MAD; SKY); Condyloma (f; JLH); Conjunctivosis (f1; APA; CAN; CEB; DEM; FNF); Constipation (f; DEM); Cough (f; APA; CEB; DEM; MAD); Cramps (1; FAD); Dementia (1; JNU); Dentition (f; DEM); Dermatosis (f; APA; KOM; MAD); Diabetes (f; KOM; MAD); Diarrhea (f12; APA; DEM; FAD;

MIC; PH2; SKY); Dropsy (f; APA; MAD; PH2); Dysentery (f1; DEM; FAD); Dysmenorrhea (f1; APA; CEB; DEM; FAD; MAD); Dysuria (f; DEM); Enterosis (1; APA; JLH; KOM; MAD; PH2); Exanthema (f; MAD); Fever (f; DEM; FEL; MAD); Fever (f; CEB; EFS; KOM); Fever Blister (f; APA); Flu (1; KOM; PED); Fungus (1; MAD); Gastrosis (f1; APA; DEM; JLH; KOM; PH2); Gleet (f; FEL); Gonorrhea (f; CEB; DEM); Gravel (f; CEB); Headache (f; DEM); Heartburn (f; DEM); Hematemesis (1; DEM; FNF); Hematuria (f; DEM); Hemoptysis (f; CEB); Hemorrhage (f1; APA; DEM; FNF); Hemorrhoid (1; APA); High Blood Pressure (f; DEM); High Cholesterol (1; JNU); Infection (f1; DEM; MAD); Inflammation (f1; APA; FAD; JLH); Lethargy (f; DEM); Leukorrhea (f; FEL; MAD); Low Blood Pressure (f; DEM); Maculitis (1; FNF); Measles (f; DEM); Metrorrhagia (f1; APA; MAD); Miscarriage (f; DEM); Morning Sickness (f; APA); Mycosis (1; MAD); Nausea (f; DEM; PED); Nephrosis (f; DEM; MAD); Ophthalmia (f 1; CEB; DEM; FNF; JNU); Parturition (1; FAD; PED); Pharyngosis (f12; APA; PHR; PH2); Phthisis (f; CEB); Pregnancy (f; APA; SKY); Proctosis (f; FEL); Prolapse (f; FEL); Prostatosis (f; APA); Pulmonosis (f; CEB; DEM; KOM); Respirosis (f; PHR; PH2); Rheumatism (f; DEM); Scab (f; MAD); Scabies (f; MAD); Sore (f1; DEM); Sore Throat (f12; APA; CEB; KOM; MIC; PH2; SKY); Stomach (f; MIC); Stomachache (f; DEM); Stomatosis (f12; APA; JLH; MAD; PHR; PH2; KOM; MIC); Stone (f; MAD); Swelling (f; APA); Tonsilosis (f1; FAD; MAD); Toothache (f; DEM); Tuberculosis (1; DEM; FNF; MAD); Ulcers (f; APA); Uterosis (f; CEB; FEL); Venereal Disease (f; CEB; DEM); Vomiting (f; DEM; FEL); Wart (f; JLH); Water Retention (f; DEM); Wound (f1; APA).

Dosages (Other *Rubus*):

FNFF = !!!

Berries widely consumed and marketed. Leaves often used as tea substitute. 1–2 tsp crushed leaf/cup water, to 6 ×/day (APA); one to three 384-mg capsules 3 ×/day (APA); two 4–8 g dry leaf, or in tea, 3 ×/day (CAN); 4–8 ml liquid extract (1:1 in 25% ethanol) 3 ×/day (CAN; SKY); 1/4–1/2 cup fresh leaf (PED); 6–12 g dry leaf (PED); 9 g dry leaf:45 ml alcohol/45 ml water (PED); 1.5 g finely cut leaf (PHR); 2–10 ml liquid leaf extract (PNC); 1–2 tsp crushed leaf/cup water, up to 6 ×/day (SKY; WIC).

Downsides (Other *Rubus*):

Class 1 (AHP, 1997). Apparently speaking of the leaves, "Should not be used during pregnancy, and, if taken during labor, should only be done under medical supervision." Still, raspberry leaf is "widely recommended to be taken during pregnancy to help facilitate easier parturition." "Unsuitable to use as an herbal remedy to treat eye infections such as conjunctivitis" (Newall et al., 1996).

Natural History (Other *Rubus*):

The insect-pollinated, bisexual flower produces the blackberry. Bumblebees (*Bombus*) are probably their most effective pollinators. Smaller solitary bees often gather pollen and aid fertilization. Despite formidable thorns, browsing mammals nip the prickly canes, and numerous fruit eaters raid the thickets. Berries are a staple in season to many birds, (bluebird, cardinal, catbird, chickadees, crow, flicker, grosbeak, grouse, jay, magpie, mockingbird, oriole, pheasant, robin, solitaire, tanager, thrasher, thrush, titmice, towhee, waxwing, woodpecker) and mammals (bear, beaver, chipmunk, deer, elk, fox, hare, marmot, mice, moose, rat, sheep, skunk, squirrel). *Rubus* twigs are relished by cottontail rabbits and white-tailed deer. Rabbits clip off the stems at an oblique angle; a ragged end indicates deer browsing. Land turtles relish low-hanging fruits (eastern box turtle, wood turtle, and Blanding's turtle) (EAS; MZN). Bright orange spots on leaf undersides, together with bunched or dwarfed shoots, indicate orange rust (*Gymnoconia peckiana*), a club fungus, is probably its most serious disease. Gall-making insects, mostly tiny wasps (*Diastrophus*) and gnats (*Lasioptera*), create characteristic swellings on stems and leaves. Curled, distorted leaves may indicate blackberry psyllids (*Trioza ripuntata*), common yellow-brown sucking insects also known as jumping plant lice. Caterpillars of the large ruby tiger moth (*Phragmatobia assimilans*) feed on blackberry. The blackberry looper caterpillar (*Chlorochlamys chloroleucaria*) forages on the fruits. A fly-catching wasp (*Hypocrabro stirpicolus*) tunnels into blackberry stems, constructing cells for eggs and stored flies. Zipper-like scars on stems are egg scars of the black-horned tree cricket (*Oecanthus nigricornis*) (EAS).

Extracts (Other *Rubus*):

Aqueous extracts anticholinesterase, myostimulant; uterotonic; fruit extracts with antiviral activity (CAN). Extract appears to relax uterus only in pregnant rats and humans, inactive on non-pregnant uterus (PNC). Anthocyanins and polyphenols in berries of several *Ribes*, *Rubus*, and *Vaccinium* spp. have *in vitro* antiradical activity on chemically generated superoxide radicals. The extracts also inhibit xanthine oxidase. All crude extracts active toward chemically generated superoxide radicals. The tannins in the leaves make leaf tea competitive with green tea for cancer and cardiopathy. I suspect that the wild strains of *Rubus sanctus* would be even better endowed with the anticancer and cardioprotective polyphenols than the cultivated blackberries and raspberries.

SHEEP SORREL (*RUMEX ACETOSELLA* L.) ++ POLYGONACEAE

Notes (Sheep Sorrel):

> **The fourteenth day of the second month at even they shall keep it, and eat it with unleavened bread and bitter herbs.**
>
> **Numbers 9:11 (KJV)**

> **In the second month on the fourteenth day in the evening they shall keep it; they shall eat it with unleavened bread and bitter herbs.**
>
> **Numbers 9:11 (RSV)**

> **In the second month on the fourteenth day between the two evenings, they should prepare it. Together with unfermented cakes and bitter greens they should eat it.**
>
> **Numbers 9:11 (NWT)**

Like so many other tart and bitter herbs, or bitter greens in the NWT, this one has been identified by some scholars as a good candidate for one of the bitter herbs of the Bible. But Israeli botanist Michael Zohary does not include it in his *Plants of the Bible*, nor is this one of more than a dozen species of *Rumex* listed in his *Flora of Palestine* (FP1). It is not even bitter; it is pleasingly tart. I have enjoyed sheep sorrel soup with unfermented corn bread on occasion. Scandinavians even add it to bread. But I consider it very unlikely to be one embraced under the "bitter herb" concept in the Bible (BIB, FP1; ZOH). Used interchangeably with *Rumex acetosa* (HHB), also not reported in the *Flora of Palestine*. But, in one NWT passage, sorrel is mentioned:

> **And the cattle and the full-grown asses cultivating the ground will eat fodder seasoned with sorrel.**
>
> **Isaiah 30:24 (NWT).**

Many of the species of *Rumex* are pleasingly tart with oxalic acid. And some tart sorrels may be among the more than a dozen species listed by Zohary in the *Flora of Palestine* (FP1).

FIGURE 1.91 Sheep Sorrel (*Rumex acetosella*). Source: Regina Hughes in Reed (1970); colored by Peggy Duke.

Common Names (Sheep Sorrel):

Acederilla (Sp.; EFS); Acetosella (It.; EFS); Acidula (JLH); Agrelleta (Cat.; KAB); Aizon (JLH); Azeda Miuda (Por.; AVP); Azedas de Ovella (Por.; AVP); Azedinha (Mad.; Por.; AVP; JAD); Azedinha Alleluia (Por.; AVP); Bodilanyana (Suto; KAB); Boksuring (Afrikan; KAB); Bread and

Cheese (Eng.; KAB); Chuk (Hindi; NAD); Chuka (Hindi; SKJ); Chuka Palam (Beng.; India; EFS; NAD); Chutrika (Sanskrit; EFS; NAD); Cizaña (Ven.; EFS); Common Sorrel (Eng.; BUR; EAS); Cow Sorrel (Eng.; BUR); Cuckoo Bread (Eng.; EFS); Cuckoo's Meat (Eng.; KAB); Feldsauramfer (Ger.; NAD); Field Sorrel (Eng.; EAS; KAB); Flora Carol Sorrel (Eng.; BUR); Horse Sorrel (Eng.; BUR); Hhummad Saghir (Arab.; Syria; HJP); Kleiner Sauer Ampfer (Ger.; EFS); Losey (Dom.; AVP); Mountain Sorrel (Eng.; BUR); Oseille (Fr.; AVP); Oseille des Brebis (Fr.; AVP; KAB); Oseille Marron (Fr.; AVP); Oseillette (Fr.; AVP); Patience (Fr.; AVP); Petite Oseille (Fr.; Reunion; EFS; KAB); Red Top Sorrel (Eng.; BUR); Red Weed (Eng.; BUR); Romasilla (Chile; AVP); Ruibarbillo (Cr.; AVP); Samhadh Caora (Irish; KAB); Sangre de Toro (Col.; AVP); Sarcille (Fr.; KAB); Schapezuring (Dutch; EFS); Sheep Sorrel (Eng.; Scn.; AH2; CR2; HJP; USN); Sorrel (Eng.; USN); Sour Dock (Eng.; EAS); Sour Sorrel (Eng.; JLH); Sourgrass (Eng.; Ocn.; AH2; EAS); Tángax Úqux (Aleutians; JLH); Vinagrerita (Sp.; KAB); Wild Sorrel (Eng.; EAS); Wilde Zuring (Dutch; EFS); Wood Sorrel (Eng.; BUR); Xiao Suan Mo (China; USN).

Activities (Sheep Sorrel):

Allergenic (1; WOI); Antiinflammatory (f; BUR); Antitumor (1; FNF); Depurative (f; EFS); Diaphoretic (f; EFS; KAB; WOI); Diuretic (f; FAD; PNC; WOI); Febrifuge (f; EFS; KAB); Hemostat (f; EFS); Laxative (1; APA); Peristaltic (f; APA); Poison (f; BIB; KAB); Purgative (f; EFS); Refrigerant (f; FAD; KAB); Styptic (f; BIB); Sudorific (f; BIB).

Indications (Sheep Sorrel):

Bleeding (f; BIB); Bruise (f; DEM); Cancer (f1; BUR; FAD; JLH; SKJ; TOM; WOI); Cancer, colon (f1; FNF; JLH); Cancer, eye (f1; FNF; JLH); Cancer, face (f1; FNF; JLH); Cancer, hand (f1; FNF; JLH); Cancer, skin (f1; FNF; JLH); Cancer, stomach (f1; FNF; JLH); Cancer, throat (f1; FNF; JLH); Dermatosis (f; EFS; JLH); Diarrhea (f; FAD); Dysmenorrhea (f; FAD); Dyspepsia (f; BIB); Dysuria (f; KAB; PNC); Epithelioma (1; FNF; JLH); Fever (f; FAD; KAB; NAD); Gastrosis (f; DEM); Inflammation (f; BUR; FAD); Jaundice (f; HJP); Metrorrhagia (f; FAD); Nephrosis (f; HJP; KAB); Pain (f; HJP); Scrofula (f; FEL); Scurvy (f1; BIB); Sore (f; BUR; DEM); Sore Throat (f; JLH; TOM); Sunburn (f; HJP); Syphilis (f; FEL); Tuberculosis (f; DEM); Tumor (f1; BUR; FNF; FAD; SKJ); Wart (f; DEM; JLH); Wen (f; JLH).

Dosages (Sheep Sorrel):

FNFF = !!

Foliage widely eaten, but not often marketed. Also used as vegetable rennet to curdle milk. Roots said to be eaten (BIB; FAC; HJP). Native Americans ate the leaves, stems, seeds, even the roots, of various species of *Rumex*. Anticosti, Bella Coola, Chehalis, Cherokee, Delaware, Hesquiat, Iroquois, Miwok, Okanagan, Saanich, and Thompson Indians reportedly eat the plant (DEM). Irishmen long ago made sorrel soup with milk. Scandinavians add it to bread. As a food pharmaceutical, a USDA 100-g serving of fresh sorrel equates to about 10 g dry sorrel (HOW).

- Alabamians suggest gargling with a strong tea for sore throat (TOM).
- Aleutian Islanders apply steamed leaves to bruises and warts (DEM; JLH).
- Asian Indians suggest the expressed juice as antiscorbutic refrigerant (NAD).
- Cherokee poultice bruised leaves and flowers on old sores (DEM).
- Europeans use the plant for renal and urinary problems (KAB).
- Lebanese consider the tea diuretic and febrifuge (HJP).
- Lebanese take decoction for dyspepsia, jaundice, kidney pain, and sunburn (HJP).
- Mohegans chew fresh leaves to alleviate stomach problems (DEM).
- North Americans apply the juice as a salve to skin cancers (JLH).

- Squaxin Indians eat raw leaves for tuberculosis (DEM).
- Tasmanians use the plant for cancer of the throat (JLH).

Downsides (Sheep Sorrel):

Class 2d (AHP). Those with kidney stones should avoid this plant. Oxalates should be avoided by endometriosis and nephrosis patients. Oxalic acid levels may attain 10 to 35% of dry matter in some species. The lowest lethal dose (LDlo) reported for humans is 700 mg/kg body weight. Weighing 100 kg, my LDlo would be 70,000 mg, or 70 g (more than 2 oz oxalic acid) (HOW). Schrader et al. (2001) reported fatal intoxication of grazing animals due to sweet clover (*Melilotus alba*), sheep's sorrel (*Rumex acetosella* L.), bracken (*Pteridium aquilinum* Kuhn), and St. John's wort (*Hypericum perforatum* L.) (X11413718).

Natural History (Sheep Sorrel):

A common fungal parasite is the leaf spot (*Cercospora acetosella*). This is the common winter host for *Botrytis cinerea* (gray mold fungus), which spreads to lowbush blueberry flowers in spring, producing blight and tip dieback. A major insect forager is the downy, slug-shaped, reddish or bright green caterpillar of the American copper butterfly (*Lycaena phlaeas*). Adult males, showing black-spotted, copper-colored forewings, defend small territories centered on sunny patches of sheep sorrel. Other *Lycaena* caterpillars may also feed. Ants frequently harvest mature seeds, dispersing them to their mounds. Bumblebees, honeybees, and some smaller butterflies visit the male plants, collecting pollen. Grouse, pheasants, prairie chickens, bobwhites, turkeys, and woodcocks consume the seeds, as do horned larks, red-winged blackbirds, bobolinks, hoary redpolls, and many sparrows. Voles and mice also eat the seeds. Poultry, rabbits, and deer readily graze the plant (EAS). Fruits are used as poultry feed. They are, like the foliage, on account of oxalic acid, reported to be poisonous to horses and sheep (BIB).

Extracts (Sheep Sorrel):

Polysaccharides show antitumor activity (ZUL).

BUTCHER'S BROOM (*RUSCUS ACULEATUS* L.) ++ RUSCACEAE

Notes (Butcher's Broom):

And there shall be no more a pricking brier unto the house of Israel, nor any grieving thorn of all that are round about them, that despised them; and they shall know that I am the Lord GOD.

Ezekiel 28:24 (KJV)

And for the house of Israel there shall be no more a brier to prick or a thorn to hurt them among all their neighbors who have treated them with contempt. Then they will know that I am the Lord GOD.

Ezekiel 28:24 (RSV)

And no more will there prove to be to the house of Israel a malignant prickle or a painful thorn out of all those round about them. Those who are treating them with scorn; and people will have to know that I am the Sovereign Lord Jehovah.

Ezekiel 28:24 (NWT)

In my first Bible Book, I accepted other authors' suggestions that this is the pricking brier of Ezekiel. So, on a nice spring day as I wrote this, I walked out to the Green Farmacy Garden to contemplate

FIGURE 1.92 Butcher's Broom (*Ruscus aculeatus*).

the differences in the words "bramble," "briar," "brier," "nettle," "prickle," "thorn," many of which connote a special thorn-like appendage. Thorny, nettle-like, and thistle-like species abound in arid lands. My *Ruscus* is very prickly, the tips of the leaves being the offender, rather than some special appendage. But there are hundreds of prickly species in the *Flora of Palestine*, and I can only say that this could be one of them. It is not, however, mentioned by Zohary in his *Flora of Palestine*, nor his later *Plants of the Bible*. I feel confident that this Mediterranean species has been introduced and survived in Israel and would survive in most Mediterranean climates. Because it shows such great promise in some of my maladies of old age, I think it belongs in any faith-based medicinal herb treatise or biblical garden, whether or not it is the thorn or the brier of Ezekiel.

Common Names (Butcher's Broom):

Acebo Menor (Sp.; EFS); Asa Barri (Arab.; BOU); Atkizounn (Ber.; BOU); Box Holly (Eng.; Ocn.; AH2; USN); Briar (Eng.; BIB); Bois Pointu (Fr.; BOU); Box Holly (Eng.; BOU; USN); Bruscolo (It.; EFS); Brusco (Sp.; EFS); Butcher's Broom (Eng.; Scn.; AH2; BOU; CR2; USN); Cobanpuskulu (Tur..; EFS); Erva dos Vasculhos (Por.; EFS); Fragon (Fr.; BOU); Fragon Epineux (Fr.; EFS); Fragon herisse (Fr.; EFS); Fragon Petit-houx (Fr.; EFS); Fragon Piquant (Fr.; BOU; USN); Gilbarbeira (Por.; EFS); Gilbardeira (Por.; EFS); Houx Frelon (Fr.; BOU; EFS); Jusbarda (Sp.; EFS); Khizana (Arab.; BOU); Kneeholm (Eng.; TAN); Knee Holly (Eng.; BIB; BOU); Muerdjel (Arab.; BOU); Myrte epineux (Fr.; EFS); Petit Houx (Fr.; BOU; USN); Prickle (Eng.; NWT); Pungitopo (It.; EFS); Rabba Bath (Arab.; Syria; HJP); Redradj (Ber.; BOU); Rusco (Sp.; Spain; EFS; VAD); Rusco Pungente (It.; EFS); Senesaq (Arab.; BOU); Shurrabet er-ra'i (Arab.; BOU); Sicilian Asparagus (Eng.; FAC); Sobhane Khallaku (Arab.; BOU); Stacheliger Mausedorn (Ger.; EFS); Stekelige Ruscus (Dutch; EFS); Unnab Barri (Arab.; BOU); Verdenace (Sp.; EFS).

Activities (Butcher's Broom):

Alpha-adrenergic (1; PNC); Alpha-adrenergic Agonist (1; X11152059); Antiaggregant (f; PED); Antiedemic (1; VAD); Anti-inflammatory (12; KOM; PHR; PH2; SKY; VAD); Aperient (f; PNC); Bitter (f; PED); Capillariprotective (1; PH2; SKY; VAD; X11152059); Cyclooxygenase Inhibitor (1; X15364641); Deobstruent (f; EFS); Diaphoretic (f;. BOU; EFS; PNC); Diuretic (2; BOU; KOM; PED; PH2); Emmenagogue (f; EFS); Expectorant (f; BIB); Febrifuge (f; BOU; EFS); Laxative (f; APA; BGB; EFS); Orexigenic (f; BOU; EFS); Tonic (1; BOU; PH2); Vasoconstrictor (1; APA; PNC; PED); Venoconstrictor (1; PNC; SKY; X11152059); Venotonic (f1; VAD; X11152059).

Indications (Butcher's Broom):

Amenorrhea (f; EFS); Anorexia (f; BOU); Arthrosis (1; APA); Atherosclerosis (f; PED; SKY); Bronchosis (f; HJP); Cancer, prostate (f; HHB; JLH); Capillary Fragility (1; PNC); Catarrh (f; BIB); Chilblain (f; BIB; X15664457); Circulosis (1; YAH); Constipation (f; DAW); Cramps (1; APA; KOM; PH2); CVI (12; APA; BGB; PH2; X14612852); Cystitis (f; VAD); Dropsy (f; BIB); Dyspnea (f; BIB); Dysuria (f; BIB); Edema (1; VAD); Erythema (1; VAD); Fever (f; BOU; EFS); Fractures (f; APA; BGB); Fungus (1; X10680445); Gout (f; VAD); Gravel (f; DAW); Hemorrhoid (12; APA; BOU; KOM; PH2; KOM; MAB; SKY); Hepatosis (f; BIB); High Blood Pressure (f; VAD); Hyperazotemia (f; VAD); Hypertriglyceridemia (1; BGB); Hyperuricemia (f; VAD); Infection (1; X10680445); Inflammation (f12; APA; KOM; VAD); Itch (1; APA; KOM; PH2); Jaundice (f; BIB; EFS); Mycosis (1; X10680445); Nephrosis (f; BIB); Obesity (f; VAD); Orthostatic Hypotension (1; X11152059); Oliguria (f; VAD); Pain (1; KOM); Phlebitis (1; PED); Pneumonia (f; HJP); Prostatosis (f; DAW; JLH); Pulmonosis (f; EFS); Respirosis (f; BIB); Retinopathy (2; BGB); Rheumatism (f; APA); Scrofula (f; BIB); Stone (f; VAD); Swelling (f1; APA; KOM; PH2; VAD); Thrombosis (1; HHB; PED); Ulcus Cruris (f; HHB); Urethrosis (f; VAD); Uterosis (f; BGB); Varicosity (1; APA; BOU; MAB; SKY; YAH); Venolymphosis (1; YAH); Water Retention (F12; BIB; BOU, PH2).

Dosages (Butcher's Broom):

FNFF = !

Young shoots cooked and eaten like asparagus, even called Sicilian Asparagus, and dressed with olive oil and lemon juice, a nice health food methinks. Scorched seed used as coffee substitute

(EFS; FAC; GMH; TAN); 300 mg tablets (APA); 7–11 mg ruscogenin (or neoruscogenin + ruscogenin)/day, or equivalent in raw extract (KOM; PH2); 1–2 Tbsp fresh root (PED); 1.5–3 g dry root (PED); 2 g dry root:10 ml alcohol:10 ml water (PED); 1 g extract, 3 ×/day (SKY); StX (with 50–100 mg ruscogenin)/day) (SKY). The VAD dosages are much higher, 60 g root/liter steeped 10 minutes; 3 cups/day; 40 g in decoction, boiled 10 minutes, 3 cups/day between meals.

- Italians treat chilblains (X15664457), perhaps self-flagellating with thorny boughs (BIB).
- Lebanese use the rhizome, sliced and dried, in decoction for catarrh, diuresis, dropsy, jaundice, kidney troubles, and respiratory difficulties (HJP).
- Middle Easterners who use this folklorically showed that its extracts inhibit *Trichophyton violaceum* (X10680445).
- North Africans use the plant for fever (BOU).
- Spanish writers hint that this may be the best of venotonic herbs (VAD).

Downsides (Butcher's Broom):

Class 1 (AHP, 1997; SKY, 1998). No health hazards or side effects known with proper therapeutic dosages (PH2). Commission E reports rhizome permitted for oral use. No contraindications or interactions. Adverse effects: rarely gastric complaints, nausea, queasiness (AEH; KOM; PHR; PH2).

Extracts (Butcher's Broom):

Ruscogenin, first isolated from this plant, is identical with Sapogenin B, which could be used as a starter material for steroids (BIB). Ruscogenins and neoruscogenins, similar to diosgenin, responsible for activities of decreasing inflammation and vascular permeability (SKY). Saponins are antiaggregant, antiinflammatory, capillarifortificant, and diuretic (PED).

FRINGED RUE (*RUTA CHALEPENSIS* L.) + RUTACEAE

Synonyms:

Ruta angustifolia Lowe; *Ruta bracteosa* DC.; *Ruta chalepensis* var. *bracteosa* (DC.) Boiss.; *Ruta graveolens* var. *angustifolia* Lowe *fide* HH2

Notes (Fringed Rue):

But woe unto you, Pharisees! for ye tithe mint and rue and all manner of herbs, and pass over judgment and the love of God: these ought ye to have done, and not to leave the other undone.

Luke 11:42 (KJV)

But woe to you Pharisees! for you tithe mint and rue and every herb, and neglect justice and the love of God; these you ought to have done, without neglecting the others.

Luke 11:42 (RSV)

But woe to you Pharisees, because you give the tenth of the mint and the rue and of every[other] vegetable, but you pass by the justice and the love of God! These things you were under obligation to do, but those other things not to omit.

Luke 11:42 (NWT)

At least the three versions above agree on mint and rue being tithed. But which rue? In my *Medicinal Plants of the Bible*, I relied on my predecessors and assumed that the biblical rue was *Ruta*

FIGURE 1.93 Fringed Rue (*Ruta chalepensis*).

graveolens. Israeli botanist Michael Zohary, however, without even indexing or mentioning *R. graveolens*, assigns it to *Ruta chalepensis*, the only species covered in the *Flora of Palestine*. Mentioned only once in the Bible, first under its Greek name *peganon*, most often post-biblically as *pigam*, closely cognate with the Arabic *fegam*. Pliny mentions honeyed wine flavored with rue, as well as 84 remedies containing rue, but I cannot be sure which species of rue he mentioned (FP2; ZOH). I feel rather certain that both could be grown in Israel but in this, my third botanical trip through the Bible, I will follow Zohary and treat *Ruta chalepensis*. The more temperate *Ruta graveolens*, thriving in Maryland in the United States, was discussed in my CRC *Handbook of Medicinal Herbs* (Edition 2).

Common Names (Fringed Rue):

Afar (Eth.; HH2); Al Shathap (Arab.; HH2); Aleppo Rue (Eng.; BOU); Aourmi (Ber.; BOU); Arouvadam Chedi ((Tam.; HH2); Arruda (Mad.; Por.; JAD); Arvada (Tam.; HH2); Bou Ghans (Arab.; BOU; HH2); Citronelle (Sp.; HH2); Citronelle Marron (Haiti; AVP); Common Rue (Eng.; ZOH); Djell (Ber.; BOU); Eastern Rue (Eng.; HH2); Egyptian Rue (Eng.; FAC); Ermul (Beng.; HH2); Fegan (Arab.; ZOH); Fidgel (Arab.; BOU; HH2); Fidjla (Arab.; BOU; HH2); Fringed Rue (Eng.; Scn.; AH2; USN); Ispunol (Beng.; HH2); Issel (Ber.; BOU); Issin (Ber.; BOU); Peganon (Greek; ZOH); Pigam

(Heb.; ZOH); Pismarum (Hindi; HH2); Red d'Alep (Fr.; BOU); Rora (Ma.; JFM); Ruda (Cr.; Peru; DAV; MDD); Ruda Antillana (Ma.; HH2; JFM); Ruda de España (Sp.; HH2); Ruda de la Tierra (Ma.; JFM; HH2); Ruda de las Antilles (Ma.; HH2; JFM); Ruda d'Espanya (Ma.; JFM); Ruda Tropical (Ma.; JFM); Rue (Eng.; Fr.; Pr.; AVP; BOU); Rue Ailee Fetide (Fr.; AVP); Rue d'Alep (Fr.; AVP); Rue d'Antilles (Fr.; AVP); Rue d'Orient (Fr.; AVP); Rura (Ma.; JFM); Ruta (Arab.; BOU); Ruta Sfangiata (It.; HH2); Rutsa (Arab.; BOU); Saadab (Arab.; ZOH); Sadab (Arab.; Hindi; GHA; HH2); Sadhab (Arab.; BOU; GHA); Satari (Hindi; HH2); Shadhab (Arab.; GHA); Shathab (Oman; Saudi; Yemen; GHA); Syrian Rue (Eng.; BOU; HH2); T'enadam (Arab.; HH2); Zent (Ber.; BOU).

Activities (Fringed Rue):

Abortifacient (f; UPH); Analgesic (1; HH2); Antiedemic (1; JEB28(3):305); Antiendotoxemic (f; JEB90:267); Antiexudative (1; HH2); Antifeedant (1; X11935899); Antifertility (1; X2748734); Antiinflammatory (f1; HH2; JEB90:267; JEB28(3):305); Antiseptic (1; HH2); Antispasmodic (f; SOU; ZOH); Aphrodisiac (f; GHA); Arachnifuge (f; BOU); Bactericide (1; X12423924); Candidicide (1; HH2); Cardiotonic (f; DAV); CNS Depressant (f1; JEB28(3):305); CNS Depressant (f; JFM); Decongestant (f; DAV; JFM); Digestive (f; DAV); Emetic (f; JFM); Embryotoxic (1; JEB69:93); Emmenagogue (f; BOU; UPH); Febrifuge (1; HH2); Fungicide (1; X10680445); Immunomodulator (f1; JEB90:267; X15013191); Insectifuge (1; PR17:202; X12672146); Molluscicide (1; FT71:308); NO Inhibitor (1; X15013191); Oxytocic (f; SOU); Phototoxic (1; DAV); Rubefacient (f; JFM); Sedative (f; BOU; DAV); Spasmodic (f; BOU); Stomachic (f; JFM); Sudorific (f; JFM); Vermifuge (f; SOU; UPH); Vulnerary (f; BOU).

Indications (Fringed Rue):

Amenorrhea (f; FP2); Arthrosis (f1; GHA; X2598777); Cold (f; BOU); Bacillus (1; X12423924); Bacteria (1; X12423924); Bronchosis (f; AHL); Candida (1; HH2); Cardiopathy (f; JFM); Childbirth (f; JFM); Cold (f; HH2); Colic (f; GHA); Congestion (f; JFM); Cough (f; HH2); Cramp (f; WOI); Dermatosis (1; X10680445); Dysmenorrhea (f; DAV; FP2; HH2); Earache (f; BOU; HH2; JFM); Edema (1; JEB28(3):305); Endotoxemia (1; X15013191); Enterosis (f; BOU; GHA); Epilepsy (f; SOU); Epistaxis (f; SOU); Escherichia (1; HH2); Fever (f1; BOU; HH2); Fungus (1; X10680445); Gingivosis (f; JFM); Headache (f; GHA; HH2); Hysteria (f; DAV; FP2); Infection (1; X12423924; HH2); Inflammation (1; HH2); Measles (f; JFM); Microsporium (1; X10680445); Myalgia (f; DAV); Mycosis (1; X10680445); Nausea (f; BOU; HH2); Nervousness (f; BOU; HH2); Neurosis (f; HH2); Ophthalmia (f; DAV; HH2; UPH); Otosis (f; DAV); Pain (f1; BOU; GHA; HH2); Palpitation (f; SOU); Paralysis (f; HH2); Pediculosis (f; DAV); Pseudomonas (1; HH2); Pulmonosis (f; HH2); Rheumatism (f1; BOU; FP2; X2598777); Rhinosis (f; BOU; HH2); Scarlet Fever (f; JFM); Shock (f; HH2); Snakebite (f; GHA); Sore (f; BOU; UPH); Soroche (f; SOU); Staphylococcus (1; HH2); Stomachache (f; GHA); Stroke (f; DAV); Swelling (f1; BOU; JEB28(3):305); Worm (f; FP2; UPH); Wound (f; BOU; GHA); Yeast (1; HH2).

Dosages (Fringed Rue):

FNFF = !
Aromatic leaves esteemed as spice by North African Jews, added to lamb and beef sausages (*merguez*); also used in Tunisian omelettes (*hajja*) (FAC).

- Arabians rub leaves on arthritic or painful areas (GHA) (watch out for photodermatitis; JAD).
- Dominicans mix juice with castor oil for bronchosis (AHL),
- Haitians use the plant as antiepileptic, emmenagogue, sudorific, vermifuge, and to treat ulcerated gums (JFM).

- Expressed juice heated and used as eardrop in earache (JFM).
- Venezuelans take the decoction to overcome shock or spasms (JFM).
- Yemeni chew leaves with sugar for stomachache (GHA).
- Vapors of the plant said to dispel eye fatigue (JFM).

Downsides (Fringed Rue):

Although widely used in Latin America, Julia Morton described the tropical rue as causing cold extremities, feeble slow pulse, gastroenterosis, salivation, swelling of the tongue, and vomiting (when taken in large doses). Overdoses taken in attempted abortion have been fatal (JFM).

Extracts (Fringed Rue):

Iauk et al. (2004) showed that the antiinflammatory biblical rue protected against murine endotoxemia (gavage at 1 g/kg per day for 7 to 14 days before injecting 0.75 mg endotoxin), There was evidence of reduced nitric oxide production. *Ruta chalepensis* has immunopharmacological properties counteracting the lethal effects of high doses of endotoxin (X15013191). Hadis et al. (2003) showed that rue (50% in coconut oil) repelled *Mansonia* mosquitoes in western Ethiopia. At 50% concentration, protection was 91.6%, 87.0%, 96.0%, 97.9% for rue, neem, pyrethrum, and deet, respectively. At 40% concentration deet, lemon eucalyptus and pyrethrum were significantly more effective than rue and neem (X12672146). Mancebo et al. (2001) demonstrated a clear antifeedant activity for rue extracts at a concentration of 0.32% (X11935899).

SUGARCANE (*SACCHARUM OFFICINARUM* L.) +++ POACEAE

Notes (Sugarcane):

Thou hast bought me no sweet cane with money, neither hast thou filled me with the fat of thy sacrifices: but thou hast made me to serve with thy sins, thou hast wearied me with thine iniquities.

Isaiah 43:24 (KJV)

You have not bought me sweet cane with money, or satisfied me with the fat of your sacrifices. But you have burdened me with your sins, you have wearied me with your iniquities.

Isaiah 43:24 (RSV)

For me you have bought no [sweet] cane with any money, and with the fat of your sacrifices you have not saturated me. In reality you have compelled me to serve because of your sins, you have made me weary with your errors.

Isaiah 43:24 (NWT)

To what purpose cometh there to me incense from Sheba, and the sweet cane from a far country? Your burnt offerings are not acceptable, nor your sacrifices sweet unto me.

Jeremiah 6:20 (KJV)

To what purpose does frankincense come to me from Sheba, or sweet cane from a distant land? Your burnt offerings are not acceptable, nor your sacrifices pleasing to me.

Jeremiah 6:20 (RSV)

FIGURE 1.94 Sugarcane (*Saccharum officinarum*). Source: BIB.

What does this matter to me that you bring in even frankincense from She'ba, and the good cane from the land far away? The whole burnt offerings of you people serve for no pleasure and your very sacrifuces have not been gratifying to me.

Jeremiah 6:20 (NWT)

Back before I read Zohary's book, my other reading had led me to conclude that the sweet cane of Isaiah was probably our sugarcane of today. But Zohary is inclined to believe that the sweet cane was more probably an aromatic grass of the genus *Cymbopogon*, or maybe even calamus, and seems to have ruled out the sugarcane and the vetiver. The sweet sugarcane is rather heavy to be carried from afar and *Cymbopogons*, *Calamus*, and today even *Vetiver* are more precious ounce for ounce than sugarcane. Still I leave sugarcane here, knowing it can be grown in Israel, as it can be in almost all tropical and subtropical countries. I have also cited many abstracts from Cuba, where sugarcane reigns supreme. I see a parallel between these good Cuban scientists trying to find more uses for King Cane, like our good United States scientists are always looking for more uses for King Soybean. The residue, bagasse, used in building materials, insulation against temperatures and sound, resins in phonograph records, mulch and litter, plastics, paper making, and in industrial chemicals, and now from Cuba, polycosanol and D-003. As the cost of fossil fuels increases, it may assume importance as an energy source. Sugarcane alcohol seems as promising as corn-based ethanol, as the price of petroleum spirals upward. Yet I do not hear of any efforts in Cuba to convert to an ethanol fuel economy.

Common Names (Sugarcane):

Adhipatra (Sanskrit; KAB); Afunu (Ada; KAB); Agbo (Cagayan; KAB); Ahleu (Korbo; KAB); Ahwerenkakraba (Fanti; KAB); Ahwereu (Twi; KAB); Ak (Beng.; KAB); Akali (Nepal; KAB); Akh (San..; KAB); Aku (Uriya; KAB); Ampeou (Cam.; KAB); Ampon (Cam.; KAB); Angarigai (Tam.; KAB); Ankhu (Majhi; NPM); Aos (Mar.; KAB); Ariva (Nc.; KAB); Arolam (Nc.; KAB); Arrake (Sokoto; KAB); Arukanupulakranuga (Tel.; KAB); Asibattiragam (Tam.; KAB); Asipatra (Sanskrit; KAB); Bhurirasa (Sanskrit; KAB); Bich (Sunwar; NPM); Bogleng (Ewe; Krebi; KAB); Boglengbiri (Krebi; KAB); Boglengfe (Krebi; KAB); Boglengyibor (Krebi; KAB); Boiepe (Nc.; KAB); Boinlioua (Nc.; KAB); Bu Ram (Tibet; NPM); Caña (Peru; Sp.; EGG); Caña de Azucar (Peru; Sp.; EGG); Cana de Asucar (Por.; AVP); Cana Doce (Por.; AVP); Cana Dolsa (Cat.; KAB); Canamelle (Fr.; KAB); Canchi (China; EFS); Canna (Brazil; KAB); Canna de Assucar (Por.; Brazil; AVP; KAB); Canna de Zucchero (It.; AVP); Canna Doce (Por.; KAB); Cannamelle (Fr.; It.; AVP; KAB); Canne (Haiti, Reunion; AHL; AVP; KAB); Canne a Sucre (Fr.; Haiti; AHL; AVP; EFS); Canne de Batavia (Fr.; KAB); Canne de la Chine (Fr.; KAB); Canne d'Haiti (Fr.; KAB); Canne Indigene (Fr.; AHL); Canne Pays (Fr.; AHL); Canya de Azucar (Dr.; Sp. AHL; EFS; IED); Canya Dulce (Sp.; EFS); Canya Miel (Sp.; EFS; KAB); Canyaduzales (Sp.; AVP); Canyuzales (Sp.; AVP); Chaku (Nepal; KAB); Cheraku (Tel.; KAB); Cherakubhedamu (Tel.; KAB); Darbheshu (Mal.; KAB); Delenole (Nc.; KAB); Dilou (Nc.; KAB); Dirghachhada (Sanskrit; KAB); Dogangueni (Nc.; KAB); Echtes Zuckerrohr (Ger.; TAN); Fary (Hova; Madagascar; KAB); Fisika (Sakalave; KAB); Fofungu (Awina; Ewe; KAB); Gadenadeboui (Nc.; KAB); Gana (Urdu; KAB); Ganda (Dec.; Hindi; KAB); Gandida (Sanskrit; KAB); Ganna (India; Nwp.; Pun.; EFS; KAN); Ghenru (Parbuttiah; KAB); Gol (Bom.; KAB); Goreate (Nc.; KAB); Gudada (Sanskrit; KAB); Gudadaru (Sanskrit; KAB); Gudakashta (Sanskrit; KAB); Gudamula (Sanskrit; KAB); Gudatrina (Sanskrit; KAB); Gura (Sanskrit; EFS); Gurdanda (Mun.; KAB); Gudodaru (Uriya; KAB); Gurkatauri (Mun.; KAB); Gurkosear (Mun.; KAB); Gursing (Sherpa; NPM); Ik (Beng.; Hindi; Nwp.; KAB); Ikhari (Nwp.; KAB); Ikhyu (Uriya; KAB); Ikku (Tam.; KAB); Ikshu (Kan.; Mal.; Sanskrit; San.; AH2; EFS; KAB); Ikshudanda (Kan.; KAB); Ikshupu (Tel.; KAB); Ikshura (Sanskrit; KAB); Impuco (Antis; EGG; RAR; SOU); Ingolu (Kan.; KAB); Inju (Tel.; KAB); Itica (Cocama; EGG; SAR); Jate (Nc.; KAB); Ka Ra (Tibet; NPM; TIB); Kabbo (Mar.; KAB); Kabbu (Kan.; KAB); Kabopolenouen (Nc.; KAB); Kajuli (Beng.; KAB); Kalai (Tam.; KAB); Kamand (Pun.; Sin.; KAB); Kan che (China; EFS; KAB); Kannal (Tam.; KAB); Kansia (Japan; KAB); Kantara (Kan.; Sanskrit; KAB); Kantaraka (Kan.; Sanskrit KAB); Kantarakam (Mal.; KAB); Kantaramu (Tel.; KAB); Kanthirikhu (Nwp.; KAB); Kanupulacheraku (Tel.; KAB); Karambu (Ceylon; KAB); Karansariki (Hausa; KAB); Karimpu (Mal.; KAB); Karkotaka (Sanskrit; KAB); Karumbu (Tam.; Tulu; KAB); Kasabishakar (Arab.; KAB); Kasibshakar (Arab.; KAB); Katari (Behar; KAB); Ketari (Behar; KAB); Keyan

(Burma; KAB); Khadgapatraka (Sanskrit; KAB); Khand (Pun.; KAB); Khum (Magar; NPM); Khunjhi (Tharu; NPM); Khusiyar (Behar; KAB); Kiaboue (Nc.; KAB); Kinemaite (Nc.; KAB); Kondimoua (Nc.; KAB); Koubala (Nc.; KAB); Koshakara (Sanskrit; KAB); Kulluar (Beng.; KAB); Kumad (Hindi; KAB); Kushiar (Beng.; KAB); Kyan (Burma; KAB); Lavucheraku (Tel.; KAB); Madhura (Kan.; KAB); Madhutrina (Sanskrit; KAB); Madhutrinam (Mal.; KAB); Madhuyashti (Sanskrit; KAB); Madudirunam (Tam.; KAB); Maharasa (Nc.; KAB); Majonana (Culina; RAR); Marakabbu (Kan.; KAB); Mebouangue (Nc.; KAB); Mengou (Nc.; KAB); Mia (Annam; Ic.; KAB); Mia co ke (Tonkin; KAB); Mia lau (Tonkin; KAB); Mia ly (Tonkin; KAB); Migao (Nc.; KAB); Misk'i wiru (Aym.; Bol.; DLZ); Misqui Huiro (Peru; EGG; SOU); Moene (Nc.; KAB); Moindiene (Nc.; KAB); Moueouete (Nc.; KAB); Mrityupushpu (Sanskrit; KAB); Nai (Iran; EFS); Ñaamura (Uvosha; EGG); Naisakar (Guj.; KAB); Naishakar (Hindi; Iran; KAB); Ngala (Nc.; KAB); Niemba (Nc.; KAB); Noble Sugarcane (Eng.; USN); Nyaamura (Uvosha; SOU); Oen mangia (Nc.; KAB); Oen ou poudendate (Nc.; KAB); Ouali (Nc.; KAB); Ouane (Nc.; KAB); Ouassab (Arab.; EFS); Oudiepe-ait (Nc.; KAB); Ouene (Nc.; KAB); Ouenebail (Nc.; KAB); Paat (Peru; EGG; SOU); Pagad (Aguaruna; RAR); Pagat (Aguaruna; Huambisa; SOU); Paiambou (Nc.; KAB); Paieme (Nc.; KAB); Pam (Lepcha; NPM); Paruvayoni (Tam.; KAB); Paunda (Pun.; KAB); Payodhara (Sanskrit; KAB); Pidiak (Nc.; KAB); Pie canne (Haiti; AVP); Pobone (Nc.; KAB); Pochoasiri (Piro; EGG; SOU); Pochwacsuru (Piro; RAR); Poilote (Nc.; KAB); Pottikamupucheraku (Tel.; KAB); Punarikhu (Nwp.; KAB); Pundaram (Tam.; KAB); Pundra (Kan.; KAB); Pundraka (Sanskrit; KAB); Puri (Beng.; KAB); Qasab al Sukkar (Arab.; GHA); Qasabussakar (Arab.; KAB); Quilaba (Vis.; KAB); Rake (Hausa; KAB); Rasadali (Kan.; KAB); Rasala (Kan.; Sanskrit; KAB); Rasalu (Sanskrit; KAB); Rastale (Kan.; KAB); Rikhu (Hindi; Kum.; Nwp.; KAB); Roseau a sucre (Fr.; KAB); Sabi (Conibo; Shipibo; EGG; RAR; SOU); Sacchar (Nepal; SUW); Sahachar (Nepal; SUW); Saharnyi trastnik (Rus.; KAB); Sastra (Sanskrit; KAB); Sato Kibi (Japan; TAN; USN); Schimate (Nc.; KAB); Seker kamizi (Tur.; EFS); Senoorr (Amuesha; SOU); Senorr (Yanesha; EGG); Serdi (Bom.; Guj.; KAB); Sha T'ang (China; KAB); Sharhara (Sanskrit; EFS); Shakarsurkh (Pun.; KAB); Sheng (Ga; KAB); Sheradi (Guj.; KAB); Sherdi (Guj.; KAB); Shih mi (China; KAB); Soo (Limbu; NPM); Sotalong (Limbu; NPM); Sthiabanghi (Nc.; KAB); Sucre de canne (Fr.; EFS); Sukumasaka (Sanskrit; KAB); Sugarcane (Eng.; Scn.; AH2; AVP; USN); Suikerriet (Dutch; EFS; KAB); Sukker (Den.; EFS); Taa vata (Amahuaca; RAR); Tacamaree (Brazil; KAB); Tacuane (Chiriguano; DLZ); Tangalite (Nc.; KAB); Tanigarbu (Kan.; KAB); Tebu (Java; KAB); Tebu gula (Malaya; EFS); Tellacheraku (Tel.; KAB); Thsiogan (Nc.; KAB); Tilibi (Nc.; KAB); Tiyyamranu (Tel.; KAB); Trestie de zahar (Rom.; KAB); Trinadhiya (Sanskrit; KAB); Trinaraja (Kan.; KAB); Tshiambo (Nc.; KAB); Tu (Newari; KAB; NPM); Tubo (Tag.; KAB); Tunta (Tel.; KAB); Uduwa (Rai; NPM); Uinkh (Mooshar; NPM); Uk (Beng.; Hindi; Nepal; Sin.; KAB); Ukgas (Sin.; KAB); Ukh (Behar; Hindi; KAB); Ukhari (Nwp.; KAB); Ukhi (Behar; KAB); Ukhu (Danuwar; Nepal; Tamang; NPM; SUW); Ukkiragandam (Tam.; KAB); Ukkiragandi (Tam.; KAB); Unkh (Bkojpuri; NPM); Uns (Guj.; KAB); Uny (Kon. KAB); Us (Bom.; Decca; Mar.; KAB); Usa (Mar.; KAB); Uss (Kon. KAB); Usyu (Gurung; NPM); Usyup (Tamang; NPM); Vamsukamu (Tel.; KAB); Vansha (Sanskrit; KAB); Velam (Tam.; KAB); Vellakarimpu (Mal.; KAB); Vengarumbu (Tam.; KAB); Viha (Brazi Zucchero (It.; EFS); Vipularasa (Sanskrit; KAB); Vrishya (Sanskrit; KAB); Xai (Cashibo; EGG; RAR; SOU); Zuckerrohr (Ger.; AVP; EFS; KAB).

Activities (Sugarcane):

Analgesic (f; X12709906); Anthelmintic (f; KAB); Antiaggregant (1; X15272645); Antiallergic (1; X15729619); Antidote (arsenic) (f; KAB); Antidote (copper) (f; KAB); Antiinflammatory (1; X15729619; X12709906); Antioxidant (1; X14756190); Antiplatelet (1; X15272645); Antiseptic (f; EFS); Antivinous (f; BIB); Aphrodisiac (f; BIB; SUW); Bactericide (f; BIB); Cardiotonic (f; BIB; EFS); Demulcent (f; EFS; SUW); Depurative (f; TIB); Diuretic (f; AHL; BIB; GHA; SUW); Emollient (f; KAB); Febrifuge (f; BIB); Hepatoprotective (1; (X14756190); Hypocholesterolemic (1; X15272645); Immunostimulant (1; X14975361); Laxative (f; AHL; BIB); Osteoprotective (1;

X15357627); Pectoral (f; BIB; KAB); Piscicide (f; BIB); Radioprotective (1; X14975361); Refrigerant (f; AHL; EFS); Stomachic (f; BIB).

Indications (Sugarcane):

Allergy (1; X15729619); Anemia (f; KAB); Arthrosis (f; BIB); Backache (f; JFM); Biliousness (f; KAB); Blenorrhagia (f; DLZ); Boil (f; BIB); Calculus (f; DLZ; KAB); Cancer (f; JLH); Cancer, breast (f; JLH); Cancer, mouth (f; JLH); Cancer, rectum (f; JLH); Cancer, stomach (f; JLH); Cancer, tonsil (f; JLH); Cancer, uterus (f; JLH); Cardiopathy (1; X15272645); Catarrh (f; BIB); Cold (f; DLZ; JFM); Colic (f; DLZ); Constipation (f; AHL); Cough (f; DLZ; GHA); Cystosis (f; DLZ); Decubitis (f; BIB); Delirium (f; KAB); Dermatosis (f1; JFM; X15729619); Diarrhea (f; JFM; KAB); Dysentery (f; JFM); Dysuria (f; JFM); Enterosis (f; KAB); Erysipelas (f; KAB); Fatigue (f; KAB); Fever (f; TIB); Fungus (f; JFM); Frambesia (f; BIB); Gastrosis (f; JLH); Gingivosis (f; BIB); Hemorrhoid (f; BIB); Hepatosis (f; DLZ); Hiccup (f; BIB); High Cholesterol (1; X15272645); Infection (f; EGG; SAR); Inflammation (f1; JFM; X15729619); Jaundice (f; EGG); Laryngosis (f; BIB); Leprosy (f; KAB); Mastosis (f; JLH); Mucososis (f; KAB); Mycosis (f; JFM); Nephrosis (f; BIB; DLZ; EGG); Neurosis (f; DLZ); Opacity (f; BIB); Ophthalmia (f; GHA; SAR); Osteoporosis (1; X15357627); Pain (f1; EGG; X12709906); Pertussis (f; BIB); Proctosis (f; JLH); Ringworm (f; JFM); Smallpox (f; BIB); Sore (f; KAB); Sore Throat (f; BIB); Splenosis (f; BIB); Splinter (f; JFM); Stomatosis (f; JLH); Thirst (f; KAB); Thrombosis (1; X15272645); Uterosis (f; JLH); Wound (f; BIB; EGG).

Dosages (Sugarcane):

FNFF = !!!.

- Arabians use cane juice as antitussive, diuretic, and ophthalmic (GHA).
- Cubans drink expressed juice as diuretic; formerly sucked roasted cane for diarrhea and dysentery (JFM).
- Curacaons make decoction of dry fallen leaves for dysuria (JFM).
- Mexicans take juice from roasted stems for colds (JFM).
- Peruvians drink fermented cane juice for liver pains (EGG).
- Peruvians put powdered sugar on wounds to prevent infection (EGG).
- Peruvians take roasted cane against jaundice and kidney pain (EGG).
- Various cultures suggest molasses for cancer of the breast, mouth, rectum, stomach, tonsils, and uterus (JLH).
- Yumbos apply a few drops of warm sap to infected eyes (SAR).
- Mashed root with vinegar poulticed onto backache (JFM).
- Ash of epidermis applied with vinegar to ringworm (JFM).

Natural History (Sugarcane):

Sugarcane is susceptible to the following viruses: cucumber mosaic, maize leaf fleck, sugarcane mosaic, tulip breaking, wheat streak mosaic, chlorotic streak, and sereh. The following fungi have been reported from sugarcane: *Allantospora radicicola*, *Alternaria* sp., *Apiospora camtospora*, *Arthrobotrys suberba*, *Aspergillus* sp., *A. flavus*, *A. fumigatus*, *A. herbariorum*, *A. nidulans*, *A. niger*, *A. penicillioides*, *A. repens*, *A. sydowii*, *A. terreus*, a form of *A. flavus* designated as *A. parasiticus* on mealybugs infesting cane, *Asterostroma cervicolor*, *Ceratostomella adiposum*, *C. paradoxa*, *Cercospora koepkei*, *C. vaginae*, *Chytridium* sp., *Cladosporium herbarum*, *Clathrus columnatus*, *Colletotrichum falcatum*, *C. graminicola*, *C. lineola*, *Corticium sasakii*, *Curvularia* sp., *Cytospora sacchari*, *Endoconidiophora adiposa*, *E. paradoxa*, *Eriosphaeria sacchari*, *Fusarium* spp., *Gibberella fujikuroi*, *Gloeocercospora sorghi*, *Gnomonia iliau*, *Graphium sacchari*, *Helminthosporium sacchari*, *H. stenospilum*, *Himantia stellifera*, *Hormiactella sacchari*, *Hypocrea gelatinosa*, *Ithyphallus rubicundis*, *Leptosphaeria sacchari*, *Ligniera vascularum*, *Lophodermium sacchari*, *Macrophoma sacchari*,

Marasmius sacchari, M. stenophyllus, Melanconium sacchari, Microdiplodia melaspora, Mycosphaerella sacchari, M. striatiformans, Myriogenospora aciculisporae, Nectria spp., *Neurospora sitophila, Nigrospora oryzae, Odontia saccharicola, Olpidium sacchari, Papularia sphaerosperma, P. vinosa, Periconia sacchari, Phyllosticta sorghina, Physalospora rhodina, P. tucumanensis, Phytophthora erythroseptica, Plectospira gemmifera, Polyporus* spp., *P. occidentalis, P. sanguineus, P. tulipiferus, Poria ambigua, Psilocybe atomatoides, Pythium* spp., *P. arrhenomanes, P. graminicola, P. aphanidermatum, P. artotrogus, P. debaryanum, P. dissotocum, P. helicoides, P. irregulare, P. mamillatum, P. monospermum, P. periilum, P. rostratum, P. splendens, P. ultimum, P. vexans, Rhizoctonia ferruginea, R. pallida, R. solani, Rosellinia paraguayensis, R. pulveracea, Saccharomyces zopfii, Schizophyllum commune, Scirrhia lophodermioides, Sclerotium rolfsii, Trichoderma lignorum, Tubercularia saccharicola, Vermicularia graminicola, Xylaria apiculata, Nectria flavociliata, and N. laurentiana.* The following nematodes have been reported on sugarcane: *Anguina spermophaga, Helicotylenchus* sp. *Heterodera* spp., *Hoplolaimus* sp., *Meloidogyne* sp., *Pratylenchus* spp., *P. pratensis, Rotylenchus* spp., *R. similes, Scutellonema spp., Trichodorus christie*, and *Tylenchorhynchus* spp. (Golden, 1984). Bacteria include *Bacillus megatherium, B. mesentericus, Xanthomonas albilineans, X. rubrilineans, X. rubrisubalbicans, and X. vasculorum* (Agriculture Handbook No. 165).

Extracts (Sugarcane):

Cuban researchers (Ledone et al. 2005) showed that a mixture of fatty acids from sugarcane (mostly palmitic, oleic, linoleic, and linolenic acids) showed antiinflammatory activity in test models for allergy, suggesting utility in allergic and inflammatory dermatosis (X15729619). Other Cuban scientists, Noa et al. (2004), looking at by-products of the sugarcane industry (like United States scientists look at soy by products) are working with a cholesterol-lowering mix called D-003. D-003 also prevents bone loss and bone resorption in ovariectomy-induced osteoporosis in rats. Compared with a sham group, prednisolone significantly reduced trabecular bone volume, while D-003 significantly and dose-dependently prevented the induced reduction of TBV. "D-003 could be useful for managing corticosteroid-induced osteoporosis" (X15357627). Gamez et al. (2004), working with beagles, reported antiplatelet and hypocholesterolemic effects. At 200 and 400 mg/kg, D-003 significantly reduced total cholesterol, inhibited platelet aggregation, and increased bleeding time, compared to controls administered D-003 for 9 months to beagles induced no signs of toxicity (X15272645).

GLASSWORT (*SALICORNIA EUROPEA* L.) ++ CHENOPODIACEAE

Synonyms:

Salicornia europaea var. *herbacea* L.; *Salicornia herbacea* (L.) L.; *Salicornia virginica* L.

Notes (Glasswort):

But who may abide the day of his coming? and who shall stand when he appeareth? for he is like a refiner's fire, and like fullers' soap.

Malachi 3:2 (KJV)

But who can endure the day of his coming, and who can stand when he appears? "For he is like a refiner's fire and like fullers' soap."

Malachi 3:2 (RSV)

But who will be putting up with the day of his coming, and who will be the one standing when he appears? "For he is like the fire of a refiner and like the lye of laundrymen."

Malachi 3:2 (NWT)

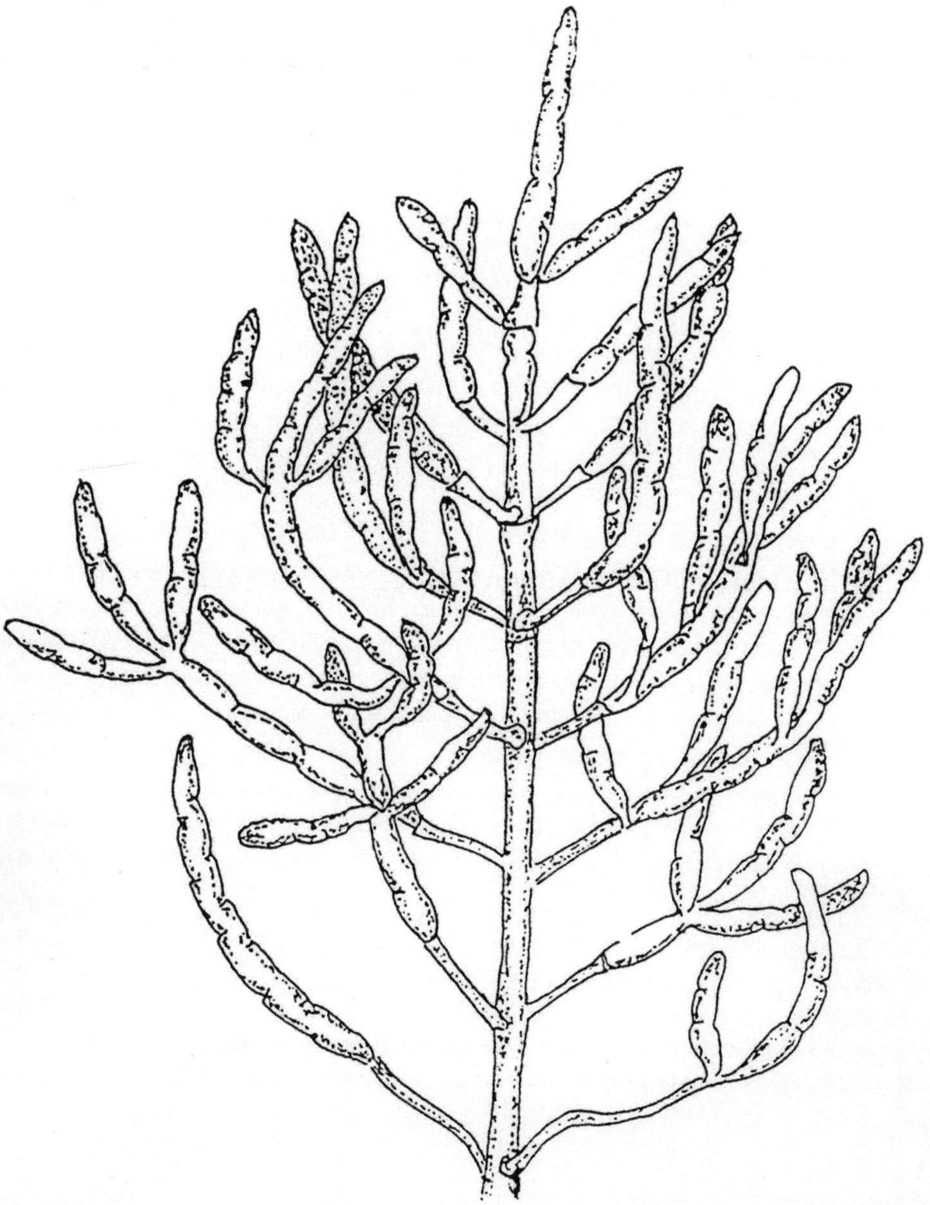

FIGURE 1.95 Glasswort (*Salicornia europea*). Source: BIB.

According to Grieve, "there are references in the Bible to the use of Glasswort for soap and for glass" (GMH). The potash, or alkaline salts used in biblical "sope-making," were derived from the ashes of glassworts and other halophytic species. The potash was then mixed with olive oil.

Hartwell lists this and other species of glasswort called *kelpwort* and *samphire* as folk remedies for tumors and superfluous flesh. As "herba salicorniae herbaceae," it is used medicinally in Palestine (BIB). Zohary lists no *Salicornias* in his *Plants of the Bible* (ZOH) but does list this species in his *Flora of Palestine* (FP1) as an edible medicinal species, a pioneer species in saline puddles drying up.

Common Names (Glasswort):

Akkeishi Sô (Japan; TAN); Chicken-claws (Eng.; USN); Common Glasswort (Eng.; USN); Glasswort (Eng.; BIB; TAN); Lye (Eng.; NWT); Marsh Samphire (Eng.; TAN; USN); Saltwort (Eng.; TAN); Soap (Eng.; BIB; KJV; RSV); Sope (Eng.; BIB); Yan Jiao Cao (China; USN).

Activities (Glasswort):

Antiscorbutic (1; EFS; FNF); Depurative (f; EFS); Digestive (1; EFS); Tonic (f; EFS).

Indications (Glasswort):

Cancer (f; JLH); Dropsy (f; EB28:315); Scurvy (1; EFS).

Dosages (Glasswort):

FNFF = !

Plant (leaves, stems and seed) is edible but ashes are more often used like lye in making soap (TAN; UPH).

WILLOW (*SALIX ACMOPHYLLA* BOISS.) ++ SALICACEAE

Synonyms:

Salix glauca Anderss.; *Salix octandra* Del. *fide* DEP

Notes (Willow):

Zohary mentions that the biblical Hebrew word refers both to this and other species of *Salix* in Israel (five species). He notes that this is a honey plant, a tannin source, used for wicker work (ZOH). I refer readers to the better-known *S. alba* and *S. babylonica* for fuller accounts. Poplars and willows belong to the same family, and are fast-growing weed trees especially when growing near water. Zohary suggests that if the willows of the Bible were native to the Holy Land, they would have been *Salix acmophylla* or *Salix alba*, which hybridize readily. I suggest that romantics like me like the idea of the weeping willow, *Salix babylonica*, in their biblical garden. These three may be keyed as follows:

- Branches erect or ascending; leaves less than 10 times as long as broad:
  - — Stamens, 4 to 5; capsule pedicels circa 1.5 mm long; branches reddish: *S. acmophylla*
  - — Stamens, 2; capsules sessile; branches yellow-green to brown: *S. alba*
- Branches pendulous; leaves more than 10 times as long as broad: *S. babylonica*

Zohary notes that, in general, *Salix alba* is found in cooler areas, *S. acmophylla* in hotter areas.

Common Names (Willow):

Aravah (Heb.; ZOH); Bada (Hindu; Pun.; DEP; NAD; SKJ); Bed (Pun.; SKJ); Bedh (Afg.; DEP); Bisu (Pun.; DEP); Budha (Bom.; Sin. DEP; NAD; SKJ); Gadphains (Up.; SKJ). Ncsn.

Activities (Willow):

Febrifuge (f; NAD; SKJ).

Indications (Willow):

Fever (f; NAD; SKJ).

FIGURE 1.96 Willow (Salix acmophylla). Source: KAB.

Dosages (Willow):

FNFF = ?

- Asian Indians suggest the bark for fever (SKL).
- Natural History (Willow):
- Leaves lopped for fodder.

WHITE WILLOW (*SALIX ALBA* L.) ++ SALICACEAE

Notes (White Willow):

And they shall spring up as among the grass, as willows by the water courses.

Isaiah 44:4 (KJV)

They shall spring up like grass amid waters, like willows by flowing streams.

Isaiah 44:4 (RSV)

And they will certainly spring up among the green grass, like poplars by the water ditches streams.

Isaiah 44:4 (NWT)

Along the Jordan River, willows frequent the fresher waters, and Euphrates poplars the more brackish waters. Willows are handsome, fast-growing trees, useful for holding banks against flooding. The wood can be used for fuel, and some willows have been recommended as energy sources. Slender willow twigs, or withes, are used in wicker work. Willow branches were used by the Jews in some of their religious rites and ceremonies. Willows were among four species the Hebrews were commanded to take at the Feast of the Tabernacles. The bark may contain as much as 13% tannin. Honey plants (BIB; FP1; ZOH). Regarded as antiperiodic, antiseptic, astringent, tonic, and vermifuge, the white willow is used in folk remedies for calluses, corns, debility, diarrhea, dysentery, dysmenorrhea, dyspepsia, fever, gout, hemotysis, malaria, rheumatism, tumors, and warts. Lebanese recognize the aspirin-like quality of the bark decoction, using it for colds, grippe, and pain, and a strong decoction for venereal disease. Even "transplanted" to America, the Lebanese used the bark of the American species for colds, flu, headache, pains, and rheumatism, even amplifying my generic belief that "all willows are the same in his medicine" (BIB; HJP).

Common Names (White Willow):

Ak Sogut (Tur.; EFS); Aravah (Heb.; ZOH); Aubier (Fr.; HH3); Bai Liu (Pin.; AH2); Bai Liu Gen (Pin.; AH2); Bai Liu Ye (Pin.; AH2); Beasa (Kas.; MKK); Bedisiah (Afg.; DEP; KAB); Bis (India; Pun.; EFS; NAD); Bushan (Pun.; KAB); Changma (Pun.; DEP); Dotterweide (Ger.; KAB); European Willow (Eng.; EFS); Falber (Ger.; HH3); Huid Piil (Den.; KAB); Huntindon Willow (Eng.; KAB); Hvit Pihl (Swe.; KAB); Isbidar (Arab.; Syria; HJP); Itea (Greek; KAB); Iva (Rus.; KAB); Kharwala (Trans-Indus; DEP; KAB); Khilaf (Arab.; BOU); Knotwilg (Dutch; EFS); Madnu (Pun.; WOI); Malchang (Pun.; SKJ); Osier Blanc (Fr.; BOU; EFS); 'Oud el Maa (Arab.; BOU); Pertiche Bianco (It.; EFS); Piletroi (Den.; EFS); Plon Blanc (Fr.; KAB); Safsaf Abiad (Arab.; BOU); Safsaf Abyad (Arab.; Syria; HJP); Salce (Sp.; EFS); Salce Blanco (Sp.; KAB); Salcie (Rom.; KAB); Salcio (It.; KAB); Salguiero (Por.; KAB); Salguiero Branco (Por.; EFS); Salicastro (It.; HHB); Salice Bianco (It.; EFS); Salice Comune (It.; HH3); Salice da Pertiche (It.; HH3); Salico da Safsaf (Arab.; Syria; HJP); Sauce Blanco (Sp.; EFS); Saudre (Fr.; HH3); Saula (Cat.; KAB); Saule Argente (Fr.; HHB); Saule Blanc (Fr.; BOU; EFS; HH3); Saule Comun (Fr.; HHB); Sausse (Fr.; HH3); Schietwilg (Dutch; EFS); Schotwilg (Dutch; EFS); Silber Weide (Ger.; EFS; HH3); Swallow-Tailed Willow (Eng.; BOU); Talezzast Amellal (Ber.; BOU); Vivir (Kas.; KAB); Vrba (Bosnia; JLH); Vuir (Kas.; NAD); Weide (Ger.; EFS); Weiss Weide (Ger.; DEP; EFS; HH3); White Willow (Eng.; AH2; CR2); Witte Wilg (Dutch; EFS; KAB); Zafzafa (Malta; KAB).

Activities (White Willow):

Analgesic (f12; HJP; KOM; PIP; PH2; WAM; PR15:344); Anaphrodisiac (f; MAD); Antiaggregant (1; VAD); Antihidrotic (f; CAN); Antiinflammatory (f12; KOM; PH2; PIP; WAM); Antiperiodic (f1; DEP; EFS); Antiprostaglandin (1; BGB); Antipyretic (f12; KOM; NAD; PIP; PH2; WAM); Antirheumatic (f12; PR15:344; CAN; EFS; FAD); Antiseptic (f1; CAN; DEP; EFS); Antispasmodic (f1; BOU; VAD); Astringent (f1; APA; EFS; FAD; PED; PH2; VAD); Bitter (1; FAD; PED); Cardioprotective (1; VAD); Contraceptive (f; MAD); Febrifuge (f1; FNF); Keratolytic (1; FNF); Sedative (f1; MAD; VAD); Tonic (f; BIB; DEM); Vulnerary (f; BOU).

FIGURE 1.97 White Willow (*Salix alba*).

Indications (White Willow):

Alopecia (f; DEM); Ankylosing Spondylitis (1; CAN); Arthrosis (f12; PR15:344; APA; FAD; MAD; SKY); Bleeding (f; BUR); Bursitis (1; SKY); Callus (f; BIB; JLH); Cancer (f; BIB; FAD; JLH); Candidiasis (f; BIB); Cardiopathy (1; FAD); Cataracts (1; FAD); Catarrh (1; CAN); Cold (f1; APA; VAD); Colic (f; MAD); Corn (f1; BIB; FAD; MAD); Cramp (f1; VAD); Debility (f; BIB; BUR); Dermatosis (1; FAD; MAD); Diabetes (f; MAD); Diarrhea (f1; APA; BIB; BUR; DEM; FAD; MAD); Dysentery (f; BIB; BUR); Dyskinesia (f; VAD); Dysmenorrhea (f; BIB; MAD; VAD); Dyspepsia (f; BIB; FEL; MAD); Dyspnea (f; DEM); Earache (1; MAD; PED); Encephalosis (f; MAD); Enterosis (f1; DEM; VAD); Fever (f1; APA; DEM; FAD; MAD; NAD; PH2; PIP); Flu (f1; BIB; FNF; VAD); Fungus (1; CEB); Gastrosis (f1; VAD); Gout (1; APA; MAD; PNC); Headache (f1; BGB; PH2; PIP; VAD; WAM); Hemoptysis (f; BIB; MAD; WOI); Hoarseness (f; DEM); Infection (f1; APA; CAN; DEP; EFS); Inflammation (f1; APA; PH2); Influenza (f1; CAN); Insomnia (f1; MAD; VAD); Malaria (f1; BIB; BUR; DEP; FEL; NAD); Myalgia (f1; APA; CAN); Neuralgia (f; MAD; VAD); Neurasthenia (f; MAD); Obesity (f; APA); Osteoarthrosis (2; SKY; PR15:344); Pain (f12; HJP; PHR; PH2; PNC; WAM; PR15:344); Pertussis (f; MAD); Poison Ivy (1; FAD); Rheumatism (f12; FAD; PHR; PH2; PIP; VAD); Scrofula (f; MAD); Sore (f; BIB; FAD; FEL; MAD); Sprains (1; APA); Stomachache

(1; VAD); Tendonitis (1; BGB); Thrombosis (1; VAD); Thrush (f; BIB); Toothache (f1; BGB; JAD); Ulcus cruris (f; MAD); Vaginosis (f; BIB); Wart (1; FNF; JLH); Worm (f; MAD).

Dosages (White Willow):

FNFF = !

Tender twigs and leaves often used for fodder, sometimes for human food. Leaves used as tea or adulterant thereof (WOI); 1–2 tsp powdered bark 1 to 3 ×/day (APA); two to three 379-mg capsules, as needed or every 3 hours, up to 18 capsules (APA); 1–2 ml bark tincture (25% alcohol) 3 ×/day (APA; SKY); 20–40 mg salicin (APA); 1–3 g dry bark, or in tea, 3 ×/day (CAN); 1–3 ml liquid extract (1:1 in 25% ethanol) 3 ×/day (CAN); 1–2 g bark (20–40 mg salicin), two or three 380-mg capsules every 3 hours (JAD); 60–120 mg salicin per day (KOM; PIP); 2–4 Tbsp fresh bark (PED); 3–6 g dry bark (PED); 4.5 g dry bark:22 ml alcohol/23 ml water (PED); 2–3 g bark in cold water, bring to boil, steep 5 minutes (PH2), 1–5 ×/day (SKY).

- Bosnians apply wine bark decoction to corns and warts (JLH).
- Cherokee take the astringent bark for alopecia, diarrhea, dyspnea, fever, and hoarseness (DEM).
- North Africans consider the leaves are calmative, antispasmodic, genital sedative, the bark useful for fever and rheumatism (BOU).

Downsides (White Willow):

Class 1. Salicylates; tannins (AHP, 1997). Commission E reports for oral use of bark, contraindications, adverse effects, and interactions: on theoretical grounds similar to those of the salicylates (AEH). (All plants contain salicylates.) In view of the lack of toxicological data, excessive use, especially during lactation and pregnancy, should be avoided. Individuals with aspirin hypersensitivity, asthma, diabetes, gastrosis, gout, hemophilia, hepatosis, hypothrombinaemia, nephrosis, and peptic ulcers should be cautious with salicylates. Alcohol, barbiturates, and oral sedatives may potentiate salicylate toxicity. Beware of salicylate interaction with oral anticoagulants, methotrexate, metoclopramide, phenytoin, pronebecid, spironolactone, and valproate. Salicylates excreted in breast milk reportedly can cause macular rashes in breast-fed babies. Salicylate toxicity may cause dermatosis, gastrosis, hematochezia, nausea, nephrosis, tinnitus, and vomiting (CAN). Excessive use of the tannin-rich bark may cause diarrhea and nausea (SKY). Still, "willow is much safer than aspirin" (SKY). Not for use during viral infections because of [remote theoretical; JAD] possibility of Reye's syndrome (WAM).

Natural History (Black Willow):

Insect pollinators include pollen collectors such as bumblebees, honeybees, and solitary ground nesting bees, and false darkling beetles (*Asclera*) and punctate leaf beetles (*Orsodacne*). In the north woods, buds and/or tender twigs are consumed by grosbeak, grouse, and ptarmigan. Beaver, deer, elk, hare, mice, moose, muskrat, rabbit, rat, and squirrel also eat twigs, foliage, and/or bark (MZN). Beaver seem to prefer this species. Sapsuckers may pit the stem seeking sap (EAS). Bugs eating poplar leaves often also eat willow, relatively immune to the salicylates. White, with brown spots, willow lacebugs (*Corythucha salicilis*) suck sap from the lower leaf surface. Adult beetles may eat the leaves, larvae stripping them; for example, imported willow or shining leaf beetle (*Plagiodera versicolora*) is metallic blue or green, the willow leaf beetles (*Chrysomela interrupta*) is yellow with black markings, and the spotted willow leaf beetle (*Lina interrupta*) is reddish with black markings. Also among the beetles consuming leaves are cottonwood leaf beetles (*Chrysomela scripta*), flea beetles, and flea weevils, not to mention the Japanese beetle. Long horn beetles may may drill weak trees; for example, the cottonwood borer may girdle leaf bases, in the process also transmitting the fungus willow scab or blight (*Pollaccia saliciperda*), which can cause leaves to blacken and die. Other fungal diseases include *Cytospora shrysosperma*, causing elongate cankers, and *Phytophora cactorum*, causing slimy weeping lesions on lower trunks. On dead or decaying willows, seek the yellowish, crown coral mushroom (*Clavicorona pyxidata*) (EAS). Many sphinx moth caterpillars eat willow; for example, big poplar or modest sphinx, one-eyed sphinx, and twin spotted sphinx. Tent moth caterpillars (*Malacosoma disstria*) and the gypsy moth are fairly common, along with several noctuiids. Among common butterfly caterpillars are mourning cloak, tortoise shells, and viceroy. Midges and sawflies may create leaf galls. Feeding en masse, willow sawfly larvae, resembling black and yellow caterpillars, may strip the plant of its leaves. Giant willow aphids may suck sap from twigs in summer. Then there are oystershell scale and willow scale (EAS).

Extracts (White Willow):

Salicylates antiaggregant, antiinflammatory, antipyretic, antiuricosuric/uricocsuric, and hyper/hypoglycemic. The pro-drug salicin, which does not irritate the stomach, is metabolized to saligenin in the GI tract and salicylic acid after absorption. "Products containing willow should preferably be standardized on their salicin content …" (CAN). "The analgesic actions of willow are typically slow-acting but last longer than standard aspirin products" (SKY). McCarty and Block (2006) note the potential in cancer and diabetes for IKKbeta Inhibitors like salicylic acid, found presumably in all willow species. IKKbeta Inhibitors may help reverse insulin resistance and control type-2 diabetes. Serving as IKKbeta Inhibitors *in vitro* are the salicylic acid, resveratrol from the biblical grape, and silybinin from milk thistle (X16880431).

WEEPING WILLOW (*SALIX BABYLONICA* L.) + SALICACEAE

Synonyms:

Salix japonica Thunb. *fide* NPM

Notes (Weeping Willow):

> **And they shall spring up as among the grass, as willows by the water courses.**
>
> **Isaiah 44:4 (KJV)**

> **They shall spring up like grass amid waters, like willows by flowing streams.**
>
> **Isaiah 44:4 (RSV)**

> **And they will certainly spring up among the green grass, like poplars by the water ditches streams.**
>
> **Isaiah 44:4 (NWT)**

Zohary suggests that if the willows of the Bible were native to the Holy Land, they would have been *Salix acmophylla* or *Salix alba*, which hybridize readily. I suggest that romantics like me like the idea of the weeping willow, *Salix babylonica*, in their biblical garden.

Common Names (Weeping Willow):

Arbol del Desmayo (Sp.; KAB); Attuppalai (Ap.; Tel.; KAB; SKJ; WOI); Babylon Weeping Willow (Eng.; USN); Bada (Pun.; DEP; WOI); Bains (Nepal; NPM); Bed (Pun.; WOI); Bedmaju (Pun.; KAB); Bes (Pun.; DEP); Besu (Pun.; DEP); Bhosi (Nepal; DEP; KAB; WOI); Bidai (Pun.; DEP); Bisa (Kas.; Pun.; DEP; SKJ; WOI); Bitsubes (Pun.; WOI); Chinese Willow (Eng.; EFS); Desmay (Cat.; KAB);

FIGURE 1.98 Weeping Willow (*Salix babylonica*).

Desmayo (Sp.; EFS; KAB); Echte Trauer Weide (Ger.; EFS); Echte Treur Wilg (Ger.; EFS); Gadhbains (Garhwal; Up.; SKJ); Garb (Fr.; KAB); Giur (Kas. DEP; KAB); Guir (Kas.; NAD; WOI); Hazomala-helo (Hova; KAB); Kashir Vir (Kas.; SKJ); Katira (Pun.; Pun.; SKJ; WOI); Laila (Pun.; DEP; WOI); Liu (China; EFS; KAB); Liu Zhi (Pin.; DAA); Lloron (Sp.; KAB); Majhinus (Kum.; WOI); Majnun (Pun.; Hindi; DEP; KAB; SKJ); Maju (Pun.; DEP); Momakha (Burma; DEP; KAB); Plakychaya Iva (Rus.; KAB); Quir (Kas.; SKJ); Salcie Pleteasa (Rom.; KAB); Salguiero (Por.; AVP); Salguiero

Chorão (Por.; AVP; EFS); Salice Piangente (It.; KAB); Salice Plangente (It.; KAB); Sauce (Peru; Sp.; EGG; ROE); Sauce de Babilonia (Sp.; USN); Sauce Llorón (Cuba; Dr.; Peru; Pr.; Sp.; AVP; EGG; ROE; USN); Saule de Babylone (Fr.; EFS); Saule Pleurier (Fr.; EFS; USN); Shidare-Yanagi (Japan; USN); Tissi (India; Nepal; EFS; NAD; WOI); Trauerweide (Ger.; KAB; USN); Treuwilg (Dutch; KAB); Wala (Pun.; DEP; WOI); Weeping Willow (Eng.; Scn.; AH2; USN).

Activities (Weeping Willow):

Anthelmintic (f; DEP; NAD); Antidote (varnish) (f; DAA); Antipyretic (f; DEP; ROE; SHJ); Antiseptic (f1; DAA; EFS; NAD; WOI); Astringent (f; DEP; WOI); Bactericide (1; WOI); Fungicide (1; LMP); Insecticide (1; LMP); Tonic (f1; DEM; DEP; EFS; NAD; WOI); Vermifuge (f; DAW).

Indications (Weeping Willow):

Abscess (f; DAA; ROE); Alopecia (f; DEM; ROE); Arthrosis (f1; DAW; ROE; SKJ; WOI); Bacillus (1; WOI); Bacteria (1; WOI); Bleeding (f; DAA); Boil (f; DAA); Cancer (f; JLH); Carbuncle (f; DAA; DAW); Dandruff (f; ROE); Dermatosis (f; ROE); Diarrhea (f; DEM); Dyspnea (f; DEM); Enterosis (f; ROE); Fever (f1; DEP; ROE; SKJ; WOI); Flu (f; ROE); Fungus (1; LMP); Gonorrhea (f; DAA; DAW); Hoarseness (f; DEM); Infection (1; ROE; WOI); Jaundice (f; DAA; DAW); Malaria (f1; DAW; ROE); Mycosis (1; LMP); Parasite (f; DAA); Rheumatism (f1; DAW; ROE; SKJ; WOI); Sore (f; DAW; ROE); Staphylococcus (1; ROE); Swelling (f; DAA); Worm (f; DEP; EFS; NAD).

Dosages (Weeping Willow):

FNFF = !
Young leaves, shoots, and flower buds parboiled and eaten; older leaves a tea adulterant; source of a manna (FAC).

- Andeans suggest the plant can whiten the teeth (ROE).
- Andeans take bark tea for fever, enterosis, and malaria, and bathe rheumatism therewith (ROE).
- Asian Indians suggest the catkins as antipyretic (SKJ), the bark as anthelmintic (DEP).
- Cherokee Indians take bark tea for alopecia, diarrhea, fever, and hoarseness (DEM).
- Chinese treat boils around the mouth with root ashes in mustard oil (LMP).
- Chinese use infusion (bark, twigs, and/or leaves) for fever, gonorrhea, jaundice, and rheumatism (LMP).
- Tippery British ash the inflorescence in ointments for burns (AAH).

Extracts (Weeping Willow):

Bark contains 3 to 4% salicin (ROE).

BRITTLE WILLOW (*SALIX FRAGILIS* L.) + SALICACEAE

Synonyms:

Salix fragilior Host.; *Salix fragilis* f. *bullata*; *Salix fragilis* var. *decipiens*; *Salix persicofolia* Host. *fide* HH3 and USN

Notes (Brittle Willow):

… the willows of the brook compass him about.

Job 40:22 (KJV)

FIGURE 1.99 Brittle Willow (*Salix fragilis*).

Zohary does not list this among the species of the Holy Land; thus, I think it highly unlikely that this is, in fact, the biblical willow (ZOH). The tree is cultivated in fuel plantations on swampy ground in India. The wood, soft, light, and even grained, is said to be more durable than other willows and is suitable for cricket bats. It is also employed in the match industry. Charcoal made from the wood is used for gunpowder. In Iran, the tree is reported to yield a sweet manna-like exudation.

Twigs are employed for basketry. This is one of the many "herbalist's aspirin." Tannin and gallic acid might explain the anticancer activity (BIB).

Common Names (Brittle Willow):

Brittle Willow (Eng.; Scn.; AH2; USN); Bruchweide (Ger.; HH3); Crack Willow (Eng.; Ocn.; AH2; BIB; DEP; HH3; USN); Kashmir. Willow (Eng.; BIB; WOI); Knackweide (Ger.; HH3; USN); Red-Wood Willow (Eng.; BIB; DEP); Saule Fragile (Fr.; HH3); Saule Rouge (Fr.; HH3); Tilhang (Lahoul; DEP); White Welsh Willow (Eng.; HH3); Zerbrechliche Weide (Ger.; HH3).

Activities (Brittle Willow):

Antiperiodic (f1; BIB; DAW); Antirheumatic (f; BIB); Astringent (f; DAW); Hemostat (f; DEM); Sedative (f1; DAW); Vulnerary (f; DEM).

Indications (Brittle Willow):

Bleeding (f; DEM); Cancer (f1; BIB; JLH); Cold (1; DAW); Fever (f1; DAW); Herpes (f; BIB); Infection (f; BIB); Insomnia (1; DAW); Malaria (f1; BIB; DAW); Rheumatism (f1; DAW); Sore (f; DEM); Tumor (f; JLH); Virus (f; BIB).

Dosages (Brittle Willow):

FNFF = !

Leaves used as manna, eaten as food (FAC).

- Iranian writers suggest the manna for herpes (BIB).

COMMON SALTWORT (*SALSOLA KALI* L.) ++ CHENOPODIACEAE

Notes (Common Saltwort):

for though thou wash thee with nitre, and take thee much sope.

Jeremiah 2:22 (KJV)

The alkaline salts used in soap manufacture during biblical times were said to be obtained by burning the plants of the saltwort. *Salsola kali* is probably the most common of 20 kinds of saltwort in the Holy Land. Soap was made by mixing such ashes with olive oil, instead of animal fat. Glass is also made therefrom, because of the high alkali content.

Common Names (Common Saltwort):

Barilla Plant (Eng.; HOC); Barrela Borda (Cat.; KAB); Barrila Borde (Sp.; USN); Barrila Pinchosa (Sp.; USN); Barrilha-Espinhosa (Por.; USN); Barrilheira (Por.; USN); Carqueja Brava (Mad.; Por.; JAD); Common Saltwort (Eng.; USN); Eestridge (Eng.; KAB); Elkali (Arab.; KAB); Erba Kali (Malta; KAB); Espinardo (Sp.; KAB); Glasswort (Eng.; HOC; JLH); Hurd (Arab.; Syria; HJP); Jaghun (Jhalawan; Kharan; KAB); Kalikraut (Ger.; USN); Kali Salzkraut (Ger.; USN); Kelpwort

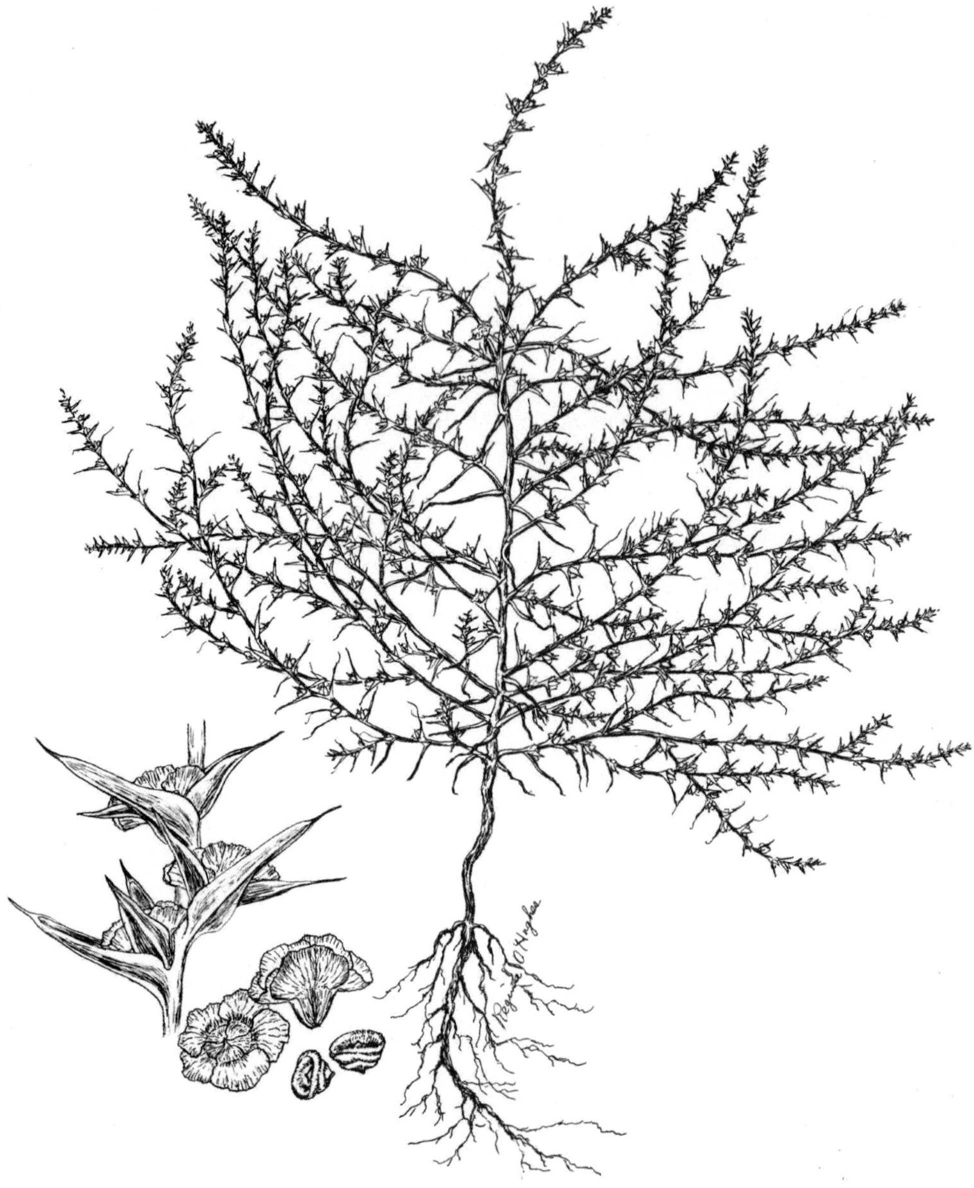

FIGURE 1.100 Common Saltwort (*Salsola kali*). Source: Regina Hughes in Reed (1970).

(Eng.; KAB); Marie Épineuse (Fr.; KAB); Marie Vulgaire (Fr.; KAB); Mata Pinchuda (Sp.; KAB); Pincho (Sp.; KAB); Prickly Saltwort (Eng.; HJP; KAB; USN); Riscolo (It.; KAB); Russian Thistle (Eng.; HOC; USN); Russian. Tumble Weed (S. Afr.; KAB); Sajji Butti (Pun.; SKJ); Sea Thrift (Eng.; KAB); Shawk Ahmar (Arab.; Syria; HJP); Soda (Malta; KAB); Sonde Commun (Fr.; KAB); Sowdwort (Eng.; KAB); Tumbleweed (Eng.; X15696348).

Activities (Common Saltwort):

Allergenic (1; X15808117); Anthelmintic (f; WOI); Antiseptic (f; BIB; HJP); Cathartic (f; WOI); Diuretic (f; WOI); Emmenagogue (f; WOI); Poison (f; DAW); Stimulant (f; WOI); Vulnerary (f; HJP).

Indications (Common Saltwort):

Amenorrhea (f; WOI); Cancer (f; JLH); Constipation (f; WOI); Dropsy (f; DAW); Excrescence (f; JLH); Infection (f; HJP); Worm (f; KAB; SKJ; WOI); Wound (f; BIB).

Dosages (Common Saltwort):

FNFF = !
Food farmacy. Young shoots edible (TAN). Roots eaten as starvation food during the Great Depression (HOC).

- East Indians use the ashes as a therapeutic antiseptic dressing (BIB).
- Lebanese apply the ash to wounds to prevent infection (HJP).

Downsides (Common Saltwort):

A major cause of hay fever (HOC).

Natural History (Common Saltwort):

Although said to be toxic in quantity, still camels, goats, and horses are said to graze it (BIB).

PEELU (*SALVADORA PERSICA* L.) ++ SALVADORACEAE

Notes (Peelu):

> **Another parable put he forth unto them, saying, The kingdom of heaven is like to a grain of mustard seed, which a man took, and sowed in his field: Which indeed is the least of all seeds: but when it is grown, it is the greatest among herbs, and becometh a tree, so that the birds of the air come and lodge in the branches thereof.**
>
> **Matthew 13:31-32 (KJV)**

> **Another parable he put before them, saying, "The kingdom of heaven is like a grain of mustard seed which a man took and sowed in his field; it is the smallest of all seeds, but when it has grown it is the greatest of shrubs and becomes a tree, so that the birds of the air come and make nests in its branches."**
>
> **Matthew 13:31-32 (RSV)**

> **Another illustration he set before them, saying, "The kingdom of the heavens is like a mustard grain which a man took and planted in his field; which is, in fact, the tiniest of all the seeds, but when it has grown it is the largest of the vegetables and becomes a tree, so that the birds of heaven come and find lodging among its branches."**
>
> **Matthew 13:31-32 (NWT)**

Zohary favors *Brassica nigra*, the black mustard, as the grain of mustard seed in the Bible. He does not consider *Moringa* or *Salvadora*, which both have seeds bigger than *Brassica* but do grow into shrubs or small trees. Both of these tropical tree species are found in the tropical vegetation found at the mouths of the Aravah and Jordan valleys (ZOH). A little late in my research, I found a "new" (actually an over 100-year-old) reference (DEP) where we read that Royle had speculated that *Salvadora persica* was indeed the Mustard Tree of the Bible (I have speculated it was the Horseradish tree (*Moringa*). Indeed, Watt (author of DEP) labeled the plant "The Mustard Tree of the Bible." He notes that the small red berries are strongly aromatic and pungent, like mustard or cress, and not much appreciated by the natives of India. The pungent shoots and leaves are eaten as a salad (DEP). According to UPW, it is popular among the Mohammedens because the prophet himself is said to have used the plant to clean his teeth and to relieve toothache (UPW). Small wonder that Saudi studies have confirmed that it works better than a toothbrush (X15560804). PubMed has more than a dozen citations relating to antiseptic and/or dental applications of the species.

Common Names (Peelu):

Abisga (Ber.; BOU); Adjou (Ber.; BOU); Arac (Fr.; BOU); Arak (Arab.; BOU; GHA); Babul (Ber.; BOU); Barir (Arab.; GHA); Irak (Arab.; GHA); Jal (Nwp.; KAB); Kabats (Arab.; BOU); Mesuak (Fr.; BOU); Miswak (Arab.; Yemen; BOU; GHA); Mustard Tree (Eng.; DEP); Peelu (Eng.; Scn.; AH2; FAC); Pilu (Ayu.; Urdu; AH2; KAB); Plewan (Pushtu; KAB); Rak (Arab.; Yemen; X15890471); Saltbush (Eng; 4PW); Siwak (Arab.; BOU); Tidjat (Ber.; BOU); Toothbrush Tree (Eng.; BOU).

Activities (Peelu):

Alexiteric (f; KAB); Analgesic (f; WOI); Antidote (f; KAB; X15890471); Antiplasmodial (1; X12426089); Antiplaque (1; X3505835); Antiscorbutic (f; WOI); Antiseptic (1; X11887585; X15890471); Antiulcer (1; PHM6:363); Aphrodisiac (1; DEP; KAB); Ascarifuge (f; WOI); Astringent (f; WOI); Biliousnesss (f; WOI); Candidicide (1; X7898373); Carminative (f1; BOU; KAB; UPW; WOI); Deobstruent (f; KAB; WOI); Diuretic (f; BOU; KAB; FP2; WOI); Emetic (f; UPW); Emmenagogue (f; WOI); Febrifuge (f; WOI); Fungicide (1; X7898373); Hepatotonic (f; WOI); Laxative (f; GHA); Litholytic (f; WOI); Orexigenic (f; BOU; KAB); Purgative (f; DEP; WOI); Sedative (1; PR16:395); Sialogogue (f; UPW); Stimulant (f; WOI); Stomachic (f; BOU; WOI); Tonic (f; BOU; WOI); Vesicant (f; WOI).

Indications (Peelu):

Ankylostomiasis (f; UPW); Anorexia (f; BOU; KAB); Asthma (f; WOI); Bacteria (1; X11887585; X14973564); Biliousness (f; KAB); Bite (f; BOU); Blennorrhea (f; UPW); Blister (f; GHA); Boil (f; BOU; UPW); Bruise (f; BOU); Cancer (f; WOI); Catarrh (f; WOI); Childbirth (f; DEP); Cold (f; UPW); Convulsion (1; PR16:395); Cough (f; WOI); Cystosis (f; UPW); Dysmenorrhea (f; GHA); Dyspepsia (1; UPW; WOI); Fever (f; WOI); Fungus (1; X7898373); Gas (f; WOI); Gastrosis (f; BOU; WOI); Gingivosis (12; BOU; UPW; WOI; X15560804); Gonorrhea (f; UPW; WOI); Headache (f; UPW); Hemorrhoid (f; WOI); Infection (1; X7898373); Infertility (f; UPW); Leukoderma (f; KAB); Malarial (1; X12426089); Mycosis (1; X7898373; X15560804); Odontosis (f1; GHA; X15890471); Ozoena (f; WOI); Pain (f; WOI); Periodontosis (f1; GHA; JAC7:405; X2239575); Plaque (f1; GHA; X15560804); Pulmonosis (f; UPW); Rheumatism (f; KAB; WOI); Rhinosis (f; WOI); Scabies (f; WOI); Snakebite (f; KAB); Sore (f; BOU; GHA); Splenosis (f; BOU; WOI); Sting (f; GHA); Stomachache (f; BOU; UPW); Stomatosis (f; UPW; X15890471); Streptococcus (1; X11887585; X14973564); Swelling (f; GHA); Syphilis (f; BOU); Toothache (f; UPW); Ulcer (1; PHM6:363); Venereal Disease (f; BOU; UPW); Worm (f; WOI); Wound (f; WOI).

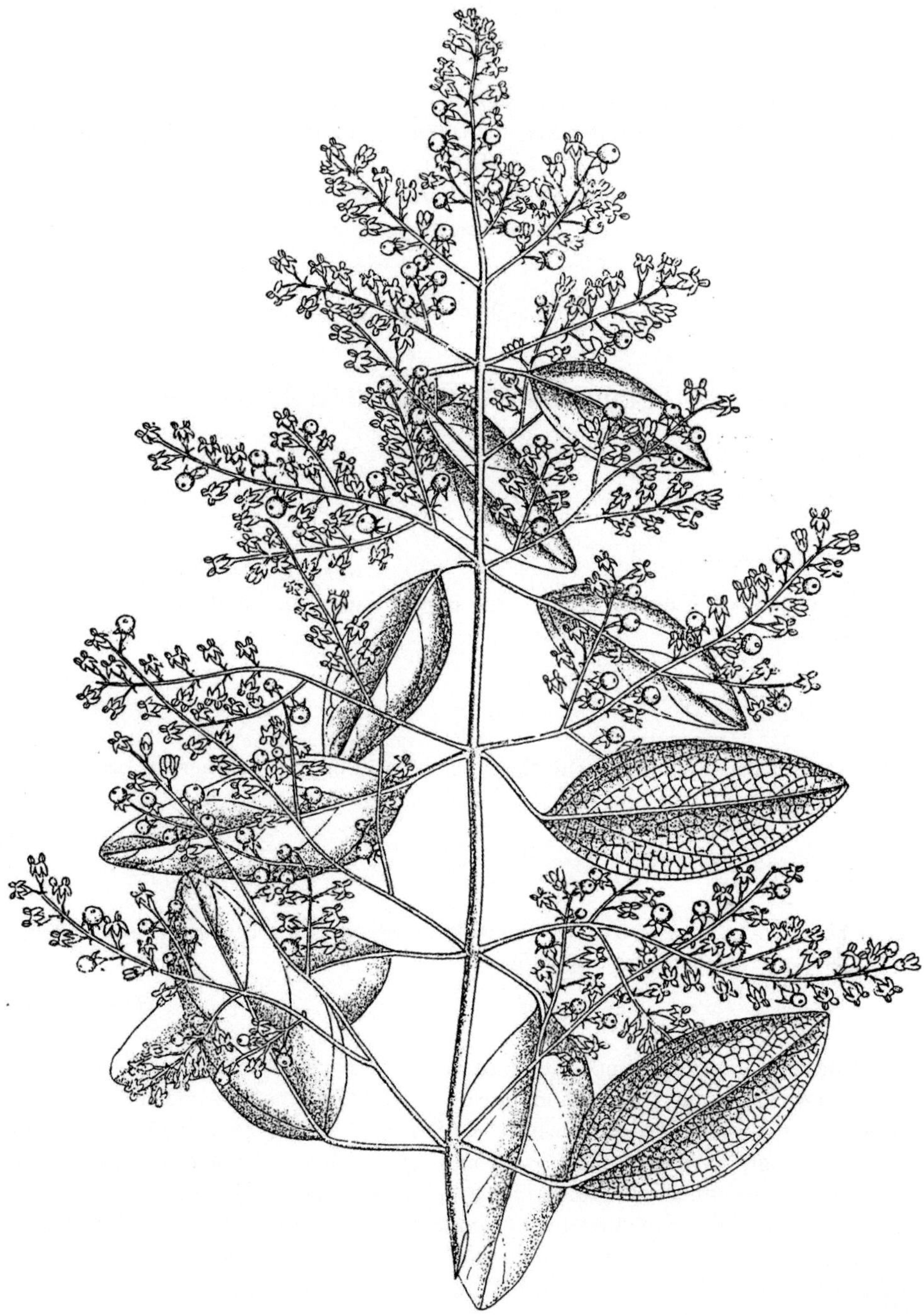

FIGURE 1.101 Peelu (*Salvadora persica*). Source: KAB.

Dosages (Peelu):

FNFF = !!

Leaves and shoots eaten raw or cooked, or made into sauce; fruits edible, fresh, dried, as a mustard substitute, or made into a beverage that can be fermented (FAC; FP2; TAN). Ash of plant made into a salt-like powder called *kegr.* Resin used for chewing gum. Seed fat used like butter (FAC).

- Arabs apply crushed leaves to swellings; dry powdered leaves to blisters and ulcers (GHA).
- Arabs take dried fruits with cloves, ginger, and sugar daily to regulate menstruation (GHA).
- North Africans consider fruits carminative, febrifuge, orexigenic, and stomachic (BOU).
- North Africans take powdered leaves, with honey and millet flour, made into balls. Taken every morning for 40 days for syphilis (BOU).
- Yemeni use frayed twigs effectively as toothbrush (on sale for a dime) (GHA).

Extracts (Peelu):

Trimethylamine is a gingival stimulant and antidyspeptic and carminative (UPW).

THORNY BURNET (*SARCOPOTERIUM SPINOSUM* (L.) SPACH ++ ROSACEAE

Synonyms:

Pimpinella spinosa Gaertn.; *Poterium spinosum* L.; *Sanguisorba spinosa* (L.) Bertol.; HH2

Notes (Thorny Burnet):

Therefore, behold, I will hedge up thy way with thorns, and make a wall, that she shall not find her paths.

Hosea 2:6 (KJV)

Therefore I will hedge up her way with thorns; and I will build a wall against her, so that she cannot find her paths.

Hosea 2:6 (RSV)

Therefore here I am hedging your way about with thorns; and I will heap up a stone wall against her, so that her own roadways she cannot find.

Hosea 2:6 (NWT)

There are multitudes of thorny plants in the desert, and Zohary has accepted this as the best candidate for the thorn in the path of the harlot in Hosea 2:6, noting that it is one of the most common dwarf shrubs, the dominant species in what the *Flora of Palestine* calls the *batha* communities. The batha(h) is a name for a vegetation type dominated by this and others such as *Gymnocarpus decander*, *Noea mucronata*, and *Traganum nudatum*. It is widely used there for fuels, for cooking and lime kilns, and for broom manufacturing and hedging, including (dare I say) "harlot hedging" (FP2). Local peasants customarily hedge their gardens and courtyards with spiny dwarf shrubs like this one. Zohary says the thorny burnet best fits the Hebrew *sir* (plural *sirim*). Abundant in Jerusalem, it might have been used to fashion the crown of thorns plaited by the Romans as recounted in Matthew, Mark, and John. In modern Hebrew, it is still called *bathah* meaning "waste" (*garigue*), a name adopted according to Zohary from Isaiah 5:6: "I will make it a waste; it shall not be pruned or hoed, and briers and thorns shall grow up."

Common Names (Thorny Burnet):

Becherblume (Ger.; HH2); Dorniger Becherstrauch (Ger.; HH2); Pimpinelle Èpineuse (Fr.; HH2); Sir (Arab.; Heb.; ZOH); Spinoporci (It.; HH2); Strauchige Becherblume (Ger.; HH2); Thir (Arab.; ZOH); Thorn Burnet (Eng.; HH2); Thorny Burnet (Eng.; ZOH).

Activities (Thorny Burnet):

Antielastase (1; HH2); Antisarcomic (1; HH2); Antitumor (1; HH2); Cardioprotective (1; HH2); Hypoglycemic (1; HH2); Hypotensive (1; HH2); Vasodilator (1; HH2).

Indications (Thorny Burnet):

Cancer (1; HH2); Cardiopathy (1; HH2); Diabetes (1; HH2); High Blood Pressure (1; HH2); Sarcoma (1; HH2).

Dosages (Thorny Burnet):

FNFF = !

Peasants in the Holy Land use the leaves as a potherb (ZOH). 5 g root bark/250 ml in decoction; 2 Tbsp after meals 3 ×/day (HH2).

- Bedouins suggest that the plant has antidiabetic activity (HH2).

COSTUS (*SAUSSUREA COSTUS* (FALC.) LIPSCH) ++ ASTERACEAE

Synonyms:

Aplotaxus lappa Decne.; *Aucklandia costus* Falc.; *Aucklandia lappa* Decne.; *Saussurea lappa* C.B. Clarke; *Theodorea costus* O. Ktze. *fide* HH2

Notes (Costus):

All thy garments smell of myrrh, and aloes, and cassia.

Psalms 45:8 (KJV)

Your robes are all fragrant with myrrh and aloes and cassia.

Psalms 45:8 (RSV)

All your garments are myrrh and aloeswood [and] cassia.

Psalms 45:8 (NWT)

Unlike Zohary, I followed Moldenke and Moldenke in my first Bible book (Duke, 1983) and concluded that the cassia of Psalms 45 was the Indian *kuth* (standardized common name "costus," widely used in perfumes and incenses). Other scholars, such as Zohary, would think that cassia in the Bible was more likely to be a species of *Cinnamomum* (which see; and I agree). Still I include the kuth here. Others might side with the Moldenkes. Although noted as an aphrodisiac, costus' chief use is as a perfume. In China and India it serves as incense in temples. The essential oil is valued in perfumery and cosmetics. The essential oil has strong antiseptic and disinfectant properties, especially against *Streptococcus* and *Staphylococcus*. The root owes its insecticidal property to its essential oil content. Roots are employed in Kashmir as insecticide to protect shawls and woolen fabrics. A process for treating costus roots or inulin obtained from them for the production of fructose has been reported. Dried stems of the plant are used as fodder in winter (BIB).

Common Names (Costus):

Agada (Sanskrit; KAB); Aucklandia (Eng.; Ocn.; AH2); Changala (Tel.; DEP); Chengulva (Tel.; NAD); Chobiqut (Kas.; WOI); Costus (Eng.; Scn.; AH2; CR2); Goda Mahanel (Sing.; DEP);

FIGURE 1.102 Costus (*Saussurea costus*).

Goshtam (Tam.; DEP); Gostan (Tam.; NAD); Indische Kostuswurzel (Ger.; HH2); Kashmirja (Sanskrit; DEP); Kashtam (Sanskrit; NAD); Koshnaha (Iran; KAB); Koshta (Kan.; WOI); Kostum (Tam.; DEP); Kostuspflanze (Ger.; HH2); Kosuta (Kan.; NAD); Kot (Hindi; Pun.; DEP; KAB); Kur (Beng.; Hindi; DEP; NAD); Kushta (Mar.; Sanskit; WOI); Kushtha (Sanskrit; Ayu.; AH2; HH2); Kust (Arab.; Iran; DEP; HH2; NAD); Kust Talkh (Pun.; DEP); Kustabeheri (Arab.; KAB); Kustam (Tel.; KAB; NAD); Kustullhalu (Arab.; KAB); Kustum (Tel.; DEP); Kut (Guj.; Hindi; Urdu; HH2; KAB); Kuth (Eng.; Hindi; Kas.; Nepal; CR2; MKK); Kutshirin (Iran; KAB); Kuttalkh (Iran; KAB); Mook Heong (Malaya; HH2); Mu Hsian (Malaya; HH2); Mu Xiang (Pin.; AH2; DAA); Muk Heung (Canton; KAB); Ouplate (Bom.; DEP); Pachak (Beng.; Hindi; DEP; KAB); Patalapadmini (Kasmir; NAD); Pokharmul (Hindi; NAD); Post Khai (Kas.; DEP); Practige Kostwurz (Ger.; NAD); Putchuk (Tam.; DEP; KAB); Rusta (Bhote; DEP); Sepuddy (Mal.; DEP; KAB); Upalet (Bom.; Guj.; NAD); Upaleta (Guj.; DEP); Yun Mu Xiang (Pin.; AH2).

Activities (Costus):

Alexipharmic (f; KAB); Alterative (f; KAB; KAP; SUW); Amebicide (1; NAD); Analgesic (f; KAB); Anthelmintic (f; HHB; KAB); Antianginal (1; PH2); Antidote (f; PH2); Antiedemic (1; X12222664); Antiinflammatory (1; X12916066); Antilambdial (f; X14497181); Antileukemic (1; X15209353); Antimycobacterial (1; PR14:303); Antioxidant (1; PR14:250); Antiperistaltic (1; WOI); Antiproliferant (1; X12916066); Antiradicular (1; PR14:250); Antiseptic (1; HH2; KAB; KAP; PH2); Antispasmodic (1; HH2; SUW; WOI; JAC7:405); Antistress (1; PH2); Antitubercular (1; PR14:303); Antiulcer (1; PH2); Aphrodisiac (f; DEP; HHB); Apoptotic (1; X15737683); Astringent (f; KAP); Bactericide (1; HH2; PR14:303; X15814268); Bronchoconstrictor (1; WOI); Bronchodilator (1; NAD; WOI; JAC7:405); Bronchospasmolytic (1; PH2); Candidicide (1; HH2); Cardiotonic (1; KAB; KAP); Carminative (f1; KAB; KAP; SUW; WOI); Caspase Inducer (1; X15209353); Cerebrotonic (f; KAB); Circulostimulant (1; WOI); CNS Depressant (1; NAD; WOI); Contraceptive (f; PH2); Cytotoxic (1; X14510592); Depurative (f; KAB); Diuretic (1; HHB; KAB; SKJ; WOI; JAC7:405); Emmenagogue (f; HHB; KAB); Expectorant (1; KAB; KAP; WOI); Fungicide (1; HH2; PH2); Hypogycemic (1; JAC7:405); Hypotensive (1; HH2; WOI; JAC7:405); iNOS Inhibitor (1; X12538000); Insecticide (1; WOI); Myorelaxant (f1; KAP; NAD); NF-kappaB Inhibitor (1; X15209353); Sedative (f1; KAP; NAD); Stimulant (f; PH2; SUW); Stomachic (f; KAB; SUW); TNF-alpha Inhibitor (1; X15209353); Tonic (f; IHB; KAB; SUW); Vasodilator (1; HH2).

Indications (Costus):

Angina (1; PH2; JAC7:405); Anorexia (f; PH2); Arthrosis (f1; KAB; X12222664); Asthma (f; DEP; IHB; PH2; SUW); Bacteria (1; HH2; PH2; X15814268); Bronchosis (f1; KAB; PH2; WOI); Cancer (f; HHB; KAP); Cancer, abdomen (f; HHB; JLH); Cancer, colon (f; JLH); Cancer, liver (f; JLH); Cancer, spleen (f; JLH); Cancer, stomach (1; X15737683); Candida (1; HH2); Cardiopathy (1; PH2; SKJ); Childbirth (f; IHB); Cholera (f; DEP; NAD; PH2; SUW); Constipation (fl IHB); Cough (f; DEP; KAB; PH2; SUW); Cramp (f; MKK); Deafness (f; KAB); Dermatosis (f; DEP; IHB; KAP; PH2; SUW); Diabetes (1; JAC7:405); Dysentery (f; MKK); Dyspepsia (f; DEP; KAP; NAD); Edema (1; X12222664); Enterosis (f; HHB); Epilepsy (f; KAB); Erysipelas (f; KAB); Fever (f; IHB; KAB); Fungus (f1; HH2; KAB); Gas (f1; KAB; PH2); Gastrosis (1; PH2; X15737683); Headache (f; KAB); Helicobacter (1; X15814268); Hepatosis (f; JLH); Hiccup (f; KAB; KAP); High Blood Pressure (f; HH2); Hysteria (f; KAB); Induration (f; JLH); Infection (1; HH2; PH2); Inflammation (f; KAB); Insomnia (f1; NAD); Itch (f; KAB); Klebsiella (1; HH2); Leprosy (f; KAB; KAP); Leukoderma (f; KAB); Malaria (f; KAB; KAP); Mycosis (f1; HH2; KAB); Nephrosis (f; KAB); Pain (f; IHB; KAB); Paralysis (f; KAB); Rheumatism (f; DEP; KAB; KAP; SUW); Ringworm (f; KAB); Scabies (f; KAB); Schistosoma (1; HH2); Shigella (1; HH2); Smallpox (f; IHB); Sore (f; DEP; KAB; KAP); Spasm (f1; NAD); Splenosis (f; JLH); Staphylococcus (1; HH2; KAP; WOI); Stomachache (f; IHB); Streptococcus (1; KAP; WOI); Stress (1; PH2); Toothache (f; DEP); Trematode (1; HH2); Tuberculosis (1; JNP61:1181); Typhus (f; KAB); Ulcer (1; PH2; X15814268); Worm (f1; HH2; KAB); Wound (f; IHB; KAP; PH2); Yeast (1; HH2).

Dosages (Costus):

FNFF = !
Root used as spice; esssential oil used to flavor baked goods, beverages, candies, frostings, gelatins, and puddings (FAC). One 500-mg capsule, twice a day for 30 days (JAC7:405).

- Ayurvedics consider the root alterative, aphrodisiac, good for blood, bronchitis, complexion, cough, epilepsy, erysipelas, headache, hysteria, itch, leukoderma, ringworm, and scabies (KAB).
- Chinese smoke the CNS-depressant root as a substitute for opium (NAD).

- Punjabi, considering the root aphrodisiac, depurative, apply powdered root to sores and wounds with worms, and to rheumatism (KAB).
- Unani consider the root alexipharmic, analgesic, anthelmintic, carminative, cerebrotonic emmenagogue, and tonic, using it for arthrosis, asthma, cough, deafness, fever, hepatosis, inflammation, nephrosis, ophthalmia, paralysis, and pulmonosis (KAB).

Downsides (Costus):

Class 1 (AHP). No health hazards or side effects known with proper therapeutic dosages (PH2) (PH2 designates no dosage!; JAD).

Extracts (Costus):

EO LD50 = 2660–4350 mg/kg orl rat (HH2).

LAKE RUSH (*SCHOENOPLECTUS LACUSTRIS* (L.) PALLA.) + CYPERACEAE

Synonyms:

Scirpus acutus.; *Scirpus lacustris* L. *fide* FAC, HOC but not necessarily USN

Notes (Lake Rush):

Therefore the LORD will cut off from Israel head and tail, branch and rush, in one day.

Isaiah 9:14 (KJV)

So the LORD cut off from Israel head and tail, palm branch and reed in one day.

Isaiah 9:14 (RSV)

And Jehovah will cut off from Israel head and tail, shoot and rush in one day.

Isaiah 9:14 (NWT)

I suspect there are no divine differences between the common name *reed* and *rush*, and concur with Zohary, suggesting this, the largest species of the six local species of *Schoenoplectus* (formerly *Scirpus*). Noting that in Isaiah 58:5 there is reference to the rush bowing down their heads, as rushes do, Zohary concludes that rush and reed are both most logical translations of the Hebrew *agmon*, and that the larger the species, the more likely the translation. It is used in construction, braiding, purifying murky waters, and most parts are consumable in emergencies. But I find little in the way of medicinal uses. DEM entries below apply to *S. acutus*. EFS entries may apply to *S. validus*.

Common Names (Lake Rush):

Agmon (Heb.; ZOH); American Great Bulrush (Eng.; HOC); Club Rush (Eng.; WOI); Great Bulrush (Eng.; EFS; FAC); Hardstem Bulrush (Eng.; DEM); Lake Rush (Eng.; ZOH); Soft Stem Bulrush (Eng.; HOC); Tule (Sp.; FAC); Nscn.

Activities (Lake Rush):

Astringent (f; BUR; SKJ); Diuretic (f; BUR; SKJ); Emetic (f; DEM); Emmenagogue (f; BUR); Hemostat (f; DEM); Poison (f; DAW; WOI); Sedative (f; EB24:86).

Indications (Lake Rush):

Bleeding (f; DEM); Insomnia (f; EB24:86).

Dosages (Lake Rush):

FNFF = !!
Facciola notes that most parts of the plants are edible; young shoots, raw or cooked; rootstocks and buds at ends of rhizomes, raw or cooked; pollen mixed with meal for breadstuffs; seeds ground for breadstuffs; Cheyenne ate inner part of stem raw (DEM; FAC).

- Chinese use the stem (with other plants) for cancer (JLH).
- Clallam Indians used them to ceremonially suck out the cause of sickness (DEM).
- Montana Indians chewed the roots to prevent thirst (DEM).
- Navajo use the plant as a ceremonial emetic (DEM).
- Thompson Indians applied ashes from burned stalks to baby's bleeding navel (DEM).
- Woodlands Cree poultice pith of stem on wounds to stop bleeding (DEM).

Downsides (Lake Rush):

Rhizomes suspected to poison grazing cattle (WOI).

GOLDEN THISTLE (*SCOLYMUS HISPANICUS* L.) ++ ASTERACEAE

Notes (Golden Thistle):

I am the rose of Sharon, and the lily of the valleys. As the lily among thorns, so is my love among the daughters.

Canticles 2:1-2 (KJV)

I am a rose of Sharon, a lily of the valleys. As a lily among brambles, so is my love among maidens.

Canticles 2:1-2 (RSV)

A mere saffron of the costal plain, I am a lily of the low plains. Like a lily among thorny weeds, so is my girl companion among the daughters.

Canticles 2:1-2 (NWT)

I was really surprised to find mention of saffron in the NWT, and even more intrigued when I read in FAC and TAN that the flowers of this species are used as a substitute or adulterant of saffron. But I suppose the KJV and RSV translation is better: rose of Sharon. However, for the *Scolymus*, I will concur with Zohary when he suggests that the quoted verse is from one of nine containing the biblical word *hoah* (plural *hohim*), apparently suggesting to him a ruderal in calcareous ruins and neglected places (*Scolymus hispanicus*) and a weed in alluvial wheat fields (*Scolymus maculatus*). The two species are keyed as follows:

- Florets orange yellow; white-hairy outside; stem wings interrupted; leaves without marginal white nerve: *Scolymus hispanicus*
- Florets lemon yellow; with long dark hairs outside; stem wings continuous; leaves with a thick marginal white nerve: *Scolymus maculatus*
- (FP3; ZOH)

After 30 years of compiling, I finally, as I recall, met the useful word *lithuretic* in PubMed abstract X9434612, but I added the word *litholytic*, knowing that was the word I had used more often. But there really is a difference: lithuretic seems to mean helping pass stones or gravel through the urethra or in the urine. Finding a lot on the food value but little on the medicinal value, I grabbed for the lithuretic new word.

Common Names (Golden Thistle):

Cardillo (Sp.; FAC; USN); Golden Thistle (Eng.; FAC; USN); Goldwurzel (Ger.; USN); Scolyme (Fr.; TAN); Scolyme d'Espagne (Fr.; USN); Spanish Oysterplant (Eng.; USN; TAN); Spanish Salsify (Eng.; FAC); Sunnariah (Eng.; Australia; USN); Tagarnina (Sp.; USN); Nscn.

Activities (Golden Thistle):

Antioxidant (1; X15800394); Antispasmodic (1; X9434612); Diuretic (f; DAW); Litholytic (f; X9434612); Lithuretic (f; X9434612); Spasmogenic (1; X9434612).

Indications (Golden Thistle):

Cramp (f; X9434612); Stone (f; X9434612).

Dosages (Golden Thistle):

FNFF = !!
Roots eaten cooked, or scorched as coffee substitute; shoots eaten boiled; tender leaves and blanched petioles cautiously consumed in salads; dressed in olive oil and vinegar in Madrid; flowers used to adulterate saffron (FAC; TAN).

Downsides (Golden Thistle):

Be careful with the thorns.

Extracts (Golden Thistle):

The ethanolic root bark extract and taraxasteryl acetate, and a fluid extract, exhibit strong biphasic antispasmodic and spasmogenic responses, on isolated rat ileum, suggesting a possible lithuretic activity (X9434612).

ALEXANDRIAN SENNA (*SENNA ALEXANDRINA* MILL.) + FABACEAE (CAESALPINIACEAE)

Synonyms:

Cassia acutifolia Delile; *Cassia angustifolia* Vahl; *Cassia lanceolata* Forssk.; *Cassia senna* L.; *Senna acutifolia* (Delile) Batka; *Senna angustifolia* (Vahl) Batka *fide* AH2.

NOTES (ALEXANDRIAN SENNA):

And the angel of the Lord appeared unto him in a flame of fire out of the midst of a bush; and he looked, and behold, the bush burned with fire, yet it was not consumed.

Exodus 3:2 (KJV)

And the angel of the Lord appeared to him in a flame of fire out of the midst of a bush; and he looked, and lo, the bush was burning, yet it was not consumed.

Exodus 3:2 (RSV)

Then Jehovah's angel appeared to him in a flame of fire in the midst of a thornbush. As he kept looking; why, here the thornbush was burning with the fire, and the thornbush was not consumed.

Exodus 3:2 (NWT)

Zohary (1982) argues convincingly that the plant called *sneh* should be translated *Senna*, not *Acacia nilotica* as Moldenke and Tristam concluded, or *Rubus* as other scholars had concluded. Why? Because *Acacia nilotica* and *Rubus* do not occur on the Sinai, (Fonck even suggests it might be a hawthorn, perhaps *Crataegus sinaica*). He concludes that "the most plausible explanation for sneh is *Cassia senna* named sene in all Arabic-speaking countries." The taxonomic name has changed to *Senna alexandrina* but sene persists in most Arabic quarters. Zohary makes one of his few medicinal observations: "It is known medically as a stimulant and purgative, under the name folia sennae." Writing concerning (ZOH) Lebanon, my friend Jane Philips says that the dried leaves and pods are a nostrum in the Middle East — the drinkers, usually elderly, sipping their unpleasant brew slowly, and sometimes complaining about their chronic constipation. The nostrum is also believed to purify the blood, alleviate hemorrhoids, and serves as a general alterative and tonic for the nervous system. Borelli et al. (2005), evaluating the carcinogenic potential of anthraquinones, found instead that senna pod extracts might exert antitumor activity on rat colon carcinogenesis (X16008128).

COMMON NAMES (ALEXANDRIAN SENNA):

Agerger (Ber.; BOU); Alexandrian Senna (Eng.; USN); Alexandrinische Senna (Ger.; USN); Bhuikhakhasa (Hindi; KAB; WO2); Bhuitarvada (Mar.; DEP; WO2); Bhumiari (Sanskrit; KAB; WO2); Bhumichari (Sanskrit; KAB); Bhumivalli (Sanskrit; KAB); Bombay. Senna (Eng.; WO2); Casse (Fr.; USN); Casse à Feuilles Étroites (Fr.; BOU); Casse Trompeuse (Fr.; BOU); Chiieh-Ming (China; EFS); Fan Hsieh Yeh (China; EFS; KAB); Fan Xie Ye (Pin.; AH2); Hemapatri (Sanskrit; KAB); Hemapatrika (Sanskrit; KAB); Hindisana (Hindi; India; DEP; EFS); Hindisana Kapat (Hindi; DEP; KAP); Indian Senna (Eng.; BOU; USN); Jalatika (Sanskrit; KAB); Jian Ye Fan Xie (Pin.; AH2); Kalyani (Sanskrit; KAB); Kattunilavirai (Tam.; KAB); Lotus Senna (Eng.; KAB); Mahaushadbi (Sanskrit; KAB); Malaharini (Sanskrit; KAB); Markandika (Sanskrit; KAB); Mecca Senna (Eng.; IHB); Medicinal Senna (Eng.; KAB); Middiawal (Guj.; WO2); Mrudirechani (Sanskrit; KAB); Mulcacha (Mah.; Mar.; DEP; NAD); Naelaponna (Tel.; NAD); Natkisana (Dec.; KAB); Natkisanapatta (Dec.; KAB); Natkisona (Dec.; DEP; NAD); Nattunelavarai (Tam.; WO2); Nattu Nila Virai (Tam.; WO2); Neelaponna (Tel.; WO2); Neelatangeedu (Tel.; WO2); Nellatangedu (Tel.; MPI); Nelavagai (Tam.; WO2); Nelavari (Sing.; Sin.; KAB; NAD); Nelavarike (Kan.; WO2); Nila Vakai (India; Tam.; EFS; MPI; NAD); Nilavaka (Mal.; KAP; NAD); Nilavaki (Mal.; WO2); Nilavari (Sindh KAB; NAD); Nilavirai (Tam.; NAD); Nilavitai (Kan.; NAD); Nubian Senna (Eng.; IHB); Pero Sama

FIGURE 1.103 Alexandrian Senna (*Senna alexandrina*).

e Hindi (Arab.; EFS); Pitapushpi (Sanskrit; KAB; WO2); Puve Kain Yoe (Burma; DEP; KAB; NAD); Rechani (Sanskrit; KAB); Sana (Arab.; GHA); Sana-e-Hindi (Arab.; Iran; DEP; NAD); Sana Hhigazf (Arab.; Syria; HJP); Sanakola (Sindh.; KAB; NAD); Sania (Aramaic; ZOH); Sanna Makki (Arab.; Beng.; GHA; WO2); Sana Sa'idi (Arab.; Syria; HJP); Sen de España (Sp.; USN); Sena (It.; Malaya; EFS); Sena Alejandrina (Sp.; USN); Senamakki (Beng.; Guj.; DEP; NAD; WO2); Sene (Arab.; ZOH); Séné (Fr.; BOU); Séné d'Egypte (Fr.; USN); Senna (Arab.; Eng.; Dutch; Ger.; EFS; USN); Senna Bush (Eng.; ZOH); Senna Haram (Arab.; BOU); Senna Hindi (Arab.; BOU); Senna

Makki (Arab.; GHA); Senna Mekki (Arab.; BOU); Senna Sa'eidi (Arab.; BOU); Senne (Den.; EFS); Shonamakhi (Mar.; WO2); Shonamukhi (Oriya; Uriya; KAB; WO2); Shonpat (Beng.; DEP); Siah Yip (China; Malaya; EFS; KAB); Sinamekia Agaci (Tur.; EFS); Sneh (Heb.; ZOH); Sona Mukhi (India; Sanskrit; EFS); Sonaehindi (Iran; KAB); Sonamakki (Mah.; NAD); Sonpat (Beng.; DEP; WO2); Soonamukhi (Kan.; WO2); Sooranatnila Avarai (Tam.; WO2); Stholotpala (Sanskrit; KAB); Sunnamukhi (Mal.; WO2); Svarnamukhi (Sanskrit; KAB); Svarnapatri (Ayu.; AH2); Svarnapatrika (Sanskrit; KAB); Svarnini (Sanskrit; KAB); Swarnamakhi (Sanskrit; WO2); Swarnapatrika (Sanskrit; WO2); Ta Hwai Yeh (China; KAB); Tinnevelly Senna (Eng.; BOU; USN); True Senna (Eng.; Ocn.; AH2; USN); Ye Fan Xie (Pin.; AH2).

Activities (Alexandrian Senna):

Alterative (f; HJP); Anthelmintic (f; WO2); Antidysenteric (f; WHO); Antihepatotoxic (1; CAN); Antiherpetic (1; WO3); Antileukemic (1; CAN); Antispasmodic (f; PED); Antiviral (1; WO3); Bactericide (1; WO2); Bitter (f; PED); Carminative (f; WHO); Cathartic (f1; SKY); Expectorant (f; WHO); Fungicide (f1; JAD); Hepatotonic (f; WO2); Hepatotoxic (1; X11324140); Laxative (f1; BGB; PH2; PIP); Mutagenic (1; BGB); Neurotonic (f; HJP); Peristaltic (1; PED; WHO); Tonic (f; WO2); Trypsin Inhibitor (1; WO3).

Indications (Alexandrian Senna):

Acne (f; WO2); Ameba (f; WO2); Anemia (f; PH2); Anorexia (f; KAB); Biliousness (f; WO2); Bronchosis (f; WO2); Burn (f; WO2); Cancer (f; JLH); Cholera (f; WO2); Constipation (f12; KOM; PH2; PIP; WHO); Cramp (F; PED); Dermatosis (f; WHO); Dysentery (f1; WHO); Dyspepsia (f; WHO); Enterosis (f; KAB; PH2); Fever (f; PH2; WHO); Fungus (1; FNF); Gas (f; CR2; WHO); Gastrosis (f; WO2); Gonorrhea (f; WHO); Gout (f; WO2); Halitosis (f; WO2); Hemorrhoid (f; BGB; HJP; PIP; WHO); Hepatosis (f; PH2); Herpes (1; WO3); Hiccup (f; WO2); Infection (f1; FNF; WO2); Jaundice (f; PH2); Leprosy (f; WO2); Leukemia (1; CAN); Mycosis; (1; JAD); Nausea (f; WO2); Nerves (f; HJP); Pimple (f; WO2); Ringworm; (f1; JAD); Splenosis (f; PH2); Syphilis (f; DEP); Typhoid (f; PH2; WO2); Venereal Disease (f; DEP; WHO); Virus (1; WO3); Worm (f; WO2); Wound (f; WHO).

Dosages (Alexandrian Senna):

FNFF = X

0.5–2 g (1/2 to 1 tsp) per cup water (APA); two 25-mg StX extracts, 1–2 ×/day (APA); 3–6 Alexandrian or 4–12 Tinnevelly pods steeped in 150 ml warm water for 6–12 hrs (CAN); 0.5–2 g dry leaflets (CAN); 0.5–2 ml liquid leaf extract (1:1 in 25% ethanol) (CAN); 0.5–2.0 ml BPC liquid extract (CAN); 20–30 mg hydroxyanthracene derivatives/day calc as sennoside B (KOM); 1 g dry leaf:5 ml alcohol/5 ml water (PED); 1–2 g dry leaf (PED; WHO); 0.5–2 g dry leaf; two 25–mg capsules/day; 10–60 mg sennosides (SKY); 1–2 g fruit (WHO).

- Algerian, Iranian, and Lebanese elders frequently sip senna tea for constipation, thinking it helps blood, nerves, and piles (HJP).
- Asian Indians sometimes chew the leaves as a laxative (DEP).
- Asian Indians use powdered leaves in secondary syphilis (DEP).
- North Africans take tea of powdered leaves and pods, with or without rose leaf, as laxative and purgative (BOU).
- Saudis use leaf decoction, alone or with other herbs, as laxative for cramps and constipation (GHA).

Downsides (Alexandrian Senna):

Class 2b, 2c, 2d (AHP, 1997). Commission E and other sources report interaction of anthranoid laxatives (AEH). Anthranoid-containing laxatives can be habit forming; some contain compounds suspected of being cytotoxic, genotoxic, mutagenic, and even tumorigenic; epidemiological studies in Germany reveal that abusers of anthranoid laxatives have a three-times-higher rate of colon carcinoma. One woman developed clubbing of her digits and hypertrophic osteoarthropathy after taking at least three tablets daily for 3 years for weight loss. She also experienced several months of secondary amenorrhea. Of senna leaf, Blumenthal et al. (1998) list abdominal pain of unknown origin, acute intestinal inflammation (e.g., Crohn's disease and colitis ulcerosa), and appendicitis as contraindications. Should not be used in lactation, during pregnancy, or with children under 12 years old (KOM). Occasional cramp-like discomfort of the GI tract may require dosage reduction (KOM). Side effects with chronic use or abuse: disturbance of electrolyte balance, especially hypokalemia (may be exacerbated by simultaneous administration of corticoadrenal steroids, licorice root, or thiazide diuretics) leading to cardiopathy and muscular weakness (especially with concurrent uses of cardiac glycosides, corticosteroids, or diuretics) (KOM). Pigmentation of the intestinal mucosa (*Pseudomelanosis coli*) is harmless and usually reverses on discontinuation of the drug. Simulating laxatives like this should not be used for more than 1 to 2 weeks without medical advice (KOM). Newall, Anderson, and Phillipson (1996) report anthraquinones are purgative and irritate the GI tract. Because of the anthraquinones, nonstandardized preparations should be avoided during pregnancy and lactation (CAN). "Anthraquinones may be secreted into breast milk CAN." Also contraindicated in hemorrohoids and nephropathy (CAN), intestinal obstruction, abdominal pain of unknown causes, any enterosis (appendicitis, colitis, Crohn's disease, irritable bowel syndrome) hemorrhoids, nephropathy, menstruation (AHP, 1997). Do not use more than eight to ten days (AHP, 1997). "Do not use this product if you have abdominal pain or diarrhea. Consult a health care provider prior to use if you are pregnant or nursing. Discontinue use in the event of diarrhea or watery stools. Do not exceed recommended dose. Not for long term use" (AHP, 1997 CAN). "Some herbal laxative preparations such as cascara and senna, for example, can cause an increase in the potency of digoxin" (D'epiro, 1997 CAN). Pedersen cautions against taking the fresh leaf (we have done that in Peru with modest laxative results). "Senna causes gripping unless taken in combination with carminative herbs such as ginger, cloves, or various mint species. Although generally recognized as safe, senna is somewhat more habit forming than cascara" (PED CAN). Lininger et al. (1998) pronounce it "safe for children over the age of six" (half the adult dose) (SKY CAN). I'd be more cautious.

Extracts (Alexandrian Senna):

Possibly representing a negative case for my synergy proposition: "The toxicity of total extracts is greater than that of of the individual sennosides. It has been proposed that the laxative and toxic components of senna could be separated" (CAN). Elsewhere I have read that the sennosides are synergic as far as their laxative action is concerned. But then anomalously, Newall et al. 1996 report that Sennosides A and B are reported to be most potent with respect to laxative action, but the least toxic compared to other anthraquinone fractions. The LD50 values of sennosides A and B are 4100 mg/kg ivn mus and of rhein-8-glycoside 400 mg/kg ivn mus. All fractions had LD50 $\geq$ 5000 mg/kg orl rat.

MILK THISTLE (*SILYBUM MARIANUM* (L.) GAERTN.) +++ ASTERACEAE

Synonyms:

Carduus marianus L.

Notes (Milk Thistle):

Thorns also and thistles shall it bring forth to thee; and thou shalt eat the herb of the field.

Genesis 3:18 (KJV)

FIGURE 1.104 Milk Thistle (*Silybum marianum*).

Thorns and thistles it shall bring forth to you; and you shall eat the plants of the field.

Genesis 3:18 (RSV)

And thorns and thistles it will grow for you; and you must eat the vegetation of the field.

Genesis 3:18 (NWT)

Being positive is not always possible. Zohary lists three species as good candidates for the thorns and briers (not for thistles), *Echinops viscosus*, *Notobasis syriaca*, and this species, *Silybum marianum*, my favorite, bearing the common names Holy *Thistle and Mary's Thistle. He confesses there* is no concrete evidence that any of these three "are the *barkanim* of Gideon", but they are more probable because they are tall plants likely used as whips. They are the most abundant of the thorns around Ophra where judge Gideon was, beating out wheat in the wine press (Judges 6: 11). I would not call milk thistle an annual, as does Zohary. Mine, like his, germinates one year, has leaves appressed to the ground that winter, bolting in the spring. Flowers will soon follow the bolting, and then come the seeds, eagerly harvested by the gleeful goldfinch. Then the biennial in midsummer is done for the year. Those few seeds that escape the goldfinches germinate in the fall. Zohary selects *Centaurea iberica* rather than *Silybum* for the thistle of the Bible (ZOH). Usually I tend to go with Zohary when such controversies arise. But I still favor the milk thistle as the biblical thistle, especially in Genesis after Eve has eaten the forbidden fruit. Not only is it edible "vegetation of the field," at least in my book, its hepatoprotective properties might even spare some of the travails of eating poisonous vegetation of the field, including mushrooms. Could it be Eve's milk rather than Mary's milk that imparted the white to the veins of the milk thistle? I do indeed doubt that, as I do the following tidbit from GMH: the tradition that the white veins making this thistle unique resulted from the milk of the Virgin falling on a thistle (GMH), leading to the instant evolution of the new milk-veined species. While agnostically doubting, I continue to tell the tale during my biblical tours of the Green Farmacy Garden.

Common Names (Milk Thistle):

Akoub (Arab.; BOU); Armurariu (Rom.; KAB); Artichaut Savage (Fr.; BOU); Baq (Arab.; BOU); Barkanim (Heb.; ZOH); Baskavkas (Tur.; EB49:406); Blessed Milk Thistle (Australia; USN); Bou Zerwal (Arab.; BOU); Cardo (Mad.; JAD); Cardo Asnal (Arg.; EFS); Cardo de María (Sp.; USN); Cardo de Santa Maria (Por.; EFS); Cardo del Late (It.; KAB); Cardo della Madonna (It.; EFS); Cardo di Maria (It.; EFS; KAB); Cardo Lechal (Sp.; EFS); Cardo Lechero (Sp.; USN); Cardo-leiteiro (Por.; USN); Cardo Mariano (It.; Sp.; EFS; KAB; USN); Cart de Maria (Cat.; KAB); Cart Gallofer (Cat.; KAB); Carthame Macule (Fr.; KAB); Chardon Argente (Fr.; BOU); Chardon Lacte (Fr.; KAB); Chardon-Marie (Fr.; USN); Chardon Notre Dame (Fr.; EFS); Chardon Tache (Fr.; KAB); Doujnilourman (Ber.; BOU); Feedistel (Ger.; KAB); Fouarek (Arab.; BOU); Frauendistel (Ger.; EFS; KAB); Froschdistel (Ger.; KAB); Gevlekt distelhruid (Dutch; EFS); Gundagai Thistle (Australia; USN); Harsharf Barri (Arab.; BOU); Hasoub (Arab.; BOU); Hedegar (Sp.; EFS); Holy Thistle (Eng.; EFS); Kocak Avas (Tur.; EB49:406); Lady's Thistle (Australia; USN); Lait de Sainte Marie (Fr.; KAB); Lekhlakh (Arab.; BOU); Lievevrouwen Destel (Dutch; EFS); Lishliash (Arab.; BOU); Marian Distel (Dutch; EFS); Marian Thistle (Eng.; EFS); Mariendistel (Dutch; Ger.; EFS; KAB; USN); Marietidsel (Den.; EFS); Melk Destel (Dutch; EFS; KAB); Meryemanadikeni (Tur.; EFS); Milk Thistle (Eng.; CR2); Ostropestro (Rus.; KAB); Sempertin (Swe.; KAB); Shouk Boulti (Arab.; BOU); Shouk el Diman (Arab.; BOU); Shouk el Gemal (Arab.; BOU); Shouk el Ghazal (Arab.; BOU); Shouk en Nasara (Arab.; BOU); Shouket el Baiidisinnari (Arab.; BOU); Shouk Sinnari (Arab.; BOU); Silberdistel (Ger.; KAB); Silybon (Greek; KAB); St. Mary's Thistle (Eng.; EFS); Stechkornerdistel (Ger.; EFS); Tataoura (Ber.; BOU); Thistle (Eng.; BIB); Thorn (Eng.; BIB; ZOH); True thistle (Eng.; EFS); Variegated Thistle

(Eng.; USN); Vrowwendistel (Dutch; KAB); Silybe de Marie (Fr.; USN); Wehedistel (Ger.; KAB); Wolldistel (Ger.; KAB); Xeuk Baghli (Malta; KAB); Zaz (Arab.; BOU).

Activities (Milk Thistle)

Adrenergic (1; WOI); Alterative (f; BIB; EFS); Antiaggregant (1; ACT9:170); Antiallergic (1; MAB); Antiangiogenic (1; X12943822; NP9(2):6); Antibilious (f; APA); Anticarcinogenic (f1; MAB; NP9(2):6); Antidepressant (f; PNC); Antidiabetic (f1; ACT9:251; Antidotal (2; SHT); Antidote (mushroom) (f1; PH2); Antiedemic (f1; MAB; NP9(2):6); Antifibrotic CGH; Antihistaminic (1; MCK); Antiinflammatory (f1; BGB; MAB; WAM); Antileukotriene (1; MAB); Antimetastatic (1; X15224346); Antioxidant (12; MAB; SHT); Antiproliferant (1; HC020444-262; NP9(2):6); Antiprostaglandin (1; MAB); Antitelomerase (1; X15076315); Antitoxic (2; SHT); Antitumor (f1; MAB); Antitumor Promoter (1; NP9(2):6); Antiulcer (1; PR14:581); Antiviral (1; PNC); Aperient (f; BIB; WOI); Apoptotic (1; JN133:3861S; X15117815); Bitter (f1; PED); cAMP-Phosphodiesterase Inhibitor (1; MAB); Caspase Activator (1; X15117815); Chemopreventive (1; NP9(2):6); Cholagogue (f2; BIB; EFS; PHR); Choleretic (1; HHB; MAB; NP9(2):6); COX-2 Inhibitor (1; NP9(2):6); Cytoprotective (1; NP9(2):6); Cytotoxic (1; NP9(2):6); Demulcent (f; KAB; PNC; WOI); Depurative (f; EFS); Dermoprotective (1; X15586237); Diaphoretic (f; EFS; WOI); Digestive (1; WAM); Diuretic (f; HC020444-262); Emmenagogue (f; BIB; EFS; PHR; PH2); Expectorant (f; BIB); Febrifuge (f; EFS); Glutathionigenic (1; MAB); HDL-genic (1; HC020444-262); Hemostat (f; BIB); Hepatoprotective (2; KOM; SHT; WAM); Hepatoregenerative (2; KOM; MAB); Hypocholesterolemic (1; MAB); Hypolipidemic (1; PNC); Hypotensive (1; MCK; ACT9:170); Hypotriglyceridemic (1; CGH); IKKbeta Inhibitor (1; X15617879); Immunostimulant (1; NP9(2):6); i-NOS Inhibitor (1; NP9(2):6); Lactogogue (f1; APA; BIB; HMM; NMH; WOI); Lipolytic (1; PNC); 5-Lipoxygenase Inhibitor (1; MAB); Litholytic (f; WOI); Nephroprotective (1; ACT9:170; NP9(2):6); Neuroprotective (1; ACT9:251); p450 Inducer (1; X12695340); p450 Inhibitor (1; X15155549); Pancreaprotective (1; MCK; X15325843); Peristaltic (1; WOI); p-Glycoprotein Inhibitor (1; X15072439); Phagocytotic (1; NP9(2):6); Photoprotective (1; NP9(2):6); PKC Inhibitor (1; X15070174); Purgative (f1; BIB; WOI); Radioprotective (1; NP9(2):6); SOD-genic (1; CGH); Stimulant (f; EFS; PHR); Sudorific (f; BIB; WOI); Sunscreen (1; MAB); Sympathicolytic (f; HHB); T-Cell Activator (1; NP9(2):6); TNF-alpha Inhibitor (1; NP9(2):6); Tonic (f1; BIB; EFS; PHR; WAM); Xanthine-oxidase Inhibitor (1; ACT9:170; X15070174; NP9(2):6).

Indications (Milk Thistle):

Acne (f; MCK); Ague (f; GMH); Allergy (f1; MCK; ACT9:251); Amenorrhea (f; MCK; NP9(2):6); Anorexia (2; FAD; PHR); Anthrax (f; BIB); Arthrosis (f; ACT9:251); Asthma (f1; BIB; ACT9:251); Atherosis (1; X15617879); Biliousness (f; GMH; WOI); Bleeding (f; BIB; HHB; NP9(2):6); Bronchosis (f; BIB; BOU; EB49:406); Calculus (f; BIB; WOI); Cancer (f; JLH; WOI; NP9(2):6); Cancer, bladder (f1; NP9(2):6); Cancer, breast (f1; HHB; JLH; MAB); Cancer, cervix (f1; HC020444-262); Cancer, colon (f1; FNF; JN133:3861S); Cancer, lung (f1; X15224346); Cancer, nose (f; HHB; JLH); Cancer, ovary (f1; HC020444-262); Cancer, prostate (f1; X15899838); Cancer, skin (f1; HC020444-262; NP9(2):6; X15586237); Cancer, tongue (f1; HC020444-262); Cardiopathy (1; X15617879); Catarrh (f; BIB; GMH); Cerebrosis (f; ACT9:251); Childbirth (f; HHB); Cholecocystosis (2; APA; BIB; PHR; PH2); Cholelithiasis (f; MCK); Cirrhosis (2; BGB; KOM; PH2; SHT); Colic (f; HH3; PH2); Colitis (f; ACT9:251); Congestion (f; MCK); Constipation (f; MCK); Cough (f; BIB); Cramp (f; BIB); CTS (f1; ACT9:251; HC020444-262); Cystosis (f; HH3); Depression (f; BIB; HC020444-262); Dermatosis (f1; BIB; MCK; PED); Diabetes (f1; MAB; WOI; HC020444-262); Diabetic Neuropathy (1; MAB); Dropsy (f; BIB; HHB; WOI); Dysmenorrhea (f; APA; HC020444-262); Dyspepsia (f12; FAD; KOM; MCK; PH2; SHT; NP9(2):6); Dyspnea (f; EB49:406); Dysuria (f; NP9(2):6); Edema (f1; ACT9:251); Enterosis (f; APA; WOI); Erysipelas (f; MCK); FAS (f; ACT9:251); Fever (f1; BIB; EFS;

HHB); Fibrosis (1; CGH); Food Allergies (1; WAM); Gallstones (1; HHB; MAB; SKY; HC020444-262; NP9(2):6); Gastrosis (f; APA); Hematuria (f; HC020444-262; NP9(2):6); Hemoptysis (f; BIB); Hemorrhage (f; KAB; MCK); Hemorrhoid (f; BIB; HHB; MAB; WOI); Hepatosis (f12; KOM; PH2; SHT; WAM); Hepatosis A (1; BGB); High Blood Pressure (1; MCK; HC020444-262); High Cholesterol (1; MAB); High Triglycerides (1; CGH; X15177299); Hydrophobia (f; BIB; GMH); Hyperemesis (f1; NP9(2):6); Hypotonia (f; HH3); Infection (f; HHB); Inflammation (f1; APA; HC020444-262; X15617879); Intoxication (1; FAD); Insulin Resistance (1; SYN); Itch (1; MAB); Jaundice (f2; BIB; HH3 MAB; PH2; PNC; WAM); Leukemia (f1; HC020444-262; NP9(2):6); Leukorrhea (f; BIB); Malaria (f1; BIB; HHB; PHR; PH2; HC020444-262); Menopause (f; HHB); Metastasis (1; X15224346); Metrorrhagia (f; HHB); Migraine (f; HH3); MS (f; ACT9:251); MS (1; HC020444-262); Mushroom Poisoning (2; FAD; SHT); Myalgia (1; HC020444-262); Nausea (f1; MAB; Cytoprotective (1; NP9(2):6); Nephrosis (f12; BGB; NP9(2):6); Neurosis (f; ACT9:251); Obesity (1; PNC); Oligolactea (f; APA); Ovarian Cancer (f1; MAB); Pain (f; NP9(2):6); Pancreatosis (1; X15325843); Peritonosis (f; BIB); Phlebitis (f; APA); Plague (f; BIB; GMH); Pleurisy (f; BIB; GMH); Psoriasis (f; PED; SKY); Pulmonosis (f; BIB); Snakebite (f; MCK); Sore (f; HHB); Splenosis (f; BGB; BIB; HH3; PHR; PH2; NP9(2):6); Steatosis (f; MCK); Swelling (f1; MCK; ACT9:251); Syndrome X (1; SYN; X15617879); Ulcer (1; PR14:581); Ulcus cruris (f; HHB; HH3); Urticaria (f; MCK); Uterosis (f; BIB; PHR; PH2; WOI); Varices (f; HHB; HH3); Worm (f; BIB).

Dosages (Milk Thistle):

FNFF = !!

Deveined leaves eaten as veggies; young shoots boiled and eaten (TAN); heads boiled and eaten like artichoke; oilseeds roasted as coffee substitute; roots also eaten like salsify (FAC; MCK); stalks, like those of most thistles, are edible and nutritious, but have caused fatalities in cattle (BIB). Seeds serve as famine food for humans (BIB), actually sold to me like sunflower seed in Pakistan. Seeds scorched as coffee substitute, seed oil used for food or lubrication, which might also serve as a famine food for humans. 1 tsp (3–5 g) mashed seed/cup water, 3–4 ×/day, 1/2 hour before meals (APA; HH3); 1 g seed HHB. 3.5–15 g seed/day (HH3). 4–9 ml/day fluid extract (1:1) (KOM); 12–15 g whole or powdered seed; an equivalent to 200–400 mg silymarin, the collective name for silybinin, silydianin, and silychristin (KOM; SHT); 4–9 g/day seed (MAB); one or two 535-mg capsules 3 ×/day; StX 420–840 mg silymarin/day (NH); one to two 540-mg capsules (StX with 175 mg certified potency seed extract with at least 80% silymarin, synergically combined in a base of turmeric and artichoke) 3 ×/day, with water (NH); 2–3 tsp fresh leaf (sic) (PED); 1–3 g dry leaf (sic) (PED); 175 mg 80% silymarin StX (PED); 420 mg silymarin/day (PNC); 200–400 mg silymarin (SHT), 200–600 mg silymarin/day for Syndrome X (SYN).

- American Eclectics use strong seed tincture for amenorrhea, bleeding, dysuria, gallstone, congestion, hematuria, hepatosis, hyperemesis, nephrosis, splenosis (NP9(2):6).
- Basque use seed decoction as antiinflammatory, antineoplastic, aperient, digestive, diuretic, hemostat, hypotensive and venotonic (NP9(2):6).
- Germans use as demulcent in catarrh and pleurisy, for jaundice and biliary disorders (GMH).
- Lebanese consider the flower infusion alterative, refrigerant, and tonic (HJP).
- Lebanese boil leaves in vinegar for skin ailments and tonics (HJP).
- Lebanese use astringent vulnerary root for hemorrhoids and worms (HJP).
- Lebanese use seed infusion for stones of gallbladder and liver, and as a hydragogue, lactogogue, litholytic, stimulant, and tonic (HJP).
- North Africans suggest the seed tincture is for cough, peritonitis, pulmonitis (or bronchitis), uteral congestion, and varicose veins (BOU).
- Turkish use ground-roasted seed decoction for shortness of breath in bronchosis (EB49:406).

Downsides (Milk Thistle):

Class 1 (AHP, 1997). No health hazards or side effects known with proper therapeutic dosages (PH2). None known (WAM); Commission E reports no contraindications or drug interactions for the fruit. Occasional mild laxative effects are reported (AEH). "The long term safety and the advisability of the use of these extracts in pregnant or women of potential childbearing remain to be established" (LRNP, March 1988). One observational study ($n \geq 2000$) reported circa 1% side effects, mostly transient GI distress (SHT). "May be used by ... pregnant and lactating women (SKY). Antagonizes phentolamine and yohimbine (PH2). One Australian report, attributed to something other than silybin in the milk thistle product, suggested abdominal cramping, diaphoresis, diarrhea, nausea, vomiting, and weakness (PH2). Silymarin may induce p450. Milk thistle decreased the trough concentrations of indinavir in humans (X15916450).

Natural History (Milk Thistle):

Birds like the seeds, goldfinches getting most of mine. GMH debunks the old birdwatcher's tale that birds line their nests with the thistle down. Most nests are built before the thistle sets seed; coltsfoot and willow down are more probable because they seed much earlier, when birds are building.

Extracts (Milk Thistle):

Milk thistle regenerates injured liver cells (SKY). Silymarin, at 100 mg/kg in rabbit diets, is reported to induce p450. Such data weakly suggest that milk thistle might detoxify (or inactivate) drugs detoxified by p450. Silymarin, the antihepatotoxic lignan, is hypolipidenic, lowering fatty deposits in the livers of experimental animals, and has been used successfully for Hepatosis B virus (PNC). Pretreatment with silymarin and silybin gives 100% protection against mushroom poisoning in experimental animals. Posttreatment? When silybin was given ivn to humans within 48 hours of ingesting death cap mushroom, it effectively prevented fatalities (PNC). Equally good results were reported in Santa Cruz, California, in January 2007. LD50 silibinin 1065 mg/kg ivn mus HH3. LDlo silymarin ≥ 20,000 mg/kg orl mus (MAB); LDlo silymarin ≥ 1000 mg/kg orl dog (MAB); LDlo Extracts >16,000 mg/kg orl mus HH3. McCarty (2005) discussed the potential of milk thistle polyphenols for reversing fat-induced insulin resistance. He noted intriguing evidence that the beta subunit of the signalsome — IKKbeta, a crucial catalyst of NF-kappaB activation — is an obligate mediator of the disruption of insulin signaling induced by excessive exposure of tissues to free fatty acids and by hypertrophy of adipocytes. He speculated that agents that safely inhibit activation of IKKbeta may help reverse insulin resistance and help control type 2 diabetes. He suggests resveratrol (from grapes) and silibinin. Dietary silibinin can inhibit the growth of certain cancers in rodents and may have potential as an IKKbeta inhibitor. Silymarin also favorably impacts glycemic and lipidemic control in type 2 diabetics with cirrhosis. IKKbeta plays a crucial role in insulin resistance, atherogenesis, inflammatory disorders, and cancer (X15617879). Katiyar (2005) elucidated the anti-inflammatory, antioxidant, and immunomodulatory potential of silymarin in preventing skin cancer. Silymarin has chemopreventive effects against chemical and photocarcinogenesis (X15586237).

CHARLOCK (*SINAPIS ARVENSIS* L.) + BRASSICACEAE

Synonyms:

Brassica arvensis Rabenh.; *Brassica kaber* (DC.) Wheeler.; *Brassica sinapis* Vis.; *Brassica sinapistrum* Boiss.; *Caulis sinapiaster* E.H.L.; *Eruca arvensis* Noulet.; *Napus agriasinapis* Schimp et Spenn.; *Raphanus arvensis* Crantz.; *Rhamphospermum arvense* Andrz.; *Sinapis kabe*r DC.; *Sinapis orientalis* L.; *Sinapis polymorhpa* Geners.; *Sinapis schkuhriana* Reichb. *fide* HH2

FIGURE 1.105 Charlock (*Sinapis arvensis*). Source: Regina Hughes in Reed (1970). Colored by Peggy Duke.

Notes (Charlock):

> **Another parable put he forth unto them, saying, The kingdom of heaven is like to a grain of mustard seed, which a man took, and sowed in his field: Which indeed is the least of all seeds: but when it is grown, it is the greatest among herbs, and becometh a tree, so that the birds of the air come and lodge in the branches thereof.**
>
> **Matthew 13:31–32 (KJV)**

> **Another parable he put before them, saying, "The kingdom of heaven is like a grain of mustard seed which a man took and sowed in his field; it is the smallest of all seeds, but when it has grown it is the greatest of shrubs and becomes a tree, so that the birds of the air come and make nests in its branches."**
>
> **Matthew 13:31–32 (RSV)**

> **Another illustration he set before them, saying, "The kingdom of the heavens is like a *mustard* grain which a man took and planted in his field; which is, in fact, the tiniest of all the seeds, but when it has grown it is the largest of the vegetables and becomes a tree, so that the birds of heaven come and find lodging among its branches."**
>
> **Matthew 13:31–32 (NWT)**

Zohary favors *Brassica nigra*, the black mustard, as the grain of mustard seed in the Bible. As an Israeli botanist, he would know better than I that *Brassica nigra* is the tallest of the local species of *Brassica* and closely related *Sinapis*, and has seed circa 1 mm. But he also admits that "The Greek sinapis is undoubtedly 'mustard (ZOH).'" No need to quibble about whether it is the least of seed or not. (I suspect a mustard seed is hundreds of times heavier than orchid seed, even my *Artemisia annua* seed.) I suspect that in biblical times, as in Maryland fields in spring, few but botanists make distinctions between the *Brassica* and *Sinapis*. Taxonomists still quibble over the genus in which to place the charlock, one candidate for the biblical mustard seed. "Mustard is not mentioned in the Old Testament but is often referred to in the Mishnah." (ZOH). *Hager's Handbuch* lists a lot of common names, phytochemicals, and synonyms, but no indications or activities (HH2). I would use it generically with black or white mustard myself. All can be weeds. More PubMed abstracts deal with weediness than medicine. But the species share the chemistry and medicine of other *Brassica* and *Sinapis*.

Common Names (Charlock):

Ackersenf (Ger.; HH2); Ager Sennep (Den.; POR); Agersennep (Den.; POR); Åkersenap (Swe.; POR); California Rape (Eng.; USN); Charlock (Eng.; GMH; USN); Collejón (Sp.; USN); Corn Kale (Ireland; AAH); Erba Falcona (It.; HH2); False Hederich (Ger.; HH2); Field Mustard (Eng.; GMH); Guelot (Fr.; HH2); Herik (Dutch; POR); Jotte (Fr.; HH2); Mostarda (Mad.; Por.; JAD); Mostarda-dos-Campos (It.; Por.; POR; USN); Mostaza de los Campos (Sp.; HH2; POR); Mostaza Silvestre (Sp.; POR); Moutarde de Champs (Fr.; HH2); Moutarde Savage (Fr.; HH2); Nabinha (Por.; POR); Rapaccina (It.; HH2); Ravanello (It.; HH2); Raveluche (Fr.; POR); Rikkasinappi (Fin.; POR); Ruche (Fr.; HH2); Sangle (Fr.; HH2); Sangle Wilder Senf (Ger.; HH2); Sanve (Fr.; POR); Saramago (Mad.; JAD); Senapa dei Campi (It.; HH2); Senape (It.; USN); Senape Selvatica (It.; POR); Senapino (It.; HH2); Sénevé, (Fr.; POR); Serapino (It.; POR); Wild Mustard (Eng.; HH2); Yang Ye Jie (China; POR).

Activities (Charlock):

Diuretic (f; EFS); Dyspepsia (f; EFS); Emetic (f; EFS); Molluscicide (1; X11775093); Ovicide (1; X11775093); Rubefacient (f; EFS); Stimulant (f; EFS); Tonic (f; AAH); Trypsin Inhibitor (1; X8070965); Vesicant (f; EFS).

Indications (Charlock):

Cancer (f; JLH); Induration (f; JLH); Jaundice (f; AAH); Scurvy (1; AAH); Splenosis (f; JLH).

Dosages (Charlock):

FFFF = !!
Seeds used as mustard substitute or oil source. Irish and Swedes boil it as a green vegetable (GMH).

- Anglicans took a floral preparation for jaundice (AAH).
- Limerick people drank the juice as a spring tonic to prevent disease the rest of the year (AAH).
- Shetland Islanders once relied on charlock (leaf with 1300 ppm ascorbic acid) and wild radish to contain scurvy (AAH).

Downsides (Charlock):

Similar to *Brassica* and *Sinapis*.

Extracts (Charlock):

As with rocket and garlic mustard, the seed oil is closely akin to Lorenzo's oil, which for awhile was promoted as a remedy for certain types of dystrophy.

VINE OF SODOM (*SOLANUM ANGUIVI* LAM.) + SOLANACEAE

Synonyms:

Solanum hermannii Dunal; *Solanum indicum* auct.; *Solanum indicum* subsp. *distichum* (Thonn.) Bitter; *Solanum scalare* C. H. Wright; *Solanum sodomeum* L.; *Solanum sodomeum* var. *hermannii* (Dun.) Dun *fide* (USN and ZUL).

Notes (Vine of Sodom):

> **For their vine is of the vine of Sodom, and of the fields of Gomorrah; their grapes are grapes of gall, their clusters are bitter.**
>
> **Deuteronomy 32:32 (KJV)**

> **For their vine comes from the vine of Sodom, and from the fields of Gomor'rah; their grapes are grapes of poison, their clusters are bitter.**
>
> **Deuteronomy 32:32 (RSV)**

> **For their vine is from the vine of Sodom, and from the terraces of Gomor'rah; their grapes are grapes of poison, their clusters are bitter.**
>
> **Deuteronomy 32:32 (NWT)**

Non-Israeli writers often identify this as the biblical vine of Sodom, but Israeli Zohary (ZOH) identifies it as *Calotropis* (which see) and does not identify this *Solanum* species or any of its synonyms in the *Flora of Palestine* (FP3). A useless weed in the past, this plant might be investigated as a source of steroids. Bad spiny weeds like this might be reduced in whole plant utilization schemes, harvesting the biomass for useful phytochemicals, producing power alcohol from the residual biomass in the process. (ZUL entries below apply to *S. sodomeum* var. *hermannii*.) The American Herbal Products Association (AH2) has designated the name *Solanum anguivi* as the standardized common name (Scn.) listing the ayurvedic name *brihati* (often spelled *brahati*), a name often associated with *Solanum indicum* L. But AH2 equates it with *Solanum indicum* auct., so I cannot necessarily be sure of which species they are talking. *Solanum indicum* L. is also an important and useful species. In the *Useful Plants of West Tropical Africa* (UPW), we read under *S. anguivi* such appealing names as children's potato and children's tomato, making them sound a bit more appealing than evil's apple and Sodom's apple. UPW adds that this is a very polymorphic aggregate of sub-shrubby plants of innumerable infraspecific taxa. Mende of Sierra Leone consider the armed (thorny) variants male, the unarmed females.

Common Names (Vine of Sodom):

Apple of Sodom (Eng.; HOC); Bitter Apple (Eng.; HOC); Bou Sikroun (Arab.; JLH); Brihati (Ayu.; AH2); Children's Potato (Eng.; UPW); Children's Tomato (Eng.; UPW) Culuta (Guinea-Bissau; UPW); Itchie (Nig.; UPW); Jakato (Sierra Leone; UPW); Katung (Togo; UPW); Kma (Liberia; UPW); Mansarin Nyateila (Gambia; UPW); Nsusuaa (Ghana; UPW); Sira Korossé (Guinea; UPW); Solanum anguivi (Scn.; AH2); Sulu Jato (Gambia; UPW); Teï (Ivo.; UPW); Vine of Sodom (Eng.; BIB).

Activities (Vine of Sodom):

Analgesic (f; BIB; ZUL); Antiseptic (f1; BIB; ZUL; X9644084); Aphrodisiac (f; BIB); Bactericide (1; X9644084); Diuretic (f; HOC); Molluscicide (1; X10925405); Orexigenic (f; UPW); Poison (f; DAW); Tonic (f; UPW); Vulnerary (f; BIB).

Indications (Vine of Sodom):

Anorexia (f; UPW); Backache (f; ZUL); Bacteria (1; X9644084); Boil (f; UPW; ZUL); Cancer (f1; JLH; X3621146); Cancer, skin (f1; JLH; X3621146); Cataract (f; UPW); Constipation (f; ZUL); Cough (f; HOC); Cystosis (f; HOC; ZUL); Dermatosis (f1; BIB; ZUL; X3621146); Dropsy (f; ZUL); Edema (f; HOC); Enterosis (f; ZUL); Fever (f; UPW); Fungus (f; BIB); Gas (f; ZUL); Impotence (f; BIB; HOC; ZUL); Infection (f1; BIB; X9644084); Infertility (f; BIB; HOC; ZUL); Itch (f; ZUL); Keratoacanthoma (1; X3621146); Keratosis (1; X3621146); Leprosy (f; UPW); Mycobacteria (1; X9644084); Mycosis (f; BIB); Ophthalmia (f; UPW); Otosis (f; UPW); Pain (f; ZUL); Proctosis (f; BIB); Ringworm (f; BIB; UPW); Sore (f; UPW); Swelling (f; HOC); Toothache (f; BIB; ZUL); Tumor (f; JLH); Wound (f; BIB; ZUL).

Dosages (Vine of Sodom):

FNFF = !?

Many species of *Solanum*, like this one, are reportedly both toxic and edible, and their primitive ancestors are more likely to be poisonous than derived cultivars. Even the picturesque Gambian names hint of doubt, *mansarin nyateila* (= the prince who causes blindness) and *sulu jato* (= hyena's bitter tomato) do not stimulate my appetite. Still, the cherry-tomato-like fruits are eaten and used in sauces and soups, and said to stimulate the appetite (UPW).

- Gambians make a collyrium for cataracts and whiteness of the eye from the plant (UPW).
- Ghanans believe the fruits are orexigenic (UPW).

- Ivory Coastals use leaves against leprosy (UPW).
- Manyika chew the root and spit the juice onto wounds as a vulnerary (BIB).
- North Africans (Morocco, Tunisia) use the plant in cataplasms for cold tumors (JLH).
- South Africans apply the fruit to ringworm in cattle and horses (BIB).
- Tonga hold the fruit onto an aching tooth (BIB).
- Xhosa apply the fruit or root juice to skin ailments (BIB).
- Zulu use the root bark for barrenness and impotency (BIB).

Downsides (Vine of Sodom):

Fatal cases of poisoning reported with children for fruits of *S. sodomeum* var. *hermannii*. Symptoms of human poisoning include colic, cramps, diarrhea, difficulty with speech and vision, dilated pupils, dizziness, exhaustion, fever, hallucinations, headache, listlessness, nausea, and sweating (ZUL). The root of even *S. anguivi* is said to be poisonous.

BITTER-APPLE (*SOLANUM INCANUM* L.) ++ SOLANACEAE

Synonyms:

Solanum sanctum L.

Notes (Bitter-Apple):

The way of the slothful man is as an hedge of thorns: but the way of the righteous is made plain.

Proverbs 15:19 (KJV)

The way of a sluggard is overgrown with thorns, but the path of the upright is a level highway.

Proverbs 15:19 (RSV)

The way of the lazy one is like a brier hedge, but the path of the upright ones is a way cast up.

Proverbs 15:19 (NWT)

This is one of the few places where the NWT was the shorter of the triad I often cover (NWT; KJV; RSV). I rather doubt that the words brier and thorn are divinely distinct so I will not question these trivial differences. Zohary comments that the Hebrew *hedek* appears only twice in the Bible (Proverbs 15:19 and Micah 7:4–5), and in either case is associated with a thorny hedge. In Arabic, *hadaq* refers specifically to *Solanum incanum*. Respecting his opinion, I have cited the Proverbs passage in its three versions above. Unlike the vine of Sodom, this *S. incanum* is reported in the *Flora of Palestine*, mostly in tropical areas of the lower Jordan Valley and the Dead Sea region (ZOH). UPW says this one species is an aggregate of more than 50 that have been recognized species, many individually recognized by African farmers. They even speculate that it came to Africa from Asia via Egypt 400 to 500 years B.C. A component of some arrow poisons and love philtres. Ethiopians use the fruits as a condiment in certain beverages. Leaves not used as a potherb. A glyco-alkaloid content up to 4.81% is found in some races, particularly the one cultivated by the Paniya tribe of Iritty, North Kerala. Bushmen use the fruit juice in arrow poison. Root and seed contain a rennet-like enzyme (BIB).

Common Names (Bitter-Apple):

'Ain al Baqar (Arab.; GHA); Akumba (Upper Volta; UPW); Angulai (Tigrinia; KAB); Angulle (Tigrinia; KAB); 'Arsam (Arab.; GHA); 'Arsan (Arab.; GHA); Atoropo (Ghana; UPW); Ba Koyo

(Guinea; UPW); Bitter Apple (Eng.; S. Afr.; KAB); Bitterappel (S. Afr.; ZUL); Bitter Tomato (Eng.; UPW); Brien (Eng.; RSV; ZOH); Culuta (Guinea-Bissau; UPW); Data (Liberia; UPW); Djang (Cemeroons; UPW); Eggplant (Eng.; UPW); Fambora (Sen.; UPW); Garden Egg (Eng.; UPW); Gifappel (S. Afr.; ZUL); Gray Nightshade (Eng.; ZOH); Grysbitterappel (S. Afr.; ZUL); Hadak (Arab.; GHA; ZOH); Háw Mòy (Niger; UPW); Hedek (Heb.; ZOH); Helkem (Dho.; GHA); Ikan (?; FAC); Kauratrim (Brahui; KAB); Koro Kaneil (Mali; UPW); Kujakataku (Gambia; UPW); Mazg (Arab.; GHA); 'Mazj (Oman; GHA); Meringam (?; FAC); Morola (Pedi; KAB); Nuqum (Yemen; GHA); Palestine Nightshade (Eng.; FAC); Tarku (Nig.; UPW); Thola (Suto; KAB); Thorn (Eng.; RSV; ZOH); Thorn Apple (Eng.; S. Afr.; ZUL); Ungulleh (Tigrinia; KAB); Yalo (Sierra Leone; UPW); Nscn.

Activities (Bitter-Apple):

Anodyne (f; BIB; UPW); Anticancer (f1; X15527763); Antiedemic (f1; HDN); Antihepatomic (1; X11108802); Antiinflammatory (f1; HDN); Antiseptic (f1; HDN; ZUL); Antitachycardic (1; HDN); Apoptotic (1; X8954074; X15527763); Bactericide (1; ZUL); Carcinogenic (f1; BIB); Cardiodepressant (1; HDN); Caspase-3 Inducer (1; X15527763); Cholagogue (1; HDN); Contraceptive (f; BIB); Cytotoxic (1; JNP53:513); Diuretic (f; HDN); Expectorant (f; HDN); Febrifuge (f1; HDN); Hepatoprotective (1; HDN; PM54:222); Hypotensive (1; HDN); Insectifuge (f; HDN); Piscicide (f1; BIB; HDN); Poison (f; BIB); Spasmolytic (f1; HDN); Stomachic (f; UPW); TNF Inducer (1; X8954074).

Indications (Bitter-Apple):

Bacteria (1; ZUL); Bleeding (f; ZUL); Bloat (f; BIB); Bronchosis (f; HDN); Bruise (f; GHA); Burn (f; UPW); Calculus (f; BIB); Cancer (f1; JLH; HDN; X15527763); Cancer, lung (f1; JLH; X15527763); Carbuncle (f; BIB); Carcinoma (f; JLH); Caries (f; UPW); Catarrh (f; UPW); Colic (f; HDN); Constipation (f; BIB); Cough (f; UPW); Cramp (f1; HDN); Craw-craw (f; HDN); Dandruff (f; HDN; ZUL); Dermatosis (f; HDN); Diarrhea (f; HDN; UPW); Dysmenorrhea (f; HDN); Dyspepsia (f; GHA; HDN; UPW); Earache (f; GHA; HDN; UPW); Edema (f1; HDN); Enterosis (f; ZUL); Epilepsy (f; HDN); Epistaxis (f; BIB); Epithelioma (f; JLH); Fever (f1; HDN); Fungus (1; HDN); Gas (f; GHA); Gastrosis (f; UPW); Gonorrhea (f; HDN); Headache (f; HDN); Hematuria (f; UPW); Hemorrhoid (f; GHA); Hepatoma (1; X11108802); Hepatosis (f1; BIB; HDN; ZUL); Herpes (f; HDN); High Blood Pressure (1; HDN); Infection (f1; HDN; ZUL); Infertility (f; BIB); Inflammation (f1; HDN); Itch (f; BIB); Laryngosis (f; UPW); Melanoma (f; JLH); Myalgia (f; HDN); Mycosis (1; HDN); Nephrosis (f; BIB); Neuralgia (f; UPW); Neurosis (f; HDN); Ophthalmia (f; UPW); Pain (f; HDN; UPW); Pleurisy (f; HDN; KAB); Pneumonia (f; HDN; KAB); Pulmonosis (f; BIB; KAB); Rhinosis (f; UPW); Ringworm (f; UPW); Ringworm (f1; BIB; HDN; ZUL); Quinsy (f; UPW); Scabies (f; BIB); Snakebite (f; HDN; ZUL); Sore (f; HDN); Sore Throat (f; HDN; KAB); Staphylococcus (1; ZUL); Stomachache (f; UPW); Swelling (f; UPW); Syphilis (f; HDN); Tachycardic (1; HDN); Tonsilosis (f; HDN); Toothache (f; GHA; HDN; KAB); Trichophytosis (1; HDN); Tumor (f; JLH); Venereal Disease (f; HDN); Vertigo (f; HDN; UPW); Wound (f; GHA; HDN).

Dosages (Bitter-Apple):

FNFF = !!
Green fruits eaten raw or diced and added to soups, fresh or dried; seeds used to curdle milk; leaves of some cultivars. edible cooked in soups; used as vegetable rennet (BIB; FAC; TAN; UPW).

- Arabians boil berries in oil to make eardrops (GHA).
- Arabians direct smoke from burning dried berries toward hemorrhoids (GHA).
- Congolese use leaf sap as expectorant in bronchosis and spasmodic coughs, a throat paint for laryngitis, quinsy, sore throat, and tonsilitis (UPW).

- Ethiopians use the leaf for bloat and epistaxis; the fruit for calculus, constipation, gonorrhea, itch, renitis, and wounds; the root for gonorrhea; and the ash for scabies (BIB).
- Europeans in South Africa use the fruit juice for dandruff (BIB).
- Guinea natives gargle the decoction or tea for sore throat or take for dyspepsia or stomachache (UPW).
- Omani insert sore fingers in fruits to draw out pus (GHA).
- Paniya women use as contraceptive, while barren Nigerians viewed as fertility symbol (BIB).
- Pedi take roasted plant for pleurisy, pneumonia, the decoction for other chest problems (KAB).
- South Africans poultice fruits onto external benign tumors (JLH).
- Sutos of South Africa use plant for sore throat and toothache (KAB).
- Taiwanese use the rhizomes for hepatitis (ZUL).
- Tanganyikans insert flower in ear for earache (UPW)
- Tanganyikans use root for abdominal pain, carbuncles, and hepatosis; the fruit for snakebite; and the gall for earache (BIB).
- Tanzanians use for dandruff, dermatosis, earache, gonorrhea, hepatosis, pneumonia, ringworm, sore, syphilis, and wounds (ZUL).
- Yemeni and Dhofari inhale smoke from burning seeds for toothache (GHA).
- Zimbabweans use roots for dysmenorrhea; gastrointestinal, respiratort, and venereal complaints; and toothache (ZUL).
- Zulu use the juice of the plant for ringworm (BIB).

Downsides (Bitter-Apple):

"Aqueous fruit extract causes skin carcinoma in animals" (HDN). Unripe fruit and other parts poisonous to animals, inducing bloat, colic, diarrhea, paralysis, polypnea, salivation, and tachycardia (ZUL).

BROOMCORN (*SORGHUM BICOLOR* (L.) MOENCH.) ++ POACEAE

Synonyms:

Andropogon sorghum Brot.; *Andropogon bicolor* Roxb.; *Holcus saccharatus*; *Holcus sorghum* L.; *Sorghum saccharatum* (L.) Moench; *Sorghum vulgare* Pers.

Notes (Broomcorn):

Now there was set a vessel full of vinegar: and they filled a sponge with vinegar, and put it upon hyssop, and put it to his mouth.

John 19:29 (KJV)

A bowl full of vinegar stood there; so they put a sponge full of the vinegar on hyssop and held it to his mouth.

John 19:29 (RSV)

A vessel was sitting there full of full of sour wine. Therefore they put a spongeful of the sour wine upon a hyssop [stalk] and brought it to his mouth.

John 19:29 (NWT)

FIGURE 1.106 Broomcorn (*Sorghum bicolor*).

While I suspect the odds are with Zohary, who identifies this biblical quote with Syrian hyssop (*Origanum syriacum*), I find it easier to visualize bunches of sorghum, much as mother used as dry floral arrangements in rural Alabama. That is, back when poor white southerners used sorghum syrup on their corn cakes and one uncle who grew the sorghum still had oxen. Yes, I can see evil guards mockingly offering a thirsty prisoner a vinegar-soaked head of sorghum. I am not the only one who offers this alternative for some of the hyssop passages in the Bible. Even Zohary admits that sorghum is better suited than common millet to the Israeli climate, thriving both in the lowlands and uplands as a very productive summer crop, even without irrigation. Zohary asserts that the species is known to have moved from East Africa via southwest Asia to India, archaeologically confirmed at circa 2000 B.C. Little will readers suspect, but sorghum is usually what was intended when corn was mentioned in the Bible (there was no corn sensu Zea mays in the old world before Columbus).

Common Names (Broomcorn):

Ägyptische Zeiskorn (Ger.; EFS); Akoko (Ga; KAB); Alboro'o (Sal.; AVP); Alcandia (Sp.; KAB); Amabele (Zulu; KAB); Ampemby (Hova; KAB); Awi (Twi; KAB); Bajrajhopanwa (Nwp.; Oudh KAB); Bakaka (Sakalave; KAB); Balai (Fr.; KAB); Balais de Jonc (Fr.; AVP); Bassiqui (Gambia; KAB); Batad (Vis.; KAB); Blé de Guinée (Fr.; AVP); Bondajanu (Tel.; KAB); Broom Corn (Eng.; DEP; EFS); Caña Dulce (Mex.; AVP); Challu (Peru; AVP); Chari (Pun.; KAB); Chavela (Ker.; Mal.; Mar.; DEP; KAB; SKJ); Chinese Sugarcane (Eng.; DEP); Cholam (Tam.; DEP; KAB; SKJ); Cholum (Sri.; KAB); Chotri Junri (Pun.; DEP); Dari (Tur.; EFS); Dari Seed (Malta; KAB); Dawa (Sudan; AVP); Deodhan (Bom.; Hindu; DEP; EFS); Devatadhanyamu (Tel.; DEP); Dhura (Arab.; DEP); Dhutar (Baghwana; KAB); Dirghamala (Sanskrit; SKJ); Djagung Roté (Malaya; EFS); Dura (It.; AVP); Durra (Eng.; Arab.; Egypt; Ven.; AVP; DEP; EFS); Durrha (Ger.; EFS); Egyptian Millet (Eng.; EFS); Escoba (Peru; EGG); Escoba Millo (Ven.; AVP); Fo (Ewe; Krepi; KAB); Ganggai (Mun.; KAB); Gao Liang (Pin.; DAA); Grand Millet (Fr.; EFS); Great Millet (Eng.; DEP; EFS); Guinea Corn (Eng.; DEP; EFS); Hirse (Ger.; AVP); Ikshupatraka (Sanskrit; KAB); Indian Millet (Eng.; DEP); Janera (Hindi; KAB); Jaoriturkimani (Afg.; KAB); Jaundri (Bom.; KAB); Jerusalem Corn (Eng.; EFS); Joar (Hindi; Kum.; DEP); Jolah (Kan.; Mysore; KAB; SKJ); Jondla (Dec.; DEP; KAB); Jovaree (India; EFS); Jowari (Las Bela; KAB); Juar (Kon.; KAB); Junali (Bhabar; Kum.; DEP; KAB); Junri (Nwp.; DEP); Kadval (Mar.; DEP; KAB); Kafferkorn (Ger.; EFS); Kafferkorn Gras (Dutch; EFS); Kaffir Corn (S. Afr.; KAB); Kangra (Bom.; DEP); Kaoliang (China; DEP; EFS); Karaliringu (Sin.; KAB); Kasajonar (Beng.; KAB); Kayda (Arab.; Egypt; DEP); Kaydi (Arab.; Egypt; KAB); Kiosagi (Afg.; DEP); Koko (Krobo; KAB); Kurbi (Beng.; DEP); Maía de Guinea (Peru; EGG); Maicillo (Cr.; AVP); Mais de Millo (Cr.; AVP); Mais Millo (Col.; AVP); Melca (Cat.; KAB); Miglio Saggina (It.; EFS); Mijo (Sp.; AVP); Mil (Fr.; AVP); Milho (Por.; KAB); Milhozaburro Vermelho (Por.; AVP); Millangue (Lan.; KAB); Millo (Cuba; Pr.; Sp.; AVP); Milo (Sp.; EFS); Milocolo (Por.; AVP); Mohammadisa (Makran; KAB); Mohrenbartgras (Ger.; EFS); Mohrhirse (Ger.; AVP; EFS); Moorhirse (Ger.; KAB); Nirgol (Kan.; DEP); Petit Mil (Haiti; AVP); Phag (India; DEP); Pti Mil (Haiti; AVP); Pyoung (Burma; DEP); Quav Ntsuas (Hmong; EB57:365); Saggina (It.; AVP; EFS); Saina (Sp.; KAB); Salu (Mar.; DEP); Serçe Darici (Tur.; EFS); Shalu (Dec.; Mar.; DEP; KAB); Shu Shu (China; KAB); Sorga a Scopa (It.; EFS); Sorghohirse (Ger.; EFS); Sorgo (Cuba; Peru; Por.; Pr.; Rus.; AVP; EFS; EGG; KAB); Sorgo Azucarado (Sp.; EFS); Sorgo Commun (Guad.; AVP); Sorgo Escobero (Peru; EGG); Sorgo Forrejero (Ven.; AVP); Sorghum (Eng.; JAD); Sundia (Guj.; DEP; KAB); Talla (Tel.; DEP); Targhar (Babian Shahrig; KAB); Tella Jonna (Tam.; Tel.; DEP; NAD); Thuthera (Cp.; KAB); Ti Chê (China; EFS); Ti Liang (China; EFS); Triguillo (Sal.; AVP); Varifemba (Betsileo; KAB); Vermelho (Por.; AVP); Zhurna (Sanskrit; DEP); Zura (Arab.; KAB); Nscn.

Activities (Broomcorn):

Anthelmintic (f; WBB); Antiabortive (f; AHL; BIB); Antioxidant (1; JAF51:6657); Aphrodisiac (f; KAB; SKJ); Cyanogenic (1; WOI); Demulcent (f; BIB; EFS; KAB; PH2; SKJ; WOI); Depurative (f; KAB); Diuretic (f; BIB; EFS; KAB; WOI); Emollient (f; BIB; EFS); Orexigenic (f; KAB); Stomachic (f; LMP).

Indications (Broomcorn):

Anorexia (f; KAB); Biliousness (f; KAB); Bleeding (f; BIB); Bronchosis (f; JFM); Burn (f; JLH; TOM); Cancer (f; JLH; KAB); Congestion (f; JFM); Cough (f; JFM); Cystosis (f; FEL; KAB);

Debility (f; BIB); Dermatosis (f; WBB); Diarrhea (f; JFM; WBB); Dyspepsia (f; PH2); Dysuria (f; BIB; KAB); Eczema (f; WBB); Epilepsy (f; BIB); Flu (f; BIB); Goiter (f; JFM); Hemorrhoid (f; KAB); Malaria (f; WBB); Mastosis (f; WBB); Measles (f; BIB; JFM); Nephrosis (f; BIB; KAB; TOM); Pulmonosis (f; JFM); Stomachache (f; BIB); Swelling (f; WBB); Tuberculosis (f; BIB; WBB); Worm (f; WBB).

Dosages (Broomcorn):

FNFF = !!!
Seeds an important cereal in many cultures. Palestinean eat the *hyssop* (or "Jerusalem corn") a main and nutritious part of their diet. The grains are gathered and ground for meal used in baking coarse bread. A single fruiting head can be large enough to feed a family. Some suggest that the "parched corn" Ruth received from Boaz was sorghum. Pearled grain cooked like rice or ground into flour. Hondurans pop the seeds like popcorn. Parched seed used as coffee substitute. Stalks used as animal feed, and the seed for bird feed. Important for silage or hay. *Sorghums* with large juicy stems are used in the manufacture of syrup, sugar, or energy alcohol. Hmong report the stems edible, and source of sugar (BIB; EFS; JFM; EB57:365). Decoct 1 Tbsp powdered toasted seed (JFM). Decoct 2 oz seed/qt water for cystosis (FEL).

- American Blacks take the seed decoction for bladder, kidney, and urinary complaints (KAB).
- Arubans poultice heated seeds in oil on backs of pulmonary patients who drink a hot tea of a diaphoretic *Pectis* (JFM).
- Brazilians drink seed decoction for bronchosis, cough, and other pulmonary complaints (JFM).
- Curacaons drink the leaf tea for measles (JFM).
- Lebanese use sorghum gruel for debility due to lung maladies like tuberculosis (HJP).
- Lebanese use hot sorghum as a poultice (HJP).
- Venezuelans decoct 1 Tbsp seed, with or without lemon, for diarrhea (JFM).

Downsides (Broomcorn):

No health hazards or side effects known with proper therapeutic dosages (PH2) (PH2 designates no specific quantified dosage!; JAD.) Cyanide poisoning can occur in grazing animals overgrazing the foliage, especially wilted foliage (WOI). Still, replacing African sorghum with American *Zea* has generated an epidemic in squamous carcinoma of the esophagus. *Fusarium* fungi grow freely on maize, producing fumonisins, which reduce nitrates to nitrites and synthesize cancer-producing nitrosamines. Nitrosamines are carcinogens. *Fusarium* fungi do not thrive on sorghum, sorghum producing two orders of magnitude less fumonisins than maize. The higher incidence of esophageal cancer in black males is ascribed to their greater consumption of traditional beer, which is produced by fermenting maize. Patients with esophageal cancer consume more beer than controls. Countries in Africa, where the staple is still sorghum, have a low incidence of squamous carcinoma of the esophagus (X15617883).

Natural History (Broomcorn):

Major diseases reported on sorghums include *Cercospora sorghi*, *Colletotrichum graminicola* (anthracnose of leaves and stems), *Helminthosporium turcicum* (leaf blight), *Macrophomina phaseoli* (charcoal rot), *Periconia circinata* (milo disease), *Phyllachora sorghi*, *Phyllosticta sorghi*, *Puccinis purpurea* (rust), *Ramulispora sorghi* (sooty strip), *Sclerospora sorghi* (downy mildew), *Sorosporium ehrenbergii*, *Sphacelia sorghi*, *Sphacelotheca sorghi* (covered smut), *Sph. cruenta* (loose smut), and *Sph. reiliana* (head smut). Plants are also severely attacked by various species of *Striga* (*S. lutea*, *S. hermontheca*, *S. senegalensis*, *S. densiflora*). Nematodes isolated from sorghum include the following species: *Helicotylenchus cavenessi*, *H. dihystera*, *H. pseudorobustus*, *Hoplolaimus pararobustus*, *Meloidogyne javanica*, *Peltamigratus nigeriensis*, *Pratylenchus brachyurus*, *P. zeae*, *Quinisulcius acutus*, *Rotylenchulus reniformis*, *Scutellonema cavenessi*, *S. clathricaudatum*, *Tylenchorhynchus acutus*, and *T. parvus* (HOE).

Extracts (Broomcorn):

Molasses, derived from the seeds, was used with sulfur as a spring tonic in rural Alabama (TOM). (I remember my father and mother talking about it, but do not know that they ever tonified me therewith but we did occasionally use it as syrup on pancakes.) Dicko et al. (2005) screened 50 sorghum varieties. Germination did not affect the content in total phenolic compounds but decreased the content of proanthocyanidins, 3-deoxyanthocyanidins, and flavan-4-ols. Sorghum grains with pigmented testa, chestnut color glumes, and reddish plants had higher contents, a larger diversity of phenolics, and more antioxidant activities. Among varieties used for *to*, *dolo*, couscous, and porridge preparation, the *dolo* (local beer) varieties had more and more diverse phenolics and the highest antioxidant activities. 3-Deoxy-anthocyanidins, namely apigenidenins and luteolinidins, are abundant in sorghum grains but rare or absent in other plants. Leucoapigenidenin (apiforol) and leucoluteolinidin (luteoforol) may confer resistance to mould (X15796598).

BENZOIN (*STYRAX BENZOIN* DRYANDER.) ++ STYRACACEAE

Synonyms:

Benzoin officinale Hayne; *Lithocarpus benzoin* Bt. *fide* HH2

Notes (Benzoin):

They sacrifice upon the tops of the mountains, and burn incense upon the hills, under oaks and poplars and elms, because the shadow thereof is good: therefore your daughters shall commit whoredom, and your spouses shall commit adultery.

Hosea 4:13 (KJV)

They sacrifice on the tops of the mountains, and make offerings upon the hills, under oak, poplar, and terebinth, because their shade is good. Therefore your daughters play the harlot, and your brides commit adultery.

Hosea 4:13 (RSV)

On the tops of the mountain they sacrifice and on the hills they make sacrificial smoke, under massive tree and storax tree and big tree because its shade is good. That is why your daughters commit fornication and your own daughters in law commit adultery.

Hosea 4:13 (NWT)

While the PH2 covers three species of the genus *Styrax* and one of the genus *Liquidambar*, few people can identify the resins of these trees, the product of commerce. Recent authorities maintain that the "incense" used in the service of the Tabernacle was a mixture, in definite proportions of frankincense, galbanum (*Ferula galbaniflua*), onycha (*Styrax benzoin*), and stacte (*Styrax officinalis*), and the use of any incense not composed of these four ingredients in the proper proportions was strictly forbidden (BIB). I certainly do not want to rock city hall, especially a biblical city hall. AH2 likewise covers three species (*S. benzoin*, *S. paralleloneurum*, and *S. tonkinensis*), giving them all the same standardized common name, that of "benzoin tree." Zohary discusses only one species, identifying it as *S. officinalis*. He notes that the Hebrew *livneh* occurs only twice in the Bible,

FIGURE 1.107 Benzoin (*Styrax benzoin*).

regarding the Hosea quote as "definitely styrax," which he identifies as *Styrax officinalis*, which is reportedly native to Israel. This is a more humid Asian species, reported from India, Cambodia, Laos, Myanmar, Thailand, Vietnam, Java, Sumatra, and Malaysia (USN). Thus, this species historically would have needed to be imported from outside Israel.

Common Names (Benzoin):

An Hsi Hsiang (China; EFS); Árbol de Benjuí (Sp.; USN); Arbre à Benjoin (Fr.; USN); Asilbent (Tur.; EFS); Bantung (Sumatra; IHB); Belzuino (It.; EFS); Benjamin Tree (Eng.; Ocn.; AH2); Benjoim de

Sumatra (Por.; HH2); Benjoin (Fr.; EFS); Benjoin de Sumatra (Fr.; HH2); Benjui (Sp.; EFS); Benjui de Sumatra (Sp.; HH2); Benzoë Baum (Ger.; EFS); Benzoë Boom (Dutch; EFS); Benzoin Tree (Eng.; Scn.; AH2); Estoraque (Sp.; EFS); Gum Benjamin Tree (Eng.; EFS); Gum Benzoin (Eng.; HH2); Kamanan (Malaya; EFS); Kamayan (Malaya; EFS); Kaminan (Malaya; EFS); Kemenjan (Malaya; EFS); Kemenyan (Indonesia; Java; Sumatra; IHB; USN); Luban (Beng.; Bom.; Hindi; India; EFS; SKJ); Manan (Malaya; EFS); Melláina (It.; EFS); Minan (Malaya; EFS); Ngan Si Hsiang (China; EFS); Onycha (Eng.; USN); Pokok Kemenyan (Malaya; IHB); Shambirani (Madras; SKJ); Shan Fan Ye (Pin.; DAA); Siam Storace Benzoino (It.; EFS); Storaxbaum (Ger.; USN); Styraxtree (Eng.; USN); Sumatra Benzoin (Eng.; Ocn.; AH2; HH2); Sumatra Benzointree (Eng.; USN); Talanan (Sumatra; IHB); Snowbell (Eng.; EFS); Wohlriechender Asant-Baum (Ger.; EFS).

Activities (Benzoin):

Allergenic (1; X6239881); Anesthetic (f1; FNF); Anticancer (f1; FNF); Antiinflammatory (f1; FNF); Antimutagenic (1; FNF); Antioxidant (1; FNF); Antiseptic (f1; CRC; DEP; EFS; SKJ); Antispasmodic (f1; FNF); Antiviral (1; FNF); Aphrodisiac (f; DEP; LMP); Bactericide (1; FNF); Carminative (f; BIB; CRC; WOI); Deodorant (f; CRC); Disinfectant (f; CRC); Diuretic (f; CRC); Expectorant (f1; CRC; PH2; SKJ); Fungicide (1; FNF); Immunostimulant (1; FNF); Insecticide (1; CRC); Insectifuge (f1; FNF); Larvicide (1; IHB); Laxative (f1; FNF); Narcotic (f1; FNF); Nematicide (1; FNF); Sedative (f1; CRC; FNF; LMP); Stimulant (f; CRC; DEP; EFS; LMP); Vermifuge (f1; FNF); Vulnerary (f; CRC; EFS).

Indications (Benzoin):

Arthrosis (f; CRC); Asthma (f; DEP); Bacteria (1; FNF); Bronchosis (f; BIB; CRC); Cancer (f1; CRC; FNF; JLH); Cardalgia (f; LMP); Catarrh (f; CRC; PH2); Circumcision (f; CRC); Cold sore (f; CRC; JFM); Colic (f; CRC); Constipation (f1; FNF); Corn (f; JLH); Coryza (f; IHB); Cough (f1; CRC); Cramp (f1; FNF); Cystosis (f; DEP); Dermatosis (f; IHB; LMP); Diarrhea (f; DAD); Earache (f1; FNF); Enteralgia (f; LMP); Fever (f; IHB); Fungus (1; FNF); Gastrosis (f; PH2); Gout (1; FNF); Heart (f; LMP); Hemorrhoid (f; LMP); Herpes (f; CRC); Infection (f1; CRC; EFS; SKJ); Inflammation (f1; FNF); Insomnia (f1; FNF); Itch (f; IHB); Labor (f; LMP); Laryngitis (f; BIB; CRC; IHB); Mastosis (f; CRC; IHB); Mucososis (f; DEP); Mycosis (1; FNF); Nipple (f; IHB); Pain (f; LMP; PH2); Pharyngosis (f; IHB); Phthisis (f; DEP); Podiatry (f; IHB); Polio (1; FNF); Polyp (f; JLH); Pulmonosis (f; PH2); Respirosis (f; PH2); Rheumatism (f; IHB; LMP); Rhinosis (f; JLH); Ringworm (f; CRC; IHB); Shingle (f; CRC); Sickle Cell Anemia (1; FNF); Spermatorrhea (f; CRC); Stomachache (f; PH2); Stroke (f; LMP; PH2); Syncope (f; LMP; PH2); Ulcer (1; FNF); Virus (1; FNF); Worm (f1; FNF); Wound (f1; CRC; EFS); Yeast (1; FNF).

Dosages (Benzoin):

FNFF = !

In the United States, *Sumatra benzoin* (*S. benzoin* and *S. paralleoneurus*) is more customarily used in pharmaceutical preparations, Siam benzoin (*S. tonkinensis* et al.) in the flavor and fragrance industries; for example, in alcoholic beverages, baked goods, beverages, candy, desserts, gelatins, and puddings, the highest average maximum level was circa 0.014% in baked goods and candies. Benzoin adds the gloss to chocolate eggs, the turbidity to syrups, and some of the flavor to baked goods, candies, chewing gums, gelatins, ice creams, puddings, and soft drinks (DAD; FAC). For topical Friar's or Turlington's Balsam (each 100 ml contains an alcohol extract from 10 g benzoin, 8 g storax, 4 g balsam of Tolu, and 2 g aloe); for cold sores, cracked skin, and indolent ulcers (BIB).

- Malayans have used a reddish root extract to kill butterfly larvae (*Parasa herbifera*) (IHB).
- Malayans use benzoin for cracked feet and circumcisions (IHB).

Downsides (Benzoin):

Class 1 (AHP).

Natural History (Benzoin):

Fungi of the genus *Pestalotia* attack the tree. Termites quickly destroy the timber. Animals eat the seeds (DAD).

STORAX (*STYRAX OFFICINALIS* L.) ++ STYRACACEAE

Notes (storax):

> **They sacrifice upon the tops of the mountains, and burn incense upon the hills, under oaks and poplars and elms, because the shadow thereof is good: therefore your daughters shall commit whoredom, and your spouses shall commit adultery.**
>
> **Hosea 4:13 (KJV)**

> **They sacrifice on the tops of the mountains, and make offerings upon the hills, under oak, poplar, and terebinth, because their shade is good. Therefore your daughters play the harlot, and your brides commit adultery.**
>
> **Hosea 4:13 (RSV)**

> **On the tops of the mountain they sacrifice and on the hills they make sacrificial smoke, under massive tree and storax tree and big tree because its shade is good. That is why your daughters commit fornication and your own daughters in law commit adultery.**
>
> **Hosea 4:13 (NWT)**

Of the three versions, only the NWT agrees with Zohary — that the poplar of Hosea 4:13 (KJV, RSV) is styrax, *livneh* in Hebrew. While the PH2 covers three species of the genus *Styrax* and one of the genus *Liquidambar*, few people can identify the resins of these trees, the product of commerce. Recent authorities maintain that the "incense" used in the service of the Tabernacle was a mixture, in definite proportions of frankincense, galbanum (*Ferula galbaniflua*), onycha (*Styrax benzoin*), and stacte (*Styrax officinalis*), and the use of any incense not composed of these four ingredients in the proper proportions was strictly forbidden (BIB). I certainly do not want to rock city hall.

EFS aggregates the common names for this and *Styrax benzoin*, but I have included them only under *Styrax benzoin*. Few, if any, could today identify the plant by its resin, and not too many could positively identify even with flowers and fruits and foliage. Hence, I doubt we will ever be sure just which species were imported into the Holy Land and mentioned in the Bible, be it as poplar (balm) or storax. I will resist the temptation to lump the *S. benzoin* and *S. officinalis*. The USDA does not. They suggest the distribution of *S. officinalis* as Cyprus, Israel, Jordan, Lebanon, Syria, Turkey in western Asia, Albania, Greece (including Crete), Italy, and Yugoslavia in Europe. Zohary adds that this species occurs in Israel's evergreen forest (*maquis*). It seems to have fewer reported activities and indications than *Styrax benzoin*. But it is the more likely candidate to be Hosea's poplar (KJV, RSV) or styrax (NWT).

Common Names (Storax):

Abhar (Arab.; ZOH); Aliboufier (Fr.; USN); Boe (Bom.; SKJ); Estoraque (Sp.; EFS); Libna (Arab.; ZOH); Livneh (Heb.; ZOH); Maiaa (Arab.; JLH); Mia (Arab.; JLH); Silajit (Beng.; SKJ); Storax

FIGURE 1.108 Storax (*Styrax officinalis*).

(Eng.; JLH; NWT; USN); Storax Tree (Eng.; FAC); Styrax (Eng.; BIB); True Storax (Eng.; DEP); Usturak (Arab.; Bom.; DEP; SKJ); Nscn.

ACTIVITIES (STORAX):

Antiseptic (f; EFS); Emmenagogue (f; HJP); Expectorant (f; EFS; SKJ); Piscicide (f; ZOH); Stimulant (f; EFS; SKJ); Vulnerary (f; EFS).

Indications (Storax):

Cold (f; HJP); Dermatosis (f; HJP); Fever (f; JHP); Hysteria (f; HJP); Infection (f; HJP); Pulmonosis (f; HJP); Venereal Disease (f; HJP).

Dosages (Storax):

FNFF = !
The balsamic gum or resin is sometimes used as a condiment (FAC).

- Arabs use the resin in sesame oil for tumors (JLH).
- Asians suggest 1/2 to 10 grains internally as a stimulant (DEP).
- Lebanese combine resin with alcohol as antiseptic or disinfectant (HJP).
- Lebanese combine the resin with butter, honey, or olive oil as skin ointment (HJP).
- Lebanese suggest the diluted resin for venereal diseases (HJP)

Downsides (Storax):

As of July 2004, the FDA Poisonous Plant Database listed this species.

SEA BLITE (*SUAEDA* SPP.) ++ CHENOPODIACEAE

Notes (Sea Blite):

There are so few ethnobotanical data on the many species of this genus (ten in Israel alone) that I have taken the generic approach here. Zohary lists two quotes of place names in 1 Chronicles that he thinks trace back to the Hebrew *shahor*, *Ashhur* (1 Chronicles 4:5) and *Sheharaih* (1 Chronicles 8:26). Saline deserts are also mentioned three times as *mlehah* but nowhere is the plant mentioned by the plant's name, although it is one of the most frequent components of this characteristic vegetation type (ZOH). Yemeni extract a black dye from stems of *S. monoica* mixed with henna (GHA). The KAB common names apply either to *Suaeda fruticosa* and/or *Suaeda monoica*. DEP entries apply to *S. fruticosa*, *S. maritima*, and *S. nudiflora*.

Common Names (Sea Blite):

Almajo (Sp.; KAB); Almajo Dulce (Sp.; KAB); Ashhur (Heb.; ZOH); Baggilana (Pun.; DEP; KAB); Bui (Delhi; WOI); Chotee Lanu (Hindi; Pun.; NAD; WOI); Choti Lani (Pun.; DEP); Common Indian Saltwort (Eng.; WOI); Dana (Pun.; DEP; KAB); Geria (Oriya; DEP; WOI); Ghobbejia Tarrmied (Malta; KAB); Hoettum (Tigre; KAB); Ila Kura (Tel.; DEP; WOI); Inkbos (Afrikan; KAB); Inkbush (Eng.; S. Afr.; KAB); Kaku (Arab.; GHA); Karuvumari; (Tam.; KAB); Khari Lani (Hindi; Sin.; DEP; WOI); Kharkhusa (Pun.; DEP; KAB); Khaskhasa (India; Pun.; DEP; KAB); Khreiza (Arab.; GHA); Kiray (Tam.; DEP); Koyyalakura (Tel.; KAB); Lana (Mar.; Pun.; DEP; WOI); Lana Lani (Guj.; Mar.; WOI); Lani (Kohlu; Pun.; Sharag; Sibi; DEP; KAB); Lano (Guj.; Mar.; WOI); Leonuk (Pun.; NAD); Lonia (Delhi; WOI); Lunak (Hindi; Pun.; DEP; KAB; NAD; WOI); Lunki (Raj.; WOI); Molleih (Arab.; GHA); Moras (Guj.; Mar.; WOI); Morasa (Bom.; Mar.; DEP; KAB; NAD); Nariumari (Tam.; WOI); Nunkhuri (Delhi; WOI); Phesaklane (Pun.; DEP; KAB); Rava Hada (Tel.; DEP); Rava Kada (Tel.; WOI); Righat (Jhalawan; Kalat; KAB); Rigilah (Arab.; GHA); Samai (Pun.; DEP); Sea Blite (Eng.; Malta; FAC; KAB); Shahor (Heb.; ZOH); Sheharaih (Heb.; ZOH); Shorag (Afg.; DEP; KAB); Shrubby Saltwort (Eng.; S. Afr.; KAB); Shrubby Seablite (Eng.; KAB); Sosa Fina de Andalucia (Sp.; KAB); Sosa Prima (Sp.; KAB); Spinazmai (Kila Saifulla; KAB); Suaid (Arab.; ZOH); Suejda (Malta; KAB); Suwaid (Arab.; GHA); Suweda (Arab.; ZOH); Suwwad (Arab.; GHA); Suwweid (Arab.; GHA); Tarteer (Arab.; GHA); Umari Nandi (S. Arcot; KAB); Uppukeerai (Tam.; WOI); Usak Lani (Pun.; DEP); Usaklana (Pun.; KAB); Ushaklan (Sin.; NAD); Ushuklani (Guj.; Sin.; DEP; KAB); Vellakeerai

FIGURE 1.109 Sea Blite (*Suaeda*). Source: KAB.

(Tam.; WOI); Veyi Kada (Tel.; DEP); White Glasswort (Eng.; Malta; KAB); Yella Kiray (Tam.; DEP); Zamai (Pun.; KAB); Zimeh (Pushtu; DEP; KAB; NAD); Zmai (Kila Saifulla; Sharig; KAB); Nscn.

Activities (Sea Blite):

Emetic (f; KAB; NAD); Laxative (f; KAB); Vulnerary (f; KAB).

Indications (Sea Blite):

Dyspnea (f; GHA); Constipation (f; KAB); Gingivosis (f; GHA); Headache (f; GHA); Hysteria (f; GHA); Nausea (f; GHA); Neurosis (f; GHA); Odontosis (f; GHA); Ophthalmia (f; GHA; KAB; NAD); Sore (f; KAB; NAD); Vertigo (f; GHA); Wound (f; KAB).

Dosages (Sea Blite):

FNFF = !

Speaking of *S. maritima*, the desalted leaves are used as a cooked vegetable; or the normally salty leaves are added to salads and soups to salt them. Young shoots also pickled. Green shoots universally eaten by Asian Indians, especially during famines. (DEP; FAC). Asian Indians eat the green leaves of *S. nudiflora* (=*fruticosa*), a source of *sajji* (DEP).

- Arabians snuff the ash for dizziness, headache, hysteria, nausea, neurosis, and poor vision (GHA).
- Arabians use stem and leaf decoctions (*S. aegyptiaca*) in gargles for gum and tooth problems (GHA).
- Arabians use the stems of *S. vermiculata* to alleviate breathing difficulties (GHA).
- Asian Indians suggest an oily application of the wooly growths of branch tips (*S. fruticosa*) for the sores of camels (KAB).
- Pakistanis use the ash in making soap (WOI).
- Sindhis apply the leaves as a poultice in ophthalmia, and the leaf infusion as an emetic (DEP).

Downsides (Sea Blite):

Said to induce a persistent black diarrhea, even death, in grazing sheep (KAB).

Natural History (Sea Blite):

Camels, goats, and sheep graze foliage of various species (DEP; WOI).

SALT CEDAR (*TAMARIX APHYLLA* (L.) H. KARST.) ++ TAMARICACEAE

Synonyms:

Tamarix articulata Vahl

Notes (Salt Cedar):

Zohary notes that Bedouins planted tamarisk trees (e.g., in the Negev Desert) for their shade and soft branches. They are indigenous in the Aravah Valley. Although originally the Hebrew *eshel* applied to the tamarisk, later in history its meaning was broadened to tree. Because of the superficial resemblance to cedar, it is sometimes called cedar. The tamarisk was used in cleansing lepers and their houses. Of the 12 species of tamarisk native to Israel, the Nile tamarisk (*T. nilotica*) is possibly most common, growing in every deep wadi, especially near water and marshes (ZOH). HJP entries apply to *Tamarix mannifera*.

Common Names (Salt Cedar):

Abiekwas-Geelhout (Afrikan; KAB); Aslularmar (Arab.; KAB); Asrelei (Sin.; KAB); Asri (Sin.; KAB); Athal (Arab.; GHA); Athel Tamarisk (Eng.; FAC; USN); Athel-pine (Eng.; USN); Atl (Arab.; ZOH); Cedar (Eng.; ZOH); Dawee (Afrikan; KAB); Erashirisaru (Ap.; India; SKJ); Eshel (Heb.; ZOH); Etel (Arab.; BOU); Ethl (Arab.; ZOH); Ettashirisaru (Tel.; KAB); Ettaverusaru (Tel.; KAB); Faras (Pun.; KAB); Farash (India; Iran; Pun.; KAB; USN); Farwa (Pun.; KAB); Gagaz (Pahrod; KAB); Gaz (Sin.; KAB); Gazesurkh (Iran; KAB); Gazlau (Sin.; KAB); Ghwa (Pun.; Pushtu; KAB); Ghuz (Pun.; KAB); Ghwaz (Pushtu; KAB); Hatab Ahmar (Arab.; Syria; HJP); Khagal (Pun.; KAB); Kharbi (Pun.; SKJ); Kharlei (Pun.; KAB); Khora Gaz (Arab.; Helmand; Iran; FAC; KAB); Kirri (Bal.; Las Bela; KAB); Laljhav (Dec.; Guj.; Hindi; KAB; SKJ); Leafless Tamarisk (Eng.; ZOH); Narlei (Pun.; KAB); Pharwan (Pun.; KAB); Raktajhav (Beng.; KAB; SKJ); Rukh

FIGURE 1.110 Salt Cedar (*Tamarix aphylla*). Source: Little & Wadsworth (1974).

(Pun.; KAB); Salt-cedar (Eng.; USN); Shakargaz (Pahrod; KAB); Siahgazz (Kharan; KAB); Sivappattuushavukku (Tam.; KAB); Sivappukottashavukku (Tam.; KAB); Sivappusirushavukku (Tam.; KAB); Tabrakat (Ber.; BOU); Takaout (Arab.; BOU); Tamaris (Fr.; USN); Tamarisk (Eng.; S. Afr.; KAB); Tamariske (Ger.; USN); Tamarix a galle (Fr.; BOU); Taray (Sp.; USN); Tarfa (Arab.; GHA); Tarfalahmar (Arab.; KAB); Tarfat Elmann (Arab.; Syria; HJP); Terfa (Arab.; Oman; GHA); Terfel (Arab.; Dho.; Oman; GHA); Ukhan (Pun.; KAB).

Activities (Salt Cedar):

Acaricide (f; BOU); Aphrodisiac (f; KAB); Astringent (f; BOU; KAB; SKJ); Pediculicide (f; BOU); Vulnerary (f; GHA).

Indications (Salt Cedar):

Childbirth (f; GHA); Dermatosis (f; GHA; KAB; WOI); Eczema (f; GHA; KAB; SKJ; WOI); Edema (f; BOU); Enterosis (f; BOU); Gastrosis (f; BOU); Impotence (f; KAB); Labor (f; GHA); Leprosy (f; ZOH); Pain (f; BOU); Pediculosis (f; BOU); Sclerosis (f; JLH); Sore (f; GHA); Splenosis (f; BOU; JLH); Stomachache (f; BOU); Swelling (f; BOU); Tumor (f; JLH); Uterosis (f; BOU); Wound (f; GHA).

Dosages (Salt Cedar):

FNFF = !
A source of a sweet manna, eaten, for example, by the Tuareg with porridge or in beverages (FAC). Resin from tree also edible; salt excreted from leaves used to salt foods (GHA). Bedouin's grazing herds are fed the foliage (ZOH).

- Asian Indians suggest the galls as astringent (KAB).
- Asian Indians combine the powdered bark with oil and kamala as an aphrodisiac (KAB).
- Baluchistani use bark for eczema and other skin disorders (GHA).
- Dhofari take tea of dried leaves to ease prolonged or difficult labor (GHA).
- Lebanese priests believe that manna from Sinai (*T. mannifera*) was very healthful for children and gently laxative for adults (HJP).
- North Africans boil bark in vinegar and water as a pediculicide (BOU).
- North Africans take astringent gall infusion for enteritis and gastralgia (BOU).
- North Africans take shoot decoction for splenomegaly, with ginger for uteral problems (BOU).
- Omani apply dried leaves to sores and wounds (GHA).
- Saudi apply green shoots of some species of *Tamarix* to head for headache and fever (GHA).
- Yemeni use tea (of *T. nilotica*) to treat diarrhea (GHA).

Natural History (Salt Cedar):

This inhabitant of the hot deserts, very wasteful of water, excretes salts from glands in the leaves.

DANDELION (*TARAXACUM OFFICINALE* G.H. WEBER EX WIGGER F.) +++ ASTERACEAE

Synonyms:

Leontodon dens-leonis Lam.; *Leontodon glaucescens* M. Bieb.; *Leontodon officinalis* With.; *Leontodon parvulum* Wall.; *Leontodon taraxacum* L.; *Leontodon vulgare* Lamn.; *Taraxacum dens-leonis* Desf.; *Taraxacum eriopodum* Schrank; *Taraxacum vulgare* Schrank; *Taraxacum wallichii* DC. *fide* BUR; DEP

Notes (Dandelion):

> **The fourteenth day of the second month at even they shall keep it, and eat it with unleavened bread and bitter herbs.**
>
> **Numbers 9:11 (KJV)**

In the second month on the fourteenth day in the evening they shall keep it; they shall eat it with unleavened bread and bitter herbs.

Numbers 9:11 (RSV)

In the second month on the fourteenth day between the two evenings, they should prepare it. Together with unfermented cakes and bitter greens they should eat it.

Numbers 9:11 (NWT)

Probably the children of Israel learned to eat bitter herbs from the Egyptians. Ancient Egyptians used to place the green herbs on the table, mixed with mustard, and then dunked their bread in the mixture. The Moldenke's believed that *Cichorium endivia*, *Cichorium intybus*, *Lactuca sativa*, *Nasturtium officinale*, *Rumex acetosella*, and *Taraxacum officinale* were among the green herbs of the Bible. By contrast, local Israeli botanist Michael Zohary lists none of these in his *Plants of the Bible* (ZOH), and only the watercress is listed as occurring in the *Flora of Palestine*. Zohary figures instead a diminutive chicory and the poppy-leaved Reichardia (which looks like dandelion) as more promising candidates. Regarding bitter herbs, Zohary says, "Many plants, especially those belonging to the Mustard and Daisy families, are frequently collected and used as potherbs and salad plants" (ZOH).

Common Names (Dandelion):

Achicoria (Sp.; JFM); Achicoria Amarga (Sp.; USN); Ackerzichorie (Ger.; KAB); Alface de Bao (Por.; AVP); Alface de Cao (Por.; JFM); Almiron (Sp.; KAB); Amargón (Sp.; KAB; USN); Apostemkraut (Ger.; KAB); Arnica (Ma.; JFM); Asnan el Saba'a (Arab.; AVP); Aspan Asad (Arab.; Syria; HJP); Augenmilch (Ger.; KAB); Baerenzahn (Ger.; KAB); Baram (Pun.; WOI); Baran; Hindi (Punj.; KAB); Barau (Hindi; WOI); Bathur (Bom.; Sin.; DEP; KAB; NAD); Bettpisser (Ger.; KAB); Bettzeiger (Ger.; KAB); Bissanliwurzel (Ger.; KAB); Bitterwort (Eng.; KAB); Blowball (Eng.; BUR); Blower (Eng.; KAB); Blowhard (Eng.; KAB); Brodawnik (Pol.; AVP); Buthur (Sin.; KAB; NAD); Butterblume (Ger.; HH2; KAB); Cais Tsearbhan (Irish.; KAB); Canker (Eng.; KAB); Cankerwort (Eng.; BUR; KAB); Capo de Frate (It.; HH2); Chicoree (Fr.; AVP); Chicoree Sauvage (Reunion; KAB); Chinita del Campo (Ma.; JFM); Chopine (Fr.; AVP; KAB); Cicueira Salvagga (Malta; KAB); Clock (Eng.; KAB); Cochet (Fr.; AVP); Common Dandelion (Eng.; USN); Consueldo (Ma.; JFM); Couronne de Moine (Fr.; KAB); Crow Parsnip (Eng.; KAB); Dandelion (Eng.; USN); Dashel Flower (Eng.; KAB); Dent de Lion (Fr.; USN); Dente-de-leão (Port, KAB; USN); Dente di Leone (It.; Fr.; Malta; AVP; HH2; KAB); Dentelion (Eng.; KAB); Dents de lleo (Cat.; KAB); Diente de León (Sp.; USN); Diente de Perro (Sp.; JFM); Dindle (Eng.; KAB); Dini Mhendo (Tamang; NPM); Doon Head Clock (Eng.; BUR; KAB); Dudal (Hindi; Pun.; KAB; NAD); Dudh-Batthal (Pun.; KAB; WOI); Dudhe Jhar (Nepal; NPM); Dudli (Hindi; Pun.; KAB; NAD; WOI); Dugdha Feni (Sanskrit;?); Dulal (Hindi; WOI); Eierblume (Ger.; KAB); Feldreis (Ger.; KAB); Florin d'or (Fr.; AVP); Florion d'Or (Fr.; HH2; KAB); Fortune Teller (Eng.; BUR; KAB); Gaddeliese (Ger.; KAB); Gemeiner Löwenzahn (Ger.; HH2); Gowan (Eng.; KAB); Groin de Porc (Fr.; AVP); Gulsagh (Kironi; KAB); Gyermeklancfu (Hun.; KAB); Habichtskraut (Ger.; KAB); Hand (Kas.; WOI); Handh (Kas.; MKK); Henbe (Arab.; Syria; HJP); Hendbe 'Elt (Arab.; Syria; HJP); Hindiba' Barri (Arab.; Syria; HJP); Horse Gowan (Eng.; BUR); Hundslattich (Ger.; KAB); Hundzahn (Ger.; KAB); Hyo Maran (Tamang; NPM); Irish Daisy (Eng.; BUR; HH2; KAB); Jamajunsha (Kas.; Lad.; MKK); Jungeblume (Ger.; KAB); Kaadu Seventhi (Kan.; WOI); Kamphul (Hindi; WOI); Kanphul (Pun.; KAB; WOI); Kara Hindiba (Tur.; AVP; EFS); Karnaphuli (Nepal; NPM); Kettenblume (Ger.; HH2); Khur Mangmang (Tibet; NPM); Khur Mong (Tibet; TIB); Khursa (Kas.; Lad.; MKK); Khuss Barri (Arab.; Syria; HJP); Kojate (Tibet; NPM); Kuhblume (Ger.; HH2; KAB); Kulattich (Ger.; KAB); Lagaina (Lanuedoc; KAB); Laiteron (Fr.; HH2); Laterneblume (Ger.; KAB); Latui de Chien (Fr.; KAB); Lechuga del Monte (Sp.; JFM); Lechuga Silvestre (Sp.; JFM); Lechuguilla (Sp.; JFM);

FIGURE 1.111 Dandelion (*Taraxacum officinale*).

Leituga (Mad.; JAD); Lejoustand (Serbia; AVP); Liondent (Fr.; KAB); Lion's-tooth (Eng.; USN); Llitso d'asa (Cat.; KAB); Lovetan (Den.; KAB); Lowenzahn (Ger.; KAB; MAD); Luchten (Ger.; KAB); Luweland (Den.; AVP); Maelkebotte (Den.; KAB); Maistoechel (Ger.; KAB); Mas'acak (Serbia; AVP); Maskros (Swe.; KAB); Maskrosoor (Serbia; AVP); Milchadistel (Ger.; KAB); Milchroedel (Ger.; KAB); Milchstoeckel (Ger.; KAB); Milk Gowan (Eng.; KAB); Mirame y No Me Toques (Sp.; JFM); Mistfinke (Ger.; KAB); Moenchsblume (Ger.; KAB); Moenchskopf (Ger.; KAB); Moisia (Dutch; EFS); Monkshead (Eng.; HH2); Monkshood (Eng.; KAB); Nedhap (Tamang; NPM); Neta Dha (Gurung; NPM); Oduvanchik (Rus.; KAB); Oduwantschiki (Rus.; HH2); One o'Clocks (Eng.; KAB); Paardenbloem (Dutch; EFS; HH2; KAB); Papadie (Rom.; KAB); Papankraut (Ger.; KAB); Papasita gainelor (Rom.; KAB); Papava Gumentse (Rus.; KAB); Pappenkruid (Fr.; AVP); Pathardi (Guj.; WOI); Pathri (Dec.; DEP; NAD; WOI); Peasant's Clock (Eng.; KAB); Pfaffendistel (Ger.; HH2; KAB); Pfaffen-Öhrlein (Ger.; AVP; HH2; KAB); Pfaffenschnell (Ger.; KAB); Pfaffenstiel (Ger.; KAB); Pfeffer-Oeslein (Ger.; KAB); Pferdeblume (Ger.; HH2; KAB); Phuli Jhar (Nepal; NPM); Pisciacane (Malta; KAB); Piscialleto (It.; KAB); Piss-abed (Eng.; AVP; KAB); Pissenlit (Fr.; Haiti; Reunion; AVP; HH2; KAB); Pissenlit Vulgaire (Fr.; USN); Pissolet (Lanuedoc; KAB); Pitachumpki (Beng.; WOI); Pixallits (Cat.; KAB); Priests' Crown (Eng.; BUR; KAB); Puffball (Eng.; BUR); Pu Gong Yeng (Pin.; DAA); P'u Kung Ying (Chinese; EFS; KAB); Pusteblume (Ger.; HH2); Radam (Pun.; DEP; KAB; WOI); Radicchiella (It. KAB); Rasuk (Lad.; DEP; KAB); Rasuke (Lad.; WOI); Ringelblume (Ger.; HH2); Salada de Toupeira (Fr.; AVP); Salade de Taupe (Fr.; AVP; KAB); Saumelke (Ger.; KAB); Saustok (Ger.; KAB); Scherkraut (Ger.; KAB); Schweineroesl (Ger.; KAB); Schwiblume (Ger.; KAB); Shag (Arab.; Syria; HJP); Shamukei (Hindi; Pun.; KAB; NAD; WOI); Smirnio (It.; KAB); Soffione (It.; EFS; HH2; KAB); Sommerdorn (Ger.; KAB); Sonnenwirbel (Ger.; KAB); Stink Davie (Eng.; KAB); Swine Snout (Eng.; BUR; KAB); Swini Miecz (Pol.; AVP); Tarakhshagun (Arab.; KAB); Tarassacio (Malta; KAB); Taraxaco (It.; Por.; AVP; HH2; KAB); Taraxacon (Sp.; AVP); Tarazacon (Sp.; KAB); Tarkhas Kun (Iran; KAB); Tete de Moine (Fr.; KAB); Teufelsrippen (Ger.; KAB); Tiefstand (Ger.; KAB); Time Table (Eng.; KAB); Tuki Phul (Nepal; NPM; SUW); Undarkarni (Mar.; WOI); Undrachekan (Kon.; NAD); Weglattich (Ger.; KAB); Wiesenlattich (Ger.; HH2; KAB); Witch Gowan (Eng.; HH2); Wolowe Oczy (Pol.; AVP); Xawiizi (Hocak; WIN); Xicoina de Burro (Cat.; KAB); Yamaghikha (Lad.; KAB; WOI); Yellow Gowan (Eng.; BUR); Zunehmkraut (Ger.; KAB).

Activities (Dandelion):

Allergenic (1; CAN); Alpha-glucosidase Inhibitor (1; X15704495); Alterative (f; CRC); Antidote (f; AKT; TIB); Antiedemic (f1; CAN); Antiinflammatory (1; CAN; SHT); Antioxidant (1; CRC; X15543940); Antirheumatic (1; CAN); Antiseptic (1; CRC; FAD); Antispasmodic (SHT); Antitumor (1; CAN); Aperient (f; NPM; SUW); Aperitif (2; KOM; PIP); Apoptotic (1; X14687655); Aquaretic (SHT); Astringent (f; PED); Bactericide (1; CRC; WOI); Bifidogenic (1; AKT; X15567259); Bitter (1; APA; PED; SHT; SUW); Candidicide (1; CRC; FAD); Carminative (2; APA; KOM); Cholagogue (f12; BIB; KOM; PH2; SHT; WAM); Cholekinetic (SHT); Choleretic (f12; APA; KOM; SHT; VAD); COX-2 Inhibitor (1; X15543940); Cytotoxic (1; X14687655); Demulcent (f; FAD); Deobstruent (f; BUR); Depurative (f; CRC; JFM; VAD); Detoxicant (f; AKT); Diaphoretic (f; KAP); Digestive (f; APA; SKY); Diuretic (f12; AKT; APA; HH3; KOM; NPM; SUW; VAD; WAM); Hepatotonic (f1; CAN; NAD; PED); Hypoglycemic (1; CAN; CRC; JAC7:405; X15704495); IL-1alpha Inducer (1; X14687655); Immunostimulant (1; JAC7:405); iNOS Inhibitor (1; X15543940); Insulinogenic (1; X14750205); Intoxicant (f; BIB); Lactagogue (f; CRC; LMP; NMH; PED; PH2); Laxative (f1; APA; BIB; CAN; SUW; VAD; WAM); Litholytic; (f1; PED); Natriuretic (1; JAD); Orexigenic (f12; AKT; BGB; JFM; KOM; VAD); Pancreaprotective (1; X15641154); Phototoxic (1; CRC); Prebiotic (1; AKT; FNF); Saluretic (12; BGB; HH3; PH2); Secretogogue (f1; PH2; X14750205); Sialogogue (1; APA); Stimulant (f; CRC); Stomachic (f1; APA; BIB; CRC; PED); Sudorific (f; CRC); TNF-alpha Inducer (1; X14687655); Tonic (f; BGB; CRC; NPM; SUW).

Indications (Dandelion):

Abscess (f; CRC; MAD); Acne (f; VAD); Adenopathy (f; JLH); Ague (f; BIB); Alactea (f; LMP; PH2); Alcoholism (f; SKY); Alzheimer's (1; FNF); Anemia (f1; AAH; DEM; JFM; WAM); Anorexia (12; APA; KOM; PH2; PIP; VAD); Arthrosis (f; BIB); Backache (f; DEM); Bacteria (1; WOI); Biliary Dyskinesia (2; PIP); Biliousness (f; BIB); Bladderstones (2; KOM); Boil (f; CRC; LMP); Bronchosis (f12; APA; BIB; LAF); Bruise (f; BIB; CRC); Cachexia (f; NAD); Cancer (f; CRC); Cancer, bladder (f; JLH); Cancer, bowel (f; JLH); Cancer, breast (f; CRC; JLH); Cancer, liver (f; JLH); Cancer, spleen (f; JLH); Caries (f; CRC; LMP); Cardiopathy (f; APA; BIB); Catarrh (f; BIB; CRC); Cellulite (1; FT71:S73); Chill (f; HJP); Cholecystosis (2; BGB; CRC; HH3; KOM; PH2); Cirrhosis (SKYf;); Cold (1; APA); Colic (1; PH2); Congestion (1; PH2); Conjunctivosis (f; AAH; AKT); Constipation (f1; FAD; SKY; FT71:S73); Consumption (f; BIB); Cough (f; MAD); Cramp (f; DEM); Cystosis (1; WAM); Dermatosis (f; APA; BGB; KAP; KOM; PH2); Diabetes (f1; BIB; CRC; JFM; KOM; MAD; PH2; X15704495; X14750205); Dropsy (f1; BGB; BIB; DEM; KAP; MAD); Dysentery (f; AKT); Dyskinesia (f1; VAD); Dysmenorrhea (f; DEM); Dyspepsia (f12; APA; KAP; KOM; PH2; PIP; VAD); Dyspnea (f; HJP); Eczema (f; CRC; HH2; KOM; PH2; VAD); Fever (f1; AKT; BIB; MAD); Flatulence (f12; BIB; KOM); Flu (f; AKT); Fracture (f; MKK); Furuncle (f; VAD); Gallstone (f; AKT; APA; CAN; MKK; PH2); Gas (2; APA; PH2); Gastrosis (f; BGB; APA; CRC; NPM); Gonorrhea (f; BIB); Gout (f; CRC; KOM; PH2); Gravel (f1; BGB; BIB HH2); Headache (f; HJP; MKK); Heart (f; CRC); Heartburn (f; BGB; CRC; SKY); Hemorrhoid (f; CRC; KOM; MKK; PH2); Hepatosis (f12; JFM; KAP; HH3; KOM; PH2; SUW; WAM); Herpes (f; VAD); High Blood Pressure (1; CRC; VAD); Hypochondria (f; MAD); Impetigo (f; BUR); Induration (f; JLH); Infection (f; AKT; APA); Inflammation (f1; BIB; PH2); Insomnia (f; BIB); Itch (f; BUR; CRC; LMP); Jaundice (f1; CAN; CRC; HJP; MAD; PH2); Kidneystone (2; KOM; PH2; X7860196); Liver Spots (f; DEM); Malaria (f; BIB); Mastosis (f; APA; PH2); Mucososis (f; MAD); Nephrosis (f1; BGB; BIB; DEM; KAB; PH2; SUW); Neurosis (f; DEM; HJP)); Obesity (f; APA; FAD); Oliguria (1; CAN; VAD); Osteoporosis (1; FNF); Pain (1; BGB; CEC; DEM; MAD); Pancreatitis (1; X15641154); Phthisis (f; JEB79:57); PMS (f; APA); Pneumonia (2; AYL; LAF); Psoriasis (f; VAD); Pulmonosis (f; DEM); Pyelosis (f1; VAD); Respirosis (2; APA; LAF); Rheumatism (f; APA; BGB; PHR); Scirrhus (f; JLH); Sclerosis (f; JLH); Scrofula (f; CRC; LMP); Scurvy (f1; BIB); Snakebite (f; CRC); Sores (1; APA; CRC); Splenosis (f; AKT; BGB; BUR); Stitch (f; MAD); Stomachache (f; AAH; DEM); Stone (f12; BIB; PHR; PH2; SHT; VAD); Swelling (f; DEM); Thrush (f; AAH); Toothache (f; AAH; DEM); Tuberculosis (f1; BIB; MAD; PH2; WOI); Tumor (f; JLH); Ulcer (f; CRC); Urethritis (f1; VAD); Urolithiasis (f1; VAD); UTI (f12 KOM; PH2; SHT; VAD); Uterosis (f; BUR); Venereal Disease (f; BIB); Vertigo (f; MKK); Vomiting (f; AKT); Wart (f; CRC); Wound (f; BIB); Yeast (f1; AAH; CRC).

Dosages (Dandelion):

FNFF = !!

Food farmacy; tender leaves valued as potherb; Winnebago make wine from the flowers when someone marries. Dandelion is sometimes eaten raw in salads, but often blanched like endive and used as a green; frequently cooked with salt pork or bacon to enhance the flavor. Roots are sometimes pickled. Ground roasted roots used for dandelion coffee, and sometimes mixed with real coffee. Dried leaves are an ingredient in many digestive or diet drinks and herb beers (BIB; NPM; WIN). 0.5 oz dry leaf/cup water (APA); 1–3 tsp powdered root/cup water (APA); 3–5 g dry root 3 ×/day (APA); 1–2 tsp tincture to 3 ×/day (APA); 1 Tbsp dandelion juice morning and evening (APA); 4–10 g dry leaf, as tea, 3 ×/day (CAN); 4–10 ml liquid leaf extract (1:1 in 25% ethanol) 3 ×/day (CAN); 2–8 g dry root, as tea, 3 ×/day (CAN); 5–10 ml root tincture (1:5 in 45% alcohol) 3 ×/day (CAN); 4–10 ml dandelion juice (CAN; PNC; SKY); 0.5–1 g powdered root (KAP); 28–56 ml root decoction (KAP); three 510-mg capsules 3 ×/day (NH); 2 tsp root juice 3 ×/day for stomach (NPM); 1/4–1/2 cup fresh root (PED); 6–12 g dry root (PED); 9 g dry root:45 ml alcohol/45 ml water (PED); 1–2 tsp (for tea) to 3–4 tsp (for decoction) cut herb/150 ml water (PH2.). 4–10 g dry leaf 3 ×/day; 1–2 tsp root/cup/morning and evening; 3–4 g powdered root (PIP); 10–15 drops root tincture (PIP); 2–8 ml liquid extract (PNC); 3–4 g/day (SHT); 2–5 ml leaf tincture3×/day (SKY).

- Asian Indians suggest 10 to 15 grains root as hepatic stimulant (NAD).
- Asian Indians, suggestive of Carter's Little Liver Pills, recommend 1 to 2 oz root (fluid extract or decoction) with podophyllin (a bit dangerous in my opinion) for dyspepsia, hepatitis, and jaundice (NAD).
- British regard this universally as a diuretic, but also use for colds, coughs, respirosis, and warts (AAH).
- Irish have even more uses for dandelion than British, adding boils, consumption, cuts, dermatosis, diabetes, fractures, hepatosis, nervousness, sore eyes, sprains, swellings, and thrush (AAH).
- Irish, because of its many "lion's teeth," believe it good for toothache (AAH).
- Italians apply dandelions to warts (X15664457).
- Lebanese extract the root in wine as a laxative or purgative, depending on the strength. Noting that "medical usage of dandelion came to western civilization through the Arabs," Philips says that gypsies use the root infusion as a depurative, and laxative, for the liver, rheumatism and sciatica, the raw leaf for a spring tonic, and the leaf tea for heavy breathing and kidney ailments (HJP).
- Limerick citizenry believe that eating a leaf with red midvein is tonic for a woman, white vein tonic for a man (AAH).
- Nepali suggest 2 tsp root juice 3 × /day for gastrosis (NPM).

Downsides (Dandelion):

Class 2d (AHP, 1997). No health hazards or side effects known with proper therapeutic dosages (PH2). Commission E reports contraindications: biliary obstruction, empyema of gallbladder, ileus; adverse effects include gastric complaints and ulcers (AEH; CAN; SKY). Other contraindications reported include biliary inflammation (AEH). Newall, Anderson, and Phillipson (1996) caution that the sesquiterpene lactones are allergenic and may cause dermatosis. May interefere with diuretic and hypoglycemic therapies (CAN). " [H]erbs with diuretic properties, such as juniper and dandelion, can cause elevations in blood levels of lithium" (D'epiro, 1997). Not for use with acute gallbladder problems (WAM). Use in cholelithiasis only under a doctor's supervision (PIP). Surprisingly, Jacobs and Burlage suggest that the root causes "mental excitement, vertigo, headache, nausea, colic, frequent urination, and gastric irritation" (BUR). Blumenthal et al. (1998) caution that, "As with all drugs containing bitter substances, discomfort due to gastric hyperacidity may occur" (KOM). Do I need to write out this caveat for all the salubrious bitter herbs of the Bible? Warning: may cause hyperacidity and gastric distress!

Natural History (Dandelion):

Birds like the seeds and pigs devour the whole plant. Goats eat the leaves, but sheep, cattle, and horses do not care for it. Omur and Handa (2005) demonstrated a priority of color over scent during flower visitation by adult *Vanessa indica* butterflies. Most flower visitors innately prefer a particular color and scent, and use them as cues for flower recognition and selection. Of colors, *V. indica* showed a color preference for yellow and blue. Aromatically, they seemed to prefer benzaldehyde, acetophenone, and (E+Z)-nerolidol. But butterflies preferred odorless yellow models to scented purple models. *V. indica* depends primarily on color and secondarily on scent during flower visitation (X15688217).

EXTRACTS (DANDELION):

LD50 herb = 28,800 ipr mus (CAN); LD50 root = 36,800 ipr mus (CAN). The leaves have a higher Vitamin A content (14,000 IU/100g) than carrots (11,000 IU/100g). Coumestrol is estrogenic. Flavonoids antiinflammatory; increase urine flow. Inulin and mucilage sooth digestive tract, absorb toxins, and regulate intestinal flora through prebiosis (help friendly flora thrive and inhibit unfriendly bacteria) and relieve muscle spasm. (PED). PH2 says the amaroids (bitter compounds) in dandelion are cholagogue (I agree) and secretolytic (I disagree; I think they are secretogogue rather than secretolytic; PH2 also says the drug is "secretion-stimulating"). Tillotson (AKT) notes that clinically the leaf is a more effective diuretic than the root and a safer alternative to Lasix. Onal et al. (2005) found anti-glucosidase activity, suggesting antidiabetic potential, in three biblical herbs: dandelion, myrtle, and stinging nettle. Agents that inhibit alpha-glucosidase can be useful oral hypoglycemics (X15704495). Proestos et al. (2005) checked the species for flavonoids and phenolics and their antioxidant and antimicrobial activity, finding circa 30 ppm caffeic acid, 20 ppm ferulic acid, eriodictyol, and 4 ppm (–)-epicatechin in dandelion (X15713039). Seo et al. (2005b) showed that dandelion protects against cholecystokinin-induced acute pancreatitis in rats. At 10 mg/kg orally, dandelion significantly decreased the pancreatic weight/body weight ratio in CCK octapeptide-induced acute pancreatitis, and IL-6 and TNF-alpha decreased, suggesting a protective effect against induced pancreatitis (X15641154). Czech scientists showed that dandelion root tea stimulated *in vitro* growth of 14 strains of bifidobacteria, the oligofructans providing an important source of carbon and energy (X15567259). Hu and Kitts (2004) found that luteolin and luteolin-7-O-glucoside from dandelion flower suppress iNOS and COX-2 at concentrations lower than 20 μM. The ethyl acetate fraction of dandelion flower extract contains 10% luteolin and luteolin-7-O-glucoside (X15543940). Hussain et al. (2004) demonstrated that dandelion increased insulin secretion, but at rather high dosage (40 μg/ml) (X14750205).

SANDARAC (*TETRACLINIS ARTICULATA* (VAHL) MAST.) + CUPRESSACEAE

SYNONYMS:

Callitris articulata (Vahl) Murb.; *Callitris quadrivalvis* Vent.; *Cupressus articulata* (Vahl) J. Forbes; *Thuja articulata* Vahl *fide* CJE; USN

NOTES (SANDARAC):

> **The merchandise of gold, and silver, and precious stones, and of pearls, and fine linen, and purple, and silk, and scarlet, and all thyine wood, and all manner vessels of ivory, and all manner vessels of most precious wood, and of brass, and iron, and marble.**
>
> **Revelation 18:12 (KJV)**

> **[C]argo of gold, silver, jewels and pearls, fine linen, purple, silk and scarlet, all kinds of scented wood, all articles of ivory, all articles of costly wood, bronze, iron and marble.**
>
> **Revelation 18:12 (RSV)**

> **A full stock of gold, and silver and precious stone and pearls and fine linen and purple and silk and scarlet, and everything in scented wood, and every sort of ivory object, and every sort of object out of most precious wood and of copper and of iron and of marble.**
>
> **Revelation 18:12 (NWT)**

Only in the KJV and only once do I find thyine. I wonder if Zohary should have been working with the KJV instead of the RSV, if he, too, would have considered including this species in his book. Although it is not listed as native to Israel, it is said to be indigenous to northern Africa (Algeria, Morocco, Tunisia), Sicily; Malta, and Spain. The passage above is talking about imports, so it is not necessary that it be indigenous to Israel or Egypt. The next stanza lists cinnamon, frankincense, myrrh, even salves, among the cargo as well. Certainly, like other members of this family, it has scented wood, as do so many other arid land species. I do not know why the KJV translated this as "thyine" instead of "scented wood."

Common Names (Sandarac):

Afrikanischer Sandarac (Ger.; HH3); Alerce (Eng.; CJE); Amelzi (Ber.; BOU); Amkouk (Ber.; BOU); Ar'ar (Arab.; BOU); Ar'ar Berboush (Arab.; BOU); Ar'ar el Eibel (Arab.; BOU); Arartree (Eng.; CJE; USN); Cyprès de l'Atlas (Fr.; USN); Gharghar (Arab.; CJE); Gliederzypresse (Ger.; HH3); Gum Juniper (Eng.; HH3); Irhkri (Ber.; BOU); Irz (Ber.; BOU); Juniper Gum Tree (Eng.; BOU); Marokkanischer Sandarac (Ger.; HH3); Mediterranean Sandarac-Cypress (Eng.; USN); Moigador Sandarak (Ger.; HH3); Sandarac (Eng.; BIB; HH3; USB); Sandarac Gum Tree (Eng.; CJE); Sandaraco (Sp.; HH3); Sandarakbaum (Ger.; HH3; USN); Sandarakharz (Ger.; HH3); Sandaraque (Fr.; HH3); Sandarus (Arab.; BOU); Shagaret el Hayat (Arab.; BOU); Tarout (Ber.; BOU); Tazout (Ber.; BOU); Thuia a la Sandaraque (Fr.; BOU); Thuia Articulé (Fr.; BOU); Thuja de Barbarie (Fr.; USN); Thyine (Eng.; BIB); Tifizza (Ber.; BOU); Tiranrat (Ber.; BOU); Tirarar (Ber.; BOU); Tuya Articulada (Sp.; USN); Vernix (Fr.; BOU); Nscn.

Activities (Sandarac):

Abortifacient (f; BIB); Apoptotic (1; X10697574); Bactericide (f1; HH3; PH2); Cytotoxic (1; X10697574).

Indications (Sandarac):

Bacteria (f1; HH3; PH2); Cancer (f1; JLH); Cancer, breast (f1; JLH; X10697574); Cancer, ovary (f1; JLH; X10697574); Carcinoma (f; JLH); Dermatosis (f; BIB; BOU); Diabetes (f; X9324004); Diarrhea (f; HH3; PH2); Edema (f; PH2); Fever (f; HH3; PH2); Gout (f; PH2); Hemorrhoid (f; DAW); Infection (f1; HH3; PH2); Mastosis (1; X10697574); Melanoma (1; X10697574); Migraine (f; BIB); Neck (f; BIB; BOU); Pain (f; BIB; BOU); Rheumatism (f; DAW; PH2); Sunburn (f; BIB; BOU); Swelling (f; PH2).

Dosages (sandarac):

FNFF = X

- Egyptians suggest the plant for diarrhea, gout, and rheumatism (DAW).
- North Africans suggest a leaf cataplasm for insolation, migraine, and headache (BOU).
- North Africans use tar from old trees for skin ailments (BOU).

Downsides (Sandarac):

No health hazards with proper administration of designated therapeutic dosages (PH2) (but PH2 designates no dosage!; JAD).

Extracts (Sandarac):

The essential oil induced apoptosis in human melanoma, breast, and ovarian cancer cell lines. Melanoma, breast, and ovarian cancer cells gave IC50 values of circa 80 μg/ml. The authors discuss the advantage of a mixture of monoterpenes over a single component (X10697574).

BEAN CAPER (*TETRAENA DUMOSA* (BOISS.) BEIER & THULIN) + ZYGOPHYLLACEAE

SYNONYMS:

Zygophyllum dumosum Boiss.

NOTES (BEAN CAPER):

And they removed from Marah, and came unto Elim: and in Elim were twelve fountains of water, and threescore and ten palm trees; and they pitched there.

Numbers 33:9 (KJV)

And they set out from Marah, and came to Elim; at Elim there were twelve springs of water and seventy palm trees, and they encamped there.

Numbers 33:9 (RSV)

Then they pulled away from Marah, and came to Elim. Now in Elim there were twelve springs of water and seventy palm trees. So they camped there.

Numbers 33:9 (NWT)

Noting that this evergreen xerophytic shrub flourishes over miles of the Israeli and Sinai deserts, Zohary notes that the inhabitants must have known it well, but equates it to the place name Elim, which shows up about a dozen times in the KJV, six times in Ruth, three times in Exodus, three in Numbers, and once in Isaiah. Persevering through rainless years on the desert, it can have more than 300 annual rings, scoring those years in which rains fell. They could be much older, as in the rainless years there are no growth rings. I suppose Zohary selected this of the five (*Zygophyllum*) species reported in the *Flora of Palestine* as it is the only conspicuously shrubby species. The other perennials are uncommon at best.

- Perennials or dwarf perennial shrubs: leaves two-foliolate:
  - —Very common dwarf shrub; fruits with five broad wings — *Z. dumosum*
- Perennials but not conspicuously shrubby; fruit wingless but sometimes five-lobed:
  - — Leaves flat; fruit 2–3 cm long — *Z. fabago*
  - — Leaves succulent, fruits much shorter:
    - – • Adults glabrous; pedicels longer than flowers — *Z. coccineum*
    - – • Adults hairy; pedicels shorter than flowers — *Z. album*
- Prostate annuals; leaves simple — *Z. simplex*

The USDA recognizes Tetraena as the correct designation for this Holy Land species, but most of my floras still call the other *Zygophyllum.* One website suggests that this species might be found on the Shroud of Turin (http://www.kensmen.com/catholic/marygardens.html). Another notes that: The Shroud of Turin, to the naked eye, is a negative image of a man with folded hands (linen 14 feet, 3 inches long and 3 feet, 7 inches wide). The shroud bears the image of a man with wounds similar to those of Jesus. Wrapped in red silk, the shroud is kept in a silver chest in the Chapel of the Holy Shroud in the Cathedral of St. John the Baptist in Turin, Italy since 1578. It is unquestionably old. It surfaced in 1357 in Lirey, France. Some carbon dating (1988) suggests that the shroud dates back only to 1260–1390. But the debate goes on. Either Carbon-14 dating is inaccurate or the shroud is a fake (says the website). Here we also read, regarding establishing the shroud's provenance, that the bean caper is most significant. Max Frei identified pollen grains of this species on adhesive tapes he studied. The northernmost limits of this species coincide with the line between Jericho and the sea-level marker on the road from Jerusalem to Jericho. Because *Zygophyllum dumosum* grows only in Israel, Jordan, and Sinai, its appearance helps to definitively limit the shroud's place of origin (source: http://www.tombofjesus.com/forums/lofiversion/index.php?t26.html).

Common Names (Bean Caper):

Bean Caper (Eng.; ZOH); Elim (Heb.; ZOH); Illam Arabic (Eng.; ZOH); Qillab (Arab.; GEP); Nscn.

Activities (Generic Tetraena):

Anthelmintic (f; KAB); Antiseptic (f; BOU); Antispasmodic (f; DAW); Depurative (f; DAW); Hemostat (f; BOU); Hypotensive (f; BOU); Poison (f; DAW); Vermifuge (f; DAW).

Indications (Generic Tetraena):

Abscess (f; BOU); Asthma (f; BOU); Bleeding (f; BOU); Boil (f; BOU); Cramp (f; DAW); Dermatosis (f; BOU); Diabetes (f; BOU); Eczema (f; BOU); Gout (f; BOU); High Blood Pressure (f; BOU); Infection (f; BOU); Leucoma (f; KAB; WOI); Ophthalmia (f; WOI); Rheumatism (f; BOU); Stiffness (f; UPW); Worm (f; BOU; WOI).

Dosages (Generic Tetraena):

FNFF = !
Facciola says fruits of *Z. fabago*, also called bean caper, are pickled and used as a substitute for capers (FAC). Tanaka says of *Z. album*: dried tips of flower clusters have the pleasant smell of tea; of *Z. coccineum*, Arabs use the seeds like black pepper; of *Z. fabago*, flowers buds used as spice; of *Z. simplex*, seeds used as food.

- Arabs, calling it balbal, bawwal, kammun karmani, and rotrayth; use *Z. coccineum* for asthma, diabetes, gout, high blood pressure, rheumatism, and worms (BOU).
- Berbers calling it aggaya, Arabs berraya; use leaf tea as antiseptic, applying dry leaf powder as a hemostat, or applying to abscesses, boils, and eczema (BOU).
- Mali citizens pound the leaves and poultice them on to stiff areas (UPW).

Downsides (Generic Tetraena):

Described as poison.

Natural History (Generic Tetraena):

Leaves browsed by camels and donkeys, but without enthusiasm (UPW).

Extracts (Generic Tetraena):

Some species may contain harmine alkaloids.

SHAGGY SPARROWWORT (*THYMELAEA HIRSUTA* (L.) ENDL.) X THYMELAEACEAE

Synonyms:

Passerina hirsuta L.

Notes (Shaggy Sparrowwort):

The princes digged the well, the nobles of the people digged it, by the direction of the lawgiver, with their staves. And from the wilderness they went to Mattanah.

Numbers 21:18 (KJV)

A well, princes dug it, the nobles of the people excavated it, with a commander's staff, with their own staffs. Then from the wilderness on to Mattanah

Numbers 21:18 (NWT)

Like Zohary, I am really stretching the point to include this one. But it is nice to have a member of a poisonous but anticancer plant family, the Thymelaeaceae, in this version of my Bible book. PubMed was amusing when I typed in *Passerina*, prompting me that there were more abstracts (hundreds on passerine birds). There was only one title for a plant species, and no abstract. I started going through my other Holy Land books; nothing on *Passerina hirsuta*. But then I stumbled onto *Thymelaea hirsuta*, a synonym preferred over *Passerina hirsuta*, in the illustrated book *Medicinal Plants of North Africa* by Loufty Boulos (BOU). What Boulos had to say about it jibed nicely with what Zohary had said, under the other name. Then I find in the *Flora of Palestine* (FP2) that Zohary, too, had called it *Thymelaea hirsuta* in that book. Here he says it is one of the most common shrubs in the Mediterranean coastal plain and also in the deserts. Bedouins use it for making rope (FP2). Justifying its inclusion, Zohary mentions that in the Aramaic version of the Pentateuch, Mattanah is translated as Matnan, a desert locality, perhaps identical with the Arabic mitnan. "Such indirect identification, which permits this important species to be included in the flora of the Bible, is supported by the fact that *matnan* or *mitnan* frequents dry wadis and other desert habitats and is so called in modern Hebrew" (ZOH).

Common Names (Shaggy Sparrowwort):

Matnan (Arab.; Aramaic; ZOH); Methnan (Arab.; Heb.; BOU; ZOH); Methnan Akhdar (Arab.; BOU); Metnan (Arab.; BOU); Mitnan (Arab.; ZOH); Passerine (Fr.; BOU); Shaggy Sparrowwort (Eng.; ZOH); Nscn.

Activities (Shaggy Sparrowwort):

Anthelmintic (f1; BOU); Cathartic (f1; BOU); Expectorant (f; BOU); Hydragogue (f1; BOU).

Indications (Shaggy Sparrowwort):

Constipation (f; BOU); Dandruff (f; BOU); Worm (f1; BOU).

Dosages (Shaggy Sparrowwort):

FNFF = X.

- Algerians use the leaf decoction as a dandruff shampoo (BOU).
- Chinese use a Chinese species for cancer; anticancer activity reported in other species (JLH).
- North Africans consider the leaves anthelmintic, cathartic, expectorant, and hydragogue (BOU).

FENUGREEK (*TRIGONELLA FOENUM-GRAECUM* L.) +++ FABACEAE

Notes (Fenugreek):

We remember the fish, which we did eat in Egypt freely; the cucumbers, and the melons, and the leeks, and the onions, and the garlick: But our soul is dried away: there is nothing at all, beside this manna, before our eyes.

Numbers 11:5–6. (KJV)

We remember the fish we ate in Egypt for nothing, the cucumbers, the melons, the leeks, the onions, and the garlic; but now our strength is dried up, and there is nothing at all but this manna to look at.

Numbers 11:5–6 (RSV)

How we remember the fish that we used to eat in Egypt for nothing, the cucumbers and the water melons, and the leeks and the onions, and the garlic; But now our soul is dried away. Our eyes are on nothing at all except the manna.

Numbers 11:5–6 (NWT)

Some scholars consider the fenugreek more likely than *Allium porrum* to be the leek of the Bible. This is not, however, included by Zohary in his *Plants of the Bible* (ZOH). Coincidentally, fenugreek shares many medicinal properties with the leek. Zohary in his flora does mention that it is widely cultivated and "subspontaneous" in Palestine, although it probably originated in southwestern Asia (FP2).

Common Names (Fenugreek):

Al Harba (Mali; UPW); Alforvas (Por.; EFS); Alholva (Sp.; Spain; EFS; VAD); Alolva (Sp.; KAB); Asaraara (Mun.; KAB); Asumodhagam (Sin.; KAB; WOI); Bahurpani (Sanskrit; KAB); Bahupatriki (Sanskrit; KAB); Bhaji (Guj.; KAB); Bockhornsklover (Swe.; TAD); Bockshorn (Ger.; EFS; KAB); Bockshornklee (Ger.; MAD); Boukeras (Greek; KAB); Boyotu (Tur.; EFS); Bukkekorn (Den.; EFS); Cemen otu (Tur.; EB54:155); Cenigrechs (Cat.; KAB); Chandrika (Sanskrit; KAB; WOI); Dipani (Sanskrit; KAB); Ervinha (Por.; EFS); Fenacho (Por.; EFS); Fenegriek (Dutch; KAB); Fenegre (Lan..; KAB); Fenegriek (Por.; EFS); Fenigrekova Trava (Rus.; KAB); Feno Grego (Por.; EFS); Fenogreco (Sp.; KAB); Fenugrec (Fr.; EFS); Fenugrechs (Cat.; KAB); Fenugreek (Eng.; CR2; KAB); Fenugrek (Fr.; BOU; KAB); Fiengreco (It.; KAB); Fieno Greco (It.; Malta; EFS; KAB); Fienu (Malta; KAB); Finegreitze (Ger.; KAB); Finmaregretjen (Ger.; KAB); Foin Grec (Fr.; KAB); Gandhabija (Sanskrit; KAB); Gandhaphala (Sanskrit; KAB); Gorog Lepkeszeg (Hun.; EFS); Graine Joyeux (Fr.; KAB); Greek Clover (Eng.; USN); Greek Hay (Eng.; KAB; USN); Gretskaya Sochevitsa (Rus.; KAB); Griechische Heu (Ger.; KAB); Griechische Heusamen (Ger.; EFS); Grieks Graszaad (Por.; EFS); Haenugraeb (Beng.; KAB); Halba (Malaya; EFS); Halna Kalabat (Malaya; TAN); Hawmar (Arab.; BOU); Helba (Arab.; Mali; GHA; UPW); Helbek (Egypt; KAB); Heno Griego (Sp.; EFS); Hilba (Arab.; GHA); Holba (Alg.; KAB); Hoornklaver (Dutch; TAD); Hornklee (Ger.; EFS); Houlba (Morocco; KAB); Hulabaha (Arab.; NAD); Hulba (Arab.; EFS; KAB); Hu Lu Ba (Pin.; DAA); Hu Lu Pa (China; KAB); Jyoti (Sanskrit; KAB); Kairavi (Sanskrit; KAB); Kelabat (Malaya; IHB); Koroha (Jap.; TAD; TAN); K'u Tou (China; EFS); Kunchika (Sanskrit; KAB); Mantha (Sanskrit; KAB); Mathi (Sin.; KAB); Medhika (Sanskrit; EFS; NAD); Menk-Palle (Kan.; WOI); Mente (Kan.; KAB); Mentepalle (Kan.; KAB); Mentesoffu (Kan.; KAB); Mentesoppu (Kan.; WOI); Menthe (Kan.; NAD); Menthya (Kan.; KAB); Mentikura (Tel.; KAB); Mentulu (Tel.; KAB; NAD); Methai (Sri.; KAB); Methi (Beng.; Guj.; Hindi; Nepal; Sanskrit; Sin.; Urdu; NAD; KAB; SUW); Methika (Sanskrit; KAB) Methini (Guj.; Sanskrit; KAB); Methishah (Beng.; KAB); Methi-Shak (Beng.; WOI); Methri (Pun.; KAB); Metthi (Tam.; KON); Methun (Pun.; KAB); Misrapushpa (Sanskrit; KAB); Munindraka (Sanskrit; KAB); Muthi (Hindu; KAB); Pagitnik (Rus.; KAB); Pazhitnik (Rus.; TAD); Penantazi (Burma; KAB); Pitabija (Sanskrit; KAB); Saine Graine (Fr.; KAB); Sainegrain (Fr.; KAB); Schabbzieberklee (Ger.; TAD); Senegrain (Fr.; KAB); Senegre (Lan.; KAB); Sennegrain (Fr.; BOU; KAB); Shamli (Afg; KAB); Shamlid (Iran; KAB); Shamlit (Iran; KAB); Shamlita (Iran; EFS; NAD); Shamliz (Iran; KAB); Shanbalid (Iran; KAB); Siebengezeit (Ger.; KAB); Tamr hindi (Arab.; BOU); Tifidas (Ber.; BOU); Tokfist (Mali; UPW); Tilis Boukeras (Greek; KAB); Treugolka (Rus.; KAB); Uluhal (Sin.; KAB); Uluva (Mal.; Sin.; Tam.; KAB; NAD);

FIGURE 1.112 Fenugreek (*Trigonella foenum-graecum*).

Uluvaarisi (Sri.; KAB); Vallari (Sanskrit; KAB); Vendayam (Sri.; Tam.; NAD); Vedhani (Sanskrit; KAB); Vendayam (Tam.; KAB); Ventayam (Mal.; Tam.; KAB); Venthiam (Mal.; KAB).

Activities (Fenugreek):

Analeptic (f; BOU); Analgesic (f1; MAD; X15374601); Anthelmintic (f; HHB); Antiadhesion (1; FNF; YAH); Antiaggregant (f; X11310527); Antianemic (f1; VAD); Antiatherosclerotic (1; BGB); Anticystitic (1; FNF; YAH); Antidiabetic (f1; BGB; PNC); Antidiuretic (1; CAN); Antihyperlipedemic (1; BGB); Antihypertensive (1; CAN); Antiinflammatory (f12; KOM; X15374601); Antiischemic (1; X16205934); Antineoplastic (f1; PR15:257); Antinociceptive (1; TAD; X15374601); Antioxidant (1; X15678722); Antiseptic (12; KOM; VAD); Antispasmodic (f1; PED; PNC); Antitumor (f1; PNC); Antiviral (f; DAA); Aperient (f; NAD); Aphrodisiac (f; BOU; CRC; HHB; KAB; SPI; SUW); Astringent (f; CRC; PED); Cardiotonic (1; APA; CAN); Carminative (f; APA; CRC; GHA; SUW; WOI); Caspase-3 Inhibitor (1; X16205934); Chemopreventive (1; X15936223); Cholagogue (f; BGB); Contraceptive (f; BOW); Demulcent (f12; CAN; PNC); Digestive (f; APA; VAD); Diuretic (f1; APA; CAN; CRC); Ecbolic (f; CRC); Emmenagogue (f1; CRC; FNF; KAB); Emollient (f12; APA; BOU; CAN; PH2; PNC); Estrogenic (1; JAD); Expectorant (f; CAN; CRC; SPI); Febrifuge (1; X15374601); Hepatoprotective (f1; VAD); Hydragogue (1; PH2); Hyperemic (2; KOM); Hypocholesterolemic (2; BGB; BRU; CRC; PNC), Hypoglycemic (f12; BRU; CAN; PHR; PH2; SKY; X15738612); Hypolipemic (f; VAD); Hypotriglyceridemic (1; BGB); Immunostimulant (1; PR15:257); Insectifuge (1; GHA; UPW; WOI); Lactagogue (f1; CRC; JAD; PHR; PH2; WOI); Laxative (f12; BOU; PNC; VAD); Lipogenic (f; BOU); Lipolytic (1; BRU; PH2; PNC); Mastogenic (f1; FNF; HAD); Mucilaginous (f1; PED); Myorelaxant (1; PNC); Nematicide (1; PR15:538); Orexigenic (12; APA; CAN; GHA; PH2; VAD); Oxytocic (1; PNC); Parturient (f; BGB); Propecic (f; MAD; NAD); Protease Inhibitor (1; HH2); Retinoprotectant (1; X16205934); Secretolytic (2; KOM); Spermicide (1; CRC); Steroid Precursor (1; JAD); Stimulant (f; BOU); Tonic (f; BOU; CRC; PED; WOI); Ureopoeitic (f; NAD); Uterotonic (f1; APA; CAN; PED); Vermifuge (1; CRC; PR15:538); Vulnerary (f1; CAN; PHR; TAD).

Indications (Fenugreek):

Abscess (f; VAD; WOI); Acne (f; VAD); Adenopathy (f; CRC; HHB); Aging (f; BOW); Alactea (f1; PH2; WOI); Allergy (f; PED); Alopecia (1; APA; KAP; MAD); Anemia (f1; BOU; GMH; SPI; VAD); Anorexia (f12; APA; CAN; KOM; PH2; JAC7:405); Aposteme (f; JLH); Arthrosis (1; KOM); Atherosclerosis (1; BGB; SKY); Backache (f; BOW); Bacteria (1; WOI; X15331344); Blepharosis (f; VAD); Boil (f; BGB; GMH; KAP); Bronchosis (f; APA; BOU; GHA; PH2); Burn (f; CRC; IHB); Calculus (1; APA); Cancer (f1; APA); Cancer, abdomen (f1; FNF; JLH); Cancer, bladder (f1; FNF; JLH); Cancer, breast (f1; FNF; JLH; X15936223); Cancer, cervix (f1; BOW); Cancer, colon (f1; FNF; JLH); Cancer, eye (f1; FNF; JLH); Cancer, gland (f1; FNF; JLH); Cancer, groin (f1; FNF; JLH); Cancer, intestine (f1; FNF; JLH); Cancer, kidney (f1; FNF; JLH); Cancer, liver (f1; FNF; JLH); Cancer, parotid (f1; FNF; JLH); Cancer, rectum (f1; FNF; JLH; MAD); Cancer, spleen (f1; FNF; JLH); Cancer, stomach (f1; FNF; JLH); Cancer, testes (f1; FNF; JLH); Cancer, throat (f1; FNF; JLH); Cancer, uterus (f1; FNF; JLH); Cancer, uvula (f1; FNF; JLH); Carbuncle (f; GMH; KAP); Cardiopathy (f1; HJP); Catarrh (f; PH2); Cellulitis (f; BOW); Cervicosis (f; BOW); Chafing (f; BIB); Chickenpox (1; DAA); Childbirth (f; BGB; BPW; GHA); Chilblain (f; CRC); Childbirth (f; BOU); Cholecystosis (f; CRC); Colic (f; GHA; KAP); Colitis (f; PH2); Conjunctivosis (f; VAD); Constipation (f1; VAD); Cough (f; APA; GHA; PED; PH2); Constipation (f1; SKY; SPI); Cystosis (f1; FNF; JLH; SKJ); Debility (f; MAD); Dermatosis (f12; APA; BOU; CRC; PHR; PH2); Diabetes (f12; APA; BRU; CAN; CRC; X15738612), Diarrhea (f1; APA; CRC; KAP); Dropsy (f; NAD); Dysentery (f; HHB; KAB; SUW); Dysgeuzia (f; KAB); Dysmenorrhea (f; BGB); Dyspepsia (f12; APA; CAN; PHR); Dystrophy (f; UPW); Eczema (f1; BGB; PHR; PH2; VAD); Edema (f; BGB; SKJ);

Enteralgia (f; APA; CRC); Edema (f; BOW); Enterosis (f; BGB; BOU; PH2; WOI); Exhaustion (f; MAD); Fever (f1; APA; BOU; CRC; PH2; X15374601); Fistula (f; CRC); Fracture (f; HJP); Furunculosis (f; BGB; HHB; PHR; VAD), Gas (f1; APA); Gastrosis (f; APA; BGB; BOU; CAN; GMH); Gonorrhea (f; UPW); Gout (f; BGB; CAN; CRC; GMH); Hay Fever (f; PED); Helicobacter (1; X15331344); Hemorrhoid (f; MAD; NAD); Hepatosis (f; CRC; JLH; KAP); Hernia (f; APA; BGB; CRC; PH2); High Blood Pressure (f1; CAN; HJP), High Cholesterol (2; APA; BRU; CAN; SKY); High Triglycerides (1; BGB; SKY); Hyperlipidemia (1; BGB); Impotence (f; APA; CRC; PH2); Impotence (f; DAA); Infection (1; APA; WOI; X15331344); Inflammation (f12; APA; BRU; KOM; PH2; X15374601); Itch (f; BOU); Ischemia (1; X16205934); Kidney stone (1; JEB26:249); Labor (f1; APA); Leprosy (f; UPW); Leukorrhea (f; KAP); Lymphadenitis (f; BGB; CAN); Mastosis (f; JLH); Muscular Dystrophy (f; UPW); Myalgia (f; BGB; CAN); Nematode (1; PR15:538); Nephrosis (f; APA; CRC; JLH); Neuralgia (f; APA; CRC); Neurasthenia (f; BOW; GMH; NAD); Ophthalmia (f; JLH; VAD); Orchosis (f; JLH); Osteomyelitis (f; HHB; MAD); Oxaluria (1; APA); Pain (f1; HJP; PH2; TAD; X15374601); Parotosis (f; JLH); Pharyngosis (f; VAD); PMS (f; BGB); Proctosis (f; JLH; MAD); Rachitis (f; MAD); Respirosis (f; APA; PH2); Retinosis (1; X16205934); Rheumatism (f; APA; CRC); Rickets (f; NAD); Sciatica (f; CRC); Scrofula (f; GMH; HHB); Smallpox (1; DAA; KAB; KAP; SUW); Sore (f; APA; BGB; MAD; PH2); Sore Throat (f1; APA; CRC; MAD); Spermatorrhea (f; BOW); Splenosis (f; HHB; KAP); Splenomegaly (f; BOU; CRC; KAB); Sprain (f; GHA); Stomatosis (f; APA); Stone (1; JEB26:249); Swelling (f; HHB; KAB; PHR); Syndrome-X (1; SYN); Syphilis (f; SKJ); Tonsilosis (f; BOU); Tuberculosis (f; APA; CRC; HHB; MAD; SPI); Tumor (f; CRC); Typhoid (f; HJP); Ulcer (f1; APA; PNC; X15331344); Ulcus cruris (f; HHB); Uterosis (f; BOU; JLH; NAD); Vaginosis (f; BGB; NAD; VAD); Venereal Disease (f; SKJ); Virus (1; DAA); Vomiting (f; PH2); Worm (1; PR15:538); Wound (f; BGB; HHB).

Dosages (Fenugreek):

FNFF = !!

Leaves and pods eaten as spice and vegetable; seeds and leaves used to make tea; seeds used to adulterate or flavor coffee; to make artificial maple syrup flavoring; ground seeds used in curry powder; sprouted seed braised in oil and eaten with parboiled cardoon; Yemeni add seeds to gravies and soups; sprouts important in Near Eastern lamb stews flavored with honey. Europeans add seed to hay, especially old hay, to make it more savory to the animals. Seeds are used as an adulterant in or substitute for coffee. Fenugreek also has wide use in the Middle and Near East today. According to Philips, the fenugreek is almost as popular in Lebanon today as the peanut is in America as a snack. In Beirut, they make a mush from the green seed after soaking, forming a fenugreek "milkshake" (FAC, GHA; HJP; TAN; EB54:155). Lydia Pinkham's Vegetable Compound contained 12 oz fenugreek seed, 8 oz unicorn root (*Aletris*), and 6 oz each of life root (*Senecio*), black cohosh (*Cimicifuga*), and pleurisy root (*Asclepias*) in enough alcohol to make 100 pints of compound. One Tbsp mashed seed/8 oz water, up to 3 ×/day as gargle (APA); 1–6 g seed, 3 ×/day (CAN); 50 g powdered seed with 1/4 l water (HH2). One 620-mg capsule 2–3 ×/day (NH); 1/4–1/2 cup seed (PED); 6–12 g dry seed (PED); 9 g dry seed, 6.3 g/day (MAD). 5–90 g seed/day (SKY).

- Arabians use seeds boiled with dry dates and figs for bronchosis and cough (GHA).
- Arabians give seeds with boiled water and egg to new mothers for 7 days after birth (GHA).
- Iranians infuse the seed for menorrhagia (BIB).
- Lebanese take fenugreek "milkshake" in Lebanon for hypertension (HJP).
- Lebanese use as a poultice, and for diabetes, dyspepsia, fever, fractures, heart trouble, lung problems, and typhoid, using the root more often than the herb for pain and rheumatism (HJP).

- Middle Easterners respect fenugreek as both preventive and panacea (HJP).
- North Africans use seed flour boiled with saltpeter and vinegar for splenomegaly (BOU).

Downsides (Fenugreek):

Class 2b (AHP, 1997). Health hazards or side effects following the proper administration of designated therapeutic dosages are not known (PH2). Contraindicated in pregnancy (f; PH2); Commission E reports no contraindications or interactions for oral use of the seed. Adverse skin reactions to repeated external use (AEH). An idiosyncratic gastroenterosis was blamed on fenugreek in LRNP (July 1987). Commission E reports no contraindications or interactions for oral use of the seed. Adverse effects: skin reactions to repeated external use (AEH). One micromastic female complained of mastogenic activity following ingestion of fenugreek sprouts. Newall, Anderson, and Phillipson (1996) caution that because of its oxytoxic and uterine stimulant actions, *in vitro*, its use in pregnancy and lactation is to be avoided. May interfere with anticoagulant, blood sugar, hormonal, and/or MAOI therapies. High mucilaginous/fiber content may interfere with absorption of other drugs. Because all herbs, spices, and food plants, like drugs, contain pharmacologically active ingredients, it seems it could go without saying that "the presence of pharmacologically active constituents would suggest that excessive ingestion is inadvisable" (CAN).

Extracts (Fenugreek):

Treating rats with seed for 4 weeks significantly decreased the quantity of calcium oxalate deposited in the kidneys, supporting Saudi folklore (JEB26:249). Leaf extract antinociceptive at 1000–2000 mg/kg ipr, but the LD50 was circa 4000 mg/kg ipr (X9406901). Alcoholic seed extract LD50 = 5000 mg/kg orl rat, 2000 der rbt (CAN); oxytocic; uterotonic. Trigonelline antihypertensive, antiinflammatory, diuretic, hypoglycemic (CAN); converted to niacin and other pyridines and pyrroles when cooked (PED). Methanol-soluble fraction showed high (>92%) nematicidal activity (PR15:538).

WHEAT (*TRITICUM AESTIVUM* L.) +++ POACEAE

Synonyms:

Triticum sativum Lam.; *Triticum vulgare* Vill.; *Triticum spelta* L.; and, although not synonymous, closely related durum wheat, *Triticum durum*, and emmer, *Triticum dicoccum*

Notes (Wheat):

> **And he slept and dreamed the second time: and, behold, seven ears of <u>corn</u> came up upon one stalk, rank and good.**
>
> **Genesis 41:5 (KJV)**

> **And he fell asleep and dreamed a second time; and behold, seven ears of <u>grain</u>, plump and good, were growing on one stalk.**
>
> **Genesis 41:5 (RSV)**

> **However he went back to sleep and dreamed a second time; and here there were seven ears of <u>grain</u>, coming up on one stalk, fat and good.**
>
> **Genesis 41:5 (NWT)**

FIGURE 1.113 Wheat (*Triticum aestivum*). Source: BIB.

The biblical term "corn" is synonymous with grain; it does not refer to Indian maize, but usually to wheat, the most common cereal. Corn in those days often included as a mixture peas, beans, lentils, cumin, barley, millet, and spelt. Egypt was the chief granary of the Roman empire. As late as 1952, Moldenke and Moldenke say, "Even today Arabia imports all its wheat from Egypt, and the caravans which leave Egypt for the Red Sea are laden with precious wheat." In my first biblical book (BIB), I acceded to the Moldenkes and figured that wheats (*Triticum aestivum*) and spelts

(*Triticum spelta*) were what was intended in the biblical passages on corn and wheat. As so often, Israeli botanist Michael Zohary rather refutes those possibilities. Yes, it was wheat, but tetraploid durum wheat, *Triticum durum*, and emmer, *Triticum dicoccum*, rather than *Tricitum aestivum* or *Triticum spelta*. The latter do not even grow in Israel (ZOH). So what? From a medicinal point of view, they are pretty much equally good or bad (if you have celiac). Zohary adds that the two tetraploids were abundantly cultivated in Israel and vicinity, the durum wheat still dominant in Israel agriculture. Not only were these free wheats (free meaning without hull) important in bread (although rich in gluten), but also offered to God in tribute as, for example, the Showbread wheat (Hebrew *hittah*), probably included in the general terms bar (Genesis 41:49), *dagan* (Numbers 18:27), and *kamah* (Judges 15:5), as well as *avur*, carmel, geresh, and omer, scattered throughout the Old Testament. Zohary concluded that the ancestral species was *Triticum dicoccoides*, also native to Israel. "Domestication of wheat … took place about 8000 years ago in one or more of the most primitive agricultural villages of the Assyrian mountains (Jarmo in Iraq), and probably in the Land of Israel as wel." (ZOH).

Common Names (Wheat):

Ble (Fr.; Haiti; AVP; EFS); Ble Ordinaire (Fr.; USN); Bread Wheat (Eng.; USN); Bugday (Tur.; EFS); Chhow (Newari; NPM); Cherbi (Sunwar; NPM); Civitella (It.; EFS); Cluster Wheat (Eng.; USN); Common Wheat (Eng.; Ocn.; AH2); Corn (Eng.; AVP; BIB); Formento (It.; AVP); Froment (Fr.; AVP; EFS; USN); Frumento (It.; AVP); Gahu (Bom.; Mah.; Kon.; NAD); Gahum (Mar.; WOI); Gahumg (Mar.; WOI); Gam (Beng.; NAD; WOI); Gahun (Bhojpuri; Chepang; Danuwar; Gurung; Magar; Mooshar; Nepal; Tharu; NPM); Ganam (Afg.; KAB); Gandam (Afg.; Iran; EFS; KAB); Gandham (Hindi; WOI); Gawn (Guj.; WOI); Gehu (Guj.; NAD); Gehun (Hindu; Nwp. EFS; KAB; NAD); Gendum (Mal.; WOI); Getreide (Ger.; AVP); Ghavum (Guj.; WOI); Giun (Beng.; Hindi; WOI); Godamba (Mal.; WOI); Godhuma (Sanskrit; EFS; NAD); Godi (Kan.; NAD); Godumai (Tam.; WOI); Godumay (Tam.; NAD); Godumbyarisi (Tam.; WOI); Godumulu (Tel.; NAD); Gom (Beng.; WOI); Goodhumalu (Tel.; NAD); Govum (Guj.; WOI); Grano (It.; AVP); Grano Grosso (It.; EFS); Gro (Tibet; NPM; TIB); Hintah (Arab.; EFS; NAD); Hsiao Mai (China; EFS); Huede (Den.; AVP); Hvete (Swe.; AVP); Kakyo (Lepcha; NPM); Kameh (Arab.; AVP); Kanak (Hindi; WOI); Kanik (Sin.; NAD); Ko Mugi (Japan; USN); Korn (Ger.; AVP); Kotanpam (Mal.; WOI); Kotampum (Mal.; NAD); Kwa (Tamang; NPM); Lumcha (Rai; NPM); Marzuolo (It.; EFS); Oun (Tur.; AVP); Pscheniza (Rus.; AVP); Pszenica (Pol.; AVP); Saatweizen (Ger.; USN); Si (Limbu; NPM); Ta (Sherpa; NPM); Tarw (Dutch; AVP); Tarwe (Dutch; EFS); Trigo (Por.; Sp.; AVP; ROE); Trigo Blandeal (Sp.; USN); Trigo Candeal (Ger.; EFS); Trigo Chamoro (Sp.; EFS); Trigo Commun (Por.; AVP); Waizen (Ger.; AVP); Weisen (Ger.; AVP); Weizen (Ger.; EFS; USN); Wheat (Eng.; Scn.; AH2; CR2); Xiao Mai (Pin.; DAA); Yava (Sanskrit; EFS; NAD).

Activities (Wheat):

Antibilious (f; BIB); Antiinflammatory (f1; VAD); Antiseptic (f; BIB); Antivinous (f; BIB); Aphrodisiac (f; KAB); Apopotic (1; X12681494); Bulking (1; SHT); Demulcent (f; BIB); Discutient (f; BIB); Emollient (f; BIB; ROE); Hypocalcemic (1; VAD); Hypocholesterolemic (1; VAD); Hypoglycemic (1; VAD); Hypolipemic (1; VAD); Lactagogue (f; ROE); Laxative (2; BGB; PH2; KAB; SHT); Lipolytic (1; PH2); Orexigenic (f; KAB); Peristaltic (1; PH2); Refrigerant (f; TIB); Sedative (f; BIB); Vulnerary (f; BIB).

Indications (Wheat):

Abscess (f; ROE); Acne (f; ROE); Adenopathy (f; JLH); Aegilops (f; PH2); Aposteme (f; JLH); Arteriosclerosis (f; VAD); Arthrosis (f; JLH); Biliousness (f; KAB); Bleeding (f; ROE); Burn (f; BIB); Callus (f; JLH); Cancer (f1; BIB; JLH); Cancer, abdomen (f1; JLH); Cancer, breast (f1; JLH); Cancer, colon (f1; JLH); Cancer, foot (f1; JLH); Cancer, joint (f1; JLH); Cancer, parotid (f1; JLH); Cancer, sinax (f1; JLH); Cancer, skin (f1; JLH); Cancer, spleen (f1; JLH); Cancer, testicle (f1; JLH); Cancer, tonsil (f1; JLH); Cancer, uterus (f1; JLH); Cardiopathy (f; BIB); Childbirth (f; ROE); Colitis (2; BGB; SHT); Condyloma (f; PH2); Constipation (f12; BGB; BIB; PH2; SHT); Corn (f; JLH); Cramp (f; ROE); Dermatosis (f1; PH2; ROE; VAD); Diabetes (f; VAD); Diarrhea (f; BIB; ROE); Diverticulitis (12; BIB; SHT; VAD); Dysentery (f; BIB); Ecchymosis (f; BIB); Enuresis (f; ROE); Epistaxis (f; BIB); Fever (f; BIB; ROE); Flu (f; ROE); Flux (f; BIB); Fracture (f; BIB; ROE); Gonorrhea (f; ROE); Gravel (f; BIB); Hematuria (f; BIB); Hemoptysis (f; BIB); Hemorrhage (f; BIB); Hemorrhoid (f; ROE); IBS (12; BGB; VAD); Impotence (f; KAB); Incontinence (f; BIB); Induration (f; JLH); Inflammation (f1; PH2; VAD); Itch (f; PH2); Leprosy (f; BIB); Leukorrhea (f; BIB); Mastosis (f; JLH); Metrorrhagia (f; BIB); Myalgia (f1; VAD); Neurasthenia (f; BIB); Night sweat (f; BIB); Obesity (f; VAD); Odontosis (f; ROE); Orchosis (f; JLH); Osteosis (f; BIB; ROE); Pain (f; JLH; ROE; VAD); Pimple (f; ROE); Scald (f; BIB); Smallpox (f; BIB); Sore (f; JLH; ROE); Sprain (f; ROE); Sting (f; BIB); Stone (f1; VAD); Sunstroke (f; BIB), Swelling (f; BIB); Syphilis (f; BIB); Tonsilosis (f; JLH); Tuberculosis (f; BIB; KAB; ROE); Ulcer (f; JLH); Urolithiosis (f1; VAD); Venereal Disease (f; BIB); Wart (f; JLH; ROE); Whitlow Cancer (f; JLH); Wound (f; BIB).

Dosages (Wheat):

FNFF =!!!
15 g bran 2 ×/day (SHT); 15–40 g bran 1–2 ×/day (PH2).

- Algerians use flour for diarrhea, fractures, metrorrhagia, and syphilis, the bran for scorpion stings.
- Ayurvedics consider the seed antibilious, aphrodisiac, laxative, orexigenic, and tonic (KAB).
- Chinese use roasted grains for sweating, especially in women with tuberculosis (KAB).
- Lebanese recommend the bran for bones, constipation, and antiseptic dressing, "claiming it was empirical penicillin" (HJP).
- Spaniards suggest the bran (salvado) and/or flour as demulcent, laxative, hypocalcemic, hypocholesterolemic, hypoglycemic, hypolipemic, satiating, and useful for arthrosis, bruises, catarrh, dermatosis, hematomas, myalgia, and tonsilitis (VAD).

Downsides (Wheat):

None at proper dosage (PH2). Bulking agents should not be taken by those with stenotic lesions of the GI tract. May lead to bowel obstruction if fluid intake is inadequate. Do not take bulking agents when lying down or at bedtime. Do not use with antiperistaltics (such as, for exasmple, loperamide) (SHT).

Natural History (Wheat):

Wheats are attacked by many fungi and other organisms. Some cultivars are resistant to the various rusts, smuts, and virus diseases. The most important fungal diseases of wheats are the following. Extension agents should be consulted concerning diseases in an area before growing wheat. Also, cultivars should be selected for growing that are resistant to such diseases. Fungal diseases of wheat include Rusts (Stem or Black rust, *Puccinia graiminis* f. sp. *tritici*; Leaf or Brown rust, *P. recondita*; Stripe or Yellow rust, *P. glumarum*); Smuts (Bunt or Covered smut, *Tilletia caries* and *T. foetida*; Dwarf Loose smut, *Ustilago tritici*); Mildews (Downy mildew, *Sclerospora macrospora*; Powdery mildew, *Erysiphe graminis* f. sp. *tritici*); Root rots (Common root rot, *Helminthosporium* spp. and *Fusarium* spp.; Take-all root rot, *Ophiobolus graminis*; Browning root rot, *Pythium* spp.); Foot rots (Eye spot, *Cercosporella herpotrichoides*; Snow mold, *Fusarium* spp.); Blights and Scabs (Head blight or scab, *Fusarium* spp.; *Rhizoctonia* blight, *Rhizoctonia* spp.; *Typhula* blight, *Typhula* spp.; Anthracose, *Colletotrichum graminicola*; Kernel smudge, *Helminthosporium* spp., *Alternaria* spp.); Blotches (Glume blotch, *Septoria nodorum*; Leaf blotch, *S. tritici*); Speckled leaf disease, *Leptosphaeria avenaria* f. sp. *triticea*; Ergot, *Claviceps purpurea*. Diseases caused by bacteria include the following: *Pseudomonas atrofaciens* (Basal glume rot or Bacterial black-tip) and *Xanthomonas transluscens* f. sp. *undulosa* (Black shaff). Diseases caused by viruses include the following: Wheat mosaic, Wheat streak mosaic, Wheat striate mosaic, and Yellow dwarf. Insect pests encountered in various areas include English grain aphid, the most common aphid affecting wheat, attacking the heads and being very damaging when populations become high prior to the late-dough stage. Other insects and cutworms, darkling beetles, hessian fly, and salt marsh caterpillars may cause damage during the seedling stage. A great number of species of nematodes have been isolated from wheats in various parts of the world. Where nematodes are a problem, the agricultural agent should be consulted (HOE).

Extracts (Wheat):

Nystrom et al. (2005) note that steryl ferulates run 300–390 ppm in wheat bran, constituting 20–25% of total sterols. Known to lower LDL cholesterol, steryl ferulates also have antiinflammatory, antioxidant, antiradicular, and antitumor activities. If metabolized more effectively than rice counterparts, steryl ferulates of wheat and rye may have more benefits than gamma-oryzanol. Alk(en)ylresorcinols, folates, lignans, phytosterols, and tocopherols tend to be localized in the outer parts of the kernel, especially the bran (X15796586). For seed, nut, and whole grain lovers, Phillips et al. analyzed the phytosterols in wheat germ, some such phytosterols medicinally important (X16302759). Delta-5-avenasterol, 161 ppm; delta-7-avenasterol, 133 ppm; campestanol, 127 ppm; campesterol, 787 ppm; phytosterols, 4130 ppm; poriferasta-7,25-dienol, 135 ppm; sitostanol, 69 ppm; beta-sitosterol, 2286 ppm; stigmasterol, 37 ppm; and delta-7-stigmastenol, 116 ppm (X16302759).

TULIP (*TULIPA* SPP.) + LILIACEAE

Notes (Generic Tulip):

The <u>flowers</u> appear on the earth; the time of the singing of birds is come, and the voice of the turtle is heard in our land.

Song of Solomon 2:12 (KJV)

The <u>flowers</u> appear on the earth, the time of singing has come, and the voice of the turtledove is heard in our land.

Song of Solomon 2:12 (RSV)

<u>Blossoms</u> themselves have appeared in the land, the very time of vine trimming, and the voice of the turtle dove itself has been heard in our land.

Song of Solomon 2:12 (NWT)

FIGURE 1.114 Tulip (*Tulipa* sp.).

Zohary offers that the mountain tulip (*Tulipa montana* Lindl.), with its wide array of colors, is probably of the Hebrew *nitzanin* (Arabic, *nissan*), the group of flowers or blossoms mentioned in Canticles. Zohary adds that even in Iraq, these groups of blossoms are recognized, as perhaps we here speak of our spring flowers, which do not bloom coincidentally but rather sequentially, In Iraq and the Holy Land, it is a group of sequential red flowers, beginning with the crown anemone and ending in the common poppy, today, as in biblical time. So we have three slightly different views of biblical phenology above. Of course there are close to 100 species of tulip, and nobody, not even Zohary, is sure that *Tulipa montana*, indeed any tulip, was intended in this passage. But he implies that the mountain tulip and/or other Israeli species have been involved in the evolved horticultural tulip.

Common Names (Mountain Tulip):

Mountain Tulip (Eng.; ZOH); Nissan (Arabic; ZOH); Nitzanin (Heb.; ZOH); Nscn.

Activities (Generic Tulip):

Allergenic (1; WOI); Alterative (f; DAW); Antibiotic (1; WOI); Antiseptic (1; WOI); Bacteristat (1; WOI); Cardiotoxic (1; WOI); Depurative (f; DAW); DNA-Synthesis Inhibitor (1; X3592627); Fungicide (1; X4850341); Hemagglutinin (1; X3595592); Mitogenic (1; X1814629); Mutagenic (1; X3088443); Poison (f; DAW); Tonic (f; DAW); Vasculoprotective (1; X10190195).

Indications (Generic Tulip):

Abscess (f; DAW); Bacillus (1; WOI); Bacteria (1; WOI); Bite (f; DAW); Bronchosis (f; HHB); Bubo (f; DAW); Cancer (f; DAW; JLH); Cancer, breast (f; DAW; JLH); Cancer, lung (f; DAW; JLH); Cancer, stomach (f; DAW; JLH); Fungus (1; X4850341); Gastrosis (f; DAW); Infection (1; WOI); Mastosis (f; DAW); Mucosis (f; HHB); Mycosis (1; X4850341); Pulmonosis (f; DAW; JLH); Scrofula (f; DAW); Struma (f; DAW); Urogenitosis (f; DAW; JLH); Varicosity (1; X10190195).

Dosages (Generic Tulip):

FNFF = !
Facciola mentions only the flowers, tasting like beans or peas, eaten as a garnish in salads, or consumed with chicken or tuna salad, cottage cheese, fruit sorbets, mixed vegetables, or mousses. Although viewed as poisonous, bulbs are eaten as food during times of scarcity (WOI); bulbs not mentioned by Facciola (FAC).

CATTAIL (*TYPHA* SPP.) +++ TYPHACEAE

Notes (Cattail):

And they smote him on the head with a reed, and did spit upon him, and bowing their knees worshipped him.

Mark 15:19 (KJV)

And they struck his head with a reed, and spat upon him, and they knelt down in homage to him.

Mark 15:19 (RSV)

Also they would hit him on the head with a reed and spit upon him, and, bending their knees, they would do obeissance to him.

Mark 15:19 (NWT)

I am once again taking the generic approach with *Typha*, but most of the entries are for Middle Eastern or tropical species. Zohary hints that the Hebrew *suf* often represents a collective word for aquatic plants, like reeds, rushes, and water weeds; but Exodus 3:3 and Isaiah 19:6 can be translated directly to *Typha*, *T. australis* being the most common species in Israel. *Typha* dominates many of the ditches and tributaries of the Nile in lower Egypt. BOU and GHA references below apply to *T. domingensis*, JLH to *T. angustifolia*, AVP and EGG to both.

Many old paintings depicting Jesus's mock trial, picture him with the cattail in his hand as a scepter. The leaves are plaited into such articles as ropes, winnowing trays, mats, and also are employed as caulking. The silky florets of spikes are used for stuffing and tinder. Ashes are sometimes used as a salt substitute. Fruiting spikes with oil serve for illumination. The pollen was used during emergency as an absorbent in surgery.

FIGURE 1.115 Cattail (*Typha* sp).

Common Names (Cattail):

Akaioud (Ber.; BOU); Barda (Arab.; GHA); Bardi (Arab.; BOU); Batbat (Tur.; EB51:195); Berdi (Arab.; BOU); Bout (Arab.; BOU); Candle Wick (Eng.; BUR; EAS); Capim de Esteira (Por.; AVP); Cat-O-Nine-Tails (Eng.; EAS); Cattail (Eng.; TAN); Cooper's Reed (Eng.; BUR); Cossack Asparagus (Eng.; EAS); Cumbungi (?; FAC); Deis (Arab.; BOU); Enea (Dr.; Peru; Sp.; AVP; EGG); Eneas (Pr.; AVP); Erva de Esteira (Por.; AVP); Espandaña (Chile; JLH); Flag Tule (Eng.; BUR); Grand Jonc (Haiti; AVP); Great Reed Mace (Eng.; AAH; BUR); Gros Jonc (Haiti; AHL); Hagla (?; FAC); Himegama (Japan; TAN); Huaricolla (Peru; Sp.; EGG); Jonc (Haiti; AHL); Junco (Dr.; AHL; AVP); Ksohi (Baby's coat) (Hocak; WIN); Léche (His.; AHL); Macios (Cuba; AVP); Marsh Beetle (Eng.; EAS); Massette (Fr.; BOU); Massette des Étangs (Fr.; BOU); Nataf (Arab.; Yemen; GHA); Pun (?; FAC); Reed (Eng.; BIB); Reed Mace (Eng.; TAN); Riche (His.; AHL); Roseau (Fr.; BOU); Small Bulrush (Eng.; BOU); Tababuia (Por.; AVP); Taboa (Por.; AVP); Tabu (Por.; AVP); Tabua (Por.; AVP; TAN); Tabuda (Ber.; BOU); Taheli (Ber.; BOU); Totora (Arg.; Chile; Peru; EGG; JLH); Tupai (Egypt; ZOH); Ugin (Ber.; BOU); Wicihu (Leaves mat) (Hocak; WIN).

Activities (Cattail):

Antiaggregant (f; DAW); Antiinflammatory (f; DEM); Antiseptic (1; DEM); Aphrodisiac (f; AHL; KAB; EB29:7); Astringent (1; BIB; EB29:20); Cicatrizant (f; EGG); Circulostimulant (f; ZUL); Cyanogenic (1; BIB); Depurative (f; ZUL); Detersive (f; EB29:20); Diuretic (f1; AHL; BIB; WBB; ZUL; EB29:20); Emetic (f; DEM; EB29:24); Febrifuge (f; EB29:7); Hemostat (f1; BIB; EB29:20); Lactagogue (f; DAW); Litholytic (f; DEM); Orexigenic (f; UPW); Poison (f; DAW; EB29:24); Purgative (f; EB29:24); Refrigerant (f; DAW); Sedative (f; DAW); Sterilizant (f; EB31:305); Stimulant (f; EB29:20); Suppurative (f; DAW); Uterocontractant (f; ZUL); Uterotonic (f; EB29:20); Vermifuge (f; DAW); Vulnerary (f; DAW).

Indications (Cattail):

Amenorrhea (f; DAW); Anodyne (f; DAW); Anorexia (f; UPW); Bite (f; EB29:7); Bleeding (1; BIB; EB29:20); Boil (f; DEM); Bruise (f; DAW); Burn (f; BUR; DEM; GHA; EB29:20); Cancer (f; DEM); Carbuncle (f; DEM); Chafing (f; DEM); Childbirth (f; DAW; ZUL); Circulosis (f; ZUL); Cramp (f; DEM); Cyanogenic (f1; EB30:400); Cyst (f; DEM); Cystosis (f; ZUL); Dermatosis (f; DEM); Diarrhea (f; DEM; ZUL; EB29:20); Dropsy (f; EB29:20); Dysentery (f; KAB; WOI; EB29:20); Dysuria (f; EB29:20); Ecchymosis (f; DAW); Enterosis (f; BUR; DAW; DEM); Epilepsy (f; AAH; BIB; EB29:7; EB24:265); Epistaxis (f; DAW); Erysipelas (f; FEL; EB29:7); Fever (f; EB29:7); Gastrosis (f; BUR; DEM); Gonorrhea (f; DEM; FEL; WOI; EB29:20); Gravel (f; DEM); Hematemesis (f; DAW); Hematochezia (f; DAW; EB29:20); Hematuria (f; DAW); Hemoptysis (f; DAW); Hemorrhoid (f; DAW); Impotence (f; DAW; EB29:20); Infection (f; DEM); Infertility (f; ZUL); Inflammation (f; DEM; EB29:20); Insanity (f; EB29:7); Kidney stone (f; DEM); Leprosy (f; DEM); Leucorrhea (f; DAW); Madness (f; BIB EB24:265); Mastosis (f; DAW; DEM); Measles (f; KAB; WOI; EB29:20); Metrorrhagia (f; DAW; ZUL); Metroxenia (f; DAW); Mucososis (f; BUR); Nephrosis (f; ZUL); Ophthalmia (f; FEL; KAB; EB29:20); Pain (f; DAW); Pertussis (f; DEM); Proctosis (f; EB29:7); Scald (f; DEM); Smallpox (f; DEM); Snakebite (f; EB29:7); Sore (f; DEM; KAB; EB29:20); Splenosis (f; KAB); Sprain (f; ZUL); Stone (f; DEM); Strangury (f; KAB); Swelling (f; DAW; FEL); Thrush (f; EB29:7; EB29:20); Toothache (f; AAH); Tumor (f; BIB; FEL; EB29:20); Ulcer (f; DAW); Urethrosis (f; ZUL); Uterosis (f; EB29:20); Vaginosis (f; DAW); Venereal Disease (f; BUR; DAW; DEM); Worm (f; DAW); Wound (f1; BIB; DAW; KAB; EB29:20); Yeast (f; EB29:7). Very few of these folk uses have proven out, but I would not hesitate to try any for any of these indications, if nothing else were available. Many folk medicines prove to have good phytochemical rationales when analyzed.

Dosages (Generic Catail):

FNFF = 1

Facciola lists five edible species of *Typha*. Young shoots, inflorescence, tender leaves, and rhizomes are eaten in various ways. Flowers and anthers are made into a sweetmeat. The sweet and soft marrow of the immature spike is considered a delicacy. Pollen is used to make bread or porridge. My first bag of cattail pollen, when taken indoors, resulted in the window pane being clouded with thousands of thrips (BIB; EAS; FAC).

- Arabians apply dried crushed flowers to cool or soothe burns (GHA).
- Chileans use decoction of *T. angustifolia* roots to wash tumors (JLH).
- Gaelic's, calling the plant what translates to "fairy wives' spindle," gathered the plant on a midsummer midnight wrapping it in a shroud, to prevent epilepsy and all other diseases (AAH).
- Hispaniolans suggest the root is aphrodisiac.
- Irish somehow use the plant for toothache (AAH).
- North Africans apply ashes of the rhizome to wounds to stop bleeding (BOU).
- Peruvians treat burns with hairs from the flower spikes (EGG)
- Peruvians make a cicatrizant pomade from cattail charcoal with oil (EGG).
- Peruvians use the buds as astringent and diuretic (EGG).

Natural History (Generic Cattail):

Most cattails are partially or entirely self-fertilized. Their own pollen sheds from above. A spider (*Clubiona riparia*) uses the leaf tip for both nursery and coffin. This sac spider folds down the leaf tip to make an enclosure fastened and lined with silk. Inside, it deposits its egg sac and simply remains there, dying inside its cage. The first meal of its young will be the mother's body. Sap-sucking leaf feeders include about a dozen species of aphids. The cattail borer moth (*Bellura obliqua*) mines downward in the leaf, eating out the transverse partitions and finally exiting through a hole at the end of its mine. Later, it bores into the stem. Related species include the white-tailed diver (*B. gortynoides*), the pickerelweed borer (*B. densa*), and the oblong sedge borer (*Archanara oblonga*). The smartweed caterpillar (*Acronicta oblinita*) feeds on the leaves. Some insects feed only on and in the flower and seed spikes. Larvae include *Dicymolomia julianalis*, a pyralid moth, which later bores into the stem. A neat, bite-sized chunk from a brown seed spike may indicate the white-veined dagger moth caterpillar, also called Henry's marsh moth (*Simyra henrici*); later, the tufted caterpillar makes a vertically aligned cocoon in a folded leaf. Syrphid flies may cluster head downward on the male spike (aphid eaters). Small, brownish adult cattail moths (*Limnaecia phragmitella*, also called the shy cosmet, a cosmopterygid) laying eggs on maturing female spikes. A dying cattail stem may signal that a stalk borer fed in the leaves or seed spike. Other common borers include snout beetles such as *Sphenophorus pertinax*, a billbug; and *Suphisellus puncticollis*, a burrowing water beetle. Red-winged blackbirds have nested in my cattail bog for years now. In cold weather, they may forage on the batons, probably more seeking large caterpillars or grubs rather than small seed. (Eastman estimates 220,000 seeds per spike. A single seed may produce a rhizomal growth some 10 feet in diameter with a hundred clonal shoots.) Few birds really eat the seed. Martin et al. list teal, geese, and sandpipers, gulf coast blue geese sometimes having as much as 25 to 50% seed and/or rootstocks. Teal may sometimes have more than a thousand seeds in their gut. Attesting to the intelligence of birds, Eastman speculates that bluejays have learned to cache corn grains in old cattail sausage tied together by silk of caterpillar, keeping their corn in relatively "dry storage" that way. Painted turtles eat the seeds and stems. Muskrats often dislodge the plants as they feed on cattail, its primary food. Many muskrats will attract the muskrat predator, mink (EAS; MZN). I once caught a mink's foot in one of my traps — no longer do I trap.

Boost your immune system and think positively if you study long in the cattail swamps. There can be significant midsummer populations of the cattail mosquito (*Coquillettidia*), the chief vector of eastern equine encephalitis. Larvae overwinter attached to roots of cattails and other aquatic plants. Adults get the virus from birds (which are unaffected) and transmit the virus when they bite humans and horses (EAS). (Sounds too much like the bird flu coming soon from Asia; Echinacea, elderberry, and garlic, here I come!). The cucumber mosaic virus has been reported from *Typha angustifolia*, the wheat streak mosiac from *T. latifolia*. Among the fungus diseases on *Typha latifolia* are *Cladosporium*, *Cryptomela typhae*, *Didymosphaeria typhae*, *Gloeosporium* sp., *Guignardia* sp., *Hendersonia typhae*, *Heterosporium maculatum*, *Hymenopsis hydrophila*, *Leptosphaeria* spp., *Leptothyrium typhina*, *Lophodermium typhinum*, *Mycosphaerella typhae*, *Ophiobolus* sp., *Phoma orthosticha*, *Phyllosticta typhina*, *Pleospora typhae*, *Pythiogeton autossytum*, *Pythium helicoides*, *Sclerotium hydrophilum*, *Scolecotrichum typhae*, *Stagonospora typhoidearum*, and *Typhula latissima*. The nematode *Meloidogyne* sp. is also reported (HOE).

HAIRY ELM (*ULMUS CANESCENS* MELV.) ++ ULMACEAE

Notes (Hairy Elm):

He heweth him down cedars, and taketh the cypress and the oak, which he strengtheneth for himself among the trees of the forest: he planteth an ash, and the rain doth nourish it.

Isaiah 44:14 (KJV)

He cuts down cedars; or he chooses a holm tree or an oak and lets it grow strong among the trees of the forest; he plants a cedar and the rain nourishes it.

Isaiah 44:14 (RSV)

There is one whose business is to cut down cedars; and he takes a certain species of tree, even a massive tree, and he lets it become strong for himself among the trees of the forest. He planted the laurel tree, and the pouring rain itself keeps making it get big.

Isaiah 44:14 (NWT)

For linguistic and ecological reasons, Zohary criticizes the translations of the Hebrew words *vegeshem yegadel* as "and the rain nourishes it" as thematically unfounded and illogical. He says geshem must also be a species of tree — one of the five species contained in the quoted passage.

This is the only elm species mentioned by Zohary, along shady creeks in Lower Galilee, Mt. Carmel, and Samaria, and Zohary suggests that this species is among those five trees (FP1; ZOH). So, as a fan of slippery elm, I include this nearly empty entry for *Ulmus canescens*, fleshing it out with a generic entry, including data below for several elm species, but mostly slippery elm data for *U. campestris* (BOU and JNP), VAD for *U. carpinifolia*. None of the activities and indications have thus far been reported for the Israeli species — to the best of my knowledge.

Common Names (Hairy Elm):

Hairy Elm (Eng.; ZOH); Geshem (Heb.; ZOH); Neshem (Arabic; ZOH); Nscn.

Activities (Generic Elm):

Antioxidant (1; CRC); Antiseptic (f; CRC); Antitussive (f1; CAN); Astringent (f1; APA; PED; WAM); Demulcent (f12; APA; PH2; PNC; WAM); Depurative (f; DEM); Diuretic (f; CRC; EFS; GMH); Emollient (f1; CRC; PH2; PNC); Expectorant (f; CRC; GMH; PED); Febrifuge (f; EFS);

Laxative (f; CRC; DEM); Stimulant (f; EFS); Sudorific (f; EFS); Tonic (f; EFS); Vermifuge (f; CRC); Vulnerary (f; CRC; GMH; PED).

Indications (Generic Elm):

Abscesses (f; CAN; FAD); Adenopathy (f; CRC; DEM; PH2); Bleeding (f; CEB; DEM); Blepharosis (f; VAD); Boils (f1; APA; CRC; GMH; PNC); Bronchosis (f; CRC); Bruise (f; FEL); Burn (f1; APA; FAD; GMH; PH2; WAM); Cancer (f; CRC; FEL; JLH); Carcinoma (f; CRC); Cardiopathy (f; GMH); Caries (f; CRC); Catarrh (f; CRC; DEM; GMH); Chilblain (f; CEB); Childbirth (f; CRC; DEM); Cholera (f; CEB); Cold (f; SKY); Cold Sore (1; APA); Colitis (f1; APA; CAN; CRC; GMH); Conjunctivosis (f; CRC; DEM); Constipation (f; CRC); Corneosis (f; VAD); Cough (12; APA; FAD; HHB; WAM); Cramp (f; CEB; CRC); Crohn's Disease (1; SKY); Cuts (f; FAD); Cystosis (f1; GMH; WAM); Dermatosis (f1; APA; PH2; PNC; VAD; WAM); Diarrhea (f1; APA; CAN; FAD); Diverticulosis (1; FNF); Duodenosis (f; PH2); Dysentery (f; CRC; FAD); Dyspepsia (f1; FAD); Dysuria (f; CRC); Eczema (f; CRC); Enterosis (12; APA; CEB; GMH); Erysipelas (f; FEL); Esophagitis (2; APA); Felon (f; CRC; JLH); Fever (f; CRC; DAW); Fistula (f; FEL); Fracture (f; CRC; DEM); Gangrene (f; CRC); Gastrosis (12; APA; GMH; PHR; PH2; SKY); Gleet (f; FEL); Gonorrhea (f; DEM); Gout (f; CRC; HH2; PH2); Heartburn (f; DEM; JAH2(2):45); Hemoptisis (f; CEB; GMH); Hemorrhoid (f; CEB); Herpes (f; CRC; FEL); Infection (f; CEB); Inflammation (12; APA; CEB; WAM); Itch (f; VAD); Labor (f; DEM); Leprosy (f; HJP); Mastosis (f; DEM); Mucososis (f; CRC); Nephrosis (f; DEM; FEL); Ophthalmia (f; DEM); Parotosis (f; FEL); Periodontosis (f; VAD); Pharyngosis (f12; APA; VAD); Pleurisy (f; CRC; FAD; GMH); Proctosis (f; FEL); Pruritis (f; VAD); Pulmonosis (f; CRC; DEM); Quinsy (f; DEM); Rheumatism (f; CRC; HH2; PH2); Scurvy (f1; DAW; FNF); Sore (f1; APA; FEL); Sore Throat (f12; APA; FAD; FEL; VAD; WAM); Stomach Distress (f1; FAD; PNC); Stomatosis (f12; APA; VAD); Strangury (f; FEL); Swelling (f; CEB; HHB); Synovitis (f; GMH); Syphilis (f; CRC; FEL); Tapeworm (f; CRC); Tenesmus (f; FEL); Toothache (f; CRC; GMH); Tuberculosis (f; CRC; DEM; GMH); Tumor (f; CRC; JLH); Typhoid (f; CRC; GMH; HJP); Ulcers (f1; APA; CAN; FAD); UTI (f; CRC; GMH); Vaginosis (f; VAD); Venereal Disease (f; CEB; CRC; DEM; FEL); Virus (f; FEL); Whitlow (f; CRC; JLH); Worm (f; CRC); Wound (f1; APA; PHR; PH2; WAM).

Dosages (Generic Elm):

FNFF = !

Oral tradition suggests that Washington's troops survived in midwinter with the inner bark of slippery elm (JAD). Facciola states that native Americans cooked it with buffalo fat, giving flavor and preventing rancidity (sounds like deep fried "elmbark chips" instead of potato chips); Kiowa Indians brewed it into a nutritive tea. Used as a meal for breading fish. Also used to fortify health beverages like smoothies (DEM; FAC), 1–3 tsp powdered herb/cup water/1–3 ×/day (APA); two 340-mg capsules as needed (APA); 4 g powdered bark in 500 ml decoction, 3 ×/day (CAN); 5 ml liquid extract (1:1 in 60% ethanol) 3 ×/day (CAN); 4–16 ml 1:8 powdered bark decoction 3 ×/day (CAN; HHB); 3 Tbsp inner bark/cup water (FAD); 1 oz powdered bark/pint water (FEL); 0.5–2 g powdered bark/cup, 2–3 ×/day; two 340–500-mg capsules as needed (JAD); 1/4–1/2 cup fresh bark (PED); 2–4 tsp dry bark (PED); 3 tsp dry bark/4 cups boiling water (PED); 1–2 g bark in tea 3–4 ×/day (SKY); 5 ml tincture 3 ×/day (SKY).

- Lebanese use the bark of *U. campestris* as we use slippery elm to make a slimy beverage, with pungent plants, for colds, dermatosis, dysentery, lungs, and throat (HJP).
- North Africans consider the bark of *U. campestris* as astringent, diuretic, emollient, resolvent, stimulant, and sudorific (BOU).
- Spaniards use bark of *U. carpinifolia* rather like we use slippery elm, internally for diarrhea and externally for blepharitis, corneal ulcers, dermatosis, erythema, inflammation, itch, periodontosis, pharyngitis, pruritis, sores, and vaginitis (VAD).

NETTLE (*URTICA DIOICA* L.) +++ URTICACEAE

Notes (Nettle):

Therefore as I live, saith the LORD of hosts, the God of Israel, Surely Moab shall be as Sodom, and the children of Ammon as Gomorrah, even the breeding of nettles, and saltpits, and a perpetual desolation: the residue of my people shall spoil them, and the remnant of my people shall possess them.

Zephaniah 2:9 (KJV)

"Therefore, as I live," says the LORD of hosts, the God of Israel, "Moab shall become like Sodom, and the Ammonites like Gomor'rah, a land possessed by nettles and salt pits, and a waste for ever. The remnant of my people shall plunder them, and the survivors of my nation shall possess them."

Zephaniah 2:9 (RSV)

"Therefore, as I am alive," is the utterance of Jehovah of armies, the God of Israel, "Mo'ab herself will become just like Sodom, and the sons of Ammon like Gomor'rah, a place possessed by nettles and a salt pit, and a desolate waste, even to time indefinite. The remaining ones of my people will plunder them, and the remnant of my nation will take possession of them."

Zephaniah 2:9 (NWT)

Yes, it is nettles in all three versions I have on hand (KJV, RSV, and NWT). So they all agree that it is nettles. As to which species, lacking voucher specimens, I cannot outguess Michael Zohary, Israeli botanist, who tends to rule out the nettle I know best, *Urtica dioica*, questionably reported from Jericho. He speculates that it is one of the Israeli species he reports in the *Flora of Palestine*.

- Stipules, 2 on each node (each of 1 connate pair), green:
  - — Pistillate racemes globular, long stalked; leaves ovate. tp 7 cm wide: *U. pilulifera*
  - — Bisexual racemes spicate; leaves narrower, lanceolate to ovate: *U. urens*
- Stipules, 4 on each node, membranous:
  - — Annuals no more than 50 cm tall; staminate rhachis winged: *U. dubia*
  - — Perennials 70 to 100 cm tall; staminate rhachis wingless: *U. hulensis*

Allen and Hatfield say that no plants, except perhaps dandelion, dock, and elder, are as important medicinally as the nettle in the British Isles. Such feelings may underline the rhyme: "Three nettles in May keeps all diseases away" (AAH). MAD entries may apply as well to *Urtica urens*, the smaller dog nettle. Ditto for EFS entries. I doubt that many of the narrators, scribes, or translators of the Bible recognized the intricate differences among the nettles, the *Urtica dioica*, or the four closely related species cited by Zohary (HP1). So I suggest that the Zohary names cited below could apply to any of the species here mentioned. The North African names are similarly merged, but were referred to as *U. urens* (Small Nettle) or *U. pilulifera* (Roman Nettle).

Common Names (Nettle):

Acalphye (?; JLH); Akalyphe (?; JLH); Andjira (Arab.; BOU); Azekdon (Ber.; BOU); Bent-en-Nar (Arab.; BOU); Bichu (India; EFS); Bichhu booti (Hindi; ?); Big String Nettle (Eng.; EAS); Bobatsi (Suso; KAB); Bou Khsas (Arab.; BOU); Bou Zeqdouf (Arab.; BOU); Braennaetsla (Swe.; KAB); Brandnekel (Afrikan; KAB); Brandnetel (Dutch; EFS); Brennessel (Ger.; EFS; MAD); Brier (Eng.; BIB; ZOH); Canada Nettle (Eng.; BUR); Caslán (Hun.; EFS); Chule (Sunwar; NPM); Chutle

FIGURE 1.116 Nettle (*Urtica dioica*).

(Magar; NPM); Cincar (Tur.; EB51:195); Cizlagan (Tur.; EB54:155); Common Nettle (Eng.; BUR); Dhyo (Magar; NPM); Dicirgin (Tur.; EB49:406); Dirik (Tur.; EB51:195); Estelkraut (Ger.; KAB); Gijirtken (Tur.; EB51:195); Great Stinging Nettle (Eng.; BUR); Groot Brandenetel (Dutch; KAB); Harrous (Ber.; BOU); Harul (Heb.; ZOH); Horreig (Arab.; Isr.; ZOH); Horreiq (Arab.; BOU); Hsieh Tzu Ts'ao (China; EFS); Hyo (Magar; NPM); Iherriquet (Ber.; BOU); Imereksin (Ber.; BOU);

Imezri (Ber.; BOU); Isirgan (Tur.; EFS; EB49:406); Isirgan otu (Tur.; EB54:155); Jhaduk (Sherpa; NPM); Jincari (Tur.; EB49:406); Kajyang (Lepcha; NPM); Ketskan (Baskir; KAB); Kirtken (Kirghiz; KAB); Krapiva (Rus.; KAB); Naetle (?; JLH); Neantog (Ire.; KAB); Nebat-en-Nar (Arab.; BOU); Naughty Man's Plaything (Eng.; EAS); Nelau (Chepang; NPM); Nesslyn (JLH); Nettle (Eng.; CR2; TAN); Nhyakan (Newari; NPM); Nockainen (Fin.; KAB); Ortica Maggiore (It.; KAB); Orticone (It.; EFS); Ortie (Fr.; EFS); Ortiga Grossa (Cat.; KAB); Ortiga Maior (Por.; KAB); Ortiga Mayor (Sp.; KAB); Ortuge (Fr.; KAB); Pokrzywa (Pol.; KAB); Polo (Gurung; Tamang; NPM); Pulu (Gurung; NPM); Qorreis (Arab.; BOU); Satu (Tibet; NPM); Seravim (Heb.; ZOH); Sha'ar el 'Agouz (Arab.; BOU); Sikya (Limbu; NPM); Sirpad (Heb.; ZOH); Sisna (Tharu; NPM); Sisnu (Bhojpuri; Danuwar; Nepal; NPM; SUW); Soi (Kas.; MKK); Singing nettle (Eng.; TAN); Sorbei (Arab.; Egypt; ZOH); Stor Broendenelde (Den.; KAB); Stor Naelde (Den.; EFS); Syak (Sherpa; NPM); Tall Nettle (Eng.; EAS); T'an Ma (China; EFS); Tezzount (Ber.; BOU); Timezrit (Ber.; BOU); Urtigão (Por.; EFS); Urzica (Rom.; KAB); Za chhag (Tibet; NPM); Zwa (Tibet; NPM).

Activities (Nettle):

Analgesic (f1; CAN; DEM; PH2; X15013182); Anesthetic (f1; PH2); Anthelmintic (f; CRC; KAB; NPM; SUW); Antiadrenaline (1; FAD); Antiaging (f; NP9(2):10); Antiallergic (f1; MAB); Antiaromatase (1; SHT); Antiarthritic (f1; PH2); Antiasthmatic (fl; CRC; DAW; NPM); Anticancer (f1; JLH; MAB); Anticomplementary (1; HH3); Anticonvulsant (1; CAN); Antiedemic (f1; FT68:387; MAB); Antiexudative (1; HH3); Antihemorrhagic (f; CAN); Antihistaminic (1; WAM); AntiHIV (1; PH2); Antihidrotic (f; MAD); Antihyperglycemic (1; FT74:677); Antiinflammatory (f1; FT68:387; MAB; PH2; X11962753); Antileukotriene (1; PP2); Antioxidant (1; X15013182); Antiperoxidant (1; X15013182); Antiproliferant (1; NP9(2):10); Antiprostatitic (2; KOM; MAB); Antiradicular (1; X15013182); Antirheumatic (f12; MAB; PH2; X11950004); Antiseptic (f1; BUR; CRC; PED); Antispasmodic (f; PED); Antitumor (f1; PED); Antiulcer (1; X15013182); Antiviral (1; MAB; FT68:387); Aphrodisiac (f; MAD); Aquaretic (1; SHT); Aromatase Inhibitor (1; HH3); Astringent (f1; CRC; MAB; PNC; SUW); Bactericide (1; FAD; MAB; WOI); Bitter (f; PED); Bradycardic (1; CAN; FT74:677); Choleretic (f; NP9(2):10); CNS Depressant (1; FAD); CVI (f1; APA; BGB); Cyclooxygenase Inhibitor (1; MAB; PH2); Cytotoxic (1; MAB); Depurative (f; BIB; FAD; MAB; PED); Diuretic (f12; CRC; PHR; PH2; PNC; SUW; NP9(2):10); Elastase Inhibitor (1; MAB); Emmenagogue (f; APA; CRC; EFS; KAB; PED; SUW); Expectorant (f; MAD; PED); Fungicide (1; HH3; MAB); Hematogenic (f1; EFS; FAD; PH2; WAM); Hemostat (f1; CAN; MAB; MAD; PED); Hepatotonic (f; BRU; NP9(2):10); Histaminic (1; FNF); Hyperglycemic (1; APA; CAN); Hypoglycemic (f1; CAN; PNC; FT74:677; NP9(2):10; EB49:406); Hypotensive (f1; CAN; NP9(2):10); FT74:677); Hypothermic (f1; CAN); Immunosupressant (1; X11950004); Impotence (f; TGP); Insecticide (f; NPM); Insectifuge (f; AAH); Insulinogenic (1; NP9(2):10); Interferonigenic (1; CAN); Lactagogue (f1; APA; CRC; HAD; MAD; NMH); Laxative (f; BGB); 5-Lipoxygenase Inhibitor (1; MAB; PH2); Litholytic (f12; MAD; NP9(2):10); Mastogenic (1; HAD); Metal Chelator (1; X15013182); Metalloproteinase Inhibitor (1; NP9(2):10; X11962753); Mitogenic (f; FAD); Myorelaxant (f; BGB); Natriuretic (1; FT74:677); Pancreatonic (1; FNF); Purgative (f; EFS); Rubefacient (f; CRC); Stimulant (f; EFS); Tonic (f; MAB; PNC); Uterotonic (1; APA; CAN); Vasoconstrictor (f; BIB; CRC); Vasorelaxant (1; NP9(2):10); Vermifuge (f; BGB; CRC; PED); Vulnerary (f; MAD).

Indications (Nettle):

Acne (f; BGB; FEL); Adenoma (1; BGB; SHT); Adenopathy (f; BIB; JLH); Ague (f; DEM; MAB); Alactia (f; CRC; MAD); Allergy (f1; BGB; HH3; MAB; WAM); Alopecia (f; APA; WOI); Amenorrhea (f; KAB); Anemia (f1; CRC; FAD; WAM); Arthrosis (f12; DEM; FAD; MAB; PH2; NP9(2):10; X10911825; X11962753; X11950004); Asthma (f1; CRC; DAW; MAB; NPM); Ataxia (f; DEM); Biliousness (f; NPM); Bladder stones (f12; PHR; PH2; NP9(2):10); Bleeding (f1; CRC; DEM; FEL);

Boil (f; NPM); BPH (root) (12; BGB; KOM; MAB; PH2; NP9(2):10); Bronchosis (f1; CRC; MAB; PED); Bug bites (1; MAB); Burns (f1; BGB; CRC; MAB); Cachexia (f; KAB); Calculus (f; CRC); Cancer (f; CRC; FAD); Cancer, breast (f1; CRC; JLH); Cancer, ear (f1; CRC; JLH); Cancer, feet (f1; JLH); Cancer, lung (f1; CRC; JLH); Cancer, mouth (f1; CRC; JLH); Cancer, prostate (f1; NP9(2):10; X15254411); Cancer, rib (f; JLH); Cancer, spleen (f1; CRC; JLH); Cancer, stomach (f1; CRC; JLH); Cancer, womb (f1; CRC; JLH); Cardiopathy (f; AAH); Carcinoma (f; BIB); Caries (f; NPM); Catarrh (f; WOI); Childbirth (f; DEM); Cholangitis (f; CRC); Cholecystosis (f; CRC; FAD; MAB; WOI); Cholera (f; FEL); Cold (f; AAH; CEB; NPM); Colic (f; CRC); Colitis (f; FEL; MAB); Congestion (f; APA); Constipation (f; CRC; WOI); Consumption (f1; BUR; MAB; SUW); Corn (f; AAH); Cough (f; AAH; NPM); Cramp (f; AAH; MAD); CVI (1; BGB); Cystosis (f; FEL); Dandruff (f; PH2; WOI); Dermatosis (f1; BGB; CAN; MAB; FT74:677); Diabetes (f1; CRC; MAD; PH2; FT74:677; EB49:406); Diarrhea (f1; BGB; BUR; FAD; FEL; MAB); Dislocation (f; NPM); Dropsy (f; AAH; BGB; CRC); Dysentery (f1; BUR; CRC; FAD; MAB); Dysmenorrhea (f; BGB; APA; MAD; PED); Dyspepsia (f; DEM; MAD); Dyspnea (f; CRC; KAB); Dysuria (2; KOM; PHR; PH2; SHT); Eczema (f; BGB; CAN; MAB; MAD); Edema (f; CRC; PH2); Endothelioma (f; BIB; JLH); Enterosis (f; FEL); Epilepsy (f; AAH); Epistaxis (f1; AAH; BGB; CAN; KAB; MAB); Epithelioma (f; BIB; JLH); Erysipelas (f; CRC); Erythema (f; CRC); Escherichia (1; WOI); Exanthema (f; MAD); Fever (f1; CAN; CEB; NPM); Flu (f; PH2); Fungus (f; AAH); Gastrosis (f; CRC); Goiter (1; MAB); Gonorrhea (f; BIB; CRC); Gout (f1; FAD; MAB; PH2; NP9(2):10); Gravel (f12; BGB; BUR; KOM; MAD; PHR); Hayfever (2; APA; MAB); Headache (f; AAH; CRC); Hematemesis (f; CEB); Hematuria (f; SUW); Hemoptysis (f; CRC); Hemorrhage (f1; PNC; FT74:677); Hemorrhoids (f; AAH; BGB; DEM; PED); Hepatosis (f; HH3); Herpes (f; BGB; X15814267); HIV (1; PH2); High Blood Pressure (f1; CAN; NP9(2):10; FT74:677); Hives (f; DEM); Hoarseness (f; CEB); Impotence (f; CEB); Infection (f; AAH); Inflammation (f1; BGB; CRC; PH2; X11962753); Itch (f; DEM); Jaundice (f; BUR; CRC; KAB; PED; SUW); Kidney stones (f12; APA; PHR; PH2; NP9(2):10); Lethargy (f; KAB); Leukorrhea (f; CRC; MAD); Malaria (f; BIB; CEB; CRC; KAB); Melaena (1; CAN); Menorrhagia (f; SUW); Mycosis (f; AAH); Myocardiopathy (1; BGB); Myosis (f; MAB); Nephrosis (f; BUR; CRC; FEL; HH3; PED; SUW); Neuralgia (f; APA; BIB; CRC); Nocturia (1; MAB); Obesity (f; BUR; CEB); Ophthalmia (f; AAH); Orchosis (f; CEB); Osteoarthrosis (f1; MAB); Osteoporosis (1; JAD); Otosis (f; MAD); Pain (f12; BUR; EB49:406; NP9(2):10; X10911825; X15013182); Palsy (f; CEB; CRC; KAB); Paralysis (f; BUR; CRC); Parotitis (f; AAH; BUB; JLH); Parturition (f; APA; BGB); Pertussis (f; BIB; CRC); Pharyngosis (f; MAB); Pimple (f; AAH); Pleurisy (f; BGB; CEB); Pollakisuria (1; BGB); Polyp (f; BIB; JLH); Pregnancy (f; SKY); Prostatosis (12; PH2; SHT; X15045190); Puerperium (f; NPM); Pulmonosis (f; CEB); Rash (f; AAH); Rheumatism (f12; FAD; KOM; PHR; PH2; NP9(2):10; EB51:195; X10911825; X11962753); Rhinosis (1; BGB; HH3; MAB); Ringworm (f; AAH); Sarcoma (f; BIB; JLH); Sciatica (f1; CRC; KAB; MAB); Seborrhea (f1; BRU); Shigella (1; WOI); Shingle (f; AAH); Snakebite (f; EB49:406); Sore (f; CEB); Sore Throat (f; CRC); Splenosis (f; CRC; FAD; JLH); Sprain (f; APA; SKJ); Sting (f; CRC); Stitch (f; MAD); Stomachache (f; DEM); Stomatosis (f; MAB); Stone (f12; KOM; MAD; PHR; PH2; SHT); Swelling (f; AAH; BIB; DEM); Tendinitis (f; APA); Toothache (f; CEB); Tuberculosis (f; CRC; KAB); Tumor (f; CRC; JLH); Ulcer (1; X15013182); Uremia (f; BIB); Urticaria (f1; MAB); Uterorosis (f; BGB; APA; CAN; KAB); UTI (2; PHR; KOM; PH2; SHT); Uvulosis (f; CEB); Vaginosis (f; APA); Venereal Disease (f; BIB; CRC); Vertigo (f; BIB; CRC); Virus (1; PH2); Worm (f; AAH; BGB; NPM); Wound (f; MAB; NPM).

Dosages (Nettle):

FNFF = !!!

Greens widely eaten, but only after cooking disarms the sting. Some make nettle pudding, adding broccoli, leeks, and rice. Some make nettle beer; British wrap their Cornish Yarg cheese in nettle leaves; leaf juice serves as rennet; dried leaves (they lose their sting in drying) used to make herb tea. In Scotland, nettles are combined with leeks or onions, broccoli or cabbage, and rice, boiled in a muslin bag and served with butter or gravy. Nettle beer and nettle tea are made by some people. Dried nettles can be fed to livestock and poultry, but few animals will eat the living plants (BIB; FAC; TAN; EB54:155). 3–4 tsp (4–6 g) shoot or leaf in 150 ml boiling water cooled, 3–4 ×/day(APA; MAD); 4–6 g/day root (APA); 3–4 tsp (circa 4 g) shoot/cup water/several ×/day (APA); 2–4 g dry herb, or in tea, 3 ×/day (CAN); 3–4 ml liquid herb extract (1:1 in 25% ethanol) 3 ×/day (CAN); 2–6 ml root tincture (1:5 in 45% ethanol) 3 ×/day (CAN); 8–12 g herb; 4–6 g root (KOM); 8–12 g dry herb/day (MAB); 9 g/day leaf for arthrosis (MAB); 4–6 g/day dry root (MAB); 3–6 ml/day fluid herb extract (1:2) (MAB); 4–9 ml/day fluid root extract (1:2) (MAB); 3–6 ml/day fluid herb extract (1:2) (MAB); 3–6 g/day root or 600–1200 mg/day 5:1 extract for BPH (MAB); 7–14 ml/day herb tincture (1:2) (MAB); 125 g juice (MAD); 3–4 tsp (circa 4.8 g) herb in hot tea (MAD); 4–6 g root/day; one or two 475-mg capsules 2 to 3 ×/day; one 450-mg StX capsules 2 ×/day (NH); 3–6 g dry leaf (PED); 4.5 g dry leaf:22 ml alcohol/23 ml water (PED); 2.5–5 ml liquid herb extract (PNC); 8–12 g dry herb/day (SHT); 4–6 g powdered root/cup water (WIC).

- Algerians mix powdered nettles with powdered jasmine for gonorrhea (BIB).
- Carolinans suggest the root for consumption, diarrhea, dysentery, gravel, hemorrhoids, jaundice, nephrosis, and pain (BUR).
- Czechs poultice the herb onto cancers (JLH).
- Devonshire locals use nettle top tea for urticaria (KAB).
- French use nettle roots steeped in vinegar for tumors of the feet and spleen, steeped in honey for tumors in lungs or ribs (JLH).
- Irish drink nettle tea to clear measle rash (AAH).
- Italians use stinging nettle (and elderberry and parietaria) for herpes zoster (X15814267).
- Russians self-urticate to energize tired muscles (KAB).
- Russians use for cholangitis, cholecystitis, constipation, dysmenorrhea, hepatitis, and jaundice (CRC; HJP).
- Herb decoction taken for cold, cough, rheumatism, and stomachache (EB51:195).
- Nettle roots crushed with vinegar for swellings of the feet or spleen (CEB).
- Nettle juice as a mouthwash for swollen uvula (CEB).
- Nettle juice boiled lightly with sugar; 2 oz taken orally for bleeding piles (CEB).
- Nettle seed with honey (or nettle juice) for cold, cough, gastrosis, orchosis, and swellings (CEB).
- 1 Tbsp seeds with jam or honey for impotence (CEB, where we read that nettle seeds in wine excite to games of love).
- Seed (and flower) tincture 1 tsp 3–4 ×/day for ague and malaria (CEB).
- Seeds boiled in wine for orchosis (CEB).
- Seeds crushed in honey for pustules on the lung, side ache, and swelling of the ribs (CEB).
- "Seed of Nettle stirreth up lust, especially drunk with Cute (thickened must) ..." (Gerarde as quoted in CEB).
- 12 to 15 seeds, 3 ×/dayfor goiter (or bigneck) (CEB).

Downsides (Nettle):

Class 1 (AHP, 1997). No health hazards or side effects known with proper therapeutic dosages (PH2). None known for herb; rare GI upsets for roots (KOM). Herbage contraindicated in fluid retention due to reduced cardiac or renal activity, rarely causing allergic reactions (PHR). Adverse effects of root: mild GI complaints (occasionally) (AEH). Occasional mild GI complaints after root ingestion. The urtication can be painful and long-lasting, in some inducing a black-and-blue reaction. No fatalities are reported in the United States. Newall, Anderson, and Phillipson (1996) caution amines are an irritant. Because it is a reputed abortifacient and to affect the menstrual cycle, its use in pregnancy and lactation should be avoided. May interfere with blood pressure, CNS, and diabetes medications (CAN). Being a nettle fan, I had never heard of it before and was reluctant when my friend Vic said the root tea almost did him in. It is almost as if he read the book, "Consumption of nettle tea has caused gastric irritation, a burning sensation of the skin, oedema, and oliguria" (CAN). Not for use in severely allergic patients, especially those with tendency toward anaphylaxis (WAM). Schulz et al. (1998) report on more than 4000 patients taking 600 to 1200 mg extract/day for 6 months. Only 35 showed side effects, 0.65% GI complaints, 9 (0.19%) dermatosis, and 2 (>0.05%) reporting hyperhidrosis (SHT). No contraindications are stated (SHT). Varro Tyler cautions against self-medication with BPH. Whenever treating BPH, a practitioner should be involved. Baseline levels of PSA should be established before considering an herbal treatment (JAD). Even JAMA (Journal of the American Medical Association) admits that there is no hard proof for any intervention in BPH; because hospitals kill 200,000 Americans a year, and prostate cancer fewer than 50,000, I will opt for nettle tea, pomegranate juice, sitosterol-rich nuts, especially pumpkin seed, and selenium-rich Brazil nuts as the foods of choice for prostate protection.

Extracts (Nettle):

Analgesic; Anticonvulsant; Bradycardic; CNS Depressant; Hemostatic; Hyperglycemic (PNC; FT68:387); Hypoglycemic (FT68:387); Hypotensive; Hypothermic; Pancreatonic (FT68:387); LD50 = 3625 mg/kg ipr mus (CAN FT68:387). Infusion LD50 = 1929 mg/kg ivn rat. HOH extract LD50 = 1721 mg/kg ivn rat. The tea was well tolerated at levels of 1310 mg/kg orally (Bombardelli and Morazzoni, 1997). LD50 infusion = 1310 orl rat (MAB); (9Z-11E)-13-Hydroxy-9,11-octadecadienoic acid, 14 octacosanol, oleanolic acid, secoisolariciresinol, and ursolic acid are listed as weak to moderate aromatase inhibitors found in the methanolic root extract. Aromatase is a key enzyme in steroid hormone metabolism, and its inhibition may partially explain the activity of the roots in BPH. The polysaccharide fraction of the aqueous root extract show prolonged antiedemic and antiinflammatory activity (40 mg/kg orl rat). Ethanolic extract also inhibits elastase, a destructive enzyme in the inflammatory process (IC50 = 68 μg/ml). The isolectin, abundant in the roots, may contribute to the antiinflammatory and antiprostatic activity of the extracts. Aqueous extracts inhibit dose dependently (0.6–10 mg/ml the binding of dihydrotestosterone to SHBG with specific receptors on human prostatic membranes). The alcoholic extract, isolectin, and stigmast-4-en-3-one were inactive. At concentrations of 0.1 mg/ml, some root extracts inhibited Na^+, K^+-ATPases 27.6 to 81.5%. Stigmast-4-en-3-one, stigmasterol, and campesterol inhibited Na^+, K^+-ATPases 23 to 67% at concentrations of 1 to 1000 μM. Such inhibition may influence prostate cell metabolism and growth (Bombardelli and Morazzoni, 1997). Root polysaccharide extracts anticomplementary (IC50 ≤ 50 μg/ml HH3). Strange that an herb should inject so many neuroactive compounds — acetylcholine, choline, formic acid, histamine, leukotrienes, serotonin (PH2) — into unsuspecting grazers. Talk about splitting hairs; *Hager's Handbuch* (1998) says that each hair of *U. dioica* contains 0.1 to 0.2 μg acetylcholine; 0.01 μg histamine; and 5 ng serotonin, while hairs of *U. urens* contain 53 ng acetylcholine, 5 ng histamine, and, if I translate correctly, circa 0.15 pg leukotriene-B4; 0.3 pg leukotriene C4 + D4. Earlier, Madaus reported secretin (MAD), but I do not think that is the same as the secretin being studied in autism. I assume that there is also some choline involved. One overindulgent physician speculated on one of my Amazonian tours that the histamine injected by the nettle sting generated an antihistaminic reaction, some of which went to the sting and some to arthritic hot spots.

VETIVER (*VETIVERIA ZIZANIOIDES* (L.) NASH EX SMALL) +++ POACEAE

Synonyms:

Andropogon muricatus Retz.; *Andropogon squarrosus* Cooke; *Andropogon zizanioides* Urb.; *Vetiveria odorata* Virey

FIGURE 1.117 Vetiver (*Vetiveria zizanioides*).

Notes (Vetiver):

Dan also and Javan going to and fro occupied in thy fairs: bright iron, cassia, and calamus, were in thy market.

Ezekiel 27:19 (KJV)

[A]nd wine from Uzal they exchanged for your wares; wrought iron, cassia, and calamus were bartered for your merchandise.

Ezekiel 27:19 (RSV)

Vedan and Javan from Uzal — for your stores they gave. Iron in wrought works, cassia and cane — for your articles of exchange they proved to be.

Ezekiel 27:19 (NWT)

Whether I was right in suggesting vetiver for this biblical calamus or cane, an association not ventured by any of my major sources, it seems to have been an aromatic cane-like vetiver, imported from afar, as the cassia in the same passage. Cassia could have been imported from the Indian subcontinent; so could vetiver; and so could *Acorus calamus*. Moldenke and Moldenke (BIB) identified it with the obscure binomial *Andropogon aromaticus* Roxb., which some authors have equated with *Vetiveria*, others with *Andropogon* (*Cymbopogon*) *schoenanthus* L, one of the "lemon-grass" assemblage; they also suggested *Andropogon muricatus*, now considered a synonym of vetiver. Zohary identifies it with *Cymbopogon*. Suggestions that it might be the calamus of today (*Acorus calamus*) have generally been rejected. *Acorus calamus* did not apparently occur in biblical Palestine (ZOH), and was less likely to have been imported than the lemon-grass or vetiver, to either of which the alternative translation "sweet cane" seems more appropriate. Oil of vetiver is described as one of the most valuable and most important perfumers' raw materials, widely used in perfumes, cosmetics, and for the scenting of soaps. In Hispaniola, the plant is cultivated as a medicinal and aromatic tea material. It serves for making awnings, bags, baskets, fans, mats, pillows, sachets, screens, and sunshades, and is used for thatch in Haiti. Young leaves, not being too aromatic, may serve as fodder (BIB).

Although originally from India, vetiver is a major money crop in Haiti, but is probably more important for holding the soil there. Although one hears a lot about it, as of Groundhog Day 2005, there were fewer than 20 abstracts on it on PubMed. But by June 15, 2005, Wilde et al. had described phytoextraction of lead from firing range soil with vetiver (X15964059). (Mark Dafforn, who for decades has been interested in vetiver, has passed on some common names tidbits that I would like to share. I have included as personal communications from Mark Dafforn, in this account only, with PER.)

Common Names (Vetiver):

Abhaya (Sanskrit; KAB); Aga Wanga (Indonesia); Akar wangi (Malaya; EFS; IHB); Akar Wangu (Malaya); Amranalam (Sanskrit; NAD); Amrinala (Sanskrit; KAB); Anias de Moras (Pamp); Anwunga-ro-gban (Sierra Leone; Temne); Arabian Kuss-Kuss (Eng.; EFS); Arabischer Kostus (Ger.; EFS); Avdaha (Sanskrit; KAB); Avurugaddiveru (Tel.; KAB); Babin (Bambara); Bala (Hindi; Sanskrit; KAB); Balah (Hindi; KAB); Barewali (Sierra Leone; Susu); Baul de Pobre (Pr.; PER); Bena (Hindi; KAB; WOI); Bhanavalo (Kon.; NAD); Bikhiwala (Iran); Birni (Sadani; KAB); Birnijono (Mun.); Botha Grass (Eng.; TAN); Capia (Arg.; Que.; PER); Capim de Boma (Por.; UPW); Capim Vetiver (Por.; UPW); Chiendent des Indes (Fr.; EFS); Chiendent Orient (Fr.; UPW); Chor'dor'de (Nig.; Fulani); Cockroach Grass (Eng.; JFM); Costus arabique (Fr.; EFS); Cus-Cus (Creole; Fr.; Haiti; EFS; VOD); Cuscus (Eng.; UK); Cuscus grass (Eng.; EFS); Dahaharana (Sanskrit; KAB); Dimi (Fulani); Diri (Songhai); Faeg (Laos; Thai); Gandhadhya (Sanskrit; KAB); Ganrar (Hindi; KAB); Giron (Bis); Grama de la India (Sp.; PER); Haripriya (Sanskrit; KAB); Ilamichamver (Tam.; KAB); Ilib (Phil. [Pamp.]); Indragupta (Sanskrit; KAB); Ivarancusa (Ger.; EFS); Izkhir (Arab.; KAB); Jalamoda (Sanskrit; KAB); Jalavasa (Sanskrit; KAB); Janur (Sunda; IHB); Jema (Hausa); Jhoor (India; SKJ); Kaadu (Kan.; WOI); Kaadu Karidappasajje Hallu (Kan.; WOI); Kabenis (Sierra

Leone; Temne; UPW); Kale (Sierra Leone; Susu; UPW); Kamare (Sarakolle); Kansh (Nepal; SUW); Karidappasajje Halu (Kannada); Katayana (Sanskrit; KAB); Khas (Ganrar; Iran; Urdu); Khas Bena (Hindi; NAD); Khas-Khas (Beng.; Hindi; India; Iran; Mar.; Nepal; EFS; KAB); Khiskhus (Eng.; UK); Khus-Khus grass (Eng.; EFS); Kieli (Fulani); Koosa (Eng.; UK); Kulikarili (Ghana); Kulkadere (Gurma); Kuruvaeru (Tel.; WOI); Kuruveeru (Tel.; WOI); Kusu Kusu (Malaya; IHB); Laamanche (Kan.; WOI); Laghubhaya (Sanskrit; KAB); Lamajjaka (Sanskrit; KAB); Lamajjakumuveru (Tel.; KAB); Laraseta (Indonesia); Lara S_tu (Java; IHB); Larawèstu (Java; IHB); Lavancha (Kan.; KAB); Magic Grass (Cameroon); Miracle Grass (Thai); Miyamoe (Burma; KAB); Mora (Phil. [Bik, Bis]); Moras (Phil. [Bik, Bis; Tag]); Moro (Pi.; Tag.; KAB) Mottenwurzel (Ger.; EFS); Muda (Cebu); Mudivala (Kan.; NAD); Naga Setu (Malaya; IHB); Nalada (Sanskrit; KAB); Nara Setu (Malaya; IHB); Nara Wastu (Malaya; Sumatra; Sunda; IHB); Narawasta (Sul; Sunda; IHB); Ngoko Ba (Bambara); Ngongon (Bambara); Ngongonari (Nig.; Fulani); Onei (Ganrar; Hindi; KAB); Pacholi (Pr.; PER); Pachuli (Peru; Por.; Sp.; EGG); Pallol (Fulani); Panni (Ganrar; Hindi; Pun.; KAB; NAD); Pindi (Sierra Leone; Mende); Raiz de Moras (Sp.); Ramaccham (Mal.; Mys.; WOI); Ramachehamver (Mal.; KAB); Ranapriya (Sanskrit; KAB); Rambhu (Sanskrit; KAB); Rarawèstu (Java; IHB); Reshira (Sanskrit; KAB); Rimodas (Phil.; Bis); Rimora (Phil.; Sbl); Rimoras (Phil.; Bik); Roudoum (Mossi); Rumput Wangi (Malaya; IHB); Saivandera (Sinh; KAB); Samagandhika (Sanskrit; KAB); Savandramul (Sin.; KAB); Seenk (India; SKJ); Sembam (Sen.; Tuk.); Sep (Sen.; Wolof); Sevya (Sanskrit; KAB); Shishira (Sanskrit; KAB); Shitamulaka (Sanskrit; KAB); Sirom (Mun.; Sant.; KAB); Sirum (Mun.; KAB); Sirum jono (Mun.; KAB); So'dornde (Nig.; Fulani); So'mayo (Nig.; Fulani); Sugandhimula (Sanskrit; KAB; WOI); Sumare (Mandinga; Sierra Leone; UPW); Swarankusa (Den.; EFS); Tiep (Sen.; Wolof); Tin (Oudh; KAB); Toul (Sen.; Falor); Tropical Sweet Grass (Eng.; JFM); Ucira (India; JLH); Usa (Sumatra; IHB); Usar (Indonesia; Sumatra; Sunda; IHB); Useeur (Sumatra; IHB); Usheera (Sanskrit; EFS); Ushira (Ayu.; Sanskrit; KAB; AH2); Usir (Arab.; KAB); Usira (India; JLH); Vairina (India; JLH); Vala (Cutch; India; Mar.; EFS; KAB; NAD); Valo (Guj.; NAD); Vattiveeru (Kan.; Tel.; KAB; WOI); Veeranam (Sanskrit; NAD); Vetive (Creole; Haiti; VOD); Vetiver (Eng.; Dutch; Tam.; AH2; EFS); Vettiveelu (Tel.; WOI); Vettiveeru (Tel.; WOI); Vettiveeru (Kan.; Mys.; Tel.); Vettiver (Tam.; WOI); Vettiveru (Mal.; WOI); Vetivert (Eng.; EFS); Vettiveru (Mysore; KAB); Vidavaliveru (Tel.; KAB); Vilhalver (Tam.); Vira (Sanskrit; KAB); Virabhadra (Sanskrit; KAB); Virana (Sanskrit; KAB); Viranam (Tam.); Virataru (Sanskrit; KAB); Vitanamulaka (Sanskrit; KAB); Vitivergras (Ger.; EFS); Xiang Geng Sao (China); Yesero mekelakeya (Eth.); Zacate Violeta (Sal.; PER); Zemako (Nig.; Fulani).

Activities (Vetiver):

Abortifacient (f; HOC; UPW; ZUL); Alexeretic (f; KAB); Analgesic (f; HOC; JFM; VOD); Anthelmintic (1; ZUL); Antifeedant (1; X12558100); Antiinflammatory (f; MAF; MHK; NAD); Antiseptic (f1; MAF; JAR12:83; X10438227); Antispasmodic (f; EFS; NAD); Astringent (f; KAB); Bactericide (1; X10438227); Cardiotonic (f; NAD); Carminative (f; JFM; WOI); Cerebrotonic (f; NAD); Diaphoretic (f; EFS; JFM; MAF; SUW; WOI); Diuretic (f; EFS); Emetic (1; MAF); Emmenagogue (f; AHL; EFS; KAB; SUW); Febrifuge (f; AHL; SUW; UPW); Fungicide (1; ZUL; JAR12:83); Hypoglycemic (1; MAF); Insecticide (f; VOD); Insectifuge (f1; JFM; MAF; ZUL; X12558100); Litholytic (f; IHB); Pulifuge (f; MAF); Refrigerant (f; EFS; SUW; WOI); Sedative (f; MAF; KAB); Stomachic (f; EFS; KAB; SUW); Stimulant (f; AHL; SKJ; SUW); Termitifuge (1; X12558100); Tonic (f; EFS; KAB; SUW); Tranquilizer (f; MAF).

Indications (Vetiver):

Amenorrhea (f; KAB); Asthma (f; ZUL); Bacteria (f1; MAF; X10438227); Biliousness (f; IHB; KAB); Bladderstones (f; IHB); Boils (f; MAF; SKJ); Bugbite (f; MAF); Burns (f; MAF; SKJ); Cancer (f; UPW); Cardiopathy (f; NAD); Cerebrosis (f; KAB); Chlamydia (f; MAF); Cholera (f; MAF;

MHK; NAD); Colic (f; WOI); Dandruff (1; JAR12:83); Dermatosis (f1; EGG; MAF; JAR12:83); Diabetes (1; MAF); Eczema (f; MAF); Enterosis (f; HOC; VOD); Epilepsy (f; SKJ); Fever (f; JFM; KAB; MAF); Flu (f; JFM; UPW; ZUL); Fungus (f1; MAF; JAR12:83); Gas (f; WOI); Gastrosis (f; IHB; MAF); Halitosis (f; KAB); Hangover (f1; JAD); Headache (f; JFM; KAB; NAD); Hematosis (f; KAB); Hepatosis (f; AHL); Infection (f1; MAF; JAR12:83); Inflammation (f; MAF; MHK); Insomnia (f; MAF); Lumbago (f; WOI); Malaria (f; MAF; SKJ; ZUL); Mycosis (f1; EGG; MAF; JAR12:83); Nausea (f; MHK); Nesseria (f; MAF); Neuralgia (f; HOC; JFM); Neurosis (f; MAF); Odontosis (f; MAF); Pain (f; MAF; VOD); Palpitation (f; NAD); Parasite (f; VOD); PID (f; MAF); Pleurisy (f; JFM; UPW; ZUL); Puerperium (f; IHB); Rheumatism (f; JFM; WOI); Septicemia (f; MAF); Snakebite (f; SKJ); Spermatorrhea (f; KAB); Sprain (f; WOI); Staphylococcus (f1; MAF; X10438227); Stings (f; SKJ); Stomatosis (f; MAF; SKJ); Toothache (f; MAF); Trichophyton (1; JAR12:83); UTI (f; MAF); Vomiting (f; WOI); Yellow Fever (f; UPW).

Dosages (Vetiver):

FNFF = ?

Khus essence and khus water used in India for flavoring sherberts, syrup sweets, and fruit drinks. Vetiver extracts added to canned asparagus to enhance the flavor (FAC). "Root yields an essential oil used in sherberts" (TAN). Grass used in cigarettes (NAD). Used in Australian barbecued rat recipe: "Fatten the rat on vetiver roots and marinate the meat in its young shoots" (Mark Dafforn, personal communication, 2006). 3.9 g powdered herb boiled in 50 ml water for UTIs and malaria (MAF). Chew 2 g root up to 4 ×/day for toothache (MAF). Two minims essence for nausea of cholera.

- Asian Indians suggest (but I do not) smoking the herb with benzoin for headache (NAD).
- Asian Indians suggest vetiver roots with those of sunflower (American) and Marsilea for malaria (SKJ).
- Asian Indians use the roots on abdominal tumors (JLH; UPW).
- Ayurvedics consider the root alexiteric, antibilious, astringent, depurative, and stomachic, suggesting it for burning sensations, fever, halitosis, strangury, and thirst (KAB).
- Americans anecdotally report that vetiver tea does wonders for tequila hangovers (Mark Dafforn, personal communication, 2006).
- Malayans failed in planting vetiver acreage to repel mosquitoes (UPW); they apply as perfume.
- Mauritians use root as abortifacient (UPW).
- Nigerians use the root in water purification (UPW).
- Peruvians suggest the decoction or tincture topically for dermatoses (e.g., fungal infections) (EGG).
- Philippinos use roots in gastrosis, and as a litholytic in bladder stones (IHB).
- Puerto Ricans use root tea as analgesic, carminative, stomachic, sudorific, for headache, neuralgia, and rheumatism (JFM).
- Trinidadans use the tea for fever, flu, pleurisy, and yellow fever (JFM; UPW).
- Unani consider the bitter root cardiotonic, cerebrotonic, depurative, soporific, using for headache, palpitations, and spermatorrhea (KAB; NAD).

Downsides (Vetiver):

Class 2b. Abortifacient emmenagogue/uterine stimulant (AHP). Not for pregnant women, neonates, or children under 12 years old. Monitor blood sugar if taking for diabetes (MAF). "To date no side effect of adverse reaction has been registered apart from cases of nausea due to overdose" (MAF, 2002).

LAURESTINUS (*VIBURNUM TINUS* L.) + CAPRIFOLIACEAE

NOTES (LAURESTINUS):

> **I will plant in the wilderness the cedar, the shittah tree, and the myrtle, and the oil tree; I will set in the desert the fir tree, and the pine, and the box tree together.**
>
> **Isaiah 41:19 (KJV)**

> **I will put in the wilderness the cedar, the acacia, the myrtle, and the olive; I will set in the desert the cypress, the plane and the pine together.**
>
> **Isaiah 41:19 (RSV)**

> **In the wilderness I shall set the cedar tree, the acacia, and the myrtle, and the oil tree. In the desert plain I shall place the juniper tree, the ash and the cypress at the same time.**
>
> **Isaiah 41:19 (NWT)**

> **The glory of Lebanon shall come unto thee, the fir tree, the pine tree, and the box together, to beautify the place of my sanctuary; and I will make the place of my feet glorious.**
>
> **Isaiah 60:13 (KJV)**

> **The glory of Lebanon shall come to you, the cypress, the plane, and the pine, to beautify the place of my sanctuary; and I will make the place of my feet glorious.**
>
> **Isaiah 60:13 (RSV)**

> **To you the very glory of Lebanon will come, the juniper tree, the ash tree and the cypress at the same time, in order to beautify the place of my sanctuary; and I shall glorify the very place of my feet.**
>
> **Isaiah 60:13 (NWT)**

The King James version calls it *box* (which suggests *Buxus* to most American botanists), the Revised Standard version calls it *plane* (which suggests *Platanus* to most American botanists), and the New World Translation calls it *ash* (which suggests *Fraxinus* to most American botanists). But Israeli botanist Michael Zohary says that only *Viburnum tinus* has the slightest linguistic support to represent the Hebrew *tidhar* in the two Isaiah passages. He is basing this on the Aramaic translation of the *Targum Yonathan*, which renders tidhar as *mornian*, which is cognate with the Arabic *murran*. Murran is the only Arabic name for *Viburnum tinus*. Having only one indication for this interesting Mediterranean species, I resort once more to the generic approach. Except for the dropsy indication, all the activities and indications below are listed for other species of the big genus *Viburnum* (e.g., those followed by VAD are from Mediterranean *V. lantana* or North American *V. prunifolium*).

COMMON NAMES (LAURESTINUS):

Ash (Eng.; NWT); Box (Eng.; KJV); Laurestinus (Eng.; ZOH); Mornian (Aramaic; ZOH); Murran (Arab.; ZOH); Plane (Eng.; KJV); Tidhar (Heb.; ZOH); Nscn.

Activities (Other Viburnums):

Alterative (f; BUR); Analgesic (f1; VAD); Antiabortive (f; FEL; TOM); Anticonvulsive (f; DEM); Antidiarrheic (f1; APA; PNC); Antiinflammatory (f; VAD); Antiseptic (f1; VAD); Antispasmodic (f1; DEM; EFS; FAD; FEL; LAF; PH2; VAD); Astringent (f1; APA; BUR; FEL; VAD); Cardiotonic (1; VAD); Cicatrizant (f; VAD); Cyanogenic (1; EB30:400); Diaphoretic (f; DEM); Diuretic (f; BUR; FEL; LAF); Emetic (f; FEL); Febrifuge (f1; VAD); Hemostat (f; VAD); Hypotensive (f; VAD); Nervine (f; APA; BUR; EFS; FAD; FEL); Sedative (f1; APA; EFS; FAD; VAD); Tonic (f; FEL; LAF); Uterorelaxant (f1; APA; BUR; FAD; LAF); Uterotonic (f; EFS; FAD; FEL); Venotonic (f1; VAD).

Indications (Other Viburnums):

Abortion (f; FEL; HH3); Ague (f; DEM); Alcoholism (f; FEL); Ameba (f; WOI); Amenorrhea (f; FEL); Arthrosis (f; TOM); Asthma (f1; APA; EFS; FAD; HOC; VAD); Bleeding (f; FEL); Blepharosis (f; VAD); Cancer (f; JLH); Cardiopathy (f; FEL); Childbirth (f; DAW); Chorea (f; FEL); Colic (f; FEL); Congestion (f; FEL); Conjunctivosis (f; VAD); Convulsions (f; DEM); Corneal Abrasians (f; VAD); Cramp (f1; APA; DEM; EFS; FAD; FEL); Dermatosis (f1; APA; FEL; LAF; PNC; VAD); Dropsy (f; DAW); Dysentery (f; FEL); Dysmenorrhea (f1; APA; FAD; HH3; LAF; PH2; TOM; VAD); Eczema (f1; VAD); Fever (f1; APA); Enterosis (f1; APA); Epilepsy (f; FEL); Erythema (f; VAD); Female Ails (f; DEM); Fever (f; DAW); Glossosis (f; DEM); Headache (f1; APA); Hemorrhoid (f1; VAD); Hiccup (f; FEL); High Blood Pressure (f1; VAD); Hot Flash (f; TOM); Hysteria (f; EFS; FEL; TOM); Infection (f; DAW); Inflammation (f; FEL; VAD); Insomnia (f1; APA; EFS; FAD); Intermittent Claudication (f; FEL; JAD); Itch (f; VAD); Jaundice (f; FEL); Lethargy (f; LAF); Malaria (f; DAW); Menorrhagia (f; FEL); Miscarriage (f; APA; FAD; LAF); Ophthalmia (f; BUR; FEL); Pain (f1; APA); Palpitation (f; FEL); Paralysis (f; FEL); Parturition (f; DEM; FAD); Periodontosis (f; VAD); Pharyngosis (f1; VAD); Phlebitis (f1; VAD); Rheumatism (f; TOM); Singultus (f; FEL); Smallpox (f; DEM); Sore (f; WOI); Spasm (f; WOI); Stomatosis (f1; VAD); Tumor (f; JLH); Uterosis (f; FEL); Uterrhagia (f; APA); Vaginosis (f; VAD); Varicosity (f1; VAD); Water Retention (f; LAF).

Dosages (Other Viburnums)

FNFF = !
Fruits sometimes edible out of hand, more often dried as "raisins," or in beverages, jams, jellies, and sauces (FAC; TAN). 2 tsp dry bark/cup water (APA); up to 2 tsp/tincture 3 ×/day (APA); 2–5 g powdered bark (HH3). 1–2 g powdered bark (PNC); 2–8 ml bark elixir (PNC); 4–8 ml liquid bark extract (PNC).

Downsides (Other Viburnums):

Class 2d. Kidney stone patients should use only with caution because of oxalates (AHP, 1997). No health hazards or side effects known with proper therapeutic dosages (PH2) (PH2 designates no specific quantified dosage, JAD). HH3 notes that there is a CNS-depressant outcome of overdose. In animals, it takes some 5 to 7 g scu to cause cardiac arrest (HH3).

FABA BEAN (*VICIA FABA* L.) ++ FABACEAE

Synonyms:

Faba vulgaris Moench.

FIGURE 1.118 Faba Bean (*Vicia faba*).

Notes (Faba Bean):

Brought beds, and basons, and earthen vessels, and wheat, and barley, and flour, and parched corn, and beans, and lentiles, and parched pulse, And honey, and butter, and sheep, and cheese of kine, for David, and for the people that were with him, to eat: for they said, The people is hungry, and weary, and thirsty, in the wilderness.

2 Samuel 17:28-9 (KJV)

brought beds, basins, and earthen vessels, wheat, barley, meal, parched grain, *beans* and lentils, honey and curds and sheep and cheese from the herd, for David and the people with him to eat; for they said, "The people are hungry and weary and thirsty in the wilderness."

2 Samuel 17:28-9 (RSV)

[brought] beds and basins, and potter's vessels, and wheat and barley and flour and roasted grain, and *broad beans* and lentils, and parched grain and honey and butter and sheep and curds of cattle they brought forward for David and the people to eat; for they said, "The people are hungry and tired and thirsty in the wilderness."

2 Samuel 17:28–29 (NWT)

Take thou also unto thee wheat, and barley, and *beans*, and lentiles, and millet and fitches, and put them in one vessel, and make thee bread thereof, according to the number of the days that thou shalt lie upon thy side, three hundred and ninety days shalt thou eat thereof.

Ezekiel 4:9 (KJV)

And you, take wheat and barley, *beans* and lentils, millet and spelt, and put them into a single vessel, and make bread of them. During the number of days that you lie upon your side, three hundred and ninety days, you shall eat it.

Ezekiel 4:9 (RSV)

And as for you, take for yourself wheat and barley, and *broad beans* and lentils, and millet and spelt, and you must put them in one utensil, and make them into bread for you, for the number of days that you are lying upon your side, three hundred and ninety days you shall eat it.

Ezekiel 4:9 (NWT)

So ancient is this vegetable that is recorded by Pliny. Even today, broadbeans are cultivated in biblical countries, especially Egypt, because of their many uses, not only as vegetable, and pulse, but also as breadstuff. Beans have been found in the Neolithic of Jericho, where they are cultivated to this day. According to Zohary, the plant is nowhere found in the wild, suggesting that the wild ancestor is extinct. In ancient days, beans were used in collecting votes from the people; a white bean signifying approval of the measure proposed; a black one meaning condemnation. Magistrates were elected by casting beans; maybe that is where we got the term "bean-counters" (BIB; ZOH).

Common Names (Faba Bean):

Aboaun (Ber.; BOU); Anhuri (Hindi; WOI); Bakala (Nepal; NPM); Bakila (Iran; EFS); Bakla (Hindi; Kum.; Tur. DEP; EFS; WOI); Bakla Sem (Delhi; WOI); Bakula (Newari; NPM); Baquela (Arabic; BOU); Bean of History (Eng.; WOI); Bean of the Bible (Eng.; WOI); Bell Bean (Eng.; USN); Bondbona (Sw.; EFS); Broad Bean (Eng.; Jam.; AVP; BOU; LEG); Buffbohne (Ger.; EFS); Can Dou (Pin.; DAA); Chas Tang (Pun.; Sutlej; DEP; WOI); Chas Tang Raiun (Him.; WOI); Djilla (Ber.; BOU); Duivenbohn (Dutch; EFS); Faba (Eng.; Sp.); Faba (It.; Por.; EFS); Faba bean (Eng.; CR2); Fava Bean (Eng.; CR2; LEG); Faveiro (Por.; EFS); Feldbohne (Ger.; EFS); Feve (Fr.; BOU); Feve Cultivee (Fr.; BOU); Feve de Marais (Fr.; BOU; EFS); Feve des Champs (Fr.; EFS); Feverolle

(Fr.; BOU); Foul (Arabic; BOU); Foul Hashadi (Arabic; BOU); Garden Bean (Eng.; NPM); Haba (Peru; Sp.; Trin.; AVP; EFS; EGG); Habas (Asa.; ROE); Habichuela (Sp.; EFS); Hende Matar (Mun.; WOI); Hestebonne (Den.; EFS); Horsebean (Eng.; LEG); Hu Tou (China; EFS); Kabli Bakla (Pun.; WOI); Kadu Huralikayee (Kan.; WOI); Kaian (Kas.; DEP); Kala Matar (Hindi; WOI); Karafasulye (Tur.; EB49:406); Katun (Kas.; WOI); Lipta (Peru; EGG; SOU); Mattz Rewari (Pun.; WOI); Mon Sran (Tibet; NPM); Nakshan (Ladak; DEP; WOI); Paardenbohn (Dutch; EFS); Pferdebohne (Ger.; EFS); Pois Blanc (Haiti; AVP); Pol (Heb.; ZOH); Raj Rawam (Pun.; Urdu; WOI); Rgya Sran (Tibet.; TIB); Saubohne (Ger.; EFS); Scotch Bean (f; NPM); Tick Bean (Eng.; LEG); Toinbohn (Dutch; EFS); T'san Tou (China; EFS); White Bean (Pr.; AVP); Windsor Bean (Eng.; LEG).

Activities (Faba Bean):

Anodyne (f; BOU); Antialcoholic (12; HAD); Antiangiogenic (1; FNF); Antidote (f; BIB); Antiparkinsonian (12; FNF); Antispasmodic (f; BOU); Aphrodisiac (f1; BIB; FNF); Astringent (f; TIB); Cholagogue (f; BOU); Cyanogenic (f; BIB); Dopaminergic (2; HAD); Diuretic (1; BIB); Estrogenic (12; BIB; HAD; WOI); Expectorant (f; BIB; TIB); Fungicide (1; X12817478); Hemolytic (1; PHR); Hypertensive (1; PH2); Litholytic (f; TIB); Mitogenic (1; X12817478); Mucolytic (f; TIB); Natriuretic (1; FNF); Nootropic (f; SOU); Reverse-Transcriptase Inhibitor (1; X12817478; Stomachic (f; BIB); Tonic (f; BIB).

Indications (Faba Bean):

Abscess (f; EB49:406); ADD (1; FNF); Addiction (1; FNF); Adenopathy (f; JLH); Asthma (f; SOU); Boil (f; SOU); Bronchosis (1; FNF); Burns (f; PHR); Callus (f; JLH); Cancer (1; FNF); Cancer, bladder (f1; FNF; JLH); Cancer, breast (f1; FNF; JLH); Cancer, eyes (f1; FNF; JLH); Cancer, eyelid (1; FNF; JLH); Cancer, foot (1; FNF; JLH); Cancer, gland (1; FNF; JLH); Cancer, liver (1; FNF; JLH); Cancer, parotid (f1; FNF; JLH); Cancer, penis (1; FNF; JLH); Cancer, spleen (1; FNF; JLH); Cancer, stomach (1; FNF; JLH); Cancer, testes (1; FNF; JLH); Corn (f; JLH); Cough (f; PHR; PH2); Cramp (f; BOU); Cystosis (f; JLH); Dermatosis (f; PHR; PH2); Drunkenness (2; BIB; FNF); Encephalitis (1; FNF); Felon (f; JLH); Flu (f; ROE); Fungus (1; WOI); Gastrosis (f; BOU; JLH); Hepatosis (f; JLH); Impotence (1; BIB; FNF); Induration (f; JLH); Leukemia (1; FNF); Mastosis (f; JLH); Melanoma (1; FNF); Mycosis (1; WOI); Nephrosis (f; BOU; PHR; PH2); Ophthalmia (f; JLH); Orchosis (f; JLH); Osteoporosis (1; FNF); Pain (f; BOU); Parkinson's (12; FNF); Pneumonia (f; BIB); Pulmonosis (f; BIB); Sclerosis (f; BIB); Smoking (1; FNF); Sore (f; EB49:406); Splenosis (f; BOU; JLH); Stomachache (f; BOU); Stomatosis (f; BIB); Swellings (f; BIB; JLH); Tumors (1; BIB), Urogenitosis (f; PH2); Wart (f; BIB; JLH; PH2); Wen (f; JLH).

Dosages (Faba Bean):

FNFF = !!!

In biblical times, broad beans, THE biblical beans, and their meal were made into bread, even as today in the biblical world. They are boiled and eaten also. Elsewhere they are cultivated as a vegetable and used green or dried, fresh or canned, and for stock feed. Broadbean has been considered a meat extender or substitute and as a skim-milk substitute. Sometimes grown for green manure, but more generally for stock feed. Large-seeded cultivars are used as a vegetable, and frequently grown as a home-garden crop for canning. Roasted seeds are eaten like peanuts in India. Beans are fed to horses and the stalks are given to camels. I once calculated that it would take a pound of faba beans, or a couple ounces of the sprouts, to give a physiological dose of L-dopa. The sprouts are also better sources of daidzein and genistein.

- North Africans nibble two parched seeds early in the morning for stomach pain (BOU).
- Iranians claim the shoots can rouse a drunkard from stupor (BIB).
- Turks apply heated leaves to abscesses (EB49:406).

Downsides (Faba Bean):

Inhaling pollen or ingesting seeds may cause favism, an especially severe hemolytic anemia, an inherited enzymatic deficiency (glucose-6-phosphate-dehydrogenase) occasional among Mediterranean people (Greeks, Italians, Semitics). Injected intravenously in rabbits, broadbean extracts have produced hemoglobinuria and death (POB). Overdosage, especially in genetically susceptible individuals, can rapidly lead to diarrhea, queasiness, vertigo, and vomiting. In more severe cases, anuria, fever, hemoglobinuria, icterus, and/or oliguria occur. Following ingestion and digestion, pyrimidine derivatives in high doses can cause hemolysis (PHR). L-Dopa may cause a rise in blood pressure (PHR). L-Dopa is contraindicated if you have heart, liver, lung, kidney, or thyroid problems; glaucoma; or are taking vitamin B_6, antidepressants (especially MAOIs), or antipsychotic drugs. L-Dopa may activate malignant melanomas, which, however, can be checked by 100 mg/day coumarin, found in the tonka bean.

Natural History (Faba Bean):

One study concluded that bees increase seed production by 15 to 20%. Honeybees were estimated to account for 80% of cross-pollination, bumblebees less than 20%, and wild bees less than 1%. A closed-flower phenotype (recessive to normal) exists which lacks the typical scent and is avoided by bees (Poulsen, 1977). Many fungi attack broadbeans, depending on the area where they grow. The following have been reported on broadbeans: *Alternaria brassicae* var. *phaseoli*, *A. tenuis*, *A. tenuissima*, *Ascochyta boltshauseri*, *A. fabae*, *A. pinodella*, *A. pinodes*, *A. pisi (A. viciae)*, *Aspergillus niger*, *Botrytis cinera*, *B. fabae*, *Cercospora fabae*, *C. viciae*, *C. zonata*, *Cladosporium cladosporioides*, *C. herbarum*, *C. pisi*, *Clonostachys araucariae*, *Colletotrichum lindemuthianum*, *Corticium rolfsii*, *C. solani*, *Cunninghamella echinulata*, *Deplosporium album*, *Dothiorella fabae*, *Erysiphe pisi*, *E. polygoni*, many species of *Fusarium*, *Gibberella fujikuroi*, *G. saubinettii*, *Gloeosporium viciae*, *Helicobasidium purpureum*, *Leveillula taurica*, *Macrophomina phaseoli*, *Melanospora papilata*, *Mycospharaella pinodes*, *Nectria anisophylla*, *Olpidium viciae*, *Peronospora fabae*, *P. lagerheimee*, *P. pisi*, *P. viciae*, *Phoma malaena*, *Phyllosticta fabae*, *Phymatotrichum omnivorum*, *Physoderma fabae*, *Phytophthora cactorum*, *Ph. cinnamoni*, *Pleospora herbarum*, *P. vulgaris*, *Pythium spp.*, *Rhizoctonia solani*, *Rhizopus nigricans*, *Sclerotinia fuckeliana*, *S. minor*, *S. sclerotiorum*, *Sclerotium rolfsii*, *Stagonospora carpathica*, *Stemphylium botryosum*, *S. consortiale*, *Trichothecium roseum*, *Uromyces appendiculatus*, *U. fabae*, *U. orobi*, *and U. viciae-fabae*. Broadbeans also attacked by the sweet pea streak, tooth-tumor swelling vein virus and broadbean wilt, red-clover vein mosaic (*Marmor trifolii*), virus 1-celery mosaic (a strain of cucumber mosaic virus: *Marmor cucumeris*), and spotted wilt (*Lethum australiensis*). Bacteria causing diseases in broadbean include *Bacterium phaseoli*, *B. viciae*, *Erwinia phytophthora*, and *Psuedomonas viciae*. Nematodes isolated from broadbean include *Ditylenchus dipsaci*, *Heterodera glycines*, *H. goettingiana*, *H. rostochiensis*, *Longidorus maximus*, *Meloidogyne arenaria*, *M. artiella*, *M. hapla*, *M. incognita*, *M. incognita acrita*, *M. javanica*, *Pratylenchus brachyurus*, *P. coffeae*, *P. goodeyi*, *P. pnetrans*, *P. pratensis*, *P. vulnus*, *P. zeae*, *Rotylenchulus reniformis*, *Tylenchorynchus dubius*, and *T. parvus*. The most serious insect pests are the broadbean weevil, *Bruchus rufimanus*, and aphids, especially the bean aphid, *Aphis fabae*. Broomrape (*Orobanche crenata*) may be a serious problem in the Middle East. Eptam, applied as a postemergence spray, was fairly effective, as was soil fumigation with dibromochloropropane, and oxak (terbutol), if deeply incorporated into the soil before sowing.

Extracts (Faba Bean):

Estrogenic activity of fresh beans equivalent to 11.4 μg estradiol monobenzoate/kg. The ethanol-ether extract of the seeds at 50-mg dose level stimulated non-pregnant uterus at diestrus (WOI). "The immature seeds exhibited appreciable antifungal activity; phytoalexins came into play to combat the infection" (WOI). The LD50 of the bean extract in mice was 19,000 mg/kg body weight. L-3,4-dihydroxyphenylalanine (L-dopa) was first synthesized as a D,L racemate in 1911. Its L-isomer was isolated from seedling, and from 1961 onward was recognized as the most efficacious drug treatment of Parkinson's disease. It is now generally recognized that L-dopa use in Parkinson's disease is a classic example of brain neurotransmitter replacement therapy. That may not be its sole action of interest. Recent evidence suggests that L-dopa may also have its own biological activity in the CNS, independent of dopamine (X15143441). Both L-Dopa and epinene have been reported from the seeds. L-Dopa is said to induce priapism in elderly males receiving it for Parkinsonianism. According to *Father Nature's Farmacy* (FNF), L-dopa has analgesic, anorexic, antidote (manganese), antiencephalopathic, antifeedant, antimorphinic, antineuroleptic, antiparkinsonian, antireserpine, aphrodisiac, arrhythmigenic, antitremor, cardiovascular, CNS-active, depressant, diuretic, dopaminergic, emetic, hallucinogenis, hypertensive, hypotensive, insectifuge, miotic, and natriuretic activities. The LD50 = 609 ppm (orl rbt); LD50 = 3650 ppm (orl mus); LD50 = 4000 ppm (orl rat) (FNF). Daidzein reportedly has antialcoholic, antiarrhythmic, anticephalalgic, antidipsomanic, antihemolytic, antiinflammatory, antileukemic, antimelanomic, antimicrobial, antimutagenic, antiosteoporotic, antioxidant, antispasmodic, coronary-dilator, estrogenic, fungicide, hypotensive, and lipase-inhibitor activities. Genistein (0–92 ppm in the stem) has abortifacient, aldose-reductase-inhibitor, alpha-reductase-inhibitor, antiaggregant, antiangiogenic, anticancer (breast), anticarcinomic (breast), antiendoccytotic, antifertility, antihemolytic, antiimplantation, antiischemic, antileukemic, antileukotrienic, antilymphomic, antimelanomic, antimicrobial, antimitogenic, antimutagenic, antineuroblastomic, antioxidant, antiproliferative, antiprostatadenomic, antiprostatitic, antispasmodic, antitumor, antitumor (GI), antitumor (ovary), antitumor (prostate), antitumor (stomach), "antiulcer?," apoptotic, cancer-preventive, catechol-O-methyl-transferase-inhibitor, cytotoxic, DOPA-decarboxylase-inhibitor, estrogenic, flatulent, fungicide, fungistat, histidine-kinase-inhibitor, lipase-inhibitor, MAO-inhibitor, peroxidase-inhibitor, phytoalexin, pituitary-sensitizer, topoisomerase-II-inhibitor, trypanosomastat, and tyrosine-kinase-inhibitor activities (FNF).

GRAPE (*VITIS VINIFERA* L.) +++ VITACEAE

Notes (Grape):

But they shall sit every man under his vine and under his fig tree; and none shall make them afraid: for the mouth of the LORD of hosts hath spoken it.

Micah 4:4 (KJV)

...but they shall sit every man under his vine and under his fig tree, and none shall make them afraid; for the mouth of the LORD of hosts has spoken.

Micah 4:4 (RSV)

And they will actually sit, each one under his vine and under his fig tree, and there will be no one making [them] tremble, for the very mouth of Jehovah of armies has spoken [it].

Micah 4:4 (NWT)

I will recount only tidbits from Zohary's interesting historical account. "From the dawn of man's history," grapes were widely cultivated in the Old World. Noah, first tiller of the soil, planted a vineyard. Viticulture was very important in the land of Israel, an image of bounty and the blessings of God in the future. The grape, one of the "seven species" with which the land was blessed, was a national emblem; but in bad years there would be little bounty. The New Testament attributes spiritual meanings to the grape, Jesus even identifying himself with the species in John. Viticulture was

FIGURE 1.119 Grape (*Vitis vinifera*).

established in Egypt, Israel, and Syria in the Early Bronze Age. Pips of grapes have been uncovered in northern Greece, circa 4500 BC (ZOH). But there seem to have been those who deemed the product of the wine (and grape juice) demotivational.

Whoredom and wine and new wine take away the heart.

Hosea 4:11 (KJV)

Wine and new wine take away the understanding.

Hosea 4:11 (RSV)

Fornication and wine and sweet wine are what take away good motive.

Hosea 4:11 (NWT)

And there were those who deem it healthy or health-giving:

Drink no longer water, but use a little <u>wine</u> for thy stomach's sake and thine often infirmities.

1 Timothy 5:23 (KJV)

No longer drink only water, but use a little <u>wine</u> for the sake of your stomach and your frequent aliments.

1 Timothy 5:23 (RSV)

Do not drink water any longer, but use a little <u>wine</u> for the sake of your stomach and your frequent cases of sickness.

1 Timothy 5:23 (RSV)

Common Names (Grape):

Abai (Bom.; KAB); Aenab (Arab.; KAB); Ampelos (Greek; KAB); Amrutaphala (Sanskrit; KAB); 'Anab (Arab.; GHA); Anavim (Heb.; ZOH); Angur (Beng.; Dec.; Hindi; India; Kan.; Nwp.; Urdu; EFS; KAP; NAD; WOI); Angura (Iran; EFS); Angurphal (Beng.; WOI); Asma (Tur.; EFS; EB54:155); Basho (Lad.; KAB); Budo (Japan; BOU); Buri (Pun.; KAB); Chagarat el'Enab (Arab.; AVP); Chawat Udi (Malaya; EFS); Common Grape (Eng.; FAC; TAN); Dakh (Hindi; WOI); Dakha (Sanskrit; EFS); Dakhnaari (Mun.; KAB); Dalia (Alg.; AVP); Darakh (Guj.; WOI); Devek (Tur.; EB49:406); Dhaku (Kon.; KAB); Dielja (Malta; KAB); Drak (Sind; KAB); Drakasha (Guj.; KAB); Drakhyaluta (Beng.; KAB; WOI); Draksha (Guj.; Kan.; Kon.; Mah.; Mar.; Sanskrit; Tam.; Tel.; EFS; NAD; WOI); Draksha Pondu (Tel.; NAD); Drakshai (India; WOI); Drakshe (Kan.; KAB); Drakshya (India; EFS); Drakya (Oriya; WOI); Edleweinrebe (Ger.; NAD); 'Enab (Arab.; GHA); Enredadera (Sp.; EFS); European Grape (Eng.; FAC; TAN); Gefen (Heb.; ZOH); Ghephen (Heb.; KAB); Gostani (Mal.; WOI); Gostanidraksha (Tam.; Tel.; KAB; WOI); Gostoni (Oriya; KAB; WOI); Grape (Eng.; CR2); Gundak Api (Malaya; EFS); Gureb (Iran; EFS); Karm (Arab.; Palestine; AVP; KAB); Kerm (Arab.; EFS); Kishamisha (India; EFS); Kishmish (Hindi; India; Iran; EFS; NAD; WOI); Kismis (India; WOI); Kmen Winny (Czech; KAB); Kodiminduri (Tam.; WOI); Kottani (Tam.; KAB); Kwar (Pushtu; KAB); Lakom (Malaya; EFS); Madhurasam (Mal.; KAB); Manakka (India; WOI); Monaangur (Bal.; KAB); Mridirka (Sanskrit; NAD); Muddrap (Sin.; KAB); Mudraka (Guj.; NAD); Mundiri (Mal.; WOI); Munteri (Malaya; EFS); Onguro (Oriya; WOI); Otrik (Arm.; KAB); Parra (Cat.; Por.; Sp.; EFS; KAB); Parreira (Por.; AVP); P'i T'ao (China; EFS); Pú Tao (Pin.; DAA); Rosinen (Ger.; NAD); Sabisi (Burma; KAB); Seben Gkak (Malaya; EFS); Sougi (India; WOI); Sur Medun (Kalmuk; KAB); Szollo (Hun.; KAB); Tak (Afg.; KAB); Trachei (Madras; KAB); Traube (Ger.; EFS); Uva (Sp.; Por.; AVP; EFS); Uzum Cibigi (Tur.; EB49:406); Vid (Sp.; EFS); Videira (Por.; EFS); Vigne (Fr.; Haiti; AVP; EFS); Vigne Cultive (Fr.; NAD); Viña (Sp.; AVP); Vinha (Por.; AVP); Vinrauka (Swe.; AVP); Vinstoock (Swe.; KAB); Vita (Rom.; KAB); Vite (It.; EFS); Vito (It.; AVP); Voaloboka (Hova; KAB); Wasia (Georgia; KAB); Weinrauke (Ger.; AVP); Weinrebe (Ger.; EFS); Weinstok (Ger.; EFS); Wijnstok (Dutch; EFS); Wine Grape (Eng.; WOI); Winna Macica (Pol.; KAB); Winograd (Rus.; KAB); Winorosl (Pol.; AVP); Wuntraee (Den.; KAB).

Activities (Grape):

Anabolic (1; X16028976); Anodyne (1; BIB; FNF); Antiaggregant (1; FNF); Antiallergic (1; FNF); Antialopecic (1; PH2); Antialzheimeran (1; COX; FNF); Antianaphylactic (1; FNF); Antiarthritic

(1; COX; FNF); Antiasthmatic (1; FNF); Antiatherosclerotric (1; FNF; PH2); Anticancer (1; COX; FNF); Anticariogenic (1; APA); Antiedemic (1; FNF); Antierythemic (1; FNF); Antihistaminic (1; FNF); AntiHIV (1; FNF); Antiinflammatory (1; COX; FNF); Antileukemic (1; X15309446); Antimelanomic (1; X15309446); Antimutagenic (1; FNF; MB); Antioxidant; (1; APA; FNF; PH2; X10820088); Antiprostaglandin (1; FNF); Antiseptic (1; FNF); Antitumor (1; PH2); Antiulcer (1; FNF); Antiviral (1; BIB; FNF); Aphrodisiac (f; BIB; KAB); Apoptotic (1; JAF51:7280); Astringent (f1; BIB; VAD); Bactericide (1; FNF); Cancer Preventive (1; FNF); Capillariprotective (1; FNF); Cardioprotective (f1; FNF; VAD); Collagen Protective (1; APA); COX-1 Inhibitor (1; FNF); COX-2 Inhibitor (1; COX; FNF); Demulcent (f; BIB; EFS); Depurative (f; BIB); Diuretic (f; BIB; EFS); Expectorant (f; BIB); Fungicide (1; FNF); Hemostat (1; EFS; FNF); Hepatoprotective (1; FNF; PH2); Hypocholesterolemic (1; FNF); Immunostimulant (1; FNF); Insulinomimetic (1; X16028976); Laxative (f; BIB); Litholytic (f; BIB; NAD); Orexigenic (f; NAD); Propecic (1; PH2); Protein-Kinase-C Inhibitor (X11823594); Radioprotective (1; X10820088); Stomachic (f; BIB); Tonic (f; BIB); Topoisomerase Inhibitor (1; X15796584); Sunscreen (1; FNF); Vasoprotective (1; APA; VAD); Venotonic (f1; VAD).

Indications (Grape):

Alopecia (1; PH2); Alzheimer's (1; COX; FNF); Anaphylaxis (1; FNF); Anemia (f; NAD; PH2); Anorexia (f; NAD); Arteriosclerosis (1; VAD); Arthrosis (1; BIB; COX; FNF); Asthma (1; BIB); Atherosclerosis(1; FNF; PH2); Biliousness (f; BIB; NAD); Bladder stone (f; BIB; NAD); Bleeding (f; BIB; FEL; HHB); Blepharosis (f1; VAD); Boil (f; SKJ); Bronchosis (f; GHA); Bruise (f; NAD); Cachexia (f; BIB; NAD); Cancer (f1; FNF; JLH; PH2); Cancer, abdomen (f1; DAD; FNF); Cancer, breast (1; JAF51:7280); Cancer, colon (1; DAD; FNF; X12935318); Cancer, ear (f1; FNF; JLH); Cancer, liver (f1; FNF; JLH); Cancer, neck (f1; FNF; JLH); Cancer, nose (f1; FNF; JLH); Cancer, testicle (f1; FNF; JLH); Cancer, throat (f1; FNF; JLH); Cancer, tonsil (f1; FNF; JLH); Cancer, uterus (f1; FNF; JLH); Cancer, uvula (f1; FNF; JLH); Candida (f; NAD); Capillary Fragility (f1; BRU; FNF; PH2; VAD); Cardiopathy (f1; BIB; FNF; VAD); Caries (1; FNF; MB); Catarrh (f; NAD); Cholera (f; BIB; DAA); Circulosis (f; PH2); Cold (f; NAD); Condyloma (f; JLH); Conjunctivosis (f1; BRU; FNF; VAD); Constipation (1; X12935318); Consumption (f; DAA; PH2); Corn (f; JLH); Cough (f; DAA; GHA); Cramp (1; BIB; PH2); CVI (1; PH2; X10356940); Cyanosis (f; PH2); Dermatosis (f; BIB; KAB; PH2); Diabetes (f1; NAD; PH2; X16028976); Diarrhea (f; FEL; PH2; WOI); Diptheria (f; BIB); Diverticulosis (1; X12935318); Dropsy (f; DAD; PH2); Dysentery (f; FEL; HHB); Dysmenorrhea (f1; VAD); Dyspnea (f; NAD); Dysuria (f; NAD; PH2); Earache (f; GHA); Ecchymosis (1; BRU); Edema (f1; FNF; PH2; VAD); Fever (f; FEL; PH2); Fibroid (f; JLH); Fibroma (f; PH2); Fungus (f1; DAD; KAD); Gastrosis (1; FNF); Gingivosis (f; EB49:406); Gonorrhea (f; PH2); Gout (1; FNF); Hangover (f; BIB); Headache (f; KAB; PH2); Hemorrhoids (f; BRU; DAD; NAD; PH2); Hepatosis (1; BIB; JLH; FNF); Herpes (1; BIB); High Blood Pressure (f1; BI2) High Cholesterol (1; FNF); HIV (1; FNF); Hoarseness (f; DAA; PH2); Hyperacidity (f; FEL); Impostume (f; JLH); Induration (f; JLH); Infection (f1; FNF; NAD); Inflammation (f1; BIB; COX; FNF; EB49:406); Jaundice (f; BIB; NAD); Maculitis (1; FNF), Mycosis (f1; DAD; NAD); Nausea (f; BIB; PH2); Nephrosis (f; DAA; PH2); Nervousness (f; PH2); Nyctalopia (1; FNF; PH2); Obesity (f1; VAD); Ophthalmia (f1; BIB; KAB; PH2); Orchosis (f; BIB; NAD); Otosis (f; GHA); Pain (f; PH2); Parathesia (1; PH2); Phlebitis (f1; VAD); Polyp (f; JLH); Pregnancy (f; DAA); Retinopathy (1; FNF; PH2); Rheumatism (f; HHB; NAD); Rhinosis (f; JLH); Scabies (f; BIB); Scirrhus (f; JLH); Smallpox (f; BIB; HJP); Sore Throat (f; KAB; PH2; WOI); Splenosis (f; DAD); Sprain (f; NAD); Sting (f; NAD); Stomachache (f; PH2); Stomatosis (f; BIB); Stone (f; BIB); Stress (f; PH2); Strangury (f; KAB); Swelling (f; PH2); Syphilis (f; BIB; KAB); Telangectasia (1; FNF; MB); Thirst (f; PH2); Thrush (f; NAD); Tonsilosis (f; JLH); Toothache (f; PH2); Tuberculosis (f; EFS; HJP; PH2); Ulcer (1; FNF); Varicosity (f1; FNF; VAD); Virus (1; BIB); Wart (f; JLH; PH2); Water Retention (f; BIB); Wound (1; FNF); Wrinkles (1; FNF).

Dosages (Grape):

FNFF = !!!

Fruits widely eaten, raw, juiced, fermented, or preserved; leaves less widely so; grape seeds are used to garnish cheeses. Italians celebrate the first pressing of the olives with noodles (pasta alla deficiera) cooked in wine instead of water. *Dibs* or *pekmez*, used to sweeten tea, consisting of evaporated grape juice or grape molasses. Unripened grape juice (verjuice or verjus) is used when tartness is needed. Grapeseed oil is preferred for cooking meat in *fondue bourguignonne*. Grape leaves (*dolmas* or *sarmas*) are often stuffed with rice, and/or currants, and/or pine nuts. Flowering shoots, tendrils and all, are tart and tender, sometimes astringent, but a good nibble in the field. Cream of tartar, scraped from the insides of wine barrels, can stabilize egg whites, and is added to candies and frostings. It is also added to baking powders to acidify. Vineyard cuttings are often burned to smoke meats (FAC; JAD; TAN; EB54:155). 75–600 mg seed extract (or OPC) daily for up to 3 weeks; maintenance dose 40–100 mg/day (APA, PH2).

- Arabians boil fruits, strain, adding honey for bronchosis and cough, as eardrops for earache (GHA).
- Asian Indians squeeze and boil dried raisins (with seeds) and take internally for diabetes (NAD).
- Asian Indians suggest that elders chew raisins for rheumatism (NAD).
- Ayurvedics regard the fruits, especially the black fruits, as aphrodisiac, diuretic, laxative, purgative, and refrigerant, and use them for asthma, biliousness, blood disorders, burning, eye ailment, fever, hangover, jaundice, sore throat, and strangury (KAB).
- Chinese use the leaf decoction for cholera, dropsy, nausea, pregnancy, and threatened abortion (DAA).
- Europeans use the juice from young branches for dermatoses and ophthalmia (KAB).
- Italians use unripe grape juice for throat afflictions (NAD).
- Lebanese have a grape "cure" for fever, liver, nervousness, smallpox, and tuberculosis. Small young leaves and/or tendrils are fed to infants to prevent scurvy and iron deficiency (the seeds and roots are ground for an anemia treatment, like wine itself). The expressed leaf juice is applied to various skin conditions, including "cancer." Lebanese use wine or brandy for cramps, stomachache, toothache, and for any pain (HJP).
- Lithuanians use raisins as a folk remedy for arthritis (JLH).
- Scientists report that two servings of raisins per day (84 g/day) can cause beneficial changes in colon function and may decrease the risk for colon cancer (X12935318).
- Unani use leaves, or leaf juice, for bleeding at the mouth, headache, nausea, piles, scabies, splenitis, and syphilis; the stem ashes for arthritis, bladder stones, orchitis, and piles; the fruit for fever; the seed ash for inflammation. They consider the seeds aphrodisiac, astringent, and refrigerant, the fruit as depurative, digestive, expectorant, and stomachic (KAB).

Downsides (Grape):

None yet known. Not in AHP (1997). No health hazards or side effects known with proper therapeutic dosages (PH2). Shahina Ghazanfar (GHA), coming from an Arabian tradition, notes that grape juice is taken with honey, as grapes are believed to increase phlegm and a tendency to jaundice (GHA).

Natural History (Grape):

Many cultivars of the *Vinifera*-type are self-unfruitful, and require another cultivar with an overlapping flowering period to be interplanted. In some cultivars, pruning affects the effectiveness of the pollen. Spraying "tame" grapes for control of insects and diseases is essential to production of fruit. However, the problems are different in different places. Each grower should study his conditions and apply only such sprays as found necessary and recommended. Infection must be prevented if clean fruit is to be produced. Thoroughness is very essential. Grapes are very sensitive to injury from 2,4-D. Grapes are affected by a great many fungi, bacteria, viruses, nematodes, insects, and mineral deficiencies. Local problems should be resolved with local agricultural agents. Eastman (1992) presents a wealth of natural history data applying to wild grapes and I excerpt here. Martin et al. note that, except for the Japanese beetle (*Popillia japonica*), wild grapes have few serious pests. Feeding on almost 300 plant species, it favors grape. I suspect that many of the pests of the wild grapes will be even worse on "tame" grapes. Conversely, I suspect that, proportionately, the wild grapes are stronger, on average, than the tame grapes. Many gamebirds (dove, duck, grouse, pheasant, pigeon, prairie chicken, quali, turkey) relish the fruit. Migrating south, Tennessee warblers sometimes descend on grapevines to feed. Martin et al. list dozens of songbirds indulging in grapes (e.g., blackbird, bluebird, catbird, finch, flicker, etc.), sometimes constituting 50% of the diet of mockingbirds, less than 25% of cardinal, fox sparrow, robin, and waxwing. Songbirds collect shreddy grapevine bark as a favorite material for nests (e.g., red-eyed vireo, gray catbird, Northern mockingbird, brown thrasher, and Northern cardinal). Raccoons often deposit scat with grape seeds at bases of trees. Other fruit feeders include black bears, coyote, fox, opossum, rabbit, raccoon, skunk, and squirrel. Deer graze the herbage (EAS; MZN). Causing diamond-shaped cankers on vines, resulting in dead branches and small angular spots with yellowish margins on leaves, is a widespread sac fungus (*Cryptosporella viticola*; dead-arm disease or branch necrosis). A "landscape scene" growing on the dead inner bark of a grapevine is a mycelial fan, probably *Armillaria*. At night, the fungus sometimes casts a faint luminescent glow. Fleshy, pea-shaped galls on leaf undersides with openings on the upper leaf surface indicate yellowish-green grape phylloxera (*Phylloxera vitifoliae*), this plant's most injurious aphid. Aphid generations migrating to the roots, feeding and forming nodules, may kill the vine. Green or red conical galls, like dunce caps on upper leaf surfaces are *Cecidomyia viticola*, the grape gall or tube midge. The grapevine tomato gall midge (*Lasioptera vitis*) may make greenish or reddish pea-size swellings on leaf veins and tendrils. Large, rear-horned sphinx caterpillars, some of which make squeaking sounds, are easily recognized. The patterns of adult wood-nymph moths (*Eudryas*) resemble bird droppings. The eight-spotted forester (*Alypia octomaculata*), a bluish-white caterpillar banded with orange, can defoliate vines. Grapevine loopers (*Eulithis*), slender, pale green inchworms, pupate in loose webs on the foliage. *Dyspteris aborivaria*, another inchworm (badwing geometer), rolls grape leaves, as does the grape plume moth (*Geina periscelidactyla*). A smoky black spotted spinulose caterpillar, the grapeleaf skeletonizer (*Harrisina americana*), feeds on upper leaf surfaces, consuming all but the veins. Roundish, white, clear-winged moth caterpillars may be grape root borers (*Vitacea polistiformis*), which drop to the ground and bore into the roots. The adults resemble paper wasps, even mimicking their behavior. In summer, an inch-long tan beetle with black spots on each side, the spotted grapevine beetle (*Pelidnota punctata*), draws leaves together making its tent.

The grape curculio (*Craponius inaequalis*) lays eggs, first feeding on the leaves. Larvae then excavate the green berry pulp beneath the skin. Caterpillars of the grape berry moth (*Paralobesia viteana*) web several ripe berries together or to leaves, making a hole in each berry. The caterpillar folds over a leaf and pupates inside the fold. Folded leaves may hang on the vine in winter. The large Virginia creeper sphinx moth, also called hog sphinx (*Darapsa myron*), punctures decaying or fermented fruit to feed, as do bees and wasps.

Extracts (Grape):

In general, agricultural selection breeds out some of the natural pesticides, like resveratrol, which also have many interesting biological activities. The recent American Chemical Society book, Wine, *Nutritional and Therapeutic Benefits* (Watkins, T.R., 1997), heaps praise on resveratrol, failing to tell us that there is 10 to 100 times more in the leaves, and I suspect seeds, than in the fruit pulp and wines. The seeds have only recently come to the market and clinical trials are few and far between. On the other hand, the fruits contain more than 30 types of anthocyanins. Small wonder that grape juice has 4 times the ORAC score of any other fruit juice studied (JNU). Resveratrol has received much press for cancer prevention. For example, Stewart et al. (2003), commenting on resveratrol as a candidate for prostate cancer prevention, comment that it may constitute 5 to 10% of grapeskin. "Resveratrol may represent the tip of the iceberg of a broad class of stilbene and related polyphenolic natural products," possibly safe and effective agents for cancer prevention. They look to resveratrol as a leading agent for prostate cancer prevention because it inhibits each stage of multistage carcinogenesis, and scavenges incipient populations of androgen-dependent and androgen-independent prostate cancer cells (X12840221). Resveratrol protects against colitis, and has antioxidant and apoptotic actvities. At levels of 5 to 10 mg/kg/day (equivalent to 1 g/day if I were the 100-kg rat), resveratrol reduced colonic injury, index of neutrophil infiltration, and levels of cytokine (X15013856). But I like to remind readers that it is a cocktail of closely related compounds, piceatannol and pterostilbene deserving almost as much praise as the resveratrol (X15309446). Many other anticancer activities are listed in the USDA database. Working with tissue culture, Jo et al. (2005) found potent topoisomerase II catalytic inhibitors: TP fractions 4 and 6 (IC50 = 0.28–0.29 μg/ml), TP-3 (IC50 = 0.74 μg/ml), and crude extract (IC50 = 1.02 μg/ml) — each significantly more potent than resveratrol (IC50 = 18.0 μg/mL) (X15796584). Preliminary data and literature searches suggest that the leaves may be a better source of resveratrol, ironically, leaves stressed by disease, insects, and physical damage (JAD). A study by Fernandez-Pachon et al. (2005) confirms what I had long heard: that red wine increases uric acid levels. That can be good in normouricemic humans, but may induce a gout crisis in hyperuricemic individuals. Maximum concentrations of maximum antioxidant capacity (and uric acid) occurred after about an hour. Uric acid, like albumin and bilirubin, is an endogenous antioxidant as well (X15941351).

SPINY COCKLEBUR (*XANTHIUM SPINOSUM* L.) + ASTERACEAE

Synonyms:

Acanthoxanthium spinosum (L.) Fourreau; *Xanthium americanum* Walter; *Xanthium canadese* Mill.; *Xanthium catharticum* H.B.K.; *Xanthium cavanillesii* Schouw; *Xanthium chinese* Mill.; *Xanthium commune* Britton; *Xanthium cylindraceum* Millsp. & Sherff; *Xanthium strumarium* var. *glabratum; Xanthium echinatum* Murray; *Xanthium echinellum* Greene & Rydb.; *Xanthium indicum* J. Koening ex Roxb.; *Xanthium inflexum* Mack. & Bush; *Xanthium italicum* Moretti; *Xanthium macrocarpum* var. *glabratum* DC.; *Xanthium occidentale* Bertol.; *Xanthium orientale* L.; *Xanthium pensylvanicum* Wallr.; *Xanthium pungens* Wallr.; *Xanthium saccharatum* Wallr.; *Xanthium speciosum* Kearney; *Xanthium strumarium* subsp. *italicum* (Moretti) D. Löve; *Xanthium vulgare* Hill

Notes (Spiny Cocklebur):

And thorns shall come up in her palaces, nettles and brambles in the fortresses thereof: and it shall be an habitation of dragons, and a court for owls.

Isaiah 34:13 (KJV)

Thorns shall grow over its strongholds, nettles and thistles in its fortresses. It shall be the haunt of jackals, an abode for ostriches.

Isaiah 34:13 (RSV)

On her dwelling towers thorns must come up, nettles and thorny weeds in her fortified places, and she must become an abiding place of jackals, the courtyard for the ostriches.

Isaiah 34:13 (NWT)

FIGURE 1.120 Spiny Cocklebur (*Xanthium spinosum*). Source: Regina Hughes in Reed (1970); Colored by Peggy Duke.

In my earlier book, I followed the Moldenkes (BIB), who concluded that the spiny cocklebur was the most likely candidate for this biblical thorn. But Zohary does not even cover this species in his Bible book, although three species of *Xanthium*, including this spiny one, are included in his *Flora of Palestine* (FP3).

- Leaves with three spines at base of leaf stalk: *Xanthium spinosum*
- Leaves spineless:
  - — Burr more than 2.5 cm, the prickles hairy at the base: *Xanthium italicum*
  - — Burr less than 2.1 cm, the prickles glandular pubescent: *Xanthium strumarium*

Common Names (Spiny Cocklebur):

Abrojillo (Arg.; MPG); Abrojito (Arg.; MPG); Abrojo (Arg.; Bol.; MPG); Alqo Khichkha (Bol.; Que.; MPG); Amor de Negro (Brazil; MPG); Amor Seco (Bol.; Ecu.; MPG); Anucháphi (Aym.; Bol.; MPG); Atulet (Arg.; MPG); Badhipjan Tariaki (Arab.; Syria; HJP); Bathurst-bur (Australia; Eng.; USN); Cachurrera Menor (Sp.; USN); Cadillos (Sp.; EFS); Carrapicho Bravo (Brazil; MPG); Casa Marutsja (Ecu.; MPG); Cepa de Caballo (Arg.; Sp.; Uru.; EFS; MPG); Choquchapi (Bol.; MPG); Chunngil (Ecu.; MPG); Clonqui (Chile; MPG); Dornige Spitzklette (Ger.; EFS; USN); Elpererin (Araucano; Arg.; MPG); Espina de Perro (Bol.; MPG); Espinho de Carneiro (Brazil; MPG); Floraria (Spain.; EFS; MPG); Gedoornde Stekelnoot (Dutch; EFS); Gratteron (Fr.; EFS); I Mop (Arg.; MPG); Juan de Alonzo (Bol.; MPG); Kachu Kawell (Arg. Vilele; MPG); Kokelin (Arg. Vilele; MPG); Lampourde aux Écrouelles (Fr.; EFS); Lampourde Épineuse (Fr.; USN); Lappolino (It.; EFS); Marucha (Ecu.; MPG); Matagallegos (Sp.; EFS); Pegotes (Sp.; EFS); Pitrak (Tur.; EFS); Prickly Burweed (Australia; Eng.; USN); Qopajchi (Bol.; Torotooro; MPG); Rat (Araucano; Arg.; MPG); Rat Ratraid (Arg.; MPG); Spino D'Asino (It.; EFS); Spiny Clotbur (Eng.; EFS); Spiny Cocklebur (Eng.; USN); Thorny Burweed (Eng.; EFS); Thorny Buttonbur (Eng.; EFS); Thorny Cocklebur (Eng.; EFS); Urusumaru (Bol.; Chiriguano; MPG); Xanthio (It.; EFS); Nscn.

Activities (Spiny Cocklebur):

Acaricide (1; MPG); Antifertility (f; VAD); Antiinflammatory (f; VAD; Antileukemic (1; MPG); Anti-MDR (1; X9364417); Antiseptic (f1; VAD; WOI); Antispasmodic (f; VAD); Antitumor (1; MPG); Astringent (f; VAD); Bactericide (1; MPG); Cicatrizant (f; VAD); CNS Depressant (1; MPG); Contraceptive (1; MPG); Depurative (f; HJP); Diaphoretic (f; DAW); Digestive (f; MPG); Diuretic (f; VAD); Emollient (f; MPG); Febrifuge (f; VAD); Hemostat (f; EFS); Hepatoprotective (f; VAD); Insulin Sparing (1; MPG); Laxative (f; VAD); Sedative (f; HJP); Sialagogue (f; EFS; VAD); Sudorific (f; VAD); Tonic (f; HJP).

Indications (Spiny Cocklebur):

Bacillus (1; MPG; X9364417); Bacteria (1; MPG); Blennorrhagia (f; MPG); Boil (f; HJP); Cancer (1; MPG); Cold (f; VAD); Cramp (f; VAD); Cystosis (f; VAD); Diarrhea (f; VAD); Dysentery (f; HJP); Dyspepsia (f; HJP); Dysuria (f; MPG); Edema (f; VAD); Epilepsy (f; HJP); Fever (f; VAD); Flu (f; VAD); Gout (f; VAD); Headache (f; MPG); Hepatosis (f; MPG); High Blood Pressure (f; VAD); Infection (f1; MPG; VAD; WOI); Inflammation (f; VAD); Insomnia (f; HJP); Klebsiella (1; X9364417); Leukemia (1; MPG); Malaria (f; EFS); Micrococcus (1; MPG); Nephrosis (f; MPG); Obesity (f; VAD); Oliguria (f; VAD); Pain (f; HJP); Pharyngosis (f; MPG); Pneumonia (1; X9364417); Pseudomonas (1; X9364417); Pyelosis (f; VAD); Rabies (f; HJP); Rheumatism (f; HJP); Salmonella (1; X9364417); Scrofula (f; EFS); Snakebite (f; DAW); Sore (f; HJP); Sore Throat (f; MPG); Splenosis (f; MPG); Staphylococcus (1; X9364417); Stress (f; HJP); Ulcer (f; HJP); Urethrosis (f; VAD); Urolithiasis (f; VAD); Uterosis (f; MPG); Wound (f; VAD).

Dosages (Spiny Cocklebur):

FNFF = X
100 g seed/liter water; 1/2 cup 2–3 ×/day (MPG); 100 g root/500 ml water; 1/2 cup 2–3 ×/day (MPG).

- Argentines take root as antimalarial, digestive, diuretic, and emollient (MPG).
- Bolivians take the leaf decoction for liver and stomach problems (MPG).

- Canary Islanders use shoots for diarrhea, dysentery, fever, hepatitis, and oliguria (MPG).
- Europeans recommend the plant for rabies (MPG).
- Lebanese reportedly feed children crushed seed to make blood and bones strong (HJP).
- Lebanese suggest a leaf tea for indigestion, a strong decoction for ulcers, internal or external (HJP).
- Lebanese use powdered seed to treat boils (HJP).
- Lebanese use the plant in steam baths to treat colds and rheumatism (HJP).
- Ukrainians use as calmative and sedative, and for dysentery, epilepsy, and hydrophobia (HJP).
- Uruguayans use the root as antiseptic, antispasmodic, diuretic, laxative, and use for hepatoses; other parts of the plant are taken in decoction or tea for malaria (MPG).

Downsides (Spiny Cocklebur):

Contraindicated in small children and lactating or pregnant women; possibly depressant, antifertility (VAD). As of November 2004, the FDA Poisonous Plant Database listed 24 titles alluding to the toxicity of this species.

SPINY ZILLA (*ZILLA SPINOSA* (L.) PRANTL.) ++ BRASSICACEAE

Synonyms:

Bunias spinosa L.; *Zilla myagriodes* Forssk *fide* FP1

Notes (Spiny Zilla):

And there shall be no more a pricking brier unto the house of Israel, nor any grieving thorn of all that are round about them, that despised them; and they shall know that I am the Lord GOD.

Ezekiel 28:24 (KJV)

And for the house of Israel there shall be no more a brier to prick or a thorn to hurt them among all their neighbors who have treated them with contempt. Then they will know that I am the Lord GOD.

Ezekiel 28:24 (RSV)

And be no will there prove to be to the house of Israel a malignant prickle or a painful thorn out of all those round about them, those who are treating them with scorn; and people will have to know that I am the sovereign Lord Jehovah.

Ezekiel 28:24 (NWT)

And there will be no more for the house of Israel a prickling brier or a painful thorn from any round about them who scorned them; then they will know that I am the Lord GOD.

Ezekiel 28:24 (NAST)

There were nearly 100 citations for *Zilla*, almost all of them for authors named Zilla, and almost no (only two) PubMed citations for the biblical plant *Zilla spinosa*. So, for a change, I went for a special Google searach <Zilla spinosa Bible> and got 15 citations, which I eagerly explored. The first two were unexpectably based on my own work, the third and fourth were a map of historical sites in the Holy Land, and the remainder were false drops. That is about the fifth time I have experienced this; most of the hot leads on the biblical plants I Google harken back to me. So I wasted 20 minutes on this Google search. The USDA nomenclature database has nothing on it. My own phytochemical database, also online at USDA, has nothing on it. And, I expect cruising through all my Mideastern books will waste even more time, but now in the Z's I am getting anxious to finish this venture. So I will paraphrase much of what Zohary has to say on his one page devoted to *Zilla*. The Arabic name *silla* for a conspicuously thorny plant (I remember it well from Egypt) was Latinized by Forsskal to *Zilla* as the generic scientific name. But quite often, as perhaps in the Ezekiel quotes above, it was reduced to a very general word for thorn rather than a specific reference to *Zilla spinosa*. Zohary offers a second quotation, "Lamech said to his wives: 'Adah and Zillah, hear my voice; you wives of Lamech, hearken to what I say: I have slain a man for wounding me, a young man for striking me.'" (Genesis 4:23; RSV). He then says, "It seems not too risky to suggest that Zillah, wife of Lamech, was named for this plant, so characteristic of the desert in the Land of Nod." (ZOH) One can think of it as spiny tumbleweed because, when mature, it may be uprooted by the wind and tumble through the desert.

Common Names (Spiny Zilla):

Chebreg (Mali; UPW); Ftozzer Oftazzen (Ber.; BOU); Hassaniya Achenfarach (Arab.; Mauritania; UPW); Oftazzen (Ber.; BOU); Roquette Épineuse (Fr.; BOU); Shabrom (Arab.; BOU); Shoubroq (Arab.; BOU); Silla (Arab.; BOU; ZOH); Silon (Heb., ZOH); Sirr (Arab.; BOU); Spiny Zilla (Eng.; ZOH); Zilla (Arab.; BOU); Nscn.

Activities (Spiny Zilla):

Antithyroid (1; X4129701); Litholytic (f; BOU).

Indications (Spiny Zilla):

Kidney stone (f; BOU); Nephrosis (f; BOU); Stone (f; BOU).

Dosages (Spiny Zilla):

FNFF = ?

- North Africans use the plant to treat kidney stones (BOU).

Natural History (Spiny Zilla):

Said to provide excellent fodder for camels and other stock, it is probably, like many Brassicaceae, quite edible and nutritious once one gets around those formidablke spines. The large leaves are near the base, getting smaller and smaller out toward the thorn tips (UPW; ZOH).

Extracts (Spiny Zilla):

Like many members of Brassicaceae, it contains goitrogenic glucosinolates and antithyroid goitrins (X4129701).

SYRIAN CHRISTTHORN (*ZIZIPHUS SPINA-CHRISTI* (L.) WILLD. ++ RHAMNACEAE

Synonyms:

Rhamnus spina-christi (L.) Desf.

Notes (Syrian Christthorn):

Then came Jesus forth, wearing the crown of thorns, and the purple robe. And Pilate saith unto them, Behold the man!

John 19:5 (KJV)

So Jesus came out, wearing the crown of thorns and the purple robe. Pilate said to them, "Behold the man!"

John 19:5 (RSV)

Accordingly Jesus came outside, wearing the thorny crown and the purple outer garment. And he said to them "Look! The man!"

John 19:5 (NWT)

Many authors view *Ziziphus* as corresponding to the crown of thorns. On the contrary, there are at least a dozen thorny species, not completely ruling out *Ziziphus*, but pushing *Sarcopoterium* as the most likely candidate. And Zohary notes that there are still a few *Ziziphus* growing on the eastern slopes of the Temple Mount (Mount Moriah).With respect to Zohary, I include both plants. The *Useful Plants of Tropical West Africa* (UPW) hints that this was the *dudaim*, the name of the plant Leah mentioned to Joseph as a possible cure for sterility. They suggest that the mandrake does not occur in the biblical arena, while the dudaim bears apples and grows along the river banks rather than in open fields. UPW further states that these fruits, the Lotiphagi ate which were *Ziziphus* rather than *Nymphaea*. Christians consider this to be the crown of thorns that Jesus wore before the crucifixion. Indian Muslims, respecting the plant highly, wash corpses in the leaf decoction. Revered by the Arabs and mentioned in the Quran (LIII: 13–18; LVI: 28–32); the "lote-tree" is revered also by Muslims throughout the Middle East. It has been esteemed both as a fruit and medicinal plant for millennia. Some believe that on the Day of Judgment, those at Allah's right hand will dwell among thornless lote-trees (UPW; ZOH).

Common Names (Syrian Christthorn):

Abaqua (Ber.; BOU); Ardj (Arab.; BOU); As Sidr (Arab.; Syria; HJP); Atad (Heb.; X16270941; ZOH); Bauyer (Arab.; Sen.; UPW); Christ Thorn (Eng.; ZOH); Christ's Thorn (Eng.; FAC; UPW); Christthorn Jujube (Eng.; X16270941; HJP); Dabi (Mali; UPW); Dawm (Arab.; Syria; HJP); Dem (Arab.; UPW); Dudaim (Heb.; UPW); Dum (Arab.; ZOH); Èpine du Christ (Fr.; UPW); Ghassel (Arab.; BOU); Jujuba de Palastine (Fr.; UPW); Kanari (Isr.; X16270941); Karnoyer (Arab.; Niger; UPW); Korna (Ber.; BOU); Lote (Eng.; UPW); Nabag (Nig.; UPW); Nabaq (Arab.; GHA); Nabas (Egypt; UPW) Nabk (Arab.; Syria; HJP); Nabq (Arab.; BOU); Ourdj (Arab.; BOU); Rimin (Isr.; X16270941); Sadr (Arab.; X16270941); Sedam (Arab.; UPW); Seder (Arab.; UPW); Sedr (Arab.; GHA); Shagar en Nabk (Arab.; HJP); Sheizaf (Heb.; X16270941); Sidr (Arab.; Syria; BOU; GHA; HJP); Syrian Christthorn (Eng.; CR2); Tsal (Arab.; X16270941); Zefzhous (Arab.; BOU); Zegzeg (Arab.; BOU).

Activities (Syrian Christthorn):

Analgesic (f1; CRC; GHA; X11395256); Anthelmintic (f; FNF); Antiinflammatory (f; BOU); Antinociceptive (f1; GHA; X11395256); Antiseptic (f1; X11167035); Antitumor (f; CRC); Astringent (f1; CRC; X12826300); Bactericide (1; X11167035); Bechic (f; UPW); CNS Depressant (1; X11744289); Collyrium (f; CRC); Demulcent (f; CRC); Depurative (f; CRC); Emollient (f; BIB; CRC; UPW); Febrifuge (f; BOU; CRC); Lactagogue (f; FNF); Laxative (f; BIB; CRC); Pectoral (f; CRC); Propecic (f; UPW); Refrigerant (f; CRC); Sedative (1; X11744289); Stomachic (f; CRC); Tonic (f; CRC; GHA); Vulnerary (f; JHP).

Indications (Syrian Christthorn):

Abscess (f; BIB; BOU); Arthrosis (f; CRC; GHA; HJP); Bacteria (1; X11167035); Blister (f; FNF); Bronchosis (f; CRC); Bruise (f; FNF); Burn (f; FNF); Cancer (1; CRC; JLH); Cardiopathy (f; FNF); Childbirth (f; GHA); Cold (f; CRC); Conjunctivosis (f; BOU); Constipation (f; BOU); Cough (f; CRC);

Dermatosis (f; BIB; UPW); Diarrhea (f1; UPW; X12826300); Dyslactea (f; FNF); Fever (f; BIB; BOU; CRC; HJP); Fungus (f; UPW); Furuncle (f; BIB; BOU); Gastrosis (f; FNF); Gingivosis (f; GHA); Headache (f; GHA); Heartburn (f; FNF); Hemorrhoid (f; FNF); Hepatosis (f; CRC; HJP); High Blood Pressure (f; CRC); Infection (f1; UPW; X11167035); Inflammation (f; BOU); Measles (f; BIB; BOU); Myalgia (f; FNF); Mycosis (f; UPW); Nervousness (f; FNF); Obesity (f; UPW); Ophthalmia (f; BIB; BOU; CRC); Pain (f1; GHA; X11395256); Proctosis (f; FNF); Pulmonosis (f; GHA); Respirosis (f; GHA); Rheumatism (f; CRC); Ringworm (f; UPW); Snakebite (f; BIB; BOU; UPW); Sore (f; BIB; UPW); Sterility (f; UPW); Toothache (f; CRC); Tuberculosis (f; CRC); Tumor (1; CRC); Venereal Disease (f; CRC; HJP); Worm (f; BOU); Wound (f; FNF).

Dosages (Syrian Christthorn):

FNFF = !!
Acid yellow drupaceous fruits eaten raw, dried, or made into breadstuffs. Sometimes in Israeli markets (FAC; TAN; ZOH).

- Arabs and Iraqi Jews apply fruit juice to wounds (X16270941).
- Bedouins drink leaf decoction as lactagogue (X16270941).
- Bedouins (Negev; Sinai) inhale leaf steam for myalgia (X16270941).
- Bedouins of Egypt put leaf cataplasm on abscesses and furuncles, and before retiring, apply to inflamed eyes (BOU).
- Camerounians plaster the plant on ringworm (UPW).
- Dhofari apply crushed leaves to arthritic joints and skin sores (GHA), seeds, fruits, and leaves to bruises (X16270941).
- Ethiopians use the plant as bechic and emollient (UPW).
- Ghanans apply the sap to cuts and wounds (UPW).
- Iraqi and Yemenite Jews use the leaves for hemorrhoids (X16270941).
- Lebanese eat boiled fruits as compote for bronchosis, cough, and tuberculosis (HJP).
- Lebanese take floral tea for fever and use as collyrium (HJP).
- Lebanese take powdered seed with lemon juice for liver ailments (HJP).
- Madagascans use the plant for diarrhea (UPW).
- Mauritanian women use powdered leaves to darken and lengthen the hair (UPW).
- Nigerians around Born take the bark as a slimming medicine (UPW).
- North Africans apply the wood ash, mixed with vinegar, to snakebite (BOU).
- Somali rub dried leaves on skin and wash, as a cleansing detergent (UPW).

Natural History (Syrian Christthorn):

Flowers said to be a favored food of the elephant in Sudan (UPW).

Extracts (Syrian Christthorn):

LD50 (aqueous root bark extract) = 2235 mg/kg ipr mus (X11395256); LD50 (methanol bark extract) = 345 mg/kg ipr mus; and LD50 (methanol bark extract) = 1200 mg/kg orl mus (X12826300).

EELGRASS (*ZOSTERA MARINA L.*) + ZOSTERACEAE

Notes (Eelgrass):

> **The waters compassed me about, even to the soul: the depth closed me round about, the weeds were wrapped about my head.**
>
> **Jonah 2:5 (KJV)**

FIGURE 1.121 Eelgrass (*Zostera marina*).

The waters closed in over me, the deep was round about me; weeds were wrapped about my head.

Jonah 2:5 (RSV)

Waters encircled me clear to the soul; the watery deep it kept enclosing me; weeds were wound about my head.

Jonah 2:5 (NWT)

In my King James Version, the word *weeds* shows up only once, in the story of Jonah. Some scholars have concluded as did I (BIB) that this weed could be it, so I include this account in my revision of biblical herbs. The leaves may prove valuable as fodder, thatching, or packing material. The foliage is an important food for some sea turtles and water fowl. An important shallow water, mud-flat stabilizer, the plant helps to sustain the productivity of estuarine areas. The plant is used by the Seri Indians of Mexico for diarrhea.

Common Names (Eelgrass):

Alva Marina (Sp.; HOC); Ama Mo (Japan; TAN); Barnacle Grass (Eng.; DAA); Eelgrass (Eng.; TAN); Grass Weed (Eng.; HOC); Grass Wrack (Eng.; BIB); Hai Dai (Pin.; DAA); Sea Ware (Eng.; HOC); Seagrass (Eng.; BIB); Seawrack (Eng.; BIB; DEM); Zostére Marine (Fr.; USN).

Activities (Eelgrass):

Antiaging (1; X15022719); Antioxidant (1; X15022719); Antiseptic (1; X1755708); ATPase (1; X12407193); Bacteria (1; X1755708); Dropsy (f; DAA); Immunomodulator (1; X1755708); Matrix Metalloproteinase Inhibitor (1; X15022719); Phagocytotic (1; X1755708); Photoprotective (1; X15022719).

Indications (Eelgrass):

Bacteria (1; X1755708); Diarrhea (f; EB28:429); Dropsy (f; DAA); Edema (f; HOC); Goiter (f; DAA; HOC); Hysteria (f; DAA; HOC); Infection (1; X1755708); Sunburn (1; X15022719); Swelling (f; HOC); Wrinkle (1; X15022719).

Dosages (Eelgrass):

FNFF = !
Bellacoola eat plants raw with eulachon grease. Cowichan use fleshy roots and leaf bases to flavor seal, porpoise, and deer meat. Hesquiat clean and eat the brownish rhizomes raw. Kwakiutl dip stems and roots in oil to eat during feasts. Nitinaht once ate the fleshy, whitish rhizomes raw. Oweekeno picked and ate the leaves with attached herring spawn. Saanich used fleshy roots and leaf bases to flavor deer, porpoise and seal. Seri Indians of Mexico ground the seeds into an edible flour. The grain, although small and relatively bland, is toasted, winnowed, and ground into flour and made into a gruel, traditionally combined with other food (e.g., sea turtle oil or honey) in Baja California. (BIB; DEM; http://www.innvista.com/health/foods/seeds/waternut.htm.)

- Seri Indians of Mexico use the plant for diarrhea (BIB).

Natural History (Eelgrass):

Brownish "roots" (actually rhizomes) eaten by black brants, Canada geese, mallard ducks, and cattle (DEM).

Extracts (Eelgrass):

Zosterin (a seagrass pectin) has antibacterial and immunomodulatory activities of therapeutic efficacy in experimental infections (X1755708).

References

Abascal, K. and Yarnell, E. 2003. The Many Faces of *Silybum marianum* (Milk Thistle). 1. Treating Cancer and Hyperlipidemia and Restoring Kidney Function. *Alternative & Complementary Therapies*, 9(4):170–175. (ACT9:170)

Abascal, K. and Yarnell, E. 2003. The Many Faces of *Silybum marianum* (Milk Thistle). 2: Clinical Uses, Safety, and Types of Preparations. *Alternative & Complementary Therapies*, 9(5):251–256. (ACT9:251)

Abe, K. and Saito, H. 2000. Effects of Saffron Extract and Its Constituent Crocin on Learning Behaviour and Long-Term Potentiation. *Phytotherapy Research* 14(3):149–152. (PR14:149)

Agriculture Handbook No. 165. 1960. *Index of Plant Diseases in the United States*. USGPO. Washington, D.C.

Akyuz, F., Kaymakoglu, S., Demir, K., Aksoy, N., Adalet, I., and Okten, A. 2005. Is There Any Medical Therapeutic Option in Hepatopulmonary Syndrome? A Case Report. *European Journal of Internal Medicine*, 16(2):126–128.

Alakbarov, F. 2003. Aromatic Herbal Baths of the Ancients. *HerbalGram* 57:40–49. (HG57:40)

Ali, B.H., Bashir, A.K., and Rasheed, R.A. 2001. Effect of the Traditional Medicinal Plants *Rhazya stricta, Balanitis aegyptiaca* and *Haplophylum tuberculatum* on Paracetamol-Induced Hepatotoxicity in Mice. *Phytotherapy Research,* 15(7):598–603. (PR15:598)

Allen, D.E. and Hatfield, G. 2004. *Medicinal Plants in Folk Tradition — An Ethnobotany of Britain and Ireland.* Timber Press, Portland, OR. 432 pp. (AAH)

Al-Qarawi, A.A., Abdel-Rahman, H.A., El-Badry, A.A., Harraz, F., Razig, NA., and Abdel-Magied, E.M. 2000. The Effect of Extracts of *Cynomorium coccineum* and *Withania somnifera* on Gonadotrophins and Ovarian Follicles of Immature Wistar Rats. *Phytotherapy Research,* 14(4):288–290. (PR14:288)

Anderson, E.F. 1986. Ethnobotany of Hill Tribes of Northern Thailand. I. Medicinal Plants of Akha. *Economic Botany,* 40:38–53. (EB40:38)

Arsene, V. Pierre-Noel. 1971. *Nomenclature Polyglotte des Plantes Haitiennes et Tropicales.* (Under supervision of Rene Laroche.) Presses Nationales d'Haiti, Port au Prince. 588 pp. (AVP)

Arvigo, R. and Balick, M. 1993. *Rain Forest Remedies — One Hundred Healing Herbs of Belize.* Lotus Press, Twin Lakes, WI. 221 pp. (AAB)

Associated Press, January 19, 2000.

Austin, D.F. 2004. *Florida Ethnobotany.* CRC Press, Boca Raton, FL. 909 pp. (AUS)

Balick, M.J., Nee, M.H., and Atha, D.E. 2000. *Checklist of the Vascular Plants of Belize.* Mem. NY Bot. Gard., Vol. 85., 246 pp. (BNA)

Bass, T. as quoted in Crellin, J.K. and Philpott, J. 1990. *Herbal Medicine Past and Present. Vol. II. A Reference Guide to Medicinal Plants.* Duke University Press, Durham, NC. 549 pp. (TOM)

Batanouny, K.H. 1981. *Ecology and Flora of Qatar.* Alden Press, Oxford. 245 pp. (BAT)

Beauvoir, M.G., DeFillips, R.A., Wolpert, B.J., with Crepin, J. 2001. *Selected Medicinal Plants of Haitian Vodou.* Typescript. Smithsonian Institution. Washington, D.C. 256 pages. (VOD)

Bejar, E., Bussmann, R., Roa, C., and Sharon, D. 2001. *Herbs of Southern Ecuador.* Latino Herbal Press, Spring Valley, CA. 350 pp. (BEJ)

Bisset, N.G. (Ed.). 1994. *Herbal Drugs and Phytopharmaceuticals.* (English translation of Wichtl, 1984, 1989). CRC Press, Boca Raton, FL. 566 pp. (BIS)

Blaschek, W., Hansel, R., Keller, K., Reichling, J., Rimpler, H., and Schneider, G. (Eds.). 1998. *Hager's Handbuch der Pharmazeutischen Praxis*, Auflage Band 2 (A–K). Springer-Verlag, Berlin. 909 pp. (HH3)

Blaschek, W., Hansel, R., Keller, K., Reichling, J., Rimpler, H., and Schneider, G. (Eds.). 1998. *Hager's Handbuch der Pharmazeutischen Praxis*, Auflage Band 3 (L–Z). Springer-Verlag, Berlin. 858 pp. (HH3)

Blumenthal, M., Busse, W.R., Goldberg, A., Gruenwald, J., Hall, T., Riggins, C.W., and Rister, R.S. (Eds.). 1998. *The Complete German Commission E Monographs. Therapeutic Guide to Herbal Medicines.* American Botanical Council, Austin, TX; and Integrative Medicine Communications, Boston, MA. 684 pp. (KOM)

Blumenthal, M., Goldberg, A., and Brinckmann, J. 2000. *Herbal Medicine — Expanded Commission E Monographs.* Integrative Medicine Communications, Newton, MA. 516 pp. (BGB)

Bnouham, M., Merhfour, F.-Z., Ziyyat, A., Mekhfi, H., Aziz, M., and Legssyer, A. 2003. Antihyperglycemic Activity of the Aqueous Extract of *Urtica dioica*. *Fitoterapia,* (74):7–8:677–681. (FT74:677)

Boik, J. 2001. *Natural Compounds in Cancer Therapy — Promising Nontoxic Antitumor Agents from Plants and Other Natural Sources*. Oregon Medical Press, Princeton, MN. 521 pp. (BO2)

Bombardelli, E. and Morazzoni, P. 1997. *Urtica dioica* L. *Fitoterapia* 68(5):387–402. (FT68:387)

Bonjar, G.H.S. 2004. Inhibition of Clotrimazole-Resistant *Candida albicans* by Plants Used in Iranian Folkloric Medicine. *Fitoterapia,* 75(1):74–76. (FT75:74)

Borek, C. 2001. Antioxidant Health Effects of Aged Garlic Extract. Journal of Nutrition, 131:1010S–1015S. (JN131:1010s)

Borrelli, F. and Izzo, AA. 2000. The Plant Kingdom as a Source of Anti-ulcer Remedies. Phytotherapy Research, 14(8):581–591. (PR14:581)

Boulos, L. 1983. *Medicinal Plants of North Africa*. Reference Publications, Algonac, MI. 286 pp. (BOU)

Bown, D. 2001. *Revised American Edition. New Encyclopedia of Herbs & Their Uses*. DK Publishing, New York. 448 pp. (BOW)

Brenner, S., Ruocco, V., Wolf, R., de Angelis, E., and Lombardi, M.L. 1995. Pemphigus and Dietary Factors. *In vitro* Acantholysis by Allyl Compounds of the Genus *Allium*. *Dermatology,* 190(3):197–202.

Bruneton, J. 1999. *Pharmacognosy, Phytochemistry, Medicinal Plants,* 2nd ed. Lavoisier Publishing, Paris. 1119 pp. (BRU)

Burits, M. and Bucar, F. 2000. Antioxidant Activity of *Nigella sativa* Essential Oil. *Phytotherapy Research,* 14(5):323–328. (PR14:323)

Burkill, H.M. 1985–1995. *The Useful Plants of West Tropical Africa*. Royal Botanical Gardens, Kew. Edition 2. Vol. 1, A–D. 1985. 960 pp., Vol. 2, E–I. 1994. 636 pp., Vol. 3, J–R. 1995. 857 pp., Vol. 4., Vol. 5, S–Z. 2000. 686 pp. (Plus Cryptogams., Addenda). (UPW)

Burkill, I.H. 1966. *A Dictionary of the Economic Products of the Malay Peninsula*. Art Printing Works, Kuala Lumpur. 2 vols. (IHB)

Bushway, R.J. 1986. Determination of Alpha- and Beta-Carotene in Some Raw Fruits and Vegetables by High-Performance Liquid Chromatography. *Journal of Agriculture and Food Chemistry,* 34(3):409–412. (JAF34:409)

Calixto, J.B., Beirith, A., Ferreira, J., Santos, A.R.S., Filho, V.C., and Yunes, R.A. 2000. Naturally Occurring Antinociceptive Substances from Plants. *Phytotherapy Research,* 14(6):401–418. (PR14:401)

Cantrell, C.L., Nunez, I.S., Castaneda-Acosta, J., Foroozesh, M., Fronczek, F.R., Fischer, N.H., and Franzblau, S.G. 1998. Antimycobacterial Activities of Dehydrocostus Lactone and Its Oxidation Products. *Journal of Natural Products,* 61(10):1181–1186. (JNP61:1181)

Castleman, M. 1996. Spice-Rack Remedies — Turn to Your Kitchen for Health Care. *Herbs for Health,* (Nov./Dec.) 22–29.

Charalambous, G. (Ed.). 1994. *Spices, Herbs and Edible Fungi*. Elsevier Science, B.V. Amsterdam. 764 pp. (SPI)

Choi, H.Y., Jhun, E.J., Lim, B.O., Chung, I.M., Kyung, S.H., and Park, D.K. 2000. Application of Flow Injection — Chemiluminescence to the Study of Radical Scavenging Activity in Plants. *Phytotherapy Research,* 14(4):250–253. (PR14:250)

Coe, F.G. and Anderson, G.J. 1999. Ethnobotany of the Sumu (Ulwa) of Southeastern Nicaragua and Comparisons with Miskitu Plant Lore. *Economic Botany*., 53(4):363–386. (ULW)

Corlett, J.L., Dean, E.A., and Grivetti, L.E. 2003. Hmong Gardens: Botanical Diversity in an Urban Setting. *Economic Botany*, 57:365–379. (EB57:365)

Crowell, P.L., Siar Ayoubi, A., and Burke, Y.D. 1996. Antitumorigenic Effects of Limonene and Perillyl Alcohol against Pancreatic and Breast Cancer. *Advances in Experimental and Medical Biology,* 401:131–136. Review.

CSIR (Council of Scientific and Industrial Research). 1948–1976. *The Wealth of India. A Dictionary of Indian Raw Materials & Industrial Products*. 11 vols. CSIR Hillside Road, New Delhi, India. (WOI)

CSIR (Council of Scientific and Industrial Research). 1985. *The Wealth of India. A Dictionary of Indian Raw Materials & Industrial Products*. First Revised Edition. Vol. 1: A. CSIR Hillside Road, New Delhi. 513 pp. + 54 p index. (WO2)

CSIR (Council of Scientific and Industrial Research). 2000. *The Wealth of India. A Dictionary of Indian Raw Materials & Industrial Products*. First Supplement Series (raw materials). Vol. 1: A-Ci. CSIR Hillside Road, New Delhi. 283 pp. + 24 p index. (WO3)

Cunnane, S.C. and Thompson, L.U. (Eds.). 1995. *Flaxseed in Human Nutrition*. AOCS Press, Champaign, IL. 384 pp.

Davies, S. and Stewart, A. 1990. *Nutritional Medicine*. Avon Books, New York. 509 pp. (DAS)

De Lucca, D.M. and Zalles, A.J. 1992. *Flor Medicinal Bolivia — Diccionario Enciclopedia*. Los Amigos del Libro, La Paz, Bolivia. 498 pp. (DLZ)

D'epiro, N.W. (Ed.). 1997. Herbal Medicine: What Works., What's Safe. *Patient Care*, 31(16):48–77.

De Pascual-Teresa, S., Gutierrez-Fernandez, Y., Rivas-Gonzalo, J.C., and Santos-Buelga, C. 1998. Characterization of Monomeric and Oligomeric Flavan-3-ols from Unripe Almond Fruits. *Phytochemical Analysis*, 9(1):21–27.

De Smet, P.A.G.M., Keller, K., Hansel, R., and Chandler, R.F. (Eds.). 1997. *Adverse Effects of Herbal Drugs 3*. Springer-Verlag, Berlin. 348 pp. (AEH)

Duke, J.A. Personal communication. (JAD)

Duke, J.A. Micmac Module of Duke's Medical Botany Syllabus. Partially Online at http://www.ars-grin.gov/duke/syllabus. (MIC)

Duke, J.A. 1981. *Handbook of Legumes of World Economic Importance*. Plenum Press, New York. 345 pp. (LEG)

Duke, J.A. 1983. *Medicinal Plants of the Bible*. Trado-Medic Books. A Division of Conch Magazine Ltd., Publishers, New York. 233 pp. (BIB)

Duke, J.A. 1983. *Handbook of Energy Species*. USDA documents, online at Purdue University. http://www.hort.purdue.edu/newcrop/duke_energy/dukeindex.html. (HOE)

Duke, J.A. 1985. *Handbook of Medicinal Herbs*. CRC Press, Boca Raton, FL. 677 pp. (CRC)

Duke, J.A. and Ayensu, E.S. 1985. *Medicinal Plants of China*. 2 vols. Reference Publications, Algonac, MI. 705 pp. (DAA)

Duke, J.A. 1986. *Isthmian Ethnobotanical Dictionary*, 3rd ed. Scientific Publishers. Jodhpur, India. 205 pp. Partially online at http://www.ars-grin.gov/duke/syllabus. (IED)

Duke, J.A. 1989. *Handbook of Nuts*. CRC Press, Boca Raton, FL. 343 pp. Illustrated by Peggy Duke. (NUT)

Duke, J.A. and Wain, K. 1991. *Medicinal Plants of the World*. USDA Computer listing of folk medicines of the world, now slightly updated and online at USDA (http://www.ars- grin.gov/duke/). (DAW)

Duke, J.A. and duCellier, J.L. 1993. *Handbook of Alternative Cash Crops*. CRC Press. Boca Raton, FL. 536 pp. (DAD)

Duke, J.A. and Vasquez Martinez, R. 1994. *Amazonian Ethnobotanical Dictionary*. CRC Press, Boca Raton, FL. 215 pp. (Partially online at http://www.ars-grin.gov/duke/syllabus.) (DAV)

Duke, J.A. 1997. *The Green Pharmacy*. Rodale Press, Emmaus, PA. 507 pp. (TGP)

Duke, J.A. 1999. *Herbs of the Bible — 2000 Years of Plant Medicine*. Interweave Press, Loveland, CO. 241 pp. (BI2)

Duke, J.A., Bogenschutz-Godwin, M.J., duCellier, J., and Duke, P.A. 2002. *Handbook of Medicinal Herbs*, 2nd ed. CRC Press, Boca Raton, FL. 936 pp. (CR2)

Duke, J.A., Bogenschutz-Godwin, M.J., duCellier, J., and Duke, P.A. 2002. *Handbook of Medicinal Spices*. CRC Press, Boca Raton, FL. 348 pp. (HOS)

Earle, C.J. 1997–2007. Gymnosperm Database. http://www.conifers.org (CJE)

Eastman, J. 1992. *The Book of Forest and Thicket*. Stackpole Books, Mechanicsburg, PA. 212 pp. (EAS)

Eastman, J. 1995. *The Book of Swamp and Bog*. Stackpole Books, Mechanicsburg, PA. 237 pp. (EAS)

Eastman, J. 2003. *The Book of Field and Roadside*. Stackpole Books. Mechanicsburg, PA. 336 pp. (EAS)

Egg, A.B. 1999. *Diccionario Enciclopédico de Plantas Útiles del Perú*. Centro de Estudios Regionales Andinos Bartolomé de Las Casas, Cuzco. 550 pp. (EGG)

Erichsen-Brown, C. 1989. *Medicinal and Other Uses of North American Plants — A Historical Survey with Special Reference to the Eastern Indian Tribes*. Dover Publications, New York. 512 pp. (CEB)

Ertugö, F. 2000. An Ethnobotanical Study in Central Anatolia. *Economic Botany* 54:155–182. (EB54:155)

Facciola, S. 1998. *Cornucopia — A Source Book of Edible Plants*. Kampong Publications, Vista, CA. 713 pp. (FAC)

Feinbrun-Dothan, N. 1978. *Flora Palaestina*. 3. Text. Ericaceae to Compositae. 481 pp. Part 3 (Plates) not paginated. 757 plates, plus indices. Israel Academy of Sciences and Humanities. (FP3)

Felter, H.W. and Lloyd, J.U. 1898. *King's American Dispensatory*. 2 vols. 18th edition, third revision. Reprinted Eclectic Medical Publications, Portland, OR. 1983. 2 vols. 2172 pp. (FEL)

Fleisher, A. and Fleisher, Z. 1988. Identification of Biblical Hyssop and Origin of the Traditional Use of Oregano-Group Herbs in the Mediterranean Region. *Economic Botany*, 42:232–241. (EB42:232)

Fleming, T. et al. 1998. *PDR for Herbal Medicine*, 1st ed. Medical Economics Co., Montvale, NJ. 1244 pp. (PHR)

Foster, S. 1996. *Herbs for Your Health — A Handy Guide for Knowing and Using 50 Common Herbs*. Interweave Press, Loveland, CO. (SF)

Foster, S. and Duke, J.A. 1990. Peterson Field Guides. *A Field Guide to Medicinal Plants. Eastern and Central North America*. Illustr. Houghton Miflin, Boston, MA, and new edition. 2000. (FAD)

Foster, S. and Hobbs, C. 2002. *Peterson Field Guides — Western Medicinal Plants and Herbs*. Houghton Mifflin, New York. 442 pp. (FAH)

Foster, S. and Yue, C.-G. 1992. *Herbal Emissaries — Bringing Chinese Herbs to the West*. Healing Arts Press, Rochester, VT. 356 pp. (FAY)

Garner-Wizard, M. 2004. *Beyond Liver Conditions: Other Uses for Milk Thistle with Clinical Guidelines*. August 16. American Botanical Council, Austin, TX. (HC020444-262)

Germosén-Robineau, L. (Ed.) 1997. *Farmacopea Vegetal Caribeña*. Enda-caribe, Tramil. Ediciones Emile Desormeaux, Martinique F.W.I. 360 pp. (TRA)

Ghazanfar, S.A. 1994. *Handbook of Arabian Medicinal Plants*. CRC Press, Boca Raton, FL. 263 pp. (GHA)

Golden, M. 1984. *Ricinus,* Morgan Golden, USDA. Personal communication.

Govere, E.M, Tonegawa, M., Bruns, M.A., Wheeler, E.F., Heinemann, P.H., Kephart, K.B., and Dec, J. 2005. Deodorization of Swine Manure Using Minced Horseradish Roots and Peroxides. *Journal of Agriculture and Food Chemistry,* 53(12):4880–4889.

Grenand, P., Moretti, C., and Jacquemin, H. 1987. *Pharmacopees Traditionnelles en Guyane*. L'orstrom, Paris. 569 pp. (GMJ)

Grieve, M. 1931. *A Modern Herbal*. Hafner Press, New York. 915 pp. Dover Reprint., 2 vols. First reprinted 1971. Dover, New York. 902 pp. (GMH)

Gruenwald, J. et al. 2000. *PDR for Herbal Medicines*. Edition 2. Medical Economics Company, Montvale, NJ. 858 pp. (PH2)

Gupta, M.P. (Ed.) 1995. *270 Plantas Medicinales Iberoamericanas*. Talleres de Editorial Presencia Ltda. Bogota. 617 pp. (MPG)

Hadis, M., Lulu, M., Mekonnen, Y., and Asfaw, T. 2003. Field Trials on the Repellent Activity of Four Plant Products against Mainly Mansonia Population in Western Ethiopia. *Phytotherapy Research,* 17(3):202–205. (PR17:202)

Hansel, R., Keller, K., Rimpler, H., and Schneider, G. (Eds.). 1992. *Hager's Handbuch der Pharmazeutischen Praxis*, Drogen (A–D). Springer-Verlag, Berlin. 1209 pp. (HH2)

Hansel, R., Keller, K., Rimpler, H., and Schneider, G. (Eds.). 1993. *Hager's Handbuch der Pharmazeutischen Praxis*, Drogen (E–O). Springer-Verlag, Berlin. 970 pp. (HH2)

Hansel, R., Keller, K., Rimpler, H., and Schneider, G. (Eds.). 1994. *Hager's Handbuch der Pharmazeutischen Praxis*, Drogen (P–Z). Springer-Verlag, Berlin. 1196 pp. (HH2)

Harborne, J.B. and Baxter, H. (Eds.). 1983. *Phytochemical Dictionary. A Handbook of Bioactive Compounds from Plants*. Taylor & Frost, London. 791 pp. (JBH)

Harris, R. 2002. Progress with Superficial Mycoses Using Essential Oils. *The International Journal of Aromatherapy,* 12(2):83–91. (JAR12:83)

Hartwell, J.L. 1982. *Plants Used against Cancer. A Survey*. Quarterman Publications, Inc. Lawrence, MA. 710 pp. (Reprinted from eleven different issues of *Lloydia,* 1967–1971). (JLH)

Hegnauer, R. 1962-1997. *Chemotaxonomie der Pflanzen* Bde 1–11. Birkhauser Verlag Basel, Boston, Berlin. (HEG)

Hill, S.R. and Duke, P.K. 1985. *100 Poisonous Plants of Maryland*. Bull. 314, University of Maryland 55 pp.

Hitchcock, A.S., and Chase, A. 1971. *Manual of the grasses of the United States*. 2nd ed., rev. by Agnes Chase. Dover Publications, New York. 2 vol. (1051 p.): ill., maps; 24cm.

Hmamouchi, M., Lahlou, M., and Agoumi, A. 2000. Molluscicidal Activity of Some Moroccan Medicinal Plants. *Fitoterapia,* 71(3):308–314. (FT71:308)

Huang, K.C. 1999. *The Pharmacology of Chinese Herbs*. Ed. 2. CRC Press, Boca Raton, FL. 512 pp. (KC2)

Humphrey, S. 2003. *The Nursing Mother's Herbal*. Fairview Press, MN. 355 pp. (NMH)

Hussein, G., Miyashiro, H., Nakamura, N., Hattori, M., Kakiuchi, N., and Shimotohno, K. 2000. Inhibitory Effects of Sudanese Medicinal Plant Extracts on Hepatitis C Virus (HCV) Protease. *Phytotherapy Research,* 14(7):510–516. (PR14:510)

Hussey, J.S. 1974. Some Useful Plants of Early New England. *Economic Botany,* 28:311–337. (EB28:316)

Hutchings, A., Scott, A.H., Lewis, G., and Cunningham, A. 1996. *Zulu Medicinal Plants. An Inventory*. University of Natal Press, Pietermaritzburg. 450 pp. (ZUL)

ICMR (Indian Council of Medical Research). 1976. *Medicinal Plants of India*. Vol. 1. Indian Council of Medical Research. Cambridge Printing Works, New Delhi. 487 pp. (MPI)

ICMR (Indian Council of Medical Research). 1987. *Medicinal Plants of India*. Vol. 2. Indian Council of Medical Research. Cambridge Printing Works, New Delhi. 600 pp. (MPI)

Ide, N. and Lau, BHS. 2001. Garlic Compounds Minimize Intracellular Oxidative Stress and Inhibit Nuclear Factor-B Activation. *Journal of Nutrition,* 131:1020S–1026S. (JN131:1020s)

Jacobs, M.L. and Burlage, H.M. 1958. *Index of Plants of North Carolina with Reputed Medicinal Uses*. Printed by H. M.Burlage, and dedicated to Marion Lee Jacobs, whose untimely death "prolonged publication of this work." 322 pp. (BUR)

Jain, S.K. and Tarafder, C.R. 1970. Medicinal Plant-Lore of the Santals. *Economic Botany,* 24:241–278. (EB24:259., EB24:265)

Jain, S.K. and deFillips, R. 1991. *Medicinal Plants of India*. 2 vols. Reference Publications, Algonac, MI. 849 pp. (SKJ)

Joseph, J., Nadeau, D., and Underwood, A. 2001. *The Color Code*. Hyperion. New York. (JNU)

Judd, A. 1995. Flax — Some Historical Considerations, pp. 1–10. In Cunnane, S.C. and Thompson, L.U. (Eds.). 1995. *Flaxseed in Human Nutrition*. AOCS Press, Champaign, IL. 384 pp.

Kalpana, C., Rajasekharan, K.N., and Menon, V.P. 2005. Modulatory Effects of Curcumin and Curcumin Analog on Circulatory Lipid Profiles during Nicotine-Induced Toxicity in Wistar Rats. *Journal of Medicinal Food,* 8(2):246–250. (JMF8:246)

Kapoor, L.D. 1990. *Handbook of Ayurvedic Medicinal Plants*. CRC Press, Boca Raton, FL. 416 pp. (KAP)

Kaul, M.K. 1997. *Medicinal Plants of Kashmir and Ladakh — Temperate and Cold Himalaya*. Indus Publishing Company, New Delhi. 173 pp. (MKK)

Khan, S. and Balick, M.J. 2001. Therapeutic Plants of Ayurveda: A Review of Selected Clinical and Other Studies for 166 Species. *The Journal of Alternative and Complementary Medicine,* 7(5):405–515. Mary Ann Liebert, Inc. (JAC7:405)

Khasbagan, S., Huai, H.-Y., and Pei, S.-J. 2000. Wild Plants in the Diet of Arhorchin Mongol Herdsmen in Inner Mongolia. *Economic Botany,* 54:528–536. (EB54:528)

Kindscher, K. and Hurlburt, D.P. 1998. Huron Smith's Ethnobotany of the Hocak (Winnebago). *Economic Botany*, 52(4):352–372. (WIN)

Kinsey, A.C., R.C. Rollins, M.L. Fernald, 1996. *Edible Wild Plants of Eastern North America*. Dover Publications, 452 pages.

Kirtikar, K.R. and Basu, B.D. Reprint 1975 (1st edition, 1918, 2nd ed. 1935 or 1938). *Indian Medicinal Plants*. 4 volumes text, 4 volumes illustrations. M/S Periodical Experts, New Delhi. (KAB)

Kletter, C. and Kriechbaum, M. (Eds.). 2001. *Tibetan Medicinal Plants*. CRC Press, Boca Raton, FL. 383 pp. (TIB)

Koch, H.P. and Lawson, L.D. 1996. *Garlic: The Science and Therapeutic Application of* Allium sativum *L. and Related Species*. Williams and Wilkins. Baltimore. (KAL)

Kuebel, K.R. and Tucker, A.O. 1988. Vietnamese Culinary Herbs in the United States. *Economic Botany,* 42:413–419. (EB42:413)

Kyo, E., Uda, N., Kasuga, S., and Itakura, Y. 2001. Immunomodulatory Effects of Aged Garlic Extract. *Journal of Nutrition,* 131:1075S–1079S. (JN131:1075s)

Lal, B., Kapoor, A.K., Agrawal, P.K., Asthana, O.P., and Srimal, R.C. 2000. Role of Curcumin in Idiopathic Inflammatory Orbital Pseudotumours. *Phytotherapy Research,* 14(6):443–447. (PR14:443)

Latorre, D.L. and Latorre, F.A. 1977. Plants Used by the Mexican Kickapoo Indians. *Economic Botany,* 31:340–357. (EB31:353)

Lawson, L.D. In Koch, H.P. and Lawson, L.D. (Eds.). 1996. *Garlic — The Science and Therapeutic Application of* Allium sativum *L. and Related Species*. Williams & Wilkins, Baltimore. 329 pp. (LAW)

Leung, A.Y. and Foster, S. 1995. *Encyclopedia of Common Natural Ingredients,* 2nd ed. John Wiley & Sons, New York. 649 pp. (LAF)

Libster, M. 2002. *Delmar's Integrative Herb Guide for Nurses*. Delmar-Thomson Learnings, Albany, NY. 931 pp. (LIB)

Lin, C.-N., Lu, C.-M., Cheng, M.-K., Gan, K.-H., and Won, S.-J. 1990. The Cytotoxic Principles of *Solanum incanum*. *Journal of Natural Products,* 53(2):513–516. (JNP53:513)

Lininger, S (Ed.)., with Wright, J., Austin, S., Brown, D., and Gaby, A. 1998. *The Natural Pharmacy*. Prima Publications, Rocklin, CA. 450 pp. (SKY)

Liogier, A.H. 1974. *Diccionario Botanico de Nombres Vulgares de la Espanola*. Impresora UNPHU, Santo Domingo, RD. 813 pp. (AHL)

List, P.H. and Hohammer, L. 1969–1979. *Hager's Handbuch der Pharmazeutischen Praxis*, Vols. 2–6, Springer-Verlag, Berlin. (HHB)

Little, E.L., Jr., Woodbury, R.O., and Wadsworth, F.H. 1974. *Trees of Puerto Rico and the Virgin Islands*. 2nd Vol. Ag. Handbook No. 449. USDA, Washington D.C. 1024 pp. (LWW)

Liu, M., Li, X.Q., Weber, C., Lee, C.Y., Brown, J., and Liu, R.H. 2002. Antioxidant and Antiproliferative Activities of Raspberries. *Journal of Agriculture and Food Chemistry,* 50(10):2926–2930. (JAF50:2926)

Mahesh, T., Balasubashini, S., and Menon, V.P. 2005. Effect of Photo-Irradiated Curcumin Treatment against Oxidative Stress in Streptozotocin-Induced Diabetic Rats. *Journal of Medicinal Food,* 8(2):251–255. (JMF8:251)

Manandhar, N.P. 2002. *Plants and People of Nepal.* Timber Press, Portland, OR. 599 pp. (NPM)

Manzan, A.C.C.M., Toniolo, F.S., Bredow, E., and Povh, N.P. 2003. Extraction of Essential Oil and Pigments from *Curcuma longa* [L.] by Steam Distillation and Extraction with Volatile Solvents. *Journal of Agriculture and Food Chemistry,* 51(23):6802–6807. (JAF51:6802)

Marczylo, T., Sugiyama, C., and Hayatsu, H. 2003. Protection against Trp-P-2 DNA Adduct Formation in C57bl6 Mice by Purpurin Is Accompanied by Induction of Cytochrome P450. *Journal of Agriculture and Food Chemistry,* 51(11):3334–3337. (JAF51:3334)

Martin, A.C., Zim, H.S., and Nelson, A.L. 1961. *American Wildlife & Plants.* Dover Publications, Inc., New York. 500 pp. (MZN)

Martin, L., Villaescusa, L., Diaz, A.M., Olliveri, E., and Delmas, F. 1997. Screening for Antiparasitic Activity of *Myrtus communis. Fitoterapia,* 68(3):276–277. (FT68:276)

Martinez, M. 1969. *Las Plantas Medinales de Mexico,* 5th ed. La Impresora Aztecas, Ediciones Botas, Mexico, Df. 657 pp. (MAX)

McGuffin, M., Hobbs, C., Upton, R., and Goldberg, A. (Eds.). 1997. *American Herbal Products Association's Botanical Safety Handbook.* CRC Press, Boca Raton, FL. 231 pp. (AHP)

McGuffin, M. (Ed.) with Kartesz, J.T., Leung, A.Y., and Tucker, A.O. 2000. *American Herbal Products Association's Herbs of Commerce.* AHPA, Washington, D.C. 421 pp. (AH2)

Miller, L.G. and Murray, W.J. (Eds.). 1998. *Herbal Medicinals — A Clinician's Guide.* Pharmaceutical Products Press, Haworth Press, Binghamton, New York. 382 pp. (MAM)

Mills, S. and Bone, K. 1999. *Principles and Practice of Phytotherapy.* Churchill Livingstone, Edinburgh. 643 pp. (MAB)

Millspaugh, C.F. 1974. *American Medicinal Plants.* Dover Publications, Inc., New York.

Milner, J.A. 2001. A Historical Perspective on Garlic and Cancer. *Journal of Nutrition,* 131:1027S–1031S. (JN131:1027s)

Mitscher, L.A. 2005. A Modern Look at Folkloric Use of Antiinfective Agents. Chap. 16, pp. 329–352. In Yaniv, Z. and Bachrach, U. (Eds.). 2005. *Handbook of Medicinal Plants.* Food Products Press: Haworth Medical Press, New York. pp. xxi, 500 p.

Moerman, D.E. 1998. *Native American Ethnobotany.* Timber Press, Portland, OR. 927 pp. (DEM)

Mohamed, A.H., Eltahir, K.E.H., Ali, M.B., Galal, M., Ayeed, I.A., Adam, S.I., and Hamid, O.A. 1999. Some Pharmacological and Toxicological Studies on *Balanites aegyptiaca* Bark. *Phytotherapy Research,* 13(5):439–441. (PR13:439)

Moldenke, H.N. and Moldenke, A.L. 1952. *Plants of the Bible.* Chronica Botanica Co., Waltham, MA. 328 pp.

Monforte, M.T., Trovato, A., Rossitto, A., Forestieri, A.M., d'Aquino, A., Miceli, N., and Galati, E.M. 2002. Anticonvulsant and Sedative Effects of *Salvadora persica* L. Stem Extracts. *Phytotherapy Research,* 16(4):395–397. (PR16:395)

Montague-Drake, B. 1997. *A Biblical Herbal — The Greatest Herbal Story Ever Told.* Earth Images, Ltd., Kendal, NSW, Australia. (BMD)

Moriguchi, T., Takasugi, N., and Itakura, Y. 2001. The Effects of Aged Garlic Extract on Lipid Peroxidation and the Deformability of Erythrocytes. *Journal of Nutrition,* 131:1016S–1019S. (JN131:1016s)

Mors, W.B, Rizzini, C.T., and Pereira, N.A. 2000. *Medicinal Plants of Brazil.* Reference Publications, Algonac, MI. 501 pp. (MPB)

Morton, J.F. 1975. Cattails (*Typha* spp.) — Weed Problem or Potential Crop?. *Economic Botany,* 29:7–29. (EB29:7., EB29:20., EB29:24)

Morton, J.F. 1977. *Major Medicinal Plants.* C.C. Thomas, Publisher, Springfield, IL. (JFM)

Morton, J.F. 1981. *Atlas of Medicinal Plants of Middle America. Bahamas to Yucatan.* C.C. Thomas, Publisher, Springfield, IL. (JFM)

Nadkarni, A.K. 1976. (Revised and Enlarged). *Dr. K. M. Nadkarni's Indian Materia Medica.* Volume One. 1319 pp. Volume Two. (Originally edited by Dr. K.M. Nadkarni, 1908.) Popular Prakashan Pvt. Ltd., Bombay. 968 pp. (NAD)

Neuwinger, H.D. 1996. *African Ethnobotany — Poisons and Drugs.* Chapman & Hall, New York. 941 pp. (HDN)

Newall, C.A., Anderson, L.A., and Phillipson, J.D. 1996. *Herbal Medicine — A Guide for Health-care Professionals.* The Pharmaceutical Press, London. 296 pp. (CAN)

Newmark, T.M. and Schulick, P. 2000. *Herbal Cox-2 Inhibition — Nature's Challenge to Arthritis, Cancer & Alzheimer's Disease.* Hohm Press, Prescott, AZ. (COX)

Newton, S.M., Lau, C., and Wright, C.W. 2000. A Review of Antimycobacterial Natural Products. *Phytotherapy Research,* 14(5):303–322. (PR14:303)

Null, G., Dean, C., Feldman, M., Rasio, D., and Smith, D. 2003. *Death by Medicine — 2.* Nutrition Institute of America, Inc., Published online.

ORAFTI 2003. Newsletter of Active Food Science Monitor, Belgium. Chicory data drawn largely from Prof. Dr. D. Jenkins, in Newsletter No. 9 (Winter).

Osborn, D.J. 1968. Notes on Medicinal and Other Uses of Plants in Egypt. *Economic Botany,* 22:165–177. (EB22:165., EB22:173)

Özcan, M. 2002. Nutrient Composition of Rose (*Rosa canina* L.) Seed and Oils. *Journal of Medicinal Food,* 5(3):137–140. (JMF5:137)

Palevitch, D. and Yaniv, Z. 2000. *Medicinal Plants of the Holy Land.* 2 vols. 358 pp. Modan Publishing House, Tel Aviv. (PAY)

Pedersen, M. 1998. *Nutritional Herbology — A Reference Guide to Health.* 4th Printing., revised and expanded edition. Wendell W. Whitman Co., Warsdaw, IN. 336 pp., plus index. (PED)

Peirce, A. 1999. *The APhA Practical Guide to Natural Medicines.* Stonesong Press Book, Wm. Morrow & Co., Inc., New York. 728 pp. (APA)

Perdue, R.E. 1958. *Arundo donax* — Source of Musical Reeds and Industrial Cellulose. *Economic Botany,* 12:368–404. (EB12:368)

Perry, L.M. 1980. *Medicinal Plants of East and Southeast Asia.* MIT Press, Cambridge, MA. 620 pp. (LMP)

Philips, H.J. 1958. *Lebanese Folk Cures. Vol. II. Some Lebanese Materia Medica.* Univ. Microfilms Internat. 457 pp. (Ph.D. thesis, Anthropology, Columbia University) (Copyright 1959). (HJP)

Pizzorno, J.E. and Murray, M.T. 1985. *A Textbook of Natural Medicine.* John Bastyr College Publications, Seattle, WA (Looseleaf). (PAM)

Press, J.R., Short, M.J., and Turland, N.J. 1994. *Flora of Madeira.* HMSO Publications Center, London. 574 pp. (PST)

Ramadan, M.F., Kroh, L.W., and Morsel, J.-T. 2003. Radical Scavenging Activity of Black Cumin (*Nigella sativa* L.), Coriander (*Coriandrum sativum* L.), and Niger (*Guizotia abyssinica* Cass.) Crude Seed Oils and Oil Fractions. *Journal of Agriculture and Food Chemistry,* 51(24):6961–6969. (JAF51:6961)

Reed, C.F. 1970. Selected Weeds of the United States. (Illustrations by Regina Hughes.) Ag. Handbook No. 366, USDA, ARS, USGPO, Washington, D.C. 463 pp.

Ridoux, O., Di Giorgio, C., Delmas, F., Elias, R., Mshvildadze, V., Dekanosidze, G., Kemertelidze, E., Balansard, G., and Timon-David, P. 2001. *In vitro* antileishmanial activity of saponins isolated from ivy, alpha-hederin, beta-hederin and hederacolchiside A_1, in association with pentamidine and amphotericin B. *Phytotherapy Research,* 15(4):298–301. (PR15:298)

Rinzler, C.A. 1990. *The Complete Book of Herbs, Spices and Condiments.* Facts on File, New York. 199 pp. (RIN)

Rizk, A.F.M. 1986. *The Phytochemistry of the Flora of Qatar.* Scientific and Applied Research Centre, University of Qatar, Kingprint, Richmond, U.K. (RIZ)

Roersch, C. 1994. *Plantas Medicinales en el Sur Andino del Peru.* 2 vols. Koeltz Scientific Books. Koenigstein. (ROE)

Roig y Mesa, J.T. 1928. *Diccionario Botanico de Nombres Vulgares Cubanos.* Imprenta y Papeleria Rambla, Bouza y Ca. Havana. 897 pp. (RyM)

Roux, K.H., Teuber, S.S., Robotham, J.M., and Sathe, S.K. 2001. Detection and Stability of the Major Almond Allergen in Foods. *Journal of Agriculture and Food Chemistry,* 49(5):2131–2136. (JAF49:2131)

Rudin, D.O. and Felix, C. 1987. *The Omega-3 Phenomenon: The Nutrition Breakthrough of the 80s.* Rawson Books, New York.

Rukkumani, R., Aruna, K., Varma, P.S., Rajasekaran, K.N., and Menon, V.P. 2005. Comparative Effects of Curcumin and Its Analog on Alcohol and Polyunsaturated Fatty Acid-Induced Alterations in Circulatory Lipid Profiles. *Journal of Medicinal Food,* 8(2):256–260. (JMF8:256)

Rutter, R.A. 1990. *Catalogo de Plantas Utiles de la Amazonia Peruana.* Instituto Linguistico de Verano. Yarinacocha, Peru. 349 pp. (RAR)

Said, S.A., El-Kasheef, H.A., El-Mazar, M.M., and Salamo, O. 1996. Phytochemical and Pharmacological Studies on *Lactuca sativa* Seed Oil. *Fitoterapia,* 67(3):215–219. (FT67:215)

Saleheen, D., Ali, S.A., and Yasinzaia, M.M. 2004. Antileishmanial Activity of Aqueous Onion Extract *in vitro. Fitoterapia,* 75(1):9–13. (FT75(1):9)

Sallal, A.K., Shraideh, Z.A., and Abu-Elteen, K.H. 1998. Ultrastructural Effects of Date Extract on *Candida albicans. Mycopathologia,* 142(3):119–123.

Samaranayake, M.D.P., Wickramasinghe, S.M.D.N., Angunawela, P., Jayasekera, S., Iwai, S., and Fukushima, S. 2000. Inhibition of Chemically Induced Liver Carcinogenesis in Wistar Rats by Garlic (*Allium sativum*). *Phytotherapy Research,* 14(7):564–567. (PR14:564)

Schilcher, H. 1997. *Phytotherapy in Paediatrics — Handbook for Physicians and Pharmacists.* 2nd German edition transl. by A.R. Meus. Med. Pharm. Publ., Stuttgart. 181 pp. (PIP)

Schinella, G.R., Tourniera, H.A., Prietoc, J.M., Ríosc, J.L., Buschiazzoa, H., and Zaidenberga, A. 2002. Inhibition of *Trypanosoma cruzi* Growth by Medical Plant Extracts. *Fitoterapia,* 73(7–8):569–575. (FT73:569)

Schmid, B., Lüdtke, R., Selbmann, H.-K., Kötter, I., Tschirdewahn, B., Schaffner, W., and Heide, L. 2001. Efficacy and Tolerability of a Standardized Willow Bark Extract in Patients with Osteoarthritis: Randomized Placebo-Controlled, Double Blind Clinical Trial. *Phytotherapy Research,* 15(4):344–350. (PR15:344)

Schultes, R.E. and Raffauf, R.F. 1990. *The Healing Forest.* Dioscorides Press, Portland, 484 pp. (SAR)

Schulz, V., Hansel, R., and Tyler, V.E. 1998. *Rational Phytotherapy — A Physician's Guide to Herbal Medicine,* 3rd ed. 1st English edition. Translated by Terry C. Telger. Springer Verlag, Heidelberg. 306 pp. (SHT)

Seigler, D.S. 1976. Plants of the Northeastern United States that Produce Cyanogenic Compounds. *Economic Botany,* 30:395–407. (EB30:400., EB30:403)

Sezik, E., Tabata, M., Yeilada, E., Honda, G., Takeda, Y., Tanaka, T., and Takaishi, Y. 1995. Traditional Medicine in Turkey. VII. Folk Medicine in Middle and West Black Sea Regions, Tetsuro Fujita. *Economic Botany,* 49:406–422. (EB49:406)

Sezik, E., Yesilada, E., Tabata, M., Honda, G., Takaishi, Y., Fujita, T., Tanaka, T., and Takeda, Y. 1997. Traditional Medicine in Turkey. VIII. Folk Medicine in East Anatolia., Erzurum, Erzincan, Agri, Kars, Igdir Provinces. *Economic Botany* 51:195–211. (EB51:195)

Shirwaikar, A., Sreenivasan, K.K., Krishnanand, R.B., and Vasanth Kumar, A. (1996). Chemical Investigation and Antihepatoxic Activity of the Root Bark of *Capparis spinosa. Fitoterapia,* 67(3):200–204. (FT67(3):200)

Shoji, T., Mutsuga, M., Nakamura, T., Kanda, T., Akiyama, H., and Goda, Y. 2003. Isolation and Structural Elucidation of Some Procyanidins from Apple by Low-Temperature Nuclear Magnetic Resonance. *Journal of Agriculture and Food Chemistry,* 51(13):3806–3813. (JAF51:3806)

Singh, B., Saxena, AK., Chandan, B.K., Gupta, D.K., Bhutani, K.K., and Anand, K.K. 2001. Adaptogenic Activity of a Novel, Withanolide-Free Aqueous Fraction from the Roots of *Withania somnifera* Dun. *Phytotherapy Research,* 15(4):311–318. (PR15:311)

Singh, R.P. Chidambara Murthy, K.N., and Jayaprakasha, G.K. 2002. Studies on the Antioxidant Activity of Pomegranate (*Punica granatum*) Peel and Seed Extracts Using *in vitro* Models. *Journal of Agriculture and Food Chemistry,* 50(1):81–86. (JAF50:81)

Skenderi, G. 2003. *Herbal Vade Mecum.* Herbacy Press, Rutherford, NJ. 480 pp. (GAZ)

Soukup, J. 1970. *Vocabulary of the Common Names of the Peruvian Flora and Catalog of the Genera.* Editorial Salesiano, Lima. 436 pp. (SOU)

Stamets, P. (and Yao, D.W.). 2002. *MycoMedicinals,* 3rd edition. Fungi Imperfecti, Olympia WA. 96 pp. (SAY)

Standley, P.C. 1920-1926. *Trees and Shrubs of Mexico.* 5 Parts. (Contr. U.S. Nat. Herb.). 1722 pp. (PCS)

Steinmetz, E.F. Circa 1957. *Codex Vegetabilis. All Botanical Drugs.* Keizersgracht 347, Amsterdam. Unpaginated. (EFS)

Suwal, P.N. 1976 (Reprint ed.). *Medicinal Plants of Nepal.* His Majesty's Government, Dept. Med. Plants, Kathmandu, Nepal. 153 pp. plus 33 appendix pages. (SUW)

Tabor, E. 1970. Plant Poisons in Shakespeare. *Economic Botany,* 24:81–94. (EB24:86)

Tackholm, V. and Drar, M. 1973. *Flora of Egype.* Vol. II. *Angiospermae, Part Monocotyledones: Cyperaceae*-Juncaceae. Cairo: Cairo University Press.

Tackholm. V. an Drar, M. 1973. *Flora of Egypt.* Vol. III. *Angiospermae, Part Monocotyledones: Liliaceae — Musaceae.* Cairo: Cairo University Press.

Tackholm. V. and Drar, M. 1969. *Flora of Egypt.* vol. IV. *Angiospermae, Part Monocotyledones; Zingiberaceae — Orchidaceae Dicotyledones: Casuarinaceae — Piperaceae.* Cairo: Cairo University Press.

Tanaka, T. 1976. *Tanaka's Cyclopedia of Edible Plants of the World.* Keigaku Publishing Co., Tokyo. 924 pp. (TAN)

Temple, A.A. 1929. *Flowers and Trees of Palestine.* Society for Promoting Christian Knowledge, London. 148 pp.

Thieret, J.W. 1996. Frankincense and Myrrh. *Lloydiana,* 1(4):6–9.

Time-Life, Editors. 1996. *The Medical Advisor. The Complete Guide to Alternative & Conventional Treatments.* Time Life, Alexandria, VA. 1152 pp. (TMA)

Tucker, A.O. and Debaggio, T. 2000. *The Big Book of Herbs.* Interweave Press, Loveland, CO. 688 pp. (TAD)

Tyler, V.E. 1985. *Hoosier Home Remedies.* Purdue University Press, West LaFayette, IN. 212 pp. (HOO)

Tyler, V.E. 1994. *Herbs of Choice — The Therapeutic Use of Phytomedicinals*. Pharmaceutical Products Press, New York. 209 pp. (HOC)

Uphof, J.C. Th. 1968. *Dictionary of Economic Plants*, 2nd ed. Verlag von J. Cramer. (UPH)

Vadmecum García, A.A. et al. 1998. *Fitoterapia Vademecum de Prescripcióón*. Masson, S.A. Barcelona, España. 1148 pp. (VAD)

Van Wyk, B.-E., Van Oudtshoorn, B., and Gerike, N. 1997. *Medicinal Plants of South Africa*. Briza Publications, Pretoria, South Africa. 304 pp. (VVG)

Verzera, A., Trozzi, A., Gazea, F., Cicciarello, G., and Cotroneo, A. 2003. Effects of Rootstock on the Composition of Bergamot (*Citrus bergamia* Risso et Poiteau) Essential Oil. *Journal of Agriculture Food Chemistry,* 51(1):206–210. (JAF51:208)

Volate, S.R., Davenport, D.M., Zander, M.E.S., Hudson, R.S., Woods, C.J., Muga, S.J., and Wargovich, M.J. 2003. Chemopreventive Effects of Herbal Supplements in Rat Colon Aberrant Crypt Foci Assay. *Journal of Nutrition,* 133(11S-I):3861S. (JN133:3861S)

Walker, W. 1957. *All the Plants of the Bible*. Harper and Bros., New York. 244 pp.

Watt, G. 1889–1892 (reprint 1972). *Dictionary of the Economic Products of India*, 6 vols. Nav Bharat Offset Process, Delhi, India. (DEP)

Watt, J.M. and Breyer-Brandwijk, M.G. 1962. *The Medicinal and Poisonous Plants of Southern Africa*. E.&S. Livingstone, Ltd., Edinburg & London. 1457 pp. (WBB)

White, L. and Mavor, S. 1998. *Kids, Herbs, Health — A Parent's Guide to Natural Remedies*. Interweave Press. 272 pp. (WAM)

White, L.B., Foster, S., et al. (staff of *Herbs for Health*). 2000. *The Herbal Drugstore*. Rodale Press. 610 pp. (WAF)

Wichtl, M. 1984. *Teedrogen. Ein Handbuch fur Apotheker und Arzte*. Wissenschaftliche Verlagsgesellscharft. mbH Stuttgart. 393 pp. (WIC)

Wichtl, M. (Ed.). 2004. *Herbal Drugs and Phytopharmaceuticals — A Handbook for Practice on a Scientific Basis*, 3rd ed. CRC Press, Boca Raton, FL. 704 pp. (WI3)

Williamson, E.M. and Evans, F.J. 1989. *Potter's New Cyclopaedia of Botanical Drugs and Preparations*, revised ed. Saffron Walden, the C.W. Daniel Co., Ltd., Essex, U.K. 362 pp, 1988, reprint 1989. (PNC)

Winston, D. 2001. Nvwoti: Cherokee Medicine and Ethnobotany. *Journal of the American Herbalists Guild.* 2(2):45–53. (JAH2(2):45)

World Health Organization. 1999. *WHO Monographs on Selected Medicinal Plants*. Vol. 1, WHO, Geneva. 289 pp. (WHO)

Wright, C.. (Ed.). 2002. *Medicinal and Aromatic Plants — Industrial Profiles. Artemisia*. Taylor & Francis, New York. 344 pp. (CWW)

Yaniv, Z., Schafferman, D., and Amar, Z. 1998. Tradition, Uses and Biodiversity of Rocket (*Eruca sativa*, Brassicaceae) in Israel. *Economic Botany,* 52:394–400. (EB52:394)

Yaniv, Z. and Bachrach, U. (Eds.). 2005. *Handbook of Medicinal Plants*. Food Products Press: Haworth Medical Press, New York. pp. 500 p + xxi. (YAB)

Yarnell, E. and Meserole, L. 1996. Topical Applications of Botanical Medicine. *Alternative & Complementary Therapies,* 2(4):241–244.

Yeh, Y.-Y. and Liu, L. 2001. Cholesterol-Lowering Effect of Garlic Extracts and Organosulfur Compounds: Human and Animal Studies. *Journal of Nutrition,* 131(3s):989S–993S. (JN131:989s)

Zhu, N., Kikuzaki, H., Sheng, S., Sang, S., Rafi, M.M., Wang, M., Nakatani, N., DiPaola, R.S., Rosen, R.T., and Ho, C.-T. 2001. Furanosesquiterpenoids of *Commiphora myrrha*. *Journal of Natural Products,* (Note) 64(11):1460–1462. (JNP64:1460)

Zia, T., Siddiqui, I.A., and Nazrul-Hasnain, S. 2001. Nematicidal Activity of *Trigonella foenum-graecum* L. *Phytotherapy Research,* 15(6):538–540. (PR15:538)

Zohary, D. 1972. The Wild Progenitor and the Place of Origin of the Cultivated Lentil: *Lens culinaris*. *Economic Botany*, 26:326.

Zohary, M. 1966. *Flora Palaestina. Part 1. Text. Equisetaceae to Moringaceae.* 364 pp. Part 1 (Plates) not paginated. 495 plates, plus indices. Israel Academy of Sciences and Humanities. (FP1)

Zohary, M. 1972. *Flora Palaestina. Part 2. Text. Platanaceae to Umbelliferae.* 489 pp. Part 2 (Plates) not paginated. 656 plates, plus indices. Israel Academy of Sciences and Humanities. (FP2)

Zohary, M. 1982. *Plants of the Bible*. Cambridge University Press, New York. 223 pp. (ZOH)

Common Name Index

I

J

L

M

N

O

P

R

S

T

V

W

Y

Scientific Name Index

A

B

C

D

E

F

N

O

P

T

U

V

X

Z